D1242723

Constants

Acceleration of gravity at sea level
  32.1739 ft/sec² = 9.807 m/s²
Avogadro's number
  1 kgmol = 6.023 × 10²⁶ molecules
Stefan-Boltzmann constant (σ)
  5.67 × 10⁻⁸ W/m²-K⁴ = 0.1714 × 10⁻⁸ Btu/hr-ft²-R⁴
Universal gas constant ($\overline{R}$)
  8.3143 kJ/kgmol-K = 1545.32 ft-lbf/pmol-R

Engineering Thermodynamics

FOURTH EDITION

Engineering Thermodynamics

FOURTH EDITION

M. David Burghardt
Hofstra University

James A. Harbach
U.S. Merchant Marine Academy

HarperCollins*CollegePublishers*

To Linda and Phyllis

Sponsoring Editor: John Lenchek
Project Editor: Randee Wire/Carol Zombo
Art Director: Julie Anderson
Cover Design: Julie Anderson
Cover Illustration: Rolin Graphics
Production Administrator: Brian Branstetter
Compositor: Progressive Typographers, Inc.
Printer and Binder: R. R. Donnelley & Sons Company
Cover Printer: The Lehigh Press, Inc.

Engineering Thermodynamics, Fourth Edition

Copyright © 1993 by HarperCollins College Publishers

All rights reserved. Printed in the United States of America. No part of this book may be used or reproduced in any manner whatsoever without written permission, except in the case of brief quotations embodied in critical articles and reviews. For information address Harper-Collins College Publishers, 10 East 53rd Street, New York, NY 10022.

Library of Congress Cataloging-in-Publication Data

Burghardt, M. David.
    Engineering thermodynamics. — 4th ed./M. David Burghardt, James A. Harbach.
        p.   cm.
    Rev. ed. of: Engineering thermodynamics with applications. 3rd ed. © 1986.
    Includes index.
    ISBN 0-06-041049-3
    1. Thermodynamics.   I. Harbach, James A.   II. Burghardt, M. David Engineering thermodynamics with applications.   III. Title
TJ265.887   1992
621.402'1 — dc20   92-30663
                                                                    CIP

95   9 8 7 6 5

Contents

Chapter 14   Gas Turbines   483

Chapter 15   Vapor Power Systems   529

Chapter 16   Refrigeration and Air Conditioning Systems   605

Answers to Selected Problems

Index

Preface

This text presents a comprehensive and comprehensible treatment of engineering thermodynamics, from its theoretical foundations to its applications in realistic situations. The thermodynamics presented will prepare students for later courses in fluid mechanics and heat transfer. In addition, practicing engineers will find the applications helpful to them in their professional work. The text is appropriate for an introductory undergraduate course in thermodynamics and for a subsequent course in thermodynamic applications.

Many features of the text distinguish it from other texts and from previous editions of this text. They include

- A systematic approach to problem solving;
- An instructor's solutions manual that includes solutions to all problems and that uses the same systematic approach found in the examples;
- Over 1500 problems in SI and English units, 90 requiring computer solution, plus 130 solved examples;
- Chapter objectives highlighted at the start of each chapter;
- Expanded and amplified development of the second law of thermodynamics, stressing availability analysis;
- The integration of the use of the personal computer for solving thermodynamics problems based on the use of TK Solver® and spreadsheet software;
- The inclusion of a disk of TK Solver® files that can be used as provided, or modified and merged into models developed to analyze new problems.

We believe that this text breaks new ground in the presentation of thermodynamics to undergraduate engineering students. The integration of the use of TK Solver® and spreadsheet software is one important aspect of this advance. In addition to including files for determining properties of steam, refrigerant 12, and air, the disk supplied with the text includes models for analyzing many thermodynamic processes and cycles. Unlike most other available software, these TK Solver® models can be easily modified to analyze new problems. This provides the student with a powerful set of tools for applications in later design courses and in their professional careers. As an indication of the software's power, although it is a long trial-and-error process to determine the percentage and effect of only one product's dissociation in a combustion process, one of the TK Solver® models supplied can analyze combustion reactions with up to three simultaneous dissociation reactions. While we believe that the integration of the computer solutions enhances the thermodynamics course, the text is also written to be used without the software.

Another key feature that elevates the text is the expanded discussion and analysis of thermodynamic fundamentals and applications. An underlying theme of the text is that understanding energy and energy utilization is vital to the well-being of an industrialized society. This is underscored when alternative energy sources such as solar, geothermal, and wind are discussed. Furthermore, the environmental effects of acid rain and global warming are discussed in the chapter on reactive systems.

A new feature of the text is a systematic problem solution methodology that is adhered to in the text and the instructor's solutions manual. This methodology guides students into thinking about the problem before proceeding with its solution. It encourages students to approach the problems logically, to state assumptions used, to detail the step-by-step analysis, to explicitly include units and conversions when numerical values are substituted, and to note in conclusion key points in the solution. We encourage faculty to require their students to follow this format in their problem solutions.

The end-of-chapter problems range in complexity from those illustrating basic concepts to more challenging ones involving judgment on the part of the student. In the applications chapters, open-ended problems allow the students to investigate alternative solutions. The provided software allows students to model complex systems, vary parameters, and undertake parameter variation in seeking optimum solutions. In the chapter on refrigeration and air conditioning, the R 12 property models and the psychrometric chart models allow solution of sophisticated HVAC problems.

Second-law analysis is ever more important in an era of heightened energy awareness and energy conservation. A thorough development of the second law of thermodynamics is provided in Chapters 7–9. The concept of entropy production is developed in Chapter 8 and used throughout the application chapters, as are the availability concepts developed in Chapter 9. The ramifications of the second law receive thorough discussion; the student not only performs calculations but understands the implications of the calculated results.

We would like to extend our gratitude to the following reviewers, who offered valuable suggestions for the fourth edition: Lynn Bellamy, Arizona State University; Alan J. Brainard, University of Pittsburgh; E. E. Brooks, University of Saskatchewan; James Bugg, University of Saskatchewan; Clinton R. Carpenter, Mohawk Valley Community College; Daniel Fairchild, Roger Williams College; Walter R. Kaminski, Central Washington University; B. I. Leidy, University of Pittsburgh; Robert A. Medrow, University of Missouri — Rolla; Edwin Pejack, University of the Pacific;

Julio A. Santander, Southern College of Technology; George Sehi, Sinclair Community College; E. M. Sparrow, University of Minnesota; Amyn S. Teja, Georgia Institute of Technology; Richard M. Wabrek, Idaho State University; and Ross Wilcoxon, South Dakota State University. We would like to acknowledge Todd Piefer of UTS for his assistance and suggestions on improving the TK Solver® models.

It is our hope that the text will form a useful part of an educational program in engineering and be a useful reference for the practicing engineer.

M. David Burghardt
James A. Harbach

1

Introduction

Thermodynamics is the science that is devoted to understanding energy in all its forms, such as mechanical, electrical, chemical, and how energy changes forms, e.g. the transformation of chemical energy into thermal energy, for instance. *Thermodynamics* is derived from the Greek words *therme,* meaning heat, and *dynamis,* meaning strength, particularly applied to motion. Literally, thermodynamics means "heat strength," implying such things as the heat liberated by the burning of wood, coal, or oil. If the word energy is substituted for heat, one can come to grips with the meaning and scope of thermodynamics. It is the science that deals with energy transformations: the conversion of heat into work, or of chemical energy into electrical energy. The power of thermodynamics lies in its ability to be used to analyze a wide range of energy systems using only a few tenets, two primary ones being the First and Second Laws of Thermodynamics. Thermodynamics applies very simple yet encompassing laws to a wide range of energy systems that have major import in our society, for example, energy use in agriculture, electric power generation, and transportation systems.

We will examine the following energy conversion systems in more detail:

- A steam power plant, fundamental to the generation of electric power.
- An internal combustion engine, used by many of us daily in our automobiles.
- Direct energy conversion, particularly photovoltaic conversion of sunlight into electricity, a projected major source of electrical power.

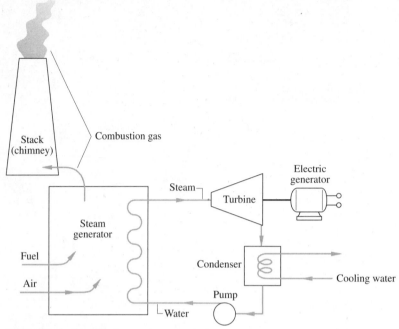

Figure 1.1 A simplified vapor power cycle.

- Cogeneration facilities where power is produced and what had formerly been waste heat is used for heating or cooling.
- Our current energy consumption patterns, and energy conservation policies that industrialized societies may follow in the future with the resultant challenges for engineers.

The Steam Power Plant

The first law of thermodynamics states that energy is conserved; it can change form, but it cannot be destroyed. This very simple, fundamental statement allows us to investigate the behavior of many devices and systems (combinations of devices). One of these is the *steam power plant,* an energy system that is essential to the industrialized world, as it often is used for the generation of electric power. Figure 1.1 depicts a simplified steam power plant. Fuel is burned to release heat in a steam generator, similar in concept to the oil- or gas-fired boilers used to provide steam or hot water for heating in homes. The process transforms the fuel's chemical energy to the thermal energy of combustion gases. The heat is used to boil water under pressure in the steam generator (boiler). The steam leaves the generator and passes through superheater tubes, where more heat is added to the steam; it then passes through the turbine, where it increases in volume, decreases in pressure, and performs work by causing the turbine rotor to rotate. The turbine is coupled to a generator, which is used to generate electric power. Thus, in the turbine another process transforms some of the

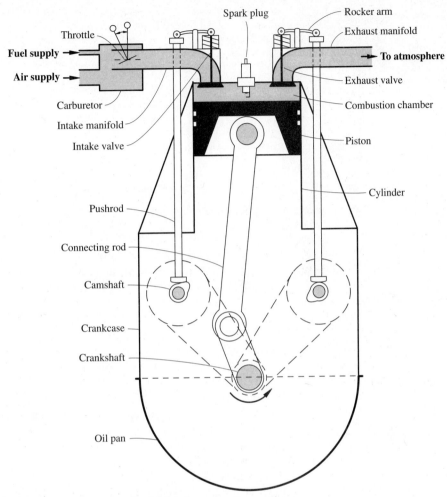

Figure 1.2 A reciprocating internal combustion engine.

thermal energy of the steam into mechanical work. The steam is then condensed, liquefied, and pumped back to the steam generator. The second law of thermodynamics tells us how much thermal energy can be converted into work. Not all of it can be.

To fully understand a steam power plant requires knowledge of substance properties: why water behaves as it does, why combustion occurs, what are the combustion products, what is the mechanism of energy transformation. Thermodynamics allows us to determine these properties, experimentally and theoretically.

Internal Combustion Engines

Another standard power plant that many of us use every day is the *internal combustion engine* used to power automobiles and many other machines (Figure 1.2). The

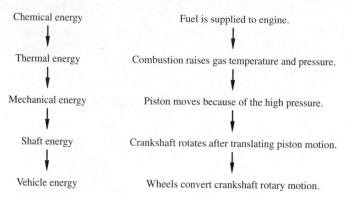

Figure 1.3 Transformation of fuel energy into vehicle motion.

engine may be viewed as a small power plant: fuel is burned, and the energy from the burning fuel is transferred to the pistons, whose gears turn the wheels, thus moving the automobile. The transformation of the fuel's chemical energy into vehicle motion is shown in Figure 1.3.

In these times of energy scarcity, we need to develop an understanding of the transformation process so we can minimize inherent losses of energy quality. Thermodynamic analysis seeks to determine ahead of time how much work we may expect from an engine and, through experiments, how efficiently the engine is performing. This is very important in minimizing the pollution from exhaust gases. A typical engine's exhaust contains unburned hydrocarbons, carbon monoxide, carbon dioxide, and nitric oxides, all of which decrease air quality. We need to minimize the total pollution, eliminating those types that are most harmful.

The *gas turbine* is another combustion engine, typically used on jet planes and for supplemental electric power generation. Air is compressed and energy added to it by burning fuel in a combustion chamber; this mixture—the products of combustion, air, and burned fuel—expands through a turbine, doing work, which drives the compressor and electric generator (Figure 1.4). On a jet engine, the work of the

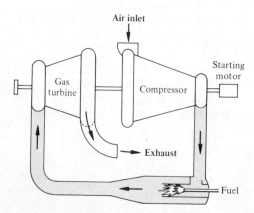

Figure 1.4 A gas-turbine unit.

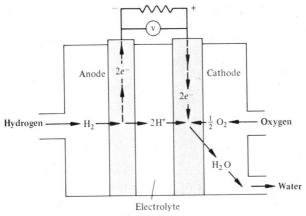

Figure 1.5 The hydrogen-oxygen fuel cell.

turbine is used only to drive the compressor, and the exhaust gases from the turbine, under a higher pressure than the outside air, expand through a nozzle, increasing in velocity. This increase in velocity generates a propulsive force on the engine and on the airplane to which it is attached.

All these analyses have a common purpose, to consider how efficiently the chemical energy of the fuel is converted into mechanical energy and with this knowledge to search for ways to improve the conversion process. The processes of converting energy are different depending on the mechanical structure (automotive engine versus steam power plant), but the laws of thermodynamics governing energy conversion remain the same.

Direct Energy Conversion

Some energy converters do not rely on intermediate devices to produce work—electric power in this case. Two familiar *direct energy conversion devices* are fuel cells, in which chemical energy is converted directly to electrical energy, and photovoltaic cells, in which the sun's radiant energy is converted directly to electrical energy.

Figure 1.5 illustrates a simplified fuel cell using hydrogen and oxygen. The hydrogen is oxidized at the anode, giving up two electrons, and the oxygen is reduced at the cathode, receiving two electrons. The load is connected to the two poles.

The half-cell reactions are

$$H_{2(g)} \rightarrow 2H^+ + 2e^- \qquad V = 0.0 \text{ volts}$$

$$\frac{1}{2} O_{2(g)} + 2H^+ + 2e^- \rightarrow H_2O \text{ (l)} \qquad V = 1.23 \text{ volts}$$

Thus the voltage of the cell has 1.23 volts, and if the load resistance is known, the current flowing through the load may be determined with Ohm's law.

One of the most promising renewable energy forms is photovoltaics, the direct conversion of sunlight into electricity. Current laboratory models can attain a con-

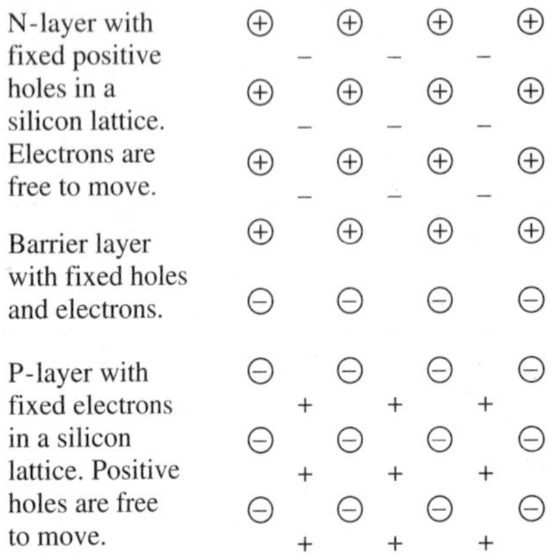

N-layer with
fixed positive
holes in a
silicon lattice.
Electrons are
free to move.

Barrier layer
with fixed holes
and electrons.

P-layer with
fixed electrons
in a silicon
lattice. Positive
holes are free
to move.

Figure 1.6 Charge distribution in a P-N photovoltaic cell.

version efficiency of 25% on small cells and an average efficiency of 13% for cells a square foot in area. Production costs are dropping, making the units more economical.

How does this conversion occur? Photovoltaic cells are a type of semiconductor involving a *P-N junction.* At a P-N junction is a voltage potential created by a positive-hole P-layer, which contains movable positive charges or "holes," and a negative N-layer, which contains movable negative electrons. When light with sufficient energy enters the crystal, electrons are released from their atomic bonds and migrate to an electrode. A wire connected from the negative electrode to a load leads back to the positive electrode, where the electrons combine with the positive holes. A barrier at the P-N junction prevents the electrons from instantly combining with the holes in the P-layer, forcing them instead to flow from the electrode through the load to the opposite electrode. No material is consumed, and the cell would theoretically last forever, except for radiation damage which limits the working life of the cell.

Figure 1.6 illustrates a P-N photovoltaic cell. The barrier layer prevents diffusion of holes or electrons from the P-layer to the N-layer, or vice versa. The crystal absorbs sunlight, producing an electron and a hole. Ordinarily these would immediately recombine, and the effect of the light absorption would be an increase in crystalline temperature. However, because of the potential barrier at the P-N junction, the electrons migrate to one electrode and the holes to the other. This produces a potential difference, and a current can flow between the electrodes when a wire connects them. Very important to the successful operation of the cell are controlled doping with selected impurities and the absence of other impurities. The effect of other impurities is to allow recombination of holes and electrons, rather than a flow of current. Because of engineering advances in refining and doping, photovoltaics is now a practical means of small-scale power generation.

Cogeneration

Another type of energy system of growing importance is a cogeneration system. *Cogeneration* means using the same energy source for more than one purpose, such

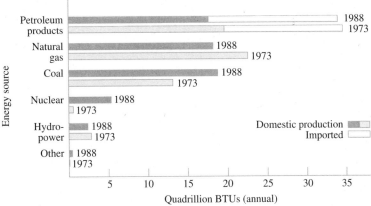

Figure 1.7 Comparison of energy sources in 1973 and 1988 in the United States.

as using the waste heat from an engine for space heating. Cogeneration facilities often generate electricity locally, such as a university using its own diesel-electric generators. The waste heat from the engine is used for additional purposes, perhaps for space heating or for air conditioning, as odd as that sounds. There are large air conditioning systems that use heat as the energy source; these are called absorption refrigeration systems. So, rather than having an engine that is 40% efficient in generating electricity, with the waste heat from the cooling water and the exhaust being dissipated to the atmosphere, a cogeneration facility uses this waste heat for other purposes. This increases the overall efficiency of the unit and decreases the total fuel or utility costs that the university has to pay.

Our Energy Consumption Lifestyle

Many studies[1,2] indicate that the world will have greatly diminished petroleum and mineral resources in the near future, resources that sustain the standard of living the industrialized world enjoys. All aspects of manufacture and living must become more energy efficient as the cost and scarcity of resources increase. A knowledge of thermodynamics is the first step. It allows engineers to create new systems and devices that will reduce our energy consumption and improve our energy utilization.

How did we get in this predicament in the first place? We love energy. Our lifestyle and our standard of living depend on a very high rate of consumption of energy. We have forgotten that all the electricity and gasoline come from a finite supply of fossil fuels. Because it is a finite supply, it will be depleted very soon, perhaps within our lifetime for domestic oil and gas reserves. Figure 1.7 graphs the sources of energy that fuel our lives. The largest source is petroleum products—gasoline and fuel oil—that we use in our cars, homes, and power plants. Notice the difference in the amount imported in 1988 versus 1973, the time of the OPEC oil embargo. The United States will become more sensitive to variations in petroleum supplies as it produces less of its total consumption, because other energy forms do not perform petroleum's function in society, for example, providing fuel for cars. Our total energy

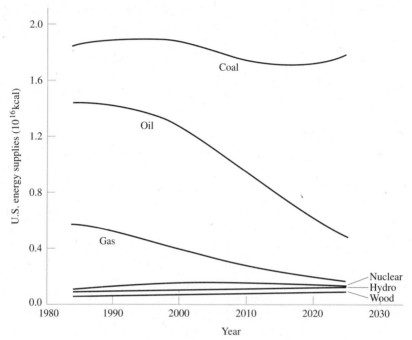

Figure 1.8 Change in U.S. energy sources.

consumption has been supplemented by increases in coal and nuclear power production. Figure 1.7 does not indicate where the fuel reserves are. The United States has a large coal supply, but the domestic petroleum and natural gas supplies are decreasing, making imports more important. Various factors are weighed to estimate when the domestic supplies will be depleted and when the United States will become totally dependent on foreign supplies. Certainly in the next ten to twenty-five years the United States will be nearing this situation. Economic and political problems will result from this increasing dependency. The problems are economic in that the cost of fuel will rise dramatically, causing inflation and social pressure, and political in that other countries will be vying for the same energy sources.

Engineers must be very aware of the social and political pressures associated with technical changes they design. Figure 1.8 represents a forecast of the changing proportions of the energy sources in the United States for the next forty years. It is apparent that oil and gas will decrease significantly and that coal will be the remaining fossil fuel, with nuclear, solar, and hydroelectric being the other energy sources.

The United States has large coal reserves that can be used for power generation, coal gasification projects, and coal liquefaction processes. Studies have indicated that there is sufficient coal to provide for our energy needs for the next four hundred years. However, there are tremendous environmental problems associated with expanding the use of coal. Much of the coal that is accessible is strip-mined, a technology that arouses environmental group action and requires water that is needed for agriculture and drinking supplies in the same geographic regions. In addition burning coal involves air pollution and air quality issues. Coal with a high sulfur content must be processed to remove as much sulfur as possible because of the direct correlation between sulfur and acid rain. Nationally 20% of the coal reserves have a high sulfur content, with the figure at 43% in coal regions east of the Mississippi River. The combustion of any hydrocarbon fuel produces carbon dioxide, heightening the probability of global warming.

Nuclear power plant construction is at a standstill in the U.S., for the legal impediments to building a plant are virtually insurmountable. However, nuclear power produces about 15% of U.S. electric power. In France, over 60% of the electric power is produced at nuclear plants. Nuclear plants create significant waste disposal problems, one of the key areas of concern. Internationally, however, nuclear plants are being constructed as an alternative power source, and the United States may see them appear again if the time and cost for construction can be decreased.

Renewable energy supplies, solar and hydro, hold some promise for future development, particularly small-scale solar power. Hydroelectric power has severe environmental impacts associated with it. Dams must be created, land flooded, water tables changed, and the habitat for certain species destroyed. All these steps are possible, but the process takes time and is sure to engender significant opposition. Certain types of solar energy utilization, particularly passive solar energy for home heating and photovoltaics for electric power generation, are increasingly being used.

During the industrial age, energy—fuel—has replaced people and animals in performing work in manufacturing, agriculture, and transportation in all areas of commerce and industry. Figure 1.9 graphs the change of energy source for work over

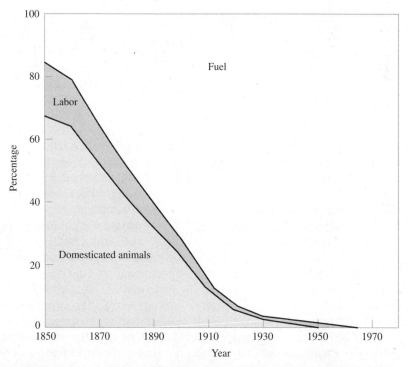

Figure 1.9 Source of energy used for work in the United States. (*Source:* C. A. S. Hall, C. J. Cleveland, and R. Kaufmann, *Energy and Resource Quality: The Ecology of the Industrial Process,* New York: John Wiley and Sons, 1986.)

Figure 1.10 Energy usage and worker productivity. *Solid line* is energy use; *dotted line* is labor productivity. Both measures are indexes, 1967 = 1.0. (*Source:* C. J. Cleveland, R. Costanza, C. A. S. Hall, R. Kaufmann, "Energy and the U.S. Economy: A Biophysical Perspective," *Science* [1984] 225:890–97. Copyright 1984 AAAS.)

time. People use machines (energy provided by fuel) for manufacturing jobs, be it a riveter or a computer assembler. Figure 1.10 illustrates the increase of productivity associated with this increased energy utilization.

The beginnings of thermodynamics in the nineteenth century coincide with the rise in energy consumption. As we reach a terminus in the nonrenewable fuel supplies, we must search for productivity gains from means other than energy. Engineering creativity is certainly needed. Figure 1.11 illustrates the near 100% dependence on nonrenewable energy sources in the United States. Using foreign supplies to replace domestic sources is no long-term solution either, as all fossil fuel supplies are finite.

Immediate actions can be taken to improve energy utilization and conservation. Because fuel costs have been very low, the manufacture of products has tended toward lowest initial cost. Operating costs were largely ignored. For instance, it is possible using current technology to build domestic refrigerators that use a maximum of 700 kilowatt-hours (kwh) per year rather than the maximum of 1900 kwh allowed in current models. California passed a law requiring that by 1993 all refrigerators achieve the greater efficiency. Engineers created models with design changes requiring more insulation and more efficient compressors, motors, condensers, and

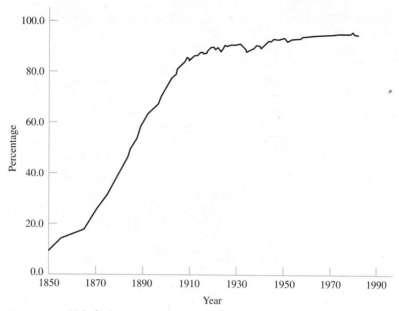

Figure 1.11 U.S. fuel consumption, percentage of nonrenewable fuel.

evaporators. These changes cost only $50 more per unit. Superefficient models are being constructed that annually consume 100 kwh, 5% of the 1900 kwh units.

Switching to natural and fluorescent lights from the predominant incandescent light bulb will save equally large amounts of electricity. A 75-W incandescent light bulb has an average lifetime of 750 hours and produces 1180 lumens of illumination; a 20-watt fluorescent light has a 7500-hour life and produces 1250 lumens of illumination. Ninety percent or more of the incandescent bulb's energy is dissipated as heat. A flow of electricity heats the bulb's filament to a high temperature, making the filament glow and produce light. Of course, this heat must be removed by a building's air-conditioning system in hot weather. Fluorescent bulbs are coated with phosphor powders that glow when excited by ultraviolet light. The ultraviolet light is created by ionizing a gas within the tube.

In the early 1970s automobiles had a comparatively low fleet average mileage, 15–16 miles per gallon (mpg). Since the energy crisis in 1973 the fleet average has risen to 25–26 mpg. This represents a tremendous savings in fuel when you consider the number of cars on the road and the miles driven. Some automobiles currently obtain 50 mpg or better. By decreasing vehicle size and weight and by using new materials, more efficient engines, and better aerodynamic design, the fleet average can continue to rise. Not only will personal transportation systems change with decreased fuel availability and/or significantly higher fuel costs, but other systems will change as well. Over the past forty years railroad transportation of goods has been replaced by trucks. The days for the mass transport of goods via trucks are numbered, however, because for long-distance hauling trains are at least five times more fuel-efficient than trucks. Compare a freight train to the number of trucks that would be needed to haul the same goods.

Populations and societies prosper because of abundant food and energy supplies. Part of the reason the United States has been able to support its standard of living at the present level is that it exports agricultural products, creating a trade surplus to partially counter the trade deficit in energy. Problems lurk in this vital sector of the economy. Farming has developed into a highly energy-intensive operation. Farms have become larger as a result of this. Agricultural engineers need to create new technologies and methodologies to support this vital economic sector. The tractors needed for tilling the soil are not the problem, although techniques in plowing dictated by large tractors cause soil erosion, the loss of a nonrenewable resource. A larger problem is chemical applications. Fertilizer, rather than naturally occurring nutrients, is used to increase crop yield per acre. Synthetic fertilizers are produced using natural gas. Pesticides and herbicides, applied to control insects and weeds, are created using fuel. Of course, the repeated applications expend energy. The pesticides, herbicides, and fertilizers pollute the water supply; much farmland is irrigated, and they become part of the runoff. The pollution of water supplies, rivers, lakes, and the ocean is caused in part by this intensive energy application to the soil in growing the U.S. food supply.

Aggravating the situation is that only 25% of U.S. crops are for human consumption. The rest are for animals. It is currently cheaper to feed animals grain rather than graze them. The meat is more tender, tastier, and higher in fat and makes up 35% of the typical American diet. As you can see, the intertwinings of various activities in support of our lives is quite complex, and energy solutions are not easy or comfortable. However, they can be accomplished.

Figure 1.12 illustrates the energy consumption and GNP per person in various countries. At one extreme is the United States and at the other is India. The energy consumption per person in the United States is about 80 kcal and in India it is about 2 kcal. However, the standard of living, indicated by the GNP, is also significantly different. Look instead at industrialized countries with a similar standard of living, but whose energy consumption is less than half that of the United States. Countries such as Norway, France, Switzerland, and the former West Germany offer energy-efficient methodologies. Engineers with thermodynamic awareness are needed to create new technologies and modify old ones so the United States may continue to prosper.

The time in which you will be a practicing engineer will be a challenging one. The energy supplies that Americans have become accustomed to will decrease dramatically. New technologies must be created to help us continue the advantages that society now offers us. A knowledge of thermodynamics is necessary for all fields of engineering. Certainly our energy habits will change. Natural and fluorescent lighting will become dominant; electricity generation will change increasingly to photovoltaic and cogeneration facilities; transportation patterns will alter; food production techniques will become less energy-intensive. All of these changes can be accomplished with talents in which engineers abound: creativity, compassion, and problem-solving ability.

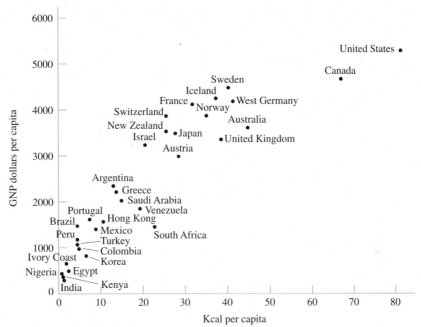

Figure 1.12 Comparison of total energy usage and GNP dollars per capita for various countries. (*Source:* C. A. S. Hall, C. J. Cleveland, and R. Kaufmann, *Energy and Resource Quality: The Ecology of the Industrial Process,* New York: John Wiley and Sons, 1986.)

PROBLEMS

1.1 Compare the operating costs of a home refrigerator that uses 700 kwh of electricity annually rather than 1900 kwh. The cost of electricity is $0.10/kwh. If this were enacted nationally, such that 10 million refrigerators were affected, what would be the total savings in kwh?

1.2 Consider the fuel savings in Problem 1.1 because less electricity would need to be generated. A power plant may be assumed 40% efficient in converting the fuel's chemical energy into electrical energy. Assuming the fuel is oil with a heating value of 43 000 kJ/kg, how many kg would be saved annually?

1.3 Recall the energy and lifetime comparisons between incandescent and fluorescent lights. Using a 75-W incandescent bulb and a 20-W fluorescent bulb, investigate the initial cost of each at a local store. Assuming that the light is used 8 hours per day annually, determine the time necessary to recover the additional cost of the fluorescent bulb.

1.4 Referring to Problem 1.3, calculate the total energy savings over the lifetime of the fluorescent bulb. Assuming that 10 million incandescent bulbs have been changed to fluorescent, calculate the fuel savings annually, using the same assumptions as in Problem 1.2.

1.5 Consider a subset of the American automotive fleet that comprises 1 million cars that are driven 10 000 miles annually. The average gasoline consumption for this fleet rises linearly from 26 mpg to 31 mpg over a 5-year period. Calculate the total fuel savings annually and cumulatively over this period.

1.6 The following data represent the proven world oil reserves as of 1989 in billions of barrels (Dept. of Commerce):

Saudia Arabia	255	Venezuela	58.5
Iraq	100	USSR	58.4
United Arab Emirates	98.1	Mexico	56.4
Kuwait	94.5	United States	25.9
Iran	92.9	China	24.9

⟹ TOTAL 864.6 BILLION BARRELS

The following data represent the consumption in millions of barrels of oil in 1989 of the largest consumers:

United States	6323	France	677
Japan	1818	Canada	643
West Germany	831	Great Britain	634
Italy	708		

Determine the number of years of oil supplies remaining, assuming the consumption remains constant. Assuming that the consumption increases at 3% per year, determine the number of years remaining.

REFERENCES

1. Lester R. Brown et al., *State of the World 1990,* Worldwatch Institute Report, New York: Norton, 1990.
2. John Gever et al., *Beyond Oil,* Cambridge, MA: Ballinger, 1986.

2

Definitions and Units

There is a certain language peculiar to any specialization; words signify certain precise concepts and once learned become a way of quickly communicating these concepts and ideas to others versed in the field. Thermodynamics is no exception, and this chapter establishes literacy in the field. These parameters will be used over and over throughout the text and are important in understanding the scope and the limitations of laws and models postulated and developed later.

In this chapter you will

- Understand the conceptualization of substances and systems;
- Discover what defines the state of a substance and how it exists;
- Differentiate between fundamental and derived units;
- Examine the limits of classical, macroscopic, thermodynamic analysis;
- Define and use specific volume and pressure in rigorous thermodynamic fashion;
- Explore equality of temperature and the zeroth law of thermodynamics and its implicit invocation in the development of temperature scales.

2.1 MACROSCOPIC AND MICROSCOPIC ANALYSIS

This text deals with macroscopic, as opposed to microscopic or statistical, thermodynamics. In microscopic thermodynamics we must look at every molecule and analyze collective molecular action by statistical methods. In macroscopic thermodynamics we concern ourselves with the overall effect of the individual molecular interactions. The macroscopic level is the level on which we live. We measure distance in meters and time in seconds. These measurements are very large compared with the measurement of events on the molecular level; because they are very large, they are called *macroscopic*. It is convenient to use this method of analysis because we have been familiar with the measurements since childhood. For instance, temperature is a macroscopic effect, and more will be said of it later in the chapter. The microscopic point of view will be used only to explain some phenomena that cannot be understood by classical (macroscopic) means.

What follows will be illustrations of the jargon of thermodynamics.

2.2 SUBSTANCES

To solve problems the parts of the problem must be enumerated. The first consideration is that there must be something performing the energy transformation. This something is a *substance*. This substance in the family automobile engine is usually the mixture of gasoline and air. In a steam turbine the substance is probably the steam. The substance may be a solid, liquid, vapor, or a mixture.

The substance may be further divided into subcategories. The substance is a pure substance if it is homogeneous in nature—that is, it must have the same chemical and physical composition at one point as it does at another point. It must not undergo chemical reaction. What would happen if it did? There would be at least two substances present, the reactant and the product, and therefore it would not meet the criteria of homogeneity. A mechanical mixture would also violate the homogeneity requirement. Oxygen by itself is a pure substance, but when mixed with another pure substance, say, nitrogen, the mixture is no longer a pure substance. Air, therefore, is a mixture, not a pure substance. It may seem that at normal temperatures and pressures the mixture of oxygen and nitrogen is the same composition everywhere. This is true. However, oxygen and nitrogen condense at different temperatures, so it is possible to have conditions under which air is not homogeneous throughout; it is a mixture and not a pure substance.

2.3 SYSTEMS—FIXED MASS AND FIXED SPACE

A substance does not exist alone; it must be contained. This brings us to the concept of a *system*. In thermodynamics a system is defined as any collection of matter or space of fixed identity, and the concept is one of the most important in thermodynamics. The selection of the system is up to the individual and requires some skill. Later we will see some problems associated with determining system boundaries.

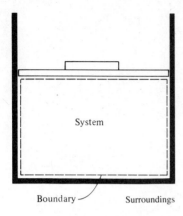

Figure 2.1 Piston-cylinder, illustrating system, boundary, and surroundings.

What does the term *boundary* mean? When a system is defined, say the fluid in a cylinder (Figure 2.1), what separates the fluid from the cylinder wall and the piston and everything external to the piston-cylinder? It is the system boundary. Everything not in the system is called the *surroundings*. Note that the piston can be raised or lowered, but the system, matter of fixed identity, is constant. Let us backtrack a moment. A system is defined as a collection of matter or space of fixed identity. You ask, "How do I know whether to choose matter or space as a system?" Consider a long pipe (Figure 2.2) filled with a flowing fluid. Pass imaginary planes through the pipe at points 1 and 2 at the same instant and consider the fluid between them as the system. This mass of fluid is our system; each molecule is identified. Now consider the problem of determining what happens to the fluid as it flows from 1 to 3. The molecules of our system become mixed with molecules from the surroundings, so we must account for the movements of each molecule. This certainly is not feasible. What if the system were defined differently for this problem? Consider our system to be a collection of space and, for this problem, consider it to be all the space in the pipe between the planes at points 1 and 3. Our system is fixed space; it is not moving, even though fluid is moving through the space. In analyzing the fluid, all we need to do is note the state of the fluid at the system boundaries at planes 1 and 3, and any changes are what has happened to the fluid in the pipe.

What conclusions can we draw from this? In the case of the piston-cylinder, where there was no flow of matter, the system was considered to be matter of fixed identity. No problem in analysis occurred even when the volume changed. This is called a *closed system,* a system closed to matter flow. When there is matter flow, the system is considered to be a volume of fixed identity, a control volume. This system is open to the flow of matter and is called an *open system.* It may seem as if the

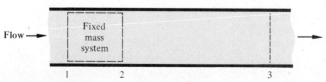

Figure 2.2 Flow through a pipe.

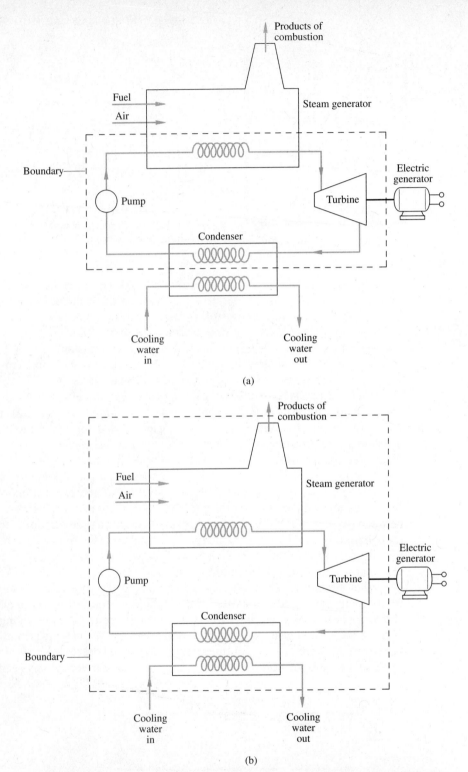

Figure 2.3 A power plant with boundary drawn for (a) a closed system and (b) an open system.

definition of a system has been greatly belabored, but a clear understanding of the two concepts eliminates a great deal of difficulty in performing analysis.

In Chapter 1 we discussed power plants for the generation of electricity. Figure 2.3a is a schematic view of a power plant with the boundary drawn so the system is closed. (No mass enters or leaves the system.) The water flows as steam to the turbine, is condensed to a liquid, and is pumped to the steam generator. The processes repeat in a cycle.

The system does not care where the energy comes from to raise its temperature and pressure in the steam generator; it may be from the combustion of coal, oil, gas, or a nuclear reaction. The water receives energy, and its temperature and pressure increase. If the boundary is drawn as in Figure 2.3b, mass flows across the system boundary; air, fuel, and cooling water enter, and products of combustion and cooling water leave. The water in the steam power plant undergoing the cycle remains a constant mass, but we may consider other aspects in the generation of power. For instance, we may be concerned about the impact on marine life when the cooling water leaving the condenser reenters at a higher temperature the river or lake from which it was drawn. In Chapter 15 we will use the system diagram shown in Figure 2.3b to coordinate the amount of fuel consumed with the products of combustion leaving the steam generator and the amount of power produced in the power plant.

2.4 PROPERTIES, INTENSIVE AND EXTENSIVE

The *state* of a substance completely describes how the substance exists. It has temperature, pressure, density, and other macroscopic properties, and by knowing these properties we can determine the state of the substance. *A property is a characteristic quality of the entire system and depends not on how the system changes state but only on the final system state.* Before looking further at the definition of a property, let us subdivide properties into two classes, *intensive* properties and *extensive* properties. An extensive property depends on the size or extent of the system, for example, mass and volume. An intensive property is independent of the size of the system, for example, temperature and pressure. Note that an extensive property per unit mass, such as specific volume (the volume divided by the mass), is an intensive property.

Getting back to the general definition of a property, we see that a property must be characteristic of the system. Consider a column of water. The temperature can be the same everywhere, and hence it can be a system property in this case. The pressure cannot be uniform throughout the system — it is greater on the bottom than on the top — so the pressure cannot be a property of this system. The definition of what constitutes a property goes further and requires that, if a system goes from one state to another state, a property must depend only on the state, not on the changes the system underwent from one state to the other. For example, let us consider the system in Figure 2.1. Let it have a temperature T_1, a pressure p_1, and a specific volume v_1. Add more weight to the piston, and the system will reach a new state characterized by a temperature T_2, a pressure p_2, and a specific volume v_2. Consider, now, that in going from 1 to 2, weights were added and subtracted, finally resulting in the same state 2. The final temperature, pressure, and specific volume must be the same as they were when just a single weight was added to achieve state 2, since these properties depend only on the system state.

System properties are defined only when they are uniform throughout a closed system; the system is then in equilibrium with respect to that property. For instance, the water in the column is in thermal equilibrium because temperature is uniform, but the water is not in mechanical equilibrium because the pressure is not uniform. The air in Figure 2.1 is in both thermal and mechanical equilibrium. There is an equilibrium associated with each intensive property, but we are primarily concerned with mechanical and thermal equilibrium. Chemical equilibrium must be considered in some instances, as delineated in Chapter 12.

From the definition of properties we know that all the states we can potentially analyze must be equilibrium states, or there would be no properties defined. When a closed system is in thermodynamic equilibrium, the properties are uniform throughout the system and may be defined. Thus thermodynamics is a misnomer, for we can define properties, the tools of analysis, only when they are at equilibrium. We shall discover, however, that this does not present a very great hardship in problem solving.

2.5 PHASES OF A SUBSTANCE

Certain groups of states of a substance may be called *phases* of that substance. Water has solid, liquid, and vapor phases; any pure substance may exist in any combination of the phases. There are specific terms to characterize phase transitions. *Melting* occurs when a solid turns to a liquid; *freezing* or *solidifying* occurs when a liquid turns to a solid. *Vaporization* occurs when a liquid turns to a gas (vapor); *condensation* occurs when a gas (vapor) turns to a liquid. *Sublimation* occurs when a solid turns to a gas.

2.6 PROCESSES AND CYCLES

A *process* is a change in the system state. Just as there are an infinite number of ways to go between two points (Figure 2.4), so too are there an infinite number of ways for a system to change from state 1 to state 2. The path describes the infinite number of system states that occur when a system undergoes a particular process from state 1 to state 2. A thermodynamic *cycle* is a collection of two or more processes for which the initial and final states are the same. The cycle in Figure 2.4 illustrates a closed system that undergoes process A and process B. The system is returned to state 1 from which it started. As an example of a simple cycle, consider an isothermal, or constant-temperature, process from state 1 to state 2. By denoting that the process is isothermal, the path the process will follow is defined. Let path A be the isothermal path as denoted on Figure 2.4. Since a cycle is two or more processes returning to the initial state, let the return process be isothermal also. The system will proceed from state 2 to state 1 along path A, but in the opposite direction. Is this the same path? No, because by definition a path is a succession of system states, and the succession from 1 to 2 differs from the succession from 2 to 1.

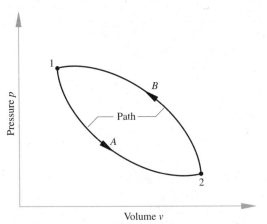

Figure 2.4 Graphical illustration of a path's being a succession of system states.

2.7 UNIT SYSTEMS

Engineers do more than work with numbers. In most instances numbers are associated with dimensions. We use these numbers to characterize systems by their length, mass, temperature, and other properties. Unit systems assign numerical values to these dimensions, and we are normally concerned with using these dimensions in conjunction with the unit systems, merging the two concepts in our minds. When they are combined into a coherent set, they form a measurement system. In this section we will discuss two measurement systems, the English engineering system and the SI metric system. These two systems are the most frequently encountered in engineering practice and in college texts. Occasionally, other systems are used; if you understand the basis for these two, however, you will understand the others.

Fundamental and Derived Units

Units that are postulated, or defined, are *fundamental units.* Once a few units are defined, others can be derived from them. Fundamental units include length, time, mass, and force.

Length (L) is the distance between two points in space.

Time (t) is the period between two events or during which something happens.

Mass (m) is the quantity of matter that a substance is composed of. It is invariant with location: the mass of a person is the same on earth and on the moon.

Force (F) is sometimes viewed as fundamental (as it is in the English engineering system) and sometimes derived (as in the SI system). Force is defined through Newton's second law, but conceptually it is associated with power or strength, the push or pull on an object. It may act directly, such as by pushing a wheelbarrow, or indirectly by fields (gravitational, electrical, or magnetic).

Other units can be derived from fundamental units. *Area* and *volume* are derived from length, L^2 and L^3, respectively. *Velocity* is distance per unit time, or L/t, and *pressure* is force per unit area, or F/L^2.

English Engineering System

The English engineering system developed in a rather haphazard fashion, with many common measures, like the yard, descended from such standards as the length from the thumb to the nose of Henry I of England. Clearly, such a system has many drawbacks, but its singular advantage is that we know it, having grown up with feet and pounds. Having force and mass as fundamental units is confusing, as they are related through Newton's second law, which states that the force acting on a body is equal to the time rate of change of momentum and is given by

$$F = \frac{d}{dt}(mv) \tag{2.1}$$

where v is the velocity and m is the mass of the body. This may be more familiar in derivative form:

$$F = v\frac{dm}{dt} + m\frac{dv}{dt}$$

Most often the mass is constant, and the change of velocity with respect to time is acceleration, *a*, so

$$F = ma \tag{2.2}$$

where F is measured in pounds force (lbf), m is measured in pounds mass (lbm), and *a* is measured in feet per second squared (ft/sec²). One lbf is defined as 1 lbm accelerated 32.174 ft/sec². When we substitute the values of the definition into equation (2.2), a problem appears in the unit balance of the equation:

$$lbf = (lbm)(ft/sec^2)$$

The units on one side do not equal the units on the other side, so let's multiply equation (2.2) by a constant with units that balance:

$$F = kma \tag{2.3}$$

or with the units included,

$$lbf = k(lbm)(ft/sec^2)$$

Thus, k must be in lbf-sec²/lbm-ft. To determine the value of the constant, we substitute the definition of 1 lbf in equation (2.3) and solve for k:

$$k = \frac{1}{32.174}\frac{lbf\text{-}sec^2}{lbm\text{-}ft} \tag{2.4}$$

Since equation (2.3) is valid for all cases, if k is determined for a given case, it must be true for all the other cases. Standard gravitational acceleration, *g*, is 32.174 ft/sec², a term with the same numeric value as $1/k$ but with different units. The term k equals $1/g_c$, where

$$g_c = 32.174 \text{ lbm-ft/lbf-sec}^2$$

and thus equation (2.3) becomes

$$F = ma/g_c \text{ lbf} \tag{2.5}$$

Example 2.1

Consider that a person seated in a car is uniformly accelerated at 5 ft/sec² and that the person has a mass of 180 lbm. Find the horizontal force acting on the person.

Solution

Given: A person is accelerated horizontally.

Find: The horizonal force as a result of this acceleration.

Sketch and Given Data:

$\frac{dv}{dt} = a = 5$ ft/sec²
$m = 180$ lbm   **Figure 2.5**

Assumptions: The acceleration is uniform and constant.

Analysis: Substitute into equation (2.5) to determine the horizontal force.

$$F = ma/g_c = \frac{(180 \text{ lbm})(5 \text{ ft/sec}^2)}{32.174 \dfrac{\text{lbm-ft}}{\text{lbf-sec}^2}} = 27.97 \text{ lbf}$$

Comments: Force acts in the direction of the applied acceleration or deceleration. Thus, if the acceleration acts in more than one dimension, there will be force components in those directions, with the resultant force being a vector in the same direction as the acceleration vector.

Thus far we have not mentioned weight. If the person in the Example 2.1 is standing on the ground, he experiences a force equal and opposite to the force he exerts on the earth. This force is his weight. He is acted upon by the local gravitational field g. If the gravitational field is located so that g has a value of 32.174 ft/sec², the

TABLE 2.1   UNITS IN THE ENGLISH ENGINEERING SYSTEM

Fundamental		Derived	
Quantity	Unit	Quantity	Unit
force (F)	lbf	acceleration (L/t^2)	ft/sec^2
length (L)	ft	area (L^2)	ft^2
mass (m)	lbm	density (m/L^3)	lbm/ft^3
temperature (T)	°R	energy (FL)	ft-lbf
time (t)	sec	power (FL/t)	(ft-lbf)/sec
		pressure (F/L^2)	lbf/ft^2
		volume (L^3)	ft^3

standard gravitational acceleration, then the force is

$$F = \frac{(180 \text{ lbm})(32.174 \text{ ft/sec}^2)}{32.174 \text{ lbm-ft/lbf-sec}^2} = 180 \text{ lbf.}$$

If the man were standing on a tall mountain where the gravitational field g had a value of 30.0 ft/sec^2, the force exerted on the ground, his weight, would be

$$F = \frac{(180 \text{ lbm})(30.0 \text{ ft/sec}^2)}{32.174 \text{ lbm-ft/lbf-sec}^2} = 167.8 \text{ lbf.}$$

In space, weightlessness results from $g = 0$; hence the force or weight of the body mass is zero.

Table 2.1 lists the fundamental units and some of the derived units found in the English engineering system.

SI Units

The abbreviation *SI* stands for Système International d'Unités, a consistent unit system developed in 1960. In 1975 the metric conversion act was signed into law in the United States, declaring in part "a national policy of coordinating the increasing use of the metric system in the United States." The process of conversion has been slow, but as the U.S. business outlook becomes more global, the United States will convert of necessity to metric sizes. It is not difficult to perform calculations in metric; in fact it is easier than in English engineering units. The difficulty in the conversion process is that standard sizes of equipment, pipes, bolts, and parts of all kinds must be changed. This requires a major investment in new manufacturing tools, which businesses must eventually pay for if they are to compete in the world marketplace, where metric is standard.

There are seven physically defined, fundamental or base SI units, and two geometrically defined supplementary ones; the rest are derived. The fundamental units are listed in Table 2.2.

TABLE 2.2   FUNDAMENTAL UNITS IN SI

Quantity	Name	Symbol
Base units		
Length	meter	m
Mass	kilogram	kg
Time	second	s
Electric current	ampere	A
Thermodynamic temperature	kelvin	K
Amount of substance	mole	mol
Luminous intensity	candela	cd
Supplementary Units		
Plane angle	radian	rad
Solid angle	steradian	sr

Force is a derived SI unit, where one unit of force is defined by Newton's second law, equation (2.2). A unit force is called a newton (N), so

$$1 \text{ N} = (1 \text{ kg})(1 \text{ m/s}^2)$$

In SI units, mass and force have not only different names but different numerical values. For instance, the force exerted on a 40-kg person by standard gravitational acceleration, 9.8 m/s², is

$$F = (40 \text{ kg})(9.8 \text{ m/s}^2) = 392 \text{ N}$$

Table 2.3 shows the prefixes that indicate magnitude of the SI units. Table 2.4 lists SI derived units.

Although SI associates an exact meaning to each symbol, this preciseness does not allow variations from the rules. For instance, in English units ft-lbf and lbf-ft have the same meaning, though the first expression is preferred. In SI N m and mN are entirely different; the former is newton meter and the latter is millinewton. Notice that a space was left between N and m in the first term, the correct way to write this combination of units. The following is a list of some important rules in using SI:

1. Unit symbols are roman letters, lower- or uppercase, as shown in Tables 2.2 and 2.4. Periods are not used after a symbol except at the end of a sentence. Thus, we write N and kg, not N. and KG. Italic letters are used for quantity symbols, thus, m is mass and m is meter. The symbol is the same in the singular and plural: 1 kg, 10 kg.
2. The prefix precedes the symbol in roman type without a space, and double prefixes are not allowed: μA, not mmA.
3. Engineering and architectural drawings use millimeter as the basic unit of length.
4. The product of two or more units is represented by a space between the units or the use of a dot between the units: N·m or N m.

5. In the division of terms, it is ambiguous to use the slash mark or the hyphen more than once on the same line. In the case of acceleration, m/s^2 or $m\text{-}s^{-2}$ are acceptable, but m/s/s is vague. It could be interpreted that the seconds cancel and only meters remain.
6. Capital and lower case letters must be used as described, for instance K means degrees kelvin and k means kilo.
7. Do not use commas in writing long numbers. In some countries commas are used in place of decimal points. Use a space after every third digit to the left and right of the decimal point. If there are only four digits, the space is optional. Thus, you would write 3 600 000 and 21.005 68, not 3,600,000 or 21.00568.
8. When writing a number less than one, always use a zero before the decimal point: 0.625, not .625.

In handwriting it may be difficult to differentiate symbols, especially roman and italic letters. Adopt a different symbol for clarity, such as a script m for mass, so your notes are not confusing.

Conversion Factors

It will remain necessary to convert between unit systems for many years, as different unit systems are in use around the world. Inside the front cover of this book is a list of conversion factors for SI and English units. All conversion factors are dimensionless;

TABLE 2.3 SI UNIT PREFIXES

Factor by which unit is multiplied	Prefix	Symbol
10^{-18}	atto	a
10^{-15}	femto	f
10^{-12}	pico	p
10^{-9}	nano	n
10^{-6}	micro	μ
10^{-3}	milli	m
10^{-2}	centi[a]	c
10^{-1}	deci[a]	d
10^{1}	deka[a]	da
10^{2}	hecto[a]	h
10^{3}	kilo	k
10^{6}	mega	M
10^{9}	giga	G
10^{12}	tera	T
10^{15}	peta	P
10^{18}	exa	E

[a] Avoid these prefixes, except for centimeter.

TABLE 2.4 SI DERIVED UNITS

Quantity	Unit	SI symbol	Formula
Acceleration	meter per second squared	—	m/s^2
Angular acceleration	radian per second squared	—	rad/s^2
Angular velocity	radian per second	—	rad/s
Area	square meter	—	m^2
Density	kilogram per cubic meter	—	kg/m^3
Electric capacitance	farad	F	$A \cdot s/V$
Electrical conductance	siemens	S	A/V
Electric field strength	volt per meter	—	V/m
Electric inductance	henry	H	$V \cdot s/A$
Electric potential difference	volt	V	W/A
Electric resistance	ohm	Ω	V/A
Electricity, quantity	coulomb	C	$A \cdot s$
Energy	joule	J	$N \cdot m$
Entropy	joule per kelvin	—	J/K
Force	newton	N	$kg \cdot m/s^2$
Frequency	hertz	Hz	$1/s$
Heat, quantity	joule	J	$N \cdot m$
Heat, specific	joule per kilogram-kelvin	—	$J/kg \cdot K$
Illuminance	lux	lx	lm/m^2
Luminance	candela per square meter	—	cd/m^2
Luminous flux	lumen	lm	$cd \cdot sr$
Magnetic field strength	ampere per meter	—	A/m
Magnetic flux	weber	Wb	$V \cdot s$
Magnetic flux density	tesla	T	Wb/m^2
Power	watt	W	J/s
Pressure	pascal	Pa	N/m^2
Radiant intensity	watt per steradian	—	W/sr
Stress	pascal	Pa	N/m^2
Thermal conductivity	watt per meter-kelvin	—	$W/m \cdot K$
Velocity	meter per second	—	m/s
Viscosity, dynamic	pascal-second	—	$Pa \cdot s$
Viscosity, kinematic	square meter per second	—	m^2/s
Volume	cubic meter	—	m^3
Work	joule	J	$N \cdot m$

hence we can multiply or divide any equation by them and not change its value. Let's examine the conversion between inches and centimeters, 1 in. = 2.54 cm. Divide this equation by 1 in., yielding 1 = 2.54 cm/in. Thus, 2.54 cm/in. is equal to a dimensionless 1.

Example 2.2

A cargo ship has tanks for carrying fuel oil. The oil is measured in barrels, while the tank dimensions are in meters, 1 m × 5 m × 15 m. How many barrels does the tank hold? How many gallons?

Solution

Given: The size of a ship's tank in meters.

Find: The number of barrels and gallons that the tank volume holds.

Sketch and Given Data:

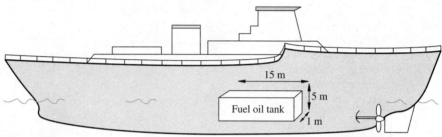

15 m

5 m

Fuel oil tank

1 m

Figure 2.6

Assumptions: None.

Analysis: Calculate the volume in cubic feet, then find in the conversion tables the value to convert from cubic meters to barrels.

$$V = 1 \times 5 \times 15 = 75 \text{ m}^3$$

The conversion table does not have a direct conversion factor, so we must convert from cubic meters to gallons and from gallons to barrels.

$$V = (75 \text{ m}^3)(2.642 \times 10^2 \text{ gal/m}^3) = 19\,815 \text{ gal}$$

Again we must calculate a new conversion factor, as we are given that 42 gal equals 1 bbl. The conversion factor is 1/42 bb/gal and the volume in barrels is

$$V = (19\,815 \text{ gal})/(42 \text{ gal/bbl}) = 471.8 \text{ bbl}$$

Comments: Often a direct conversion factor is not available, and we must perform intermediate steps.

Unit Conversions in TK Solver

TK Solver is a general-purpose equation-solving software package. Its ability to solve sets of nonlinear simultaneous equations makes it an excellent tool for solving thermodynamics problems. TK Solver also conveniently converts the units of variables used in a problem solution. Conversions are entered by selecting the Unit Sheet [= U] and entering the conversion factors and offsets (if required). TK Solver will automatically combine two conversions to produce a third. As long as the required conversions are entered, TK Solver will combine them as needed.

   Contained on the disk included with your text is the UNITS.TK file. See Figure 2.7 for a listing. It contains the most commonly used units conversions for thermo-

```
═══════════════════════ UNIT SHEET ═══════════════════════
From ─────── To ─────── Multiply by ─ Add Offset ───
degF          degR                       459.67
degC          degK                       273.15
degK          degR          1.8
psia          kPa           6.8948
psia          inHg          2.0359
MPa           kPa           1000
kPa           Pa            1000
ft3/lbm       m3/kg         .62428
BTU/lbm       kj/kg         2.3261
B/lbm-R       kj/kg-K       4.1868
kW            W             1000
hp            ft-lbf/s      550
hp            W             745.7
BTU/sec       kW            1.0551
BTU/sec       BTU/hr        3600
BTU           ft-lbf        778.169
kJ            J             1000
kg            lbm           2.20462
slug          lbm           32.1739
ft            m             .3048
ft/sec        m/sec         .3048
ft            in            12
lbf           N             4.4482
N             dyne          100000
m2            ft2           10.764
ft2           in2           144
m2            cm2           10000
m3            ft3           35.3147
m3            l             1000
gal           in3           231
ft3           in3           1728
l             cm3           1000
```

Figure 2.7 Listing of UNITS.TK.

dynamics problems. UNITS.TK is included as part of all the property models included on the disk. It can be added to a model by loading it "on top" of the model already in the computer.

Once information has been entered into the Unit Sheet of TK Solver, typing over the units listed with the ones desired involves a conversion. Try the following:

1. Load UNITS.TK into TK Solver.
2. Change to the Variable Sheet [= V].
3. Enter "Power" as the variable name, "1" as the input value, and "kW" as the unit.
4. Successively change the units of "Power" by typing the following one after another in the Unit column: W, hp, BTU/sec, BTU/hr, ft-lbf/s. You should get results like those in Figure 2.8.

For further information on units in TK Solver, see Appendix D and your TK Solver manual.

```
═══════════════════════════ VARIABLE SHEET ═══════════════════════════
St Input ──── Name ──── Output ──── Unit ──── Comment ─────────────
   1          Power                 kW


═══════════════════════════ VARIABLE SHEET ═══════════════════════════
St Input ──── Name ──── Output ──── Unit ──── Comment ─────────────
   1000       Power                 W


═══════════════════════════ VARIABLE SHEET ═══════════════════════════
St Input ──── Name ──── Output ──── Unit ──── Comment ─────────────
   1.3410219  Power                 hp


═══════════════════════════ VARIABLE SHEET ═══════════════════════════
St Input ──── Name ──── Output ──── Unit ──── Comment ─────────────
   .94781699  Power                 BTU/sec


═══════════════════════════ VARIABLE SHEET ═══════════════════════════
St Input ──── Name ──── Output ──── Unit ──── Comment ─────────────
   3412.1412  Power                 BTU/hr


═══════════════════════════ VARIABLE SHEET ═══════════════════════════
St Input ──── Name ──── Output ──── Unit ──── Comment ─────────────
   737.56202  Power                 ft-lbf/s
```

Figure 2.8 Units conversion with TK Solver.

2.8 SPECIFIC VOLUME

There are several properties that are familiar to us but must be rigorously defined so we know exactly what is meant. *Specific volume* is the volume of a substance divided by its mass. But is there a point at which this is no longer true? Would specific volume have a meaning if we were to select a single molecule as the substance we want to measure? No. First, specific volume is a macroscopic phenomenon; second, for a property to be macroscopic, a continuum must exist. That is to say, macroscopic properties must vary continuously from region to region without a discontinuity. We say region instead of point because a point is too small (it is infinitesimal) to contain enough particles of substance; the smallest space a macroscopic substance can occupy and still have macroscopic properties is a region with a characteristic volume, $\delta V'$. If the specific volume is v, and δV is a small substance volume with mass δm, then

$$v = \lim_{\delta V \to \delta V'} \frac{\delta V}{\delta m} \tag{2.6}$$

Equation (2.6) is precise, but it could prove to be a bit tedious to work with. If the system substance of which the specific volume is desired is of finite size and homogeneous, then the specific volume at one part of the system is the same as at another part of the system, and for a system volume V and mass m, the specific volume is

$$v = \frac{V}{m} \ \text{m}^3/\text{kg} \quad [\text{ft}^3/\text{lbm}] \tag{2.7}$$

There is also a property known as *density* (ρ), which may be defined as the mass divided by the volume, or the reciprocal of the specific volume; that is,

$$\rho = \frac{1}{v} \ \text{kg/m}^3 \quad [\text{lbm/ft}^3] \tag{2.8}$$

2.9 PRESSURE

The second property to be rigorously defined is *pressure*. If we consider a small cubical volume $\delta V'$ of a substance, there will be forces acting normally in the x, y, and z directions as well as forces acting at angles to these directions. In solids these pressures are known as the stress of the solid. In fluids at equilibrium the nonnormal forces are zero, and the pressure that is measured is the average of pressures, that is, forces per unit area, acting in the three directions. If the average normal force is δF_n and the area that is characteristic of $\delta V'$ is $\delta A'$, then the pressure p is

$$p = \lim_{\delta A \to \delta A'} \frac{\delta F_n}{\delta A} \tag{2.9}$$

As in the case of specific volume, if the substance in the system is homogeneous and at equilibrium, then the pressure acting on a finite area A with normal force F is

$$p = \frac{F}{A} \ \text{N/m}^2 \quad [\text{lbf/in.}^2] \tag{2.10}$$

The units of pressure in the English system are normally pounds force per square inch (psi); but there are instances where pounds force per square foot (psf) are used, so the conversion of square inches to square feet is necessary:

$$(10 \ \text{lbf/in.}^2)(144 \ \text{in.}^2/\text{ft}^2) = 1440 \ \text{lbf/ft}^2$$

The unit of pressure in SI is defined as one unit force per unit area, or 1 pascal (Pa). Thus,

$$1 \ \text{Pa} = 1 \ \text{N/m}^2 \tag{2.11}$$

The pressure that is measured most often is the difference between the pressure of the surroundings and that of the system. Consider, for instance, the Bourdon tube pressure gage, shown schematically in Figure 2.9. The pressure of the fluid acts against the inside tube surface, causing the tube to move. This movement is relayed through linkages to a pointer, which, when a calibrated dial is attached, indicates the difference in pressure between the system fluid inside the tube and the surrounding fluid outside the tube. This pressure is called *gage pressure*. The absolute pressure of the system is the sum of the system gage pressure and the surrounding absolute pressure:

$$p_{\text{abs}} = p_{\text{gage}} + p_{\text{surr}} \tag{2.12}$$

Most often the pressure gage is located in a room, and the pressure acting on it is atmospheric pressure. The air in the atmosphere has mass that is acted upon by

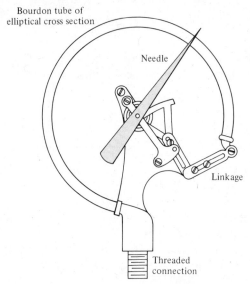

Bourdon tube of
elliptical cross section

Needle

Linkage

Threaded
connection

Figure 2.9 Bourdon tube pressure gage.

gravity, and this force per unit area has an average value of 101 325 Pa at sea level. Because this is a very large number to work with, we make use of the prefixes in Table 2.3; thus

$$101\ 325\ \text{Pa} = 101.325\ \text{kPa (kilopascals)}$$
$$= 0.101\ 325\ \text{MPa (megapascals)}$$

Very often atmospheric pressure is assumed to be 100 kPa for simplicity. In English units atmospheric pressure is equal to 14.696 psia.

What happens when the gage pressure is zero? Then the absolute system pressure is equal to that of the surroundings. Should the gage pressure be less than the surrounding pressure, it is considered negative, in that the reading is subtracted from the surrounding absolute pressure to obtain the system absolute pressure. A pressure less than atmospheric pressure is called a *vacuum pressure,* or vacuum.

Let us consider the sketch in Figure 2.10, where gage A reads 200 kPa (gage), gage B reads 120 kPa (gage), and the atmospheric pressure is 101.3 kPa. Find the absolute pressure of the air in box B and box A.

$$p_{B_{abs}} = p_{B_{gage}} + p_{surr}$$
$$= 120\ \text{kPa (gage)} + 101.3\ \text{kPa} = 221.3\ \text{kPa}$$
$$p_{A_{abs}} = p_{A_{gage}} + p_{surr}$$
$$= 200\ \text{kPa (gage)} + 221.3\ \text{kPa} = 421.3\ \text{kPa}$$

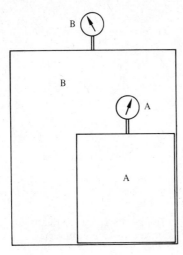

Figure 2.10 Two containers, one within the other, at different pressures.

Figure 2.11 depicts the various pressures. The negative pressure, or vacuum, is often measured by a manometer, which has readings of millimeters of mercury or whatever other fluid the manometer may contain. This pressure may also be called kPa (vacuum).

Consider Figure 2.12, in which a manometer is used to measure the system pressure in the container. The system has a pressure p, the fluid in the manometer has a density ρ, and the surroundings are atmospheric with pressure p_{atm}. The difference in pressure between the system and the atmosphere is able to support the fluid in the manometer with density ρ for a distance L. This may be expressed by

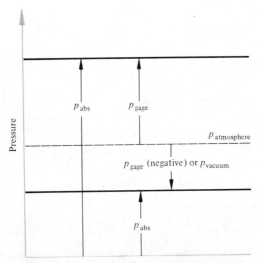

Figure 2.11 Graphical representation of pressure.

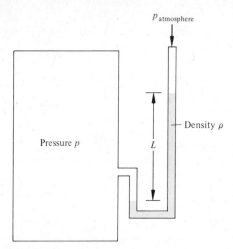

Figure 2.12 Diagram relating pressure and density in a manometer.

$$p - p_{atm} = \Delta p = (\rho \text{ kg/m}^3)(L \text{ m})(g \text{ m/s}^2) = \text{N/m}^2$$

$$\Delta p = \rho L g \text{ Pa}$$

Note that the units in the equation balance; the checking of units is a necessary step in all engineering practice, especially in thermodynamics.

In English units

$$\Delta p = (\rho \text{ lbm/ft}^3) \, (L \text{ ft}) \, \frac{g \text{ ft/sec}^2}{g_c \text{ lbm-ft/lbf-sec}^2}$$

$$\Delta p = \rho L \frac{g}{g_c} \text{ lbf/ft}^2$$

There are several units for pressure, and it is necessary to know the equivalence between them. For instance,

$$14.696 \text{ lbf/in.}^2 = 1 \text{ atmosphere (atm)} = 760 \text{ mm Hg} = 29.92 \text{ in. Hg}$$

One atmosphere of pressure exerts a force of 101.325 kPa, or supports a column of mercury 760 mm high. A pressure less than atmospheric could not support so high a column of mercury, and zero pressure absolute could not support the column at all. This is said to be a perfect vacuum. Looking at Figure 2.13, we see that when the system has zero pressure, the height L to which the atmospheric pressure can push the mercury is 760 mm. The system pressure would be

$$0 \text{ kPa} = 760 \text{ mm Hg (vacuum)} = 101.325 \text{ kPa (vacuum)}$$

$$= 29.92 \text{ in. Hg (vacuum)}$$

Pressure is often measured by means other than a manometer or a Bourdon tube gage. Two of the most frequently encountered devices measure a change in an

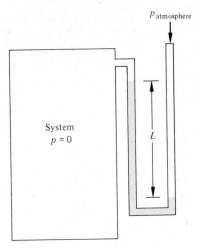

Figure 2.13 Schematic diagram illustrating the concept of inches of mercury vacuum.

electrical property, such as induced voltage or resistance. Certain solid materials exhibit a piezoelectric effect when subjected to a force, such as pressure, that causes the material to deform. This piezoelectric effect is manifested by a change in the crystal's charge, or voltage, proportional to the pressure applied. Other sensors relate the force applied to changes in electrical resistance, inductance, or capacitance. Having a voltage signal corresponding to pressure fits ideally with computerized data acquisition systems, where the signals must be in the form of voltages.

2.10 EQUALITY OF TEMPERATURE

Temperature cannot be defined at this point, but equality of temperature can. Consider two blocks of material, say, iron; if these two blocks are brought together and there is no change in any observable property, then the two blocks are said to be in thermal equilibrium and their temperature is the same. But if one block has a higher temperature than the other, it will also have a different stress, a different electrical resistivity, and a different density. When the blocks are brought together, heat will flow from the hotter to the colder body, causing a change in these three properties; when thermal equilibrium is reached, no more property changes occur, and the temperatures are equal. Note that temperature equality is measured by changes in other properties. This will have greater significance when we discuss thermometers.

2.11 ZEROTH LAW OF THERMODYNAMICS

The zeroth law of thermodynamics states that when two bodies are in thermal equilibrium with a third body, they are in thermal equilibrium with each other and hence are at the same temperature. Figure 2.14 illustrates this law. In this case the third body is a thermometer. As a result of the zeroth law, a third body, the thermometer, may be used for relating temperatures of two bodies without bringing them in contact with each other, since they are both related to the thermometer. It is the thermometer's scale that permits such a comparison.

2.14 Illustration of the equality of temperature, the zeroth law of thermodynamics.

2.12 TEMPERATURE SCALES

Many devices, from hands to glass bubbles, have been used for measuring temperature, but the most common thermometers are sealed glass capillary tubes filled with colored alcohol or mercury. Note that the thermometer relies on changes in another property, density, to indicate a temperature. There have been many researchers in the field of thermometry, but for the most part they have based the temperature scales on two points: the triple point of water (where solid, liquid, and vapor coexist) and the boiling point of water at 1 atm. (See Figure 2.15.)

One of the common temperature scales, the Fahrenheit scale, has a slightly different beginning. Daniel Gabriel Fahrenheit, a German physicist, was interested in thermometry and visited an astronomer friend, Rømer, who devised a temperature scale of 60 degrees. There are 60 seconds to a minute, 60 minutes in an hour, so why not 60 degrees for a temperature scale? The lower limit was an ice, salt, and water mixture that corresponded to 0 degrees Fahrenheit. Anyway, Fahrenheit thought the scale too rough, and increased the number of divisions to 240; why not 360 divisions remains a mystery. He originally set the human body temperature to be 90 but later decided it should be 96; the scale's temperature having been fixed by body temperature, and an ice-salt-water mixture, the triple point became 32 degrees, and water boiled at 212 degrees. To eliminate these inconvenient numbers, a Swedish astronomer by the name of Anders Celsius devised a scale that started at 100 (triple point) and went to 0 (boiling point). A friend suggested he reverse them, which resulted in the current centigrade or Celsius temperature scale, with 0 at the triple point and 100 at the boiling point.

A Frenchman named Guillaume Amontons first proposed that there was an absolute zero temperature. He was rather ignored, but in 1851 Lord Kelvin developed the idea further and devised an absolute temperature scale, the Kelvin scale. A second absolute temperature scale is the Rankine scale. The Kelvin scale has the same

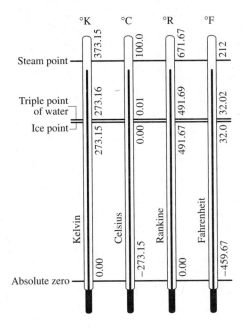

Figure 2.15 Graphical illustration of four temperature scales.

degree division as the Celsius scale, whereas the Rankine scale uses the Fahrenheit degree divisions. The relationship between the scales is

$$T_K = T_C + 273.15$$

$$T_R = T_F + 459.67$$

where T_K is degrees Kelvin, T_C is degrees Celsius, T_R is degrees Rankine, and T_F is degrees Fahrenheit. The Celsius and Fahrenheit temperature scales are related by

$$T_F = \tfrac{9}{5}(T_C + 32)$$

$$T_C = \tfrac{5}{9}(T_F - 32)$$

The symbol T denotes temperature throughout this text.

A thermometer relies on at least one property that will change with temperature. The familiar liquid-in-glass thermometer (Figure 2.16) consists of a sealed glass capillary tube with one end connected to a liquid-filled bulb and the space above the liquid filled with an inert gas or the liquid's vapor. Specific volume is the thermometric property in this instance. As the temperature of the liquid increases, its volume increases, causing the liquid to rise in the capillary tube. The distance L corresponds to a given temperature. The scale on the thermometer is calibrated using two points, typically 0°C (32°F) and 100°C (212°F).

Another temperature-measuring device is the thermocouple, which develops a voltage as a function of temperature. Thermocouples are formed by joining two dissimilar metal wires such as copper and constantan (a copper-nickel alloy) or iron and constantan. An electromotive force (emf) produced at the metal junction is primarily a function of temperature. Thus, the voltage variation corresponds directly to temperature, and by measuring the voltage output of the thermocouple, the

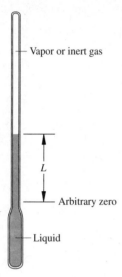

Vapor or inert gas

L

Arbitrary zero

Liquid

Figure 2.16 Schematic diagram of a liquid-in-glass thermometer.

temperature of the thermocouple's junction can be ascertained. Again, having voltage as an output signal is useful in data acquisition systems.

Thermistors and resistance temperature detectors (RTDs) use electrical resistance as the thermometric property. In this situation the variation of electrical resistance with temperature of conductors or semiconductors is known, and by measuring the resistance, we can deduce the temperature. When extremely high temperatures or temperatures of moving objects need to be measured, radiation and optical pyrometers are used. These devices measure the thermal radiation emitted by the object and correlate that to temperature.

When extremely accurate temperature measurements are required, such as in setting national or international standards, gas thermometers are used. In these the pressure of a gas, such as hydrogen or helium, corresponds directly to temperature. By accurately measuring the pressure using a manometer, the temperature is determined. It should be noted, while very accurate, gas thermometers are slow in responding to variations in temperature and require elaborate test procedures.

Example 2.3
A thermocouple has the following characteristics:

T (°F)	mV
50	0.397
150	2.666
250	4.964
350	7.205

Assuming the thermocouple is represented by a cubic equation of the form $mV = A + B*T + C*T^2 + D*T^3$, compute the values of the four coefficients, A, B, C, and D.

Solution

Given: The temperature/voltage characteristics of a thermocouple.

Find: The coefficients of the cubic equation representing the thermocouple behavior.

Assumptions: None.

Analysis: Use TK Solver to solve the set of four simultaneous equations written from the data given. Enter the four equations in the Rule Sheet and the given values of voltage and temperature in the Input column of the Variable Sheet. Since the Iterative Solver is required, enter "G" in the Status column for the variables A, B, C, and D. Press **F9** and the values of A, B, C, and D will appear in the Output column of the Variable Sheet.

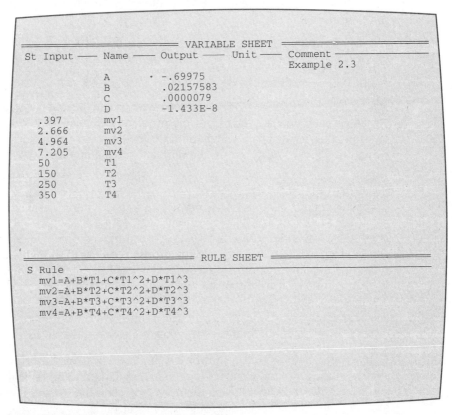

Figure 2.17 Variable Sheet for Example 2.3.

Comments:

1. The voltage characteristic of a thermocouple can be represented by a polynomial equation.
2. TK Solver is a convenient tool for calculating the coefficients of the equation.

2.13 GUIDELINES FOR THERMODYNAMIC PROBLEM SOLUTION

A primary reward for studying thermodynamics is being able to successfully solve engineering problems requiring thermodynamic analysis. This text has been written with this in mind, and the examples and end-of-chapter problems assist in this effort. There is a procedure that helps in mastering the solution of all thermodynamic problems, though true mastery comes with actual practice in solving many problems.

The examples use a methodology for solving thermodynamic problems; it is not a "cookbook" approach, but rather it forces one to pay attention to the underlying thermodynamic principles, conservation of mass, conservation of energy, and the second law of thermodynamics, of the physical situation being analyzed. While it may seem cumbersome to use at first, particularly with elementary problems, the following approach can develop your confidence and ability to solve quite complicated problems.

Given: State in your own words what the given information is. This requires that you think about the problem statement and not simply restate it.

Find: State what must be determined.

Sketch and Given Data: Draw a sketch of the system, deciding whether it is open or closed; include relevant data from the problem statement. The purpose here is to conceptualize what is happening and translate the word problem into sketches and diagrams. This is a key point in the problem-solving methodology.

Assumptions: Engineering analyses use models of physical systems; list the assumptions made in the modeling of this particular problem. Sometimes this information should be included on the sketch as well.

Analysis: Apply the assumptions to the governing equations and relationships. You should endeavor to work with the equations as long as possible before substituting numerical data in them. Substitute the data when the equations have been reduced to their final form. It is important to make sure the equations are conceptually correct and dimensionally consistent. Dimensional inconsistency is often the cause of unsatisfactory problem solution. Ask yourself if the answer seems reasonable: Is the sign correct? Is the magnitude of the term consistent with other terms?

Comments: Discuss your results briefly. You will be using the problem solution in studying for exams and in reviewing the material at other times. What insight have you gained that will help during the review process? This might include principles and concepts you had to consider in the solution and an interpretation of the results.

CONCEPT QUESTIONS

1. What is a property?
2. Describe the difference between closed and open systems.
3. Weight and mass are often used interchangeably. Is this wise? Why?
4. We have seen pictures of astronauts floating in space in a condition of "weightlessness." What causes this condition?
5. Describe absolute and gage pressures.
6. What are the standard units of pressure in the SI and English unit systems?
7. What are the four common temperature scales? Derive the conversion formulas between them.
8. The concept of a continuum was used to describe the limit of macroscopic thermodynamic analysis. Describe this in your own words.
9. A can of beer is placed in a refrigerator to cool. Would this be analyzed as an open or closed system? Would your analysis change if an entire six-pack were placed in the refrigerator?
10. Is a closed system closed to energy flow as well as mass flow? What about an open system?
11. You are watching a candle burn. What energy transformations occur during this process?
12. What defines thermodynamic equilibrium of a system?
13. What is the difference between extensive and intensive properties?
14. Explain how a liquid-filled thermometer operates in determining temperature.
15. How can a thermocouple be used to measure temperature?

PROBLEMS (SI)

2.1 Referring to Figure 2.10, the atmospheric pressure is 100 kPa, and the pressure gages A and B read 210 kPa (gage). Determine the absolute pressures in boxes A and B in (a) kPa; (b) mm Hg absolute.

2.2 A manometer is used to measure pressure as indicated in Figure 2.12. The fluid in the column is mercury with a density 13.6 times that of water. If atmospheric pressure is 95 kPa and the height of the column is 1.5 m, determine the pressure of the system.

2.3 During takeoff in a space ship, an 80-kg astronaut is subjected to an acceleration equal to 5 times the pull of the earth's standard gravity. If the takeoff is vertical, what force does he exert on the seat?

2.4 A system has a mass of 20 kg. Determine the force necessary to accelerate it 10 m/s^{-2} (a) horizontally, assuming no friction; (b) vertically, where $g = 9.6$ m/s^{-2}.

2.5 A pump discharges into a 3-m-per-side cubical tank. The flow rate is 300 liters/min, and the fluid has a density 1.2 times that of water (density of water = 1000.0 kg/m^3). Determine (a) the flow rate in kg/s; (b) the time it takes to fill the tank.

2.6 Someone proposes a new absolute temperature scale in which the boiling and freezing points of water at atmospheric pressure are 500°X and 100°X, respectively. Develop a relation to convert this scale to degrees Celsius.

2.7 A vertical column of water will be supported to what height by standard atmospheric pressure?

2.8 A tank contains a mixture of 20 kg of nitrogen and 20 kg of carbon monoxide. The total tank volume is 20 m³. Determine the density and specific volume of the mixture.

2.9 An automobile has a 1200-kg mass and is accelerated to 7 m/s². Determine the force required to perform this acceleration.

2.10 A hiker is carrying a barometer that measures 101.3 kPa at the base of the mountain. The barometer reads 85.0 kPa at the top of the mountain. The average air density is 1.21 kg/m³. Determine the height of the mountain.

2.11 Convert 225 kPa into (a) atmospheres; (b) mm Hg absolute.

2.12 A skin diver wants to determine the pressure exerted by the water on her body after a descent of 35 m to a sunken ship. The specific gravity of seawater is 1.02 times that of pure water (1000 kg/m³). Determine the pressure.

2.13 A new temperature scale is desired with freezing of water at 0°X and boiling at atmospheric pressure occurring at 1000°X. Derive a conversion between degrees Celsius and degrees X. What is absolute zero in degrees X?

2.14 A 50-kg frictionless piston fits inside a 20-cm-diameter vertical pipe and is pulled up to a height of 6.1 m. The pipe's lower end is immersed in water, and atmospheric pressure is 100 kPa. The gravitational acceleration is 9.45 m/s². Determine (a) the force required to hold the piston at the 6.1-m mark; (b) the pressure of the water on the underside of the piston.

2.15 A spring scale is used to measure force and to determine the mass of a sample of moon rocks on the moon's surface. The springs were calibrated for the earth's gravitational acceleration of 9.8 m/s². The scale reads 4.5 kg, and the moon's gravitational attraction is 1.8 m/s². Determine the sample mass. What would the reading be on a beam balance scale?

2.16 Determine the pressure at points A and B in the figure shown if the density of mercury is 13 590.0 kg/m³ and that of water is 1000.0 kg/m³.

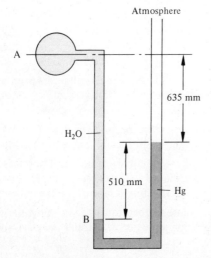

Problem 2.16

2.17 For the situation sketched below, the following information is known: $\rho_{H_2O} = 1000.0$ kg/m³, $\rho_{H_g} = 13\ 590.0$ kg/m³, $g = 9.8$ m/s^{-2}, $p_1 = 500$ kPa. Determine p_{II}.

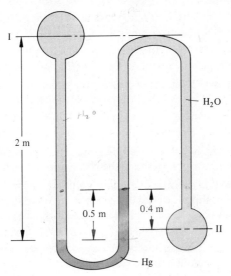

Problem 2.17

2.18 The piston shown above is held in equilibrium by the pressure of the gas flowing through the pipe. The piston has a mass of 21 kg; $p_I = 600$ kPa; $p_{II} = 170$ kPa. Determine the pressure of the gas in the pipe, p_{III}.

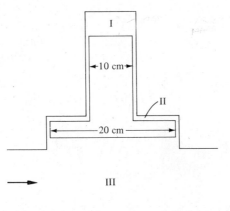

Problem 2.18

2.19 Referring to Problem 2.18, $p_I = 350$ kPa, $p_{II} = 130$ kPa, $p_{III} = 210$ kPa. Find the mass of the piston.

2.20 A new temperature scale is proposed where the boiling and freezing points of water at atmospheric pressure are 500°X and 100°X, respectively. Determine absolute zero for this new temperature scale.

2.21 A diver descends 100 m to a sunken ship. A container is found with a pressure gage reading 100 kPa (gage). Atmospheric pressure is 100 kPa. What is the absolute pressure of the gas in the container? (The density of water is 1000 kg/m³.)

2.22 A room is 10 m by 10 m by 3 m. Determine the mass and weight of air in the room if the air's average density is 1.18 kg/m³.

2.23 A beer barrel has a mass of 10 kg and a volume of 20 liters. Assuming the density of beer is 1000 kg/m³, determine the total mass and weight of the barrel when it is filled with beer.

2.24 A tank's pressure gage indicates that the pressure in the tank is 300 kPa (gage) where the atmospheric pressure is 745 mm Hg. Determine the tank's absolute pressure.

2.25 A tank has a vacuum gage attached to it indicating 20 kPa (vacuum) where the atmospheric pressure is 100 kPa. Determine the absolute pressure in the tank.

2.26 A tank has a vacuum gage attached to it indicating 35 kPa (vacuum) where the atmospheric pressure is 740 mm Hg. Determine the absolute pressure in the tank.

2.27 A barometer can be used to measure an airplane's altitude by comparing the barometric pressure at a given flying altitude to that on the ground. Determine an airplane's altitude if the pilot measures the barometric pressure at 700 mm Hg, the ground reports it at 758 mm Hg, and the average air density is 1.19 kg/m³. $g = 9.8$ m/s².

2.28 The pressures in Problem 2.27 are now 90 kPa aboard the plane and 101 kPa at ground level. What is the plane's altitude if the average air density is 1.19 kg/m³ and $g = 9.8$ m/s²?

2.29 A submarine is cruising 200 m below the ocean's surface. Determine the pressure on the submarine's surface if atmospheric pressure is 101 kPa and the density of seawater is 1030 kg/m³. $g = 9.8$ m/s².

2.30 A vertical, frictionless piston cylinder contains air at a pressure of 300 kPa with atmospheric pressure of 100 kPa. The diameter of the piston is 0.25 m, and $g = 9.8$ m/s². Determine the piston's mass.

2.31 The vertical, frictionless piston-cylinder shown below contains a gas at an unknown pressure. The piston has a mass of 10 kg and a cross-sectional area of 75 cm². The spring exerts a downward force of 100 N on the piston, and atmospheric pressure is 100 kPa. Determine the pressure of the gas.

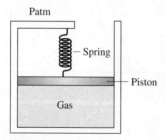

Problem 2.31

2.32 For the configuration in Problem 2.31, let the pressure in the cylinder be 200 kPa, atmospheric pressure be 100 kPa, and the mass of the piston be 12 kg. Determine the spring force on the piston.

2.33 A pressure cooker operates by cooking food at a higher pressure and temperature than is possible at atmospheric conditions. Steam is contained in the sealed pot, with a small vent hole in the middle of the cover, allowing steam to escape. The pressure is regulated by covering the vent hole with a small weight, which is displaced slightly by the escaping steam. Atmospheric pressure is 100 kPa, the vent hole area is 7 mm², and the pressure inside should be 250 kPa. What is the mass of the weight?

2.34 Determine the temperature at which readings in degrees Fahrenheit and Celsius are the same.

2.35 One February morning the outside temperature measured 2°C. If the absolute value of this temperature is doubled and decreased by 239 degrees, what would be the final temperature and what season of the year would this indicate?

2.36 A thermocouple produces the following readings:

T (°C)	Voltage (mV)
25	1.000
100	4.095
175	7.139
250	10.151

What is a first-order equation that relates temperature to mV? Determine the mV that would correspond to a temperature of 350°C.

PROBLEMS (English Units)

*2.1 Referring to Figure 2.10 in the text, the atmospheric pressure is 100 kPa and the pressure gages A and B read 25 psig. Determine the absolute pressure in boxes A and B in (a) psia; (b) in. Hg absolute.

*2.2 Determine the temperature for which a thermometer with degrees Fahrenheit is numerically twice the reading of the temperature in degrees Celsius.

*2.3 During takeoff in a space ship, an astronaut is subjected to an acceleration equal to 5 times the pull of the earth's standard gravity. If the astronaut is 180 lbm and the takeoff is vertical, what force does he exert on the seat?

*2.4 Convert 20 in. Hg vacuum to (a) in. Hg absolute; (b) psia; (c) kPa.

*2.5 A spring scale is used to measure force and to determine the mass of a sample of moon rocks on the moon's surface. The springs were calibrated for g_c. The scale reads 10 lbf, and the moon's gravitational attraction is 5.40 ft/sec². Determine the sample mass. What would be the reading on a beam balance scale?

*2.6 Determine the pressure at points A and B if the density of mercury is 724.4 lbm/ft³ and that of water is 62.4 lbm/ft³. Refer to sketch for Problem 2.16.

*2.7 The following information is known: $\rho_{H_2O} = 62.4$ lbm/ft³, $\rho_{Hg} = 724.4$ lbm/ft³, $g = 30$ ft/sec², $p_I = 100$ psia. Determine p_{II}. Refer to sketch for Problem 2.17.

*2.8 A room is 30 ft by 30 ft by 10 ft. Determine the mass and weight of air in the room if the air's average density is 0.074 lbm/ft³.

*2.9 A beer barrel has a mass of 20 lbm and a volume of 5 gal. Beer's density is 62.4 lbm/ft³. Determine the total mass and weight of the barrel when it is filled with beer.

*2.10 A tank's pressure gage indicates that the pressure in the tank is 50 psig and the atmospheric pressure is 28.72 in. Hg. Determine the tank's absolute pressure.

*2.11 A tank has a vacuum gage attached to it indicating 25 in. Hg (vacuum) where atmospheric pressure is 14.5 psia. Determine the absolute pressure in the tank.

*2.12 A tank has a vacuum gage attached to it indicating 28.6 in. Hg (vacuum) where atmospheric pressure is 29.5 in. Hg absolute. Determine the absolute pressure in the tank.

*2.13 A barometer can be used to measure an airplane's altitude by comparing the barometric pressure at a given flying altitude to that on the ground. Determine an airplane's altitude if the pilot measures the barometric pressure at 27.55 in. Hg absolute, the ground reports it to be 29.92 in. Hg absolute, and the average air density if 0.077 lbm/ft³. $g =$ 32.174 ft/sec².

*2.14 A hiker is carrying a barometer that measures 29.92 in. Hg absolute at the base of the mountain. The barometer reads 25.5 in. Hg at the top of the mountain. The average air density is 0.076 lbm/ft³, and gravitational acceleration remains constant at 32.174 ft/sec². What is the mountain's height?

*2.15 A submarine is cruising 600 ft below the ocean's surface. Determine the pressure on the submarine's surface if atmospheric pressure is 14.7 psia, the density of seawater is 64.2 lbm/ft³, and $g =$ 32.174 ft/sec².

*2.16 A vertical, frictionless piston-cylinder contains air at a pressure of 50 psia where the atmospheric pressure is 14.7 psia. The diameter of the piston is 10 in., and $g =$ 32.174 ft/sec². Determine the piston's mass.

*2.17 A vertical, frictionless piston-cylinder, similar to the one illustrated in Problem 2.31, contains a gas at an unknown pressure. The piston has a mass of 20 lbm and a cross-sectional area of 50 in². In addition the spring exerts a force of 15 lbf on the piston, and atmospheric pressure is 14.7 psia. Determine the pressure of the gas.

*2.18 Referring to Problem *2.19, let the pressure in the cylinder be 40 psia, atmospheric pressure 14.7 psia, and the mass of the piston 20 lbm. Determine the spring force on the piston.

*2.19 A pressure cooker operates by cooking food at a higher pressure and temperature than is possible at atmospheric conditions. Steam is contained in the sealed pot, with a small vent hole in the middle of the cover, allowing steam to escape. The pressure is regulated by covering the vent hole with a small weight, which is displaced slightly by the escaping steam. Atmospheric pressure is 14.7 psia, the vent hole area is 0.015 in.², and the pressure inside should be 30 psig. What is the mass of the weight?

*2.20 A piston-cylinder contains 2 lbm of water. The initial volume is 0.1 ft³. The piston rises, causing the volume to double. Determine the final specific volume of the water.

*2.21 A cylindrical tank is 50 in. long, has a diameter of 16 in., and contains 1.65 lbm of water. Calculate the specific volume and density of the water.

*2.22 Steam is held in two compartments separated by a membrane as shown in the sketch shown. The total volume is 777 ft³, and the volume of compartment B is 280 ft³. The specific volume of the steam in B is 9.5 ft³/lbm. The membrane breaks, and the resulting specific volume is 12.75 ft³/lbm. Determine the original specific volume in compartment A.

Problem *2.22

*2.23 Referring to the sketch for Problem *2.22, let the volume of A be 7 ft³ and that of B 17.5 ft³. The specific volumes of A and B are, respectively, 275 ft³/lbm and 55 ft³/lbm. If the membrane breaks, determine the resulting specific volume.

*2.24 The device shown below contains a free-moving, frictionless piston. The total volume of the cylinder is 40 ft³ evenly divided between A and B. The initial specific volumes in A and B are 1000 ft³/lbm and 500 ft³/lbm, respectively. The piston moves so x is one-quarter of the distance L. Determine the final specific volumes in A and B.

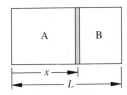

Problem *2.24

COMPUTER PROBLEMS

C2.1 Compute the weight of a 50-kg mass at different heights above the earth's surface. At sea level $g = 9.8$ m/s² and decreases by 0.000913 m/s² for each 300 m of ascent. Consider a total ascent of 2.5 km with increments of 100 m.

C2.2 Compute the pressure on a diver as she descends in the ocean to a depth of 200 m at 10 m increments. The pressure at sea level is 101.3 kPa. The density of the seawater is 1025 kg/m³, and g remains constant at 9.8 m/s².

C2.3 Using a spreadsheet program, prepare a table converting temperatures between $-100°$ and 300°C in increments of 10° into Kelvin, Fahrenheit, and Rankine.

C2.4 A thermistor is to be used to measure temperatures between $-75°$ and 150°C. The manufacturer quotes resistances of 1071K, 7355, and 41.9 ohms at $-75°, 0°$, and 150°C respectively. A common equation for the characteristic of a thermistor is $1/T = A + B\ln R + C(\ln R)^3$. Compute the values of the coefficients A, B, and C and the resistance of the thermistor at 50°C.

3

Conservation of Mass and Energy

Not only is energy the lifeblood of industrialized society, it is the fundamental focus of thermodynamic analysis. In this chapter we will develop the equations describing the conservation of energy and mass for open and closed systems. To understand how energy is conserved, we must conceptualize the various energy forms that are intrinsic to total energy. In this chapter you will

- Analyze the conservation of mass and determine the conditions necessary for steady-state and steady-flow conditions.
- Develop expressions for the first law of thermodynamics for open and closed systems;
- Develop a thermodynamic understanding of heat and work;
- Derive expressions for kinetic and potential energies;
- Conceptualize the internal energy of a substance;
- Apply first-law analysis to a variety of real-life situations;
- Expand your knowledge of thermodynamic cycles.

3.1 CONSERVATION OF MASS

The law of conservation of mass states that the total mass is a constant. As with all laws, this is a deduction of experimental evidence; it can be demonstrated, but not proved, to be true.

In thermodynamics there are two types of systems, open and closed. For a closed system, a mass of fixed identity, the conservation of mass is true. No equation is necessary to demonstrate this; it is self-evident in the definition of the system. However, in the case of an open system, a system of fixed space, we can develop an expression for the conservation of mass. Consider Figure 3.1, which shows a control volume in a stream of moving fluid. At some time t a given mass will occupy the control volume. This will be a mass of fixed identity. We follow this mass of fixed identity as it moves across the boundaries of the control volume, and from this we can determine the conservation of mass, or continuity equation, for the control volume.

After a time Δt, some of the mass in the control volume has moved across the control surface. This is represented as the system boundary at time $t + \Delta t$. The control surface boundary and the system boundary are coincident at time t.

The regions of space that the mass occupies at times t and $t + \Delta t$ may be visualized as three regions, I, II, and III. Since the system mass is constant with time, we may write

$$m_{It} + m_{IIt} = m_{IIt+\Delta t} + m_{IIIt+\Delta t} \tag{3.1}$$

where the subscripts on each mass term represent first the spatial region and then the time. This equation may be rearranged as follows:

$$m_{IIt+\Delta t} - m_{IIt} - m_{It} + m_{IIIt+\Delta t} = 0 \tag{3.2}$$

Let us add zero to equation (3.2), but zero expressed in a mathematically convenient form. What is the mass in region III at time t? Zero. What is the mass in region I at

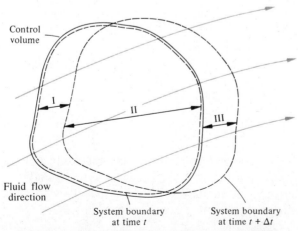

Figure 3.1 A control volume located in a moving fluid.

time $t + \Delta t$? Zero. Add these quantities to equation (3.2) and divide by Δt.

$$\frac{m_{\mathrm{II}t+\Delta t} - m_{\mathrm{II}t}}{\Delta t} + \frac{m_{\mathrm{I}t+\Delta t} - m_{\mathrm{I}t}}{\Delta t} + \frac{m_{\mathrm{III}t+\Delta t} - m_{\mathrm{III}t}}{\Delta t} = 0 \tag{3.3}$$

The first term of equation (3.3) represents the change in mass of region II. As Δt approaches zero, region II coincides with the control volume, so the change in mass of region II is identical to the change in mass of the control volume.

$$\lim_{\Delta t \to 0} \frac{m_{\mathrm{II}t+\Delta t} - m_{\mathrm{II}t}}{\Delta t} = \frac{dm_{\mathrm{II}}}{dt} = \frac{dm_{cv}}{dt}$$

If the limit is taken for the other two terms in equation (3.3), the result is

$$\frac{dm_{cv}}{dt} - \frac{dm_{\mathrm{I}}}{dt} + \frac{dm_{\mathrm{III}}}{dt} = 0 \tag{3.4}$$

where the negative sign is due to the fact that $m_{\mathrm{I}t+\Delta t} - m_{\mathrm{I}t}$ is a decreasing, or negative, term physically. The mass in region I is decreasing, not increasing, with time. As Δt approaches zero, the change of mass of region I represents the mass entering the control volume, and the change of mass of region III represents the mass leaving the control volume. Let $dm/dt \equiv \dot{m}$; then $dm_{\mathrm{I}}/dt = \dot{m}_{\mathrm{in}}$ and $dm_{\mathrm{III}}/dt = \dot{m}_{\mathrm{out}}$. Equation (3.4) becomes

$$\frac{dm_{cv}}{dt} = \dot{m}_{\mathrm{in}} - \dot{m}_{\mathrm{out}} \tag{3.5}$$

Equation (3.5) may be interpreted physically this way: the change of the mass within the control volume is equal to the difference between the mass entering and the mass leaving. We would probably have deduced this, but it is comforting to have a rigorous development as support.

It is possible to express the conservation of mass, equation (3.5), in terms of area, velocity, and density, all measurable quantities. The volume will be denoted as $\mathscr{V}$ in the next few paragraphs to distinguish it from the mass velocity, v. Let $d\mathscr{V}$ be a differential volume in the total control volume. The fluid in $d\mathscr{V}$ has a density ρ. Thus the total mass in the control volume is the volume integral of $\rho \, d\mathscr{V}$:

$$m_{cv} = \int_{cv} \rho \, d\mathscr{V} \tag{3.6}$$

The time rate of change of m_{cv} is

$$\frac{dm_{cv}}{dt} = \frac{d}{dt} \int_{cv} \rho \, d\mathscr{V} \tag{3.7}$$

The volume flow across a differential surface is the product of the area, dA, times the fluid velocity normal to the area, v_n. The mass flow is found by multiplying this term by the fluid density, ρ, at area dA. The total mass flow across the entire area is found by integrating over the entire area, A. Thus

$$\dot{m} = \int_A \rho v_n \, dA \tag{3.8}$$

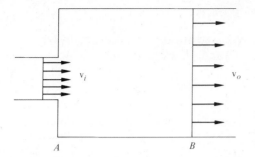

Figure 3.2 One-dimensional slug flow in a pipe.

Equation (3.5) may now be written as

$$\frac{d}{dt}\int_{cv}\rho\, d\mathcal{V} = \int_A \rho v_n\, dA_{\text{in}} - \int_A \rho v_n\, dA_{\text{out}} \tag{3.9}$$

For steady flow, the change in mass within the control volume is zero; thus the mass flow rates in and out are identical:

$$\int_A \rho v_n\, dA_{\text{in}} = \int_A \rho v_n\, dA_{\text{out}} \tag{3.10}$$

For steady, one-dimensional flow, where the fluid enters at state i and leaves at state o, as illustrated by Figure 3.2, the conservation of mass becomes

$$\frac{A_i v_i}{v_i} = \frac{A_o v_o}{v_o} \tag{3.11}$$

Thus the conservation of mass expression for steady, one-dimensional flow, frequently referred to as the continuity equation, is

$$\dot{m} = \frac{A v}{v} = \rho A v = \text{constant} \; \frac{\text{kg}}{\text{s}} \; \left[\frac{\text{lbm}}{\text{sec}}\right] \tag{3.12}$$

Equation (3.12) expresses the conservation of mass for open systems in terms of readily measured or determined properties.

3.2 ENERGY FORMS

We have mentioned the energy of a system, but we have not defined energy. This was no accident. The total energy, E, cannot be simply defined, although parts of the total energy will be later in the chapter. There is a conservation law for energy — the first law of thermodynamics. It states that whenever energy transfer occurs, energy must be conserved. This is stated simply enough, but before we develop the first law for open and closed systems, let us examine some of the energy forms that we will be using in thermodynamics.

Work and heat are two of the most fundamental energy forms with which we will deal, and it is essential that we understand them thoroughly.

Work

The work, W, done on or by a system may be expressed in terms of the product between an observable force, F, and a displacement, ds. Thus,

$$W = \int_{s_1}^{s_2} F \cdot ds \qquad (3.13)$$

In the mechanical sense, work may be viewed as the expansion or compression of a fluid or the stretching of a bar. Thermodynamics is not limited to mechanics; the force could be that necessary to move a charged particle in an electric field. In general, a system is viewed as performing work on the surroundings if the sole effect on everything external to the system could have been the raising of a weight. Raising of weight, a force, through a distance is mechanical work. The system need not perform mechanical work, however, but only be capable of doing so.

When work is considered in thermodynamics, a system is involved. Either the system will be performing work on the surroundings or the surroundings will be performing work on the system. Conventionally, work done by a system on the surroundings is positive, and work done by the surroundings on the system is negative.

Consider the system in Figure 3.3. What is the direction of the work in these three cases? Referring to the definition of work, we note that in case (a) the falling mass in the gravitational field provides the energy for work to be done on the system. Note that the work crosses the system boundary, going into the system, and thus the work is negative. In case (b) the mass and pulley are considered the system, and we note that the work is done by the system to the surroundings. In this case the work is positive as it goes from the system to the surroundings, the box with paddle wheels. In case (c) the

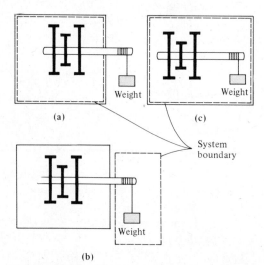

Figure 3.3 The effect of a system's boundary on the work.

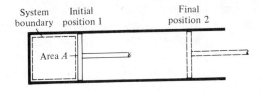

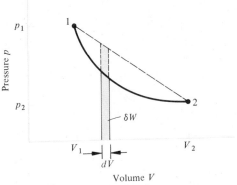

Figure 3.4 The relationship between pressure and volume on a p-V diagram and in a piston-cylinder system.

paddle wheel and the mass and pulley are considered the system. Is any work done on the system by the surroundings? No. Does the system do any work on the surroundings? No. So the work is zero. What happens within the system does not matter. It becomes important, then, to draw the system boundaries judiciously, or parts of a problem may not be apparent. In the reverse situation unnecessary complications may also occur. Both types of difficulties are eliminated with practice in problem solving.

Let us find the work done by a closed system, a system in which there is no mass flow. Again the piston-cylinder provides a ready tool for our analysis (Figure 3.4). There is a piston of uniform area, A, acted upon by a pressure, p_1; the pressure pushes the piston to the final position, where the pressure has dropped to p_2. The system mass has remained constant, but the volume has changed from V_1 to V_2, where $V_2 > V_1$. From our previous experience we note that the system has done work on the surroundings and that the work is positive.

We would find this qualitative description somewhat lacking if we were asked how much work was done. We know that work was performed by the system; now we must attempt to model, with a mathematical equation, what is actually happening. We must remember that this is a model, not a description.

Looking back to Chapter 2 we find that properties such as pressure are defined only for equilibrium states. The pressure must be uniform throughout the system. This is useful, because the pressure acting on the piston face will be uniform, and a pressure multiplied by an area yields a force. Consider what happens when the piston moves a distance, dL. First, there will be a drop in pressure, and then, according to our model, the pressure will drop uniformly throughout the system for mechanical equilibrium. A process is called a *quasi-equilibrium process* when the system is in equilibrium (e.g., mechanical, thermal, chemical) at each point going from its initial to final state. Our model will be valid or not depending on whether or not this assumption can be made. It turns out that this assumption is reasonable for many processes, because the time the substance requires to establish equilibrium is usually much less than the time of the macroscopic event, the work done.

Using the quasi-equilibrium model, the force, $F = pA$, and the work done by moving a distance dL are

$$\delta W = pA \, dL \tag{3.14}$$

however,

$$A \, dL = dV$$

so

$$\delta W = p \, dV \tag{3.15}$$

The total work is found by integrating between positions 1 and 2:

$$W = \int_1^2 \delta W = \int_1^2 p \, dV \tag{3.16}$$

The reason for the inexactness of the work differential is that work is a path function. This may be seen graphically in Figure 3.4, where δW is denoted by the shaded area on the diagram. This, however, is for a given distribution of pressure with volume. If the distribution were changed to that of a straight line between states 1 and 2, the δW would be greater, indicating that work is a function of how the pressure varies in going from state 1 to state 2. The δ operator indicates an inexact differential, which approaches an infinitesimal limit, rather than the exact differential, which approaches zero.

Example 3.1

The pressure of a gas in a piston-cylinder varies with volume according to (a) $pV = C$; (b) $pV^2 = C$. The initial pressure is 400 kPa, the initial volume is 0.02 m³, and the final volume is 0.08 m³. Determine the work for both processes.

Solution

Given: Gas pressure in a piston-cylinder varies as a function of volume.

Find: The work done in the expansion process.

Sketch and Given Data:

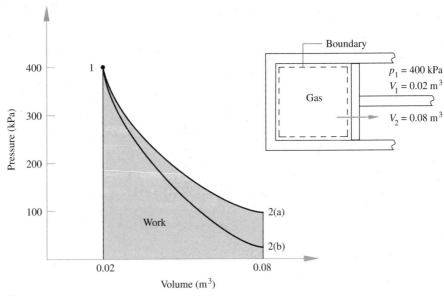

Figure 3.5

Assumptions:

1. The gas undergoing the expansion is a closed system.
2. The expansion is in quasi-equilibrium as the pressure is continuously defined during the expansion process.

Analysis: The work is determined by integrating equation (3.16) for the functional relationship for $p(V)$ for cases (a) and (b).

Case (a):

$$W = \int_1^2 p\,dV = C \int_1^2 \frac{dV}{V} = C \ln\left(\frac{V_2}{V_1}\right)$$

$$C = pV = p_1 V_1$$

$$W = p_1 V_1 \ln\left(\frac{V_2}{V_1}\right) = \left(400\,\frac{\text{kN}}{\text{m}^2}\right)(0.02\ \text{m}^3)\ln\left(\frac{0.08}{0.02}\right) = 11.09\ \text{kJ}$$

Case (b):

$$W = \int_1^2 p\,dV = C \int_1^2 \frac{dV}{V^2} = C\left[\frac{1}{V_1} - \frac{1}{V_2}\right]$$

$$C = p_1 V_1{}^2 = p_2 V_2{}^2$$

$$W = p_1 V_1 - p_2 V_2 = (400\ \text{kN/m}^2)(0.02\ \text{m}^3) - (25\ \text{kN/m}^2)(0.08\ \text{m}^3)$$

$$W = 6\ \text{kJ}$$

Comments:

1. The area under the $p(V)$ curve represents the work done by the gas in the expansion process. The magnitude of the work terms in cases (a) and (b) correspond to the graphical areas.
2. The sign on the work is positive, indicating the system did work on the piston, moving it outward.
3. The gaseous substance did not have to be specified as long as the pressure's functional relationship was known.
4. The assumption of a quasi-equilibrium process is key to the solution, as pressure must be continuously defined throughout the expansion process to integrate the work function.

Example 3.2

Solve Example 3.1 numerically on a computer using a spreadsheet program.

Solution

Given: Same as Example 3.1.

Find: Same as Example 3.1.

Assumptions: Same as Example 3.1.

Analysis: The work can be computed by summing the work over small increments of volume change. The work in each volume increment is approximately the average pressure multiplied by the volume change.

Case (a):

Enter the following equations and data into the cells of the spreadsheet:

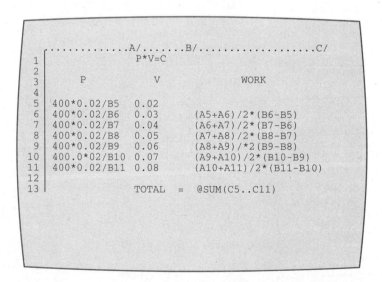

```
.............A/.......B/...................C/
1                    P*V=C
2
3        P              V              WORK
4
5   400*0.02/B5     0.02
6   400*0.02/B6     0.03        (A5+A6)/2*(B6-B5)
7   400*0.02/B7     0.04        (A6+A7)/2*(B7-B6)
8   400*0.02/B8     0.05        (A7+A8)/2*(B8-B7)
9   400*0.02/B9     0.06        (A8+A9)/*2(B9-B8)
10  400.0*02/B10    0.07        (A9+A10)/2*(B10-B9)
11  400*0.02/B11    0.08        (A10+A11)/2*(B11-B10)
12
13                  TOTAL   =   @SUM(C5..C11)
```

The spreadsheet yields the following results:

```
..........A/......B/..........C/
1              P*V=C
2
3        P           V          WORK
4
5          400      0.02
6     266.6666      0.03       3.333333
7          200      0.04       2.333333
8          160      0.05            1.8
9     133.3333      0.06       1.466666
10    114.2857      0.07       1.238095
11         100      0.08       1.071428
12
13                 TOTAL  =    11.24285
```

Case (b):

Enter the following equations and data:

```
............................E/....F/.............G/
1                            P*V^2=C
2
3              P             V            WORK
4
5  400*0.02*0.02/(F5*F5)    0.02
6  400*0.02*0.02/(F6*F6)    0.03   (E5+E6)/2*(F6-F5)
7  400*0.02*0.02/(F7*F7)    0.04   (E6+E7)/2*(F7-F6)
8  400*0.02*0.02/(F8*F8)    0.05   (E7+E8)/2*(F8-F7)
9  400*0.02*0.02/(F9*F9)    0.06   (E8+E9)/*2(F9-F8)
10 400.0*02*0.02/(F10*F10)  0.07   (E9+E10)/2*(F10-F9)
11 400*0.02*0.02/(F11*F11)  0.08   (E10+E11)/2*(F11-F10)
12
13                          TOTAL = @SUM(G6..G11)
```

The spreadsheet yields the following results:

```
........E/.......F/.........G/
1              P*V^2=C
2
3       P              V              WORK
4
5       400            0.02
6       177.7777       0.03           2.888888
7       100            0.04           1.388888
8       64             0.05           0.82
9       44.44444       0.06           0.542222
10      32.65306       0.07           0.385487
11      25             0.08           0.288265
12
13                     TOTAL =        6.313752
```

Comments:

1. The numerical solutions agree reasonably well with the analytic solutions in Example 3.1.
2. To improve the accuracy, reduce the size of the volume increment, increasing the number of steps over which the work is summed. ▬

Let us now consider the quasi-equilibrium process in more detail. Consider the piston-cylinder arrangement in Figure 3.6(a). The piston is held in place by a mass acted upon by the gravitational field, hence a force. If we move one block from the piston as in Figure 3.6(b), the piston will shoot upward and oscillate up and down before reaching an equilibrium position. Clearly the pressure is not uniform during such a process, and the process cannot be called a quasi-equilibrium process. The next step toward reaching a quasi-equilibrium process is to divide the mass into smaller and smaller quantities. Figure 3.7 illustrates the mass being replaced by a stack of cards. As each card is removed, the piston moves upward a slight amount and after a very slight oscillation reaches its equilibrium state. To improve this process even further, we must make the mass of each card infinitesimal, so that the change in the piston height would occur in differential steps. This process is imaginary, but it represents the ideal, that is, the maximum work possible: all the energy goes into moving the piston and is not dissipated by the piston's oscillation. Note also that the addition of an infinitesimal card to the piston would start the process in reverse. This process is called an *equilibrium* or *reversible* process. In Figure 3.6 we could not slide the mass onto the piston to start the process over. This process is called an *irreversible* process; it cannot return to its original state along the same path. There are many

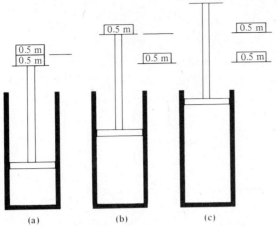

Figure 3.6 Piston movement when large weight increments are removed.

factors that render a process irreversible. Nonuniform pressure in the system causes the mass to move within the system, using energy that then will not be available for work. Frictional effects, too, are apparent irreversibilities. The energy used to overcome mechanical friction is lost for useful purposes. Also, fluid viscous forces, that is, fluid friction, dissipates useful energy. There are other causes of irreversibilities, but the effect is always to decrease the useful energy output or increase the energy required.

Measuring how quickly work is accomplished gives us another quantity—

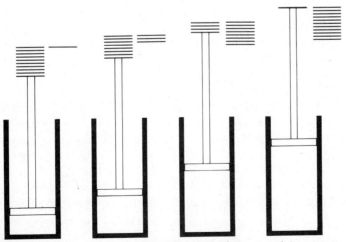

Figure 3.7 Piston movement when infinitesimal weights are removed, modeling a quasi-equilibrium process.

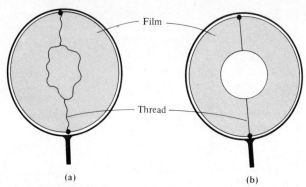

(a)   (b)

Figure 3.8 A hoop with a soap film. This demonstrates the effect of surface tension.

power, $\dot{W}$.

$$\dot{W} = \lim_{\delta t \to 0} \frac{\delta W}{\delta t} \text{ energy/unit time} \tag{3.17}$$

One horsepower is defined as the ability to perform 33,000 ft-lbf of work in 1 min.

$$1 \text{ hp} = 33{,}000 \text{ ft-lbf/min}$$

Other common power equalities are

$$1 \text{ hp} = 2545 \text{ Btu/hr}$$

$$1 \text{ hp} = 0.746 \text{ kW}$$

The SI unit of power is the watt. Thus

$$1 \text{ J/s} = 1 \text{ N} \cdot \text{m} \cdot \text{s}^{-1} = 1 \text{ W}$$

Associated with work and power is torque. *Torque* is the turning moment exerted by a tangential force acting at a distance from the axis of rotation. Torque represents the capacity to do work, whereas power represents the rate at which work may be done. Section 13.9 explores torque, work, and power in more detail.

Forms of work other than mechanical work are seldom dominant, but neglecting them can lead to error. For instance, a film on the surface of a liquid has surface tension, which is a property of the liquid and the surroundings. The surface tension of a film is a function of the surrounding medium. A simple experiment illustrates the surface tension action. In Figure 3.8(a), the entire loop is covered with a soap film. The film is punctured within the thread loop, and the surface tension of the film acts to make the remaining area a minimum. The surface tension, σ, has units of force per unit length

$$W = -\int_{1}^{2} \sigma \, dA \text{(force/length)} \times \text{area} = \text{force} \times \text{length} \tag{3.18}$$

The minus sign indicates that an increase in area results in work being done on the system. The units of work are correct. Thus, we have seen that when pressure with units of force per unit area acts on a volume, work is performed; and now we see that surface tension acting on an area is work. Surface tension is an area phenomenon, whereas pressure is a volume phenomenon.

To demonstrate another form of work, let us consider a wire as the system. Stretch the wire within the elastic limits, and the work is done on the wire by the surroundings; F is the tension, and

$$W = - \int_1^2 F \, dL \qquad (3.19)$$

where the minus sign indicates that a positive displacement results from work being supplied to the system. If we limit the problem to within the elastic limit, where E is the modulus of elasticity, s is the stress, ϵ is the strain, and A is the cross-sectional area, then

$$F = sA = E\epsilon A$$

$$d\epsilon = dL/L_1$$

$$\delta W = -F \, dL = -E\epsilon AL_1 \, d\epsilon$$

$$W = -AEL_1 \int_1^2 \epsilon \, d\epsilon = -\frac{AEL_1}{2} (\epsilon_2^2 - \epsilon_1^2)$$

To develop expressions for electrical work and electrical power, we note that current flow i in amps is

$$i = \frac{dc}{dt}$$

where c is the charge in coulombs and t is time. Thus dc is the charge crossing a boundary during time dt. If $\mathcal{V}$ is the voltage potential, the work is

$$\delta W = \mathcal{V} \, dc$$

$$\delta W = \mathcal{V} i \, dt$$

$$W = \int_1^2 \mathcal{V} i \, dt$$

One needs to know how i and $\mathcal{V}$ vary with time to integrate the expression. The electric power is

$$\dot{W} = \lim_{\delta t \to 0} \frac{\delta W}{\delta t} = \mathcal{V} i$$

Shaft work is another work form, such as the rotating shaft in Figure 3.3. The torque, τ, is necessary to rotate the shaft $d\theta$ degrees, the shaft work is

$$W = \int_1^2 \tau \, d\theta$$

and the power is

$$\dot{W} = \int_1^2 \tau \frac{d\theta}{d\tau} = \tau\omega$$

where ω is the angular velocity and τ is a constant in this case.

There are other, less visible modes of work, such as magnetic work, which do not play a significant role in this course in thermodynamics. The following equations summarize some work forms:

Compressible fluid $\qquad W = \int_1^2 p\, dV$

Surface film $\qquad W = -\int_1^2 \sigma\, dA$

Stretched wire $\qquad W = -\int_1^2 F\, dL$ $\qquad$ (3.20)

Electrical $\qquad W = \int_1^2 \mathcal{V}\, dc = \int_1^2 \mathcal{V}i\, dt$

Shaft $\qquad W = \int_1^2 Fr\, d\theta = \int_1^2 \tau\, d\theta$

What is common to all these work expressions? That work is equal to an intensive property times the change in its extensive property. In advanced thermodynamics the intensive properties are defined in terms of the partial derivatives of the extensive property on which they act. If $\overline{F}$ is an intensive property and $d\overline{X}$ is the change in the related extensive property, then

$$\delta W = \overline{F} \cdot d\overline{X} \qquad (3.21)$$

Equation (3.21) is the general work expression for any reversible work. It might be possible to have a system in which many different forms of work are occurring; thus

$$\delta W = \sum_i \overline{F}_i \cdot d\overline{X}_i \qquad (3.22)$$

and for equation (3.20), this becomes

$$\delta W = p\, dV - \sigma\, dA - F\, dL + \mathcal{V}i\, dt + \tau\, d\theta \qquad (3.23)$$

Example 3.3
A battery charger produces 3 A at 12 V and charges a battery for 2 h. What is the work done to the battery?

Solution

Given: The flow of electrical charge from a battery charger flows into a battery.

Find: The electrical work done in the charging process.

Sketch and Given Data:

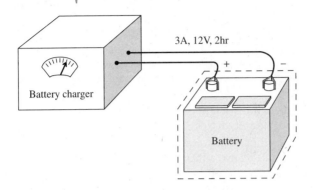

Figure 3.9

Assumptions:

1. The battery is a closed system.
2. Voltage and current are constant for the 2-hour charging period.

Analysis: The electrical work may be determined by integrating the expression

$$W = \int_1^2 \mathcal{V}i \, dt$$

In this case voltage and current are constant with time, thus

$$W = \mathcal{V}_i \int_1^2 dt = \mathcal{V}_i \int_0^{7200} dt$$

The current is flowing into the system and hence has a negative value. Notice if the charger were assumed to be the system the value of i would be positive.

$$W = (12 \text{ V})(-3 \text{ A})(7200 \text{ s})$$

$$W = -259.2 \text{ kJ}$$

Comments:

1. It is important to identify the system to determine whether the current flow is viewed as positive or negative.
2. If voltage and current vary with time, the integration is more complicated. Do not assume it is direct current unless it is correct to do so.
3. Unit balance is very important. In this case $1 \text{ W} = (1 \text{ V})(1 \text{ A}) = 1 \text{ J/s}$. Time must be in seconds to give the dimensionally correct value of work.

Heat

Having looked at work in some of its many forms, let us consider now a transfer of thermal energy between a system and its surroundings. *Heat* is defined as the energy crossing a system's boundary because of a temperature gradient between the system and its surroundings. Heat and work are similar in that they both are energy fluxes and must cross a system's boundary to have any meaning. This definition differs from the colloquial usage of heat, in that bodies do not have any property called "heat." Heat exists only as energy crossing a boundary; once inside the system it has no meaning.

Heat has conventions attached to the direction in which it crosses a system's boundary. *Heat flow into the system is positive; heat flow from the system is negative.* Heat is represented by the symbol Q. A process in which $Q = 0$ is *adiabatic.* For instance, if a pipe is well insulated from its surroundings, then any heat flow between the system and its surroundings is very small, in fact negligible in most cases. If it were zero, the pipe would be adiabatically insulated. If we were to model the pipe as the system, it might be possible to assume that the walls were adiabatically insulated. This simplifies the analysis to be made without introducing errors of any great magnitude. Of course, if there actually were substantial heat flow through the walls, then to assume that the walls were adiabatic would be extremely inaccurate.

Heat is a function of the method in which thermal energy is transferred between a system and its surroundings. This means that heat is a path function; in other words, it depends on how the energy is transferred.

$$Q = \int_1^2 \delta Q \text{ kJ } \text{ [Btu]} \qquad (3.24)$$

Q is the heat transferred when a system goes from state 1 to state 2 for any given process. Another convenient notation is heat per unit mass, q.

$$q = \frac{Q}{m} \text{ kJ/kg } \text{ [Btu/lbm]} \qquad (3.25)$$

where m is the system mass. The integral of δQ may be evaluated only when the process is reversible (continuous)—the only time a functional relationship may be determined. Note that the δ operator must be used.

Modes of Heat Transfer

In equilibrium thermodynamics we deduce the heat flow by knowing the other energy terms in the equation for conservation of energy. It is possible to determine the flow of heat directly, however, and a course in heat transfer will yield this information in detail, as will Chapter 18. Here we will merely survey the variety of modes of heat transfer.

Many heat transfer problems require the inclusion of time, so their analyses involve more than investigation of equilibrium states. The laws of heat transfer obey the first and second laws of thermodynamics: energy is conserved, and heat must flow from hot to cold. There are three modes of heat transfer—conduction, radiation, and convection.

Conduction is heat transfer within a medium. In solids, particularly metals, conduction is due to (a) the drift of free electrons and (b) phonon vibration. At low temperatures phonon vibration, the vibration of the crystalline structure, is the primary mechanism for conduction, and at higher temperatures electron drift is the primary mechanism. Regardless of the mechanism, energy is transferred from one atom or molecule to another, resulting in a flow of energy within a medium. In a gas

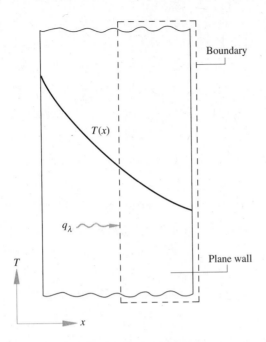

Figure 3.10 Illustration of Fourier's law of heat conduction.

the mechanism for conduction is primarily molecular collision. The conduction is dependent on pressure and temperature, which act in obvious ways to increase the chance of molecular collisions. In liquids the mechanism for conduction is the combination of electron drift and molecular collision. Conduction in liquids most commonly depends on temperature, not pressure.

Conduction heat transfer follows Fourier's law, which states that the conductive heat flow, $\dot{Q}_\lambda$, is a product of the thermal conductivity of the material, λ, the area normal to the heat flow, A, and the temperature gradient, dT/dx, across the area. Thus,

$$\dot{Q}_\lambda = -\lambda A \frac{dT}{dx}$$

Figure 3.10 illustrates Fourier's law. The reason for the inclusion of the minus sign in Fourier's law is that we want the heat flow to be positive when the temperature gradient is negative (hot to cold) in the $+x$ direction, and the minus sign corrects for this. Notice that the temperature gradient is the potential that causes heat flow through a material. The values of the thermal conductivity vary from high for metal conductors to low for nonmetallic insulators.

Radiation is the flow of thermal energy, via electromagnetic waves, between two bodies separated by a distance. Electromagnetic waves, which are a function of body-surface temperature, transfer heat and thus constitute thermal radiation. The

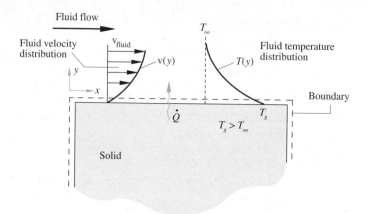

Figure 3.11 Illustration of Newton's law of cooling, showing velocity and temperature distributions.

expression for radiative heat transfer from a surface is given by the expression

$$\dot{Q} = \epsilon \sigma A T^4$$

where ϵ is the surface's emissivity, σ is the Stefan-Boltzmann constant, A is the surface area, and T is the surface absolute temperature. This is a highly nonlinear expression and represents only the radiation from a surface; it does not include the effect of radiation from other surfaces and any surrounding fluid. The final equations describing radiative heat transfer include the effect of geometrical orientation of the various surfaces, surface material properties, and the effect of any intervening fluids.

Convection is heat transfer between a solid surface and a moving fluid. At the solid-fluid interface heat is transferred by conduction, energy transfer resulting from molecular collisions between the solid and fluid molecules. These collisions cause a temperature change in the fluid, a density variation is produced, and bulk fluid motion occurs. There is a mixing of the high-and-low temperature fluid elements, and heat is transferred between the solid and fluid by convection. This type of convection is called *natural convection* in that the fluid motion results from fluid density variations. In *forced convection* an external device such as a pump or fan creates the fluid motion.

The empirical expression that describes this motion is Newton's law of cooling:

$$\dot{Q}_c = hA(I_s - I_\infty)$$

Figure 3.11 illustrates the fluid parameters implicit in Newton's law: A is the solid-fluid surface area, h is the unit coefficient for convective heat transfer, T_s is the solid surface temperature, and T_∞ is the fluid temperature far from the surface. The coefficient for convective heat transfer is not a thermodynamic property but a parameter that includes such effects as fluid properties, flow pattern, and surface geometry.

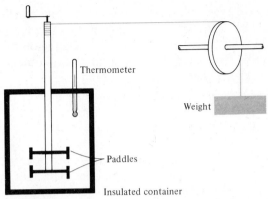

Figure 3.12 Joule's experiment on the equivalence of heat and work.

Convertibility of Energy

We have seen that both foot-pound force (ft-lbf) and the British thermal unit (Btu) are energy forms, but how are they related? Starting in 1845, an Englishman, James Prescott Joule, initiated experiments to find the equivalences between heat (Btu) and mechanical work (ft-lbf). Figure 3.12 illustrates a simplified version of his experiment. A closed, insulated container is fitted with a spindle to which paddles are attached. A wire is wound around the external spindle shaft and led over a pulley to a weight (force). The weight is dropped at constant velocity, eliminating all acceleration but local gravity. The temperature increases in the water after the weight is dropped. From the definition of a British thermal unit—at the time, the heat required to raise 1 lbm of water from 59.5° to 60.5°F—Joule was able to calculate the equivalence between mechanical energy and thermal energy. This has been refined up until today, when

$$778.169 \text{ ft-lbf} = 1 \text{ Btu}$$

Actually, the British thermal unit is defined in terms of joules, but the foregoing definition is quite adequate. In SI units the joule is the basic unit of energy.

$$1 \text{ N} \cdot \text{m} = 1 \text{ J}$$

The joule, the British thermal unit, and foot-pound force are related as follows:

$$1 \text{ Btu} = 1055 \text{ J} = 1.055 \text{ kJ}$$

$$1 \text{ ft-lbf} = 1.355 \text{ J}$$

Potential, Kinetic, and Internal Energies

There are many types of energy forms, but we will discuss only three: gravitational potential, kinetic, and internal energies—the ones that occur in the typical situations we will be analyzing.

The potential energy of a system mass depends on its position in the gravitational force field. There must be a reference datum and a distance from the datum to the system mass; let this distance be z. Let the system be acted upon by a force F, and assume that this force raises the system. The change in potential energy is equal to the work (force through a distance) necessary to move the system.

$$d(\text{P.E.}) = F\,dz = mg\,dz$$

$$\int_1^2 d(\text{P.E.}) = \int_1^2 mg\,dz = m\int_1^2 g\,dz \qquad (3.26)$$

The limits are from the initial to the final position. If the distance from 1 to 2 is not too great, so that the gravitational acceleration, g, is essentially constant, equation (3.26) becomes

$$(\text{P.E.})_2 - (\text{P.E.})_1 = mg(z_2 - z_1)\,\text{N}\cdot\text{m} = \frac{mg}{g_c}(z_2 - z_1)\,\text{ft-lbf} \qquad (3.27)$$

Thus, we see that our assumption about the form of this energy equation is realistic: the potential energy of a system increases as the height increases.

The kinetic energy of a system is developed in an analogous manner. Again let us consider a system of mass m. Let a horizontal force act on the system and move it a distance dx. Since it moves horizontally, there is no change in potential energy. The change in kinetic energy is defined as the work in moving the system a distance dx.

$$d(\text{K.E.}) = F\,dx$$

$$F = \frac{d(mv)}{dt} = m\frac{dv}{dt} + v\frac{dm}{dt}^{\,0}$$

$$\frac{dv}{dt} = \frac{dv}{dx}\frac{dx}{dt} = v\frac{dv}{dx}$$

$$d(\text{K.E.}) = mv\frac{dv}{dx}dx = mv\,dv \qquad (3.28)$$

Integrating

$$\int_1^2 d(\text{K.E.}) = m\int_1^2 v\,dv$$

$$(3.29)$$

$$(\text{K.E.})_2 - (\text{K.E.})_1 = \frac{m}{2}(v_2^2 - v_1^2)\,\text{N}\cdot\text{m} = \frac{m}{2g_c}(v_2^2 - v_1^2)\,\text{ft-lbf}$$

Note that this is the translational velocity of the system, and if the velocity of the system increases, there is a positive change of kinetic energy. If a car is moving at 40 km/h and is accelerated (force through a distance) to 80 km/h, the kinetic energy increases. If the car's velocity is zero, its kinetic energy is zero.

One of the less tangible forms of energy of a substance is its internal energy. This is the energy associated with the substance's molecular structure. Although we cannot measure internal energy, we can measure changes of internal energy. We are

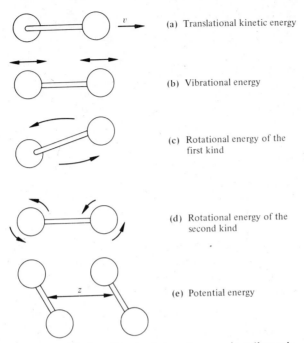

(a) Translational kinetic energy

(b) Vibrational energy

(c) Rotational energy of the first kind

(d) Rotational energy of the second kind

(e) Potential energy

Figure 3.13 Modes of storing internal energy in a diatomic molecule.

aware of internal energy from everyday experiences—bending a paper clip, compressing a gas, or compressing a spring. For instance, when a spring is compressed, its final energy is greater than its initial value, as a result of the compression. The spring's—the system's—energy change is not a result of changes in kinetic or potential energies, but of changes in the system's internal energy.

The internal energy of a diatomic molecule can be visualized as a function of four energies associated with the molecule's motion, rotation, and vibration, and the energy of position. Energy is stored in the molecular structure by these five modes. The more complex the molecular structure, the more modes are available for storing energy. A monatomic gas has less than five modes of energy storage, but we will take the general case first. In Figure 3.13 the different forms of a diatomic molecule are illustrated. The energy associated with the translational motion of the molecule is *translational kinetic energy*. The back-and-forth movement of the atoms in the molecule, toward and away from one another, is a *vibrational energy*. The atoms, as a pair, are visualized as rotating around their center of mass; this is the "first" kind of *rotational energy*. An atom may rotate around its own center of mass, and this is the "second" kind of rotational energy. As the number of atoms in the molecule increases, the complexity of the rotational energy increases. The last term in the internal energy model is the potential energy of attraction between adjacent molecules. This force is quite large, as can be realized by the energy it takes to convert water to steam by overcoming the attractive force of the liquid molecules.

The symbol for specific internal energy is u, for total internal energy, U.

$$u = \text{specific internal energy, J/kg} \quad [\text{Btu/lbm}] \qquad (3.30a)$$

$$U = mu, \text{total internal energy, J} \quad [\text{Btu}] \qquad (3.30b)$$

Many other energies actually contribute to the internal energy, such as electron spin, but we are more interested in using internal energy than in building models of it.

Several system energy properties have now been defined and discussed: kinetic energy, potential energy, and internal energy. Heat and work, which are energy forms that cross a system's boundary but are not properties, have also been analyzed. Chemical energy will be discussed in detail in Chapter 12. It may be possible that for a given system all work types, energies, and heat must be included, but that is very rare. Most often, only a few are important for each process.

3.3 FIRST COROLLARY OF THE FIRST LAW

There are two types of systems: fixed mass (closed) and fixed space (open). The first corollary of the first law of thermodynamics is the application of the conservation of energy to closed systems; the second corollary is the application of the conservation of energy to open systems. We will consider the closed system first.

$$\text{Final energy} - \text{initial energy} = \text{energy added to system}$$

$$E_2 \qquad - \qquad E_1 \qquad = \qquad Q - W \qquad (3.31)$$

where heat has been added $(+Q)$ and work has been done on the system $(-W)$.

Let us examine the components of the initial energy of the system, E_1. What energy forms cannot be in E_1? Since work and heat exist only when crossing the boundary, they cannot exist as energy at a system state, unlike kinetic, potential, and internal energies. Thus

$$E = \left(U + \frac{mv^2}{2} + mgz \right) \text{J} = \left(U + \frac{mv^2}{2g_c \mathcal{J}} + \frac{mg}{g_c \mathcal{J}} z \right) \text{Btu} \qquad (3.32)$$

or dividing by the system mass, m,

$$e = \left(u + \frac{v^2}{2} + gz \right) \text{J/kg} = \left(u + \frac{v^2}{2g_c \mathcal{J}} + \frac{gz}{\mathcal{J} g_c} \right) \frac{\text{Btu}}{\text{lbm}} \qquad (3.33)$$

The symbol $\mathcal{J}$ denotes the conversion from foot-pound force to British thermal units.

$$\mathcal{J} = 778.169 \text{ ft-lbf/Btu}$$

In solving problems it is often easier to find the solution per unit mass and multiply by the system mass at the conclusion.

Let us introduce heat and work into the energy equation by having a change occur. Assume the system does work W and receives heat Q, as in Figure 3.14. Let the piston expand, doing work, and the system receives heat from the surroundings. If the system operates between initial and final states, then the first law of thermodynamics,

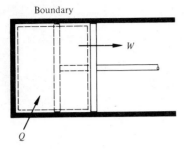

Boundary

Figure 3.14 A piston-cylinder system receives heat and does work.

equation (3.31), becomes

$$Q = E_2 - E_1 + W \qquad (3.34)$$

Q and W are both positive in the thermodynamic sense. Solve equation (3.34) for Q and divide by the system mass, m:

$$q = (e_2 - e_1) + w \qquad (3.35)$$

$$q = \left[(u_2 - u_1) + \frac{v_2^2 - v_1^2}{2} + g(z_2 - z_1) + w \right] J/kg \qquad (3.36)$$

Usually the kinetic and potential energies are neglected, especially for closed systems, so

$$q = (u_2 - u_1) + w \qquad (3.37)$$

or

$$q = \Delta u + w$$

If the derivative of equation (3.37) is taken,

$$\delta q = du + \delta w \qquad (3.38)$$

the total system change is

$$\delta Q = dU + \delta W \qquad (3.39)$$

Equations (3.38) and (3.39) are commonly referred to as the first-law equations for a closed system. You have seen they are nothing more than an energy balance.

Example 3.4
A piston-cylinder contains 2 kg of steam, which expands from state 1 ($u_1 = 2700$ kJ/kg) to state 2 ($u_2 = 2650$ kJ/kg). During the expansion process the system receives 30 kJ of heat. Determine the system work.

Solution

Given: A piston-cylinder expands from state 1 to state 2 while receiving heat. The change in specific internal energy is given.

Find: The piston work done during the expansion process.

Sketch and Given Data:

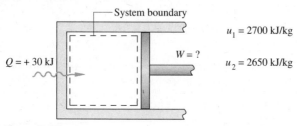

Figure 3.15

Assumptions:

1. The steam is a closed system.
2. Changes in kinetic and potential energies can be neglected.

Analysis: The first law for a closed system is

$$Q = \Delta U + \Delta K\!.E. ^{0} + \Delta P\!.E.^{0} + W$$

Write the change of system internal energy in terms of specific internal energy.

$$Q = m(u_2 - u_1) + W$$

Substitute numerical values.

$$30 \text{ kJ} = (2 \text{ kg})(2650 - 2700) \text{ kJ/kg} + W$$

$$30 \text{ kJ} = -100 \text{ kJ} + W$$

$$W = +70 \text{ kJ}$$

Comments:

1. The system work is positive, indicating the steam did work on the piston and expands during the change of state. System volume increased.
2. No information is known about the expansion process; it may be reversible or irreversible, as the first law is independent of the process. ▬

Example 3.5

An adiabatic tank similar to the one that Joule used in determining the mechanical-thermal energy equalities contains 10 kg of water. A 110-V, 0.1-A motor drives a paddle wheel and runs for 1 h. Determine the change of specific and total internal energy of the water.

Solution

Given: An adiabatic tank containing water receives work from a motor-driven paddle.

Find: The change of the water's specific and total internal energy.

Sketch and Given Data:

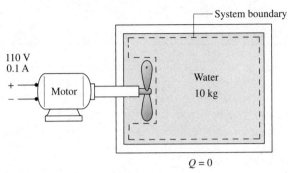

Figure 3.16

Assumptions:

1. The water is a closed system.
2. The process is adiabatic, so the heat transfer is zero.
3. All the electrical work is converted into paddle work.
4. Changes in system kinetic and potential energies are zero.

Analysis: The first law for a closed system is

$$Q = \Delta U + \Delta K.E.^{\;0} + \Delta P.E.^{\;0} + W$$
$$Q = \Delta U + W$$

The work term is only paddle work, W_p, which is done on the system (negative), and the system is adiabatic ($Q = 0$).

$$0 = \Delta U - W_p$$
$$\Delta U = m(u_2 - u_1) = W_p$$

The paddle work needs to be determined from the power consumption data.

$$\dot{W}_{electrical} = \mathscr{V}_i = (110 \text{ V})(0.1 \text{ A}) = 11 \text{ W} = 11 \text{ J/s}$$

$$W_{electrical} = (\dot{W}_{electrical})(t) = (11 \text{ J/s})(3600 \text{ s}) = 39.6 \text{ kJ}$$

$$\Delta U = 39.6 \text{ kJ}$$

$$\Delta u = \Delta U/m = 39.6 \text{ kJ}/10 \text{ kg} = 3.96 \text{ kJ/kg}$$

Comments:

1. Understanding the term *adiabatic,* meaning the heat transfer is zero, is pivotal to the problem solution.
2. Paddle work only goes into a system and hence is negative.

3. It is important to distinguish between specific and total internal energy, making sure the first-law equation is dimensionally correct.   ▬

3.4 ENERGY AS A PROPERTY

In the preceding analysis we have tacitly assumed that the system energy is a property. The use of an exact differential in describing energy, for instance, implies this. In the following development we will prove that this assumption is valid. First, a cycle analysis of a closed system will be made. Referring to the cycle definition we see that the system state is the same before and after a cyclic process. We assume that both heat and work enter and leave the system as it completes the cyclic process. From equation (3.35), noting that $E_1 = E_2$, we may write

$$Q_{in} + W_{in} = Q_{out} + W_{out}$$

$$Q_{in} - Q_{out} = W_{out} - W_{in}$$

$$\Sigma Q = \Sigma W \tag{3.40}$$

This may be written as a cyclic integral

$$\oint \delta Q = \oint \delta W \tag{3.41}$$

or for any cyclic process

$$\oint \delta Q - \oint \delta W = 0 \tag{3.42}$$

If

$$\delta Q = dE + \delta W \qquad \text{or} \qquad \delta Q = \delta W = dE$$

then

$$\oint \delta Q - \oint \delta W = \oint dE = 0$$

Hence, the system energy depends only on the system state and not on the path followed to arrive at the state; it is a property. Also, the components of the total energy are properties, so

$$\oint dU = \oint d(\text{K.E.}) = \oint d(\text{P.E.}) = 0 \tag{3.43}$$

It is also important to note from equation (3.40) that $\Sigma Q = \Sigma W$ for any system undergoing a cyclic process. The individual work and heat terms may be positive or negative, depending on direction. The algebraic sum of these terms, then, is the net heat and work, the conclusion being that the net heat is equal to the net work for a closed system operating on a cycle.

3.5 SECOND COROLLARY OF THE FIRST LAW

We will consider now the system in which there is a mass flow, namely, the open system. Here the volume is fixed, and the mass flows into and out of this fixed space, or control volume. This is a very common situation: in a jet engine, air enters and leaves the engine; in a steam turbine, the steam enters and leaves the turbine. In both cases work is done, heat is exchanged, and energy is transferred.

The second corollary of the first law is the conservation of energy principle applied to open systems. Consider the control volume shown in Figure 3.17 to be located in a stream of moving fluid. Heat may be transferred and work may be done, as illustrated.

At some time t there is a control mass, which is the sum of the masses in regions I and II. At some time $t + \Delta t$, some of the mass has moved into region III, and heat and work interactions may have occurred to the control mass. An energy balance on the control mass for the time interval Δt is

$$Q - W = \Delta E_{CM} = E_{CMt+\Delta t} - E_{CMt} \tag{3.44}$$

where Q and W are, respectively, the heat and work done on the control mass.

The energy of the control mass may be expressed in terms of the control volume:

$$E_{CMt} = E_{IIt} + E_{It} \tag{3.45}$$

$$E_{CMt+\Delta t} = E_{IIt+\Delta t} + E_{IIIt+\Delta t} \tag{3.46}$$

Subtract equation (3.45) from equation (3.46) and add zero to the right-hand side. In

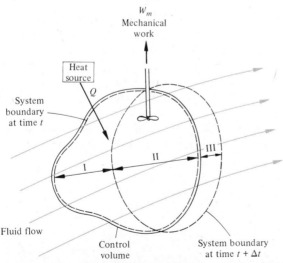

Figure 3.17 Control volume located in a moving fluid with heat and work being transferred.

this case $E_{\mathrm{III}t} = 0$ and $E_{\mathrm{I}t+\Delta t} = 0$.

$$E_{\mathrm{CM}t+\Delta t} - E_{\mathrm{CM}t} = E_{\mathrm{II}t+\Delta t} - E_{\mathrm{II}t} + E_{\mathrm{I}t+\Delta t} - E_{\mathrm{I}t} + E_{\mathrm{III}t+\Delta t} - E_{\mathrm{III}t}$$

Divide by Δt and take the limit as $\Delta t \to 0$. Again region II coincides with the control volume as $\Delta t \to 0$, hence $dE_{\mathrm{II}}/dt = dE_{cv}/dt$.

$$\frac{dE_{\mathrm{CM}}}{dt} = \frac{dE_{cv}}{dt} - \frac{dE_{\mathrm{I}}}{dt} + \frac{dE_{\mathrm{III}}}{dt} \tag{3.47}$$

The negative sign on dE_1/dt is due to the difference $(E_{\mathrm{I}t+\Delta t} - E_{\mathrm{I}t})$ being negative in the physical sense: the energy flow in region I is decreasing. Furthermore, we let

$$\lim_{\Delta t \to 0} \frac{Q}{\Delta t} = \frac{\delta Q}{dt} = \dot{Q} \tag{3.48}$$

and

$$\lim_{\Delta t \to 0} \frac{W}{\Delta t} = \frac{\delta W}{dt} = \dot{W} \tag{3.49}$$

The derivative dE_1/dt represents the energy entering the control volume, dE_{in}/dt; and dE_{III}/dt represents the energy leaving, dE_{out}/dt. Thus equation (3.44) becomes

$$\frac{dE_{cv}}{dt} = \dot{Q} - \dot{W} + \frac{dE_{\mathrm{in}}}{dt} - \frac{dE_{\mathrm{out}}}{dt} \tag{3.50}$$

The total energy may be represented as $E = em$, and $dE/dt = e\dot{m}$; hence

$$\frac{dE_{cv}}{dt} = \dot{Q} - \dot{W} + e_{\mathrm{in}}\dot{m}_{\mathrm{in}} - e_{\mathrm{out}}\dot{m}_{\mathrm{out}} \tag{3.51}$$

The development is almost complete, but we must account for the different work terms. There is mechanical work done by the fluid on the machinery (turbine, compressor); there is also flow work, the energy transmitted across the system boundary as a result of a pumping process occurring outside the system, causing the fluid to enter the system. This is the work by the fluid to overcome the normal stress, that is, pressure, at the boundary. Thus there is a flow-work term entering and leaving the system boundary.

Let us consider a differential element on the control surface with area A. The fluid has a pressure p at area A, and the fluid moves a distance dx normal to the area. The flow work is

$$\delta W_{\mathrm{flow}} = pA\,dx$$
$$A\,dx = v\,\mathrm{dm}$$
$$\delta W_{\mathrm{flow}} = pv\,\mathrm{dm}$$
$$\dot{W}_{\mathrm{flow}} = pv\,\dot{m}$$

The net flow work is the difference between the flow work entering and leaving.

Equation (3.51) becomes

$$\frac{dE_{cv}}{dt} = \dot{Q} - \dot{W}_m + (e + pv)_{in}\,\dot{m}_{in} - (e + pv)_{out}\,\dot{m}_{out} \qquad (3.52)$$

For the steady state, $dE_{cv}/dt = 0$, and the first law may be stated as an energy balance:

$$\text{Energy in} = \text{Energy out}$$

$$\dot{Q} + (e + pv)_{in}\,\dot{m}_{in} = (e + pv)_{out}\,\dot{m}_{out} + \dot{W}_m \qquad (3.53)$$

and for steady-flow conditions, $\dot{m}_{in} = \dot{m}_{out} = \dot{m}$,

$$\dot{Q} + (e + pv)_{in}\,\dot{m} = (e + pv)_{out}\,\dot{m} + \dot{W}_m \qquad (3.54)$$

Separating the total energy into three components—internal, kinetic, and potential energies—yields for the energy equation on the unit mass basis

$$q + u_1 + p_1 v_1 + (\text{k.e.})_1 + (\text{p.e.})_1 = u_2 + p_2 v_2 + (\text{k.e.})_2 + (\text{p.e.})_2 + w \qquad (3.55a)$$

$$q = \Delta u + \Delta(pv) + \Delta(\text{k.e.}) + \Delta(\text{p.e.}) + w \qquad (3.55b)$$

$$\delta q = du + d(pv) + d(\text{k.e.}) + d(\text{p.e.}) + \delta w \qquad (3.55c)$$

and so

$$\dot{Q} + \dot{m}[u_1 + p_1 v_1 + (\text{k.e.})_1 + (\text{p.e.})_1]$$
$$= \dot{m}[u_2 + p_2 v_2 + (\text{k.e.})_2 + (\text{p.e.})_2] + \dot{W} \qquad (3.56)$$

Two energy properties u and pv appear on both sides of the equation in the energy term. Combining them creates a new property, h.

$$h \equiv u + pv$$
$$H \equiv U + pV \qquad (3.57)$$

The sum of these two properties, h, is called *enthalpy*. Note that u and pv should have compatible units. At times there may be a temptation to associate enthalpy with heat or work, but enthalpy is a property having no function physically other than being the sum $U + pV$.

Equations (3.55b) and (3.56) become, respectively,

$$q = \Delta h + \Delta(\text{k.e.}) + \Delta(\text{p.e.}) + w \qquad (3.58a)$$

and

$$\dot{Q} + \dot{m}[h_1 + (\text{k.e.})_1 + (\text{p.e.})_1] = \dot{m}[h_2 + (\text{k.e.})_2 + (\text{p.e.})_2] + \dot{W} \qquad (3.58b)$$

The first law may be written on a differential basis as

$$\delta q = dh + d(\text{k.e.}) + d(\text{p.e.}) + \delta w \qquad (3.59)$$

Equation (3.59) has been developed with regard to the control volume. We will consider the unit mass of the system; the first law for the mass is

$$\delta q = du + p\,dv$$

From the definition of enthalpy,

$$dh = du + p\,dv + v\,dp$$

Therefore,

$$dh = \delta q + v\,dp \qquad (3.60)$$

Substituting equation (3.60) into equation (3.59),

$$\delta w = -v\,dp - d(\text{k.e.}) - d(\text{p.e.}) \qquad (3.61)$$

If the kinetic and potential energy variations are zero,

$$\delta w = -v\,dp \qquad (3.62\text{a})$$

$$\dot{W} = -\dot{m}\int v\,dp \qquad (3.62\text{b})$$

The work in equation (3.61) is valid for reversible processes, since that assumption was used in expressing $\delta w = p\,dv$ for constant mass. The expression for power in equation (3.62b) is the power transferred to or from the fluid to a rotating shaft in the control volume. Examples include a turbine's rotor and a pump's impeller.

Example 3.6

A ship's steam turbine receives steam at a pressure of 900 psia and a velocity of 100 ft/sec. The steam leaves the turbine at 27 in. Hg vacuum with a velocity of 900 ft/sec. The turbine entrance is 10 ft above the exit level. Find the power produced if the mass flow rate of steam is 100,000 lbm/hr and the heat loss from the turbine is 50,000 Btu/hr. The steam has the following properties.

	Inlet	Exit
Pressure	900 psia	27 in. Hg vacuum
Temperature	1000°F	114°F
Velocity	100 ft/sec	900 ft/sec
Specific internal energy	1354.5 Btu/lbm	951 Btu/lbm
Specific volume	0.9273 ft³/lbm	214 ft³/lbm

Solution

Given: A turbine receives steam at a high pressure and discharges it at a low pressure, doing work in the expansion process.

Find: The power produced by the steam turbine.

Sketch and Given Data: See Figure 3.18 on the following page.

Assumptions:

1. The turbine is an open system with steam flowing through it.
2. The sign on the heat flow is negative as it flows from the system to the surroundings.
3. The flow is steady-state, so the flow rate in equals the flow rate out.

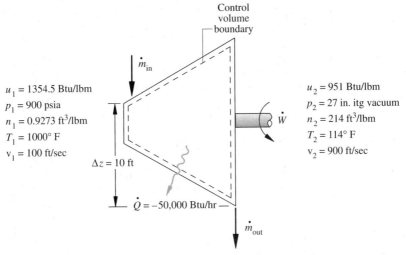

$u_1 = 1354.5$ Btu/lbm
$p_1 = 900$ psia
$n_1 = 0.9273$ ft^3/lbm
$T_1 = 1000°$ F
$v_1 = 100$ ft/sec

$\Delta z = 10$ ft

$\dot{Q} = -50,000$ Btu/hr

$u_2 = 951$ Btu/lbm
$p_2 = 27$ in. itg vacuum
$n_2 = 214$ ft^3/lbm
$T_2 = 114°$ F
$v_2 = 900$ ft/sec

Figure 3.18

Analysis: The first law for an open system is

$$\dot{Q} + \dot{m}[u + pv + (\text{k.e.}) + (\text{p.e.})]_{\text{in}} = \dot{W} + \dot{m}[u + pv + (\text{k.e.}) + (\text{p.e.})]_{\text{out}}$$

In this equation all the terms are known except the power, W. However, care must be taken to assure that all the terms are dimensionally correct. First, the pressure must be converted to absolute pressure. (Subscript 1 refers to inlet state, 2 to outlet state.)

$$p = 27 \text{ in. Hg vacuum} = 2.92 \text{ in. Hg absolute} = 1.43 \text{ psia}$$

$$\dot{W} = \dot{m}\{(u_1 - u_2) + (p_1 v_1 - p_2 v_2) + [(\text{k.e.})_1 - (\text{k.e.})_2] + [(\text{p.e.})_1$$
$$- (\text{p.e.})_2\} + \dot{Q}$$

$$u_1 - u_2 = 1354.5 - 951 = +403.5 \text{ Btu/lbm}$$

$$p_1 v_1 - p_2 v_2 = \{(900 \text{ lbf/in.}^2)(144 \text{ in.}^2/\text{ft}^2)(0.9273 \text{ ft}^3/\text{lbm})$$
$$- (1.43 \text{ lbf/in.}^2)(144 \text{ in.}^2/\text{ft}^2)(214 \text{ ft}^3/\text{lbm})\}/(778.16 \text{ ft-lbf/Btu})$$

$$p_1 v_1 - p_2 v_2 = 97.81 \text{ Btu/lbm}$$

$$(\text{k.e.})_1 - (\text{k.e.})_2 = \left(\frac{v_1^2 - v_2^2}{2 g_c \mathscr{J}}\right)$$

$$= [(100^2 - 900^2)\text{ft}^2/\text{sec}^2]/[(2)(32.174 \text{ lbm-ft}/\text{lbf-sec}^2)(778.16 \text{ ft-lbf/Btu})]$$

$$(\text{k.e.})_1 - (\text{k.e.})_2 = -15.98 \text{ Btu/lbm}$$

$$(\text{p.e.})_1 - (\text{p.e.})_2 = g(z_1 - z_2)/g_c \mathscr{J} = (32.174 \text{ ft/sec}^2)(10 \text{ ft})/$$
$$\{(32.174 \text{ lbm-ft}/\text{lbf-sec}^2)(778.16 \text{ ft-lbf/Btu})\}$$

$$(\text{p.e.})_1 - (\text{p.e.})_2 = +0.01 \text{ Btu/lbm}$$

The sum of the terms in the brackets is $+485.34$ Btu/lbm. The power is

$$\dot{W} = -50,000 \text{ Btu/hr} + (100,000 \text{ lbm/hr})(485.34 \text{ Btu/lbm})$$

$$\dot{W} = 4.848 \times 10^7 \text{ Btu/hr} = 19,051 \text{ hp}$$

Comments:

1. The implication of the minus sign on the kinetic energy term indicates this portion of the steam's energy was not converted into work, as the steam left the turbine with a high exit velocity. Ideally the kinetic energy would be converted to work. Note this is possible from the original derivation of kinetic energy.
2. It is very important to maintain dimensionally correct equations. Attention is particularly needed when performing problems in the English unit system.
3. The turbine's power is calculated by determining the power imparted to the turbine rotor by the steam and assuming they are the same.
4. Note the magnitude of the potential energy term. Very often in problems the change of potential energy is neglected as being numerically insignificant compared to the other energy terms.   ▬

Example 3.7

A nozzle is a device that converts fluid thermal energy, enthalpy, into kinetic energy. The kinetic energy may be used to drive a mechanical device such as a turbine wheel, thus converting the fluid energy into mechanical work. A nozzle receives 0.5 kg/s of air at a pressure of 2700 kPa and a velocity of 30 m/s and with an enthalpy of 923.0 kJ/kg, and the air leaves at a pressure of 700 kPa and with an enthalpy of 660.0 kJ/kg. Determine the exit velocity from the nozzle for (a) adiabatic flow; (b) flow where the heat loss is 1.3 kJ/kg.

Solution

Given: Air flows through a nozzle, a device that converts fluid thermal energy into kinetic energy.

Find: The air's exit velocity from the nozzle.

Sketch and Given Data:

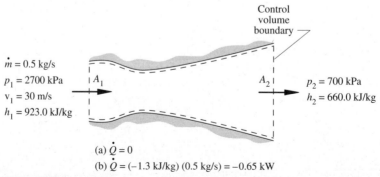

$\dot{m} = 0.5$ kg/s
$p_1 = 2700$ kPa
$v_1 = 30$ m/s
$h_1 = 923.0$ kJ/kg

$p_2 = 700$ kPa
$h_2 = 660.0$ kJ/kg

(a) $\dot{Q} = 0$
(b) $\dot{Q} = (-1.3 \text{ kJ/kg})(0.5 \text{ kg/s}) = -0.65 \text{ kW}$

Figure 3.19

Assumptions:

1. The flow rate is steady, the system open.
2. The work crossing the system control volume is zero.
3. The change in potential energy from inlet to exit can be neglected.

Analysis: Case (a): The first law for an open system with steady flow (subscript 1 refers to inlet state and 2 refers to exit state) is

$$\dot{Q} + \dot{m}[h + \text{k.e.} + \text{p.e.}]_{\text{in}} = \dot{W} + \dot{m}[h + \text{k.e.} + \text{p.e.}]_{\text{out}}$$

$$(\text{k.e.})_2 = (h_1 - h_2) + (\text{k.e.})_1 + [(\text{p.e.})_1 - (\text{p.e.})_2]$$

$$\frac{v_2^2}{2} = (h_1 - h_2) + (\text{k.e.})_1$$

where $\dot{W}$ was eliminated by assumption 2, Δp.e. by assumption 3, and $\dot{Q}$ by adiabatic flow for case (a).

$$\frac{v^2 \text{m}^2/\text{s}^2}{2(1000)\, \text{J}/\text{kJ}} = (923.0 - 660)\, \text{kJ}/\text{kg} + \frac{30^2 \text{m}^2/\text{s}^2}{2000\, \text{J}/\text{kJ}}$$

$$v_2 = 725.9 \text{ m/s}$$

Case (b): The heat loss $q = -1.3$ kJ/kg. The first law for a steady open system is

$$\dot{Q} + \dot{m}[h + \text{k.e.} + \text{p.e.}]_{\text{in}} = \dot{W} + \dot{m}[h + \text{k.e.} + \text{p.e.}]_{\text{out}}$$

Divide by $\dot{m}$ and solve for the kinetic energy out.

$$(\text{k.e.})_2 = q + (h_1 - h_2) + (\text{k.e.})_1 + [(\text{p.e.})_1 - (\text{p.e.})_2]$$

$$\frac{v_2^2}{2} = q + (h_1 - h_2) + (\text{k.e.})_1$$

$$\frac{v_2^2 \text{m}^2/\text{s}^2}{2(1000)\, \text{J}/\text{kJ}} = -1.3\, \text{kJ}/\text{kg} + (923.0 - 660)\, \text{kJ}/\text{kg} + \frac{30^2 \text{m}^2/\text{s}^2}{2000\, \text{J}/\text{kJ}}$$

$$v_2 = 724.1 \text{ m/s}$$

Comments:

1. The effect of heat transfer from the nozzle reduces the velocity of the air exiting the nozzle, as that energy cannot be converted into kinetic energy.
2. It is important to maintain correct unit balances in converting specific kinetic energy to kilojoules/kilograms.
3. To obtain work to or from a control volume, shaft rotation is required. There will never be work from a nozzle, as no shaft is present.   ▬

Example 3.8

A windmill has a blade diameter D and receives air with a velocity v and a density ρ. The exit velocity from the blades can be considered negligible. Derive an expression for the maximum power that the windmill can develop.

Solution

Given: A windmill receives air at a given velocity and uses the kinetic energy of the air to produce power.

Find: An expression for the maximum power that a windmill can produce.

Sketch and Given Data:

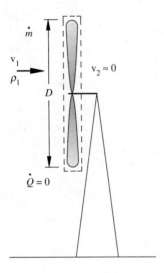

Figure 3.20

Assumptions:

1. The control volume encompassing the windmill's blades is open.
2. Steady-state, steady-flow conditions exist.
3. The exit velocity from the blades is zero.
4. The system is adiabatic, with no temperature difference between the surroundings and the system.
5. The change of potential energy is zero.
6. The change of enthalpy is zero, as there is no change of temperature or pressure across the control volume.

Analysis: The first law for an open system is

$$\dot{Q} + \dot{m}[h + \text{k.e.} + \text{p.e.}]_{\text{in}} = \dot{W} + \dot{m}[h + \text{k.e.} + \text{p.e.}]_{\text{out}}$$

$$\dot{W} = \dot{m}\{(h_1 - h_2) + [(\text{k.e.})_1 - (\text{k.e.})_2] + [(\text{p.e.})_1 - (\text{p.e.})_2]\}$$

$$\dot{W} = \dot{m}\,\frac{\text{v}^2}{2}$$

This is not sufficiently reduced to determine the power; the conservation of mass can be used to write the mass flow rate in terms of measurable properties. Thus

$$\dot{m} = \rho A v$$

and the area may be expressed in terms of the diameter as

$$A = \frac{\pi}{4} D^2$$

Substituting these into the expression for power yields

$$\dot{W} = \frac{\pi}{8} \rho D^2 v^3$$

Comments:

1. The windmill power is a function of the velocity cubed, so windmill power is very sensitive to velocity changes.
2. In actual windmills air has an exit velocity from the windmill blading, which reduces the theoretical maximum to 59.3% of the value determined in this problem. This is developed in Chapter 17.
3. Enthalpy is the sum of internal energy and pressure divided by density. In this case the pressure and density are constant and the internal energy of a air, a function of temperature at normal atmospheric conditions, is constant. Thus, their change and the consequent change of enthalpy is zero.

3.6 FURTHER EXAMPLES OF ENERGY ANALYSIS

One of the most useful features of the first law is the ease with which it can be applied to a wide variety of energy-consumption situations. In the variety of energy-consumption problems facing society, thermodynamic analysis, even a simplified first-law analysis, provides great insight.

Example 3.9
A hydroelectric power-generating facility is to be created at a location that has a change of elevation of 45 m and where the river flow rate averages 227 m³/s. Determine the maximum power that can be generated in adiabatic hydraulic turbines located at the base of the dam. The density of water is 1000 kg/m³.

Solution

Given: Water flows through a hydraulic turbine located at the base of a dam. The water decreases in elevation as it passes across the dam.

Find: The power produced by the water flowing through the turbine.

Sketch and Given Data:

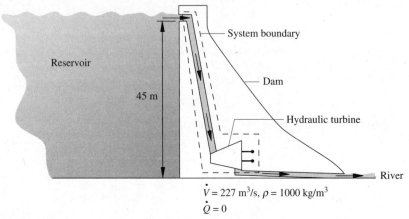

$\dot{V} = 227 \text{ m}^3/\text{s}, \rho = 1000 \text{ kg/m}^3$

$\dot{Q} = 0$

Figure 3.21

Assumptions:

1. The process is adiabatic, so the heat loss is zero.
2. The only energy property that will change is the potential energy. The water's velocity into and out of the system will be the same, hence the change of kinetic energy will be zero. In addition the internal energy, pressure, and density of the water into and out of the system will be constant; hence their change will be zero.

Analysis: Perform a first-law analysis on the control volume shown in Figure 3.21.

$$\dot{Q} + \dot{m}[u + p + \text{k.e.} + \text{p.e.}]_{\text{in}} = \dot{W} + \dot{m}[u + p + \text{k.e.} + \text{p.e.}]_{\text{out}}$$

Rearranging the terms yields

$$\dot{W} = \dot{m} \{(u_1 \overset{0}{\cancel{-u_2}}) + (p_1 v_1 \overset{0}{\cancel{-p_2 v_2}}) + [(\text{k.e.})_1 \overset{0}{\cancel{-(\text{k.e.})_2}}] + [(\text{p.e.})_1 - (\text{p.e.})_2]\}$$

$$\dot{W} = \dot{m} [(\text{p.e.})_1 - (\text{p.e.})_2]$$

Solve for the mass flow rate from the volume flow rate, using the density relating the two.

$$\dot{m} = \dot{V}\rho = (227 \text{ m}^3/\text{s})(1000 \text{ kg/m}^3) = 227\,000 \text{ kg/s}$$

$$(\text{p.e.})_1 - (\text{p.e.})_2 = g(z_1 - z_2) = (9.8 \text{ m/s}^2)(45 \text{ m}) = 441 \text{ J/kg}$$

$$\dot{W} = (227\,000 \text{ kg/s})(441 \text{ J/kg}) = 100.1 \text{ MW}$$

Comments:

1. By drawing the control boundary around the piping and the turbine, we eliminate the need to determine the pressure differential across the turbine and consider only the change of potential energy across the control surface.
2. The mass flow rate is quite large. Comparing the work performed per kilogram in this example, 0.441 kJ/kg, to that in Example 3.5, 485 Btu/lbm = 1128.9 kJ/kg,

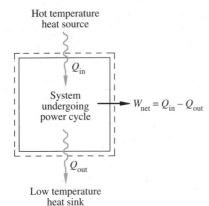

Hot temperature
heat source

Low temperature
heat sink

Figure 3.22 Schematic diagram of a power-producing cycle.

we see an order of magnitude difference in the thousands. This is typical of most renewable energy sources; the work that can be performed per unit mass is low, so it takes enormous numbers of units to reach values useful in society. The initial costs and sizes are quite large compared to conventional energy sources.

A thermodynamic cycle has special usefulness when we consider the generation of power, the transformation of heat into work on a continuous basis. When we were developing the concept of energy being a property in Section 3.4, the derivation led us to the conclusion that the net heat added to a cycle was equal to the net work produced, or

$$Q_{\text{net cycle}} = W_{\text{net cycle}} \tag{3.63}$$

These terms represent the algebraic summation of heat and work around the cycle, respectively. Figure 3.22 represents the schematic of a possible power-producing cycle. Notice that we do not need to know about all the processes within the cycle to perform an overall analysis, just the flow of work and heat energy crossing the system boundary.

The net-cycle heat term is the algebraic sum of the heat supplied, which is positive, and the heat rejected, which is negative. Thus, the net-cycle work becomes

$$W_{\text{net cycle}} = Q_{\text{in}} - Q_{\text{out}} \tag{3.64}$$

Physically these terms represent the net work produced by a power cycle, such as the electrical energy generated or the horsepower produced. The heat is supplied by such energy sources as the combustion of oil or coal, or a nuclear reaction. For an automobile's engine, a power plant, the heat rejection occurs primarily from the radiator to the surrounding atmosphere. For larger power plants used in the generation of electric power, typically the heat rejected is discharged into a body of water.

Often we are interested in how efficiently power is produced, how well we are converting heat into work. The efficiency, η, is the output, or desired effect, divided by the input, or cost of achieving that effect, or

$$\eta = \frac{W_{\text{net}}}{Q_{\text{in}}} \tag{3.65}$$

Since the net-cycle work can be expressed in terms of heat from equation (3.64), the efficiency may also be expressed in terms of only heat terms:

$$\eta = \frac{Q_{in} - Q_{out}}{Q_{in}} = 1 - \frac{Q_{out}}{Q_{in}} \tag{3.66}$$

The efficiency and cycle relationships that have been derived above apply to energy flows as well as to energy; thus power and heat flux, $\dot{W}$ and $\dot{Q}$, terms may also be used.

The efficiency of a cycle can never exceed unity and is usually much less than that value. When we discuss the second law of thermodynamics starting in Chapter 7, we will begin to see the theoretical limits imposed on efficiency.

Example 3.10

A student investigates the space requirements for generating power using solar heat. The average incident solar radiation is 1.36 W/m², or 430 Btu/hr-ft², and there are on average 3000 h/yr of sunshine in the area she lives in. The total power desired is 20 MW, and the average thermal efficiency of the power plant is 0.25. Determine the total energy that must be supplied and the collector surface area required. The power is to be delivered over a 24-h day. While it is impossible to generate power during the night and on overcast days, the assumption is that excess power may be sold during the hours of operation and repurchased at no net cost during the other hours.

Solution

Given: The incident sunshine intensity for a given locality and the net power that a solar power plant is to produce given the efficiency of that plant.

Find: The net heat flow that is required and the land area that is required to collect the solar energy for the plant.

Sketch and Given Data:

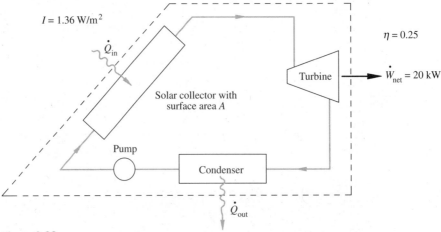

Figure 3.23

Assumption: The power produced in 3000 h can be averaged over all the hours of the year.

Analysis: The total heat flow into the cycle can be determined from equation (3.65).

$$\eta = \frac{\dot{W}_{net}}{\dot{Q}_{in}} \qquad 0.25 = \frac{20\ 000\ \text{kW}}{\dot{Q}_{in}}$$

Thus,

$$\dot{Q}_{in} = 80\ 000\ \text{kW}$$

This value represents the heat flux at any moment. We now must find the total heat supply (subject to our assumption) required for the year.

$$\dot{Q}_{in} = (80\ 000\ \text{kW})(24\ \text{h/day})(365\ \text{day/yr}) = 7.008 \times 10^8\ \text{kwh/yr}$$

This energy comes from the 3000 h of incident solar radiation, which has an intensity of 1.36 kW/m². Using the relationship between radiation intensity and heat in, namely

$$\dot{Q}\ \text{kwh/yr} = (I\ \text{kW/m}^2)(3000\ \text{h/yr})(A\ \text{m}^2)$$

yields

$$(80\ 000\ \text{kwh/yr}) = (1.36\ \text{kW/m}^2)(3000\ \text{h/yr})(A\ \text{m}^2)$$
$$A = 171\ 765\ \text{m}^2 = 42.4\ \text{acres}$$

Comment: This is certainly a staggering land area to cover with collector surface, an inherent problem with large-scale solar electric power generation plants. For water-heating systems, however, partial coverage of a house roof with collectors will often provide sufficient area for the domestic heating requirements.   ▬

Example 3.11
A power plant produces 1000 MW of electricity and uses residual oil as the fuel. The oil has a heating value, the maximum energy liberated in combustion, of 43 000 kJ/kg. The overall efficiency of the power plant is 52%. Determine the instantaneous heat input, the daily fuel consumption, and the dimensions of a cylindrical storage tank ($L = D$) to hold a 3-day supply of oil. The density of the oil is 990 kg/m³. Determine in addition the amount of heat rejected to the environment.

Solution

Given: A power plant operating on a thermodynamic cycle produces power while consuming energy, heat. The cycle efficiency and the net power produced are given.

Find: The heat input required for the cycle, the fuel necessary in kilograms per second, the size of a tank to hold a 3-day fuel supply, and the heat rejected to the environment.

Sketch and Given Data:

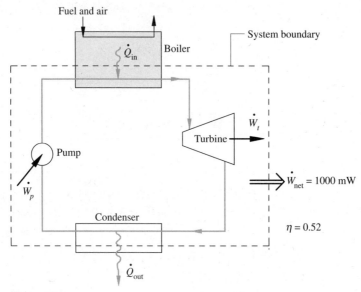

Figure 3.24

Assumptions:

1. The power plant operates on a thermodynamic cycle.
2. All the energy liberated by burning the fuel is converted to heat into the system.

Analysis: From the expression for the cycle efficiency, equation (3.65), the heat input may be determined.

$$0.52 = \frac{1000}{\dot{Q}_{in}}$$

$$\dot{Q}_{in} = 1923 \text{ MW}$$

From equation (3.64) we can determine the heat out.

$$\dot{W}_{net} = \dot{Q}_{in} - \dot{Q}_{out}$$

$$\dot{Q}_{out} = 1923 - 1000 = 923 \text{ MW}$$

The heat input is created by burning fuel, thus

$$\dot{Q}_{in} = \dot{m}_f h_{RP}$$

where $\dot{m}_f$ is the fuel flow rate in kilograms/second and h_{RP} is the heating value of the fuel. Thus, the fuel flow rate may be solved for.

$$1\,923\,000 \text{ kW} = (\dot{m}_f \text{ kg/s})(43\,000 \text{ kJ/kg})$$

$$\dot{m}_f = 44.72 \text{ kg/s}$$

The total mass for a 3-day supply is

$$m_f = (44.72 \text{ kg/s})(3600 \text{ s/h})(24 \text{ h/day})(3 \text{ day}) = 1.159 \times 10^7 \text{ kg}$$

The density relates volume and mass, hence the total volume required is

$$(V_f \text{ m}^3) = (m_f \text{ kg})/(\rho \text{ kg/m}^3)$$

$$V_f = (1.159 \times 10^7)/(990) = 11\ 708 \text{ m}^3$$

Lastly, the volume of a cylinder is $V = \pi L D^2/4$, thus for the oil tank with $L = D$,

$$11\ 708 = \pi D^3/4 \qquad D = L = 24.6 \text{ m}$$

Comments:

1. The fuel and air are not part of the system; they only provide a means to obtain the heat input.
2. This example illustrates the tremendous mass of fuel that must be transported, stored, and consumed in a power plant with excellent overall efficiency.
3. We do not need to know how the cycle produces the power to determine some of the overall effects, such as total fuel consumption and the heat rejected to the environment.
4. The heat rejected to the environment is usually to a water source, but may be directly to the atmosphere via cooling towers.

CONCEPT QUESTIONS

1. What is the meaning of *flow work?*
2. What does *adiabatic* mean?
3. Why are kinetic and potential energies directly convertible into work?
4. What is the difference between steady flow and steady state?
5. In your own words define *heat* and *work.*
6. Describe what is meant by the internal energy of a substance.
7. Explain the sign convention for heat and work.
8. Describe power and how it differs from work.
9. Why are heat and work not system properties?
10. What does the area under the curve on a *p-V* diagram represent?
11. What is a quasi-equilibrium process?
12. Why is the work greater for a reversible process than for an actual process?
13. Describe conductive heat transfer.
14. What is Newton's law of cooling?
15. Describe radiative heat transfer.
16. An automobile accelerates to 60 mph in 12 sec. Is the work different if it accelerated to 60 mph in only 8 sec.? Why?
17. It is a hot summer day. A student leaves his dormitory room with a fan on and with the door and windows closed. When he returns in the afternoon, will the room be warmer or cooler than an identical room without the fan?

$Ke = \frac{1}{2}mv^2$ $Pe = mgh$
$W = F \cdot d$
$F = m \cdot a$

PROBLEMS (SI)

$Ke = \frac{1}{2}mv^2$ $Pe = mgh$

3.1 The weight of a bridge crane plus its load equals 100 metric tons (1 metric ton = 1000 kg). It is driven by a motor and travels at 1.17 m/s along the crane rails. Determine the energy that must be absorbed by the brakes in stopping the crane.

$1 \cdot 10$

3.2 Determine the work required to accelerate a unit mass between the following velocity limits: (a) 10 m/s to 110 m/s; (b) 50 m/s to 150 m/s; (c) 100 m/s to 200 m/s.

3.3 Five people must be lifted on an elevator a distance of 100 m. The work is found to be 341.2 kJ, and the gravitational acceleration is 9.75 m/s². Determine the average mass per person.

3.4 In Problem 3.3 the initial potential energy of the elevator was 68.2 kJ vis-à-vis the earth's surface. What is the height above the ground that the people were lifted?

3.5 A student is watching pilings being driven into the ground. From the size of the pile driver the student calculates the mass to be 500 kg. The distance that the pile driver is raised is measured to be 3 m. Determine the potential energy of the pile driver at its greatest height (the piling is considered the datum). Find the driver velocity just prior to impact with the piling.

3.6 A child (25-kg mass) is swinging on a play set. The swing's rope is 2.3 m long. Determine (a) the change in potential energy from 0° with the vertical when the swing goes from 0° to an angle of 45° with the vertical; (b) the velocity when the swing reaches the 0° angle.

3.7 A piston-cylinder contains air at a pressure of 500 kPa. The piston movement is resisted by a spring and atmospheric pressure of 100 kPa. The air moves the piston and the volume changes from 0.15 m³ to 0.60 m³. Determine the work when (a) the force of the spring is directly proportional to the displacement; (b) the force of the spring is proportional to the square root of the displacement.

3.8 An automobile produces 20 kW when moving with a velocity of 50 km/h. Determine (a) the resisting force; (b) the resisting (drag) force if it is proportional to the velocity cubed and the automobile travels at 100 km/h.

3.9 An elastic sphere of 0.5-m diameter contains a gas at 115 kPa. Heating of the sphere causes it to increase to 0.62 m, and during this process the pressure is proportional to the sphere diameter. Determine the work done by the gas.

3.10 A force F is proportional to x^2 and has a value of 133 N when $x = 2$. Determine the work done as it moves an object from $x = 1$ m to $x = 4$ m.

3.11 A fluid at 700 kPa, with a specific volume of 0.25 m³/kg and a velocity of 175 m/s, enters a device. Heat loss from the device by radiation is 23 kJ/kg. The work done by the fluid is 465 kJ/kg. The fluid exists at 136 kPa, 0.94 m³/kg, and 335 m/s. Determine the change in internal energy.

3.12 An air compressor handles 8.5 m³/min of air with a density of 1.26 kg/m³ and a pressure of 1 atm, and it discharges at 445 kPa (gage) with a density of 4.86 kg/m³. The change in specific internal energy across the compressor is 82 kJ/kg, and the heat loss by cooling is 24 kJ/kg. Neglecting changes in kinetic and potential energies, find the power in kW.

3.13 A centrifugal pump compresses 3000 liters/min of water from 98 kPa to 300 kPa. The inlet and outlet temperatures are 25°C. The inlet and discharge piping are on the same level, but the diameter of the inlet piping is 15 cm, whereas that of the discharge piping is 10 cm. Determine the pump power in kW.

3.14 Two gaseous streams containing the same fluid enter a mixing chamber and leave as a single stream. For the first gas the entrance conditions are $A_1 = 500$ cm^2, $v_1 = 130$ m/s, $\rho_1 = 1.60$ kg/m^3. For the second gas the entrance conditions are $A_2 = 400$ cm^2, $\dot{m}_2 = 8.84$ kg/s, $v_2 = 0.502$ m^3/kg. The exit stream condition is $v_3 = 130$ m/s and $v_3 = 0.437$ m^3/kg. Determine (a) the total mass flow leaving the chamber; (b) the velocity of gas 2.

3.15 An insulated 2-kg box falls from a balloon 3.5 km above the earth. What is the change in internal energy of the box after it has hit the earth's surface?

3.16 A 4-mm-diameter steel wire, with the Young's modulus (E) of the material equal to 2.067×10^8 kPa, has a length of 4 m and is gradually subjected to an axial force of 5000 N. Determine the work done.

3.17 A soap bubble with a 15-cm radius is formed by blowing through a 2.5-cm-diameter wire loop. Assume that all the soap film goes into making the bubble. The surface tension of the film is 0.02 N/m. Find the total surface work required to make the bubble.

3.18 A closed system containing a gas expands slowly in a piston-cylinder from 600 kPa and 0.10 m^3 to a final volume of 0.50 m^3. Determine the work done if the pressure distribution is determined to be (a) $p = C$; (b) $pV = C$; (c) $pV^{1.4} = C$; (d) $p = -300V + 630$, where V is in m^3 and p in kPa.

3.19 An elevator is on the thirtieth floor of an office building when the supporting cable shears. The elevator drops vertically to the ground, where several large springs absorb the impact of the elevator. The elevator mass is 2500 kg, and it is 100 m above the ground. Determine (a) the potential energy of the elevator before its fall; (b) the velocity and kinetic energy the instant before impact; (c) the energy change of the springs when they are fully compressed.

3.20 A closed gaseous system undergoes a reversible process in which 30 kJ of heat is rejected and the volume changes from 0.14 m^3 to 0.055 m^3. The pressure is constant at 150 kPa. Determine (a) the change in internal energy of the system; (b) the work done.

3.21 A fluid enters with a steady flow of 3.7 kg/s and an initial pressure of 690 kPa, an initial density of 3.2 kg/m^3, an initial velocity of 60 m/s, and an initial internal energy of 2000 kJ/kg. It leaves at 172 kPa, $\rho = 0.64$ kg/m^3, v = 160 m/s, and $u = 1950$ kJ/kg. The heat loss is found to be 18.6 kJ/kg. Find the power in kW.

3.22 A 32-cm cube of ice at 0°C melts while being used to cool beer and soda at the beach. The specific volume of liquid water at 0°C is 1.002 cm^3/g and that of ice at 0°C is 1.094 cm^3/g. Is there any work done by the surroundings, namely, the atmosphere, on the ice?

3.23 Air and fuel enter a furnace used for home heating. The air has an enthalpy of 302 kJ/kg and the fuel an enthalpy of 43 027 kJ/kg. The gases leaving the furnace have an enthalpy of 616 kJ/kg. There is 17 kg air/kg fuel. Water circulates through the furnace wall receiving heat. The house requires 17.6 kW of heat. What is the fuel consumption per day?

3.24 An air compressor compresses air with an enthalpy of 96.5 kJ/kg to a pressure and temperature that have an enthalpy of 175 kJ/kg. There are 35 kJ/kg of heat lost from the compressor as the air passes through it. Neglecting kinetic and potential energies, determine the power required for an air mass flow of 0.4 kg/s.

3.25 A steam condenser receives 9.47 kg/s of steam with an enthalpy of 2570 kJ/kg. The steam condenses to a liquid and leaves with an enthalpy of 160.5 kJ/kg. (a) Find the total heat transferred from the steam. (b) Cooling water passes through the condenser with an unknown flow rate; however, the water temperature increases from 13°C to 24°C. Also, it is known that 1 kg of water will absorb 4.2 kJ of energy per degree temperature rise. Find the cooling water flow rate.

3.26 Steam enters a turbine with a pressure of 4826 kPa, $u = 2958$ kJ/kg, $h = 3263$ kJ/kg, and a flow of 6.3 kg/s. Steam leaves with $h = 2232$ kJ/kg, $u = 2102$ kJ/kg, and $p = 20.7$ kPa. There is radiative heat loss equal to 23.3 kJ/kg of steam. Determine (a) the power produced; (b) the adiabatic work; (c) the inlet specific volume; (d) the exit velocity if the exit area is 0.464 m².

3.27 Steam with a flow rate of 1360 kg/h enters an adiabatic nozzle at 1378 kPa, 3.05 m/s, with a specific volume of 0.147 m³/kg and a specific internal energy of 2510 kJ/kg. The exit conditions are $p = 137.8$ kPa, specific volume $= 1.099$ m³/kg, and internal energy $= 2263$ kJ/kg. Determine the exit velocity.

3.28 An injector was used on steam locomotives as a means of providing water to the boiler at a pressure higher than the steam pressure. The injector looks like a T with steam entering horizontally and water entering vertically from the bottom. The fluids mix and discharge horizontally; 18.2 kg/min of steam enters at 1378 kPa with an internal energy of 2590 kJ/kg and a specific volume of 0.143 m³/kg. The water enters 1.52 m below the horizontal with an enthalpy of 42 kJ/kg. The mixture has a pressure of 1550 kPa, an internal energy of 283.8 kJ/kg, and a specific volume of 0.001024 m³/kg. The ratio of water to steam is 10.5 to 1. Neglect changes in kinetic energy. Determine the rate of heat transfer in kW.

3.29 A steady-flow system receives 1 kg/s of a substance with $\mu_1 = 1000$ kJ/kg, $p_1 = 500$ kPa, $v_1 = 1.2$ m³/kg, and $v_1 = 50$ m/s. There is a heat loss of 100 kJ/kg, and the fluid exits at $v_2 = 150$ m/s, $v_2 = 0.8$ m³/kg, $p_2 = 100$ kPa, and $h_2 = 1000$ kJ/kg. Determine the power and the exit specific internal energy.

3.30 Determine the power delivered by a shaft rotating at 200 revolutions per second against a constant torque of 10 J.

3.31 The torque of an engine is found to be $\tau = 200 \sin(\tau\omega/2000)$ J, when ω varies between 500 and 1000 rpm and $(\pi\omega/2000)$ is in degrees. Determine the power at these two rpm's.

3.32 A battery charger can produce 3 A at 12 V and charges a battery for 2 h. What is the work?

3.33 The current used by a device at a constant voltage of 120 V varies with time according to $i = 6e^{-t/60}$, where i is in amperes and t is in seconds. Calculate work for the first minute.

3.34 A gas is compressed according to the relationship $p = aV + b$, where $a = -1000$ kPa/m³ and $b = 500$ kPa. The initial volume is 0.4 m³, and the final volume is 0.1 m³. Determine the work done in this process by integrating the expression for pressure, plotting the curve on a p-V diagram, and finding the area under it.

3.35 A gas expands in a piston from an initial pressure of 1000 kPa and an initial volume of 0.15 m³ to a final pressure of 200 kPa while following the process described by $p = aV + b$, where $a = 1000$ kPa/m³ and b is a constant. Calculate the work performed.

3.36 A spherical balloon contains air at 150 kPa and 300°K. Furthermore, the pressure is proportional to the square of the diameter. Heat is added to the balloon, and the volume doubles. Determine the work done by the balloon. Assume $d_1 = 1$ m.

3.37 Determine the work required to accelerate a 1000 kg car from 10 to 80 km/h on a hill where the elevation increases by 35 m.

3.38 Determine the work necessary to compress a linear spring with a spring constant of 90 kN/m a distance of 30 cm from its extended, rest position.

3.39 Determine the time to accelerate a 1000-kg automobile from rest to 80 km/h if it has an engine rated at 90 kW.

3.40 A 1500-kg automobile is stopped at a traffic light proceeding in a direction up a 20° hill. The length of road is 400 m, and the car accelerates from rest to 60 km/h in this distance. Determine the minimum power required to accelerate the car.

3.41 While standing in line for a ski lift, you decide to determine the minimum power required to operate the lift. The distance from the bottom of the hill where the lift begins to the top is 0.9 km, and the elevation increases 200 m. You time the lift and find its steady operating speed is 3.0 m/s. You estimate that the total mass of the chairs plus their two occupants is 18000 kg. Because you are at higher elevations, the gravitational acceleration is 9.7 m/s². Determine (a) the power required to operate the lift; (b) using this value of power, the time required to reach the operating velocity when the lift is first started.

3.42 A vertical piston-cylinder assembly containing water is being heated on a stove. During this process, 100 kJ of heat is transferred to the water. In the water's expansion process (because of the heat addition), 10 kJ of work is done. Also, there is 20-kJ heat loss from the assembly. What is the water's change of energy as a result of the heat addition process?

3.43 Fill in the missing data in the table below for a closed system changing from state 1 to state 2.

Q (kJ)	W (kJ)	E_1 (kJ)	E_2 (kJ)	ΔE (kJ)
−20			10	−15
23	−7		37	
	15	18		21
25	10	15		
40		19		16

3.44 A system operates on a two-process cycle. During the first process 50 kJ of heat is added to the system, and the system performs 70 kJ of work. In the second process 50 kJ of work is added to the system. Determine (a) the heat transfer during the second process; (b) the net work and net heat transfer for the cycle.

3.45 A system operates on a three-process cycle. The first process is adiabatic, and 60 kJ of work is done on the system. In the second process no work occurs, but 240 kJ of heat is added. In the third process the system rejects 160 kJ of heat. Determine (a) the work done for the last process; (b) the net work and net heat transfer for the cycle.

3.46 A large adiabatic room containing 300 kg of air at 15°C is heated to 25°C in 15 min by an electric heater that consists of a 200-W fan pushing 1 kg/s of air across electric heating coils. The enthalpy of the air at 15°C is 289 kJ/kg and at 25°C is 300 kJ/kg. What is the rate of heat transfer in kW?

3.47 The heater in Problem 3.46 is now located in a nonadiabatic room of the same size. The heat loss from the room is 120 kJ/min. Determine the time for the room to reach 25°C from the initial 15°C.

3.48 A 1500 W electric hair dryer is essentially an adiabatic duct and consists of a small fan that blows air over a heating element, increasing the temperature of the air from its inlet temperature of 292°K to an exit temperature of 320°K. The air density at inlet conditions is 1.205 kg/m³ and at outlet conditions is 1.100 kg/m³. The specific internal energy changes from 209.5 kJ/kg at inlet to 229.6 kJ/kg at outlet. The pressure remains constant at 101.3 kPa throughout the hair dryer. The exit cross-sectional area of the hair dryer when the nozzle is in place is 5 cm². Determine (a) the mass flow rate of air through the dryer; (b) the volume flow rate of air at inlet conditions; (c) the velocity of the air leaving the nozzle.

3.49 An electric hot-water heater consists of a 4-cm-diameter pipe containing a resistance electric heater. Cold water enters the pipe at 10°C with an enthalpy of 42 kJ/kg and leaves at 50°C with an enthalpy of 209 kJ/kg. The water flow rate is 20 liters/min, and its density is 1000 kg/m³. Determine the rating of electric resistance in kW and the velocity of water in the pipe.

3.50 A 12-kW pump draws water from a well 15 m below ground level and discharges it into a water tower atop a building 100 m above ground. There is no change in the water's internal energy, temperature, pressure, or specific volume during the process. What is the maximum flow rate possible? The pump may be considered adiabatic, and the density of water is 1000 kg/m³.

3.51 A 10-kg mass slides down a ramp inclined at 30° from the horizontal a total vertical distance of 3 m. Determine the velocity of mass when it reaches the bottom, neglecting friction and air resistance.

3.52 A force applied to a 10-kg mass initially at rest for 15 s causes it to be accelerated at 3 m/s² during this time period. Determine the work in kJ.

3.53 Air contained in a piston-cylinder undergoes two processes in series. In the first the air expands according to $pV = C$ from 300 kPa and a specific volume of 0.021 m³/kg to a pressure of 140 kPa. The second process is a constant pressure compression until specific volume 3 equals specific volume 1. Sketch the processes on a p-V diagram and determine the work per unit mass.

3.54 Air contained in a piston-cylinder undergoes two processes in series. In the first the volume remains constant while the pressure decreases from 300 kPa to 60 kPa. At this point the second process, $pV^{1.3} = C$, occurs, pressure increases to 300 kPa, and the final volume is 0.1 m³. Sketch the processes on a p-V diagram and determine the work in kJ.

3.55 The following table illustrates the variation of pressure and volume in the cylinder of an internal combustion engine during the expansion process.

Data point	Pressure (kPa)	Volume (cm³)
1	2000	400
2	1600	490
3	1200	620
4	990	730
5	600	1120
6	300	1930

Plot the data on a p-V diagram and determine the work done in kJ. Is this exact or an estimate? Why?

3.56 An industrial furnace has a 10-cm-thick, 15-m^2 brick wall. The bricks have a thermal conductivity of 0.2 W/m K and have a uniform and steady temperature on one surface of 1000°K and on the other of 1300°K. Determine the heat transfer through the wall.

3.57 A surface of 1.5 m^2 with an emissivity of 0.85 emits thermal radiation. Find the radiant heat emitted for surface temperatures of 300°, 500°, and 700°K.

3.58 Refrigerant flows through a 1-m, 1-cm-diameter tube and evaporates at a constant temperature of 260°K. The tube's surface temperature is constant at 268°K, and the unit convective coefficient is 2000 W/m^2 K. Determine the rate of convective heat transfer from the refrigerant to the tube surface.

3.59 A windmill produces on average 6 kW of electrical power over an 8-h period. The electricity is used to charge storage batteries. In the charging process the batteries increase in temperature, causing them to lose heat to the surroundings at a rate of 0.5 kW. Determine the total energy stored in the batteries during this 8-h period.

3.60 A piston-cylinder contains gas initially at 3500 kPa with a volume of 0.03 m^3. The gas is compressed during a process where $pV^{1.25} = C$ to a pressure of 8500 kPa. The heat transfer from the gas is 2.5 kJ. Determine the change in internal energy, neglecting changes in kinetic and potential energies.

3.61 An adiabatic tank with a volume of 0.25 m^3 receives paddle work at a rate of 4.3 W for 30 min. The gas in the tank has an initial density of 1.25 kg/m^3. Determine the specific volume at the final state and the change of specific internal energy.

3.62 A nonadiabatic tank receives paddle work for 30 min. The power to the paddle varies with time according to $\dot{W}_p = -8t$, where $\dot{W}_p$ is in watts and t is in minutes. In addition there is heat transfer from the tank at a constant rate of 40 W. Determine the net change of the gas's energy after 30 min.

3.63 A closed system containing a gas undergoes a cycle composed of three processes. The system's initial state is at 100 kPa, 1.5 m^3, and an internal energy of 510 kJ. The gas is compressed according to $pV = C$ until the pressure is 2000 kPa and the internal energy is 685 kJ. The second process has constant volume, and the heat loss is 150 kJ. In the final process returning the system to the initial state, the work is 50 kJ. Determine the heat transfer for the first and last processes.

3.64 A heat power cycle with a thermal efficiency of 0.4 produces 12 000 kJ of net work. Determine the heat added and heat rejected for the cycle.

3.65 A heat power cycle with a thermal efficiency of 35% receives 1500 MW of heat added. Determine the net power produced in MW.

3.66 A power plant with an efficiency of 35% produces 250 MW of power. It uses coal as the fuel supply which has a heating value of 12 000 kJ/kg. Determine the fuel required per day. Every 1000 kg of fuel is considered a metric ton. How many metric tons of coal must be transported to the power plant to have 1 week's supply of coal on hand?

3.67 A power plant produces 750 MW of electric power while operating with an efficiency of 42%. The heat rejected from the cycle goes into cooling water supplied from an adjacent river. The water's enthalpy increases by 45 kJ/kg as it receives the heat rejected. Determine the mass flow rate of water required.

3.68 A south-facing roof of a home measures 10 m by 17 m and receives on average 2500 h/yr of sunshine with an average solar radiation of 1.36 kW/m^2. Determine the total annual energy received by the roof.

3.69 A student living in the home described in Problem 3.68 decides to invest in a solar collector with an efficiency of 60% that occupies one-half of the roof area. If the home annually uses 3000 kg of oil with a heating value of 43 000 kJ/kg for heating, what will be the annual savings in kg of oil?

3.70 In hiking along a river an engineering student notices the site of a former sawmill that used hydro power. Remnants of the original dam are still visible, and the stream is still flowing. The student decides to make some measurements and determine the maximum power that could be generated if the site were redeveloped. The stream flow is found to be 25 m³/s, and the change in elevation is 4 m. Determine the maximum power that could be produced.

PROBLEMS (English Units)

*3.1 A system undergoes a cycle where 10 Btu of heat is removed and 15,000 ft-lbf of work is done by the system during the first process. In the second process 15 Btu of heat is added. What is the work necessary to complete the cycle?

*3.2 A closed system expands from 0.5 to 2.5 ft³; the pressure varies according to $p = 50V + 10$ psia. Determine the work.

*3.3 Calculate the kinetic energy of a 3000-lbm automobile moving at 60 mph.

*3.4 A fluid at 100 psia, with a specific volume of 4 ft³/lbm and a velocity of 600 ft/sec, enters a device. Heat loss from the device by radiation is 10 Btu/lbm. The work done by the fluid is 200 Btu/lbm. The fluid exits at 20 psia, 15 ft³/lbm, and 1100 ft/sec. Determine the change in specific internal energy.

*3.5 An air compressor handles 300 ft³/min of air with a density of 0.079 lbm/ft³ and a pressure of 14.7 psia, and it discharges at a pressure of 75 psig with a density of 0.305 lbm/ft³. The change in specific internal energy across the compressor is 35 Btu/lbm, and the heat loss by cooling is 10 Btu/lbm. Neglecting changes in kinetic and potential energies, find the power in Btu/h, hp, and kW.

*3.6 Two gaseous streams containing the same fluid enter a mixing chamber and leave as a single stream. For the first gas the entrance conditions are $A_1 = 80$ in.², $v_1 = 400$ ft/sec, and $\rho_1 = 0.10$ lbm/ft³. For the second gas the entrance conditions are $A_2 = 60$ in.², $\dot{m}_2 = 70{,}000$ lbm/hr, and $v_2 = 8.04$ ft³/lbm. The exit stream condition is $v_3 = 400$ ft/sec and $v_3 = 7$ ft³/lbm. Determine (a) the total mass flow leaving the chamber; (b) the velocity of gas 2.

*3.7 Steam with a flow rate of 3000 lbm/hr enters an adiabatic nozzle at 200 psia, 600 ft/min, with a specific volume of 2.36 ft³/lbm, and with a specific internal energy of 1122.7 Btu/lbm. The exit conditions are $p = 20$ psia, specific volume = 17.6 ft³/lbm, and internal energy = 973 Btu/lbm. Determine the exit velocity.

*3.8 A gas is compressed according to the relationship $p = aV + b$, where $a = -1000$ psia/ft³ and $b = 500$ psia. The initial volume is 0.4 ft³, and the final volume is 0.1 ft³. Determine the work done in this process by integrating the expression for pressure, plotting the curve on a p-V diagram, and finding the area under it.

*3.9 A gas expands in a piston from an initial pressure of 1000 psia and an initial volume of

0.15 ft^3 to a final pressure of 200 psia, while following the process described by $p = aV + b$, where $a = 1000$ psia/ft^3 and b is a constant. Calculate the work performed.

*3.10 A spherical balloon contains air at 18 psia and 535°R. Furthermore, the pressure is proportional to the square of the diameter. Heat is added to the balloon, and the volume doubles. Determine the work done by the balloon. Assume $d_1 = 1$ foot.

*3.11 Determine the work required to accelerate a 2000 lbm car from 10 to 80 mph on a hill where the elevation increases by 100 ft.

*3.12 Determine the work necessary to compress a linear spring with a spring constant of 9 lbf/in. a distance of 10 in. from its extended, rest position.

*3.13 Determine the time to accelerate a 1-ton automobile from rest to 60 mph if it has an engine rated at 60 hp.

*3.14 A 3000-lb automobile is stopped at a traffic light proceeding in a direction up a 20° hill. The length of road is 1200 ft, and the car accelerates from rest to 40 mph in this distance. Determine the minimum power required to accelerate the car.

*3.15 While standing in line for a ski lift, you decide to determine the minimum power required to operate the lift. The distance from the bottom of the hill where the lift begins to the top is 0.5 mi, and the elevation increases 600 ft. You time the lift and find its steady operating speed is 10 ft/sec. You estimate that the total mass of each chair plus its two occupants are 40 000 lbm. Because you are at higher elevations, the gravitational acceleration is 31.2 ft/sec^2. Determine (a) the power required to operate the lift; (b) using this value of power, the time required to reach the operating velocity when the lift is first started.

*3.16 A vertical piston-cylinder assembly containing water is being heated on a stove. During this process, 100 Btu of heat is transferred to the water. In the water's expansion process (because of the heat addition), 8000 ft-lbf of work is done. Also, there is 20-Btu heat loss from the assembly. What is the water's change of energy as a result of the heat addition process?

*3.17 Fill in the missing data in the table below for a closed system changing from state 1 to state 2.

Q (Btu)	W (Btu)	E_1 (Btu)	E_2 (Btu)	ΔE (Btu)
	20		10	−15
30	−7	37		
15		18		21
25	10		15	
40		19		16

*3.18 A system operates on a two-process cycle. During the first process 60 Btu of heat is added to the system, and the system performs 80 Btu of work. In the second process 60 Btu of work is added to the system. Determine (a) the heat transfer during the second process; (b) the net work and net heat transfer for the cycle.

*3.19 A system operates on a three-process cycle. The first process is adiabatic, and 60 Btu of work is done on the system. In the second process no work occurs, but 240 Btu of heat is added. In the third process the system rejects 160 Btu of heat. Determine (a) the work done for the last process; (b) the net work and net heat transfer for the cycle.

*3.20 A large adiabatic room containing 700 lbm of air at 60°F is heated to 80°F in 15 min by an electric heater that consists of a 200-W fan pushing 2 lbm/sec of air across electric heating coils. The enthalpy of the air at 60°F is 124 Btu/lbm and at 80°F is 129 Btu/lbm. What is the rate of heat transfer in Btu/min? In kW?

*3.21 The heater in Problem *3.20 is now located in a nonadiabatic room of the same size. The heat loss from the room is found to be 120 Btu/min. Determine the time for the room to reach 80°F from the initial 60°F.

*3.22 A 1500 W electric hair dryer is essentially an adiabatic duct and consists of a small fan that blows air over a heating element, increasing the temperature of the air from its inlet temperature of 77°F to an exit temperature of 127°F. The air density at inlet conditions is 0.0739 lbm/ft³ and at outlet conditions is 0.0709 lbm/ft³. The specific internal energy changes from 91.5 Btu/lbm at inlet to 101.1 Btu/lbm at outlet. The pressure remains constant at 14.7 psia throughout the hair dryer. The exit cross-sectional area of the hair dryer when the nozzle is in place is 1 in.². Determine (a) the mass flow rate of air through the dryer; (b) the volume flow rate of air at inlet conditions; (c) the velocity of the air leaving the nozzle.

*3.23 An electric hot-water heater consists of a 2-in.-diameter pipe containing a resistance electric heater. Cold water enters the pipe at 40°F with an enthalpy of 8 Btu/lbm and leaves at 140°F with an enthalpy of 108 Btu/lbm. The water flow rate is 5 gal/min, and its density is 62.4 lbm/ft³. Determine the rating of electric resistance in kW and the velocity of water in ft/sec in the pipe.

*3.24 A 10-hp pump draws water from a well 30 ft below ground level and discharges it into a water tower atop a building 300 ft above ground. There is no change in the water's internal energy, temperature, pressure, or specific volume during the process. What is the maximum flow rate possible? The pump may be considered adiabatic, and the density of water is 62.4 lbm/ft³.

*3.25 A 10-lb mass slides down a ramp inclined at 45° from the horizontal a total vertical distance of 10 ft. Determine the velocity of mass when it reaches the bottom, neglecting friction and air resistance.

*3.26 A force applied to a 10-lb mass initially at rest for 15 sec causes it to accelerate 10 ft/sec² during this time period. Determine the work in ft-lbf.

*3.27 Air contained in a piston-cylinder undergoes two processes in series. In the first the air expands according to $pV = C$ from 500 psia and a specific volume of 0.444 ft³/lbm to a pressure of 20 psia. The second process is a constant-pressure compression until specific volume 3 equals specific volume 1. Sketch the processes on a p-V diagram and determine the work per unit mass.

*3.28 Air contained in a piston-cylinder undergoes two processes in series. In the first the volume remains constant, while the pressure decreases from 45 psia to 9 psia. At this point the second process, $pV^{1.3} = C$, occurs, pressure increases to 45 psia, and the final volume is 3.0 ft³. Sketch the processes on a p-V diagram and determine the work in Btu.

*3.29 The following table illustrates the variation of pressure and volume in the cylinder of an internal combustion engine during the expansion process.

Data point	Pressure (psia)	Volume (in.^{3})
1	350	25
2	280	31
3	210	39
4	175	50
5	110	70
6	50	120

Plot the data on a p-V diagram and determine the work done in ft-lbf. Is this exact or an estimate? Why?

*3.30 An industrial furnace has a 5-in.-thick, 100-ft^2 brick wall. The bricks have a thermal conductivity of 0.2 Btu/hr-ft-F and have a uniform and steady temperature on one surface of 750°F and on the other of 1200°F. Determine the heat transfer through the wall.

*3.31 A surface of 5 ft^2 with an emissivity of 0.85 emits thermal radiation. Find the radiant heat emitted for surface temperatures of 77°, 277°, and 577°F.

*3.32 Refrigerant flows through a 6-ft, 0.5-in.-diameter tube and evaporates at a constant temperature of 0°F. The tube's surface temperature is constant at 10°F, and the unit convective coefficient is 350 Btu/hr-ft^2-F. Determine the rate of convective heat transfer from the refrigerant to the tube surface.

*3.33 A windmill produces on average 6 kW of electrical power over an 8-hr period. The electricity is used to charge storage batteries. In the charging process the batteries increase in temperature, causing them to lose heat to the surroundings at a rate of 500 Btu/hr. Determine the total energy stored in the batteries during this 8-hr period.

*3.34 A piston-cylinder contains gas initially at 510 psia with a volume of 1.06 ft^3. The gas is compressed during a process where $pV^{1.25} = C$ to a pressure of 1235 psia. The heat transfer from the gas is 3.0 Btu. Determine the change in internal energy, neglecting changes in kinetic and potential energies.

*3.35 An adiabatic tank with a volume of 8.8 ft^3 receives paddle work at a rate of 4.5 W for 30 min. The gas in the tank has an initial density of 0.078 lbm/ft^3. Determine the specific volume at the final state and the change of specific internal energy.

*3.36 A nonadiabatic tank receives paddle work for 30 min. The power to the paddle varies with time according to $\dot{W}_p = -12t$, where $\dot{W}_p$ is in watts and t is in minutes. In addition there is heat transfer from the tank at a constant rate of 170 Btu/hr. Determine the net change of the gas's energy after the 30 min.

*3.37 A closed system containing a gas undergoes a cycle composed of three processes. The system's initial state is at 14.7 psia, 53.0 ft^3, and an internal energy of 540 Btu. The gas is compressed according to $pV = C$ until the pressure is 290 psia and the internal energy is 723 Btu. The second process has constant volume, and the heat loss is 160 Btu. In the final process returning the system to the initial state, the work is 41,300 ft-lbf. Determine the heat transfer for the first and last processes.

*3.38 A heat power cycle with a thermal efficiency of 0.4 produces 8×10^6 ft-lbf of net work. Determine the heat added and heat rejected for the cycle.

*3.39 A heat power cycle with a thermal efficiency of 35% receives 2.0×10^6 Btu/sec of heat added. Determine the net power produced in hp and MW.

*3.40 A power plant with an efficiency of 35% produces 250 MW of power. It uses coal as the fuel supply which has a heating value of 5200 Btu/lbm. Determine the fuel required per day. How many tons of coal must be transported to the power plant to have 1 week's supply of coal on hand?

*3.41 A power plant produces 500 MW of electric power while operating with an efficiency of 45%. The heat rejected from the cycle goes into cooling water supplied from an adjacent river. The water's enthalpy increases by 20 Btu/lbm as it receives the heat rejected. Determine the mass flow rate of water required.

*3.42 A south-facing roof of a home measures 25 ft by 40 ft and receives on average 2800 hr/yr of sunshine with an average solar radiation of 430 Btu/hr-ft²-F. Determine the total energy received by the roof annually.

*3.43 A student living in the home described in Problem *3.42 decides to invest in a solar collector with an efficiency of 65% that occupies one-half of the roof area. If the home annually uses 1000 gal of oil with a heating value of 18,000 Btu/lbm and a specific gravity of 0.9 (based on water's density of 62.4 lbm/ft³) for heating, what will be the annual savings in gallons of oil?

*3.44 In hiking along a river an engineering student notices the site of a former sawmill that used hydro power. Remnants of the original dam are still visible, and the stream is still flowing. The student decides to make some measurements and determine the maximum power that could be generated if the site were redeveloped. The stream flow is found to be 900 ft³/sec, and the change in elevation is 12 ft. Determine the maximum power that could be produced in hp.

COMPUTER PROBLEMS

C3.1 Develop a spreadsheet template or computer program that will determine the exit conditions for the adiabatic mixing of up to five inlet streams characterized by area, mass flow rate, velocity, density, pressure, and specific internal energy. The outlet stream should have the same characteristics. Test it using the information from Problem 3.14.

C3.2 The engine from Problem 3.31 accelerates from 500 to 1000 rpm in 5 s. Compute the power at increments of 1 s and the average power during the period of acceleration.

C3.3 Compute the energy and power used by the device in Problem 3.33 over the first minute. Plot the results in increments of 5 s.

C3.4 Solve Problem 3.18 (except for part [a]) using a spreadsheet program. Compute the work numerically by summing the average pressure over small volume increments. Compare the results for volume increments of 0.1 m³ and 0.025 m³ with the results obtained by calculus.

4

Properties of Pure Substances

In Chapter 3 the concepts of conservation of mass and energy were introduced. The system energy transformation had an unknown substance in it whose properties were given. In this chapter you will learn about pure substances, their behavior, and their properties. This will include

- Understanding the various phases, solid, liquid, and vapor;
- Learning the conditions for phase equilibria;
- Visualizing the three-dimensional pvT structure of a substance and its implications;
- Developing the ability to use the tables of properties on the computer as well as in the appendix;
- Investigating how the properties of two phase mixtures are determined theoretically and experimentally.

4.1 THE STATE PRINCIPLE

The state of an equilibrium system can be defined by its thermodynamic properties. How many properties are needed to uniquely define the state? We can answer this question using the state principle, a rule that has been developed from many years of observations. A closed system may have a variety of heat and work interactions; for instance, there could be mechanical work of compression, electrical work, magnetic work, and heat transfer to the surroundings. Each of the work modes is characterized by an extensive property and a related intensive property. For example, volume (extensive) and pressure (intensive) characterize mechanical work. The intensive property describing heat transfer is temperature. In systems undergoing equilibrium processes or that can be approximated by equilibrium processes, the number of independent properties necessary to define the system is one more than the number of relevant work interactions. Thus, for simple compressible systems (mechanical work) in the absence of potential and kinetic energies that are included in the total energy term, two independent properties can define the state.

The predominant system used in this text is that of simple compressible systems, where work is equal to $\int p\, dV$. Should we write the first law for such a system

$$Q = \Delta E + \int p\, dV$$

there would be no way to determine electrical potential, for example. Information concerning a system having electrical work interactions as well as compressible work interactions would require an additional independent property.

Returning to the simple compressible system undergoing equilibrium processes and assuming it contains a unit mass for simplicity, we can define all the other properties of the system, using two independent intensive properties. Thus, if we know the temperature, T, and pressure, p, other properties can be determined: $h = h(T, p)$; $u = u(T, p)$; $v = v(T, p)$.

In this chapter we consider simple compressible systems using pure substances. These substances are of constant chemical composition, regardless of phase. Let us pick water as the substance in the system. The water can be liquid, solid, or vapor. If we sample the system now, the samples will be distinguishable by the various phases of water. Water is an example of a pure substance, which may exist as a liquid, a solid, or a vapor, phases by which all pure substances are characterized. Recall that the definition of a pure substance is that it is homogeneous by nature, does not undergo chemical reactions, and is not a mechanical mixture of different species. The substances in our systems considered heretofore have been phases of a substance. The phases are physically homogeneous. There is usually a sharp distinction between phases, inasmuch as the properties of one phase are decidedly different from those of another phase. There is a difference between ice and liquid water, for example.

4.2 LIQUID-VAPOR EQUILIBRIUM

Where do these phase changes occur and how can these changes be quantified and used? To answer this, let us consider the piston-cylinder arrangement in Figure 4.1(a). This is just a scientific way of considering a pot of water on a kitchen stove with the weightless top sitting on the water instead of the edge of the pot. Heat is added at constant pressure, $q = \Delta h$, to the liquid water at 40°C initially. As heat is added, the water temperature rises until it reaches 100°C; see Figure 4.1(b). Then what happens? The water boils, as shown in Figure 4.1(c). What is the boiling process? It is a phase transition. The water is going from liquid to vapor. In Figure 4.1(d) all the liquid has just changed to vapor and finally, as more heat is added (Figure 4.1[e]), the temperature of the vapor increases.

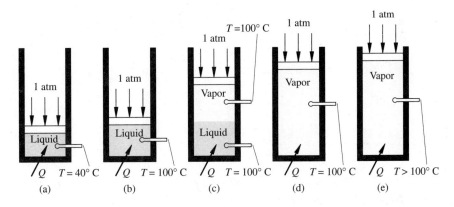

Figure 4.1 The change of water from a subcooled liquid to a superheated vapor by constant-pressure heat addition.

4.3 SATURATED PROPERTIES

Let us plot a T-v diagram showing this process for 1 atm of pressure (Figure 4.2). When the water is at point b, it is a *saturated liquid*. This means that it is at the highest temperature at which, for this pressure, it can remain liquid. If more heat is added, some of the liquid changes to vapor, and a mixture of vapor and liquid occurs, such as at point c. This vapor is a *saturated vapor*. At point d, all the water exists as a saturated vapor. Any addition of heat results in the vapor's being superheated, such as the vapor at point e. This is a *superheated vapor*. The vapor has a temperature greater than the saturation temperature, that is, the temperature of the water when it is saturated liquid and vapor, for a given pressure. Note that the temperature of the water does not

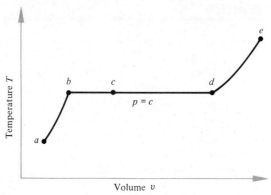

Figure 4.2 Constant-pressure heat addition to water.

begin to change again until it has all changed phase. All the heat added during this phase transition goes into changing the water in the liquid phase to the vapor phase.

What about point *a*? The water at point *a* is a *subcooled,* or compressed, *liquid* because its temperature is less than the saturated temperature for this pressure.

4.4 CRITICAL PROPERTIES

Now let us run the same test at different pressures, and plot a family of constant-pressure lines on a *T-v* diagram (Figure 4.3). Furthermore, let us connect the locus of point *b* and the locus of point *d*. These are the *saturated liquid line* and the *saturated vapor line,* respectively. The point of inflection, point *f* on Figure 4.3, is the *critical*

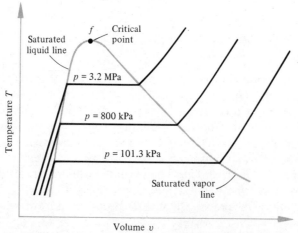

Figure 4.3 The *T-v* diagram for water at various pressures.

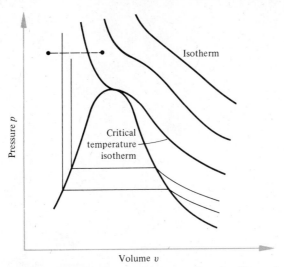

Figure 4.4 A p-v diagram for water, showing lines of constant temperature.

point. This point has a unique temperature and pressure known as the *critical temperature* and *critical pressure.* At pressures higher than the critical pressure, the liquid could be heated from a low temperature to a high temperature without a phase transition occurring. This is illustrated by the dotted line in Figure 4.4, which illustrates the p-v diagram for the liquid vapor phases of water. At temperatures that are greater than the critical temperature, the pressure may be increased to very high values, and no liquefaction will occur.

4.5 SOLID-LIQUID-VAPOR EQUILIBRIUM

Let us consider the solid phase of water—ice. We take a piece of ice at $-17.7°C$ and put it in a vacuum chamber until the pressure is 348 Pa. We now heat the ice. The temperature of the ice will rise to $-6.67°C$, and then further addition of heat will cause the solid water to go directly to water vapor at the same temperature and pressure. This is called *sublimation.* The pressure of the system for the next case is raised to 610.8 Pa. Again we start with the ice at $-17.7°C$ and heat it. The temperature will rise to $0.01°C$. Further heating causes some of the solid water to turn to liquid and some to vapor. This point, characterized by this temperature and pressure, is the *triple point* for water. It is the only point at which all three phases may coexist. If the pressure is further increased from the initial setting to 1.0 kPa and heat is supplied, the ice will rise to $0.01°C$ and change to liquid. This is called *melting.* These processes are illustrated in Figure 4.5.

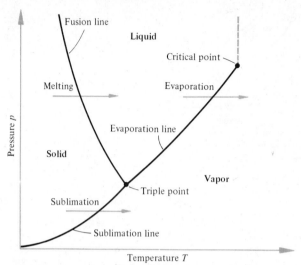

Figure 4.5 A *p-T* diagram showing phase equilibrium lines, the triple point, and the critical point.

The amount of heat added to effect the various phase changes is equal to the change in enthalpy, as noted earlier. These various enthalpy differences across the phase boundaries have certain names. The change of enthalpy between a solid and liquid phase is the *latent heat of fusion.* This is somewhat of a misnomer since heat refers only to thermal energy crossing a boundary, but the term was developed and adopted before this classification. The change of enthalpy between a liquid and vapor phase is the *latent heat of vaporization.* Finally, the change of enthalpy in going from a solid to a vapor phase is the *latent heat of sublimation.*

Referring to Figure 4.2, it is possible to denote the state of water by knowing the temperature and pressure if the water is a subcooled liquid or superheated vapor. This means the lines of constant temperature and constant pressure cross at some unique point. This point is the state of the system at this pressure and temperature. However, what if the state lies somewhere between points *b* and *d*, in the two-phase region? The lines of temperature and pressure are coincident, so they do not uniquely locate the system state. How could the state be located? Knowing the temperature and pressure gives us one line. If the fraction of vapor were known, then we could find out how much water had been evaporated and where along the *a-b* line the system existed. It is important to realize that we need two independent intensive properties to determine the state of a pure substance. In the saturated mixture region, temperature and pressure are not independent and thus do not define a state. This is demonstrated graphically by their being coincident and not intersecting.

4.6 QUALITY

We now define a quantity, x, called the *quality*, as the mass of vapor in the system divided by the total system mass (the mass of vapor plus the mass of liquid). Note that the definition of quality presumes a homogeneous mixture of vapor and liquid.

$$x = \frac{\text{mass vapor}}{\text{mass vapor} + \text{mass liquid}} \qquad (4.1)$$

Let us see then whether we can find the value of some extensive property on a unit mass basis. The specific enthalpy, h, is desired at point c on Figure 4.2, and point c is characterized by a quality x. Let the enthalpy at state b be h_b, and at state d let it be h_d. The change of enthalpy in going from b to d is $h_{bd} = h_b - h_d$. To find the enthalpy of state c, we consider first that all the water was initially at state b, with an enthalpy h_b. Then a fraction of the water was evaporated, increasing its enthalpy by h_{bd}. Thus, the total enthalpy of the mixture of state c is

$$h_c = h_b + x h_{bd} \qquad (4.2a)$$

or

$$h_c = h_d - (1 - x)h_{bd} \qquad (4.2b)$$

This has been illustrated for specific enthalpy, but it is valid for all specific extensive properties. Equations (4.2a) and (4.2b) assume that the enthalpy of each phase remains constant when determining the mixture enthalpy.

4.7 THREE-DIMENSIONAL SURFACE

Water is not the only pure substance with which thermodynamics concerns itself. Another pure substance is carbon dioxide, which is used in refrigeration cycles and has a T-p diagram different from that of water. One should not expect all phase diagrams to be identical to those of water. The various phases and phase transitions are present, and that is the similarity.

Any two-dimensional diagram is really a projection of a three-dimensional surface on that plane. Figure 4.6(a) illustrates the three-dimensional surface from which the pressure-temperature and pressure-volume diagrams are projected.

The surface and projections in Figure 4.6(a) are for a substance whose volume decreases on freezing, which means that the freezing temperature increases as the pressure increases. This is not true with water, as we know; the volume of solid water is greater than that of the same mass of liquid. The result of this is that ice floats, and we can enjoy ice skating. Furthermore, all aquatic life would be destroyed if the solid phase were denser than the liquid phase. The three-dimensional diagram for water is illustrated in Figure 4.6(b). Note the difference in the solid-liquid interface.

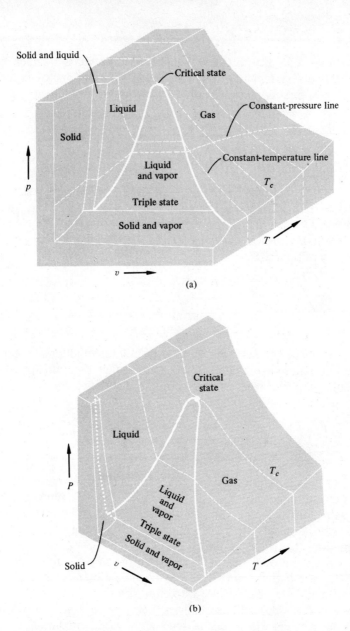

Figure 4.6 A pvT surface for a substance that (a) contracts and (b) expands on freezing.

4.8 TABLES OF THERMODYNAMIC PROPERTIES

In the appendixes of the text several tables and diagrams list the values of the thermodynamic functions for various substances. This same information is provided in computerized form as TK Solver files.

Saturated Steam

We shall consider the steam tables first (appendix Tables A.5, A.6, A.14, and A.15). How are these tables organized? Tables A.5 and A.6 are concerned only with the properties in the saturated region, region II, in Figure 4.7. Table A.5 gives data using temperature as the independent variable while Table A.6 gives data using pressure; data are given for even increments of each. The TK Solver model SATSTM.TK can provide saturation properties as a function of any property in both SI and English units. Let us consider an example using Table A.6. The specific volume, specific internal energy, specific enthalpy, and specific entropy are tabulated. Let the symbol r stand for any of these properties: r_f is the value of the property as a saturated liquid, denoted by point f in Figure 4.7; r_g is the value of the property as a saturated vapor, denoted by point g in Figure 4.7; r_{fg} is the difference between r_g and r_f, $r_{fg} = r_g - r_f$. This is sometimes called the change in the property due to evaporation.

Example 4.1

Find the enthalpy and specific volume of steam at the following states, using the tables and the TK Solver model SATSTM.TK: (a) 250 kPa and $x = 50\%$; (b) 100 psia and $x = 75\%$.

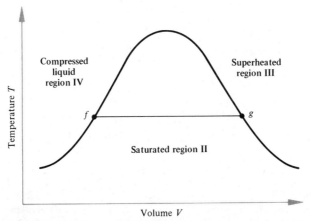

Figure 4.7 A T-v diagram illustrating three regions included in the steam tables.

Solution

Given: The values of pressure and quality of steam.

Find: The values of enthalpy and specific volume at these states.

Sketch and Given Data:

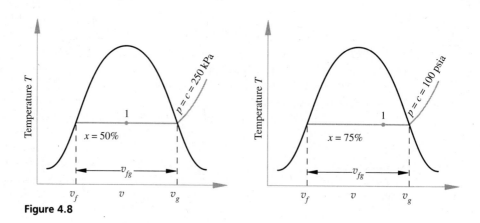

Figure 4.8

Assumption: The state is an equilibrium state, allowing use of the property tables.

Analysis: Case (a): Go to Table A.6 and determine the saturated liquid and vapor values for the enthalpy and specific volume at 250 kPa: $h_f = 535.23$ kJ/kg, $h_g = 2717.1$ kJ/kg, $h_{fg} = 2181.9$ kJ/kg, $v_f = 0.001\ 067\ 5$ m³/kg, $v_g = 0.718\ 77$ m³/kg, $v_{fg} = 0.717$ m³/kg.

$$h = h_f + xh_{fg}$$
$$= 535.23 + (0.5)(2181.9)$$
$$= 1626.18 \text{ kJ/kg}$$

$$v = v_f + xv_{fg}$$
$$= 0.001\ 067\ 5 + (0.5)(0.7177)$$
$$= 0.3599 \text{ m}^3/\text{kg}$$

To solve using TK Solver, load the model SATSTM.TK and enter the equations for enthalpy and specific volume in the mixture region. In the Input column of the Variable Sheet, enter the values of pressure (250) and quality (.5) and press **F9**. The Iterative Solver will be called, and the following results should appear.

```
========================== VARIABLE SHEET ==========================
St Input—Name—Output——— Unit ——— Comment————————————————————————————
                                  ***Saturated Steam Properties***


        Tsat   127.404    degC     Temperature (degK, degC, degR, degF)
        Psat              kPa      Pressure (kPa, MPa, psia)
   250
        vf     .00106748  m3/kg    Specific volume (m3/kg, ft3/lbm)
        vg     .718769    m3/kg    Specific volume (m3/kg, ft3/lbm)

        hf     535.226    kJ/kg    Enthalpy (kJ/kg, BTU/lbm)
        hg     2717.1     kJ/kg    Enthalpy (kJ/kg, BTU/lbm)

        sf     1.60539    kJ/kg-K  Entropy (kJ/kg—K, B/lbm—R)
        sg     7.05253    kJ/kg-K  Entropy (kJ/kg—K, B/lbm—R)

  .5    x                          Quality
        h      1626.16             Enthalpy
        v      .359918             Specific volume

============================= RULE SHEET =============================
S Rule ——————————————————————————————————————————————————————————————
  "Calculation Units are SI: degK, kPa, m3/kg, kJ/kg, kJ/kg—K
  "Call Saturated Steam Property Functions
  Psat=Psat(Tsat)
  hf=Hf(Tsat)
  hg=Hg(Tsat)
  vf=Vf(Tsat)
  vg=Vg(Tsat)
  sf=Sf(Tsat)
  sg=Sg(Tsat)

  h=hf+x*(hg—hf)
  v=vf+x*(vg—vf)
```

Case (b): Go to Appendix Table A.15 and determine the saturated liquid and vapor values for the enthalpy and specific volume at 100 psia: $h_f = 298.41$ Btu/lbm, $h_g = 1187.9$ Btu/lbm, $h_{fg} = 889.47$ Btu/lbm, $v_f = 0.017\,738$ ft^3/lbm, $v_g = 4.4339$ ft^3/lbm, $v_{fg} = 4.416\,16$ ft^3/lbm.

$$h = h_f + xh_{fg}$$

$$= 298.41 + (0.75)(889.47)$$

$$= 965.51 \text{ Btu/lbm}$$

$$v = v_f + xv_{fg}$$

$$= 0.017\,738 + (0.75)(4.416\,16)$$

$$= 3.329\,85 \text{ ft}^3/\text{lbm}$$

To solve using TK Solver, change the units in the Unit column to English, enter the new values for pressure and quality, and press **F9.** Your monitor should display the following results.

```
================= VARIABLE SHEET =================
St Input— Name— Output ——  Unit ——  Comment—
                                     ***Saturated Steam Properties***


         Tsat   327.76       degF      Temperature (degK, degC, degR, degF)
   100   Psat                psia      Pressure (kPa, MPa, psia)

         vf     .017738      ft3/lbm   Specific volume (m3/kg, ft3/lbm)
         vg     4.43391      ft3/lbm   Specific volume (m3/kg, ft3/lbm)

         hf     298.407      BTU/lbm   Enthalpy (kJ/kg, BTU/lbm)
         hg     1187.88      BTU/lbm   Enthalpy (kJ/kg, BTU/lbm)

         sf     .473842      B/lbm—R   Entropy (kJ/kg-K, B/lbm—R)
         sg     1.60338      B/lbm—R   Entropy (kJ/kg-K, B/lbm—R)
   .75
         x                             Quality
         h      965.51       BTU/lbm   Enthalpy
         v      3.32987      ft3/lbm   Specific volume

==================== RULE SHEET ====================
S Rule
  "Calculation Units are SI: degK, kPa, m3/kg, kJ/kg, kJ/kg—K
  "Call Saturated Steam Property Functions
  Psat=Psat(Tsat)
  hf=Hf(Tsat)
  hg=Hg(Tsat)
  vf=Vf(Tsat)
  vg=Vg(Tsat)
  sf=Sf(Tsat)
  sg=Sg(Tsat)

  h=hf+x*(hg—hf)
  v=vf+x*(vg—vf)
```

Comment: Sketches of the *T-v* or *p-v* diagrams are useful in visualizing what region the state is in, liquid, mixture, or superheated. ▬

Superheated Steam

Now that we have a certain confidence in finding the values of properties in the saturated region, region II, let us consider properties in the superheated region, region III. Water vapor existing in this region must be defined by two independent properties before the state can be determined. Usually one of these properties is the pressure, and frequently the other is the temperature. These properties are tabulated in Tables A.7 and A.16 and as TK Solver model SHTSTM.TK.

Example 4.2
Determine the enthalpy and specific volume of steam at 500 kPa and (a) 300°C; (b) $u = 2800$ kJ/kg.

Solution

Given: Steam pressure and temperature.

Find: Steam's enthalpy and specific volume at this state.

Sketch and Given Data:

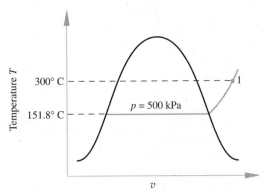

Figure 4.9

Assumption: The state is an equilibrium state, allowing use of the property tables.

Analysis: Case (a): We find steam with these properties in Table A.7, because the temperature of the steam, 300°C, is greater than the saturated steam temperature, 151.8°C, for the pressure of 500 kPa. Proceed to the 500 kPa columns and find a temperature of 300°C. At this line the values of h and v are 3064.2 kJ/kg and 0.522 72 m³/kg.

   Case (b): The internal energy of steam is not listed. To determine if the steam state is in the superheated region, first calculate the saturated vapor value of u_g.

$$u_g = h_g - pv_g = 2748.96 \text{ kJ/kg} - (500 \text{ kN/m}^2)(0.374\ 894 \text{ m}^3/\text{kg}) = 2561.5 \text{ kJ/kg}$$

Since the value of internal energy, 2800 kJ/kg, is greater than the saturated vapor value, the steam state is in the superheated region. To determine the enthalpy, specific volume, and temperature in the superheated region requires creating a column for internal energy for different temperature values, say at 250°C and 300°C, and interpolating between these, yielding a temperature of 298.2°C for $u = 2800$ kJ/kg. The values for enthalpy and specific volume would be found in a similar fashion.

   Using TK Solver simplifies the solution of this problem. Load SHTSTM.TK, and enter "h = u + P*v" into the Rule Sheet and the values for P and u in the Input column of the Variable Sheet. Press **F9,** and the following results should appear.

```
================================ VARIABLE SHEET ================================
St Input—Name—Output——   Unit ——   Comment—
                                           ***Superheated Steam Properties***

   500    P                kPa        Pressure (kPa, MPa, psia)
          T      298.196    degC       Temperature (degK, degC, degR, degF)
          v      .520993    m3/kg      Specific Volume (m3/kg, ft3/lbm)
          h      3060.5     kJ/kg      Enthalpy (kJ,kg, BTU/lbm)
          s      7.45273    kJ/kg—K  Entropy (kJ/kg—K, B/lbm—R)

  2800    u                kJ/kg      Internal Energy


===================================== RULE SHEET =====================================
S Rule ──────────────────────────────────────────────────────────────────────────
   "Calculation Units are SI: degK, kPa, m3/kg, kJ/kg, kJ/kg—K
   "Call Superheated Steam Property Functions
   v=Vsuper(T,P)
   h=Hsuper(T,P)
   s=Ssuper(T,P)

   h=u+P*v
```

Comment: Using problem-solving software readily allows the determination of property values other than those tabulated.

Subcooled Liquid Water

This time let us consider water at 10 MPa and 60°C. We want to find the enthalpy and specific volume at this condition. We might suspect that the water is a subcooled liquid in region IV on Figure 4.7, which has property values located in Table A.8. Why? Again we check the saturated water temperature at 10 MPa. It is 311.06°C. If the temperature is less than this saturated value for water with a pressure of 10 MPa, the water is subcooled and its properties may be found in Table A.8. Since the conditions in the problem meet this criterion, Table A.8 will be used for the solution. We go to the column for 10 MPa and proceed vertically down until it intersects the 60°C line. At this level we read the property values as

$$h = 259.49 \text{ kJ/kg}$$

$$v = 0.001\ 012\ 7 \text{ m}^3/\text{kg}$$

Measuring the Saturated Vapor State

There are tabulated values for vapor properties, but it is sometimes very difficult to measure the vapor state. Let us consider a situation in which steam is driving a turbine and we want to know the steam state as it enters the turbine.

This means finding the state of a flowing vapor. If the vapor is superheated, measuring the temperature and pressure would determine the state. What happens if the steam is a saturated mixture? The temperature and pressure are not independent in the saturated region, so more information is needed to determine the state. There is no ready means to measure other properties, such as specific volume, enthalpy, or internal energy.

There are several methods to resolve this problem, one of the most common being a throttling calorimeter, schematically illustrated in Figure 4.10. Note that the steam supply is sampled from a vertical pipe. If the pipe were horizontal, separation of liquid and vapor would be more apt to occur. In a throttling process the pressure is decreased adiabatically by use of a valve. This is a totally irreversible process. A first-law analysis across the adiabatic valve shows that the initial and final enthalpies have the same value. If there is a sufficient decrease in pressure, the steam will be superheated at a lower pressure and temperature.

Temperature and pressure measurements can be made for the steam leaving the valve (in the cup), and the enthalpy can be determined. It is important that there be negligible heat loss and negligible velocity at the point of temperature and pressure measurement. By knowing the enthalpy and the initial pressure or temperature, the initial steam state may be determined. This is the average value of the steam state, since it is based on a representative sample.

Other types of calorimeters operate with varying degrees of success. A problem with the throttling calorimeter is that the initial steam state must not be very wet or throttling will not produce superheated steam. A separating calorimeter mechanically separates the liquid and vapor, volumetrically measuring the liquid and measuring the remaining vapor flow rate. An electric calorimeter, through resistance heating, superheats a sample of steam that is steadily withdrawn through a sampling

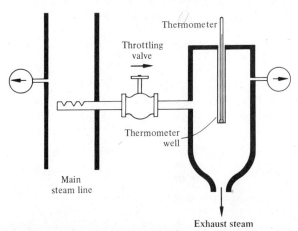

Figure 4.10 Schematic diagram of a throttling steam calorimeter.

tube. By measuring the electrical energy added and the superheated steam state and steam flow rate, the steam quality is determined.

Example 4.3
A throttling calorimeter is connected to a desuperheated steam line supplying steam to the auxiliary feed pump in a steam power plant. The line pressure is 400 psia, the calorimeter pressure is 14.7 psia, and the temperature is 250°F. Determine the enthalpy and quality of the desuperheated steam.

Solution

Given: Steam at a given pressure is expanded through a throttling calorimeter. The pressure and temperature of the steam in the calorimeter are known.

Find: The enthalpy and quality of the steam before entering the calorimeter.

Sketch and Given Data:

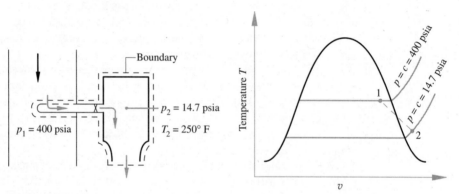

Figure 4.11

Assumptions:

1. The throttling process is adiabatic.
2. Equilibrium exists at the described steam states, so tables of properties can be used.
3. There is no work crossing the control volume.
4. The changes in potential and kinetic energies across the control volume are negligible.
5. Steady-state conditions apply.

Analysis: Noting that $T_2 = 250°F$ is greater than T_{sat} at 14.7 psia (212°), look for the value of enthalpy at state 2 in the superheated steam tables. Thus, $h_2 = 1168.5$ Btu/lbm.

The first law for an open system with steady flow is

$$\dot{Q} + \dot{m}[h + \text{k.e.} + \text{p.e.}]_{in} = \dot{W} + \dot{m}[h + \text{k.e.} + \text{p.e.}]_{out}$$

Applying the assumptions, this reduces to

$$h_1 = h_2$$

Thus, the enthalpy at state 1 is the same as at state 2. In addition we can determine the quality at state 1 as follows.

$$h_1 = 1168.5 = h_f + x_1 h_{fg}$$

$$1168.5 = 424.35 + (x_1)(781.08)$$

$$x = 0.953 \text{ or } 95.3\%$$

SATSTM.TK can be used to solve the problem. Input the state 1 enthalpy value and the pressure; the program determines the quality. You still need to perform the first-law analysis to realize that the enthalpies at state 1 and state 2 are equal based on the assumptions listed.

Comment: The throttling calorimeter is useful in determining steam qualities at moderate to high pressures but only at quality values about 94% or greater. ■

Ammonia and Refrigerant 12 Tables

Two other substances have property tables that are listed in the appendix. These are ammonia and refrigerant 12 (R 12), dichlorodifluoromethane. Both are used as refrigerants, especially R 12. Although it is not necessary to repeat all the examples that can be done with these tables, since the techniques developed in using the steam tables are equally valid, it is worthwhile to do some. The thermodynamic properties of ammonia are listed in Tables A.9 and A.10, and the thermodynamic properties of refrigerant 12 are listed in Tables A.11 and A.12 for SI. The English unit tables begin with Tables A.18 and A.19 for ammonia and Tables A.20 and A.21 for R 12. Refrigerant 12 properties are also available in the TK Solver models R12SAT.TK and R12SHT.TK.

Example 4.4
Refrigerant 12 is used in a refrigeration unit. At one point in the refrigeration process, the refrigerant leaves the compressor at 125 psia and 180°F and enters the high-pressure tubing with a flow rate of 5 lbm/min. Determine the velocity of the refrigerant in the tubing in feet/minute if the tubing diameter is 1 in.

Solution

Given: R 12 flowing through 1-in.-diameter tubing at a certain pressure and temperature and with a constant flow rate.

Find: The velocity of the refrigerant in the tubing.

Sketch and Given Data:

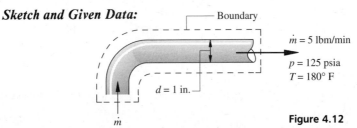

Figure 4.12

Assumptions:

1. The flow is steady.
2. The state is an equilibrium state, allowing use of the property tables.

Analysis: There is no mention of energy in the problem statement or in the information sought in the answers, so a first-law analysis is not immediately indicated. What relates the flow rate, diameter, and velocity? The conservation of mass. Thus,

$$\dot m = \rho A v$$

but

$$\rho = 1/v$$

and

$$A = \pi d^2/4$$

Next determine the specific volume of R 12. The R 12 is superheated, because its temperature, 180°F, is greater than the saturation temperature at 125 psia, 97°F. Looking in the superheated tables yields a value of the specific volume of 0.40857 ft³/lbm.

Solving the conservation of mass equation for velocity and substituting in the variable values yields

$$v = \frac{\dot m v}{A} = \frac{(5 \text{ lbm/min})(0.40857 \text{ ft}^3/\text{lbm})(4)(144 \text{ in.}^2/\text{ft}^2)}{\pi(1 \text{ in.}^2)}$$

$$v = 374.6 \text{ ft/min}$$

Comment: While the conservation of energy is very important, many problems must use conservation of mass or other property relationships. In this problem, even though velocity was asked for, information about kinetic energy does not provide the connection. Look for connections between the parameters given in the problem statement to provide initial direction to the problem analysis.   ▬

Example 4.5
In the evaporator of a refrigeration system the cold refrigerant receives heat from the refrigerated space. The flow through a given evaporator is 50 lbm/hr, and the R 12 enters the evaporator as a saturated liquid at −20°F and leaves this constant-pressure process at +20°F. Determine the heat transfer during this process.

Solution

Given: A refrigerant flows through an evaporator at constant pressure, entering as a saturated liquid and leaving as a superheated vapor.

Find: The heat transfer to the refrigerant in this process.

Sketch and Given Data:

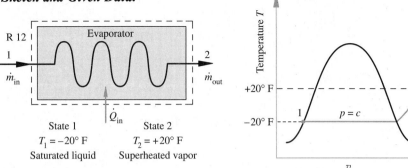

State 1
$T_1 = -20°$ F
Saturated liquid

State 2
$T_2 = +20°$ F
Superheated vapor

Figure 4.13

Assumptions:

1. The flow is steady.
2. There is no work crossing the control volume.
3. The changes in kinetic and potential energies are negligible.
4. States 1 and 2 are equilibrium states, so tables of properties can be used.

Analysis: This problem involves energy, so the first-law analysis is an appropriate starting point.

$$\dot{Q} + \dot{m}[h + \text{k.e.} + \text{p.e.}]_{\text{in}} = \dot{W} + \dot{m}[h + \text{k.e.} + \text{p.e.}]_{\text{out}}$$

Employing the assumptions yields

$$\dot{Q} = \dot{m}(h_2 - h_1)$$

From Table A.20 the enthalpy for a saturated liquid at $-20°$F is

$$h_1 = h_f = 4.2357 \text{ Btu/lbm}$$

Since the pressure at state 2 is the same as at state 1, and the pressure at state 1 is the saturated pressure corresponding to a temperature of $-20°$F,

$$p_1 = p_{\text{sat}} = 15.27 \text{ psia}$$

$$p_2 = p_1 = 15.27 \text{ psia}$$

Using the superheated tables, find the enthalpy.

$$h_2 = 80.712 \text{ Btu/lbm}$$

Substituting in the first-law equation and solving for the heat transfer yields

$$\dot{Q} = \dot{m}(h_2 - h_1) = (50 \text{ lbm/hr})(80.712 - 4.236 \text{ Btu/lbm})$$

$$\dot{Q} = 3823.8 \text{ Btu/hr}$$

Instead of the R 12 property tables, the TK Solver models R12SAT.TK and R12SHT.TK can be used to determine the refrigerant enthalpy. Simply load R12SAT.TK, change the units to English, enter -20 for Tsat, and press **F9**. Load R12SHT.TK, change the units to English, enter the temperature and pressure, and press **F9**.

```
================================== VARIABLE SHEET ==================================
St Input─ Name─ Output ──  Unit ──  Comment ─────────────────────────────────
                                    ***Saturated R 12 Properties***


        Psat  15.2669      psia     Pressure (kPa, MPa, psia)
   -20  Tsat               degF     Temperature (degK, degC, degR, degF)

        vf    .0107879     ft3/lbm  Liquid Specific Volume (m3/kg, ft3/lbm)
        vg    2.44344      ft3/lbm  Vapor Specific Volume (m3/kg, ft3/lbm)

        hf    4.2248       BTU/lbm  Liquid Enthalpy (kJ/kg, BTU/lbm)
        hg    75.111       BTU/lbm  Vapor Enthalpy (kJ/kg, BTU/lbm)

        sf    .00980607    B/lbm-R  Liquid Entropy (kJ/kg-K, B/lbm-R)
        sg    .171013      B/lbm-R  Vapor Entropy (kJ/kg-K, B/lbm-R)


================================== VARIABLE SHEET ==================================
St Input─ Name─ Output ──  Unit ──  Comment ─────────────────────────────────
                                    ***Superheated R 12 Properties***


15.2669  P                 psia     Pressure (kPa, MPa, psia)
20       T                 degF     Temperature (degK, degC, degR, degF)
         v    2.69985       ft3/lbm  Specific Volume (m3/kg, ft3/lbm)
         h    80.6953       BTU/lbm  Enthalpy (kJ/kg, BTU/lbm)
         s    .183166       B/lbm-R  Entropy (kJ/kg-K, B/lbm-R)
```

Comments:

1. When a state is that of a saturated liquid or vapor, knowing one additional condition, such as temperature, enables one to determine all the properties at this state.
2. Often in open-system heat transfer problems, the assumption of a constant-pressure process is applicable.

CONCEPT QUESTIONS

1. You are drinking ice water. Is this a pure substance? Explain.
2. You are drinking ice tea with sugar added. Is this a pure substance? Explain.
3. What is the difference between saturated and compressed liquids?
4. What is the difference between superheated vapor and saturated vapor?
5. During a heating process when a liquid is boiling, the pressure is increased. Will the temperature change or not?
6. What is the difference between the critical point and the triple point?
7. Can water vapor exist at $-20°C$?
8. Can liquid water exist at 0.08 psia?

9. In the phase transition process from compressed liquid to superheated vapor, what difference do subcritical and supercritical pressures make?

10. You are given a substance's specific internal energy, specific volume, and pressure. Can you determine its enthalpy?

11. What does h_{fg} mean physically?

12. Can quality be expressed in terms of volume rather than mass?

13. What is quality? Where is it defined?

14. Consider two cases of vaporization of a saturated liquid to a saturated vapor at constant pressures of 100 kPa and 500 kPa. Which case requires more energy?

15. Consider two cases of vaporization of a saturated liquid to a saturated vapor at constant temperatures of 100°C and 200°C. Which case requires more energy?

16. Without compressed liquid tables, how would you determine the liquid's properties given its pressure and temperature?

17. How many independent properties are needed to define the state of a pure substance?

18. Are liquid properties primarily dependent on temperature or on pressure?

19. Why do foods cook faster in a pressure cooker?

20. When a block of dry ice, solid CO_2, is placed on a table at room temperature and pressure, no liquid is observed. Why?

PROBLEMS (SI)

4.1   A 2-m³ tank contains a saturated vapor at 40°C. Determine the pressure and mass in the tank if the substance is (a) steam; (b) ammonia; (c) R 12.

4.2   Determine for R 12 the following:
(a) h if $T = 85°C$ and $p = 1000$ kPa.
(b) x if $h = 100$ kJ/kg and $T = 0°C$.
(c) u if $T = 100°C$ and $p = 800$ kPa.
(d) p if $T = 20°C$ and $v = 0.001\ 020$ m³/kg.

$M = PAV$

$V = M \cdot v$

4.3   Complete the following table for ammonia.

	T (°C)	p (kPa)	x (%)	h (kJ/kg)	u (kJ/kg)	v (m³/kg)	State
(a)	10			1225.5			
(b)	50	700					
(c)		752.79	80				
(d)	12	1000					
(e)	0		90				
(f)		1554.3	80				

Indicate for each state whether it is subcooled liquid, saturated liquid, mixture, saturated vapor, or superheated vapor.

4.4   Determine the volume occupied by 2 kg of steam at 1000 kPa and 500°C.

4.5 Complete the following table for water.

	T (°C)	p (kPa)	x (%)	h (kJ/kg)	u (kJ/kg)	v (m³/kg)	State
(a)	200			852.4			
(b)		150			1000.0		
(c)	300	1000					
(d)	200	5000					
(e)		250				0.8500	
(f)	300		80				
(g)		1000	90				

Indicate for each state whether it is subcooled liquid, saturated liquid, mixture, saturated vapor, or superheated vapor.

4.6 R 12 is contained in a storage bottle with a diameter of 20 cm and a length of 120 cm. The weight of the R 12 is 370 N ($g = 9.8$ m/s²) and the temperature is 20°C. Determine (a) the ratio of mass vapor to mass liquid in the cylinder; (b) the height of the liquid-vapor line if the bottle is standing upright.

4.7 Steam has a quality of 90% at 200°C. Determine (a) the enthalpy; (b) the specific volume.

4.8 A refrigeration system uses R 12 as the refrigerant. The system is evacuated, then charged with refrigerant at a constant 20°C temperature. The system volume is 0.018 m³. Determine (a) the pressure and quality when the system holds 0.8 kg of R 12; (b) the mass of R 12 in the system when the pressure is 200 kPa.

4.9 A rigid steel tank contains a mixture of vapor and liquid water at a temperature of 65°C. The tank has a volume of 0.5 m³, the liquid phase occupying 30% of the volume. Determine the amount of heat added to the system to raise the pressure to 3.5 MPa.

4.10 Steam enters an isothermal compressor at 400°C and 100.0 kPa. The exit pressure is 10 MPa. Determine the change of enthalpy.

4.11 Steam enters an adiabatic turbine at 300°C and 400 kPa. It exits as a saturated vapor at 30 kPa. Determine (a) the change of enthalpy; (b) the work; (c) the change of internal energy.

4.12 A 0.5-m³ tank contains saturated steam at 300 kPa. Heat is transferred until the pressure reaches 100 kPa. Determine (a) the heat transferred; (b) the final temperature; (c) the final steam quality; (d) the process on a T-v diagram.

4.13 A 500-liter tank contains a saturated mixture of steam and water at 300°C. Determine (a) the mass of each phase if their volumes are equal; (b) the volume occupied by each phase if their masses are equal.

4.14 A 1-kg steam-water mixture at 1.0 MPa is contained in an inflexible tank. Heat is added until the pressure rises to 3.5 MPa and the temperature to 400°C. Determine the heat added.

4.15 A rigid vessel contains 5 kg of wet steam at 0.4 MPa. After the addition of 9585 kJ the steam has a pressure of 2.0 MPa and a temperature of 700°C. Determine the initial internal energy and specific volume of the steam.

4.16 Two kg/min of ammonia at 800 kPa and 70°C are condensed at constant pressure to a saturated liquid. There is no change in kinetic or potential energy across the device.

Determine (a) the heat; (b) the work; (c) the change in volume; (d) the change in internal energy.

4.17 Three kg of steam initially at 2.5 MPa and a temperature of 350°C have 2460 kJ of heat removed at constant temperature until the quality is 90%. Determine (a) T-v and p-v diagrams; (b) pressure when dry saturated steam exists; (c) work.

4.18 You want 400 liters/min of water at 80°C. Cold water is available at 10°C and dry saturated steam at 200 kPa (gage). They are to be mixed directly. Determine (a) steam and water flow rates required; (b) the pipe diameters, if the velocity is not to exceed 2 m/s.

4.19 Steam condensate at 1.8 MPa leaves a heat exchanger trap and flows at 3.8 kg/s to an adjacent flash tank. Some of the condensate is flashed to steam at 180 kPa, and the remaining condensate is pumped back to the boiler. There is no subcooling. Determine (a) the condensate flow returned at 180 kPa; (b) the steam flow produced at 180 kPa.

4.20 A chemical process requires 2000 kg/h of hot water at 85°C and 150 kPa. Steam is available at 600 kPa and 90% quality, and water is available at 600 kPa and 20°C. The steam and water are mixed in an adiabatic chamber, with the hot water exiting. Determine (a) the steam flow rate; (b) the steam-line diameter if the velocity is not to exceed 70 m/s.

4.21 The main steam turbine of a ship is supplied by two steam generators. One generator delivers steam at 6.0 MPa and 500°C, and the other delivers steam at 6.0 MPa and 550°C. Determine the steam enthalpy and temperature at the entrance to the turbine.

4.22 An adiabatic feed pump in a steam cycle delivers water to the steam generator at a temperature of 200°C and a pressure of 10 MPa. The water enters the pump as a saturated liquid at 180°C. The power supplied to the pump is 75 kW. Determine (a) the mass flow rate; (b) the volume flow rate leaving the pump; (c) the percentage of error if the exit conditions are assumed to be a saturated liquid at 200°C.

4.23 An adiabatic rigid tank has two equal sections of 50 liters separated by a partition. The first section contains steam at 2.0 MPa and 95% quality. The second section contains steam at 3.5 MPa and 350°C. Determine the equilibrium temperature and pressure when the partition is removed.

4.24 A throttling calorimeter is attached to a steam line where the steam temperature reads 210°C. In the calorimeter the pressure is 100 kPa, and the temperature is 125°C. Determine the quality of the steam, using the steam tables.

4.25 A throttling calorimeter is connected to a main steam line where the pressure is 1750 kPa. The calorimeter pressure is 100 mm Hg vacuum and 105°C. Determine the main steam quality.

4.26 A piston-cylinder containing steam at 700 kPa and 250°C undergoes a constant-pressure process until the quality is 70%. Determine per kg (a) the work done; (b) the heat transferred; (c) the change of internal energy; (d) the change of enthalpy.

4.27 An electric calorimeter samples steam with a line pressure of 0.175 MPa. The calorimeter adds 200 W of electricity to sampled steam having a resultant pressure of 100 kPa, temperature of 140°C, and flow rate of 11 kg/h. Determine the main steam quality.

4.28 Three kg of ammonia are expanded isothermally in a piston-cylinder from 1400 kPa and 80°C to 100 kPa. The heat loss is 495 kJ/kg. Determine (a) the system work; (b) the change of enthalpy; (c) the change of internal energy.

4.29 Refrigerant 12 is expanded steadily in an isothermal process. The flow rate is 13.6 kg/min with an inlet state of wet saturated vapor with an 80% quality to a final state of 70°C and

200 kPa. The change of kinetic energy across the device is 3.5 kJ/kg, and the heat added is 21.81 kW. Determine the system power.

4.30 A tank contains 0.2 kg of a steam-water mixture at 100 kPa. Heat is added until the substance is a saturated vapor at 1.0 MPa. Determine (a) the T-v diagram; (b) the heat added; (c) the tank's volume.

4.31 A piston-cylinder contains R 12 as a saturated vapor at 100 kPa and compresses it to 600 kPa and 40°C. The work done is 35 kJ/kg. Determine the heat loss or gain in kJ/kg.

4.32 A pressure cooker has a volume of 4 liters and contains 0.75 kg of water at 20°C. The lid is secured and heat is added, with the air being vented in the process, until the pressure is 200 kPa. Determine the heat added and the final quality.

4.33 An adiabatic steam turbine receives 5 kg/s of steam at 1.0 MPa and 400°C, and the steam exits at 50 kPa and 100% quality. Determine (a) the power produced; (b) the exit area in m² if the exit velocity is 250 m/s.

4.34 Plot the p-v diagram for water on log-log coordinates from the triple point to the critical point, denoting the saturated liquid and vapor values. The unit of pressure is kPa and of specific volume is m³/kg.

4.35 Plot the T-v diagram for water on log-log coordinates from the triple point to the critical point, denoting the saturated liquid and vapor values. The unit of temperature is °C and of specific volume is m³/kg.

4.36 Plot the T-v diagram for R 12 on linear-log coordinates from −50°C to 112°C, denoting the saturated liquid and vapor values. The unit of specific volume is m³/kg.

4.37 Determine the quality of a two-phase mixture of (a) water at 180°C and a specific volume of 0.15 m³/kg; (b) R 12 at 745 kPa and a specific volume of 0.020 m³/kg.

4.38 An R 12 tank has a volume of 54 liters and contains 3.6 kg at 1000 kPa. What is the temperature?

4.39 Determine the volume occupied by 3 kg of water at 1000 kPa and temperatures of 100°C and 1000°C.

4.40 Determine the volume occupied by 2 kg of steam with a quality of 75% and a pressure of 500 kPa.

4.41 Refrigeration tubing is 2 cm in diameter and 3 m long and contains R 12 as a saturated vapor at 0°C. What is the mass of R 12 in the tubing?

4.42 A plastic container holds 8 liters of water of 25°C. The plastic itself has a mass of 50 g. What is the total weight of the filled container?

4.43 A 0.5-m³ tank contains ammonia at a temperature of 0°C and a quality of 85%. Determine the volume occupied by the vapor and by the liquid and the percentage of the total volume each represents.

4.44 A 10-m³ tank is used to hold high-pressure steam from a steam generator during emergency conditions when the turbine fails. The tank contains steam at 500°C and 10 MPa. The steam cools until the temperature is 200°C. What is the pressure and how much, if any, liquid is present in the tank?

4.45 A rigid tank contains 3 kg of saturated steam at a pressure of 3000 kPa. Because of heat transfer to the surroundings, the pressure decreases to 1000 kPa. Determine the tank's volume and the quality of steam at the final state.

4.46 Steam in a rigid tank is at a pressure of 5 MPa and a temperature of 400°C. As a result of heat transfer, the temperature decreases to that of the surroundings, 20°C. Determine (a) the final pressure in kPa; (b) the percentage of the total mass that is liquid in the final state; (c) the percentage of volume occupied by the liquid and vapor at the final state.

4.47 A 20-liter tank contains a saturated mixture of water and vapor at 100 kPa. The volume occupied by the liquid is 20 cm³. Heat is added until all the water evaporates and the tank contains only saturated vapor. Determine the pressure at this final state.

4.48 Water expands at constant temperature from a saturated vapor at 250°C until the specific volume is 1.0 m³/kg. Determine the final pressure.

4.49 Two kg of steam is compressed at constant pressure in a piston-cylinder from an initial state of 500 kPa and 300°C to a saturated vapor. Determine the work for the process.

4.50 A rigid adiabatic tank contains 1.5 kg of water at a quality of 90% and at a pressure of 200 kPa. Paddle work is applied until the water becomes a saturated vapor. Determine the paddle work, neglecting changes in kinetic and potential energies.

4.51 A rigid nonadiabatic tank contains 2 kg of R 12 with a quality of 85% and a temperature of 40°C. Fifty kJ of heat is added. Determine the final state.

4.52 A rigid adiabatic tank has a 50-W electric resistance heater and contains 2 kg of R 12 at 30°C and 90% quality. The heater is turned on, and the temperature of the tank is measured at 100°C. How long a time interval was required for the heating?

4.53 Refrigerant 12 initially a saturated vapor at 10°C is compressed adiabatically to a pressure of 2.5 MPa and 100°C. Determine the work per unit mass, neglecting changes in potential and kinetic energies.

4.54 A rigid adiabatic tank contains two compartments. One is filled with 1.5 kg of steam at a pressure of 500 kPa and 50% quality, and the other is completely evacuated. The partition between them is removed, and the steam expands to fill the entire tank. The final pressure is 200 kPa. Determine the volume of the evacuated compartment.

PROBLEMS (English Units)

*4.1 Fill in the data omitted in the following table for water.

Pressure (psia)	Temperature (°F)	Specific volume (ft³/lbm)	Enthalpy (Btu/lbm)	Quality x (%)	State
500		0.650			
	250		1000		
600	700				
800			1399.1		
	300			90	
1000	200				

Indicate for each state whether it is subcooled liquid, saturated liquid, mixture, saturated vapor, or superheated vapor.

*4.2   Fill in the data omitted in the following table for R 12.

Pressure (psia)	Temperature (°F)	Specific volume (ft³/lbm)	Enthalpy (Btu/lbm)	Internal energy (Btu/lbm)	Quality x (%)	State
90			53.6			
	100			33.2		
	120	0.01317				
160					80	
	0	1.6089				
	20			50.0		

Indicate for each state whether it is subcooled liquid, saturated liquid, mixture, saturated vapor, or superheated vapor.

*4.3   Determine the correct proportions of water and steam at 100 psia in a rigid tank that would allow the mixture to pass through the critical point when heated. Determine the proportion on the mass basis.

*4.4   A piston-cylinder contains steam at 500 psia and 900°F and has a volume of 4 ft³. The piston compresses the steam until the volume is one-half its initial value. The pressure remains constant. Find the work.

*4.5   One lbm of a steam-water mixture at 160 psia is contained in an inflexible tank. Heat is added until the pressure rises to 600 psia and the temperature to 600°F. Determine the heat added.

*4.6   A rigid vessel contains 10 lbm of wet steam at 60 psia. After adding 6997 Btu, the steam has a pressure of 300 psia and a temperature of 540°F. Determine the initial internal energy and specific volume of the steam.

*4.7   Three lb of ammonia is expanded isothermally in a piston-cylinder from 200 psia and 200°F to 15 psia. The heat loss is 212.7 Btu/lbm. Determine (a) the system work; (b) the change of enthalpy; (c) the change of internal energy.

*4.8   Refrigerant 12 is expanded steadily in an isothermal process. The flow rate is 30 lbm/min with an inlet state of wet saturated vapor with an 80% quality to a final state of 160°F and 25 psia. The change of kinetic energy across the device is 1.5 Btu/lbm and the heat added is 21.81 kW. Determine the system power.

*4.9   A rigid steel tank contains a mixture of vapor and liquid water at a temperature of 150°F. The tank has a volume of 15 ft³, the liquid phase occupying 30% of the volume. Determine the amount of heat added to the system to raise the pressure to 500 psia.

*4.10   Plot the p-v diagram for water on log-log coordinates from the triple point to the critical point, denoting the saturated liquid and vapor values. The unit of pressure is psia and of specific volume is ft³/lbm.

*4.11   Plot the T-v diagram for water on log-log coordinates from the triple point to the critical point, denoting the saturated liquid and vapor values. The unit of temperature is °F and of specific volume is ft³/lbm.

*4.12   Plot the p-v diagram for R 12 on log-log coordinates from 15 psia to the critical point, denoting the saturated liquid and vapor values. The unit of specific volume is ft³/lbm.

*4.13 Determine the quality of a two-phase mixture of (a) water at 400°F and a specific volume of 0.55 ft³/lbm; (b) R 12 at 350 psia and a specific volume of 0.025 ft³/lbm.

*4.14 An R 12 tank has a volume of 15 gal and contains 10 lbm at 175 psia. What is the temperature?

*4.15 Determine the volume occupied by 5 lbm of water at 100 psia and temperatures of 100°F and 500°F.

*4.16 Determine the volume occupied by 2 lbm of a water with a quality of 75% and a pressure of 500 psia.

*4.17 Refrigeration tubing is 2 in. in diameter and 10 ft long and contains R 12 as a saturated vapor at 0°F. What is the mass of R 12 in the tubing?

*4.18 A plastic container holds 2.5 gal of water at 70°F. The plastic itself weighs 4 oz. What is the total weight of the filled container?

*4.19 A 10-ft³ tank contains ammonia at a temperature of 0°F and a quality of 85%. Determine the volume occupied by the vapor and by the liquid and the percentage of the total volume each represents.

*4.20 A 300-ft³ tank is used to hold high-pressure steam from a steam generator during emergency conditions when the turbine fails. The tank contains steam at 1000°F and 1000 psia. The steam cools until its temperature is 400°F. What is the pressure and what, if any, liquid is present in the tank?

*4.21 A rigid tank contains 3 lb of saturated steam at a pressure of 500 psia. Because of heat transfer to the surroundings, the pressure decreases to 100 psia. Determine the tank's volume and the quality of steam at the final state.

*4.22 Steam in a rigid tank is at a pressure of 400 psia and a temperature of 600°F. As a result of heat transfer, the temperature decreases to that of the surroundings, 70°F. Determine (a) the final pressure in psia; (b) the percentage of the total mass that is liquid in the final state; (c) the percentage of volume occupied by the liquid and vapor at the final state.

*4.23 A 1-ft³ tank contains a saturated mixture of water and vapor at 1 atm. The volume occupied by the liquid is 20 in.³ Heat is added until all the water evaporates and the tank contains only saturated vapor. Determine the pressure at this final state.

*4.24 Water expands at constant temperature from a saturated vapor at 600°F until the specific volume is 1.2 ft³/lbm. Determine the final pressure.

*4.25 Two lbm of steam are compressed at constant pressure in a piston-cylinder from an initial state of 400 psia and 500°F to a saturated vapor. Determine the work for the process.

*4.26 A rigid adiabatic tank contains 1.5 lbm of water at a quality of 90% and a pressure of 50 psia. Paddle work is applied to the water until it becomes a saturated vapor. Determine the paddle work, neglecting changes in kinetic and potential energies.

*4.27 A rigid nonadiabatic tank contains 3 lbm of R 12 with a quality of 85% and a temperature 40°F. Fifty Btu of heat is added. Determine the final state.

*4.28 A rigid adiabatic tank has a 50-W electric resistance heater and contains 2 lbm of R 12 at 30°F and 90% quality. The heater is turned on, and the temperature of the tank is measured at 160°F. How long a time interval was required for the heating to occur?

*4.29 Ammonia initially a saturated vapor at 30 psia is compressed adiabatically to a pressure of 200 psia and 280°F. Determine the work per unit mass, neglecting changes in potential and kinetic energies.

*4.30 A rigid adiabatic tank contains two compartments. One is filled with 1.5 lbm of steam at a pressure of 500 psia and 50% quality, and the other is completely evacuated. The partition between them is removed, and the steam expands to fill the entire tank. The final pressure is 200 psia. Determine the volume of the evacuated compartment.

COMPUTER PROBLEMS

C4.1 Using the TK Solver model SHTSTM.TK, determine the pressure and enthalpy of steam at 700°C and 2.0 m³/kg.

C4.2 Using the TK Solver model SHTSTM.TK, determine the enthalpy, specific volume, and temperature of steam at 95 psia and an internal energy of 1325 BTU/lbm.

C4.3 Using the TK Solver model R12SAT.TK, determine the pressure of a mixture of R 12 with a quality of 50% and an enthalpy of 125 kJ/kg.

C4.4 Using TK Solver or a spreadsheet program, plot using log-log format the properties of water for saturated liquid and vapor from 25°C to the critical point as follows:
(a) pressure versus specific volume.
(b) pressure versus temperature.
(c) pressure versus enthalpy.
(d) temperature versus specific volume.
(e) temperature versus enthalpy.
(f) enthalpy versus specific volume.

C4.5 Solve Problem 4.23 using TK Solver and the models SATSTM.TK and SHTSTM.TK.

C4.6 Develop a TK Solver model that will calculate the quality of steam entering a throttling calorimeter when the line pressure and calorimeter temperature and pressure are entered.

C4.7 Use the model developed for Problem C4.6 to determine the maximum moisture that can be measured by a throttling calorimeter exhausting to the atmosphere for line pressures of 200 kPa, 2000 kPa, and 20 000 kPa. Assume a minimum superheat of 3°C in the calorimeter.

C4.8 A cylinder contains R 12 at 70 psig, half saturated liquid and half saturated vapor by mass. It is set too close to a space heater and is heated to 300°F. What is the pressure in the cylinder?

C4.9 The pressure in a well-insulated vessel with a volume of 20 ft³ is 100 psia. The vessel is filled 80% with saturated liquid water and the rest with saturated vapor. Vapor is allowed to escape slowly through a small valve. Using SATSTM.TK, plot the pressure in the vessel as the fluid fraction is reduced to 60%. At what fluid fraction does the pressure in the vessel drop to atmospheric?

1 MOLE = 6.023 × 10²³ MOLECULES.

$P = PRESS.$ (P_a)
$V = VOL.$ (M^3)
$M = MASS$ (Kg)
$R = INDIV. GAS CONS. (APPEND A)$
$T = ABS. TEMP. (°K)$

$$\frac{P_1 V_1}{T_1} = \frac{P_2 V_2}{T_2}$$

$P \bar{v} = R T × n.$ $P V = n R T$

5

Ideal and Actual Gases

$\bar{R} = UNIV. GAS CONSTANT.$
$\bar{v} = VOL OF 1 MOL OF GAS$

$$P \bar{v} = M R T$$

IDEAL GAS LAW

$$\frac{P_1 V_1}{T_1} = \frac{P_2 V_2}{T_2}$$

NON - IDEAL GAS EQUATION.

$$P = \frac{\bar{R} T}{\bar{v} - b} - \frac{a}{\bar{v}^2} (VAN DER WAALS \ eq)$$

$$VOL \ SPHERE = \frac{4}{3} \pi r^3$$

Chapter 4 addressed the property determination for all phases of a pure substance. The vapor phase, particularly a highly superheated vapor, or gas, occurs in a tremendous number of situations. For instance, air at atmospheric pressure is a mixture of gases, though it is often modeled as one substance. In this chapter you will learn about the behavior and properties of ideal and actual gases, including

- The ideal-gas equation and other equations of state;
- The kinetic theory underlying the ideal-gas laws;
- The development of specific heats of gases;
- Use of gas tables to determine actual gas properties.

5.1 IDEAL-GAS EQUATION OF STATE

There are several equations of state for gases, the most common being the *ideal-gas equation of state,* which relates the dependence of pressure, volume, temperature, and mass at a state. Not all states are allowed; consider Figure 5.1, a simplified version of the equilibrium surfaces in Chapter 4.

States *A, B,* and *C* are shown. States *A* and *B* are on the surface; these represent all possible equilibrium states for a substance. Remember that properties are defined

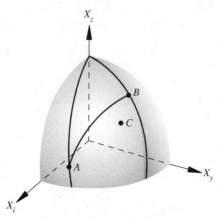

Figure 5.1 Diagram of an equilibrium surface with points A and B on the surface and point C below the surface.

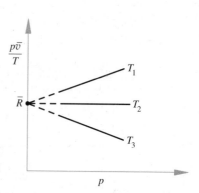

Figure 5.2 A plot of a gas's isotherms used in determining $\bar{R}$.

only at the equilibrium states. State C is not an equilibrium state; it is not possible for a substance to remain in that state for any period of time. The system could pass through state C in going from A to B, but it could not exist there. The ideal-gas equation of state represents all the states on the AB surface, relating the various properties to each other. The ideal-gas equation of state, often called the *ideal-gas law,* is

$$pV = mRT \qquad (5.1)$$

where p is the absolute pressure, V is the total volume, m is the mass, T is the absolute temperature, and R is the individual gas constant. This equation comes in many forms. Dividing by the mass yields

$$pv = RT \qquad (5.2)$$

Equations (5.1) and (5.2) are the most frequently used forms of the ideal-gas equation of state. The value of R is tabulated in Appendix Table A.1 for several gases.

The gas constant R may be calculated for any gas if its molecular weight is known. Avogadro's law states that equal volumes of an ideal gas at the same temperature and pressure have equal numbers of molecules. If M is the molecular mass, $Mv = \bar{v}$; multiply equation (5.2) by it:

$$p\bar{v} = MRT$$
$$p\bar{v} = \bar{R}T \qquad (5.3)$$

where $\bar{R} = MR$.

The value of $\bar{R}$ may be determined independently by considering the following experiment. A piston-cylinder contains a gas, and the entire unit may be maintained at constant temperature as the piston is moved to various positions. The pressure and specific volume at each of the states is determined, with the resulting isotherm plotted in Figure 5.2. Several of these isotherms are plotted and extrapolated to $p \rightarrow 0$,

resulting in the following

$$\lim_{p \to 0} \frac{p\bar{v}}{T} = \bar{R}$$

$\bar{R}$ is a common value for all temperatures. Additional experiments for different gases show that $\bar{R}$ is the same for all gases. Thus, it is called the *universal gas constant*.
  The constant $\bar{R}$ has a value of

$$\bar{R} = 8.3143 \text{ kJ/kgmol-K} = 1545.32 \text{ ft-lbf/pmol-R} = 1.986 \text{ Btu/pmol-R} (5.4)$$

Note that $n\bar{v} = V$, where n is the number of moles, and equation (5.3) becomes

$$pV = n\bar{R}T \qquad (5.5)$$

Equations (5.1) and (5.2) are the ones most often used, but equation (5.5) is used when dealing with chemical reactions. The abbreviation *kgmol* stands for the molecular weight expressed in kilograms; the abbreviation *pmol* stands for the molecular weight expressed in pounds mass.

Example 5.1
Two kg of air at 280°K are contained in a 0.2-m³ tank. Consider the air to be an ideal gas. Determine the pressure, the number of moles, and the specific volume on a mass and molal basis.

Solution

Given: 2 kg of air at 280°K and 0.2 m³.

Find: The pressure, moles, and specific volume on a mass and mole basis.

Sketch and Given Data:

Air
2 kg
280° K
0.2 m³

Figure 5.3

$PV = MRT$

$\frac{PV}{RT} = M'$

Assumption: The air is in an equilibrium state.

Analysis: From Table A.1, $R = 0.287$ kJ/kg-K and $M = 28.97$ kg/kgmol. Substituting in the ideal-gas law yields

$$p = \frac{mRT}{V} = \frac{(2 \text{ kg})(0.287 \text{ kJ/kg-K})(280°\text{K})}{(0.2 \text{ m}^3)} = 803.6 \text{ kPa}$$

$$n = \frac{m}{M} = \frac{(2 \text{ kg})}{(28.97 \text{ kg/kgmol})} = 0.069 \text{ kgmol}$$

$$v = \frac{V}{m} = \frac{(0.2 \ \text{m}^3)}{(2.0 \ \text{kg})} = 0.1 \ \text{m}^3/\text{kg}$$

$$\bar{v} = Mv = (28.97 \ \text{kg/kgmol})(0.1 \ \text{m}^3/\text{kg}) = 2.897 \ \text{m}^3/\text{kgmol}$$

Comments:

1. The ideal-gas law provides a simple method of determining ideal-gas state properties.
2. It is important to be able to convert from the mass to the mole system.

5.2 NONIDEAL-GAS EQUATIONS OF STATE

In actual gases the molecular collisions are inelastic; at high densities in particular there are intermolecular forces that the simplified equations of state do not account for. There are many gas equations of state that attempt to correct for the nonideal behavior of gases. The disadvantage of all methods is that the equations are more complex and require the use of experimental coefficients.

One of the equations is the *van der Waals equation of state* for a gas, which was developed in 1873 as an improvement on the ideal-gas law. The van der Waals equation of state is

$$p = \frac{\bar{R}T}{\bar{v} - b} - \frac{a}{\bar{v}^2} \tag{5.6}$$

The coefficients a and b compensate for the nonideal characteristics of the gas. The constant b accounts for the finite volume occupied by the gas molecules, and the $a/\bar{v}^2$ term accounts for intermolecular forces. Constants for selected gases are given in Table 5.1.

TABLE 5.1 VAN DER WAALS CONSTANTS

Substance	a (kPa [m³/kgmol]²)	b (m³/kgmol)
Air	135.8	0.0364
Ammonia (NH_3)	423.3	0.0373
Carbon dioxide (CO_2)	364.3	0.0427
Carbon monoxide (CO)	146.3	0.0394
Helium (He)	3.41	0.0234
Hydrogen (H_2)	24.7	0.0265
Methane (CH_4)	228.5	0.0427
Nitrogen (N_2)	136.1	0.0385
Oxygen (O_2)	136.9	0.0315

TABLE 5.2 CONSTANTS FOR THE BEATTIE-BRIDGEMAN EQUATION OF STATE

Substance	A_0	a	B_0	b	$10^{-4}c$
Air	131.8441	0.019 31	0.046 11	−0.001 101	4.34
Argon (Ar)	130.7802	0.023 28	0.039 31	0.0	5.99
Carbon dioxide (CO_2)	507.2836	0.071 32	0.104 76	0.072 35	66.00
Helium (He)	2.1886	0.059 84	0.014 00	0.0	0.0040
Hydrogen (H_2)	20.0117	−0.005 06	0.020 96	−0.043 59	0.0504
Nitrogen (N_2)	136.2315	0.026 17	0.050 46	−0.006 91	4.20
Oxygen (O_2)	151.0857	0.025 62	0.046 24	0.004 208	4.80

Pressure in kPa; specific volume in m^3/kgmol; temperature in K; $\bar{R} = 8.314\ 34$ kJ/kgmol-K.

A second equation of state is the *Beattie-Bridgeman equation of state* for a gas:

$$p = \frac{\bar{R}T(1-\epsilon)}{\bar{v}^2}(\bar{v}+B) - \frac{A}{\bar{v}^2} \tag{5.7}$$

where

$$A = A_0(1 - a/\bar{v})$$

$$B = B_0(1 - b/\bar{v})$$

$$\epsilon = \frac{c}{\bar{v}T^3}$$

and the constants A_0, B_0, a, b, and c are determined for individual gases. Table 5.2 gives these for certain gases.

A simple but accurate two-constant equation is that proposed by Redlich and Kwong[1] in 1949:

$$p = \frac{RT}{v-b} - \frac{a}{T^{1/2}v(v+b)}$$

The value of the constants a and b can be determined by noting that the first and second derivatives of pressure with respect to specific volume at the critical point are zero (refer to Figure 4.4).

$$\left.\frac{\delta p}{\delta v}\right|_{T_c} = 0 \qquad \left.\frac{\delta p^2}{\delta v^2}\right|_{T_c} = 0$$

---

[1] O. Redlich and J.N.S. Kwong, "On the Thermodynamic of Solutions, V-An Equation of State," *Chemical Review,* vol. 44, 233–244, 1949.

When these derivatives are taken, we find

$$a = 0.42748 \frac{R^2 T_c^{5/2}}{p_c}$$

$$b = 0.08664 \frac{R T_c}{p_c}$$

This same approach can be applied to the van der Waals equation. When the partial derivatives are taken, the following equations for the constants result:

$$a = \frac{27 R^2 T_c^2}{64 p_c}$$

$$b = \frac{R T_c}{8 p_c}$$

These equations can be used to calculate the van der Waals constants rather than using tabulated values.

Virial Equation of State

Gas equations of state attempt to approximate the behavior of gases over a wide range of conditions and thus become quite complex. An equation of state based on an infinite series is called the *virial equation of state* and is of the form

$$p\bar{v} = \bar{R}T \left(1 + \frac{B(T)}{\bar{v}} + \frac{C(T)}{\bar{v}^2} + \frac{D(T)}{\bar{v}^3} + \ldots \right) \tag{5.8}$$

The coefficients in Equation (5.8) are called *virial coefficients* and are functions of temperature. The coefficients may be found experimentally, through curve fitting of data, or theoretically from statistical mechanics. Even with this complex expression, as the pressure approaches zero the virial coefficients vanish and equation (5.8) reduces to the ideal-gas law. All the more complex equations of state reduce to the ideal-gas law as the limiting case.

Example 5.2
Two kgmol of air at 400°K is contained in a 0.5 m³ piston-cylinder. A change of state occurs, and the final pressure is 15 MPa and the final temperature 350°K. Determine the initial pressure and the final value of specific volume, using the ideal-gas law and the van der Waals equation of state.

Solution

Given: 2 kgmol of air initially at 400°K and 0.5 m³ and finally at 15 mPa and 350°K.

Find: The initial pressure and the final specific volume using the ideal-gas and van der Waals equations of state.

Sketch and Given Data:

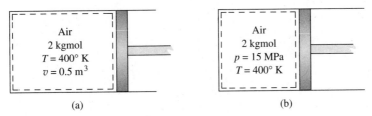

(a)   (b)

Figure 5.4

Assumption: Air is in equilibrium state initially and finally.

Analysis:

$$\bar{v} = \frac{V}{n} = \frac{0.5 \text{ m}^3}{2 \text{ kgmol}} = 0.25 \text{ m}^3/\text{kgmol}$$

Find the values of the constants a and b from Table 5.1.

$$a = 135.8 \text{ kPa (m}^3/\text{kgmol})^2$$
$$b = 0.0364 \text{ m}^3/\text{kgmol}$$

Substituting in the van der Waals equation of state yields

$$p = \frac{(8.3143 \text{ kJ/kgmol-K})(400°\text{K})}{(0.25 \text{ m}^3/\text{kgmol} - 0.0364 \text{ m}^3/\text{kgmol})} - \frac{135.8 \text{ kPa (m}^3/\text{kgmol})^2}{(0.25 \text{ m}^3/\text{kgmol})^2}$$

$$p = 13.4 \text{ MPa}$$

Using the ideal-gas equation of state yields

$$p = \frac{n\bar{R}T}{V} = \frac{(2.0 \text{ kgmol})(8.3143 \text{ kJ/kgmol-K})(400°\text{K})}{(0.5 \text{ m}^3)}$$

$$p = 13.3 \text{ MPa}$$

For the final state, substitute into the ideal-gas law, yielding

$$\bar{v} = \frac{RT}{p} = \frac{(8.3143 \text{ kJ/kgmol-K})(350°\text{K})}{(15\ 000 \text{ kPa})}$$

$$\bar{v} = 0.194 \text{ m}^3/\text{kgmol}$$

Using the van der Waals equation of state requires a trial-and-error solution. This is easily done by TK Solver. The van der Waals equation is entered into the Rule Sheet, a guess value for specific volume is entered into the Input Column of the Variable Sheet, and an iterative solution is invoked by pressing **F9**.

```
============================ VARIABLE SHEET ============================
St Input— Name— Output——  Unit ——  Comment——————————————————————
                                    Example 5.2

   15000  P                 kPa
   350    T                 degK
          vbar  .19222164   m3/kgmol
   135.8  a
   .0364  b


============================ RULE SHEET ============================
S Rule —————————————————————————————————————————————————
 * p=8.31434*T/(vbar—b)—a/vbar^2    "van der Waals Equation
```

Comment: At the initial state the difference between the values of pressure is only 0.7%. Hence the ideal-gas law is accurate in predicting air behavior in many instances, except in those cases of very high pressure and/or low temperature. ▪

Compressibility Factor

The ideal-gas law works well for gases at low densities. As the pressure of the gas is increased for a given temperature, the molecules are packed closer and closer together. This brings about nonideal behavior due to additional forces acting on the

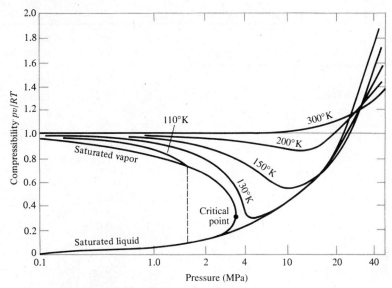

Figure 5.5 Compressibility diagram for nitrogen.

molecule. The other equations of state account for this deviation by introducing empirical constants. The form of equation (5.2) is very convenient to work with, and a method to account for nonideal-gas behavior using this form is available. If we divide equation (5.2) by RT, it yields

$$\frac{pv}{RT} = 1$$

for an ideal gas. If the gas is not ideal, this will equal some other number, Z.

$$\frac{pv}{RT} = Z$$

The symbol Z is the *compressibility factor*. It is equal to one for an ideal gas, but will have a value other than one for an actual gas. Figure 5.5 illustrates how the compressibility factor for nitrogen varies with pressure along different isotherms. At room temperature, 300°K, the compressibility factor is unity up to 7000 kPa. Note that as the temperature rises, it requires greater and greater pressures to cause nonideal-gas behavior. In many application areas the temperature and pressure limits allow us to use the ideal-gas equation of state.

Generalized Compressibility Factor

The form of the equation of state using the compressibility factor is simple. At this point the only difficulty lies in acquiring charts for all the gases. Fortunately, this task may be reduced to that of developing only one chart.

This is accomplished by using reduced equations of state. The critical pressure,

TABLE 5.3   CRITICAL CONSTANTS

Substance	Molecular weight	Temp (K)	Pressure (MPa)	Volume (m³/kgmol)
Air	28.97	133	3.76	0.0828
Ammonia (NH_3)	17.03	405.5	11.28	0.0724
Argon (Ar)	39.948	151	4.86	0.0749
Carbon dioxide (CO_2)	44.01	304.2	7.39	0.0943
Carbon monoxide (CO)	28.011	133	3.50	0.0930
Ethane (C_2H_6)	30.07	305.5	4.88	0.1480
Ethylene (C_2H_4)	28.054	282.4	5.12	0.1242
Helium (He)	4.00	38.4	1.66	—
Hydrogen (H_2)	2.016	33.3	1.30	0.649
Methane (CH_4)	16.043	191.1	4.64	0.0993
Methyl alcohol (CH_3OH)	32.0042	513.2	7.95	0.1180
Nitrogen (N_2)	28.013	126.2	3.39	0.0899
Oxygen (O_2)	32.0	154.8	5.08	0.0780
Propane (C_3H_8)	44.097	370	4.26	0.1998
Water (H_2O)	18.015	647.3	22.09	0.568

temperature, and specific volume are unique at the critical point for each gas. Table 5.3 lists the critical properties of some substances. The reduced coordinates are

Reduced pressure $p_r = p/p_c$, where p_c = critical pressure

Reduced temperature $T_r = T/T_c$, where T_c = critical temperature

Reduced specific volume $v_r = v/v_c$, where v_c = critical specific volume

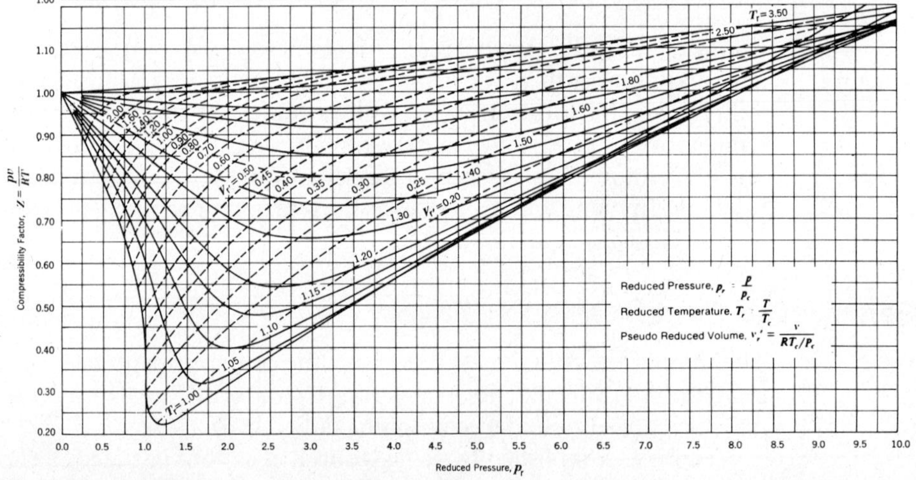

Figure 5.6 A generalized compressibility chart.

A generalized chart of compressibility factors for reduced temperatures and pressures, Figure 5.6, was developed by Nelson and Obert. This may be used to find the properties of a gas if we know the critical properties. The reduced volume, V'_r, is defined as

$$V'_r = \frac{V}{RT_c/p_c} = \frac{ZRT/p}{RT_c/p_c} = Z\frac{T_r}{p_r}$$

by Nelson and Obert.[2]

Example 5.3

Ammonia is contained in a 2.0-m³ tank at 2000 kPa and 160°C. Determine the specific volume of the ammonia using the generalized compressibility chart, and compare this value to that found in Table A.10. In addition determine the mass of ammonia contained in the tank.

Solution

Given: Ammonia is contained in a 2.0-m³ tank at 2000 kPa and 160°C.

Find: The specific volume using the generalized compressibility chart and compared to tabulated data, and the mass of ammonia in the tank.

Sketch and Given Data:

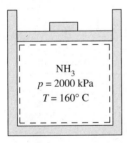

NH$_3$
$p = 2000$ kPa
$T = 160°$ C

Figure 5.7

Assumptions:

1. The ammonia is a closed system.
2. The state is an equilibrium state.

---

2 L. C. Nelson and E. F. Obert, "General p-V-T Properties of Gases," Trans. ASME, Oct. 1954, 1057–1066.

Analysis: From Table 5.3 find the critical pressure and critical temperature of ammonia, 11 280 kPa and 405.5°K. The reduced pressure and temperature are

$$p_{r1} = \frac{2000 \text{ kPa}}{11\ 280 \text{ kPa}} = 0.177$$

$$T_{r1} = \frac{433°\text{K}}{405.5°\text{K}} = 1.068$$

From Figure 5.6, the value of Z is 0.94. From Table A.1, $R = 0.4882$ kJ/kg-K, thus

$$v = Z\frac{RT}{p} = \frac{(0.94)(0.4882 \text{ kJ/kg-K})(433°\text{K})}{(2000 \text{ kN/m}^2)}$$

$$v = 0.0993 \text{ m}^3/\text{kg}$$

This compares very well with the value of specific volume found in Table A.10, 0.0999 m³/kg.

The mass of ammonia in the tank is found from $mv = V$; thus

$$m = \frac{(2.0 \text{ m}^3)}{(0.0993 \text{ m}^3/\text{kg})} = 20.14 \text{ kg}$$

Comments:

1. The absolute temperature and pressure must be used in calculating the reduced pressure and temperature values.
2. The compressibility factor is dimensionless.

5.3 SPECIFIC HEAT

Specific Heat at Constant Volume

Specific heat at constant volume, c_v, is defined for any substance as

$$c_v \equiv \left(\frac{\partial u}{\partial T}\right)_v \text{ kJ/kg-K } [\text{Btu/lbm-R}] \tag{5.9}$$

An experiment by Joule, Figure 5.8, verified the equation's premise that thermal energy (heat) transfer at constant volume is a function only of temperature. In Figure 5.8 a vessel with two chambers, I and II, each containing the same gas at different pressures and connected by a valve, was placed in a constant-temperature bath. When the system achieved thermal equilibrium with the bath, the valve was opened. No temperature change occurred in the bath, the system work was zero, and hence

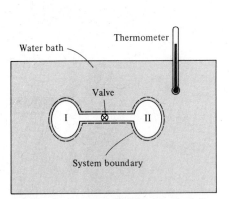

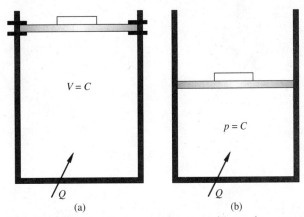

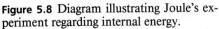

Figure 5.8 Diagram illustrating Joule's experiment regarding internal energy.

Figure 5.9 Piston-cylinder arrangements illustrating constant-volume and constant-pressure processes.

the internal energy change was zero. This led to the conclusion that specific internal energy is not a function of volume for an ideal gas.

Mathematically, we may say that

$$u = u(v, T)$$

$$du = \left(\frac{\partial u}{\partial v}\right)_T dv + \left(\frac{\partial u}{\partial T}\right)_v dT$$

Since $dv > 0$ from Joule's experiment and the internal energy did not change, we may conclude that $u = u(T)$ only.

From the definition of specific heat at constant volume, equation (5.9), and from the fact that the specific internal energy of an ideal gas is not a function of specific volume (Joule's experiment), the partial derivative in equation (5.9) becomes a total derivative; hence

$$c_v = \frac{du}{dT}$$

$$du = c_v \, dT \tag{5.10}$$

$$dU = mc_v \, dT \tag{5.11}$$

If we consider a constant-volume heating process, as in Figure 5.9(a), the first law states

$$\delta Q = dU + \delta W$$

but

$$\delta W = \int p \, dV = 0$$

so

$$\delta Q = dU$$

$$Q = \int_1^2 \delta Q = \int_1^2 dU = U_2 - U_1$$

for an ideal gas,

$$U_2 - U_1 = \int_1^2 mc_v \, dT \qquad (5.12)$$

For constant-volume processes, with constant mass and constant specific heat,

$$Q = \Delta U = mc_v(T_2 - T_1)$$

The change of internal energy for an ideal gas with constant specific heat is always

$$U_2 - U_1 = \Delta U = mc_v(T_2 - T_1) \qquad (5.13)$$

The change of internal energy for an ideal gas is always denoted by equation (5.11). The process may be reversible, irreversible, constant-pressure, constant-volume, or any other process. Internal energy, specific heat at constant volume, temperature, and mass are properties and do not depend on path. The equation of state for internal energy, equation (5.11), follows from the definition of specific heat at constant volume.

Specific Heat at Constant Pressure

The coefficient c_p is the specific heat at constant pressure, and is rigorously defined as

$$c_p \equiv \left(\frac{\partial h}{\partial T}\right)_p \quad \text{kJ/kg-K} \quad \text{[Btu/lbm-R]} \qquad (5.14)$$

From the definition of specific heat at constant pressure and from $h = u + pv = u + RT$ for an ideal gas,

$$dh = c_p \, dT \qquad (5.15a)$$

$$dH = mc_p \, dT \qquad (5.15b)$$

We now consider heat addition at constant pressure, as in Figure 5.9(b). Let us analyze the first-law equation for this system.

$$\delta Q = dU + p \, dV = dU + d(pV) \qquad (5.16)$$

$$\delta Q = dH \qquad \text{for } p = c \qquad (5.17)$$

hence

$$\delta Q = dH = mc_p \, dT \qquad (5.18)$$

for ideal-gas constant-pressure processes.

For an ideal gas, the changes in internal energy and enthalpy depend only on the temperature; if the initial and final temperatures are known, the enthalpy and internal energy changes may be calculated. Thus, equations (5.15a) and (5.11) are true only for ideal gases, by the definitions of specific heat. We have therefore a method of calculating the internal energy and enthalpy for any process if we know the end states. The work may be calculated if we know the process path, that is, how the system changes state.

Further Considerations

An important ratio that we shall use extensively is that of the specific heats, k.

$$k = \frac{c_p}{c_v} \qquad (5.19)$$

Heretofore the specific heat has been written on a mass basis; it can also be written on the mole basis.

We denote the molal constant-volume specific heat as

$$\bar{c}_v \equiv \left(\frac{\partial \bar{u}}{\partial T}\right)_{\bar{v}} \text{ kJ/kgmol-K } \text{ [Btu/pmol-R]} \qquad (5.20)$$

and the molal constant-pressure specific heat as

$$\bar{c}_p \equiv \left(\frac{\partial \bar{h}}{\partial T}\right)_p \text{ kJ/kgmol-K } \text{ [Btu/pmol-R]} \qquad (5.21)$$

The bar over any property denotes that it is on the mole basis; hence both enthalpy, $\bar{h}$, and internal energy, $\bar{u}$, have units of kJ/kgmol-K. The property on the mass basis is multiplied by the substance's molecular weight, M, to obtain the property on the mole basis, that is, $\bar{h} = Mh$.

Another relation between the specific heats and the gas constant may be developed by considering the enthalpy of an ideal gas.

$$h = u + pv$$

$$pv = RT$$

$$h = u + RT$$

We now differentiate, and substitute equations (5.11) and (5.15a) into the resulting equation.

$$dh = du + R\, dT$$

$$c_p\, dT = c_v\, dT + R\, dT$$

$$c_p - c_v = R \qquad (5.22)$$

Thus even though the specific heats may be a function of temperature, for an ideal gas their difference is always constant.

The specific heats in Table A.1 are measured experimentally. Specific heats may be developed mathematically, but using experimental values adds to the accuracy with which we may calculate, and hence predict, system changes.

It is interesting to note that the idea of heat capacity developed from the days when the caloric theory of heat was in vogue. A body, when heated, was considered to be gaining "caloric." Heat capacity was the caloric necessary to raise the substance one degree in temperature. To make heat capacity an intensive property, we divide by the system mass, yielding specific heat capacity.

Variation of Specific Heat with Temperature

It may have been inferred that the specific heats are constant for all temperatures of a gas. This is not the case. The functional relationships denoting this variation are determined from experimental tests. Table 5.4 lists several formulas that predict the specific heat of a gas at a given temperature. Figure 5.10 shows a diagram of the specific heat variation with temperature for five substances. If increased accuracy is desired, an average value of specific heat can be used for the temperature range under consideration, or the specific heat equation for Table 5.4 can be integrated. The specific heats vary with temperature because the energy associated with each vibrational mode becomes greater, especially at high temperatures. This is illustrated in Figure 5.10 for water and carbon dioxide, where there are more vibrational modes available in the atomic structure than in the simpler diatomic molecules hydrogen and oxygen. Hence the specific heat variation is more pronounced.

TABLE 5.4 FORMULAS FOR SPECIFIC HEAT VARIATION WITH TEMPERATURE

Substance	Temperature range	c_p(kJ/kg-K)
Air	280–1500°K	$0.9167 + 2.577 \times 10^{-4}T - 3.974 \times 10^{-8}T^2$
Ammonia (NH_3)	300–1000°K	$1.5194 + 1.936 \times 10^{-3}T - 1.789 \times 10^{-7}T^2$
Sulfur dioxide (SO_2)	300–1000°K	$0.7848 + 0.7113 \times 10^{-4}T - 1.73 \times 10^4 T^{-2}$
Hydrogen (H_2)	300–2200°K	$11.959 + 2.160 \times 10^{-3}T + 30.95 T^{-1/2}$
Oxygen (O_2)	300–2750°K	$1.507 - 16.77 T^{-1/2} + 111.1 T^{-1}$
Nitrogen (N_2)	300–5000°K	$1.415 - 287.9 T^{-1} + 5.35 \times 10^4 T^{-2}$
Carbon monoxide (CO)	300–5000°K	$1.415 - 273.3 T^{-1} + 4.96 \times 10^4 T^{-2}$
Water (H_2O)	300–3000°K	$4.613 - 103.3 T^{-1/2} + 967.5 T^{-1}$
Carbon dioxide (CO_2)	300–3500°K	$1.540 - 345.1 T^{-1} + 4.13 \times 10^4 T^{-2}$
Methane (CH_4)	300–1500°K	$0.8832 + 4.71 \times 10^{-3}T - 1.123 \times 10^{-6}T^2$
Ethylene (C_2H_4)	300–1500°K	$0.4039 + 4.35 \times 10^{-3}T - 1.35 \times 10^{-6}T^2$
Ethane (C_2H_6)	300–1500°K	$0.306 + 5.34 \times 10^{-3}T - 1.53 \times 10^{-6}T^2$
n-butane (C_4H_{10})	300–1500°K	$0.314 + 5.23 \times 10^{-3}T - 1.60 \times 10^{-6}T^2$
Propane (C_3H_8)	300–1500°K	$0.214 + 5.48 \times 10^{-3}T - 1.67 \times 10^{-6}T^2$
Acetylene (C_2H_2)	280–1250°K	$1.921 + 7.06 \times 10^{-4}T - 3.73 \times 10^4 T^{-2}$
Octane (C_8H_8)	225–610°K	$0.290 + 3.97 \times 10^{-3}T$

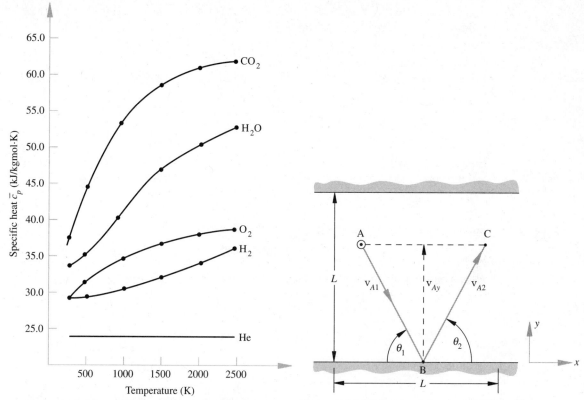

Figure 5.10 The variation of constant-pressure specific heat with temperatures for four gases, plotted from equations in Table 5.4, and for helium.

Figure 5.11 Molecule A having elastic collision on wall B.

5.4 KINETIC THEORY—PRESSURE AND SPECIFIC HEAT OF AN IDEAL GAS

The pressure that a gas creates is caused by the impact of the gas molecules against a surface. The force created by the impact is due to the change of momentum of a gas particle as it hits and bounces from the surface. Kinetic theory offers us an insight into how an ideal gas behaves. We must invoke a variety of assumptions, such as the volume of the molecule is negligible, the forces between the molecules are negligible, the molecules can be treated as rigid spheres that have elastic collisions with themselves and with any surface.

Consider molecule A in Figure 5.11; it has a incident velocity v_{A1} and an angle of incidence Θ_1. The molecule rebounds from the wall at an angle of incidence Θ_2, where $\Theta_2 = \Theta_1$. Since the initial kinetic energy equals the final kinetic energy, the absolute value of the initial velocity must equal the absolute value of the final velocity, $|v_{A2}| = |v_{A1}|$. The velocity vector of the molecule changes because the direction changes, but the magnitude of the velocity vector is the same. Thus, the change of momentum for molecule A in the y direction is

$$m_A(v_{A1y} - (-v_{A2y})) = 2m_A v_{Ay} \tag{5.23}$$

For any molecule this may be written without the A subscript. From Newton's second law

$$F = \frac{d(mv)}{dt} \approx \frac{\Delta(mv)}{(\Delta t)} \tag{5.24}$$

Equation (5.23) represents the change of momentum. The time for a molecule to leave B, reach the opposite wall a distance L away, and return is

$$\Delta t = 2L/v_y$$

Substituting these values in equation (5.24) yields

$$F = mv_y^2/L$$

Assuming that all the molecules have the same mass and summing over the total number of molecules present, N, yields

$$F = \frac{m}{L} \sum_{i=1}^{N} v_{yi}^2$$

Multiplying and dividing by N yields

$$F = \frac{Nm}{L} \sum_{i=1}^{N} \frac{v_{yi}}{N}$$

Noting that

$$\overline{v}_y^2 = \sum_{i=1}^{N} \frac{v_{yi}^2}{N}$$

which represents the average of the square of the velocity in the y direction, we may determine pressure from its definition of force per unit area, noting that the area is L^2.

$$p = \frac{Nm}{L^3} \overline{v}_y^2 = \frac{Nm}{V} \overline{v}_y^2 \tag{5.25}$$

where $L^3 = V$, the volume containing N molecules.

Molecules are in random motion, so the pressure is a result of forces in the x, y, and z directions. Furthermore, because molecular motion is random, the kinetic energies in each direction must be the same. This has been experimentally verified. Thus

$$\overline{v}_x^2 = \overline{v}_y^2 = \overline{v}_z^2$$

and the magnitude of the total velocity vector for the molecule is

$$\overline{v}^2 = \overline{v}_x^2 + \overline{v}_y^2 + \overline{v}_z^2 = 3\,\overline{v}_y^2$$

and the pressure may be written as

$$p = \frac{Nm}{3V} \bar{v}^2 \tag{5.26}$$

Consider one mole of a substance; it will have a volume and contain N_A molecules, where N_A is Avogadro's number. Equation (5.26) becomes

$$p\bar{v} = \frac{N_A m}{3} \bar{v}^2 = \frac{2}{3} \left[N_A \frac{m\bar{v}^2}{2} \right] \tag{5.27}$$

The term in the brackets represents the total translational kinetic energy of all the molecules in $\bar{v}$. Denoting this expression as $\bar{u}_t$ and using the ideal-gas law, $p\bar{v} = \overline{RT}$, yields

$$\bar{u}_t = \frac{3}{2} \overline{R}T \tag{5.28}$$

We have noted that the velocity vector has components in each direction whose average velocity is the same. This is consistent with Boltzmann's principle of equipartition of energy, wherein the total translational kinetic energy has three equal components (the kinetic energy in each direction). Thus the molecular mass has three degrees of freedom, $f = 3$, corresponding to the x, y, and z directions.

The internal energy for a monatomic gas molecule, $f = 3$, is described by equation (5.28). The equipartition principle may be generalized to include energy forms other than translational kinetic energy, such as rotational and vibrational modes. In this case the internal energy may be written as

$$\bar{u} = \bar{c}_v T = f \frac{\overline{R}T}{2} \tag{5.29}$$

Dividing equation (5.29) by the molecular weight of the gas yields

$$u = c_v T = f \frac{RT}{2} \quad \text{or} \quad c_v = \frac{fR}{2}$$

Since $c_p = c_v + R$,

$$c_p = R(1 + f/2) \tag{5.30}$$

For a monatomic gas such as helium, $f = 3$. Substituting into equation (5.30) yields

$$c_p = 2.077(1 + 3/2) = 5.1925 \text{ kJ/kg-K}$$

which corresponds very well with the value found in Table A.1 (5.1954 kJ/kg-K). As the structure of the molecule becomes more complicated, however, even a diatomic molecule such as hydrogen, H_2, illustrates the limitation of the kinetic theory model and the influence of other modes of energy storage within the molecular structure. For a diatomic gas there are three translational modes, two rotational modes, and two

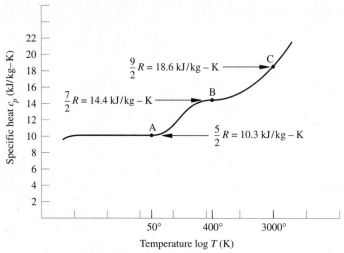

Figure 5.12 Variation of c_p of hydrogen with temperature.

vibrational modes of energy storage, or degrees of freedom ($f = 7$). Thus for hydrogen,

$$c_p = 4.125(1 + 7/2) = 18.56 \text{ kJ/kg-K}$$

which is much greater than the tabulated value of 14.31 kJ/kg-K.

Figure 5.12 illustrates the variation of specific heat with temperature for hydrogen. The molecule uses more modes of energy storage, exhibiting more degrees of freedom, as the temperature increases. Thus at low temperatures only translational kinetic energy modes are used to store energy, whereas at high temperatures all modes are used.

5.5 GAS TABLES

We have examined two methods of analyzing gas behavior: the first is the use of an ideal-gas equation of state and the second is the use of the compressibility chart for real-gas behavior. A third resource is available to us, the gas tables by Keenan and Kaye. Tables A.2–A.4 are adapted from these gas tables, and Table A.13 is an abridgement for air. These tables account for the variation of specific heat with temperature for an ideal gas. Temperature is the independent variable, and tabulated values for internal energy and enthalpy are given on the same line. Some other symbols are also given values, and we will learn about them when we deal with process paths, that is, the way in which the system changes state. Note that equations using constant specific heats cannot be used, since the substance is now a gas with variable specific heats.

The TK Solver model AIR.TK contains information like that in the tables for air. It permits the convenient determination of the pressure, temperature, and specific volume at two state points, and the change of internal energy and enthalpy between them.

Example 5.4

Air with a temperature of 27°C receives heat at constant volume until the temperature is 927°C. Find the heat added per kilogram using (a) specific heat from Table A.1; (b) specific heat using the equation from Table 5.4 and an average temperature; (c) the air tables in Table A.2; (d) TK Solver model AIR.TK.

Solution

Given: Air is heated at constant volume from 27°C to 927°C.

Find: The heat added per kilogram.

Sketch and Given Data:

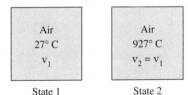

State 1   State 2   **Figure 5.13**

Assumptions:

1. The system is closed with no change in volume.
2. The initial and final states are equilibrium states.
3. There are no changes in kinetic and potential energies.

Analysis: The first-law equation is

$$q = \Delta u + w$$

$w = 0$ for constant volume, so

$$q = \Delta u$$

Case (a): For an ideal gas the equation for internal energy is

$$\Delta u = c_v(T_2 - T_1) = (0.7176 \text{ kJ/kg-K})(1200 - 300\text{K}) = 645.84 \text{ kJ/kg}$$

Case (b): From Table 5.4, the specific heat at constant volume is

$$c_v = c_p - R = (0.9167 + 2.577 \times 10^{-4}T - 3.974 \times 10^{-8}T^2) - 0.287$$

$$T = \frac{(300.15 + 1200.15)}{2} = 750.15°K$$

$$c_v = 0.800\ 65 \text{ kJ/kg-K}$$

$$u = (0.800\ 65)(1200 - 300K) = 720.6 \text{ kJ/kg}$$

Case (c): From Table A.2, at $300°K$, $u_1 = 214.09$ kJ/kg; at $1200°K$, $u_2 = 933.40$ kJ/kg; thus, $\Delta u = 719.31$ kJ/kg.

Case (d): Using AIR.TK, enter the equation vl = v2 into the Rule Sheet and the two temperatures and an assumed initial pressure of 100 kPa into the Variable Sheet. Solve by pressing **F9**.

```
============================= VARIABLE SHEET =============================
St Input—Name—Output —— Unit —— Comment
                                    ***Air Properties***

                                       POINT 1
    100    P1                kPa      Pressure (kPa, MPa, psia)
     27    T1                degC     Temperature (degK, degC, degR, degF)
           v1    .86116      m3/kg    Specific Volume (m3/kg, ft3/lbm)

                                       POINT 2
           P2    400.37      kPa      Pressure (kPa, MPa, psia)
    927    T2                degC     Temperature (degK, degC, degR, degF)
           v2    .86116      m3/kg    Specific Volume (m3/kg, ft3/lbm)

                                    CHANGE (POINT 2—POINT 1)
           DELh  977.21      kJ/kg    Enthalpy (kJ,kg, BTU/lbm)
           DELu  718.88      kJ/kg    Internal Energy (kJ/kg, BTU/lbm)
           DELs  1.0837      kJ/kg-K  Entropy (kJ/kg-K, B/lbm-R)
```

Comments:

1. Because of the wide variation in specific heat over the range of temperature, using the specific heat value from Table A.1 results in significant error.
2. Using the air tables (Tables A.2), the TK Solver model AIR.TK, or a specific heat from Table 5.4 based on an average temperature produces similar results. ■

Example 5.5

A piston-cylinder contains 2 kg of helium at 300°K and 100 kPa. The helium is compressed irreversibly to 600 kPa and 450°K, and 100 kJ of heat is transferred to the surroundings in the process. Determine the final volume and the work done, considering helium to be an ideal gas.

Solution

Given: Helium, an ideal gas, is compressed irreversibly from 300° K and 100 kPa to 450°K and 600 kPa. The process transfers 100 kJ of heat to the surroundings.

Find: The final volume and the work done in the process.

Sketch and Given Data:

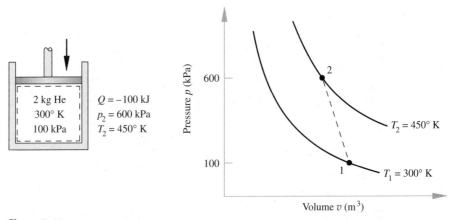

Figure 5.14

Assumptions:

1. The helium is a closed system.
2. The initial and final states are equilibrium states.
3. Helium may be considered an ideal gas.
4. The changes in kinetic and potential energies are zero.

Analysis: The first law for a closed system is

$$Q = \Delta U + \Delta \cancel{K.E.}^{0} + \Delta \cancel{P.E.}^{0} + W$$

The changes in kinetic and potential energies are assumed to be zero. Solving for the work yields

$$W = Q - \Delta U$$

For an ideal gas the equation of state for internal energy is

$$\Delta U = mc_v(T_2 - T_1) = (2 \text{ kg})(3.1189 \text{ kJ/kg-K})(450^\circ\text{K} - 300^\circ\text{K})$$
$$\Delta U = 935.7 \text{ kJ}$$

Solving for work yields

$$W = -100 \text{ kJ} - 935.7 \text{ kJ} = -1035.7 \text{ kJ}$$

The minus sign on the answer indicates that work was added to the system. The ideal-gas law may be used to determine the final volume at state 2:

$$V = \frac{mRT}{p} = \frac{(2 \text{ kg})(2.077 \text{ kJ/kg-K})(450^\circ\text{K})}{(600 \text{ kN/m}^2)} = 3.12 \text{ m}^3$$

Comments:

1. The ideal-gas law may be used to evaluate properties at a given state, regardless of whether the system's process in reaching that state is reversible or not.
2. It is important to remember the sign convention for heat and work.
3. The first law is valid for all processes; hence the work may be determined from it for irreversible processes. The $\int p \, dV$ can be evaluated only for reversible processes.
4. When a process is irreversible, use a dashed line to indicate that the system's succession of states is not a succession of equilibrium states.   ▅

Example 5.6
Two adiabatic tanks contain air at different temperatures and pressures and are connected by a valve. Tank A holds 3 ft³ of air at 500°R and 100 psia and tank B holds 2 lbm of air at 750°R and 50 psia. The valve is opened, and the pressure and temperature reach equilibrium values. What are they?

Solution

Given: Two adiabatic tanks contain air at different temperatures and pressures. The valve connecting the tanks is opened, and the air reaches an equilibrium temperature and pressure.

Find: The final temperature and pressure.

Sketch and Given Data:

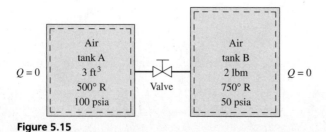

Figure 5.15

Assumptions:

1. The total amount of air can be considered a closed system.
2. The initial and final states are equilibrium states.
3. Air may be modeled as an ideal gas.
4. There is no change in kinetic and potential energies.
5. There is no work done in the process.

Analysis: The first law for a closed system, assuming no work done and no change in potential or kinetic energy, is

$$Q = \Delta U_A + \Delta U_B = m_A c_v(T_f - T_A) + m_B c_v(T_f - T_B)$$

Since the tanks are adiabatic and considered as one tank at the final state with temperature T_f, the first law simply indicates that the energy gained by one subsystem is given up by the other subsystem. The mass for tank A may be found by using the ideal-gas law

$$m_A = \frac{pV}{RT} = \frac{(100 \text{ lbf/in.}^2)(144 \text{ in.}^2/\text{ft}^2)(3 \text{ ft}^3)}{(53.34 \text{ ft-lbf/lbm-R})(500 \,^\circ R)} = 1.62 \text{ lbm}$$

Substituting into the first law equation, noting that $Q = 0$ and the specific heats cancel, yields

$$(1.62)(T_f - 500) = (2)(750 - T_f)$$

$$T_f = 638.1 \,^\circ R$$

The final pressure may be determined from the ideal-gas law, but first we must find the volume of tank B.

$$V_B = \frac{mRT}{p} = \frac{(2 \text{ lbm})(53.34 \text{ ft-lbf/lbm-R})(750 \,^\circ R)}{(50 \text{ lbf/in.}^2)(144 \text{ in.}^2/\text{ft}^2)}$$

$$V_B = 11.1 \text{ ft}^3$$

The final pressure is

$$p_f = \frac{mRT}{V} = \frac{(3.62 \text{ lbm})(53.34 \text{ ft-lbf/lbm-R})(638.1 \,^\circ R)}{(14.1 \text{ ft}^3)(144 \text{ in.}^2/\text{ft}^2)}$$

$$p_f = 60.7 \text{ psia}$$

Comments:

1. If the variation in temperature within the system were large, the assumption of constant values for specific heat would not be valid. The problem could be solved using the tabulated values for internal energy, which account for variation in specific heat with temperature.
2. Use of the generalized compressibility chart shows that $Z = 1$ for all gas states in this problem; thus the usefulness of ideal-gas equations of state is a good assumption.

CONCEPT QUESTIONS

1. Why is the term "specific heat" a misnomer?
2. Upon what ideal-gas property are internal energy and enthalpy dependent?
3. How would you define an ideal gas?
4. Can actual or ideal-gas laws be used to predict nonequilibrium property values? Why?
5. Are the specific heats of an ideal gas constant?
6. What does the generalized compressibility factor indicate?
7. Why does the specific heat of a gas vary with temperature?
8. Can the ideal-gas equation of state be used at the initial and final equilibrium states for an ideal gas undergoing an irreversible process?

PROBLEMS (SI)

5.1 An unknown gas has a mass of 1.5 kg and occupies 2.5 m³ while at a temperature of 300°K and a pressure of 200 kPa. Determine the ideal-gas constant for the gas.

5.2 A motorist equips her automobile tires with a relief-type valve so that the pressure inside the tire never will exceed 240 kPa (gage). She starts a trip with a pressure of 200 kPa (gage) and a temperature of 23°C in the tires. During the long drive the temperature of the air in the tires reaches 83°C. Each tire contains 0.11 kg of air. Determine (a) the mass of air escaping each tire; (b) the pressure of the tire when the temperature returns to 23°C.

5.3 A 6-m³ tank contains helium at 400°K and is evacuated from atmospheric pressure to a pressure of 740 mm Hg vacuum. Determine (a) the mass of helium remaining in the tank; (b) the mass of helium pumped out. (c) The temperature of the remaining helium falls to 10°C. What is the pressure in kPa?

5.4 A 1.5-kg quantity of ethane is cooled at constant pressure from 170°C to 65°C. Determine (a) the change of enthalpy; (b) the change of internal energy; (c) the heat transferred; (d) the work done.

5.5 A 5-m³ tank contains chlorine at 300 kPa and 300°K after 3 kg of chlorine has been used. Determine the original mass and pressure if the original temperature was 315°K.

5.6 Carbon dioxide at 25°C and 101.3 kPa has a density of 1.799 kg/m³. Determine (a) the gas constant; (b) the molecular weight based on the gas constant.

5.7 Given that a carbon monoxide gas has a temperature of 500°K and a specific volume of 0.4 m³/kg, determine the pressure using the van der Waals equation of state and the ideal-gas equation of state.

5.8 Determine the specific volume of helium at 200 kPa and 300°K using the van der Waals equation of state and the ideal-gas equation of state.

5.9 Helium is assumed to obey the Beattie-Bridgeman equation of state. Determine the pressure for a temperature of 500°C and a specific volume of 5.2 m³/kg. Compare with the ideal-gas equation of state.

5.10 Given a pressure of 500 kPa and a temperature of 500°K for carbon dioxide, compute the specific volume using the Beattie-Bridgeman and ideal-gas equations of state.

5.11 Determine the specific volume of the following gases using the generalized compressibility chart:
(a) Methyl alcohol at 6000 kPa and 600°K.
(b) Carbon dioxide at 15 MPa and 300°C.
(c) Helium at 500 kPa and 60°C.

5.12 For a certain ideal gas, the value of R is 0.280 kJ/kg-K and the value of k is 1.375. Determine the values of c_p and c_v.

5.13 For a certain ideal gas, $R = 0.270$ kJ/kg-K and $k = 1.25$. Determine (a) c_p; (b) c_v; (c) M.

5.14 For a certain ideal gas, $c_p = 1.1$ kJ/kg-K and $k = 1.3$. Determine (a) M; (b) R; (c) c_v.

5.15 The specific heat of carbon dioxide may be found in Table 5.4 as a function of temperature. Find the change of enthalpy of carbon dioxide when its temperature is increased from 325°K to 1100°K. Compare this value with that calculated for constant specific heat found in Table A.1.

5.16 A mountain is to be measured by finding the change in pressure at constant temperature. A barometer at the base of the mountain reads 730 mm Hg, while at the top it reads 470 mm Hg. The local gravitational acceleration is 9.6 m/s^{-2}. Find the mountain's height. Assume $T = 298$°K.

5.17 An empty, opened can is 30 cm high with a 10-cm diameter. The can, with the open end down, is pushed under water with a density of 1000 kg/m^3. Find the water level in the can when the top of the can is 50 cm below the surface. Thermal equilibrium exists at all times.

5.18 Compute the average specific heat at constant pressure using the equations found in Table 5.4 between temperature limits of 350°K and 1200°K for (a) nitrogen; (b) methane; (c) propane.

5.19 Determine the specific heats, c_p and c_v, of steam at 7 MPa and 500°C. Compare this to the specific heats calculated at 7 MPa and 350°C.

5.20 An insulated, constant-volume system containing 1.36 kg of air receives 53 kJ of paddle work. The initial temperature is 27°C. Determine (a) the final temperature; (b) the change of internal energy.

5.21 A 1-kg gaseous system in a piston-cylinder receives heat at a constant pressure of 350 kPa. The internal energy increases 200 kJ, and the temperature increases 70°K. If the work done is 100 kJ, determine (a) c_p; (b) the change in volume.

5.22 Using Table A.2, calculate the specific heats at constant pressure and constant volume for air at 1000°K. (*Hint:* Use $c_p \approx \Delta h/\Delta T$.)

5.23 A spherical balloon measures 10 m in diameter and is filled with helium at 101 kPa and 325°K. The balloon is surrounded by air at 101 kPa and 320°K. Determine the lifting force.

5.24 Repeat Problem 5.23, except that now the balloon contains hydrogen. Note that the molecular weight of hydrogen is one-half that of helium. Does this halve the lifting force? Why?

5.25 Determine the size of a spherical balloon required to lift a payload of 1360 kg. The gas to be used is helium at 101.3 kPa and 23°C. The surrounding air is 101.3 kPa and 10°C.

5.26 A nitrogen cylinder of 0.1 m^3 originally has a pressure of 17.25 MPa and a temperature of 20°C. The nitrogen is gradually used until the pressure is 2.75 MPa, the temperature remaining at 20°C. What is the mass of nitrogen used? Use the compressibility chart and ideal-gas law and compare results.

5.27 A hot air balloonist wishes to operate at an altitude where the pressure is 94 kPa and the temperature is 274°K. The balloon diameter is 10 m, and the payload is 200 kg. What temperature must the air in the balloon be to achieve this?

5.28 A 0.1-m³ scuba tank holds air at 20 MPa and 10°C. At a depth of 100 m in water ($\rho =$ 1000 kg/m³) a diver uses 0.04 m³/min of air at a pressure equal to that of the water surrounding her. Assume the temperature of the air remains at 10°C. Determine the time until the air is consumed.

5.29 Given the equation $pv = RT$, show that the following equations may be derived from it: (a) $pV = mRT$; (b) $pV = n\bar{R}T$; (c) $p\bar{v} = \bar{R}T$; (d) $pv = \bar{R}T/M$.

5.30 A weather balloon is 3 m in diameter and contains helium at 27°C and 101 kPa. The balloon rises to an altitude where the temperature is −17°C and the pressure is 15 kPa. Calculate the mass of helium in the balloon and the change of balloon volume at the high altitude.

5.31 A rigid 4.0-m³ tank contains 40 kgmol of an ideal gas at 50°C with a molecular weight of 25 kg/kgmol. (a) Determine the gas pressure. (b) Heat transfer occurs, and the temperature decreases to 20°C. What is the pressure?

5.32 A 5.0-m³ tank contains 1.0 kgmol of an ideal gas at 400 kPa with a molecular weight of 31 kg/kgmol. (a) Determine the gas temperature. (b) Gas is removed from the tank, temperature remaining constant, until the pressure decreases to 100 kPa. What mass of gas was removed?

5.33 A typical adult breathes 0.5 liters of air with each breath and has 25 breaths per minute. At 101.3 kPa and 22°C, determine the mass of air per hour entering a person's lungs. This person now is skiing on a mountain where the air is −10°C and the pressure is 89 kPa. How many breaths per minute are required if the mass of air per hour entering the lungs is to be constant?

5.34 Determine the compressibility factor for water vapor at 10 MPa and 600°K using (a) data from the generalized compressibility chart; (b) data from the steam tables.

5.35 Determine the pressure range for air in kPa for $0.95 < Z < 1.05$ at temperatures of $2T_c$, $3T_c$, and $4T_c$, where T_c is the critical temperature.

5.36 The same as Problem 5.35 except the gas is argon.

5.37 An adiabatic tank has an internal partition that separates two gases. On one side of the partition is air—1.5 kg at 500 kPa and 350°K; on the other side is ammonia—3.0 kg at 200 kPa and 500°K. Imagine now that the partition is free to move and allows the conduction of heat from one side to the other. Determine the final temperature and pressure of each gas, assuming that the ideal-gas laws, constant specific heats, may be used.

5.38 Two well-insulated tanks contain air at different conditions. Tank 1 contains 2 kg of air at 800°K and tank 2 contains 1 kg of air at an unknown temperature. A valve in the line connecting the tanks is opened, and the equilibrium temperature is found to be 600°K. Determine the initial temperature of the air in tank 2.

5.39 Air expands in a piston-cylinder from 200 kPa, 0.2 m³, and 300°K to a final state of 0.4 m³ and 400°K. The pressure varies linearly with volume during the process. Determine the work and the heat transfer.

PROBLEMS (English Units)

*5.1 Determine the change of enthalpy for air and for carbon dioxide when the temperature changes from 70°F to 1000°F. Use equations from Table 5.4.

*5.2 One lbm of air is heated in a 0.5-ft³ tank from 70°F to 200°F. Determine (a) the heat transferred; (b) the final pressure.

*5.3 Ammonia has a specific volume of 1.496 at 300 psia and 300°F. Determine the specific volume, using the ideal-gas law and the compressibility factor.

*5.4 An ideal gas with molecular weight of 45 expands at constant pressure from 150°F to 350°F. During the process 5.0 Btu are added to the gas, whose mass is 0.2 lbm. Determine c_p and c_v.

*5.5 Carbon dioxide at 537°R and 14.7 psia has a density of 0.1123 lbm/ft³. Determine (a) the gas constant; (b) the molecular weight based on the gas constant.

*5.6 For a certain ideal gas, $R = 43.0$ ft-lbf/lbm-R and $k = 1.25$. Determine (a) c_p; (b) c_v; (c) M.

*5.7 For a certain ideal gas, $c_p = 0.255$ Btu/lbm-R and $k = 1.3$. Determine (a) M; (b) R; (c) c_v.

*5.8 Determine the specific heats, c_p and c_v, of steam at 1000 psia and 900°F. Compare to specific heats calculated at 1000 psia and 600°F.

*5.9 A mountain is measured by finding the change in pressure at constant temperature. A barometer at the base of the mountain reads 28.74 in. Hg, while at the top it reads 19.14 in. Hg. The average local gravitational acceleration is 31.2 ft/sec². Determine the height if the temperature may be assumed constant at 70°F.

*5.10 An empty, opened can is 10 in. high and 4 in. in diameter. The can, with the open end down, is pushed under water with a density of 62.4 lbm/ft³. Find the water level in the can when the top of the can is 1 ft below the surface. Thermal equilibrium exists at all times.

*5.11 A 500-ft³ tank contains chlorine at 720°R and atmospheric pressure. It is evacuated to a pressure of 29 in. Hg vacuum. Determine the mass of chlorine pumped out of the tank, temperature remaining constant.

*5.12 An oxygen cylinder of 2 ft³ originally has a pressure of 3000 psia and a temperature of 60°F. The oxygen is gradually used until the pressure is 400 psia, temperature remaining constant. What mass of oxygen is used?

*5.13 An unknown gas has a mass of 3.3 lbm and occupies 25 ft³ while at 540°R and 30 psia. Determine the gas constant.

*5.14 An insulated, constant-volume system containing 3 lbm of air receives 39,000 ft-lbf of paddle work. The initial temperature is 85°F. Determine the change of internal energy of the system and the air's final temperature.

*5.15 A piston-cylinder contains 2.5 lbm of a gas and receives heat at a constant pressure of 55 psia. The internal energy increases by 240 Btu, and the temperature increases by 120°F. If the work done is 120 Btu, determine the specific heat at constant pressure for the gas and the change in volume of system.

*5.16 A weather balloon is 10 ft in diameter and contains helium at 77°F and 14.7 psia. The balloon rises to an altitude where the temperature is 0°F and the pressure is 2.2 psia.

Calculate the mass of helium in the balloon and the change of balloon volume at the high altitude.

*5.17 A rigid 40-ft^3 tank contains 30 pmol of an ideal gas at 120°F with a molecular weight of 24 lbm/pmol. (a) Determine the gas pressure. (b) Heat transfer occurs, and the temperature decreases to 60°F. What is the pressure?

*5.18 A 50-ft^3 tank contains 1.0 pmol of an ideal gas at 75 psia with a molecular weight of 31 lbm/pmol. (a) Determine the gas temperature. (b) Gas is removed from the tank, temperature remaining constant, until the pressure decreases to 14.7 psia. What mass of gas was removed?

*5.19 A typical adult breathes 30 in.^{3} of air with each breath and has 25 breaths per minute. At 14.7 psia and 70°F, determine the mass of air per hour entering a person's lungs. This person now is skiing on a mountain where the air is 10°F and the pressure is 13.1 psia. How many breaths per minute are required if the mass of air per hour entering the lungs is to be constant?

*5.20 Determine the compressibility factor for water vapor at 2000 psia and 1000°F using (a) data from the generalized compressibility chart; (b) data from the steam tables.

*5.21 Determine the pressure range for air in psia for $0.95 < Z < 1.05$ at temperatures of $2T_c$, $3T_c$, and $4T_c$, where T_c is the critical temperature.

*5.22 The same as Problem *5.21 except the gas is chlorine.

*5.23 An adiabatic tank has an internal partition that separates two gases. On one side of the partition is air—3.3 lbm at 75 psia and 170°F; on the other side is ammonia—6.6 lbm at 30 psia and 440°F. Imagine now that the partition is free to move and allows the conduction of heat from one side to the other. Determine the final temperature and pressure of each gas, assuming that the ideal-gas laws, constant specific heats, may be used.

*5.24 Two well-insulated tanks contain air at different conditions. Tank 1 contains 4 lbm of air at 1440°R, and tank 2 contains 2 lbm of air at an unknown temperature. A valve in the line connecting the tanks is opened, and the equilibrium temperature is found to be 1080°R. Determine the initial temperature of the air in tank 2.

*5.25 Air expands in a piston-cylinder from 30 psia, 2.0 ft^3, and 540°R to a final state of 4.0 ft^3 and 720°R. The pressure varies linearly with volume during the process. Determine the work and the heat transfer.

COMPUTER PROBLEMS

C5.1 Compute the pressure of air at a temperature of 100°C and a specific volume of 0.2 m^3/kg, using (a) the ideal-gas law; (b) the van der Waals equation; (c) the Beattie-Bridgeman equation; (d) the Redlich-Kwong equation.

C5.2 Solve Problem 5.8 using TK Solver.

C5.3 Solve Problem 5.10 using TK Solver.

C5.4 Compute the specific volume of air at 5 MPa and 200°K, using (a) the ideal-gas law; (b) the van der Waals equation; (c) the Beattie-Bridgeman equation; (d) the Redlich-Kwong equation.

C5.5 Compute the compressibility factor for nitrogen using the Redlich-Kwong equation for temperatures of 300°K and 150°K and a range of pressures between 100 kPa and 30 MPa. Plot the results and compare them to Figure 5.5.

C5.6 Compute the specific volume of superheated steam using the ideal-gas law, the van der Waals equation, and the Redlich-Kwong equation for (a) 10 MPa and 350°C; (b) 100 kPa and 150°C. (c) Compare the results to the tabulated values in the steam tables.

C5.7 Compute the lifting force of a helium-filled balloon from sea level to an altitude of 10,000 ft in increments of 500 ft. Assume atmospheric pressure and temperature vary linearly from 14.7 psia and 60°F at sea level to 10.1 psia and 30°F at 10,000 ft. The balloon has a diameter of 30 ft, and its pressure and temperature are at equilibrium with the surrounding air. As the balloon rises, excess helium is vented to the atmosphere.

C5.8 Compute a table of the internal energy and enthalpy of air as a function of temperature from 300°K to 1500°K. Use the equation for specific heat variation in Table 5.4. Assume the enthalpy is 300 kJ/kg and the internal energy is 214 kJ/kg at 330°K.

C5.9 Using the equations in Table 5.4, compute and plot curves of specific heat ratio (k) versus temperature in the range of 300°K to 1500°K for (a) methane; (b) ethane; (c) propane.

6

Energy Analysis of Open and Closed Systems

At this point you have the knowledge and the techniques to undertake a variety of energy analyses for a variety of open and closed systems undergoing reversible or irreversible energy transfers. Industrial plants are designed in part by combining individual thermodynamic processes; a thermodynamic analysis typically precedes the mechanical design, an analysis of the type that you can begin. In this chapter you will

- Further understand equilibrium and nonequilibrium processes;
- Investigate constant-property processes for open and closed systems;
- Consider generalized reversible processes;
- Explore transient analysis of open systems;
- Develop a greater understanding of processes combined to form cycles;
- Extend your perceptions of energy applications and conservation.

6.1 EQUILIBRIUM AND NONEQUILIBRIUM PROCESSES

We know that a process occurs whenever a system changes from one state to another state. It would be convenient if we also knew the path this process follows. However, as soon as the process path is mentioned, an equilibrium of reversible process is assumed to occur. The equilibrium surface illustrated in Figure 6.1 shows the path of

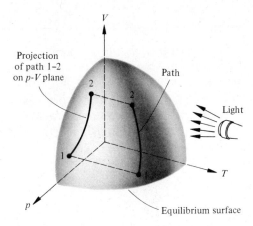

Figure 6.1 Projection of an equilibrium path on a p-V plane.

the process from state 1 to state 2. Imagine that a light is shown on the surface; a shadow, or projection, of the path will be made on the plane surface. This is denoted by the solid line on the p-V plane. Often a two-dimensional surface is used in drawing a system path. It may be helpful, however, to remember how this projection occurs; a projection of an equilibrium process is implied as soon as a continuous function is drawn.

What if the process were an irreversible one, going again between states 1 and 2? Let us assume that as it crosses the surface in Figure 6.1, a dot is marked on the surface of each crossing. Figure 6.2 might denote such a process, with its projection on the p-V plane.

From mathematics we know that a dotted line is a discontinuous function. To find work, we must be able to integrate $p\,dV$ along a continuous path from state 1 to state 2. Hence, when the process is irreversible, the path is discontinuous, and we cannot integrate to find the work; the work must be found from first-law analysis in these cases.

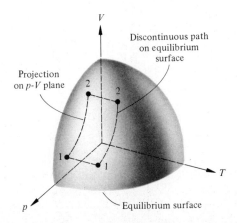

Figure 6.2 Projection of a discontinuous, or irreversible, path on a p-V plane.

6.2 CLOSED SYSTEMS

Science evolves by taking individual cases and relating their common aspects to form a general case. The temptation is to present the general case and assume the special cases will be understood. However, experience indicates that people grasp the more complex or general case when they first understand the simpler cases.

Before we proceed in solving some problems, we must first decide whether an open- or a closed-system analysis is to be used. Although here the sample problems and discussion will indicate the system classification, in homework problems and in real-world situations you must decide.

Constant-Pressure Process

We consider first a constant-pressure process for a closed system, illustrated in Figure 6.3(a). The system receives an amount of heat Q, performs work W, and experiences changes in internal energy, $\Delta U = U_2 - U_1$. The first law of thermodynamics for a closed system at rest, valid for all processes, is

$$Q = (U_2 - U_1) + W \qquad (6.1)$$

For reversible processes we may write the first law as

$$Q = (U_2 - U_1) + \int_1^2 p \, dV$$

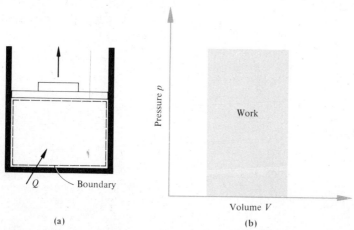

(a)   (b)

Figure 6.3 (a) A constant-pressure process with heat addition. (b) Graphical interpretation of mechanical work for a constant-pressure process.

or

$$\delta Q = dU + p\,dV \tag{6.2}$$

Also,

$$H = U + pV$$

$$dH = dU + p\,dV + V\,dp \tag{6.3}$$

$$dp = 0 \quad \text{for } p = C$$

$$\delta Q = dH \quad \text{for } p = C \tag{6.4}$$

For an ideal gas with constant specific heats,

$$\Delta H = mc_p(T_2 - T_1)\,\text{kJ} \tag{6.5a}$$

$$\Delta U = mc_v(T_2 - T_1)\,\text{kJ} \tag{6.5b}$$

For pure substances, the values of enthalpy and internal energy may be looked up in tables.

For a reversible constant-pressure process in a closed system, the work is

$$W = \int_1^2 p\,dV = p(V_2 - V_1) \tag{6.6a}$$

$$W = p_2 V_2 - p_1 V_1$$

Using the ideal-gas equation of state, $pV = mRT$, yields

$$W = mR(T_2 - T_1) \tag{6.6b}$$

Examining the system in Figure 6.3(a) more closely, we note that when the piston expands, it does work against the surroundings, displacing them as well as delivering additional or net work available for other purposes. Very often the surroundings' pressure is simply atmospheric pressure. The work expended in displacing the surroundings, W_{surr}, is

$$W_{surr} = \int_1^2 p_{surr}\,dV = p_{surr}(V_2 - V_1) \tag{6.7}$$

Does the work against the surroundings' pressure depend on the system process? No. This pressure is almost always constant, typically atmospheric pressure, so the work against the surroundings is a function of the initial and final system volume. Thus we may find the system work from the first-law analysis of the system, but to find the net work available to run a piece of equipment, we must subtract the work against the surroundings from the system work.

$$W_{net} = W_{sys} - W_{surr} \tag{6.8}$$

The net work does not influence the system work. It provides us with additional information, but in most problems we are interested only in system work.

Example 6.1

A piston-cylinder containing air expands at a constant pressure of 150 kPa from a temperature of 285°K to a temperature of 550°K. The mass of air in the cylinder is 0.05 kg. Determine the system heat and work for the process as well as the net work available if the surroundings' pressure acting on the piston is 101.3 kPa.

Solution

Given: A cylinder containing air expands at constant pressure from an initial state to a final state. The states are specified, as is the surroundings' pressure.

Find: The system heat and work and the net work available.

Sketch and Given Data:

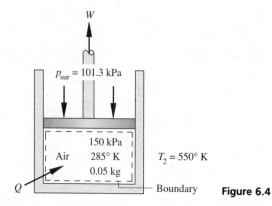

Figure 6.4

Assumptions:

1. The process is constant-pressure and reversible.
2. The changes in kinetic and potential energies may be neglected.
3. This is a closed system.
4. Air is an ideal gas.

Analysis: For a closed system undergoing a constant-pressure process, the first law may be reduced as follows:

$$\delta Q = dU + p\, dV$$

$$\delta Q = dH - V\, dp$$

$$dp = 0$$

$$\therefore \delta Q = dH$$

or

$$Q = \Delta H$$

Air is an ideal gas at the temperature and pressures in this problem, thus

$$\Delta H = mc_p \, \Delta T = (0.05 \text{ kg})(1.0047 \text{ kJ/kg-K})(550 - 285°\text{K})$$

$$\Delta H = 13.3 \text{ kJ}$$

From the first law for a closed system, $Q = \Delta U + W$, we note that by finding the change in internal energy for air, the work may be determined.

$$\Delta U = mc_v \, \Delta T = (0.05 \text{ kg})(0.7176 \text{ kJ/kg-K})(550 - 285°\text{K})$$

$$\Delta U = 9.5 \text{ kJ}$$

Substituting in the first law yields

$$W = Q - \Delta U = 13.3 - 9.5 = 3.8 \text{ kJ}$$

This is the work that the system did in moving the piston from state 1 to state 2. Some of this work is not available for a useful purpose, however, but must be expended in displacing the surroundings. The work against the surroundings is found from equation (6.7).

$$W_{\text{surr}} = p_{\text{surr}} \, (V_2 - V_1)$$

From the ideal-gas equation of state

$$V_1 = mRT_1/p_1 = (0.05 \text{ kg})(0.287 \text{ kJ/kg-K})(285°\text{K})/(150 \text{ kPa})$$

$$V_1 = 0.0273 \text{ m}^3$$

$$V_2 = mRT_2/p_2 = (0.05 \text{ kg})(0.287 \text{ kJ/kg-K})(550°\text{K})/(150 \text{ kPa})$$

$$V_2 = 0.0526 \text{ m}^3$$

$$W_{\text{surr}} = (101.3 \text{ kPa})(0.0526 - 0.0273 \text{ m}^3) = 2.56 \text{ kJ}$$

$$W_{\text{net}} = W_{\text{sys}} - W_{\text{surr}} = 3.8 - 2.56 = 1.23 \text{ kJ}$$

Comments:

1. There are other ways to solve for the system work, integrating $p \, dV$, for instance. Regardless of the methodology, the results will be the same.
2. The work against the surroundings can always be determined using equation (6.7), whether the process the system undergoes is reversible or not.
3. The heat transfer is equal to the change of system enthalpy only for reversible, constant-pressure processes.

Example 6.2
A rigid 1-m³ tank receives 500 kJ of heat and paddle work delivered for 1 h with a shaft torque of 1 J and a speed of 300 rpm. The tank contains steam initially at 300 kPa and 90% quality. Determine the final system temperature.

Solution

Given: A tank contains wet steam and receives heat and paddle work, increasing the system's energy.

Find: The final system temperature.

Sketch and Given Data:

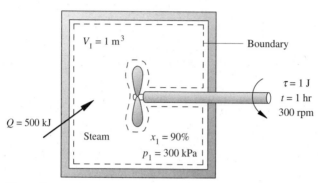

Figure 6.5

Assumptions:

1. The system is constant-volume and closed.
2. The system mechanical work is zero in light of assumption 1.
3. The initial and final states are equilibrium states.
4. The changes in kinetic and potential energies are zero.

Analysis: The first law for a closed system, invoking assumptions 2 and 4, may be written as

$$Q = \Delta U + W_p$$

where W_p is the paddle work added to the system by the rotating propeller.
The properties of steam at state 1 are found by using Table A.6, yielding

$$u_1 = u_f + x_1 u_{fg} = 560.9 + (0.90)(1982.9) = 2345.5 \text{ kJ/kg}$$

$$v_1 = v_f + x_1 v_{fg} = 0.001\ 073 + (0.90)(0.604\ 78) = 0.5454 \text{ m}^3/\text{kg}$$

$$m = V_1/v_1 = (1 \text{ m}^3)/(0.5454 \text{ m}^3/\text{kg}) = 1.833 \text{ kg}$$

The paddle work may be found from equation (3.20).

$$W_p = \tau \omega\ t = (1.0 \text{ J/rad})(5 \text{ rev/s})(2\pi \text{ rad/rev})(1 \text{ h})(3600 \text{ s/h})$$

$$W_p = 113\ 100 \text{ J} = 113.1 \text{ kJ}$$

Substituting into the first-law equation, noting that work into a system is negative and heat into a system is positive, yields

$$500 \text{ kJ} = \Delta U - 113.1 \text{ kJ}$$

$$\Delta U = 613.1 \text{ kJ} = m(u_2 - u_1) = (1.833 \text{ kg})(u_2 - 2345.5 \text{ kJ/kg})$$

$$u_2 = 2680.0 \text{ kJ/kg}$$

In this situation we know the final state is defined by its internal energy and specific volume, $v_2 = v_1$, for a constant-volume process.

From the superheated steam table we find that the final state's temperature is approximately 220°C. This is found by performing a double interpolation matching both the specific volume and internal energy, a tedious manual task, but one readily accomplished by TK Solver. SATSTM.TK computes a temperature of 221.1°C at a pressure of 410.8 kPa.

Comments:

1. Paddle work is always negative and irreversible: it can only be added to the system, hence negative and the system cannot cause the paddle to rotate, hence irreversible.
2. Double interpolations in the steam table are quite time-consuming when done manually. Solution on the computer is much easier and faster.
3. Mechanical work for a constant-volume process is zero, regardless of any other energy interactions.

Constant-Temperature Processes

The third type of special process we can envision is a reversible constant-temperature process. Because an ideal-gas equation of state is used in this discussion, the results must be limited to ideal gases. It is, of course, possible to have constant-temperature processes using tables of properties, but the following equations do not apply.

Consider again a piston-cylinder in Figure 6.6(a), which receives heat and ex-

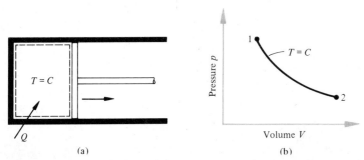

(a)   (b)

Figure 6.6 A constant-temperature process in a piston-cylinder and illustrated on a p-V diagram.

pands, doing work at constant temperature. Figure 6.6(b) illustrates the process on a p-V diagram. The first law for a closed system is

$$Q = \Delta U + W = \Delta U + \int p \, dV$$

We need p as a function of V for a constant-temperature process. The ideal-gas law shows that

$$p_1 V_1 = mRT_1 = mRT_2 = p_2 V_2 = C$$

or

$$pV = C \quad \text{for } T = C$$

$$W = \int_1^2 p \, dV = \int_1^2 C \frac{dV}{V} = C \ln\left(\frac{V_2}{V_1}\right) \tag{6.9}$$

$$W = p_1 V_1 \ln\left(\frac{V_2}{V_1}\right)$$

and since $U = U(T)$, then $\Delta U = 0$. This may be readily seen for an ideal gas:

$$\Delta U = mc_v(T_2 - T_1) = 0 \quad \text{for } T = C$$

and

$$Q = W \quad \text{for reversible } T = C \tag{6.10}$$

6.3 OPEN SYSTEMS

When dealing with open systems in which there are steady-state, steady-flow conditions, as discussed in Chapter 3, the analysis is straightforward. One simply equates all the energy entering the system to all the energy leaving the system and solves that equation for the unknown, usually the heat or work of the system. Figure 6.7 illustrates a typical open system.

In Chapter 3 we learned that heat for reversible processes may be expressed as

$$\delta q = dh - v \, dp \tag{6.11}$$

Figure 6.7 An open system with various energy forms.

and work may be expressed as

$$\delta w = -v\,dp - d(\text{k.e.}) - d(\text{p.e.}) \tag{6.12}$$

Equation (6.12) reduces to

$$\delta w = -v\,dp \tag{6.13}$$

$$\dot{W} = -\dot{m}\int v\,dp \tag{6.14}$$

for negligible changes in potential and kinetic energies.

At this point let us consider the various processes. Of course, constant volume is meaningless for open systems, but constant specific volume may be considered.

Constant-Pressure Process

Referring to equation (6.13), we note that in the absence of changes in kinetic and potential energies, the work is zero. The heat transfer, determined from the first-law equation, becomes equal to the change in enthalpy of the fluid. Heat exchangers are modeled as constant-pressure open systems: no work is done, ideally no pressure drop occurs, and heat is transferred from one fluid to another. Figure 6.8 has schematic diagrams for four basic types of heat exchangers. Figure 6.8(a) illustrates a tube-in-tube heat exchanger where fluid A flows in the inside tube and fluid B in the outside tube. Two flow configurations are possible: the fluids flow in the same direction, parallel flow; and the fluids flow opposite to one another, counterflow. Figure 6.6(b) illustrates a shell-and-tube heat exchanger, where fluid A flows through the tubes and baffling forces fluid B to flow over the tubes. Figure 6.6(c) illustrates a cross-flow heat exchanger, typified by automotive radiators. Figure 6.6(d) illustrates a direct-contact heat exchanger, where two phases of the same substance are brought together, in this case heating cold water by mixing it with steam. Often heat exchangers are considered adiabatic in that no heat is lost to the surroundings; the heat transfer occurs within the boundary of the heat exchanger from one fluid to another.

Example 6.3
An adiabatic shell-and-tube heat exchanger receives 200 lbm/min of steam at 1 psia and 90% quality on the shell side, and the steam condenses to a saturated liquid. Cooling water enters the tubes at 70°F and exits at 85°F. Determine (a) the mass flow rate of the cooling water; (b) the rate of energy transfer from the steam to the cooling water.

Solution

Given: A shell-and-tube heat exchanger with steam on the shell side and water flowing through the tubes.

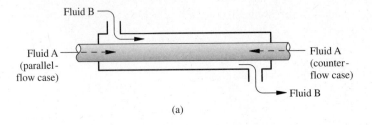

Fluid B

Fluid A
(parallel-
flow case)

Fluid A
(counter-
flow case)

Fluid B

(a)

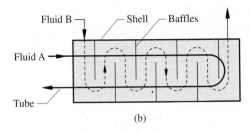

Fluid B   Shell   Baffles

Fluid A

Tube

(b)

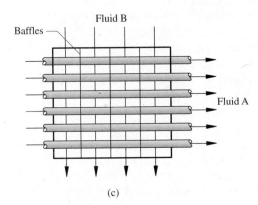

Fluid B

Baffles

Fluid A

(c)

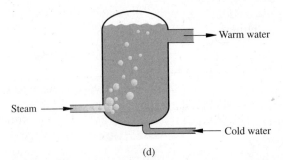

Warm water

Steam

Cold water

(d)

Figure 6.8 Various heat exchanger configurations. (a) Parallel and counterflow. (b) Shell-and-tube. (c) Cross-flow. (d) Direct-contact.

Find: The mass flow rate of cooling water required and the energy-transfer rate between the fluids.

Sketch and Given Data:

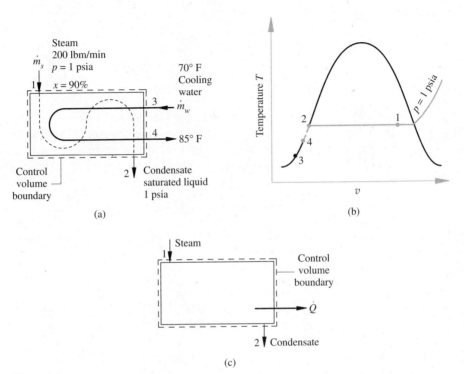

(a)

(b)

(c)

Figure 6.9

Assumptions:

1. The heat loss from the heat exchanger to the surroundings is zero.
2. The flow of both fluids is steady-state.
3. The changes in kinetic and potential energies for both fluids are zero.
4. The work across the control volume is zero.

Analysis: Part (a): Consider the heat exchanger to be an open system with two fluids entering and leaving. The first law becomes

$$\dot{Q} + \dot{m}_s[h_1 + (\text{k.e.})_1 + (\text{p.e.})_1 + \dot{m}_w[h_3 + (\text{k.e.})_3 + (\text{p.e.})_3]$$
$$= \dot{W} + \dot{m}_s[h_2 + (\text{k.e.})_2 + (\text{p.e.})_2] + \dot{m}_w[h_4 + (\text{k.e.})_4 + (\text{p.e.})_4$$
$$\dot{m}_s(h_1 - h_2) = \dot{m}_w(h_4 - h_3)$$

Using the values of enthalpy found in the steam tables,

$h_1 = h_f + x_1 h_{fg} = 69.58 + (0.9)(1036.2) + 1002.16$ Btu/lbm

$h_2 = h_f = 69.58$ Btu/lbm

$h_3 = h_f$ at $70°F = 37.68$ Btu/lbm

$h_4 = h_f$ at $85°F = 52.79$ Btu/lbm

$\dot{m}_w = (200$ lbm/min$)(1002.16 - 69.58$ Btu/lbm$)/(52.79 - 37.68$ Btu/lbm$)$

$\dot{m}_w = 12{,}345$ lbm/min of cooling water

Part (b): The control volume for the shell side of the heat exchanger is illustrated in Figure 6.9(c). In this case we are considering the control volume only in terms of the steam. It enters with an initial quality, condenses to a saturated liquid, and gives heat to the surroundings in the process. The first law for an open system with steady flow is

$$\dot{Q} + \dot{m}_s[h_1 + (\text{k.e.})_1 + (\text{p.e.})_1] = \dot{W} + m_s[h_2 + (\text{k.e.})_2 + (\text{p.e.})_2]$$

From the assumptions, the power and changes in kinetic and potential energies are zero; hence

$$\dot{Q} + \dot{m}_s h_1 = \dot{m}_s h_2$$
$$\dot{Q} = \dot{m}_s(h_2 - h_1) = (200 \text{ lbm/min})(69.58 - 1002.16 \text{ Btu/lbm})$$
$$\dot{Q} = -186{,}516 \text{ Btu/min}$$

The heat is negative, indicating that it leaves the shell control volume and enters the cooling-water control volume.

Comments:

1. At low pressures the property values of liquids are primarily a function of temperature and secondarily a function of pressure.
2. The entire heat exchanger is adiabatic, as illustrated in part (a). In part (b) the adiabatic feature was implicitly invoked, as no heat entered from the surroundings—the only heat transfer was a result of the steam condensing.

Constant-Temperature Processes—Ideal Gases

Let us calculate the power produced by a constant-temperature expansion process. This is found by integrating equation (6.14). In this situation we assume negligible changes in kinetic and potential energies. For an ideal gas

$$pv = RT = C$$

thus

$$v = C/p$$

$$\dot{W} = -\dot{m} \int v \, dp = -\dot{m} \int C \, dp/p = -\dot{m}_1 v_1 \ln (p_2/p_1)$$

Note that the assumption of reversibility was incorporated into the development when the work was expressed as $v \, dp$. This is true only for reversible flow processes, which is not a serious limitation: since a constant-temperature process is not physically realizable but is used in some instances as the theoretical minimum work of compression, assuming constant temperature is at least as great a restriction as assuming reversibility. The heat flow is found from equation (6.11); note that it is equal to the power when $dh = 0$, which for an ideal gas, $dh = c_p \, dT$, occurs in isothermal processes.

Constant – Specific Volume Process

A process has constant specific volume when it is under the constraint of incompressibility. The specific volume, hence density, remains constant. Very often in the pumping of liquids and sometimes in compressing air slightly, the assumption of incompressibility is realistic.

Figure 6.10 diagrams rotative devices, turbines that produce power and compressors and pumps that consume power. Figure 6.10(a) shows an axial-flow turbine; the flow of the fluid is parallel to the rotor, or main axis, of the turbine. This type of turbine is used in gas and steam turbine applications. The gas or steam enters the turbine at a high pressure and temperature and exits at a lower pressure and temperature. Usually the change in potential energy is negligible, as the elevation change across the turbine is quite small. Very often the change in kinetic energy is small enough to be neglected, particularly when contrasted with the much larger changes in thermal energy, denoted by the integral of $v \, dp$ in equation (6.14). The section on turbomachinery in Chapter 17 explains the fluid mechanics of the fluid-blade energy transfer. Because the fluid entering the turbine is often at high temperatures, the control volume can lose heat to the surroundings if the turbine casing is not well insulated. Engineers have noted that losing heat to the surroundings precludes its being converted into work, and in most instances the turbine casing is very well insulated.

In both pumps and compressors as illustrated in Figures 6.10(b) and 6.10(c), the changes in kinetic and potential energies across the device are quite small compared to the change in thermal energy. The centrifugal compressor or pump has fluid entering the rotating impeller axially. The fluid has a component of velocity changed to the radial direction because of centrifugal force, while it develops an angular velocity approaching the speed of the impeller. As the fluid moves outward, more gas flows into the impeller, creating a continuous flow. As the fluid flows through the

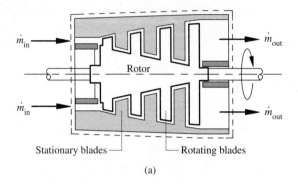

Stationary blades ⎯⎦   ⎣⎯ Rotating blades

(a)

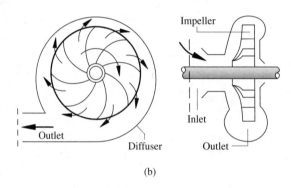

(b)

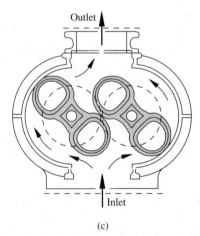

(c)

Figure 6.10 Schematics of relative devices. (a) An axial-flow turbine. (b) A centrifugal pump or compressor. (c) A Roots-type compressor.

diffuser, whose area increases with radial distance, the kinetic energy decreases; this energy is manifested primarily by an increase in the pressure component of the enthalpy.

The Roots-type compressor traps a gas between the lobes and the casing of the compressor and pushes it up to the pressure of the discharge line. The clearance between the lobes and the casing and between the lobes themselves is very close, to minimize leakage. Note that the lobes are rotating in opposite directions and that a steady supply of gas is pushed through the compressor. In many applications neither the pressure rise nor the temperature change is great, so the density variation is minimal, and the fluid is considered to be incompressible.

Example 6.4

A compressor receives 0.75 m^3/s of air at 290°K and 101 kPa. The compressor discharges the air at 707 kPa and 435°K. The heat transfer from the control volume is determined to be 2.1 kW. Determine the power required to drive the compressor.

Solution

Given: An air compressor receives air at known inlet and exit conditions and with a known rate of heat transfer from the control volume.

Find: The power to drive the compressor.

Sketch and Given Data:

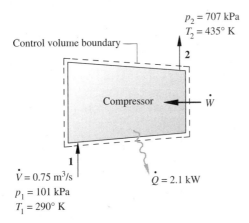

$p_2 = 707$ kPa
$T_2 = 435°$ K

Control volume boundary

2

Compressor ← $\dot{W}$

1
$\dot{V} = 0.75$ m^3/s
$p_1 = 101$ kPa
$T_1 = 290°$ K

$\dot{Q} = 2.1$ kW

Figure 6.11

Assumptions:

1. Air at these temperatures and pressures may be considered an ideal gas.
2. The flow is steady-state.
3. The changes in kinetic and potential energies may be neglected.

Analysis: The first law for a steady-flow open system is

$$\dot{Q} + \dot{m}[h_1 + (\text{k.e.})_1 + (\text{p.e.})_1] = \dot{W} + \dot{m}[h_2 + (\text{k.e.})_2 + (\text{p.e.})_2]$$

The mass flow rate of the air is not known but may be calculated from the ideal-gas equation of state written as

$$p\dot{V} = \dot{m}RT$$

thus,

$$\dot{m} = (101 \text{ kPa})(0.75 \text{ m}^3/\text{s})/[(0.287 \text{ kJ/kg-K})(290°\text{K})]$$

$$\dot{m} = 0.91 \text{ kg/s}$$

Using assumptions 1 and 3 the first law becomes

$$\dot{W} = \dot{Q} + \dot{m}(h_1 - h_2) = \dot{Q} + \dot{m}c_p(T_1 - T_2)$$

where the ideal-gas equation of state for enthalpy has been used.

$$\dot{W} = -2.1 \text{ kW} + (0.91 \text{ kg/s})(1.0047 \text{ kJ/kg-K})(290 - 435°\text{K})$$

$$\dot{W} = -134.7 \text{ kW}$$

Comments:

1. The heat and work flux terms are negative, indicating heat leaving the control volume and power being required to compress the air.
2. One can verify the use of the ideal-gas model by checking that the generalized compressibility factor is unity for the conditions in this problem. ▬

Example 6.5

A pump in a municipality's water-supply system receives water from the filtration beds and pumps it up to the top of a water tower. The tower's height is 35 m, and the inlet piping to the pump is 2 m below the pump's intake. The water temperature is 20°C, measured at both the inlet and the discharge from the pump. The mass flow rate through the pump is 100 kg/s, the diameter of the inlet piping is 25 cm, and the diameter of the discharge piping is 15 cm. Determine the power required by the pump.

Solution

Given: Water flows steadily through a pump that discharges it to a higher elevation. The flow rate and pipe diameters are given.

Find: The power required by the pump.

Sketch and Given Data:

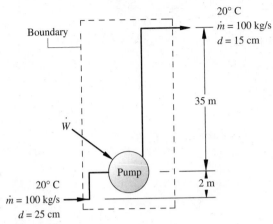

Figure 6.12

Assumptions:

1. The water flows steadily through the control volume.
2. The heat loss to the surroundings is negligible.
3. The property values of water are those of a saturated liquid at 20°C.

Analysis: The first law for a steady, open system is

$$\dot{Q} + \dot{m}[h_1 + (\text{k.e.})_1 + (\text{p.e.})_1] = \dot{W} + \dot{m}[h_2 + (\text{k.e.})_2 + (\text{p.e.})_2]$$

Since $h_1 = h_2 = h_f$ at 20°C, these two terms add out. Furthermore, the heat transfer is zero from assumption 2, reducing the first-law equation to terms involving kinetic energy and potential energy. The velocities of the water into and out of the control volume are not given but can be determined from the conservation of mass equation.

$$\dot{m} = A\text{v}/v$$

where $v = v_f$ at 20°C = 0.001 002 m³/kg. Thus, for the inlet conditions

$$(0.001\ 002\ \text{m}^3/\text{kg})(100\ \text{kg/s}) = \pi(0.25\ \text{m})^2(\text{v}_1\ \text{m/s})/4$$

$$\text{v}_1 = 2.0\ \text{m/s}$$

and for discharge conditions

$$(0.001\ 002\ \text{m}^3/\text{kg})(100\ \text{kg/s}) = \pi(0.15\ \text{m})^2(\text{v}_2\ \text{m/s})/4$$

$$\text{v}_2 = 5.7\ \text{m/s}$$

Solving the first-law equation for power yields

$$\dot{W} = \dot{m}\{[(\text{k.e.})_1 - (\text{k.e.})_2] + [(\text{p.e.})_1 - (\text{p.e.})_2]\}$$

$$[(\text{k.e.})_1 - (\text{k.e.})_2] = \frac{v_1^2 - v_2^2}{2} = \frac{4.0 - 32.49}{2} = -14.2 \text{ J/kg}$$

$$[(\text{p.e.})_1 - (\text{p.e.})_2] = g(z_1 - z_2) = (9.8 \text{ m/s}^2)(-2 - 35 \text{ m}) = -362.6 \text{ J/kg}$$

$$\dot{W} = (100 \text{ kg/s})(-14.2 - 362.6 \text{ J/kg}) = -37.7 \text{ kW}$$

Comments:

1. The power is negative, indicating that it had to be supplied to the control volume.
2. The heat loss is negligible because there is little temperature difference between the system and the surroundings in situations like this.
3. The major component of the work term is the potential energy change across the control volume.

Throttling Devices

In Chapter 4 we discussed a throttling calorimeter used to determine the quality of steam. The throttling process occurs whenever adiabatic pressure decreases in the flow of a fluid with no work in or out. Such a pressure decrease occurs most frequently through partially opened valves, illustrated in Figure 6.13(a), or through a porous material located in the flow line, such as a filter, illustrated in Figure 6.13(b). The first law for an open-system, steady-state condition is

$$\dot{Q} + \dot{m}[h_1 + (\text{k.e.})_1 + (\text{p.e.})_1] = \dot{W} + \dot{m}[h_2 + (\text{k.e.})_2 + (\text{p.e.})_2]$$

The control volume is adiabatic, no work is done, and the change in potential energy is zero. If the velocities are measured upstream and downstream from the valve itself, where velocities across the valve seat may be high, the change in velocity

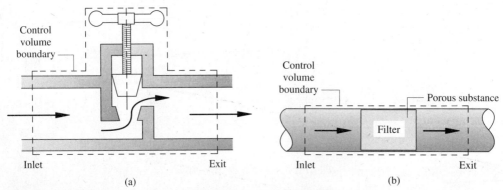

Figure 6.13 Throttling processes. (a) Partially open valve. (b) Filter.

between the inlet and outlet conditions for steady flow is usually quite small, yielding for the first law that

$$h_1 = h_2$$

6.4 POLYTROPIC PROCESS

Closed System

The previous process types for gases in open and closed systems are special cases of a more general type of process, called the *polytropic process,*

$$pV^n = C \tag{6.15a}$$

$$pv^n = C \tag{6.15b}$$

This process is illustrated in Figure 6.14 on a *p-V* diagram and exemplified by the special constant-property processes.

The constant-temperature process is a case when $n = 1$, the constant-pressure process when $n = 0$, and the constant-volume process when $n = \infty$. The polytropic processes are all assumed to be reversible. Note that $pV^n = C$ is the equation of the line projected from the equilibrium surface.

Relationships may be developed between the temperature, pressure, and volume for a polytropic process between state 1 and state 2. Using the ideal-gas law,

$$pv = RT$$

$$p_1 v_1^n = p_2 v_2^n \tag{6.16}$$

$$p_1 = \frac{RT_1}{v_1} \qquad p_2 = \frac{RT_2}{v_2}$$

$$\frac{p_1}{p_2} = \left(\frac{T_1}{T_2}\right)\left(\frac{v_2}{v_1}\right)$$

For a polytropic process between states 1 and 2,

$$\frac{p_1}{p_2} = \left(\frac{v_2}{v_1}\right)^n \tag{6.17a}$$

We substitute equation (6.16) into equation (6.17a):

$$\left(\frac{T_1}{T_2}\right)\left(\frac{v_2}{v_1}\right) = \left(\frac{v_2}{v_1}\right)^n$$

$$\frac{T_1}{T_2} = \left(\frac{v_2}{v_1}\right)^{n-1} = \left(\frac{V_2}{V_1}\right)^{n-1} \tag{6.17b}$$

$$\frac{V_2}{V_1} = \left(\frac{T_1}{T_2}\right)^{1/(n-1)} \tag{6.17c}$$

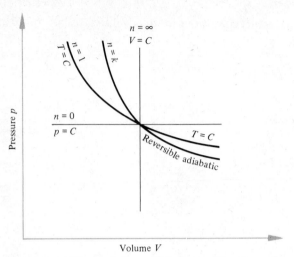

Figure 6.14 The variation of pressure with volume for different values of n.

On eliminating the volumes the result is

$$\frac{T_1}{T_2} = \left(\frac{p_2}{p_1}\right)^{(1-n)/n} = \left(\frac{p_1}{p_2}\right)^{(n-1)/n} \tag{6.18a}$$

$$\frac{p_1}{p_2} = \left(\frac{T_1}{T_2}\right)^{n/(n-1)} \tag{6.18b}$$

The work may also be calculated from the pressure-volume functional relationship, as follows:

$$W = \int_1^2 p \, dV \qquad p = CV^{-n}$$

$$W = \int_1^2 CV^{-n} \, dV = \frac{CV_2^{1-n} - CV_1^{1-n}}{1-n} \tag{6.19a}$$

$$C = p_2 V_2^n = p_1 V_1^n$$

$$W = \frac{p_2 V_2 - p_1 V_1}{1-n}$$

Using the ideal-gas law,

$$p_2 V_2 = mRT_2$$

$$p_1 v_1 = mRT_1 \tag{6.19b}$$

$$W = \frac{mR(T_2 - T_1)}{1-n}$$

To derive equation (6.19a) the ideal-gas equation of state is not required; however, it is required to derive equation (6.19b). Reversibility is common to both, since the integral of $p\,dV$ is evaluated by a continuously defined pressure.

One type of process that is frequently used as a standard of comparison for any actual process is the reversible adiabatic process. For a closed system the first law of thermodynamics for a reversible change may be written as

$$\delta q = du + p\,dv \tag{6.20a}$$

$$\delta q = dh - v\,dp \tag{6.20b}$$

Since $\delta q = 0$ for a reversible adiabatic process, defining du and dh in terms of an ideal gas yields

$$c_v\,dT = -p\,dv \tag{6.20c}$$

$$c_p\,dT = v\,dp \tag{6.20d}$$

Dividing equation (6.20d) by equation (6.20c),

$$\frac{c_p}{c_v} = k = -\frac{v}{p}\frac{dp}{dv}$$

$$k\frac{dv}{v} = -\frac{dp}{p} \tag{6.20e}$$

Integrating equation (6.20e),

$$k\ln\left(\frac{v_2}{v_1}\right) = \ln\left(\frac{p_1}{p_2}\right)$$

We then take antilogs

$$\left(\frac{v_2}{v_1}\right)^k = \left(\frac{p_1}{p_2}\right)$$

or

$$p_2 v_2^k = p_1 v_1^k = pv^k = C \tag{6.21a}$$

$$pV^k = C \tag{6.21b}$$

Equation (6.21b) is the same as equation (6.15a) except that $n = k$; hence, the relationships among temperature, pressure, and volume are

$$\frac{V_2}{V_1} = \left(\frac{T_1}{T_2}\right)^{1/(k-1)} \tag{6.22a}$$

$$\frac{p_1}{p_2} = \left(\frac{T_1}{T_2}\right)^{k/(k-1)} \tag{6.22b}$$

Open System (Steady-State, Steady-Flow)

In the case of open systems, an expression for the power may be determined because the process is known. Neglecting variations in kinetic and potential energies, from equation (6.14) and the process $pv^n = C$,

$$\dot{W} = -\dot{m}\int_1^2 v\,dp = -\dot{m}\int_1^2 C^{1/n}p^{-1/n}\,dp$$

$$C^{1/n} = p_1^{1/n}v_1 = p_2^{1/n}v_2$$

$$\dot{W} = \frac{n}{n-1}\dot{m}(p_1v_1 - p_2v_2)$$

Note that

$$\frac{p_2v_2}{p_1v_1} = \left(\frac{p_2}{p_1}\right)^{(n-1)/n}$$

$$\dot{W} = \frac{n}{n-1}\dot{m}RT_1\left[1 - \left(\frac{p_2}{p_1}\right)^{(n-1)/n}\right]$$

(6.23)

Example 6.6

A piston-cylinder compresses 0.11 kg of helium polytropically from 1 atm and 0°C to 10 atm and 165°C. Determine the work and heat for the process.

Solution

Given: Helium is compressed polytropically between two states.

Find: The work and heat for the compression process.

Sketch and Given Data:

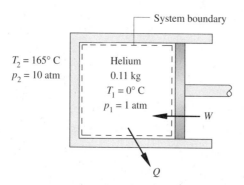

Figure 6.15

Assumptions:

1. Helium is an ideal gas with constant specific heats.
2. The process is reversible.
3. The changes in kinetic and potential energies may be neglected.

Analysis: The first law for a closed system, neglecting changes in kinetic and potential energies, is

$$Q = \Delta U + W$$

The change of internal energy of an ideal gas with constant specific heats is

$$\Delta U = mc_v(T_2 - T_1) = (0.11 \text{ kg})(3.1189 \text{ kJ/kg-K})(438 - 273°\text{K}) = 56.6 \text{ kJ}$$

To evaluate the work for a polytropic compression, we must first evaluate the polytropic exponent, n. Substituting into equation (6.18b)

$$\left(\frac{1}{10}\right) = \left(\frac{273}{438}\right)^{n/(n-1)}$$

Taking logs of both sides and solving for n yields

$$n = 1.259$$

The work may be evaluated from equation (6.19b).

$$W = \frac{mR(T_2 - T_1)}{1 - n} = \frac{(0.11 \text{ kg})(2.077 \text{ kJ/kg-K})(438 - 273°\text{K})}{(1 - 1.259)} = -145.6 \text{ kJ}$$

The heat may be found by substituting into the first-law equation

$$Q = 56.6 - 145.6 = -89.0 \text{ kJ}$$

This problem can be easily solved using TK Solver. The basic relationships of the polytropic equation, the ideal-gas law, the first law, and polytropic work for a closed system, are entered in the Rule Sheet. Entering the Input values and a guess value for n in the Variable Sheet produces the following results:

```
======================= VARIABLE SHEET =======================
St Input—Name—Output——  Unit —— Comment ————————————
                                Polytropic Process - Closed System

   100     P1                  kPa
           v1     .62406581    m3/kg
   0       T1                  degC

   1000    P2                  kPa
           v2     .10010413    m3/kg
   165     T2                  degC

           n      1.2582124
   2.077   R                   kJ/kg-K
   3.1189  Cv                  kJ/kg-K
   .11     m                   kg

           Q      -89.38633    kJ
           delU   56.608035    kJ
           W      -145.9944    kJ


========================= RULE SHEET =========================
S  Rule ——————————————————————————————————————————————
   "Calculation Units are SI: degK, kPa, m3/kg, kJ/kg-K
   P1*v1^n=P2*v2^n
   P1*v1=R*T1
   P2*v2=R*T2
   IF n<>1 THEN W=m*R*(T2-T1)/(1-n)
*  IF n=1 THEN W=m*P1*v1*LN(v2/v1)
   delU=m*Cv*(T2-T1)
   Q=delU+W
```

Comments:

1. The negative sign shows that work was done on the helium, compressing it.
2. The negative sign shows that heat was rejected from the system.
3. The energy terms indicate that work was done on the gas; some of the energy raised the helium's internal energy, and the remainder was rejected as heat to the surroundings.   ▬

Example 6.7

A turbine receives 150 lbm/sec of air at 63 psia and 2450°R and expands it polytropically to 14.7 psia. The exponent n is equal to 1.45 for the process. Determine the power and heat flux.

Solution

Given: A turbine receives air and expands it between two states polytropically.

Find: The heat and work fluxes across the control volume.

Sketch and Given Data:

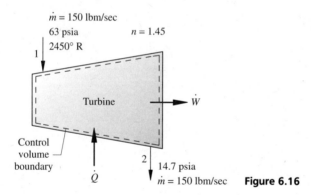

Figure 6.16

Assumptions:

1. Air may be considered an ideal gas with constant specific heats.
2. The polytropic process is reversible.
3. The changes in kinetic and potential energies may be neglected.
4. The air flows steadily through the turbine.

Analysis: The first law for an open, steady-flow system is

$$\dot{Q} + \dot{m}[h_1 + (\text{k.e.})_1 + (\text{p.e.})_1] = \dot{W} + \dot{m}[h_2 + (\text{k.e.})_2 + (\text{p.e.})_2]$$

The power may be evaluated from equation (6.23).

$$\dot{W} = \frac{n}{n-1} \dot{m}RT \left[1 - \left(\frac{p_2}{p_1} \right)^{(n-1)/n} \right]$$

$$\dot{W} = \left(\frac{1.45}{0.45} \right) \frac{(150 \text{ lbm/sec})(53.34 \text{ ft-lbf/lbm-R})(2450°R)}{778.16 \text{ ft-lbf/Btu}}$$

$$\left[1 - \left(\frac{14.7}{63.0} \right)^{0.45/1.45} \right]$$

$$\dot{W} = 29{,}499 \text{ Btu/sec} = 37{,}492 \text{ hp}$$

The change of enthalpy for an ideal gas is

$$(h_2 - h_1) = c_p(T_2 - T_1)$$

The final temperature may be found from equation (6.18a).

$$T_2 = (2450°R) \left(\frac{14.7}{63} \right)^{0.45/1.45} = 1560°R$$

The heat flux, subject to assumptions, is

$$\dot{Q} = \dot{m}(h_2 - h_1) + \dot{W}$$

$$= (150 \text{ lbm/sec})(0.24 \text{ Btu/lbm-R})(1560 - 2450°R) + 29{,}499 \text{ Btu/sec}$$

$$\dot{Q} = -2541 \text{ Btu/sec}$$

Comment: The negative sign on the heat transfer indicates that it is leaving the control volume.

6.5 THREE-PROCESS CYCLES

By combining two or more processes it is possible to construct a cycle. In later chapters we will analyze several applications of cycles, but for now let us consider three processes, joined together, forming a cycle. Three is the minimum number of processes to create a practical cycle. In Chapter 3 we noted some interesting behavior for cycles, namely, that $\oint \delta Q = \oint \delta W$ or $\Sigma Q = \Sigma W$. With this in mind, let us consider the following example.

Example 6.8

A three-process cycle uses 3 kg of air and undergoes the following processes: polytropic compression from state 1 to state 2, where $p_1 = 150$ kPa, $T_1 = 360°$K, $p_2 = 750$ kPa, and $n = 1.2$; constant-pressure cooling from state 2 to state 3; and constant-temperature heating from state 3 to state 1, completing the cycle. Find the temperatures, pressure, and volumes at each state and determine the process and cycle heat and work.

Solution

Given: A three-process cycle using air, with details of the various processes that constitute the cycle.

Find: The temperature, pressure, and volume for each state in the cycle and the heat and work for each process as well as the net heat and work for the cycle.

Sketch and Given Data:

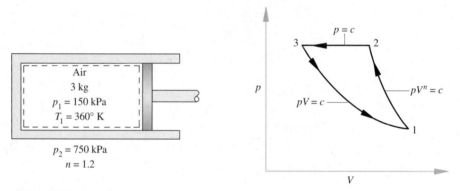

Figure 6.17

Assumptions:

1. Air behaves as an ideal gas with constant specific heats.
2. The processes are reversible.
3. The changes in kinetic and potential energies may be neglected.

Analysis: The key to solving three-process-cycle problems is to draw a sketch. Once that is done, follow the processes around the cycle.

State 1:

$$V_1 = \frac{mRT_1}{p_1} = \frac{(3.0 \text{ kg})(0.287 \text{ kJ/kg-K})(360°\text{K})}{150 \text{ kPa}} = 2.066 \text{ m}^3$$

Process 1-2, $pV^n = C$:

$$\frac{T_2}{T_1} = \left(\frac{p_2}{p_1}\right)^{(n-1)/n} \qquad T_2 = (360°\text{K})\left(\frac{750}{150}\right)^{0.2/1.2} = 470.7°\text{K}$$

$$V_2 = \frac{mRT_2}{p_2} \qquad V_2 = \frac{(3.0 \text{ kg})(0.287 \text{ kJ/kg-K})(470.7°\text{K})}{750 \text{ kPa}} = 0.540 \text{ m}^3$$

Process 2-3, $p = C$ and $T_3 = T_1 = 360°$K, since the process 3-1 is at a constant temperature:

$$p_3 = p_2 = 750 \text{ kPa}$$

$$V_3 = \frac{mRT_3}{p_3} = 0.413 \text{ m}^3$$

$$T_3 = 360°\text{K}$$

At this point let us calculate Q and W for each process.

$$Q_{1\text{-}2} = U_2 - U_1 + W_{1\text{-}2}$$

$$W_{1\text{-}2} = \frac{mR(T_2 - T_1)}{1 - n} = -476.5 \text{ kJ}$$

$$U_2 - U_1 = mc_v(T_2 - T_1) = 238.3 \text{ kJ}$$

$$Q_{1\text{-}2} = -238.2 \text{ kJ}$$

$$Q_{2\text{-}3} = H_3 - H_2 = mc_p(T_3 - T_2) = -333.6 \text{ kJ}$$

$$Q_{2\text{-}3} = U_3 - U_2 + W_{2\text{-}3}$$

$$U_3 - U_2 = mc_v(T_3 - T_2) = -238.3 \text{ kJ}$$

$$W_{2\text{-}3} = -95.3 \text{ kJ}$$

The work could also be evaluated by $\int p \, dV$. For constant-temperature processes,

$$Q_{3\text{-}1} = W_{3\text{-}1} = p_3 V_3 \ln\left(\frac{V_1}{V_3}\right) = mRT_3 \ln\left(\frac{V_1}{V_3}\right)$$

$$Q_{3\text{-}1} = W_{3\text{-}1} = (3.0 \text{ kg})(0.287 \text{ kJ/kg-K})(360°\text{K}) \ln\left(\frac{2.066}{0.413}\right) = 499 \text{ kJ}$$

$$\Sigma Q = -72.8 \text{ kJ} \qquad \Sigma W = -72.8 \text{ kJ}$$

Comments:

1. The net work is equal to the net heat.
2. The cyclic work is negative, indicating that the cycle requires a net work input to operate.

6.6 TRANSIENT FLOW

The typical situations studied in undergraduate thermodynamics are closed systems and steady-state open systems. However, the unsteady open system must be considered in transient flow; the two cases that we will consider are charging and discharging a tank.

Consider equation (3.52), which may be rewritten as

$$dE_{cv} = \delta Q - \delta W + \left(h + \frac{v^2}{2} + gz\right)_{in} dm_{in} - \left(h + \frac{v^2}{2} + gz\right)_{out} dm_{out} \qquad (6.24)$$

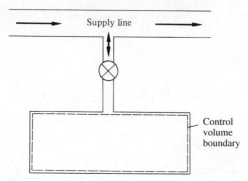

Figure 6.18 Sketch for charging and discharging a tank.

where $e = u + v^2/2 + gz$. Equation (6.24) may be integrated over the control volume for any instant of time, yielding

$$\Delta E_{cv} = Q - W + \int \left(h + \frac{v^2}{2} + gz \right)_{in} dm_{in} - \int \left(h + \frac{v^2}{2} + gz \right)_{out} dm_{out} \qquad (6.25)$$

Furthermore, the change of energy within the control volume is equal to the change of internal energy. Thus, equations (6.24) and (6.25) become

$$dU_{cv} = \delta Q - \delta W + \left(h + \frac{v^2}{2} + gz \right)_{in} dm_{in} - \left(h + \frac{v^2}{2} + gz \right)_{out} dm_{out} \qquad (6.26)$$

eeq.and

$$\Delta U_{cv} = Q - W + \int \left(h + \frac{v^2}{2} + gz \right)_{in} dm_{in} - \int \left(h + \frac{v^2}{2} + gz \right)_{out} dm_{out} \qquad (6.27)$$

Discharging a Tank

Consider Figure 6.18, which illustrates a tank connected to a supply line. The volume of the connection is considered negligible. Apply the first law to the control volume (equation [6.26]).

$$dU_{cv} = \delta Q + \left(h + \frac{v^2}{2} + gz \right)_{out} dm_{out} \qquad (6.28)$$

Note that $\delta W = 0$ and $dm_{in} = 0$. Furthermore, let us restrict the velocity of the exit stream to less than 50 m/s and the change of z to be small. Thus, kinetic and potential energies may be neglected. This yields

$$dU_{cv} = \delta Q + (h\, dm)_{out} \qquad (6.29)$$

The typical assumption at this point is that the tank is adiabatic; if it is not, δQ must be evaluated by other means (heat-transfer analysis, experimental data). Thus, for $\delta Q = 0$, equation (6.29) becomes

$$d(mu) = h\, dm \qquad (6.30)$$

However, $d(mu) = u\,dm + m\,du$, and upon substituting into equation (6.30),

$$u\,dm + m\,du = h\,dm$$

$$\frac{dm}{m} = \frac{du}{h-u}$$

For any substance, $h = u + pv$; hence

$$\frac{dm}{m} = \frac{du}{pv} \tag{6.31}$$

A separate relationship between specific volume and mass may be written by considering the total control volume, $V = vm = \text{const}$; hence

$$dV = 0 = v\,dm + m\,dv$$

or

$$\frac{dm}{m} = -\frac{dv}{v} \tag{6.32}$$

Substituting equation (6.32) into equation (6.31) yields

$$\frac{du}{pv} = -\frac{dv}{v}$$

or

$$du + p\,dv = 0$$

Since $du + p\,dv = \delta q = 0$, the process is reversible adiabatic. In order to evaluate properties at any instant of time, the substance in the tank is assumed to be in equilibrium at that instant.

By using the reversible adiabatic properties for an ideal gas,

$$\frac{T_2}{T_1} = \left(\frac{v_1}{v_2}\right)^{k-1}$$

$$\ln\left(\frac{T_2}{T_1}\right) = (1-k)\ln\left(\frac{v_2}{v_1}\right)$$

$$\frac{dT}{T} = (1-k)\frac{dv}{v}$$

Substituting in equation (6.32) yields

$$\left(\frac{1}{k-1}\right)\frac{dT}{T} = \frac{dm}{m}$$

and integrating yields

$$\frac{m_2}{m_1} = \left(\frac{T_2}{T_1}\right)^{1/(k-1)} \tag{6.33}$$

In a similar fashion

$$\frac{m_2}{m_1} = \left(\frac{p_2}{p_1}\right)^{1/k} = \frac{v_1}{v_2} \tag{6.34}$$

Charging a Tank

Let us consider the opposite situation, that of charging a tank. Figure 6.18 still applies, but the flow is into the control volume. As before, we will neglect changes in kinetic and potential energies, note that the work is zero, and let the tank be adiabatic. Equation (6.27) becomes

$$\int (h\, dm)_{in} = \Delta U_{cv} = m_2 u_2 - m_1 u_1 \tag{6.35}$$

At this point, let us assume that the properties in the line are constant with time (charging the tank does not affect them). Integrate from 0 to $m_L(m_L = m_{line})$, with $h_L = $ const. Thus

$$m_L h_L = m_2 u_2 - m_1 u_1 \tag{6.36}$$

Let us apply this to the situation where the tank is initially evacuated ($m_1 = 0$); thus, $m_L = m_2$ and

$$m_2 h_L = m_2 u_2$$

or

$$h_L = u_2$$

Consider an ideal gas as the substance entering the tank ($h = c_p T$, $u = c_v T$); thus,

$$T_2 = k T_L$$

Hence, when charging a tank, the final equilibrium temperature in the tank is k times the line temperature. Again equilibrium at any instant is assumed to exist in the tank, so properties are defined.

Example 6.9
An adiabatic tank containing air is used to power an air turbine during times of peak power demand. The tank has a volume of 500 m³ and contains air at 1000 kPa and 500°K. Determine (a) the mass remaining when the pressure reaches 100 kPa; (b) the temperature at this instant; (c) the total work done by the turbine, considering that it is also adiabatic and that all the air exiting the tank also exits the turbine.

Solution

Given: A tank containing air at given conditions is discharged to a final pressure. The tank is also connected to an air turbine, and the discharged air flows through the turbine.

Find: The mass and temperature of the air remaining in the tank at a certain pressure and the work done by the discharged air in the turbine.

Sketch and Given Data:

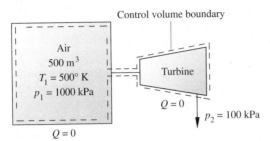

Figure 6.19

Assumptions:

1. Air is an ideal gas with constant specific heats.
2. The heat loss from the control volume is zero.
3. All the air discharging from the tank also discharges from the turbine. The turbine is viewed as initially having no air in it.
4. The changes in kinetic and potential energies are zero.

Analysis: Part (a): The mass remaining in the tank when the pressure reaches 100 kPa is determined from equation (6.34) with the value of $k = 1.4$ for air, found from Table A.1.

$$\frac{m_2}{m_1} = \left(\frac{p_2}{p_1}\right)^{1/k} = \left(\frac{100}{1000}\right)^{1/1.4} = 0.193$$

$$m_2 = 0.193 \, m_1$$

The initial mass in the tank may be evaluated using the ideal-gas law.

$$m_1 = \frac{p_1 V_1}{R T_1} = \frac{(1000 \text{ kPa})(500 \text{ m}^3)}{(0.287 \text{ kJ/kg-K})(500°\text{K})} = 3484.3 \text{ kg}$$

$$m_2 = (0.193)(3484.3 \text{ kg}) = 672.5 \text{ kg}$$

Part (b): The temperature may be calculated in several ways, using the ideal-gas equation of state or equation (6.33).

$$T_2 = \frac{p_2 V_2}{m_2 R} = \frac{(100 \text{ kPa})(500 \text{ m}^3)}{(672.5 \text{ kg})(0.287 \text{ kJ/kg-K})} = 259°\text{K}$$

or

$$T_2 = T_1 \left(\frac{p_2}{p_1}\right)^{(k-1)/k} = (500°\text{K})\left(\frac{100}{1000}\right)^{0.4/1.4} = 259°\text{K}$$

or

$$T_2 = T_1 \left(\frac{m_2}{m_1}\right)^{k-1} = (500°\text{K})\left(\frac{672.5}{3484.3}\right)^{0.4} = 259°\text{K}$$

```
 ..A/....B/....C/............D/............E/............F/...........G/
1
2  Example 6.9
3  P1   P2  Mstart   Mend                    delM       T2                       W
4
5  1000 900 3484.3   +C5*(B5/A5)^(1/1.4)     +C5-D5    500*(B5/A5)^(0.4/1.4)    (1.4/0.4)*E5*0.287*F5*(1-(100/((A5+B5)/2))^(0.4/1.4))
6  900  800 +D5      +C6*(B6/A6)^(1/1.4)     +C6-D6    +F5*(B6/A6)^(0.4/1.4)    (1.4/0.4)*E6*0.287*F6*(1-(100/((A6+B6)/2))^(0.4/1.4))
7  800  700 +D6      +C7*(B7/A7)^(1/1.4)     +C7-D7    +F6*(B7/A7)^(0.4/1.4)    (1.4/0.4)*E7*0.287*F7*(1-(100/((A7+B7)/2))^(0.4/1.4))
8  700  600 +D7      +C8*(B8/A8)^(1/1.4)     +C8-D8    +F7*(B8/A8)^(0.4/1.4)    (1.4/0.4)*E8*0.287*F8*(1-(100/((A8+B8)/2))^(0.4/1.4))
9  600  500 +D8      +C9*(B9/A9)^(1/1.4)     +C9-D9    +F8*(B9/A9)^(0.4/1.4)    (1.4/0.4)*E9*0.287*F9*(1-(100/((A9+B9)/2))^(0.4/1.4))
10 500  400 +D9      +C10*(B10/A10)^(1/1.4)  +C10-D10  +F9*(B10/A10)^(0.4/1.4)  (1.4/0.4)*E10*0.287*F10*(1-(100/((A10+B10)/2))^(0.4/1.4))
11 400  300 +D10     +C11*(B11/A11)^(1/1.4)  +C11-D11  +F10*(B11/A11)^(0.4/1.4) (1.4/0.4)*E11*0.287*F11*(1-(100/((A11+B11)/2))^(0.4/1.4))
12 300  200 +D11     +C12*(B12/A12)^(1/1.4)  +C12-D12  +F11*(B12/A12)^(0.4/1.4) (1.4/0.4)*E12*0.287*F12*(1-(100/((A12+B12)/2))^(0.4/1.4))
13 200  100 +D12     +C13*(B13/A13)^(1/1.4)  +C13-D13  +F12*(B13/A13)^(0.4/1.4) (1.4/0.4)*E13*0.287*F13*(1-(100/((A13+B13)/2))^(0.4/1.4))
14 _____
15                           Total M= @SUM(E5..E13)                              Work= @SUM(G5..G13)
16
```

Part (c): Let's solve for the turbine work numerically, using a spreadsheet. The turbine work will be calculated using equation (6.23) for polytropic open-system work. For each increment of tank pressure, the mass of air removed from the tank is determined, and the work done by the mass is calculated. The average tank pressure for each increment will be used to calculate the turbine work. The sum of the work for all the increments represents the total turbine work for the expansion from 1000 kPa to 100 kPa. The data and equations are entered into the spreadsheet as shown on the previous page.

The results are as follows:

```
........A/......B/ ......C/........D/......E/ ......F/........G/
 1   Example 6.9
 2
 3     P1        P2     Mstart     Mend      delM     T2          W
 4
 5   1000.00   900.00  3484.30   3231.70   252.60   485.17     58401.29
 6    900.00   800.00  3231.70   2970.94   260.76   469.12     56209.08
 7    800.00   700.00  2970.94   2700.67   270.27   451.56     53656.57
 8    700.00   600.00  2700.67   2419.09   281.58   432.10     50624.15
 9    600.00   500.00  2419.09   2123.70   295.39   410.17     46925.83
10    500.00   400.00  2123.70   1810.81   312.90   384.83     42251.76
11    400.00   300.00  1810.81   1474.45   336.36   354.47     36034.54
12    300.00   200.00  1474.45   1103.70   370.75   315.69     27080.31
13    200.00   100.00  1103.70    672.71   430.99   258.97     12264.30
14
15
16                             Total M= 2811.59   Work=    383447.84
```

Comments:

1. The ideal-gas equation of state may be used at any instant as long as the properties are equilibrium ones.
2. Should the turbine and tank not be adiabatic, other means than the first law must be used to determine the heat flow.

3. The turbine is viewed as being within the control volume with the tank and contains no initial or final mass, which allows the analysis to be direct. ▬

CONCEPT QUESTIONS

1. What is a polytropic process?

2. What is the difference between a polytropic process for an ideal gas and a reversible adiabatic one?

3. How can an adiabatic heat exchanger transfer heat?

4. Air is compressed adiabatically. Will the temperature rise? Why?

5. Will a pressure drop across the tube side of a shell-and-tube heat exchanger affect the amount of heat transferred?

6. A tank, initially evacuated, is charged with air at 300°K. When the tank is charged, the temperature of the air in the tank is measured and found to be 400°K. Why?

7. In unsteady flow must the energy entering the control volume equal that leaving? Does the mass flow entering equal the mass flow leaving?

8. An adiabatic storage tank containing air at a high pressure develops a slight leak, allowing some air to escape to the surroundings. Will the temperature in the tank change? Why?

9. In a throttling process the enthalpies into and out of the control volume are equal, but a temperature change can occur. Explain how the energy transformation occurs that allows this.

PROBLEMS (SI)

6.1   An insulated box containing carbon dioxide gas falls from a balloon 3.5 km above the earth's surface. Determine the temperature rise of the carbon dioxide when the box hits the ground.

6.2   Air from the discharge of a compressor enters a 1-m^3 storage tank. The initial air pressure in the tank is 500 kPa, and the temperature is 600°K. The tank cools, and the internal energy decreases by 213 kJ/kg. Determine (a) the work done; (b) the heat loss; (c) the change of enthalpy; (d) the final temperature.

6.3   A rigid, perfectly insulated system contains 0.53 m^3 of helium at 1000 kPa. The system receives 1000 kJ of paddle work. Determine the final pressure.

6.4   A constant-pressure, adiabatic system contains 0.15 kg of air at 150 kPa. The system receives 20.79 kJ of paddle work. The temperature of the air is initially 278°K and finally 416°K. Find the mechanical work and the changes of internal energy and enthalpy.

6.5   A closed rigid container has a volume of 1 m^3 and holds air at 344.8 kPa and 273°K. Heat is added until the temperature is 600°K. Determine the heat added and the final pressure.

6.6   A piston-cylinder containing air receives heat at a constant temperature of 500°K and an initial pressure of 200 kPa. The initial volume is 0.01 m^3, and the final volume is 0.07 m^3. Determine the heat and work.

6.7 Steam enters an adiabatic turbine at 1 MPa and 600°C and exits at 60 kPa and 150°C. Determine the work per kg of steam.

6.8 A steam turbine receives 5 kg/s of steam at 2.0 MPa and 600°C and discharges the steam as a saturated vapor at 100 kPa. The heat loss through the turbine is 6 kW. Determine the power.

6.9 A nozzle receives 5 kg/s of steam at 0.6 MPa and 350°C and discharges it at 100 kPa and 200°C. The inlet velocity is negligible, and the heat loss is 250 kJ/kg. Determine the exit velocity.

6.10 Air is compressed polytropically from 101 kPa and 23°C and delivered to a tank at 1500 kPa and 175°C. Determine per kg of air (a) the heat removed during compression; (b) the reversible work.

6.11 A tank with a volume of 50 m^3 is being filled with air, an ideal gas. At a particular instant, the air in the tank has a temperature of 400°K and 1380 kPa. At this instant the pressure is increasing at a rate of 138 kPa/s, and the temperature is increasing at a rate of 25 K/s. Calculate the air flow into the tank at this instant in kg/s.

6.12 Air at a pressure of 100 kPa has a volume of 0.32 m^3. The air is compressed in a reversible adiabatic manner until the temperature is 190°C. The reversible work is -63 kJ. Determine (a) the initial temperature; (b) the air mass; (c) the change of internal energy.

6.13 Air in a piston-cylinder occupies 0.12 m^3 at 552 kPa. The air expands in a reversible adiabatic process, doing work on the piston until the volume is 0.24 m^3. Determine (a) the work of the system; (b) the net work if the atmospheric pressure is 101 kPa.

6.14 Three mol of oxygen is compressed in a piston-cylinder in a reversible adiabatic process from a temperature of 300°K and a pressure of 102 kPa until the final volume is one-tenth the initial volume. Determine (a) the final temperature; (b) the final pressure; (c) the system work.

6.15 A scuba tank contains 1.5 kg of air. The air in the tank is initially at 15°C. The tank is left near an engine exhaust line, and the tank's pressure doubles. Determine (a) the final temperature; (b) the change in internal energy; (c) the heat added.

6.16 In a reversible adiabatic manner, 17.6 m^3/min of air are compressed from 277°K and 101 kPa to 700 kPa. Determine (a) the final temperature; (b) the change of enthalpy; (c) the mass flow rate; (d) the power required.

6.17 An adiabatic device looks like an inverted T with 3.03 kg/s of steam at 4 MPa and 600°C entering from the top, and two streams exiting horizontally, one at 0.5 kg/s, 0.2 MPa, and 6°C, and the other at 0.2 MPa and an unknown temperature. Determine the unknown temperature.

6.18 A control volume receives three streams of steam (1, 2, 3). Power is produced or consumed. Eight-hundred kW of heat leaves the control volume, and two streams (4, 5) leave. Determine the power, given that $\dot{m}_1 = 0.5$ kg/s, $v_1 = 400$ m/s, $p_1 = 1.5$ MPa, dry saturated vapor; $\dot{m}_2 = 5.0$ kg/s, $v_2 \approx 0$ m/s, $p_2 = 1.0$ MPa, $v_2 = 0.100$ m^3/kg; $\dot{m}_3 = 3.0$ kg/s, $v_3 \approx 0$ m/s, $p_3 = 600$ kPa, $T_3 = 450$°C; $\dot{m}_4 = 4.0$ kg/s, $v_4 = 500$ m/s, $p_4 = 2.0$ MPa, $T_4 = 300$°C; $\dot{m}_5 = 4.5$ kg/s, $v_5 \approx 0$ m/s, $p_5 = 800$ kPa, $T_5 = 500$°C.

6.19 An adabatic condenser receives 100 kg/s of steam at 92% quality and 60 kPa. The steam leaves at 60 kPa and 60°C. The cooling water enters at atmospheric pressure and 40°C and discharges at 60°C. Determine (a) the heat transferred; (b) the cooling water flow rate.

6.20 An adiabatic rigid container, holding water initially at 100 kPa and 20°C, receives 209 kJ/kg of paddle work. Determine the final temperature.

6.21 A pneumatic lift system is being demonstrated at a sales show. The total load is 70 kg, and the lift piston is 15.2 cm in diameter and has a 20.2-cm stroke. A portable air bottle with an initial pressure of 20 MPa and a temperature of 23°C is to be used as the pneumatic supply. A regulator reduces the pressure from the bottle to the lift system. Neglecting all volume in the lines from the bottle to the piston, determine the number of times the piston can operate per air bottle if the air in the bottle remains at 23°C and the volume of the bottle is 0.05 m³.

6.22 One kg/s of air initially at 101 kPa and 300°K is compressed polytropically according to the process $pv^{1.3} = C$. Calculate the power necessary to compress the air to 1380 kPa. An aftercooler removes 100 kW from the air before it enters a storage tank. Determine the change in internal energy of the air across the aftercooler.

6.23 Oxygen expands in a reversible adiabatic manner through a nozzle from an initial pressure and initial temperature and with an initial velocity of 50 m/s. There is a decrease of 38°K in temperature across the nozzle. Determine (a) the exit velocity; (b) the exit pressure for inlet conditions of 410 kPa and 320°K.

6.24 An ideal gas having a mass of 2 kg at 465°K and 415 kPa expands in a reversible adiabatic process to 138 kPa. The ideal-gas constant is 242 J/kg-K and $k = 1.40$. Determine (a) the final temperature; (b) the change in internal energy; (c) the work; (d) c_p, c_v.

6.25 An ideal gas with a molecular weight of 6.5 kg/kgmol is compressed in a reversible manner from 690 kPa and 277°K to a final specific volume of 0.47 m³/kg according to $p = 561 + 200v + 100v^2$, where p is the pressure in kPa and v is the specific volume in m³/kg. The specific heat at constant volume is 0.837 kJ/kg-K. Determine (a) the work; (b) the heat; (c) the final temperature; (d) the initial specific volume.

6.26 Three kg of neon is in a constant-volume system. The neon is initially at a pressure of 550 kPa and a temperature of 350°K. Its pressure is increased to 2000 kPa by paddle work plus 210 kJ of heat addition. Determine (a) the final temperature; (b) the change of internal energy; (c) the work input.

6.27 A 2-kg mass of oxygen expands at a constant pressure of 172 kPa in a piston-cylinder system from a temperature of 32°C to a final temperature of 182°C. Determine (a) the heat required; (b) the work; (c) the change of enthalpy; (d) the change of internal energy.

6.28 One kg of carbon dioxide is contained in a constant-pressure piston-cylinder system. The carbon dioxide receives paddle work and rejects heat while changing from an initial temperature of 417°K to a final temperature of 277°K. The heat rejected is found to be three times the work input. Determine (a) the heat; (b) the paddle work; (c) the change of enthalpy; (d) the net work if the surroundings are at 101 kPa.

6.29 One kg of air expands at a constant temperature from a pressure of 800 kPa and a volume of 2 m³ to a pressure of 200 kPa. Determine (a) the work; (b) the heat; (c) the change of internal energy; (d) the change of enthalpy.

6.30 An ideal compressor compresses 12 kg/min of air isothermally from 99 kPa and a specific volume of 0.81 m³/kg to a final pressure of 600 kPa. Determine (a) the power in kW; (b) the heat loss in kW.

6.31 A vertical piston-cylinder system is constructed so the piston may travel between two stops. The system is surrounded by air at 100 kPa. The enclosed volume is 0.05 m³ at the lower stop and 0.12 m³ at the upper stop. Carbon dioxide is contained in the system at a temperature of 300°K and a pressure of 200 kPa. Heat is added, and the piston rises, the system changing at constant pressure, until the piston reaches the upper stop. Heat continues to be added until the temperature is 900°K. Determine (a) the heat added; (b) the change of enthalpy; (c) the system work; (d) the net work; (e) the final pressure.

6.32 Air is contained in a piston-cylinder and is compressed in a reversible adiabatic manner from a temperature of 300°K and a pressure of 120 kPa to a final pressure of 480 kPa. Determine (a) the final temperature; (b) the work per kg.

6.33 Two m³ per second of helium at 277°K and 101 kPa is compressed to 404 kPa in a reversible adiabatic manner. Determine (a) the final temperature; (b) the power required.

6.34 In a natural gas–pipeline compressor 110 m³/min propane is compressed polytropically. The inlet pressure is 101 kPa, and the temperature is 38°C. The process follows $pV^{1.08} = C$. The exit pressure is 510 kPa. Determine (a) the exit temperature; (b) the heat loss; (c) the power required; (d) the mass flow rate.

6.35 Air is compressed polytropically in a cylinder according to $pV^2 = C$. The work required is 180 kJ. Determine (a) the change of internal energy; (b) the heat transferred.

6.36 A piston-cylinder contains 2 kg of steam. The piston is frictionless and rests on two stops. The steam inside is initially at 1000 kPa and 75% quality. Heat is added until the temperature is 500°C. However, the piston does not move until the system pressure reaches 2 MPa and then it moves at constant pressure. Determine (a) the system work; (b) the total heat transferred.

6.37 Helium expands polytropically through a turbine according to the process $pV^{1.5} = C$. The inlet temperature is 1000°K, the inlet pressure is 1000 kPa, and the exit pressure is 150 kPa. The turbine produces 1×10^5 kW. Determine (a) the exit temperature; (b) the heat transferred (kW); (c) the mass flow rate.

6.38 Carbon dioxide flows steadily at 1 kg/s through a device where the pressure, specific volume, and velocity of the gas are tripled according to $pV^n = C$. The inlet conditions are $p_1 = 200$ kPa, $v_1 = 0.4$ m³/kg, and $v_1 = 100$ m/s. Determine (a) n; (b) the initial and final temperatures; (c) the change of enthalpy; (d) the power; (e) the heat.

6.39 Nitrogen expands through a turbine in a reversible adiabatic process. The nitrogen enters at 1100°K and 550 kPa and exits at 100 kPa. The turbine produces 37 MW. Using Table A.1 determine (a) the exit temperature; (b) the exit enthalpy (kJ/kg); (c) the flow rate required.

6.40 Three kg of dry saturated steam expands at a constant pressure of 300 kPa to a final temperature of 400°C. Determine (a) the heat required; (b) the work done; (c) the change of internal energy.

6.41 In Problem 6.40 the atmospheric pressure is 100 kPa. Determine the net system work.

6.42 In a piston-cylinder, 0.5 kg of air expands polytropically, $n = 1.8$, from an initial pressure of 5000 kPa and an initial volume of 0.07 m³ to a final pressure of 500 kPa. Calculate the system work and heat.

6.43 A cycle, composed of three processes, is polytropic compression ($n = 1.5$) from 137 kPa and 38°C to state 2, constant pressure from state 2 to state 3, and constant volume from state 3 and to state 1. The heat rejected is 1560 kJ/kg, and the substance is air. Determine (a) the pressures, temperatures, and specific volumes around the cycle; (b) the heat added; (c) the heat rejected; (d) the work for each process.

6.44 A three-process cycle operating with nitrogen as the working substance has constant-temperature compression 1-2 ($T_1 = 40°C$, $p_1 = 110$ kPa); constant-volume heating 2-3; and polytropic expansion 3-1 ($n = 1.35$). The isothermal compression requires -67 kJ/kg of work. Determine (a) p, T, and v around the cycle; (b) the heats in and out; (c) the net work.

6.45 Two kg of helium operates on a three-process cycle where the processes are constant volume (1-2); constant pressure (2-3); and constant temperature (3-1). Given that $p_1 = 100$ kPa, $T_1 = 300°K$, and $v_1/v_3 = 5$, determine (a) the pressure, specific volume, and temperature around the cycle; (b) the work for each process; (c) the heat added.

6.46 An insulated tank initially evacuated with a volume of 3 m³ is charged with helium from a constant-temperature supply line at $300°K$ and 1000 kPa. If the final tank pressure is 900 kPa, determine the final helium temperature.

6.47 The tank in Problem 6.46 has an initial charge of helium at $300°K$ and 100 kPa. For the same final pressure of 900 kPa, determine the final helium temperature.

6.48 An adiabatic tank contains 2 kg of air at 3000 kPa and $325°K$. The tank is discharged until the pressure reaches 500 kPa. Determine (a) the mass remaining; (b) the final tank temperature.

6.49 The tank in Problem 6.48 is now heated so that the temperature remains constant at $325°K$. Determine the heat added.

6.50 Ethylene is compressed according to $pV^{1.15} = C$ from 1.5 MPa and $500°K$ to 2.8 MPa. The mass of ethylene is 3 kg. Determine the final temperature, work, and heat for the process.

6.51 Methane undergoes a constant pressure process from 6 MPa and $600°K$ to $400°K$. Determine the heat transfer and the work done per unit mass.

6.52 Acetylene is compressed according to $pV^n = C$ from an initial pressure of 200 kPa and $300°K$ to a final pressure of 1200 kPa. Determine the work and heat for $n = 1, 1.1, 1.232,$ 1.3, and 1.4. What conclusions can you draw?

6.53 Air, initially at 120 kPa and $320°K$, occupies 0.11 m³. It is compressed isothermally until the volume is halved and then compressed at constant pressure until the volume decreases to one-quarter the initial volume. Sketch the processes on a p-V diagram and determine the total heat and total work for the two processes.

6.54 A system holds 2 kg of neon in a piston-cylinder where the initial pressure is 200 kPa and the initial temperature is $300°K$. The system operates on a three-process cycle comprised of the following processes: from state 1 to state 2 constant-volume heating until the pressure is 800 kPa; from state 2 to state 3 expansion according to $pV^{1.67} = C$; from state 3 to 1 constant-pressure compression. Sketch the cycle on a p-V diagram and determine the net work and heat added.

6.55 Air occupies a 0.3-m³ piston-cylinder at 150 kPa and $278°K$. The piston-cylinder operates on a three-process cycle where from state 1 to state 2 the pressure is constant and $V_2 = 1.4 \, V_1$. From state 2 to 3 the compression is reversible adiabatic until $V_3 = V_1$, and the temperature at state 3 is $450°K$. A constant-volume process completes the cycle. Sketch the cycle on a p-V diagram and determine the net work and heat flows for the cycle.

6.56 1 kg of water vapor initially at 1000 kPa is cooled at a constant volume, $V_1 = V_2 = 0.25$ m³. The temperature at state 2 is $150°C$. The system expands at constant pressure from state 2 to state 3 until the volume is 0.75 m³. Determine the specific volumes and internal energies at each state and the heat and work for each process.

6.57 A piston-cylinder containing 1.5 kg of water vapor operates on a three-process cycle. At state 1 the water is a saturated vapor at 10 MPA, which expands adiabatically to a pressure of 1000 kPa and a quality of 0.7817. At this point a constant-pressure process occurs until the specific volume at state 3 equals that at state 1. Finally, from state 3 to state 1 constant-volume heating occurs. Sketch the cycle on a T-v diagram. Determine the work and heat for each process as well as the net work for the cycle.

6.58 A four-process cycle using 2 kg of water operates with the following processes: from state 1 to state 2 constant-volume heating from 500 kPa and 200°C to 1000 kPa; from state 2 to 3 constant-pressure cooling until the water is a saturated vapor; from state 3 to 4 constant-volume cooling; from state 4 to 1 isothermal expansion where $Q_{4-1} = 447.8$ kJ. Sketch the cycle on a T-v diagram. Determine the specific volume and internal energy at each state and the net work and total heat added for the cycle.

6.59 An adiabatic steam turbine receives 10 kg/s of steam at 5000 kPa and 500°C. The steam expands through the turbine to a pressure of 1000 kPa and a temperature of 300°C where 2 kg/s is removed. The remainder expands to 5 kPa and a quality of 90%. Determine the power produced by the turbine.

6.60 An air compressor receives 2 m³/min of air at 100 kPa and 300°K and compresses it to a pressure of 700 kPa in a reversible adiabatic process. Heat transfer occurs, and the air's temperature is reduced by 40°K before it leaves the compressor. Determine the power required to compress the air and the total heat transfer in the aftercooler. What circulating water flow rate is required in the aftercooler if the temperature rise is from 25° to 40°C?

6.61 An adiabatic compressor receives 1.5 m³/s of air at 30°C and 101 kPa. The discharge pressure is 505 kPa, and the power supplied is 325 kW. What is the discharge temperature?

6.62 A refrigeration compressor receives R 12 at 200 kPa and 0°C and discharges it at 1200 kPa and 50°C. The refrigerant flow rate at inlet conditions is 0.01 m³/s. The heat transfer from the compressor to the surroundings is 2.5 kJ/kg of refrigerant. Determine (a) the power required; (b) the volume flow rate at exit conditions.

6.63 A compressor receives carbon dioxide gas at 140 kPa with a specific volume of 0.37 m³/kg and compresses it to a temperature of 325°K. The work per unit mass for compression is 80 kJ/kg. The gas enters through a 15-cm-diameter line with a velocity of 10 m/s and leaves with a velocity of 25 m/s. Determine the heat transfer in kW.

6.64 A hydroelectric power plant has the hydraulic turbine located 60 m below the elevation of the intake pipe. The 1-m-diameter pipe delivers water with a velocity of 3 m/s at a temperature of 15°C to the turbine. The heat transfer between the turbine and the surroundings is negligible. Determine the power developed in kW.

6.65 A pump delivers 50 liters/s of water. The intake to the pump is 75 m below the final discharge. The inlet and discharge pressure is essentially atmospheric, and the temperature of the water remains constant at 20°C during the process. Determine the power required by the pump.

6.66 An oil-transfer pump uses 1 kW of power in transferring 5 kg/s of oil through a 3-cm-diameter pipe. The changes in kinetic and potential energies are essentially zero, and the process is adiabatic. Additionally, there is no appreciable temperature change in the oil, which has a density of 1500 kg/m³. Determine the change in pressure from inlet to exit.

6.67 An adiabatic heat exchanger receives 15 kg/min of R 12 at 1000 kPa and 60°C and condenses it to a saturated liquid at constant pressure. This is accomplished by using air available at 30°C with a flow rate of 2 kg/s. Determine the air exit temperature from the heat exchanger.

6.68 Air is heated as it flows steadily through a constant-area duct of 175 cm². It enters at 200 kPa, 300°K, and a velocity of 90 m/s and leaves at 150 kPa and 360 m/s. Determine the heat transfer required in kW.

6.69 An adiabatic counterflow heat exchanger receives 0.3 m³/s of R 12 at 2000 kPa and 90°C and discharges it as a saturated liquid at 2000 kPa. Water enters at 25°C and leaves at 40°C. Determine (a) the water flow rate in kg/s; (b) the heat transfer from the refrigerant to the water in kW.

6.70 The cooling unit of an air conditioning system is a heat exchanger that passes air over coils that have refrigerant flowing through them. In one home 50 m³/min of air enters the heat exchanger at 40°C and atmospheric pressure. It leaves at 20°C and atmospheric pressure. The cooling is accomplished by R 12 evaporating at a temperature of 10°C and an initial quality of 50% to a saturated vapor. Determine the refrigerant flow rate and the heat transfer between the air and refrigerant.

6.71 A solar collector has a surface area of 4 m². It receives solar radiation with an intensity of 1.3 kW/m². Of the incident radiation 65% is retained; the rest is lost from the collector. The energy that is retained is used to heat water from 40° to 60°C. What is the maximum flow rate of water that can be heated between these temperature limits in a 30-min interval?

6.72 Often a power-generating facility has superheated steam available, but a process needs saturated steam; a desuperheater can combine superheated steam with water to produce saturated steam. In the desuperheater 12 kg/s of steam at 10 MPa and 500°C is desuperheated to dry saturated steam at 10 MPa with water at 10 MPa and 100°C. What flow rate of water is required, assuming the desuperheater is adiabatic?

6.73 A direct-contact heat exchanger operates by combining 20 kg/s of water at 5 MPa and 100°C with saturated steam at 5 MPa to produce a saturated liquid at 5 MPa. Determine the total mass flow rate leaving the heat exchanger.

6.74 The steam entering a steam turbine flows through a control valve that regulates the flow. The steam entering the valve has a pressure of 20 MPa and a temperature of 500°C. The pressure downstream of the valve is 15 MPa. What is the steam temperature downstream of the valve?

6.75 The thermostatic expansion valve in a refrigeration system receives R 12 at 2.0 MPa and 40°C and discharges it at 308 kPa. What is the quality of the R 12 leaving the valve, assuming the device is adiabatic?

6.76 Steam enters a valve with a quality of 98% and an unknown pressure. The pressure of the steam leaving the adiabatic valve is 150 kPa, and the steam is a dry saturated vapor. What is the inlet pressure?

6.77 An adiabatic pressure-reducing valve has equal inlet and exit diameters of 4 cm, receives 5 kg/s of saturated steam at 10 MPa, and reduces the pressure to 1000 kPa. Determine the exit velocity and temperature of the steam leaving the valve.

6.78 An axial-flow compressor receives 2.0 m³/s of air at 100 kPa and 289°K and compresses it to 550 kPa and 520°K. The air leaves the compressor and enters an aftercooler, where the temperature decreases to 350°K. Determine the compressor power and heat removed in the aftercooler in kW.

6.79 A gas turbine compressor unit receives 32.6 m³/min of air at 98 kPa and 295°K and compresses it in a reversible adiabatic process to 450 kPa. Determine the power required to do this.

6.80 A simplified gas turbine unit may be visualized as an adiabatic compressor receiving intake air and discharging it at a higher pressure to a heat exchanger. In the heat exchanger heat is added to the air, raising its temperature at constant pressure and discharging the air to an adiabatic turbine. The air expands through the turbine, doing work. Some of this work returns to drive the compressor, and the remainder is the net work produced. A particular gas turbine unit operating on this cycle receives 0.7 m³/s of air at 100 kPa and 300°K and compresses the air at 500 kPa and 500°K. The heat exchanger raises the air's temperature to 800°K, and the turbine discharges the air at 100 kPa and 550°K. Determine (a) the power consumed in the compressor; (b) the total power produced by the turbine; (c) the net power available; (d) the total heat added.

6.81 Dry saturated steam is produced by heating a liquid-vapor mixture in a 50-liter tank. The mixture with an initial quality of 25% is maintained at constant pressure by a pressure regulator attached to the discharge line from the tank. The pressure is 2 MPa. Determine (a) the heat required to evaporate all the liquid; (b) the mass of steam that leaves the tank.

6.82 A 0.1-m³ adiabatic tank contains air at 2500 kPa and 100°C. The tank develops a small leak and air escapes. Determine (a) the mass remaining when the pressure is 250 kPa; (b) the temperature of the remaining air.

6.83 An initially evacuated 250-m³ adiabatic tank is charged with dry saturated steam at 5000 kPa until the pressure in the tank is 5000 kPa. Determine the mass of steam in the tank and its temperature when the pressure is 5000 kPa.

6.84 A room is heated by a portable heater that has a fan blowing air over heating coils. The fan will circulate 1 kg/s of air at a pressure of 100 kPa. The air in the 10 m × 4 m × 8 m room is initially at 15°C and is heated to 27°C in 20 min. The room loses 5 kW to the surroundings during this time. Determine (a) the power rating of the heater; (b) the temperature rise of the air as it passes through the heater.

PROBLEMS (English Units)

*6.1 A constant-pressure, insulated closed system receives paddle work. The pressure is 100 psia, and the paddle turns 10,000 revolutions with an average torque of 0.5 ft-lbf. The piston moves 0.2 ft³. Find the change of internal energy of the fluid in Btu.

*6.2 A four-process cycle has the following states for an unknown gas: $T_1 = 530°R$, $p_1 = 14$ psia, $V_1 = 0.2$ ft³; process 1-2, reversible adiabatic, $p_2 = 352$ psia, $V_2 = 0.02$ ft³, $T_2 = 1330°R$; process 2-3, constant pressure, $V_3 = 3 V_2$, $T_3 = 3990°R$; process 3-4, reversible adiabatic, $p_4 = 65$ psia, $T_4 = 2460°R$; process 4-1, constant volume. Additionally, $U_1 = 1.39$, $U_2 = 3.48$, $U_3 = 10.45$, $U_4 = 6.44$ Btu. Determine the work and heat for each process and the net work and net heat for the cycle.

*6.3 One lbm of air is compressed at a constant temperature of 75°F from 15 to 100 psia. Determine (a) the change of internal energy; (b) the work in ft-lbf; (c) the heat in Btu.

*6.4 A rigid adiabatic cylinder is separated into two compartments, one containing 2 lbm of air at 70 psia and 100°F, and the other 1 lbm of air at 15 psia and 200°F. The division between the compartments is removed. Determine the equilibrium temperature and pressure.

*6.5 A wet steam mixture is throttled so the final state is 14.7 psia and 250°F. Determine the initial quality if the initial pressure is 450 psia.

*6.6 Refrigerant 12 undergoes a constant-temperature process from 125 psia to 50 psia. The initial specific volume is 0.35 ft^3/lbm. Determine (a) the final specific volume; (b) the final internal energy.

*6.7 A steel tank has a volume of 0.6 ft^3 and is filled with saturated steam at 14.7 psia. The tank is cooled to 100°F. Determine the final pressure and heat transfer.

*6.8 A constant-volume container holds 3 lbm of dry saturated steam at 60 psia. 150 Btu of heat is added, and paddle work is done on the steam until the pressure is 100 psia. Calculate the paddle work.

*6.9 Determine the specific volume of water vapor at 760°F and 1000 psia using steam tables, the ideal-gas law, and the compressibility factor.

*6.10 A tank contains 2 lbm of air at 50 psia and 140°F. A valve is opened, and air leaves the tank until the pressure is 25 psia. Determine the heat required to keep the temperature constant.

*6.11 Ethylene is compressed according to $pV^{1.15} = C$ from 200 psia and 900°R to 450 psia. The mass of ethylene is 3 lbm. Determine the final temperature and the work and heat for the process.

*6.12 Methane undergoes a constant-pressure process from 850 psia and 600°F to 300°F. Determine the heat transfer and the work done per unit mass.

*6.13 Acetylene is compressed according to $pV^n = C$ from an initial pressure of 30 psia and 77°F to a final pressure of 200 psia. Determine the work and heat for $n = 1, 1.1, 1.232,$ 1.3, and 1.4. What conclusions can you draw?

*6.14 Air, initially at 20 psia and 100°F, occupies 6.0 ft^3. It is compressed isothermally until the volume is halved and then compressed at constant pressure until the volume decreases to one-quarter the initial volume. Sketch the processes on a p-V diagram and determine the total heat and total work for the two processes.

*6.15 A system holds 2 lbm of neon in a piston-cylinder where the initial pressure is 25 psia and the initial temperature is 80°F. The system operates on a three-process cycle comprised of the following processes: from state 1 to state 2 constant-volume heating until the pressure is 100 psia; from state 2 to state 3 expansion according to $pV^{1.67} = C$; from state 3 to 1 constant-pressure compression. Sketch the cycle on a p-V diagram and determine the net work and heat added.

*6.16 Air occupies a 15-ft^3 piston-cylinder at 20 psia and 40°F. The piston-cylinder operates on a three-process cycle where from state 1 to state 2 the pressure is constant and $V_2 = 1.4 V_1$. From state 2 to 3 the compression is reversible adiabatic until $V_3 = V_1$ and the temperature at state 3 is 350°F. A constant-volume process completes the cycle. Sketch the cycle on a p-V diagram and determine the net work and heat flows for the cycle.

*6.17 A cycle operates on 1 lbm of water vapor initially at 150 psia and cooled at a constant volume, $V_1 = V_2 = 6$ ft^3. The temperature at state 2 is 300°F. The system expands isothermally from state 2 to state 3 until the volume is 18 ft³. Determine the specific volumes and internal energies at each state and the heat and work for each process in the cycle.

*6.18 A piston-cylinder containing 3.0 lbm of water vapor operates on a three-process cycle. At state 1 the water is a saturated vapor at 1500 psia, which expands adiabatically to a pressure of 150 psia and a quality of 0.78. At this point a constant-pressure process occurs until the specific volume at state 3 equals that at state 1. Finally, from state 3 to state 1 constant-volume heating occurs. Sketch the cycle on a T-v diagram. Determine the work and heat for each process as well as the net work for the cycle.

*6.19 A four-process cycle using 2 lbm of water operates with the following processes: from state 1 to state 2 constant-volume heating from 80 psia and 320°F to 400 psia; from state 2 to 3 constant-pressure cooling until the water is a saturated vapor; from state 3 to 4 constant-volume cooling; from state 4 to 1 isothermal expansion where $Q_{4-1} = 1395$ Btu. Sketch the cycle on the T-v diagram. Determine the specific volume and internal energy at each state and the net work and total heat added for the cycle.

*6.20 An adiabatic steam turbine receives 1500 lbm/min of steam at 800 psia and 900°F. The steam expands through the turbine to a pressure of 160 psia and a temperature of 550°F where 300 lbm/min is removed. The remainder expands to 1 psia and a quality of 90%. Determine the power produced by the turbine.

*6.21 An air compressor receives 100 ft³/min of air at 14.7 psia and 80°F and compresses it to a pressure of 100 psia in a reversible adiabatic process. Heat transfer occurs, and the air's temperature is reduced by 70°F before it leaves the compressor. Determine the power required to compress the air and the total heat transfer in the aftercooler. What circulating water flow rate is required in the aftercooler if the temperature rise is from 70° to 90°F?

*6.22 An adiabatic compressor receives 3000 ft³/min of air at 80°F and 14.6 psia. The discharge pressure is 75 psia, and the power supplied is 425 hp. What is the discharge temperature?

*6.23 A refrigeration compressor receives R 12 at 20 psia and 20°F and discharges it at 150 psia and 140°F. The refrigerant flow rate at inlet conditions is 20 ft³/min. The heat transfer from the compressor to the surroundings is 2.5 Btu/lbm of refrigerant. Determine (a) the power required; (b) the volume flow rate at exit conditions.

*6.24 A compressor receives carbon dioxide gas at 21 psia with a specific volume of 5.7 ft³/lbm and compresses it to a temperature of 125°F. The work per unit mass for compression is 34 Btu/lbm. The gas enters through a 6-in.-diameter line with a velocity of 30 ft/sec and leaves with a velocity of 75 ft/sec. Determine the heat transfer in Btu/min.

*6.25 A hydroelectric power plant has the hydraulic turbine located 200 ft below the elevation of the intake pipe. The 3-ft-diameter pipe delivers water with a velocity of 10 ft/sec and a temperature 60°F to the turbine. The heat transfer between the turbine and the surroundings is negligible. Determine the power developed in hp.

*6.26 A pump delivers 700 gal/min of water. The intake to the pump is 230 ft below the final discharge. The inlet and discharge pressure is essentially atmospheric, and the temperature of the water remains constant at 68°F during the process. Determine the power required by the pump.

*6.27 An oil-transfer pump uses 1 hp of power in transferring 13.75 lbm/sec of oil through a 1.5-in.-diameter pipe. The changes in kinetic and potential energies are essentially zero, and the process is adiabatic. Additionally, there is no appreciable temperature change in the oil, which has a density of 93 lbm/ft³. Determine the change in pressure from inlet to exit.

*6.28 An adiabatic heat exchanger receives 33 lbm/min of R 12 at 150 psia and 200°F and condenses it to a saturated liquid at constant pressure. This is accomplished by using air available at 85°F with a flow rate of 275 lbm/min. Determine the air exit temperature from the heat exchanger.

*6.29 Air is heated as it flows steadily through a constant area duct of 25 in.². It enters at 20 psia, 80°F, and a velocity of 200 ft/sec and leaves at 16 psia and 800 ft/sec. Determine the heat transfer required in Btu/sec.

*6.30 An adiabatic counterflow heat exchanger receives 624 ft³/min of R 12 at 300 psia and 220°F and discharges it as a saturated liquid at 300 psia. Water enters at 80°F and leaves at 100°F. Determine (a) the water flow rate in lbm/min; (b) the heat transfer from the refrigerant to the water in Btu/min.

*6.31 The cooling unit of an air conditioning system is a heat exchanger that passes air over coils that have refrigerant flowing through them. In one home 1700 ft³/min of air enters the heat exchanger at 100°F and atmospheric pressure. It leaves at 68°F and atmospheric pressure. The cooling is accomplished by R 12 evaporating at a temperature of 40°F and an initial quality of 50% to a saturated vapor. Determine (a) the refrigerant flow rate; (b) the heat transfer between the air and refrigerant.

*6.32 A solar collector has a surface area of 50 ft². It receives solar radiation with an intensity of 420 Btu/hr-ft². Of the incident radiation 65% is retained; the rest is lost from the collector. The energy that is retained is used to heat water from 100° to 150°F. What is the maximum flow rate of water that can be heated between these temperature limits in a 30-min interval?

*6.33 Often a power-generating facility has superheated steam available, but a process needs saturated steam; a desuperheater can combine superheated steam with water to produce saturated steam. In the desuperheater 1600 lbm/min of steam at 1500 psia and 1000°F is desuperheated to dry saturated steam at 1500 psia with water at 1500 psia and 200°F. What flow rate of water is required, assuming the desuperheater is adiabatic?

*6.34 A direct-contact heat exchanger operates by combining 2500 lbm/min of water at 700 psia and 200°F with saturated steam at 700 psia to produce a saturated liquid at 700 psia. Determine the total mass flow rate leaving the heat exchanger.

*6.35 The steam entering a steam turbine flows through a control valve that regulates the flow. The steam entering the valve has a pressure of 3000 psia and a temperature of 1000°F. The pressure downstream of the valve is 2000 psia. What is the steam temperature downstream of the valve?

*6.36 The thermostatic expansion valve in a refrigeration system receives R 12 at 300 psia and 100°F and discharges it at 35.7 psia. What is the quality of the R 12 leaving the valve, assuming the device is adiabatic?

*6.37 Steam enters a valve with a quality of 98% and an unknown pressure. The pressure of the steam leaving the adiabatic valve is 20 psia, and the steam is a dry saturated vapor. What is the inlet pressure?

*6.38 An adiabatic pressure-reducing valve has equal inlet and exit diameters of 2 in., receives 700 lbm/min of saturated steam at 1500 psia, and reduces the pressure to 150 psia. Determine the exit velocity and temperature of the steam leaving the valve.

*6.39 An adiabatic axial-flow compressor receives 4250 ft³/min of air at 14.6 psia and 65°F and compresses it to 80 psia and 435°F. The air leaves the compressor and enters an aftercooler, where the temperature decreases to 170°F. Determine (a) the compressor power in hp; (b) the heat removed in the aftercooler in Btu/min.

*6.40 A gas turbine compressor unit receives 1200 ft³/min of air at 14.5 psia and 70°F and compresses it in a reversible adiabatic process to 70 psia. Determine the power required to do this.

*6.41 A simplified gas turbine unit may be visualized as an adiabatic compressor receiving intake air and discharging it at a higher pressure to a heat exchanger. In the heat exchanger heat is added to the air, raising its temperature at constant pressure and discharging the air to an adiabatic turbine. The air expands through the turbine, doing work. Some of this work returns to drive the compressor, and the remainder is the net work produced. A particular gas turbine unit operating on this cycle receives 1500 ft³/min of air at 14.7 psia and 80°F and compresses the air at 75 psia and 430°F; the heat exchanger raises the air's temperature to 1000°F, and the turbine discharges the air at 14.7 psia and 480°F. Determine (a) the power consumed in the compressor; (b) the total power produced by the turbine; (c) the net power available; (d) the total heat added.

*6.42 Dry saturated steam is produced by heating a liquid-vapor mixture in a 2-ft³ tank. The mixture with an initial quality of 25% is maintained at constant pressure by a pressure regulator attached to the discharge line from the tank. The pressure is 300 psia. Determine (a) the heat required to evaporate all the liquid; (b) the mass of steam that leaves the tank.

*6.43 A 3.5-ft adiabatic tank contains air at 375 psia and 200°F. The tank develops a small leak, and air escapes. Determine (a) the mass remaining when the pressure is 35 psia; (b) the temperature of the remaining air.

*6.44 An initially evacuated 9000-ft³ adiabatic tank is charged with dry saturated steam at 800 psia until the pressure in the tank is 800 psia. Determine the mass of steam in the tank and its temperature when the pressure is 800 psia.

*6.45 A room is heated by a portable heater that has a fan blowing air over heating coils. The fan will circulate 130 lbm/min of air at a pressure of 14.7 psia. The air in the 30 ft × 13 ft × 25 ft room is initially at 60°F and is heated to 80°F in 20 min. The room loses 290 Btu/min to the surroundings during this time. Determine (a) the power rating of the heater; (b) the temperature rise of the air as it passes through the heater.

COMPUTER PROBLEMS

C6.1 Air at 100 kPa and 300°K is compressed in a piston-cylinder to 10% of its initial volume. Develop a spreadsheet or TK Solver model to compute the final temperature and pressure for a range of polytropic coefficients between 0 and 2.0. Plot the final pressure and temperature as a function of n, first with linear scales and then with log-linear scales.

C6.2 Air at 100 kPa and 300°K is compressed in a piston-cylinder in a reversible adiabatic manner. The compression ratio (V_1/V_2) varies from 1 to 10. Compute and plot the final pressure and temperature as a function of compression ratio, first with linear scales and then with log-log scales.

C6.3 Carbon dioxide is expanded in a turbine from 1000 psia and 540°R to 14.7 psia. The flow rate is 5 lb/sec. For a range of outlet temperatures from 100°R to 540°R, compute and plot the polytropic coefficient, the turbine work, and the heat transferred.

C6.4 A tank with a volume of 0.4 m³ contains saturated steam at 350 kPa. It is connected to a pipe with steam flowing at 1.4 MPa and 300°C. A valve isolating the tank from the line is opened, and steam flows into the tank until the tank pressure is 1.4 MPa. Using TK Solver, compute the final tank temperature and the mass of steam that flows into the tank.

7

The Second Law of Thermodynamics and the Carnot Cycle

The second law of thermodynamics is for many the most intriguing law of natural science: it can tell us whether an event may happen, but not that it will happen. Whereas most natural laws are expressed mathematically as equalities, the second law is expressed mathematically as an inequality. Chapters 7–9 each introduce new concepts and considerations, developing greater understanding of the second law of thermodynamics. In this chapter you will

- Relate the second law to energy value;
- Delineate in more detail the differences between reversible and irreversible processes;
- Develop greater understanding of the direction of processes;
- Investigate different expressions of the second law;
- Explore the Carnot cycle;
- Expand upon the second law through corollaries applicable for all reversible heat engines;
- Analyze the thermodynamic temperature scale.

7.1 INTRODUCTION AND OVERVIEW

One of the philosophically interesting and upsetting aspects of the second law is that it is not a conservation law, as in the case of energy and mass. A property related to second-law analysis, entropy, is not conserved, and thus an inequality enters the picture. We will be able to determine factors that affect the magnitude of this inequality through our understanding of the second law.

How can this understanding be useful to us in the practice of engineering? In several ways. It establishes an energy quality level, a value system for energy. Essentially, the more useful energy is in its ability to help us, the greater its value. Thus, energy at a high temperature is more valuable than energy at a low temperature. For instance, 1000 kJ of energy at 35°C will warm us, but 1000 kJ of energy in the ocean at 4°C will not. It is less valuable. Additionally, the second law will determine the direction of energy flow, from a higher energy value to a lesser energy value. Thus, energy flows from a high temperature to a low temperature. In addition to processes, when a device produces or uses work in a cyclic manner, the second law will establish the ideal limit. Very importantly, we will also be able to determine whether a system will change state spontaneously. For example, can a chemical reaction occur, and what will be its final equilibrium state? The second law is our guide.

7.2 THE SECOND LAW OF THERMODYNAMICS

The second law of thermodynamics may be written in several ways. Regardless of the terminology used, however, the purpose of the second law is to give a sense of direction to energy-transfer processes. Combining this with the first law gives us the information necessary to analyze energy-transfer processes. The second law of thermodynamics states:

> Whenever energy is transferred, the value of energy cannot be conserved and some energy must be permanently reduced to a lower value.

When this is combined with the first law of thermodynamics, the law of energy conservation, the following results:

> Whenever energy is transferred, energy must be conserved, but the value of energy cannot be conserved and some energy must be permanently reduced to a lower value.

We understand the conservation of energy, but what is the energy value? What follows is a wholly qualitative description of the combined first and second laws of thermodynamics. It is intended to give a feeling and a sense of appreciation for the laws and their implications.

Let us consider two identical blocks of material with different temperatures. Let the temperature of one block of 1000°K and that of the other be 500°K. The block at

1000°K has a higher thermal energy value than the one at 500°K. The internal energy of the blocks is

$$U_1 = m_1 c T_1 \text{ kJ}$$

$$U_2 = m_2 c T_2 \text{ kJ}$$

If these blocks are brought together and reach thermal equilibrium at temperature T_3, the energy of the two blocks is

$$U_3 = (m_1 + m_2) c T_3$$

However, by the first law,

$$U_3 = U_1 + U_2$$

$$(m_1 + m_2) c T_3 = m_1 c T_1 + m_2 c T_2$$

Let

$$m_1 = m_2$$

$$T_3 = \frac{T_1 + T_2}{2} = 750°K$$

Thus the final temperature is less than one of the initial temperatures. We assume the energy value to be proportional to temperature, so we see that there has been an energy-value decrease. There has also been an energy-value increase, but the second law states only that there will be a decrease. The energy value finally is less than the highest energy value initially. Note that the total energy of either block has nothing to do with the energy value. The mass of block 2 could be 10 times larger than that of block 1, but there will still be an energy-value decrease. Check this and you will find that $T_3 = 545.4°K$.

Actually, the energy value is more than the temperature level of a system. A block of material of a small mass at a distance above a zero datum would have a higher energy value than a block of large mass of the same material resting at the zero datum. In this case the energy value is indicated by the potential energy. The potential energy of the small mass is greater than that of the larger mass. The energy value is a function of all the thermodynamic forces in a system that will cause energy transfer from the system. The energy value of the system is lowered when the values of these thermodynamic forces within the system decrease. The direction of energy transport is from a high energy value to a low energy value. In a similar fashion imagine a tank containing air at a pressure greater than atmospheric. The air in the tank because of its pressure has a greater value than the surrounding air, a higher energy value.

What gives the energy in certain instances a greater value? The opportunity to obtain work from the energy as it goes from a higher value to a lower value. For instance, when the temperature of the system is greater than that of the surroundings, a heat engine producing work and operating on a cycle can be placed between the system and the surroundings. In the case of pressure differential between the system and surroundings, one could use an air turbine to convert part of the air's energy into work as it expanded to the surroundings' pressure. By controlling the lowering of the mass in a gravitational field, one could use the energy to rotate a shaft, lift another object, or turn a gear, thus doing work.

In Chapter 2 we noted that an equilibrium is associated with every intensive property; a system will spontaneously change to establish this equilibrium. For instance, if a tank contains air at a pressure higher than that of the surroundings and we open a valve connecting it to the atmosphere, air would immediately flow from the tank until the pressure was in equilibrium with the surroundings. The energy flow is from a higher value to a lower value; the potential driving the flow is a nonequilibrium condition sensed by the system. Second-law analysis allows us to determine what the conditions of equilibrium are for a system. By being cognizant of these conditions, we can determine the direction of energy flow as well as the maximum work that can be obtained from the energy flow.

Reversible and Irreversible Processes

When considering the maximum work that can be produced from a given energy flow, we must understand the concepts of reversibility and irreversibility. All actual processes are irreversible. Consider a system and its surroundings and imagine a process. The process is reversible only if both the system and the surroundings can be returned to their initial states. Let a block, the system, slide down an inclined plane. The block can be returned to the same position on the plane, but it takes work from the surroundings to accomplish this. Thus, the surroundings are not returned to their initial state. If friction occurs between the block and plane, the system and surroundings are slightly warmed; thus, the initial state cannot be returned to, and the process is irreversible.

Although actual processes are irreversible, it does not follow that studying reversible processes is pointless. To the contrary, understanding what needs to be accomplished to make processes more reversible has the potential to increase the work that can be derived from the irreversible processes. Consider a simple pendulum. Two irreversibilities prevent the pendulum from returning to its initial position once set in motion—friction at the pivot and friction of the pendulum moving through air. As the friction is reduced, the process becomes more reversible.

Consider a special case of internally reversible processes: situations in which the system undergoes reversible processes, but the surroundings do not. (If the process is internally and externally reversible, both the system and the surroundings undergo reversible processes.) Internally reversible processes are used in this text unless otherwise indicated. This allows us to model internal combustion engines, power plants, and a variety of thermodynamic systems quite simply and effectively.

7.3 THE SECOND LAW FOR A CYCLE

The first law of thermodynamics gives us a technique for energy analysis, but it does not describe how the energy will flow. The second law of thermodynamics gives direction to the energy flow.

Carnot first observed that in steam engines no work could be produced unless there was heat flow from a high-temperature reservoir to a low-temperature reservoir.

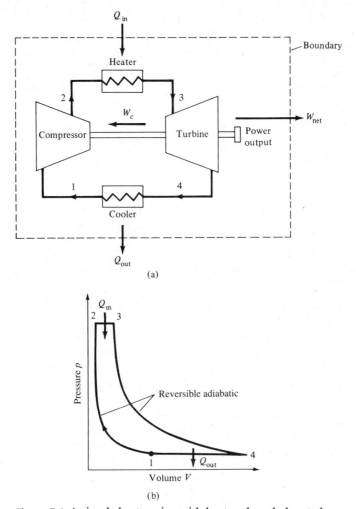

Figure 7.1 A simple heat engine with heat and work denoted.

He further noted that the work produced was a function of the temperature difference between the reservoirs and was greater for a greater temperature difference.

Let us consider the simple heat engine in Figure 7.1 (a), which models a closed gas turbine cycle. In this figure heat, Q_{in}, is added to a high-pressure gas in the high-temperature heat reservoir, the heater. The gas expands through the turbine, producing work. Some of the work, W_c, is used to operate the compressor, and the remainder, W_{net}, can be used for useful purposes, for example, driving an electric generator. The gas exits the turbine at a low pressure and at a lesser temperature than state 3, but higher than state 1. A low-temperature heat exchanger removes heat, Q_{out}, reducing the gas's temperature to that of state 1, and thus the cycle is completed.

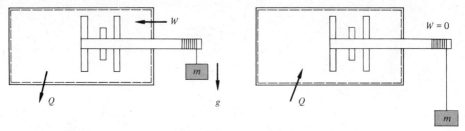

Figure 7.2 Demonstration of the effect of the second law.

The Kelvin-Planck statement of the second law is

> No cyclic process is possible whose sole result is the flow of heat from a single heat reservoir and the performance of an equivalent amount of work.

In relating this statement of the second law to the simple power cycle shown in Figure 7.1(a), could we eliminate the cooler from the system and maintain the cycle? No. Inspection of Figure 7.1(b), the p-V diagram for the cycle, indicates there must be heat rejection for the gas to go from state 4 to state 1. It is impossible to complete the cycle without this process.

Consider a constant-volume container that receives paddle work, as shown in Figure 7.2(a). The substance will increase in temperature, but cooling will return the system to its initial condition. Let us reverse the situation and add heat to the system. Can the weight be raised? No. This is analogous to trying to produce work with only a single heat reservoir, which violates the second law. Notice that all the work can be converted into heat, but not all the heat can be converted into work.

Thus, the second law tells us that not all heat may be converted into work, but it does not tell us the maximum amount of work that can be produced for heat flow between two temperatures. That is the purpose of the next section. The concept of the second law will be more completely understood after the next chapters, which discuss entropy and available energy.

7.4 CARNOT CYCLE

A thermodynamic cycle occurs when a system undergoes two or more processes and returns to its initial state. Many machines are analyzed using cyclic investigation.

The first person to recognize the energy-transfer processes in machines was Sadi Carnot. A retired army corporal at the age of 24, he had the time and inclination to think about these transfer processes. In 1824 he published *Reflections on the Motive Power of Fire,* the document that established the basis for the second law of thermodynamics and introduced the concepts of "cycle" and "reversible processes." It remained for Lord Kelvin and Rudolf Clausius to amplify the principles in modern thermodynamic form.

Necessity is the mother not only of invention but also of discovery. The basis for many scientific discoveries has been a practical need to understand some physical phenomenon. Although the steam engine was becoming the motive force of the

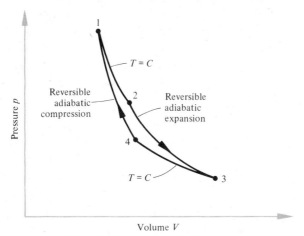

Figure 7.3 The *p-V* diagram for the Carnot cycle.

industrial revolution, no one had the understanding necessary to advance its design, and its manufacture proceeded on a hit-or-miss basis.

Carnot Engine

Carnot noted that the higher the temperature of the steam entering the engine and the lower the temperature of the steam leaving the engine, the greater the engine's work output. He devised a theoretical engine that would operate on a closed cycle—it would receive heat at a constant temperature and reject heat at a constant temperature. The boiling and condensing of water takes place at a nearly constant temperature. The engine would be perfectly insulated, the work would be done reversibly. Hence there would be a reversible adiabatic expansion by the engine to produce work, and to complete the cycle there would be a reversible adiabatic compression. Figure 7.3 illustrates the Carnot cycle on pressure-volume coordinates.

How does the Carnot engine produce work? We can see that in the *p-V* diagram a net area is equal to the net work produced, but this does not always give a physical appreciation of the work performed. If the engine could turn a shaft, as an internal-combustion engine turns a crankshaft, then a physical appreciation could be attached to the use of the *p-V* diagram.

Consider a Carnot heat engine, as depicted in Figure 7.4. This engine can provide work to turn the shaft as shown. Let us follow the engine through a cycle and compare the piston movement to the points in Figure 7.3. The cylinder arrangement starts with the fluid in the cylinder at state 1. A perfect conductor is connected between the engine and the heat source, and heat is transferred at constant temperature until the fluid reaches state 2, shown in Figure 7.4(b). A perfect insulator is now placed between the heat reservoirs, and the engine and the piston expand in a reversible adiabatic process until state 3 is reached, shown in Figure 7.4(c). A perfect conductor is now inserted between the engine and the heat sink, and heat is rejected from the engine to the heat sink at constant temperature. This brings the engine to state 4,

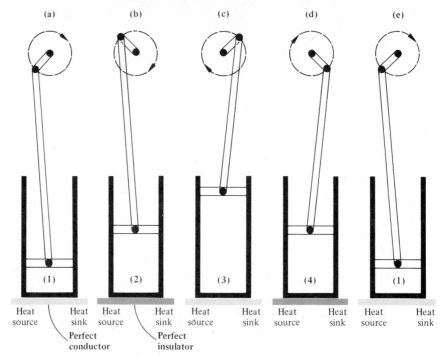

Figure 7.4 A visualization of the Carnot engine.

shown in Figure 7.4(d). In order to return the fluid to state 1, some of the energy developed as work in going from state 1 to state 3 must be returned to the engine, compressing the fluid from state 4 to state 1. This compression must be reversible adiabatic, so a perfect insulator is inserted between the engine and the heat reservoirs, and the fluid is compressed to state 1, shown in Figure 7.4(e). The cycle is completed as the engine returns to its starting point, and work has been produced.

The Carnot engine is important because it converts thermal energy (heat) into the maximum amount of mechanical energy (work). No other engine or device works more efficiently between two temperature reservoirs, although it may be possible to develop engines with the same efficiency as the Carnot engine. It is desirable to have an expression for the thermal efficiency of the engine and to see what factors influence the efficiency. Later we will be able to compare other engines to the Carnot engine and determine the relative merit of the particular engine.

Carnot Engine Efficiency

The thermal efficiency, η_{th}, of any engine is the net work produced divided by the energy supplied to produce that work; that is,

$$\eta_{th} = \frac{W_{net}}{Q_{in}} \tag{7.1}$$

The net work can be evaluated by using a cyclic integral. Ideal-gas relationships will be used for this analysis, but in Chapter 8 it will be shown that the results are valid for any substance as the working fluid.

$$W_{net} = \oint \delta W = \oint p \, dV = \int_1^2 p \, dV + \int_2^3 p \, dV + \int_3^4 p \, dV + \int_4^1 p \, dV$$

$$\int_1^2 p \, dV = p_1 V_1 \ln\left(\frac{V_2}{V_1}\right) = mRT_H \ln\left(\frac{V_2}{V_1}\right)$$

$$\int_2^3 p \, dV = \frac{p_3 V_3 - p_2 V_2}{1-k} = \frac{mRT_C - mRT_H}{1-k} \tag{7.2}$$

$$\int_3^4 p \, dV = p_3 V_3 \ln\left(\frac{V_4}{V_3}\right) = mRT_C \ln\left(\frac{V_4}{V_3}\right)$$

$$\int_4^1 p \, dV = \frac{p_1 V_1 - p_4 V_4}{1-k} = \frac{mRT_H - mRT_C}{1-k}$$

where T_H is the temperature of the high-temperature heat source and T_C is the temperature of the low-temperature heat sink.

If these terms are added, the net work reduces to

$$W_{net} = mRT_H \ln\left(\frac{V_2}{V_1}\right) + mRT_C \ln\left(\frac{V_4}{V_3}\right) \tag{7.3}$$

The expression for the heat supplied must be deduced. Heat is supplied at constant temperature from state 1 to state 2. This is a closed system, so the first law of thermodynamics tells us

$$Q_{1-2} = U_2 - U_1 + W_{1-2}$$

$$U_2 - U_1 = 0 \quad \text{since } T = C$$

$$Q_{1-2} = W_{1-2} = mRT_H \ln\left(\frac{V_2}{V_1}\right) \tag{7.4}$$

The heat rejected may be found in a similar manner to be

$$Q_{3-4} = mRT_C \ln\left(\frac{V_4}{V_3}\right) \tag{7.5}$$

An important conclusion regarding theoretical heat engines is demonstrated when equations (7.5), (7.4), and (7.3) are compared. That is,

$$W_{net} = \Sigma Q = Q_{in} + Q_{out} \tag{7.6}$$

Substituting equations (7.3) and (7.4) into equation (7.1) yields

$$\eta_{th} = \frac{mRT_H \ln (V_2/V_1) + mRT_C \ln (V_4/V_3)}{mRT_H \ln (V_2/V_1)} \tag{7.7}$$

Using the fact that processes 2-3 and 4-1 are reversible adiabatic processes,

$$\frac{T_2}{T_3} = \left(\frac{V_3}{V_2}\right)^{k-1} = \frac{T_H}{T_C} \qquad \frac{T_1}{T_4} = \left(\frac{V_4}{V_1}\right)^{k-1} = \frac{T_H}{T_C}$$

$$\frac{V_2}{V_1} = \frac{V_3}{V_4}$$

Equation (7.7) may now be written

$$\eta_{th} = \frac{mRT_H \ln (V_2/V_1) - mRT_C \ln (V_2/V_1)}{mRT_H \ln (V_2/V_1)}$$

$$\eta_{th} = \frac{T_H - T_C}{T_H} \tag{7.8}$$

Thus, we discover that the thermal efficiency of a Carnot engine with ideal gas as the working fluid is dependent only on the absolute temperature of the heat supplied and the heat rejected. The greater the difference between the two, the more efficient the engine. In this development the heat was transferred at constant temperature. This is ideal; usually a temperature difference exists between a heat reservoir and an actual heat engine.

7.5 MEAN EFFECTIVE PRESSURE

The work for the Carnot engine may be evaluated by equation (7.3). However, it is sometimes convenient to have another means of evaluating an engine. One such indicator is the mean effective pressure, an imaginary pressure developed by equating the cycle work to an equivalent work:

$$W_{net} = \int_1^3 p \, dV = p_m \int_1^3 dV \tag{7.9}$$

$$W_{net} = p_m \times \text{piston displacement volume} \tag{7.10}$$

$$p_m = \frac{W_{net}}{\text{piston displacement volume}}$$

The mean effective pressure, p_m, is equal to the net cycle work divided by the piston displacement volume. Figure 7.5 illustrates the p-V diagram for the Carnot engine. The shaded rectangular area is the same area as that enclosed by the Carnot engine, but it is characterized by a mean effective pressure. The higher the mean effective pressure is, the greater will be the force per cycle and the work output per cycle.

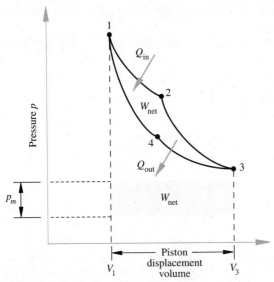

Figure 7.5 A graphical illustration of mean effective pressure p_m.

Example 7.1

A heat engine operates on the Carnot cycle. It produces 50 kW of power while operating between temperature limits of 800°C and 100°C. Determine the engine efficiency and the amount of heat added.

Solution

Given: A heat engine operating on the Carnot cycle, its net work, and the temperature limits of the cycle.

Find: The efficiency and the heat added.

Sketch and Given Data:

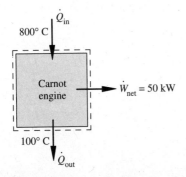

Figure 7.6

Assumption: The system is shown on the sketch and follows the Carnot cycle.

Analysis: The efficiency of the Carnot cycle can be expressed in terms of absolute temperature:

$$\eta_{th} = \frac{T_H - T_C}{T_H} = \frac{1073 - 373}{1073} = 0.652$$

Knowing the efficiency and the power produced, the heat added may be determined.

$$\dot{Q}_{in} = \frac{\dot{W}_{net}}{\eta_{th}} = \frac{50}{0.652} = 76.6 \text{ kW}$$

Comments:

1. Absolute temperatures must be used in the efficiency equation.
2. The heat and work in the efficiency equation may be on a unit mass basis, on a total energy basis, or in terms of power. The units of heat and work must be the same. ▬

Example 7.2

A six-cylinder engine with a 4 × 4 – in. bore and stroke operates on the Carnot cycle. It receives 51 Btu/cycle of heat at 1040°F and rejects heat at 540°F while running at 300 rpm. Determine the mean effective pressure, hp, and heat flow out of the engine.

Solution

Given: An engine operating on the Carnot cycle between fixed temperature limits and receiving a certain amount of heat.

Find: The mean effective pressure and power produced.

Sketch and Given Data:

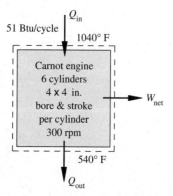

Figure 7.7

Assumption: The system is shown on the sketch and operates on the Carnot cycle.

Analysis: The temperature limits and heat added per cycle allow us to calculate the work produced per cycle. Knowing the work and information about the engine displacement volume allows calculation of the mean effective pressure. The heat rejected is the difference between the heat added and work produced per cycle. The efficiency for this engine is

$$\eta_{th} = \frac{T_H - T_C}{T_H} = \frac{1500 - 1000}{1500} = 0.333$$

The work per cycle may be determined from the efficiency to be

$$W_{net} = (0.333)Q_{in} = (0.333)(51 \text{ Btu/cycle}) = 17 \text{ Btu/cycle}$$

The total piston displacement volume is equal to the displacement of each cylinder multiplied by the number of cylinders.

$$V_{PD} = (6 \text{ cylinders}) \left(\frac{\pi}{4}\right) \left(\frac{4 \text{ in.}}{12 \text{ in./ft}}\right)^2 \left(\frac{4 \text{ in.}}{12 \text{ in./ft}}\right) = 0.1745 \text{ ft}^3$$

The mean effective pressure is

$$p_m = \frac{W_{net}}{V_{PD}} = \frac{(17 \text{ Btu})(778.16 \text{ ft-lbf/Btu})}{(0.1745 \text{ ft}^3)(144 \text{ in.}^2/\text{ft}^2)}$$

$$p_m = 526 \text{ lbf/in.}^2$$

The power produced is found by multiplying the net work per cycle by the number of cycles per minute.

$$\dot{W}_{net} = (17 \text{ Btu/cycle}) (300 \text{ cycle/min})(0.02358 \text{ min-hp/Btu})$$

$$\dot{W}_{net} = 120.5 \text{ hp}$$

The heat flow from the engine is found in an analogous manner. The heat rejected per cycle is

$$Q_{out} = W_{net} - Q_{in} = 17 - 51 = -34 \text{ Btu}$$

The heat flow rate out is

$$\dot{Q}_{out} = (-34 \text{ Btu/cycle})(300 \text{ cycles/min}) = -10,200 \text{ Btu/min}$$

Comments:

1. The net work of 17 Btu/cycle represents the total work of all six cylinders completed in one cycle or revolution of the engine.
2. Conversion between ft-lbf and Btu is important.
3. The heat rejected represents the energy that would be dissipated to the atmosphere by an actual automobile's radiator and exhaust.   ▬

Example 7.3

A Carnot engine uses 0.05 kg of air as the working substance. The temperature limits of the cycle are 300°K and 940°K, the maximum pressure is 8.4 MPa, and the heat added per cycle is 4.2 kJ. Determine the temperature, pressure, and volume at each state of the cycle.

Solution

Given: A Carnot engine, the mass of air it operates on, temperature limits, heat added, and maximum pressure.

Find: The pressure, temperature, and volume at each state point around the cycle.

Sketch and Given Data:

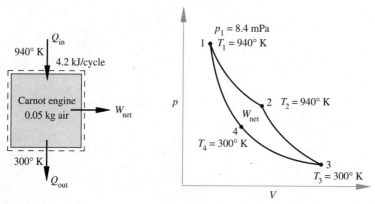

Figure 7.8

Assumption: Air is an ideal gas.

Analysis: Determine the cycle state points by using the ideal-gas equations of state and the expressions for the processes forming the cycle.

State 1:

$$T_1 = 940°K \qquad p_1 = 8.4 \text{ MPa} \qquad V_1 = \frac{mRT_1}{p_1}$$

$$V_1 = (0.05 \text{ kg})(0.287 \text{ kJ/kg-K})(940°K)/(8400 \text{ kN/m}^2) = 0.001\ 606 \text{ m}^3$$

State 2: The process from state 1 to state 2 is constant-temperature heat addition; thus,

$$Q_{1\text{-}2} = p_1 V_1 \ln \{V_2/V_1\}$$

$$4.2 \text{ kJ} = (8400 \text{ kN/m}^2)(0.001\ 606 \text{ m}^3)(\ln \{V_2/V_1\})$$

$$\ln \{V_2/V_1\} = 0.3113 \qquad V_2 = 0.002\ 193 \text{ m}^3$$

$$p_2 = \frac{mRT_2}{V_2} = (0.05 \text{ kg})(0.287 \text{ kJ/kg-K})(940°\text{K})/(0.002\ 193 \text{ m}^3)$$

$$p_2 = 6150 \text{ kPa} \qquad T_2 = 940°\text{K}$$

State 3: Using the relationships for a reversible adiabatic process from state 2 to state 3, the values of pressure and volume at state 3 may be determined.

$$\frac{V_3}{V_2} = \left(\frac{T_2}{T_3}\right)^{1/(k-1)} = \left(\frac{940}{300}\right)^{2.5}$$

$$V_3 = 0.038\ 11 \text{ m}^3$$

Similarly for pressure

$$\frac{p_3}{p_2} = \left(\frac{T_3}{T_2}\right)^{k(k-1)} = \left(\frac{300}{940}\right)^{3.5}$$

$$p_3 = 112.9 \text{ kPa}$$

State 4: The heat out may be calculated from the thermal efficiency. Knowing this and the process, constant-temperature heat rejection, the values at state 4 may be determined. However, it is also possible to use the process relationships from state 1 to state 4 in a manner analogous to going from state 2 to state 3.

$$\left(\frac{V_4}{V_1}\right) = \left(\frac{T_1}{T_4}\right)^{1/(k-1)} = \left(\frac{940}{300}\right)^{2.5}$$

$$V_4 = 0.027\ 91 \text{ m}^3$$

$$p_4 = \frac{mRT_4}{V_4} = (0.05 \text{ kg})(0.287 \text{ kJ/kg-K})(300°\text{K})/(0.027\ 91 \text{ m}^3)$$

$$p_4 = 154.2 \text{ kPa}$$

Comment: One need not proceed in one direction around the cycle to find the cycle state points.

Example 7.4
A heat engine operating on the Carnot cycle uses solar energy as the source of high-temperature heat input. The solar irradiation, averaged over the day, has a value of 0.51 kW/m². This provides energy to the cycle at a uniform temperature of 450°K, and the cycle rejects heat to the environment at a temperature of 300°K. The engine produces 2000 kW of power. Determine the minimum area needed to provide this power.

Solution

Given: A Carnot engine receiving heat from solar energy, operating between fixed temperature limits, and producing power.

Find: The land area required for the collector to supply the heat for the power required.

Sketch and Given Data:

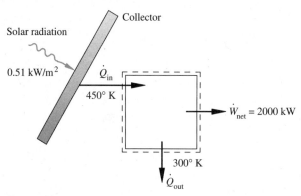

Figure 7.9

Assumption: No energy is lost from the collector.

Analysis: The heat flux in at any moment is found from the thermal efficiency of the cycle and the net power required from the cycle. Thus,

$$\eta_{th} = 1 - \frac{T_C}{T_H} = 1 - \frac{300}{450} = 0.333$$

$$0.333 = \frac{\dot{W}_{net}}{\dot{Q}_{in}} = \frac{2000 \text{ kW}}{\dot{Q}_{in}}$$

$$\dot{Q}_{in} = 6000 \text{ kW}$$

This heat flux comes from the solar radiation incident on the collector. An energy balance on the collector yields

$$6000 \text{ kW} = (0.51 \text{ kW/m}^2)(A \text{ m}^2)$$

$$A = 11\ 765 \text{ m}^2 \qquad \text{or about 2.9 acres}$$

Comments:

1. Collectors with large areas are required to generate power using solar energy.
2. The given average value of incident solar radiation was for when the sun was shining. During periods of darkness no power can be generated.

7.6 REVERSED CARNOT ENGINE

The Carnot engine is a power-producing engine that uses heat supplied as an energy source and delivers mechanical work as an energy output. When the Carnot cycle is reversed, mechanical work is supplied as an energy input and heat may be moved from one energy value (temperature) to another energy value. Why do this? Consider the manner in which refrigerators operate. Energy is supplied by an electric motor, driving a compressor, and the refrigerant (the working substance) picks up heat from inside the refrigerator at a low temperature and discharges it at a high temperature to the condensing coils on the outside of the refrigerator. We will discuss actual refrigeration systems later, but for now let us limit the discussion to the reversed Carnot cycle.

The reversed Carnot cycle has all the same processes as the power-producing Carnot cycle, but the cycle operates in the counterclockwise direction. The p-V diagram for the reversed Carnot engine is given in Figure 7.10.

The purpose of the reversed Carnot engine is to remove a quantity of heat at a low temperature Q_{in} by supplying work. This result in a rejection of heat at a high temperature Q_{out}. The performance of reversed engines is denoted by the coefficient of performance (COP), rather than an efficiency term. This has, however, the same purpose as efficiency—the desired effect divided by the cost of achieving the desired effect. For a reversed Carnot engine using an ideal gas, the working substance, the coefficient of performance is

$$(COP)_C = \frac{Q_{in}}{W_{net}} = \frac{mRT_C \ln (V_3/V_2)}{-mRT_C \ln (V_3/V_2) - mRT_H \ln (V_1/V_4)}$$

$$(COP)_C = \frac{T_C}{T_H - T_C} \tag{7.11a}$$

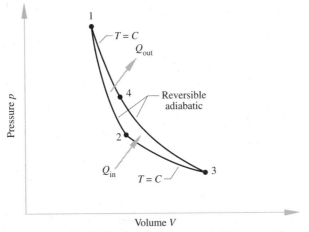

Figure 7.10 A p-V diagram for the reversed Carnot cycle.

where the net work has been written as a positive quantity. Thus, the coefficient of a performance of a reversed Carnot engine depends only on the absolute temperatures of the heat reservoirs. The coefficient of performance of most reversed cycles is greater than unity.

The reversed Carnot engine can also be used to provide heat to warm a space. The actual device is called a *heat pump*, where the low-temperature heat source is the surroundings, work is added, and the heat out of the cycle is discharged to the space at a higher temperature and used for heating. Thus, the desired effect is heat out. The coefficient of performance of a heat pump is

$$(\text{COP})_h = \frac{Q_{\text{out}}}{W_{\text{net}}} = \frac{T_H}{T_H - T_C} \tag{7.11b}$$

where the net work is written as a positive quantity.

Example 7.5

A home requires 5.0 kW to maintain an indoor temperature of 21°C when the outside temperature is 0°C. A heat pump is used to provide the energy. Determine the minimum power required.

Solution

Given: A heat pump providing energy for home heating with temperature limits and heat required noted.

Find: The minimum power required to operate the heat pump.

Sketch and Given Data:

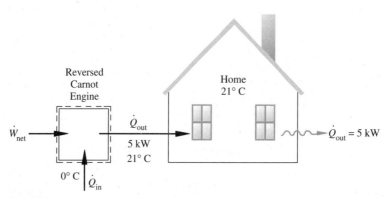

Figure 7.11

Assumption: The surroundings and the house are the low- and high-temperature heat reservoirs, respectively.

Analysis: The reversed Carnot cycle requires the least power to move heat from one temperature to another. The house requires 5.0 kW of heat out of the cycle to maintain itself at 21°C. If this heat flow differs from 5.0 kW, the house's temperature will increase or decrease. The reversed cycle is the system being analyzed; thus, the heat out of the cycle is the desired effect even though this becomes the heat into the house. The $(COP)_h$ for the reversed Carnot cycle is

$$(COP)_h = \frac{\dot{Q}_{out}}{\dot{W}_{net}} = \frac{T_H}{T_H - T_C}$$

$$(COP)_h = \frac{5 \text{ kW}}{\dot{W}_{net}} = \frac{294}{294 - 273} = 14$$

$$\dot{W}_{net} = 5/14 = 0.357 \text{ kW}$$

Comments:

1. Absolute temperatures must be used in the calculations.
2. The power required in an actual heat pump will be greater due to irreversibilities.

7.7 FIRST COROLLARY OF THE SECOND LAW

The second law gives a sense of direction to a process, whereas the first law does not. It also tells us that it is impossible to have a perpetual-motion machine of the second kind, one that violates the second law. The first of two corollaries to this law states:

> It is impossible to construct an engine to operate between two heat reservoirs, each having a fixed and uniform temperature, that will exceed the efficiency of a reversible engine operating between the same reservoirs.

Figure 7.12 illustrates such a situation. Note that the reversible engine is a Carnot engine, whereas the other engine may have irreversibilities associated with it.

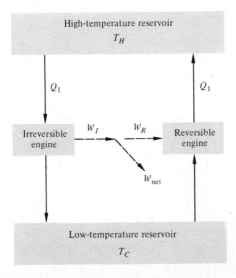

Figure 7.12 Two heat engines, one reversible and one irreversible, operating between the same temperature limits.

Assume that the irreversible engine has an efficiency greater than that of the reversible engine. Let W_I be the work produced by the irreversible engine, while receiving heat Q_1, and let W_R be the work produced by the reversible engine (Carnot) when receiving the same amount of heat Q_1. The assumption tells us that $W_I > W_R$.

Now run the reversible engine as a reversed Carnot engine so it will supply the amount of heat Q_1 to the high-temperature reservoir. It would be possible to run the reversible engine by supplying work from the irreversible engine and still have some net work remaining. This is a perpetual-motion machine of the second kind and cannot exist because it violates the second law of thermodynamics. Thus, the assumption that $W_I > W_R$ is not valid, and the first corollary must be valid.

7.8 SECOND COROLLARY OF THE SECOND LAW

The second corollary of the second law states:

> All reversible engines have the same efficiency when working between the same two constant-temperature heat reservoirs.

The proof of the proposition follows that of the first corollary. Assume that the proposition is not true in a particular case, and then show that a perpetual-motion machine of the second kind is a consequence. In this case the irreversible engine is replaced by a Carnot engine of higher efficiency than the other reversible engine.

7.9 THERMODYNAMIC TEMPERATURE SCALE

When discussing the zeroth law of thermodynamics, we discovered that temperature could be measured by observing a change in the magnitude of a property. Usually this is the volume change of mercury or another liquid in a thermometer. Temperature scales have been developed using various properties of different substances. The scales may be standardized at the boiling and ice points of water, but they will differ at other temperature levels. It is possible to develop a temperature scale that is independent of substance properties but that relies on a reversible engine.

We are at the point where we can discuss a thermodynamic temperature scale by considering reversible engines operating between constant-temperature heat reservoirs. Figure 7.13 illustrates this. Three reversible engines that produce work, W_{12}, W_{23}, and W_{13}, are placed between constant-temperature reservoirs at temperatures T_1, T_2, T_3, respectively. Engine 12 receives heat Q_1 at temperature T_1 and rejects heat Q_2 at temperature T_2 while producing work W_{12}. Engine 23 receives heat Q_2 at temperature T_2 and rejects heat Q_3 at temperature T_3 while producing work W_{23}. Engine 13 receives heat Q_1 at temperature T_1 and rejects heat Q_3 at temperature T_3 while producing work W_{13}. Heat Q_3 is the same for engine 23 and engine 13, as demonstrated by the first corollary.

The thermal efficiency of a reversible engine operating between two constant-temperature reservoirs has been shown to be a function of only the temperatures of the reservoirs. The thermal efficiency for engine 12, η_{12}, is then

$$\eta_{12} = f(T_1, T_2) \tag{7.12}$$

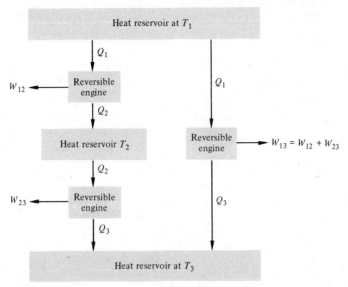

Figure 7.13 Reversible engines operating between two temperature reservoirs.

and the thermal efficiency is also

$$\eta_{12} = 1 - \frac{Q_2}{Q_1} \tag{7.13}$$

and hence a new functional relationship ϕ is shown to exist between the heat ratio and the temperatures of the heat reservoir by comparing equations (7.12) and (7.13).

$$\frac{Q_1}{Q_2} = \phi(T_1, T_2)$$

In the same manner, relationships for the other heat ratios and temperatures may be developed or deduced as

$$\frac{Q_2}{Q_3} = \phi(T_2, T_3)$$

$$\frac{Q_1}{Q_3} = \phi(T_1, T_3)$$

A relationship between the three heat ratios is

$$\frac{Q_1}{Q_2} = \frac{Q_1/Q_3}{Q_2/Q_3}$$

and substitution of the temperature equivalents yields

$$\phi(T_1, T_2) = \frac{\phi(T_1, T_3)}{\phi(T_2, T_3)}$$

Temperature T_3 must cancel from the right-hand side if the equation is to maintain a functional identity. This yields a new relationship

$$\phi(T_1, T_2) = \frac{\psi(T_1)}{\psi(T_2)} \tag{7.14}$$

where ψ denotes the functional relationship.

There are many functional relationships that will satisfy equation (7.14). The simplest, and the one used throughout this text, was proposed by Kelvin:

$$\psi(T) = T \tag{7.15}$$

Thus,

$$\frac{Q_1}{Q_2} = \frac{T_1}{T_2} \tag{7.16}$$

and equation (7.12) becomes

$$\eta_{12} = 1 - \frac{Q_2}{Q_1} = 1 - \frac{T_2}{T_1}$$

This is the Kelvin temperature scale with which the reader is familiar. The limits are from 0 to $+\infty$.

Kelvin proposed another functional relationship that would permit the lower temperature limit to be $-\infty$, rather than zero. To do this he introduced a logarithmic scale with temperature θ. The relationship between the scales may be shown as

$$\psi(\theta) = e^{\theta} \tag{7.17}$$

$$\frac{Q_1}{Q_2} = \frac{e^{\theta_1}}{e^{\theta_2}} \tag{7.18}$$

Equating equations (7.16) and (7.18) yields

$$\frac{e^{\theta_1}}{e^{\theta_2}} = \frac{T_1}{T_2}$$

$$\theta_1 - \theta_2 = \ln(T_1) - \ln(T_2)$$

and therefore

$$\theta = \ln(T) + C$$

where C is a constant that determines the level of temperature corresponding to zero on the logarithmic temperature scale.

The Kelvin temperature scale showed that it was possible to have an absolute-zero temperature that is independent of the properties of any substance. We must resort, in the final analysis, to a "thermometer" type instrument to measure temperature, as it is impossible to design a heat engine running on the Carnot cycle.

CONCEPT QUESTIONS

1. Is satisfying the first law of thermodynamics sufficient to know whether a process can occur?
2. A home refrigerator may be viewed as receiving work from the surroundings and rejecting heat to the surroundings. Does this violate the second law of thermodynamics?
3. What makes some energy more valuable than other energy?
4. What is a thermal (heat) energy reservoir?
5. Can a heat engine operate with only a high-temperature heat reservoir?
6. Explain the concept of thermal efficiency of a heat engine.
7. An inventor creates a heat engine with a thermal efficiency of 100%. Does this violate the first law of thermodynamics? The second law of thermodynamics? Why?
8. Would the thermal efficiency of a heat engine reach 100% if all irreversibilities were eliminated?
9. Electric heaters are advertised as having an efficiency of 100%. Is this possible? Explain.
10. Is there a conceptual difference between efficiency and coefficient of performance? Explain.
11. Why are there two coefficients of performance for reversed heat engines?
12. Why does an actual process require more work than a corresponding equilibrium process?
13. Describe the differences between internally, externally, and totally reversible processes.
14. A hot cup of coffee left on a table cools to room temperature. Is the process reversible? Why?
15. A can of cold soda left on a table in a warm room reaches room temperature. Is the process reversible? Why?
16. A house is heated by a heat pump. The owner turns down the house's thermostat. Will this affect the performance of the heat pump? Explain.
17. A refrigerator is cooled by a reversed Carnot engine. The thermostat is raised in the refrigerated space. Does this affect the performance of the reversed Carnot cycle? Explain.
18. Solar energy is to be the energy source for a power plant. Two plans are being considered. In one system the solar collectors gather energy at 100°C and in the other at 300°C. The total energy collected is the same in both cases. Is one likely to have a higher efficiency? Explain.

PROBLEMS (SI)

7.1 A Carnot engine operates with 0.136 kg of air as the working substance. The pressure and volume at the beginning of isothermal expansion are 2.1 MPa and 9.6 liters. The air behaves as an ideal gas, the sink temperature is 50°C, and the heat added is 32 kJ.

Determine (a) the source temperature; (b) the cycle efficiency; (c) the pressure at the end of isothermal expansion; (d) the heat rejected to the sink per cycle.

7.2 A Carnot engine produces 25 kW while operating between temperature limits of 1000° K and 300°K. Determine (a) the heat supplied per second; (b) the heat rejection per second.

7.3 A Carnot engine uses nitrogen as the working fluid. The heat supplied is 53 kJ, and the adiabatic expansion ratio is 16 : 1. The receiver temperature is 295°K. Determine (a) the thermal efficiency; (b) the heat rejected; (c) the work.

7.4 A Carnot engine uses air as the working substance, receives heat at a temperature of 315° C, and rejects it at 65°C. The maximum possible cycle pressure is 6.0 MPa, and the minimum volume is 0.95 liter. When heat is added, the volume increases by 250%. Determine the pressure and volume at each state in the cycle.

7.5 A Carnot engine operates between temperatures of 1000°K and 300°K. The engine operates at 2000 rpm and develops 200 kW. The total engine displacement is such that the mean effective pressure is 300 kPa. Determine (a) the cycle efficiency; (b) the heat supplied (kW); (c) the total engine displacement (m^3).

7.6 A Carnot cycle uses nitrogen as the working substance. The heat supplied is 54 kJ. The temperature of the heat rejected is 21°C, and $V_3/V_2 = 10$. Determine (a) the cycle efficiency; (b) the temperature of heat added; (c) the work.

7.7 Helium is used in a Carnot engine where the volumes beginning with the constant-temperature heat addition are $V_1 = 0.3565$ m^3, $V_2 = 0.5130$ m^3, $V_3 = 8.0$ m^3, and $V_4 = 5.57$ m^3. Determine the thermal efficiency.

7.8 Air is used in a Carnot engine where 22 kJ of heat is received at 560°K. Heat is rejected at 270°K. The displacement volume is 0.127 m^3. Determine (a) the work; (b) the mean effective pressure.

7.9 A Carnot engine operates between temperature limits of 1200°K and 400°K, using 0.4 kg of air and running at 500 rpm. The pressure at the beginning of heat addition is 1500 kPa and at the end of heat addition is 750 kPa. Determine (a) the heat added per cycle; (b) the heat rejected; (c) the power; (d) the volume at the end of heat addition; (e) the mean effective pressure; (f) the thermal efficiency.

7.10 A reversed Carnot engine receives 316 kJ of heat. The reversible adiabatic compression process increases by 50% the absolute temperature of heat addition. Determine (a) the COP; (b) the work.

7.11 Two reversible engines operate in series between a high-temperature (T_H) and a low-temperature (T_C) reservoir. Engine A rejects heat to engine B, which in turn rejects heat to the low-temperature reservoir. The high-temperature reservoir supplies heat to engine A. Let $T_H = 1000$°K and $T_C = 400$°K, and the engine thermal efficiencies are equal. The heat received by engine A is 500 kJ. Determine (a) the temperature of heat rejection by engine A; (b) the works of engine A and engine B; (c) the heat rejected by engine B.

7.12 Use the engine arrangement described in Problem 7.11 with engines of equal thermal efficiency. Let the work of engine B be equal to 527.5 kJ, the low temperature be equal to 305°K, and the heat received by engine A be equal to 2110 kJ. Determine (a) the thermal efficiency of each engine; (b) the temperature of heat supplied to engine A; (c) the work of engine A; (d) the heat rejected by engine B; (e) the temperature of heat supplied to engine B.

7.13 A nonpolluting power plant can be constructed using the temperature difference in the ocean. At the surface of the ocean in tropical climates, the aveage water temperature year-round is 30°C. At a depth of 305 m, the temperature is 4.5°C. Determine the maximum thermal efficiency of such a power plant.

7.14 A heat pump is used to heat a house in the winter months. When the average outside temperature is 0°C and the indoor temperature is 23°C, the heat loss from the house is 20 kW. Determine the minimum power required to operate the heat pump.

7.15 A heat pump is used for cooling in summer and heating in winter. The house is maintained at 24°C year-round. The heat loss is 0.44 kW per degree difference between outside and inside temperatures. The average outside temperature is 32°C in the summer and −4°C in the winter. Determine (a) the power requirements for both heating and cooling; (b) the condition that determines the unit size that must be purchased.

7.16 The Carnot efficiency of a heat engine is a function of the high-temperature heat source, T_H, and the low-temperature heat sink, T_C. It is possible to change one heat reservoir by an amount ΔT to improve the engine efficiency. Which reservoir should be changed?

7.17 A Carnot refrigerator rejects 2500 kJ of heat at 80°C while using 1100 kJ of work. Find (a) the cycle low temperature; (b) the COP; (c) the heat absorbed.

7.18 A Carnot heat engine rejects 230 kJ of heat at 25°C. The net cycle work is 375 kJ. Determine the thermal efficiency and the cycle high temperature.

7.19 A Carnot refrigerator operates between temperature limits of −5°C and 30°C. The power consumed is 4 kW, and the heat absorbed is 30 kJ/kg. Determine (a) the COP; (b) the refrigerant flow rate.

7.20 The Novel Air Conditioning Company claims to have developed a new air conditioner that will maintain a room at 22°C while rejecting 105 kJ of heat at 42°C. The unit requires 10 kJ of work. Is this device believable?

7.21 A Carnot heat pump is being considered for home heating in a location where the outside temperature may be as low as −35°C. The expected COP for the heat pump is 1.50. To what temperature could this unit provide heat?

7.22 A refrigerator operating on the reversed Carnot cycle provides a refrigerated space at 4°C while in a room at 25°C. A wattmeter supplying power to the refrigerator reads 3.0 kW. What is the amount of cooling being provided?

7.23 Determine the minimum power required to provide 3 kW of cooling at −10°C while the surrounding air is at 24°C.

7.24 A Carnot refrigerator, consuming 3.0 kW, is located in a room where the air temperature is 25°C. The COP is found to be 5.0. Determine the temperature of and the amount of heat removed from the refrigerated space.

7.25 Derive the expression for the $(COP)_h$ for a heat pump operating on the reversed Carnot cycle.

7.26 A reversed Carnot cycle is used to remove 1 kJ of heat at a variety of low temperatures with a fixed high temperature of 300°K. Determine the work required for low temperatures of 273°K, 200°K, 150°K, 100°K, and 50°K.

7.27 An experiment is being conducted on a new refrigerator design. During one test that lasts 1 h, the following measurements were made: 2.1 kW of power consumed, and 90 000 kJ of heat removed from refrigerated space at a temperature of 250°K. Are these measurements reasonable?

7.28 The inventor of a new refrigerator claims to have maintained a cooled space at $0°C$ in surrounding air of $25°C$ while maintaining a $(COP)_c$ of 7.0. Is this reasonable?

7.29 A Carnot engine receives 1000 kJ of heat from a heat reservoir at an unknown temperature and rejects 400 kJ of heat to a low-temperature reservoir at $25°C$. Determine (a) the high temperature; (b) the thermal efficiency.

7.30 An inventor claims that a heat engine receives 1000 kJ of heat at $400°K$ and rejects 500 kJ of heat at $300°K$. Is this a reasonable assertion?

7.31 It is desired to double the efficiency of a Carnot engine from 30% by raising its temperature of heat addition, while keeping the temperature of heat rejection constant. What percentage of increase in high temperature is required?

7.32 It is desired to double the $(COP)_c$ of a reversed Carnot engine from 5.0 by raising the temperature of heat addition for a fixed high temperature. By what percentage must the low temperature be raised?

7.33 An 8-kW heat pump is designed to maintain a house at $21°C$ when the outside temperature is $-5°C$. The heat loss from the house is estimated to be 85 000 kJ/h for these temperature conditions. Can the heat pump provide the necessary heat?

7.34 A building is heated by a heat pump and must be maintained at $20°C$. The power to the heat pump is 8 kW, and the heat loss from the building is 2900 kJ/h for each degree in temperature difference between the inside and outside. Determine the lowest outside temperature possible, subject to the heating requirements.

7.35 A house is maintained at $22°C$ by a heat pump. The outside temperature is $-5°C$, and the heat loss is estimated to be 90 000 kJ/h. Find the minimum power required when the heat supplied to the heat pump is (a) at the surrounding air temperature; (b) at the temperature of pond water at $4°C$.

7.36 A Carnot engine using 1 kg air has the following conditions: heat addition beginning at 15 MPa and $1200°K$ and continuing until the pressure is 10 MPa; isothermal compression from 100 kPa and $300°K$, continuing until the pressure is 150 kPa. Determine (a) the heat transfer into the cycle; (b) the heat transfer from the cycle; (c) the work for each of the processes; (d) the cycle efficiency.

7.37 Show for a Carnot engine using an ideal gas as the working substance that $V_4 V_2 = V_1 V_3$.

7.38 A Carnot cycle heat pump provides 90 000 kJ/h to a house maintained at $21°C$ when the outside temperature averages $0°C$. When operating, the heat pump uses 11 kW of power. The cost of electricity is $0.11/kwh. Determine (a) the number of hours the heat pump runs in one day; (b) the cost of electricity for the day; (c) the cost of electricity if the heat pump were replaced by electric baseboard heat.

7.39 A house is cooled by a reversed Carnot engine where the indoor temperature is maintained at $21°C$, and the average outside temperature is $32°C$. The heat gain by the house is 43 000 kJ/h. When the reversed Carnot engine operates, it uses 5 kW of power. Determine (a) the number of hours per day the reversed Carnot engine runs; (b) the cost per day of electricity if the price is $0.11/kwh.

7.40 A Carnot engine receives 100 kW at $1000°K$ and rejects heat to the surroundings at $300°K$. The work from the engine is used to drive a reversed Carnot engine that operates between $-20°C$ and $300°K$. Determine (a) the heat input to the reversed engine; (b) the total heat from both engines rejected to the environment at $300°K$.

7.41 A reversed Carnot engine operates between $250°K$ and $300°K$ and receives 100 kW of heat at the lower temperature. The power to drive the reversed engine comes from a

Carnot engine operating between 900°K and 300°K. Determine the heat input to the Carnot engine.

7.42 A reversed Carnot engine operating as a heat pump consumes 6 kW and has a $(COP)_h$ of 3.0. A house containing 1800 kg of air is initially at 12°C, and the heat raises the air temperature to 21°C. How long will this take?

PROBLEMS (English Units)

*7.1 A Carnot engine rejects 1000 Btu/min at 50°F and produces 40 hp. Determine the temperature of heat addition and the amount of heat flow into the engine.

*7.2 A Carnot engine uses air as the working substance. There is 10 Btu/lbm of heat added, the pressure at the beginning of adiabatic expansion is 200 psia, and the specific volume is 1.3 ft³/lbm. The sink temperature is 40°F. Determine the engine efficiency and the net work per unit mass.

*7.3 A Carnot engine uses air and operates between temperature limits of 740°F and 40°F. The pressure at state 1 is 200 psia, and the specific volume doubles during the heat addition process. Determine the net work per unit mass.

*7.4 A Carnot engine operates between 1200°F and 100°F and rejects 40 Btu to the low-temperature heat reservoir. Find the net work.

*7.5 The engine in Problem *7.4 is used to drive a heat pump that receives 250 Btu from the low-temperature heat reservoir. It rejects heat at 100°F. Determine the temperature of the heat that is added.

*7.6 A Carnot engine uses air as the working substance, receives heat at a temperature of 600°F, and rejects it at 150°F. The maximum possible cycle pressure is 1000 psia, and minimum volume is 0.03 ft³. When heat is added, the volume increases by a factor of 2.5. Determine the pressure and volume at each state in the cycle.

*7.7 A Carnot cycle uses air as the working substance. The heat supplied is 50 Btu. The temperature of the heat rejected is 70°F, and $V_4/V_1 = 10$. Determine (a) the cycle efficiency; (b) the temperature of heat added; (c) the work.

*7.8 A refrigerator operating on the reversed Carnot cycle provides a refrigerated space at 40°F while in a room at 80°F. A wattmeter supplying power to the refrigerator reads 3.0 kW. What is the amount of cooling being provided?

*7.9 Determine the minimum power required to provide 180 Btu/min of cooling at 0°F while the surrounding air is at 75°F.

*7.10 A Carnot refrigerator, consuming 3.0 kW, is located in a room where the air temperature is 80°F. The COP is found to be 5.0. Determine the temperature of and the amount of heat removed from the refrigerated space.

*7.11 A reversed Carnot cycle is used to remove 1 Btu of heat at a variety of low temperatures with a fixed high temperature of 540°R. Determine the work required for low temperatures of 460°R, 360°R, 260°R, 160°R, and 60°R.

*7.12 An experiment is being conducted on a new refrigerator design. During one test that lasts 1 h, the following measurements were made: 2.1 kW of power consumed and 85,000 Btu of heat removed from refrigerated space at a temperature of 450°R. Are these measurements reasonable?

*7.13 The inventor of a new refrigerator claims to have maintained a cooled space at 32°F in surrounding air of 77°F while maintaining a (COP)$_c$ of 7.0. Is this reasonable?

*7.14 A Carnot engine receives 1000 Btu of heat from a heat reservoir at an unknown temperature and rejects 400 Btu of heat to a low-temperature reservoir at 77°F. Determine (a) the high temperature; (b) the thermal efficiency.

*7.15 An inventor claims that a heat engine receives 1000 Btu of heat at 720°R and rejects 500 Btu of heat at 540°R. Is this a reasonable assertion?

*7.16 It is desired to double the efficiency of a Carnot engine from 40% by raising its temperature of heat addition, while keeping the temperature of heat rejection constant. What percentage of increase in high temperature is required?

*7.17 It is desired to double the (COP)$_c$ of a reversed Carnot engine from 5.0 by raising the temperature of heat addition for a fixed high temperature. By what percentage must the low temperature be raised?

*7.18 A 10-hp heat pump is designed to maintain a house at 70°F when the outside temperature is 23°F. The heat loss from the house is estimated to be 85,000 Btu/hr for these temperature conditions. Can the heat pump provide the necessary heat?

*7.19 A building is heated by a heat pump and must be maintained at 68°F. The power to the heat pump is 8 kW, and the heat loss from the building is 2900 Btu/hr for each degree in temperature difference between the inside and outside. Determine the lowest outside temperature possible, subject to the heating requirements.

*7.20 A house is maintained at 72°F by a heat pump. The outside temperature is 20°F, and the heat loss is estimated to be 90,000 Btu/hr. Find the minimum power required when the heat supplied to the heat pump is (a) at the surrounding air temperature; (b) at the temperature of pond water at 40°F.

*7.21 A Carnot engine using 1 lbm air has the following conditions: heat addition beginning at 2200 psia and 2200°R and continuing until the pressure is 1400 psia; isothermal compression from 14.7 psia and 540°R and continuing until the pressure is 23.1 psia. Determine (a) the heat transfer into the cycle; (b) the heat transfer from the cycle; (c) the work for each of the processes; (d) the cycle efficiency.

*7.22 A Carnot cycle heat pump provides 90,000 Btu/hr to a house maintained at 70°F when the outside temperature averages 32°F. When operating, the heat pump uses 15 hp. The cost of electricity is $0.11/kwh. Determine (a) the number of hours the heat pump runs in one day; (b) the cost of electricity for the day; (c) the cost of electricity if the heat pump were replaced by electric baseboard heat.

*7.23 A house is cooled by a reversed Carnot engine where the indoor temperature is maintained at 70°F and the average outside temperature is 92°F. The heat gain by the house is 43,000 Btu/hr. When the reversed Carnot engine operates it uses 7 hp. Determine (a) the number of hours per day the reversed Carnot engine runs; (b) the cost per day of electricity if the price is $0.11/kwh.

*7.24 A Carnot engine receives 600 Btu/min of heat at 1800°R and rejects heat to the surroundings at 540°R. The work from the engine is used to drive a reversed Carnot engine that operates between −4°F and 540°R. Determine (a) the heat input to the reversed engine; (b) the total heat from both engines rejected to the environment at 540°R.

*7.25 A reversed Carnot engine operates between 440°R and 540°R and receives 500 Btu/ min of heat at the lower temperature. The power to drive the reversed engine comes from a Carnot engine operating between 1600°R and 540°R. Determine the heat input to the Carnot engine.

*7.26 A reversed Carnot engine operating as a heat pump consumes 6 kW and has a $(COP)_h$ of 3.0. A house containing 3000 lbm of air is initially at 54°F, and the heat raises the air temperature to 70°F. How long will this take?

COMPUTER PROBLEMS

C7.1 Develop a TK Solver or spreadsheet model to analyze the Carnot cycle. For the cycle conditions in Example 7.3, vary the heat supplied to the cycle between 25 and 300 kJ/kg and plot the pressures at the end of the heat addition and expansion processes.

C7.2 Repeat Problem C7.1 but use carbon dioxide as the working fluid in the Carnot cycle. Compare your results with those from Problem C7.1.

C7.3 A solar collector heats a 100-m³ insulated tank of water to 350°K. The surroundings are at 300°K. A Carnot engine will be used to extract work from the stored thermal energy. Note that, as heat is extracted from the tank, the temperature will drop and the engine efficiency will decrease. Using the average tank temperature, compute the maximum engine work, using temperature intervals of 50°K, 10°K, and 5°K.

8

Entropy

Understanding entropy is essential to understanding the ramifications of the second law of thermodynamics. In this chapter we will develop

- A further understanding of system equilibrium conditions;
- A derivation of the Clausius inequality and of an expression for entropy;
- Expressions for the change of entropy of gases and pure and incompressible substances;
- An understanding as to why entropy production occurs in both open and closed systems;
- A relationship between entropy and disorder;
- A philosophical view of entropy.

In the late 1800s, Clausius was investigating the equilibrium conditions for an isolated system. He knew that the total energy of the system was a constant, and he wanted to determine whether or not there would be a change of state in the isolated system. Experience had shown that a change of state might occur, but could a method be devised by which the change could be predicted? If the system were in equilibrium, no change would occur. It would be especially helpful if there were a system property that would denote whether or not the system was in equilibrium.

Let us consider the types of mechanical equilibrium a ball can exhibit. If we were to place the ball in a bowl, it would be in *stable equilibrium;* in this state it would return to its original position when subjected to a light disturbance. Now let us place the ball in a very slight depression, for example, a marble on a partially burned candlestick. This marble exists in *metastable equilibrium.* It can stay in this configuration indefinitely, but a slight disturbance of the system can cause a tremendous change of the system state. Now, place the ball at the apex of a convex surface. It may exist in this state, but the smallest disturbance will cause a dramatic change; hence, the state is *unstable.* A fourth possibility exists when the ball is lying on a horizontal surface; this is *neutral equilibrium.* A slight disturbance will cause it to change state slightly. The quest of Clausius, which led to the property entropy, was to find whether a system was in stable equilibrium. Of course, one is not usually interested in marbles and the like, but the same types of equilibrium exist in thermal systems, such as steam in an equilibrium or nonequilibrium state, which affects equipment design.

8.1 CLAUSIUS INEQUALITY

Clausius developed a method to evaluate a closed system operating as a heat engine, as shown in Figure 8.1. The system receives heat Q_{S1} and Q_{S2} at temperatures T_1 and T_2, respectively. The system produces work W_S and rejects heat Q_{S3} at temperature T_3. All the processes are assumed to be reversible. The work from the system drives two Carnot engines, A and B, which act as heat pumps. What Clausius did, and we shall do, is assume that the system develops a slight irreversibility. What happens provides the answer to his original question: Does a property exist to determine system equilibrium?

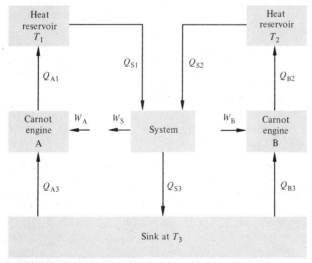

Figure 8.1 A system used to derive two reversed Carnot engines.

The energy into the heat reservoirs is equal to the energy out; hence

$$Q_{A1} = Q_{S1} \quad \text{and} \quad Q_{B2} = Q_{S2}$$

Since the Carnot engines are reversible, all the work leaving the system is used as work by them:

$$W_S = W_A + W_B$$

The system rejects heat Q_{S3}. Since all processes are reversible,

$$Q_{S3} = Q_{A3} + Q_{B3} \tag{8.1}$$

We let the system act in an irreversible manner, causing rejected heat to be greater than the reversible heat supplied to the Carnot engines.

$$Q_{S3} > Q_{A3} + Q_{B3} \tag{8.2}$$

Equations (8.1) and (8.2) may be combined as

$$Q_{S3} \geq Q_{A3} + Q_{B3} \tag{8.3}$$

Analyzing Carnot engine A, we find

$$Q_{A3} + W_A = Q_{A1}$$

From the Carnot engine efficiency,

$$W_A = Q_{A1}\left(1 - \frac{T_3}{T_1}\right)$$

hence,

$$Q_{A3} = Q_{A1}\left(\frac{T_3}{T_1}\right) = Q_{S1}\left(\frac{T_3}{T_1}\right) \tag{8.4}$$

and similarly, for Carnot engine B,

$$Q_{B3} = Q_{B2}\left(\frac{T_3}{T_2}\right) = Q_{S2}\left(\frac{T_3}{T_2}\right) \tag{8.5}$$

Consider the algebraic summation for the heat flowing into and out of the sink. Let the heat Q_{S3} entering the sink be negative (as it is leaving the system), and let the heats Q_{A3} and Q_{B3} be positive (as they are when they enter the system). The summation of these heats will be

$$Q_{A3} + Q_{B3} + Q_{S3} \leq 0 \tag{8.6}$$

Substituting equation (8.4) and equation (8.5) in equation (8.6) yields

$$\frac{Q_{S1}}{T_1} + \frac{Q_{S2}}{T_2} + \frac{Q_{S3}}{T_3} \leq 0 \tag{8.7}$$

Equation (8.7) was developed for three heat reservoirs and could be expanded for n heat reservoirs. This means that there would be $n - 1$ temperatures at which heat could be added to the system:

$$\sum_n \frac{Q}{T} \leq 0$$

As the number of heat reservoirs becomes large, the heat transferred at any given reservoir becomes small. Taking the limit as n becomes infinitely large, Q becomes infinitesimally small, and

$$\lim_{n \to \infty} \sum_n \frac{Q}{T} = \int \frac{\delta Q}{T} \leq 0$$

$$\oint \frac{\delta Q}{T} = \leq 0$$

(8.8)

Equation (8.8) is the *Clausius inequality.*

8.2 DERIVATION OF ENTROPY

The definition of entropy was developed by applying the Clausius inequality to a cyclic process. Figure 8.2 illustrates a two-process cycle, composed of reversible paths A and B. Since the cycle is reversible, it is possible to proceed in the clockwise or counterclockwise direction.

Point P is located on path A. If the system proceeds an infinitesimal distance from P in the clockwise direction, an amount of heat δQ_{cl} will be added. If the direction is reversed — and the system returns that infinitesimal distance to point P — an infinitesimal amount of heat δQ_{cc} is removed.

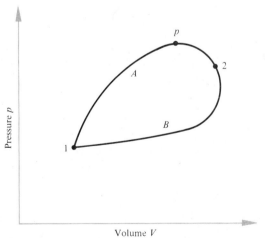

Figure 8.2 Two paths, A and B, joining states 1 and 2.

Heat added is positive, and heat removed is negative; so

$$\delta Q_{cl} = -\delta Q_{cc}$$

Applying the Clausius inequality to the cycle,

$$\oint \frac{\delta Q_{cl}}{T} \leq 0 \qquad (8.9)$$

or

$$\oint \frac{\delta Q_{cc}}{T} \leq 0 \qquad (8.10)$$

but

$$-\delta Q_{cl} = +\delta Q_{cc}$$

hence

$$-\oint \frac{\delta Q_{cl}}{T} \leq 0 \qquad (8.11)$$

There is only one solution that satisfies both equations (8.9) and (8.10):

$$\oint \frac{\delta Q_{cl}}{T} = 0$$

Since it would be possible to obtain the same result for the counterclockwise direction, the expression may be written in general as

$$\oint \frac{\delta Q}{T} = 0 \qquad (8.12)$$

We apply equation (8.12) to the cycle illustrated in Figure 8.2.

$$\oint \frac{\delta Q}{T} = \int_1^2 \left(\frac{\delta Q}{T}\right)_A + \int_2^1 \left(\frac{\delta Q}{T}\right)_B = 0$$

$$\int_1^2 \left(\frac{\delta Q}{T}\right)_A = \int_1^2 \left(\frac{\delta Q}{T}\right)_B$$

Since the paths A and B are arbitrary paths, for any path between the points 1 and 2

$$\int_1^2 \frac{\delta Q}{T} = \int_1^2 \left(\frac{\delta Q}{T}\right)_A = \int_1^2 \left(\frac{\delta Q}{T}\right)_B = C$$

The integral must be equal to some constant value because it does not depend on a particular system path. This indicates that the quantity $(\delta Q/T)_{rev}$ is an exact differential. In the thermodynamic sense the quantity is not a function of the system but depends only on the system states 1 and 2. This is the definition of property. Clausius called this property "entropy," S. Hence

$$S_2 - S_1 = \int_1^2 \left(\frac{\delta Q}{T}\right)_{rev} \qquad (8.13)$$

Although the quantity $(\delta Q/T)_{rev}$ may denote a differential change in the property S, the property difference is described by equation (8.13). This poses no problem in undergraduate thermodynamics, since the analysis of problems uses the entropy difference. However, the resolution of this dilemma remains a point of interest.

Let us now apply the Clausius inequality to a system operating on a cycle composed of a reversible path A and an irreversible path B.

$$\oint \frac{\delta Q}{T} = \int_1^2 \left(\frac{\delta Q}{T}\right)_B + \int_2^1 \left(\frac{\delta Q}{T}\right)_A \le 0$$

$$\int_1^2 \left(\frac{\delta Q}{T}\right)_A = -(S_2 - S_1)$$

However, for an irreversible path $\delta Q/T$ cannot be evaluated. The entropy change for an irreversible process between states 1 and 2 is

$$S_2 - S_1 > \int_1^2 \left(\frac{\delta Q}{T}\right)_B$$

The equality sign may be included if we write the previous equation for both reversible and irreversible situations.

$$S_2 - S_1 \ge \int_1^2 \left(\frac{\delta Q}{T}\right)_{rev} \tag{8.14a}$$

or

$$dS \ge \left(\frac{\delta Q}{T}\right)_{rev} \tag{8.14b}$$

We may reach several conclusions regarding the entropy change of a system. For adiabatic closed systems, the entropy will increase due to internal irreversibilities. Only for reversible adiabatic processes will $dS = 0$.

8.3 CALCULATION OF ENTROPY CHANGE FOR IDEAL GASES

Thus far we have not calculated the value of entropy, or the difference in entropy between two states for a closed system. We will first calculate the entropy of gases, using the ideal-gas laws and the gas tables, and then calculate the entropy of pure substances.

The first law for a closed system is

$$\delta Q = dU + \delta W$$

For a reversible process, and considering only mechanical work, this becomes

$$T\,dS = dU + p\,dV \tag{8.15}$$

Equation (8.15) is a very important thermodynamic relation. A second important relation is

$$dH = dU + p \, dV + V \, dp$$

and for reversible processes, with only mechanical work, this becomes

$$T \, dS = dH - V \, dp \tag{8.16}$$

If the ideal-gas equation of state is used, $pV = mRT$, and the change of enthalpy and internal energy is written for an ideal gas, equations (8.15) and (8.16) become, respectively,

$$dS = mc_v \frac{dT}{T} + mR \frac{dV}{V} \tag{8.17}$$

and

$$dS = mc_p \frac{dT}{T} - mR \frac{dp}{p} \tag{8.18}$$

Integrating between states 1 and 2 with constant specific heats and mass yields

$$S_2 - S_1 = mc_v \ln \left(\frac{T_2}{T_1} \right) + mR \ln \left(\frac{V_2}{V_1} \right) \tag{8.19}$$

$$S_2 - S_1 = mc_p \ln \left(\frac{T_2}{T_1} \right) - mR \ln \left(\frac{p_2}{p_1} \right) \tag{8.20}$$

Thus, there are two equations describing the change of entropy for an ideal gas in a closed system. The particular equation to use is the one that will make the solution simplest.

The gas tables account for the variation of specific heat with temperature. If equation (8.18) is partially integrated as follows,

$$S_2 - S_1 = m \int_1^2 c_p \frac{dT}{T} - mR \ln \left(\frac{p_2}{p_1} \right)$$

and from the gas tables,

$$\int_1^2 c_p \frac{dT}{T} = \phi_2 - \phi_1$$

then

$$S_2 - S_1 = m(\phi_2 - \phi_1) - mR \ln \left(\frac{p_2}{p_1} \right) \tag{8.21}$$

For convenience, values of the temperature-dependent integral are listed in the gas tables under the heading ϕ, where

$$\phi = \int_{T_{ref}}^{T} c_p \frac{dT}{T}$$

Example 8.1

Calculate the change of entropy of 3 kg of air that changes state from 300°K and 100 kPa to 800°K and 500 kPa.

Solution

Given: Air changing state with both states completely identified.

Find: The change of entropy of the air.

Sketch and Given Data:

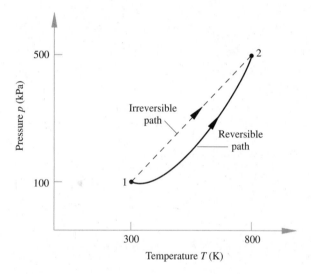

Figure 8.3

Assumption: Air behaves as an ideal gas.

Analysis: The change of entropy of an ideal gas is given by equations (8.19), (8.20), and (8.21). In equation (8.21) the variation of specific heats with temperature is included. Solve for the air volume at each state using the ideal-gas equation of state.

$$V_1 = \frac{mRT_1}{p_1} = (3 \text{ kg})(0.287 \text{ kJ/kg-K})(300°K)/(100 \text{ kN/m}^2)$$

$$V_1 = 2.583 \text{ m}^3$$

$$V_2 = \frac{mRT_2}{p_2} = (3 \text{ kg})(0.287 \text{ kJ/kg-K})(800°K)/(500 \text{ kN/m}^2)$$

$$V_2 = 1.378 \text{ m}^3$$

The change of entropy using equation (8.19) yields

$$S_2 - S_1 = mc_v \ln\left(\frac{T_2}{T_1}\right) + mR \ln\left(\frac{V_2}{V_1}\right)$$

$$S_2 - S_1 = (3 \text{ kg})(0.7176 \text{ kJ/kg-K}) \ln\left(\frac{800}{300}\right) + (3 \text{ kg})(0.287 \text{ kJ/kg-K}) \ln\left(\frac{1.378}{2.583}\right)$$

$$S_2 - S_1 = 1.57 \text{ kJ/K}$$

From equation (8.20)

$$S_2 - S_1 = mc_p \ln\left(\frac{T_2}{T_1}\right) - mR \ln\left(\frac{p_2}{p_1}\right)$$

$$S_2 - S_1 = (3 \text{ kg})(1.0047 \text{ kJ/kg-K}) \ln\left(\frac{800}{300}\right) - (3 \text{ kg})(0.287 \text{ kJ/kg-K}) \ln\left(\frac{500}{100}\right)$$

$$S_2 - S_1 = 1.57 \text{ kJ/K}$$

Using equation (8.21) requires use of Table A.2.

At $300°K$, $\phi_1 = 2.5153$ kJ/kg-K; at $800°K$, $\phi_2 = 3.5312$ kJ/kg-K.

$$S_2 - S_1 = (3 \text{ kg})(3.5312 - 2.5153 \text{ kJ/kg-K}) - (3 \text{ kg})(0.287 \text{ kJ/kg-K}) \ln (5)$$

$$S_2 - S_1 = 1.66 \text{ kJ/K}$$

Using TK Solver and the model AIR.TK, the change in entropy can be easily calculated.

```
========================= VARIABLE SHEET =========================
 St Input—Name—Output —— Unit —— Comment
                                   ***Air Properties***

                                    POINT 1
    100    P1              kPa      Pressure (kPa, MPa, psia)
    300    T1              degK     Temperature (degK, degC, degR, degF)
           v1    .86073    m3/kg    Specific Volume (m3/kg, ft3/lbm)

                                    POINT 2
    500    P2              kPa      Pressure (kPa, MPa, psia)
    800    T2              degK     Temperature (degK, degC, degR, degF)
           v2    .45985    m3/kg    Specific Volume (m3/kg, ft3/lbm)

                                    CHANGE (POINT 2—POINT 1)
           DELh  521.67    kJ/kg    Enthalpy (kJ,kg, BTU/lbm)
           DELu  378.15    kJ/kg    Internal Energy (kJ/kg, BTU/lbm)
           DELs  .55616    kJ/kg—K  Entropy (kJ/kg-K, B/lbm—R)
```

The change in entropy for the 3 kg is thus

$$S_2 - S_1 = (3 \text{ kg})(0.55616 \text{ kJ/kg-K}) = 1.66848 \text{ kJ/K}$$

Comments:

1. In problems where there is a significant temperature variation, as there is in this example, the variation of specific heat with temperature has a significant effect in calculating the change of entropy. In this case the difference is nearly 6%.
2. When the initial and final states of a system are given, the entropy change may be computed whether or not the process the system underwent was reversible. Entropy is a system property and as such does not depend on the path taken by the system in changing state, and a reversible path may be chosen. This is implicit in the derivation of the expression for change of entropy.

8.4 RELATIVE PRESSURE AND RELATIVE SPECIFIC VOLUME

The gas tables also give the means of calculating an *isentropic process* from state 1 to state 2. For an isentropic process, $\Delta S = 0$; hence

$$\ln \left(\frac{p_2}{p_1}\right) = \frac{\phi_2 - \phi_1}{R}$$

The relative pressure, p_r, is defined as

$$\ln (p_r) = \ln \left(\frac{p}{p_0}\right) = \frac{\phi}{R}$$

where p_0 is a reference state. Then

$$\ln \left(\frac{p_{r2}}{p_{r1}}\right) = \frac{\phi_2 - \phi_1}{R} \qquad (8.22)$$

and hence,

$$\frac{p_{r2}}{p_{r1}} = \left(\frac{p_2}{p_1}\right)_s \qquad (8.23)$$

There is also a relative specific volume v_r

$$v_r = \frac{RT}{p_r}$$

Hence,

$$\left(\frac{v_2}{v_1}\right)_s = \frac{v_{r2}}{v_{r1}} \qquad (8.24)$$

Example 8.2

Three lbm/sec of air expands isentropically through a nozzle from 1000°R, 200 psia, and negligible velocity to 20 psia. Determine the velocity exiting the nozzle, using the gas tables as appropriate.

Solution

Given: Air flowing isentropically through a nozzle from given inlet conditions to an exit pressure.

Find: The air velocity exiting the nozzle.

Sketch and Given Data:

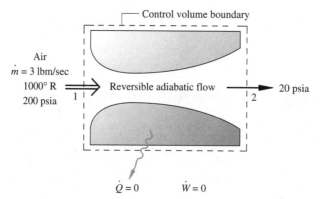

Figure 8.4

Assumptions:

1. Air behaves as an ideal gas with variable specific heats.
2. The negligible inlet velocity allows us to neglect inlet kinetic energy.
3. Inlet and exit potential energy changes are negligible.
4. There is no heat flow, as *isentropic* denotes reversible adiabatic conditions.
5. The flow is steady, and steady-state conditions apply.
6. There is no work crossing the control volume.

Analysis: The first law for an open system yields

$$\dot{Q} + \dot{m}[h_1 + (\text{k.e.})_1 + (\text{p.e.})_1] = \dot{W} + \dot{m}[h_2 + (\text{k.e.})_2 + (\text{p.e.})_2]$$

Invoking the stated assumptions yields

$$(\text{k.e.})_2 = (h_1 - h_2)$$

The substance is an ideal gas with variable specific heats. The enthalpy and relative pressure at state 1 may be read directly from the tables, yielding

$$h_1 = 240.98 \text{ Btu/lbm} \qquad p_{r1} = 12.298$$

The process from state 1 to state 2 is isentropic, thus

$$p_{r2} = p_{r1} \frac{p_2}{p_1} = 12.298(20/200) = 1.2298$$

$$h_2 = 124.75 \text{ Btu/lbm} \qquad T_2 = 522°\text{R}$$

Substituting into the first-law equation yields

$$\frac{v_2^2}{2g\mathcal{J}} = (240.98 - 124.75 \text{ Btu/lbm})$$

$$v_2 = 2412 \text{ ft/sec}$$

Solve the problem using TK Solver and AIR.TK.

```
================ VARIABLE SHEET ================
St Input─Name─Output ── Unit ── Comment ─────────
                              ***Air Properties***

                                       POINT 1
     200   P1                psia    Pressure (kPa, MPa, psia)
    1000   T1                degR    Temperature (degK, degC, degR, degF)
           v1      1.8585    ft3/lbm Specific Volume (m3/kg, ft3/lbm)

                                       POINT 2
      20   P2                psia    Pressure (kPa, MPa, psia)
           T2      525.24    degR    Temperature (degK, degC, degR, degF)
           v2      9.7226    ft3/lbm Specific Volume (m3/kg, ft3/lbm)

                                  CHANGE (POINT 2─POINT 1)
           DELh    -115.41   BTU/lbm Enthalpy (kJ/kg, BTU/lbm)
           DELu    -82.862   BTU/lbm Internal Energy (kJ/kg, BTU/lbm)
       0   DELs              B/lbm─R Entropy (kJ/kg─K, B/lbm─R)
```

Therefore, since DELh is -115.41 Btu/lbm,

$$v_2 = 2405 \text{ ft/sec}$$

Comments:

1. The relative pressure and relative specific volume ratios may be used only for isentropic processes.
2. The temperature leaving the nozzle is quite cool compared to the inlet conditions. The entropy increase caused by the decreasing pressure is balanced by the entropy decrease associated with lower temperature. ■

8.5 ENTROPY OF A PURE SUBSTANCE

Entropy has been defined as a property of a system, and the specific entropy is tabulated in the tables of thermodynamic properties. Since entropy is a property, it may be calculated for any quality within the saturated region. It is listed in the compressed liquid and superheated vapor tables. Figure 8.5 illustrates a *T-s* diagram for steam. Another frequently used chart for steam is the *h-s* diagram, called the *Mollier diagram,* illustrated in Figure 8.6. Inspection shows a disadvantage of the Mollier diagram is that it cannot be used for steam with a quality of 50% or less. As we will see in later chapters, the Mollier diagram is the traditional means available for the solution of steam turbine problems. The *T-s* diagram usually does not list specific volume, a disadvantage in its use. However, both charts are very useful for particular problems and will be used for them. The charts are listed in Appendix Tables B.1 and B.2.

Entropy Change for Incompressible Substances

The internal energy of a substance is a function of temperature and specific volume, $u(T, v)$. If the specific volume is constant, the internal energy is a function of temper-

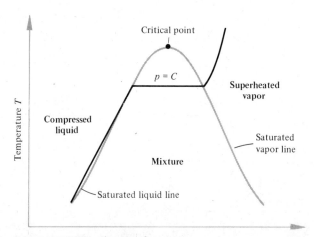

Figure 8.5 A *T-s* diagram for water.

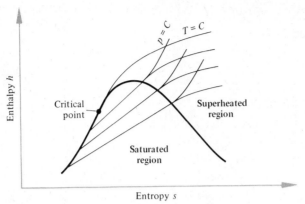

Figure 8.6 An *h-s*, or Mollier, diagram for water.

ature only, $u = u(T)$, as is the specific heat, $c = c(T)$. The change of internal energy is

$$du = c(T)\, dT$$

From the first law for closed systems,

$$T\, ds = du + p\, dv$$
$$dv = 0 \quad \text{for } v = c$$

thus,

$$ds = c(T)\, \frac{dT}{T}$$

Integrating this between states 1 and 2 yields

$$s_2 - s_1 = \int_1^2 \frac{c(T)\, dT}{T}$$

and if $c(T)$ is a constant (an average value over the temperature range is often used), the change of entropy for an incompressible substance becomes

$$s_2 - s_1 = c \ln [T_2/T_1] \tag{8.25}$$

Example 8.3

Twelve kg/min of R 12 flows through an evaporator coil at constant pressure. The inlet conditions are 308.6 kPa and 30% quality, and the refrigerant exits as a saturated vapor. Determine the change of entropy across the evaporator and the heat flow rate to the evaporator.

Solution

Given: Refrigerant flowing through an evaporator, receiving heat, and evaporating from a given initial state to a given final state at constant pressure.

Find: The change in entropy across the evaporator and the heat flow rate.

Sketch and Given Data:

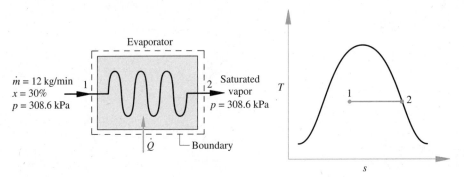

Figure 8.7

Assumptions:

1. The conditions are steady-state.
2. Refrigerant 12 is a pure substance, and the states are equilibrium ones, so tables of properties may be used.
3. The changes in potential and kinetic energies may be neglected.
4. The work done at constant pressure in an open system is zero.

Analysis: The first law for an open system is

$$\dot{Q} + \dot{m}[h_1 + (\text{k.e.})_1 + (\text{p.e.})_1] = \dot{W} + \dot{m}[h_2 + (\text{k.e.})_2 + (\text{p.e.})_2]$$

Solving for the heat flow subject to the assumptions above yields

$$\dot{Q} = \dot{m}(h_2 - h_1)$$

The properties for R 12 found from Table A.11 are

$$h_1 = 36.022 \text{ kJ/kg} + (0.3)(151.376 \text{ kJ/kg}) = 81.43 \text{ kJ/kg}$$

$$s_1 = 0.1418 + (0.3)(0.5542) = 0.3081 \text{ kJ/kg-K}$$

$$h_2 = h_g = 187.397 \text{ kJ/kg}$$

$$s_2 = s_g = 0.6960 \text{ kJ/kg-K}$$

The heat flow is

$$\dot{Q} = \frac{(12 \text{ kg/min})(187.397 - 81.43 \text{ kJ/kg})}{(60 \text{ s/min})} = 21.19 \text{ kW}$$

The change of entropy is

$$m(s_2 - s_1) = \frac{(12 \text{ kg/min})(0.6960 - 0.3081 \text{ kJ/kg-K})}{(60 \text{ s/min})} = 0.0776 \text{ kW/K}$$

Because R 12 is a pure substance, the process is not only constant-pressure but constant-temperature. In this case the equation for heat in terms of entropy may be readily integrated.

$$\int \delta q = \int T \, ds$$

for $T = C$

$$q = T(s_2 - s_1)$$

In the case where flow is occurring, $\dot{Q} = \dot{m}q$. Checking the results in this case, the temperature of evaporation is 0°C or 273.15°K, hence

$$\dot{Q} = \dot{m}T(s_2 - s_1) = (273.15°\text{K})(0.776 \text{ kW}/\text{K}) = 21.19 \text{ kW}$$

Comments:

1. When a reversible process is constant-temperature, the heat flow may be found from the $T \, ds$ equation. If this process were not reversible, the equation would not be applicable. In actual processes a pressure drop occurs; hence the process would be neither constant-pressure nor constant-temperature.
2. Absolute temperatures must always be used in entropy calculations. ▇

8.6 FURTHER DISCUSSION OF THE SECOND LAW FOR CLOSED SYSTEMS

Rather than allowing the Clausius inequality, equation (8.8), to remain an inequality, it can be changed to an equality by including an entropy production term ΔS_{prod}. Equation (8.8) becomes

$$\oint \frac{\delta Q}{T} = -\Delta S_{\text{prod}} \tag{8.26}$$

The entropy production is caused by internal irreversibilities. This term will disappear only when the internal irreversibilities become zero. The entropy production term is process-dependent. It depends on how the system changes state and, as such, is not a property. In Equation (8.26) ΔS_{prod} is zero when there are no irreversibilities present. ΔS_{prod} is positive when there are irreversibilities contributing to entropy production. It can never be less than zero.

Equation (8.26) can be expanded similarly to the previous development by considering a reversible and an irreversible path going from state 1 to state 2 and from state 2 to state 1, respectively. Thus,

$$\int_1^2 \left(\frac{\delta Q}{T}\right)_I + \int_2^1 \left(\frac{\delta Q}{T}\right)_R = -\Delta S_{\text{prod}}$$

This reduces to

$$S_2 - S_1 = \int_1^2 \left(\frac{\delta Q}{T}\right)_I + \Delta S_{\text{prod}} \tag{8.27}$$

Let us evaluate the irreversible heat term. Assume that the system receives heat Q_i at temperature T_i. This temperature is the system temperature at the boundary receiving the heat. For one heat source at a fixed temperature, equation (8.27) becomes

$$S_2 - S_1 = \frac{Q_i}{T_i} + \Delta S_{prod} \tag{8.28}$$

It is possible to extend this by considering a system that has several heat flows, each at a different but constant temperature. Equation (8.28) becomes

$$S_2 - S_1 = \sum_{i=1}^{n} \frac{Q_i}{T_i} + \Delta S_{prod} \tag{8.29}$$

If equation (8.29) is differentiated with respect to time, putting it on a time-rate basis yields

$$\frac{dS}{dt} = \sum_{i=1}^{n} \frac{\dot{Q}_i}{T_i} + \Delta \dot{S}_{prod} \tag{8.30}$$

Let us apply equation (8.28) to a reservoir supplying heat Q_i at temperature T_i to a system. In this situation we wish to determine the entropy change of the reservoir. The entropy production within the reservoir is zero; there are no internal irreversibilities, for it is at a fixed and constant temperature.

$$\Delta S_{res} = -\frac{Q_i}{T_i} \tag{8.31}$$

Consider that the reservoir and the system are two subsystems within an isolated, adiabatic, constant-mass system. Comparing equations (8.28) and (8.31) we see that the entropy-production term is due to internal system irreversibilities.

Let us consider that an isolated system, one that is closed, is adiabatic and has no energy flow across its boundary. We know from equation (8.14b) that

$$dS_{isol} \geq 0 \tag{8.32a}$$

or

$$\Delta S_{isol} \geq 0 \tag{8.32b}$$

The system may have several parts to it, that is, i subsystems, so if we let equation (8.32b) apply to the overall system, then

$$\Delta S_{total} = \sum_i \Delta S_i \geq 0 = \Delta S_{prod} \tag{8.33}$$

Example 8.4

A 5-kg steel billet is heated to 1200°K and then quenched by placing it in 50 kg of water, initially at 20°C. Determine the entropy change of the water and of the billet and the entropy-production term. The average specific heat of steel is 0.5 kJ/kg-K.

Solution

Given: A heated steel billet of known mass and temperature is placed in cool water with a known initial temperature.

Find: The change of entropy of the water and metal, as well as the entropy production due to irreversibilities in heat transfer.

Sketch and Given Data:

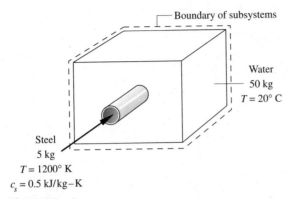

Boundary of subsystems

Water
50 kg
$T = 20°$ C

Steel
5 kg
$T = 1200°$ K
$c_s = 0.5$ kJ/kg–K

Figure 8.8

Assumptions:

1. The metal and water form two subsystems within an isolated system.
2. The changes in kinetic and potential energies are zero.
3. No work is done.
4. No heat flux crosses the boundary formed by these two subsystems.

Analysis: To determine the entropy change of the subsystem, the final temperature must be found. The first law for a closed system is

$$Q = \Delta U + \Delta \text{K.E.} + \Delta \text{P.E.} + W$$

The heat, work, and changes in kinetic and potential energies are zero, making ΔU zero. The total change of internal energy is the sum of the subsystem ΔU's. Thus,

$$\Delta U_{water} + \Delta U_{steel} = 0$$

$$u_1 = u_f \text{ at } 20°C = 82.9 \text{ kJ/kg}$$

$$s_1 = s_f \text{ at } 20°C = 0.2914 \text{ kJ/kg-K}$$

Substitute these values into the equation

$$m(u_2 - u_1) + mc_s(T_2 - T_1) = 0$$

Solve by trial and error for T_2.

$$T_2 = 303.7°K \qquad u_2 = 128.0 \text{ kJ/kg}$$

$$s_2 = s_f \text{ at } 30.7°C = 0.4424 \text{ kJ/kg-K}$$

The entropy changes may be calculated as

$$\Delta S_{\text{water}} = (50 \text{ kg})(0.4424 - 0.2914 \text{ kJ/kg-K}) = 7.55 \text{ kJ/K}$$

$$\Delta S_{\text{steel}} = (5 \text{ kg})(0.5 \text{ kJ/kg-K}) \ln [303.7/1200] = -3.44 \text{ kJ/K}$$

There is no heat transfer crossing the boundary, so the entropy production is the sum of the individual entropy-change terms of the subsystems, yielding

$$\Delta S_{\text{prod}} = 7.55 - 3.44 = 4.11 \text{ kJ/K}$$

The entropy change of the water can also be calculated using TK Solver and the model SATSTM.TK as follows:

```
================================ VARIABLE SHEET ================================
St Input—Name— Output —— Unit —— Comment————————————————————————————————————————
                                   Example 8.4

   50     Mw                  kg
   5      Ms                  kg
   .5     Cs                  kJ/kg—K
   20     T1w                 degC
   1200   T1s                 degK
          T2     303.781      degK
          u1     82.9067      kJ/kg
          u2     127.718      kJ/kg
          delSw  7.50655      kJ/K
          P1     2.34141      kPa
          P2     4.04695      kPa

==================================== RULE SHEET ================================
S Rule ————————————————————————————————————————————————————————————————————————
  "Calculation Units are SI: degK, kPa, m3/kg, kJ/kg, kJ/kg—K
  "Call Saturated Steam Property Functions
  P1=Psat(T1w)
  P2=Psat(T2)
  u1=Hf(T1w)−P1*Vf(T1w)              "Definition of Enthalpy
  u2=Hf(T2)−P2*Vf(T2)               "Definition of Enthalpy
  Mw*(u2−u1)+Ms*Cs*(T2−T1s)=0       "First Law
  delSw=Mw*(Sf(T2)−Sf(T1w))
```

Comments:

1. The steel decreases in entropy because the temperature drops, but the water increase exceeds it by the amount of the entropy-production term.

2. For liquid water at relatively low temperatures such as in this example, the internal energy may be represented as $c_w T$, where $c_w = 4.186$ kJ/kg-K. Even at elevated temperatures an average specific heat may be used, but it will have a different numerical value than the one given. ▬

8.7 EQUILIBRIUM STATE

Clausius started his investigation of systems and developed, or discovered, the property of entropy in order to determine whether an isolated system was in equilibrium. Since $dS = (\delta Q/T)_{rev}$, there are two methods by which dS will be nonzero. One method can be an exchange of heat δQ between the system and the surroundings. The system is isolated, however, and so there can be no energy transfer between the system and the surroundings. The other method requires a temperature differential within the system. This would cause an irreversible energy flow within the system and result in a change in the system temperature. Thus, it is only by an irreversible process that the entropy of the system will increase.

The following conclusions about the change in entropy and system equilibrium may be made.

1. When the entropy of an isolated system is at its maximum value, no change in state can occur.
2. When it is possible for the entropy of an isolated system to increase, the system cannot be in a state of equilibrium, and it is possible for a change of system state to occur.

Figure 8.9 illustrates an entropy equilibrium surface. The system is characterized by temperature T_1 and pressure p_1. If the system is in equilibrium, it is denoted by point A on the equilibrium surface. If the system is not in equilibrium, however, it has to be below the equilibrium surface, as illustrated by point B. A change may occur that would bring the system to the equilibrium surface; however, we cannot determine whether or not it will occur. Note that the entropy is a maximum for its state; if the

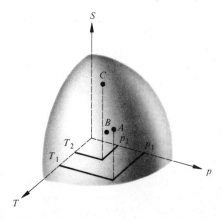

Figure 8.9 An equilibrium surface with states A and C on the surface and state B below the surface.

temperature or pressure were different, another maximum value of entropy for that state would be the equilibrium value, such as point *C*.

The following statements summarize entropy changes of a closed system:

1. The entropy will decrease when heat is removed from the system, all processes being reversible ones.
2. The entropy will remain constant when reversible adiabatic processes occur within the system.
3. The entropy will increase when heat is added to the system, reversibly or irreversibly.
4. The entropy of an isolated system will increase when irreversible processes occur within it.

Statement 1 requires the restriction of reversibility because irreversibilities within the system may cause an entropy increase, and although the heat removal results in an entropy decrease, the net entropy change could be positive or negative, depending on the magnitude of each term.

Let us examine in more detail what constitute irreversibilities. In Chapter 3 we examined a quasi-equilibrium process and found this to be an example of a reversible process, with an irreversible process being one where mechanical, thermal, or chemical equilibrium does not exist as the system changes state. This is one type of irreversibility. A second type is caused by the natural dissipative effects of a substance. For instance, friction dissipates some useful energy (work) into a less useful form (heat). This characterizes all dissipative effects. Many common phenomena demonstrate this dissipative effect — springs are not truly elastic, inductors contain some resistance, viscous effects appear in flowing fluids.

All actual processes contain some of these or similar irreversibilities. Thus, for a process to be truly reversible, it must meet two general criteria: it must be quasi-equilibrium and it must be nondissipative. This is not possible; thus, reversible processes become a limit that we cannot reach and can only strive for.

8.8 CARNOT CYCLE USING *T-S* COORDINATES

One of the great advantages of a *T-S* diagram is that reversible heat transfer may be represented. Since $Q_{rev} = \int T \, dS$, these are the natural coordinates for reversible heat flow. When talking about the Carnot cycle, we had to invoke the ideal-gas law to determine the efficiency. This is no longer necessary. Figure 8.10 illustrates the Carnot cycle on the *T-S* coordinates. Notice that the reversible adiabatic process, a constant-entropy or isentropic process, is a straight vertical line.

From the equation for reversible heat transfer,

$$\delta Q = T \, dS$$

$$Q = T \int_1^2 dS \quad \text{for } T = C$$

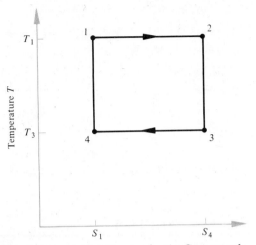

Figure 8.10 A T-S diagram for the Carnot cycle.

Hence

$$Q_{1\text{-}2} = T_H(S_2 - S_1) \tag{8.34}$$

and

$$Q_{3\text{-}4} = T_C(S_4 - S_3) \tag{8.35}$$

but

$$S_4 - S_3 = -(S_2 - S_1)$$

and recalling that

$$\eta_{th} = \frac{Q_{in} + Q_{out}}{Q_{in}}$$

$$\eta_{th} = \frac{T_H - T_C}{T_H} \tag{8.36}$$

Not surprisingly, the result is obtained more directly when using the proper thermo-dynamic coordinates. In this case the ideal-gas law did not have to be invoked, thus demonstrating that equation (8.36) is valid regardless of the working fluid in the Carnot engine.

8.9 HEAT AND WORK AS AREAS

That the work is equal to the algebraic sum of the heat supplied and heat rejected may be demonstrated graphically. The enclosed area in Figure 8.11(a) represents the heat supplied. The enclosed area in Figure 8.11(b) represents the heat rejected. Since heat in is positive and heat out is negative, the two areas, when added, yield the net work, shown in Figure 8.11(c).

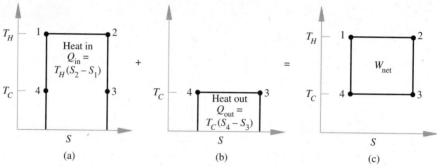

Figure 8.11 A graphical illustration that the net work is the sum of the heats.

8.10 THE SECOND LAW FOR OPEN SYSTEMS

We have considered the second law of thermodynamics for closed systems; let us extend this to open systems. Figure 8.12 illustrates a general open system, a control volume located in a moving fluid with heat and work being done.

The second law may be stated for constant mass as

$$\frac{dS_{CM}}{dt} = \sum_{i=1}^{n} \frac{\dot{Q}_i}{T_i} + \Delta\dot{S}_{prod} \tag{8.37}$$

The mass in the control volume at time t, that is, the control mass, is constant. We will develop another expression for the entropy of the control mass and from this develop the expression of the second law for open systems.

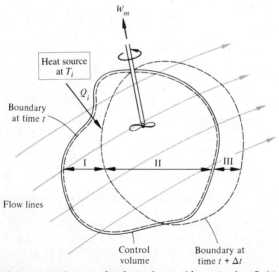

Figure 8.12 A control volume located in a moving fluid.

The entropy of the control mass at time t is

$$S_{CM,t} = S_{II,t} + S_{I,t} + S_{III,t} \tag{8.38}$$

The term $S_{III,t}$ is zero, so it may be included. At time $t + \Delta t$,

$$S_{CM,t+\Delta t} = S_{II,t+\Delta t} + S_{I,t+\Delta t} + S_{III,t+\Delta t} \tag{8.39}$$

Again, $S_{I,t+\Delta t}$ is a convenient zero to have in the equation. Take the difference between equations (8.38) and (8.39) and divide by Δt, taking the limit as Δt approaches zero. This yields

$$\frac{dS_{CM}}{dt} = \frac{dS_{II}}{dt} - \frac{dS_{I}}{dt} + \frac{dS_{III}}{dt} \tag{8.40}$$

The negative sign on dS_1/dt is due to a physical decrease in the quantity $(S_{I,t+\Delta t} - S_{I,t})$. Let us adopt the following convention:

$$S = ms$$

$$\frac{dS}{dt} = \frac{dm}{dt} s = \dot{m}s$$

As Δt approaches zero, region II coincides with the control volume, and regions I and III represent flow in and out of the control volume, respectively.

Equation (8.37) becomes

$$\frac{dS_{CV}}{dt} = \sum_{i=1}^{n} \frac{\dot{Q}_i}{T_i} + \dot{m}_{in} s_{in} - \dot{m}_{out} s_{out} + \Delta \dot{S}_{prod} \tag{8.41}$$

For steady, one-dimensional flow, where $\dot{m} = \dot{m}_{in} = \dot{m}_{out}$, equation (8.41) reduces to

$$\dot{m}(s_2 - s_1) = \sum_{i=1}^{n} \frac{\dot{Q}_i}{T_i} + \Delta \dot{S}_{prod} \tag{8.42}$$

For no heat transfer across the control surface, the surface of the control volume, equation (8.42) becomes

$$\dot{m}(s_2 - s_1) = \Delta \dot{S}_{prod} \tag{8.43a}$$

or

$$s_2 \geq s_1 \tag{8.43b}$$

Thus for steady, one-dimensional, adiabatic flow, the second law tells us that the entropy increases or remains constant; it cannot decrease.

Example 8.5

Steam enters a 6-in.-diameter pipe as a saturated vapor at 20 psia with a velocity of 30 ft/sec. The steam exits the pipe at 14.7 psia with a quality of 95%. Heat loss from the pipe, surface temperature 220°F, occurs by convection to the surroundings at 80°F. Determine (a) the entropy production across the pipe; (b) the entropy change of the surroundings.

Solution

Given: A pipe has steam enter and exit at known conditions. The pipe loses heat.

Find: Entropy production due to irreversibilities, considering the pipe as an open system, and the entropy change of the surroundings due to heat transfer from the pipe.

Sketch and Given Data:

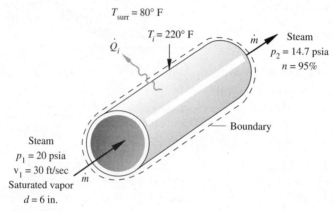

Figure 8.13

Assumptions:

1. Equilibrium steam property values may be used.
2. The flow is steady, and changes in kinetic and potential energies may be neglected.
3. There is no work flow crossing the system's boundary.
4. The heat flow from the system occurs at a constant and uniform temperature.

Analysis: The change of entropy across the pipe may be calculated using equation (8.42):

$$\dot{m}(s_2 - s_1) = \frac{\dot{Q}_i}{T_i} + \Delta\dot{S}_{prod}$$

The heat flux can be determined from an open system, first-law analysis across the control volume for steady-state conditions.

$$\dot{Q} + \dot{m}[h_1 + (k.e.)_1 + (p.e.)_1] = \dot{W} + \dot{m}[h_2 + (k.e.)_2 + (p.e.)_2]$$

Applying the assumptions reduces this to

$$\dot{Q} = \dot{m}(h_2 - h_1)$$

The mass flow rate may be determined from the equation for conservation of mass, since we know the velocity, specific volume, and cross-sectional area. The steam properties entering and leaving the pipe are

$h_1 = h_g$ at 20 psia $= 1156.4$ Btu/lbm

$s_1 = s_g$ at 20 psia $= 1.7318$ Btu/lbm-R

$v_1 = v_g$ at 20 psia $= 20.094$ ft³/lbm

$h_2 = h_f + xh_{fg}$ at 14.7 psia $= 180.35 + 0.95(970.12) = 1102.0$ Btu/lbm

$s_2 = s_f + xs_{fg}$ at 14.7 psia $= 0.31192 + 0.95(1.4445) = 1.6842$ Btu/lbm-R

The cross-sectional area is

$$A = \frac{\pi}{4} d^2 = \frac{\pi}{4} (0.5)^2 = 0.1963 \text{ ft}^2$$

The mass flow rate is

$$\dot{m} = \frac{Av}{v} = \frac{(0.1963 \text{ ft}^2)(30 \text{ ft/sec})}{(20.094 \text{ ft}^3/\text{lbm})} = 0.293 \text{ lbm/sec}$$

and the heat flux is determined by

$\dot{Q} = \dot{m}(h_2 - h_1) = (0.293 \text{ lbm/sec})(1102.0 - 1156.4 \text{ Btu/lbm}) = -15.94 \text{ Btu/sec}$

and the change of entropy across the pipe is

$\dot{m}(s_2 - s_1) = (0.293 \text{ lbm/sec})(1.6842 - 1.7318 \text{ Btu/lbm-R}) = -0.0139 \text{ Btu/sec-R}$

Applying equation (8.42) yields the solution for the entropy production:

$$\dot{m}(s_2 - s_1) = \frac{\dot{Q}_i}{T_i} + \Delta\dot{S}_{\text{prod}}$$

$$-0.0139 \text{ Btu/sec-R} = -\frac{15.94 \text{ Btu/sec}}{680°\text{R}} + \Delta\dot{S}_{\text{prod}}$$

$$\Delta\dot{S}_{\text{prod}} = +0.0095 \text{ Btu/sec-R}$$

The surroundings are a constant-mass system; hence applying equation (8.31), written with a heat flux, it becomes

$$\Delta\dot{S}_{\text{surr}} = -\frac{\dot{Q}_i}{T_i}$$

$$\Delta\dot{S}_{\text{surr}} = +\frac{15.94 \text{ Btu/sec}}{540°\text{R}} = +0.0295 \text{ Btu/sec-R}$$

Comments:

1. The entropy change of the surroundings is always positive because of heat flow into the surroundings.
2. If the pipe were insulated—adiabatic—the enthalpy of the steam leaving the pipe

would be the same as the inlet conditions, but the steam would be superheated at the lower exit pressure. A glance at the steam tables shows that the entropy increases; thus, entropy production always occurs, except for reversible adiabatic flow conditions. ∎

8.11 THIRD LAW OF THERMODYNAMICS

The third law of thermodynamics allows the calculation of absolute entropy. The Nernst postulate of the third law is

> The absolute entropy of a pure crystalline substance in complete internal equilibrium is zero at zero degrees absolute.

Although a complete discussion of the third law and its implications is beyond the scope of this text, it can be seen that the absolute entropy of a substance at some state 2 may be determined as follows:

$$S_2 - S_1 = \int_{T=T_1}^{T=T_2} \frac{\delta Q}{T} \tag{8.44}$$

However,

$$S_1 - S_0 = \int_{T=0}^{T=T_1} \frac{\delta Q}{T} \tag{8.45}$$

but $S_0 = 0$, so the absolute value of S_1 may be determined by equation (8.45) and the absolute value of S_2 may be determined from equation (8.44). If S_0 is a constant, the absolute value of S_2 may be found in the same manner.

8.12 FURTHER CONSIDERATIONS

Heretofore we have considered a system to be an amount of fixed mass or space, most likely an engine of a relatively small size. Consider now that the system is the universe, which, being of fixed mass, satisfies the definition of a system. If in general all the processes within any system and the universe were isentropic or reversible cycles, there would be no increase in entropy. Actual processes are irreversible, however, and so there is a continual increase in the entropy of the universe. This means that the entropy must have been originally constrained, or it could not be increasing now. People are still speculating about how the entropy of the universe was minimized originally. Although it is beyond the scope of this text, it remains a point from which many interesting discussions and speculations may result.

Entropy and Disorder

It is possible to consider entropy from a probabilistic viewpoint, which leads to viewing it as a measure of disorder or chaos. To do this let's start with an insulated or adiabatic container with 1 g mol of some gas inside it. Furthermore, let us divide the container into two equal volumes and let the gas be equally divided between the two volumes. The division between the two volumes will be a membrane with a small hole in it. Furthermore, let us imagine that we paint a molecule, located in the left-hand volume, white. It will bounce around the container, colliding with the other molecules and the container walls. We assume there are no external forces, so all locations are equally possible, and the molecule will occupy all these locations eventually. It can also go through the hole, and, since the volumes are equal, we can expect to find the molecule half of the time on the right side and half of the time on the left side of the container. Thus, the probability that it is on the right side is $1:2$.

Now let us imagine that another molecule is painted a different color and that it too will migrate between both volumes and have a probability of being on the right-hand side of $1:2$. The probability that both molecules will be on the right-hand side is $\frac{1}{2} \times \frac{1}{2} = \frac{1}{4}$. If we consider more molecules, say three identifiable molecules, the probability of finding all three on the same side is $(\frac{1}{2})^3 = \frac{1}{8}$. If we now extend this to n molecules, the probability is $(\frac{1}{2})^n$. The probability of finding the entire gram mole of molecules on one side is extremely small: $(\frac{1}{2})^{6.02 \times 10^{23}}$.

Since there are no leaks, the probability of finding all the molecules in the container is certain, or $1:1$. It is possible to show that for two volumes defining the enclosure, V_1 and V_2, the probability of finding all the molecules on the right-hand side is $(V_1/V_2)^n$.

Let us now consider that we observe the expansion of the gas from volume V_1 to volume V_2 in this adiabatic container. The gas is going from a state of low probability, $\mathcal{P}_1$, to a state of high probability, $\mathcal{P}_2 = 1$.

$$\mathcal{P}_1 = \left(\frac{V_1}{V_2}\right)^n$$

$$\ln(\mathcal{P}_1) = n \ln\left(\frac{V_1}{V_2}\right) \qquad \ln(\mathcal{P}_2) = 0$$

$$\ln \mathcal{P}_2 - \ln \mathcal{P}_1 = -n \ln\left(\frac{V_1}{V_2}\right) = n \ln\left(\frac{V_2}{V_1}\right)$$

$$\frac{R}{n} \ln\left(\frac{\mathcal{P}_2}{\mathcal{P}_1}\right) = R \ln\left(\frac{V_2}{V_1}\right)$$

We have seen that the entropy increase of an ideal gas during adiabatic expansion when no work is done is also

$$S_2 - S_1 = R \ln\left(\frac{V_2}{V_1}\right)$$

Let the ratio R/n be equal to a constant k, called the *Boltzmann constant*. Thus, the entropy of a state may be related to its probability by

$$S_2 - S_1 = k \ln\left(\frac{\mathcal{P}_2}{\mathcal{P}_1}\right)$$

Entropy increases for any spontaneous process, since it goes from a state of lower probability to one of higher probability.

How does this relate to chaos? The answer lies in understanding entropy's relationship to probability. Let's examine disorder. Have you noticed that spaces tend to get messy if we don't make an effort to keep them neat? Consider a classroom with desks and chairs in neat rows at the beginning of a class period. Now look at the room when the period is over and the students have left. The disorder has increased, the chairs and desks are no longer in neat rows, some are turned at angles, some are pushed together.

How do we measure this disorder? Consider the following situation. You neatly place ten dimes on your desk, all facing heads up. They are then knocked off the desk. What is the probability of finding them all heads up on the floor? The answer is $(\frac{1}{2})^{10}$, or about one in a thousand. Thus, the probability for disorder is much greater than that for order.

In comparing order and entropy, we see that when a system changes state spontaneously it goes from a state of a lower probability of existence to a state of a higher probability. Also, the state of higher probability is more disordered. Thus, we find that, as the entropy of a system increases, so does its disorder or chaos.

CONCEPT QUESTIONS

1. Explain stable, unstable, and metastable equilibrium.
2. How is the Clausius inequality related to entropy?
3. Is entropy a property? Why?
4. What is an isolated system?
5. Explain entropy to a layperson.
6. Is there a conservation of entropy principle for isolated systems?
7. What effect does work have on entropy production?
8. What are the causes of entropy production?
9. Is it true that the change of entropy for a closed system is identical for all processes operating between the same initial and final states?
10. Is it true that the entropy change of a closed system cannot be negative?
11. Is it possible to conceive of a process that would violate the second law of thermodynamics but not the first law? Explain.
12. Is it true that the change of entropy a system experiences resulting from heat transfer to the surroundings is equal and opposite to the entropy change of surroundings resulting from this heat transfer?
13. Write the second law for closed systems in terms of entropy change for adiabatic processes.
14. Write the second law for steady-state, adiabatic, open systems in terms of entropy.
15. Is the expression $T \, ds = du + p \, dv$ restricted to certain types of processes? Explain.
16. A process is internally reversible and adiabatic. Is this the same as isentropic? Explain.
17. Two solid blocks of the same material and identical mass but at different temperatures are brought together in an adiabatic container until they reach thermal equilibrium with one another. Does the entropy decrease of one block equal the entropy increase of the other?

18. Using the relationship between specific heats, show that equations (8.19) and (8.20) are identical.

19. What does the property ϕ mean in the gas tables for air?

20. What are the limitations on using p_r and v_r in the gas tables for air?

21. Consider an ideal gas with constant specific heats undergoing an isothermal process. Does the entropy change?

PROBLEMS (SI)

8.1 Two kg of a gas is cooled from 500°C to 200°C at constant pressure in a heat exchanger. Determine the change of entropy for (a) air; (b) carbon dioxide; (c) helium.

8.2 Calculate the change in entropy per kg between 250°K and 75 kPa and 750°K and 300 kPa using the ideal-gas law and the gas tables for air.

8.3 Isentropic compression of a gas occurs. Determine the temperature ratio that allows this if the pressure ratio is six and the gas is (a) air; (b) propane; (c) helium.

8.4 An isothermal expansion of air occurs in a piston-cylinder from 298°K and 800 kPa to 225 kPa. Determine the change of entropy per kg.

8.5 A piston-cylinder arrangement has been developed to compress air adiabatically from 30°C and 100 kPa to 450 kPa with 120 kJ/kg of work. Is this possible?

8.6 An isothermal compression process of 50 kg/s of nitrogen from 305°K and 120 kPa to 480 kPa occurs in a large water-cooled compressor. Determine (a) the change of entropy; (b) the power required.

8.7 A piston-cylinder containing 0.2 kg of air at 800°K and 2.1 MPa expands isentropically until the pressure is 210 kPa. Determine (a) the work done (use gas tables); (b) the initial and final specific volumes.

8.8 A closed system undergoes a polytropic process according to $pV^{1.3} = C$. The initial temperature and pressure are 600°K and 200 kPa, respectively. Three kg of air is the working substance. Determine the change of entropy for the process if the final temperature is 900°K.

8.9 When air is throttled, entropy increases. For 2 kg/s of air the entropy increases by 0.06 kW/K. Determine the pressure ratio of final to initial for this to occur.

8.10 In a constant-volume process 1.5 kg of air is cooled from 1200°K. The change of entropy is −1.492 kJ/K. Determine the final temperature.

8.11 A cooling process at constant pressure involves the removal of 3.5 kW from the system. Compute the mass flow required and the change in entropy of the following substances: (a) air decreasing in temperature from 0°C to −20°C; (b) R 12 condensing at −20°C; (c) ammonia condensing at −20°C.

8.12 A Carnot engine operates between 4°C and 280°C. If the engine produces 310 kJ of work, determine the entropy change during heat addition.

8.13 Two kg of an ideal gas, $R = 317$ J/kg-K and $k = 1.26$, is contained in a rigid cylinder; 21.1 kJ of heat is added to the gas, which has an initial temperature of 305°K. Determine (a) the final temperature; (b) the change of entropy; (c) the change of enthalpy; (d) the change of internal energy.

8.14 A gas turbine expands 50 kg/s of helium polytropically, $pV^{1.8} = C$, from 1100°K and 500 kPa to 350°K. Determine (a) the power produced; (b) the entropy change; (c) the final pressure; (d) the heat loss; (e) the power produced by isentropic expansion to the same temperature.

8.15 A steam turbine receives steam as a dry saturated vapor at 7.0 MPa and expands adiabatically to atmospheric pressure, 100 kPa. The turbine uses 24.3 kg/h of steam for each kW. What is the entropy of the steam leaving the turbine?

8.16 Hydrogen is compressed isentropically from $p_1 = 750$ kPa and $T_1 = 305°$K to a pressure of 2.25 MPa. Determine per kg (a) the final temperature; (b) the work; (c) the change of enthalpy and internal energy.

8.17 If 2.06 kg of steam expands adiabatically from a volume of 0.234 m³ and a temperature of 300°C to a pressure of 125 kPa, (a) for reversible adiabatic expansion determine the work, the initial pressure, and the final quality; (b) for irreversible expansion — where the final quality is 100% — find the work, the initial pressure, and the change of entropy.

8.18 A closed system contains an ideal gas, $R = 269$ J/kg-K, $c_v = 0.502$ kJ/kg-K. The system undergoes the following cycle: at state 1 the temperature is 444°K and the pressure is 448 kPa; heat is transferred at constant pressure until the temperature is 889°K — state 2; the gas is compressed at constant temperatures until the value of the entropy equals that at state 1 — state 3; and finally there is an isentropic expansion from state 3 to state 1. Determine (a) the p-V and T-s diagrams; (b) the pressure at state 3; (c) the heat transferred from state 1 to 2 and from state 2 to 3; (d) the cycle work.

8.19 Steam, contained in a 22.6-liter piston-cylinder at 1.4 MPa and 250°C, expands isentropically until it becomes a saturated vapor. Determine (a) the final pressure; (b) the work required; (c) the final volume.

8.20 A refrigeration system uses ammonia as the working substance. The compressor receives ammonia as a dry saturated vapor at $-22°$C and discharges it at 1.4 MPa. The compression process is isentropic. Determine (a) the discharge temperature; (b) the work of compression per kg.

8.21 A natural gas pipeline distributes gas throughout the country, pumped by large gas turbine–driven centrifugal compressors. Assume that the natural gas is methane, the pipe diameter is 0.2 m, and the gas enters the compressor at 300°K and 105 KPa. The velocity of the methane entering the compressor is 4 m/s. The compression process is isentropic, and the discharge pressure is 700 kPa. Determine (a) the discharge temperature; (b) the mass flow rate; (c) the power required.

8.22 Two insulated tanks, one containing 1.7 m³ of argon at 1.0 MPa and 38°C, the other completely evacuated and having a volume of 3.4 m³, are connected by a short pipe and valve. The valve is opened and the argon pressure is equalized between the two tanks. Determine (a) the final pressure; (b) the final temperature; (c) the change of entropy; (d) the change of enthalpy; (e) the work.

8.23 Develop the expression for the Clausius inequality, starting with the premise that the thermal efficiency of a Carnot engine is greater than or equal to the thermal efficiency of a heat engine.

8.24 A capillary tube is used in lieu of a throttling valve in home refrigerators. Liquid R 12 enters saturated at 40°C and leaves at a pressure of 362 kPa. Determine the entropy change per kg across the capillary tube.

8.25 Two kg/s of saturated steam at 200°C is condensed to a saturated liquid. The coolant is R 12, which is vaporized at 30°C. Determine (a) the mass flow rate of vaporized R 12; (b) the change of entropy of the steam and of R 12.

8.26 A thermodynamic cycle is composed of the following reversible processes: isothermal expansion, state 1 to 2; isentropic compression, state 2 to 3; and constant pressure cooling, state 3 to 1. The cycle operates on 1.5 kg of steam. State 1 is dry saturated steam at 200°C; state 2 has a pressure of 100 kPa. Determine (a) the heat addition; (b) the net work; (c) the entropy change from state 1 to 2.

8.27 An isothermal compressor consumes 8 kW while compressing air from p_1 to p_2 at a constant temperature of 300°K. The surroundings are at a temperature of 280°K. Determine (a) the rate of entropy change of the air; (b) the rate of entropy change of the surroundings.

8.28 A constant-pressure system contains 1 kg of steam as a saturated liquid at 300 kPa. The system receives 500 kJ of heat from a reservoir at 600°K. Considering the system and the reservoir as an isolated system, calculate the entropy production.

8.29 A Carnot cycle receives 1000 kJ of heat while operating between temperature limits of 1000°K and 500°K. Determine the entropy change during heat addition.

8.30 In a home refrigerator, 2 kg/min of R 12 enters the evaporator coil as a saturated liquid at 200 kPa and leaves as a saturated vapor at the same pressure. The refrigerated space is maintained at a constant temperature of 5°C. Determine the rate of entropy change of the refrigerant and of the refrigerated space.

8.31 A steam radiator has a volume of 0.05 m³ and contains saturated vapor at 100 kPa. As a result of heat transfer to the room, the steam condenses and the final temperature of the liquid before the steam trap opens is 50°C. Determine the entropy change of the steam for this process.

8.32 An adiabatic tank contains 0.75 kg of saturated liquid water and 0.25 kg of saturated steam at 100 kPa. An electric heater is inserted in the tank and turned on until all the liquid is vaporized. Determine the final steam pressure and the entropy change of the steam.

8.33 An adiabatic tank is partitioned into two equal volumes, one containing 0.5 kg of saturated steam at 200 kPa and the other totally evacuated. The partition is removed. What is the entropy change of the steam?

8.34 An insulated R 12 container develops a leak. The refrigerant before the leak is at 500 kPa and 20°C. Determine the percent mass of refrigerant remaining when the pressure is 200 kPa.

8.35 A 0.3-m³ adiabatic tank containing steam at 1000 kPa and 300°C is connected via a line and valve to a piston-cylinder. The valve connecting the tank to the piston is opened, and the piston rises at a constant pressure of 100 kPa. This process continues until the tank pressure reaches 100 kPa. Assume the processes to be reversible. Determine (a) the final steam state in the tank; (b) the work done in the piston-cylinder.

8.36 An adiabatic heat exchanger receives 2 kg/s of saturated steam on the shell side at 200 kPa, which condenses to a saturated liquid. Water enters the tubes at 25°C and leaves at 40°C. Determine the entropy change for each substance and the entropy production.

8.37 A direct-contact heat exchanger receives saturated steam at 300 kPa and 20 kg/s of water at 300 kPa and 80°C. Water leaves the heat exchanger as a saturated liquid at 300 kPa. Determine the entropy production.

8.38 A 10-kg copper ingot, $c_p = 0.39$ kJ/kg-K, is heated to 500°C and dropped into an adiabatic tank holding 60 kg of water, initially at 25°C. Determine the entropy change for the water and for the copper, and the total entropy production.

8.39 An insulated piston-cylinder contains 0.02 m³ of carbon dioxide at 300°K and 120 kPa. A 500-W electric heater is turned on for five minutes and the piston moves, maintaining a constant pressure. Determine the entropy change of the carbon dioxide.

8.40 Air is compressed polytropically according to $pV^{1.3} = C$ from 300°K and 100 kPa to 750 kPa. The surroundings are at 25°C. Determine the entropy change for the air and for the surroundings, and the entropy production per unit mass.

8.41 A nonadiabatic nozzle accelerates R 12 from negligible inlet velocity to an exit velocity of 250 m/s. The inlet pressure is 600 kPa, and the inlet temperature is 100°C. The exit pressure is 100 kPa. The exit temperature of the refrigerant is 50°C. What heat must be added to the nozzle during the expansion process?

8.42 Derive an expression for the change of entropy of an ideal gas with constant specific heats in terms of specific volume and pressure change.

8.43 A room with dimensions of 4 m × 3 m × 6 m contains air at 15°C and 1 atm pressure. A 1-kW electric heater is placed in the room and turned on for 1 h. (a) What is the final air temperature, assuming the room is adiabatic? (b) What is the air's change of entropy?

8.44 A heater at a constant temperature of 370°K is used to heat 100 liters of water from 10°C to 30°C. Determine the entropy production for the water and the heater.

8.45 A 2-kg adiabatic container has an average specific heat of 3.5 kJ/kg-J and an initial temperature of 300°K and is dropped 2 km from a balloon. Determine the change of entropy of the container.

8.46 An adiabatic tank has a partition, creating two equal compartments of 0.05 m³ each. Compartment A contains steam at 100 kPa and 240°C, and compartment B contains steam at 1000 kPa and 300°C. The partition is removed, and the steam reaches a new equilibrium state. Determine the entropy production.

8.47 A 100-Ω resistor has 5 A of current flow through it for 5 s. During this time the resistor is maintained at a constant temperature of 300°K by a reservoir at 285°K. Determine the change of entropy of the resistor and of the reservoir, and the total entropy production.

8.48 The resistor in Problem 8.47 is now adiabatically insulated. The resistor's mass is 25 g and its specific heat 0.71 kJ/kg-K. The current flows for the same time period. What is the entropy change in the resistor?

8.49 A 7.5-kW compressor handles 2.4 kg/min of air from 100 kPa and 290°K to 600 kPa and 440°K. The surroundings' temperature is 290°K. Determine (a) the entropy change of the air in the compressor; (b) the entropy change of the surroundings; (c) the entropy production.

8.50 An adiabatic container has two compartments, one containing 1 kg of oxygen at 300°K and 5 MPa and the other containing 0.5 kg of oxygen at 500°K and 10 MPa. The partition is removed. What is the change of entropy of the gas?

8.51 Air enters an adiabatic nozzle at 300 kPa and 400°K with an initial velocity of 50 m/s. The exit pressure is 100 kPa, and the process is reversible. Determine the exit velocity using (a) constant specific heats; (b) the gas tables for air.

8.52 Methane is compressed isothermally from 325°K and 100 kPa to 200 kPa. The surroundings are at 298°K. Is the process internally reversible? Is it externally and internally reversible?

8.53 A piston-cylinder contains 0.1 kg of steam at 1000 kPa and 300°C and expands adiabatically to 100 kPa. What is the maximum work that the steam can produce in the expansion process?

8.54 A piston-cylinder contains 2 kg of ammonia, initially at 40% quality and a pressure of 400 kPa and finally as a saturated liquid at 16°C. What is the change of entropy of the ammonia? Can this process occur adiabatically?

8.55 A piston-cylinder contains 0.2 kg of R 12 as a saturated vapor at −20°C. The refrigerant is compressed adiabatically until the final volume is 0.00418 m³. Is it possible for the discharge pressure to be 700 kPa? 900 kPa?

8.56 An electric motor produces 4 kW of shaft power while using 18 A at 240 V. The motor's surface temperature is measured at 50°C. Determine the rate of entropy production for the motor.

8.57 During the heat-addition process of a Carnot engine, the cycle temperature is 800°K and the reservoir supplying the heat is constant at 1000°K. Fifteen hundred kJ of heat is transferred. Determine the entropy change of the system and of the reservoir, and the entropy production.

8.58 A tank contains 3 kg of air at 100 kPa and 500°K. Heat is transferred from a constant-temperature heat reservoir at 1000°K until the air temperature is 800°K. The system's boundary has a constant temperature of 850°K during the heat-transfer process. Determine the system entropy production.

8.59 A constant-volume system contains neon at 200 kPa and 320°K. The system receives energy until the temperature rises to 440°K. Determine the entropy production if (a) the system receives adiabatic paddle work to raise the temperature; (b) the system receives heat from a constant-temperature heat reservoir at 500°K and the system boundary is 500°K during the heat-addition process.

8.60 An adiabatic cylinder contains a frictionless, thermally conducting piston initially held in place by a pin. One side of the piston contains 0.2 m³ of air at 500°K and 300 kPa, and the other side contains 0.2 m of neon at 300°K and 500 kPa. The pin is removed. Determine (a) the equilibrium temperature and pressure of the gases; (b) the entropy change of each gas.

8.61 Air is flowing through an adiabatic, horizontal duct. Measurements at the A end indicate a temperature of 340°K, a pressure of 105 kPa, and a velocity of 75 m/s. At the B end of the duct the temperature is 300°K, the pressure is 90 kPa, and the velocity is 305 m/s. What is the flow direction, A to B or B to A?

8.62 Three kg/s of helium at 500 kPa and 300°K enters an insulated device where the work performed is zero. The fluid divides into two equal streams leaving the device, both at 100 kPa and one at 85°C and the other at an unknown temperature. Neglecting changes in kinetic and potential energies, what is the exit temperature of the second stream? Is it possible for the device to operate?

8.63 Steam flows steadily through a turbine from inlet conditions of 5 MPa, 400°C, and negligible velocity to exit conditions of 100 kPa, saturated vapor, and a velocity of 100 m/s. Heat transfer, 25 kJ/kg, from the turbine casing occurs at an average temperature of 100°C. Determine (a) the work per unit mass; (b) the rate of entropy production in the turbine per unit mass of steam flowing through the turbine.

8.64 Consider the turbine in Problem 8.63; let the control also envelop part of the power plant, which is at a temperature of 37°C. What is the entropy production per unit mass of steam in the control volume under this circumstance?

8.65 A compressor receives 0.2 m³/s of air at 27°C and 100 kPa and compresses it to 700 kPa and 290°C. Heat loss per unit mass from the compressor surface at 100°C is 20 kJ/kg. (a)

Determine the power required, neglecting changes in kinetic and potential energies. (b) Determine the entropy production for the compressor.

8.66 A 0.5-m³ tank is initially evacuated. A valve connecting it to a very large supply of steam at 500 kPa and 300°C is opened, and steam flows into the tank until the pressure is 500 kPa. If the process is adiabatic, determine (a) the final steam temperature in the tank; (b) the entropy produced in the tank.

8.67 A 0.2-m³ tank is to be charged with air from a large supply that is at 320°K and 1000 kPa. The tank is charged adiabatically until the pressure in the tank equals that of the supply. Determine (a) the temperature of the air in the tank; (b) the entropy produced in the process.

PROBLEMS (English Units)

*8.1 Oxygen is heated at constant volume from 50 psia and 100°F to 500°F. Determine the change of entropy per unit mass.

*8.2 Air is cooled at constant pressure from 400°F to 100°F. Determine the change of entropy per unit mass.

*8.3 A three-process cycle operates with 1 lbm of air on the following cycle: $p_1 = 15$ psia, $T_1 = 500$°R, process 1-2, $V = C$ heating; $p_2 = 60$ psia, process 2-3, isentropic expansion; $p_3 = 15$ psia, $T_3 = 1400$ R, process 3-1, $p = C$ cooling. Determine the T-s diagram and the cycle thermal efficiency.

*8.4 Which cycle, indicated below, has the higher thermal efficiency?

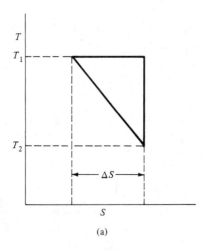

(a)

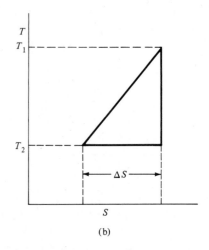

(b)

*8.5 Air is contained in a 1-ft³ tank at 2000 psia and 200°F. It is cooled by the surroundings until it reaches the surrounding temperature of 70°F. Considering the tank and the surroundings as an isolated system, what is the net entropy change?

*8.6 Air is contained in a 10-ft^3 tank at 75°F and 25 psia. Paddle work is added, 24 ft-lbf torque for 100 revolutions. If 9.1 Btu of heat is removed, determine the final temperature and change of entropy of the air.

*8.7 If 2 lbm/sec steam is throttled from 1000 psia and 1000°F to 100 psia, determine the change of entropy.

*8.8 Steam condensate at 250 psia leaves a heat-exchanger trap and flows at 30,000 lbm/hr to an adjacent flash tank. Some of the condensate is flashed to steam at 25 psia, and the remaining condensate is pumped back to the boiler. There is no subcooling. Determine the net entropy change.

*8.9 A chemical process requires 5000 lbm/hr of hot water at 180°F and 25 psia. Steam is available at 100 psia and 90% quality, and water is available at 100 psia and 70°F. The steam and water are mixed in an adiabatic chamber, with the hot water exiting. Determine the net entropy change.

*8.10 An isothermal compressor consumes 10 hp while compressing air from p_1 to p_2 at a constant temperature of 77°F. The surroundings are at a temperature of 55°F. Determine (a) the rate of entropy change of the air; (b) the rate of entropy change of the surroundings.

*8.11 A constant-pressure system contains 1 lbm of steam as a saturated liquid at 50 psia. The system receives 200 Btu of heat from a reservoir at 1000°R. Considering the system and the reservoir as an isolated system, calculate the entropy production.

*8.12 A Carnot cycle receives 1000 Btu of heat while operating between temperature limits of 1000°R and 500°R. Determine the entropy change during heat addition.

*8.13 In a home refrigerator, 5 lbm/min of R 12 enters the evaporator coil as a saturated liquid at 35.7 psia and leaves as a saturated vapor at the same pressure. The refrigerated space is maintained at a constant temperature of 40°F. Determine the rate of entropy change of the refrigerant and of the refrigerated space.

*8.14 A steam radiator has a volume of 2.0 ft^3 and contains saturated vapor at 14.7 psia. As a result of heat transfer to the room, the steam condenses and the final temperature of the liquid before the steam trap opens is 120°F. Determine the entropy change of the steam for this process.

*8.15 An adiabatic tank contains 0.75 lbm of saturated liquid water and 0.25 lbm of saturated steam at 14.7 psia. An electric heater is inserted in the tank and turned on until all the liquid is vaporized. Determine the final steam pressure and the entropy change of the steam.

*8.16 An adiabatic tank is partitioned into two equal volumes, one containing 0.5 lbm of saturated steam at 30 psia and the other totally evacuated. The partition is removed. What is the entropy change of the steam?

*8.17 An insulated R 12 container develops a leak. The refrigerant before the leak is at 70 psia and 60°F. Determine the mass of refrigerant remaining when the pressure is 30 psia.

*8.18 A 10-ft^3 tank containing steam at 160 psia and 550°F is connected via a line and valve to a piston-cylinder. The valve connecting the tank to the piston is opened, and the piston rises at a constant pressure of 14.7 psia. This process continues until the tank pressure reaches 14.7 psia. Assume the processes to be reversible. Determine (a) the final steam state in the tank; (b) the work done in the piston-cylinder.

*8.19 An adiabatic heat exchanger receives 300 lbm/min of saturated steam on the shell side at 25 psia, which condenses to a saturated liquid. Water enters the tubes at 80°F and leaves at 95°F. Determine the entropy change for each substance and the entropy production.

*8.20 A direct-contact heat exchanger receives saturated steam at 75 psia and 2500 lbm/min of water at 75 psia and 170°F. Water leaves the heat exchanger as a saturated liquid at 75 psia. Determine the entropy production.

*8.21 A 10 lbm copper ingot, $c_p = 0.10$ Btu/lbm-F, is heated to 900°F and dropped into an adiabatic tank holding 125 lbm of water, initially at 70°F. Determine the entropy change for the water and for the copper, and the total entropy production.

*8.22 An insulated piston-cylinder contains 0.75 ft³ of carbon dioxide at 540°R and 18 psia. A 500-W electric heater is turned on for five minutes and the piston moves, maintaining a constant pressure. Determine the entropy change of the carbon dioxide.

*8.23 Air is compressed polytropically according to $pV^{1.3} = C$ from 80°F and 14.7 psia to 115 psia. The surroundings are at 70°F. Determine the entropy change for the air and for the surroundings, and the entropy production per unit mass of air.

*8.24 A nonadiabatic nozzle accelerates R 12 from negligible inlet velocity to an exit velocity of 825 ft/sec. The inlet pressure is 100 psia, and the inlet temperature is 200°F. The exit pressure is 15 psia. The exit temperature of the refrigerant is 90°F. What heat must be added to the nozzle during the expansion process?

*8.25 A room with dimensions of 13 ft × 10 ft × 20 ft contains air at 55°F and 1 atm pressure. A 1-kW electric heater is placed in the room and turned on for 1 hr. (a) What is the final air temperature, assuming the room is adiabatic? (b) What is the air's change of entropy?

*8.26 A heater at a constant temperature of 200°F is used to heat 25 gal of water from 50°F to 100°F. Determine the entropy production for the water and the heater.

*8.27 A 2-lbm adiabatic container has an average specific heat of 0.83 Btu/lbm-F and with an initial temperature of 70° F and is dropped 1.5 m from a balloon. Determine the change of entropy of the container.

*8.28 An adiabatic tank has a partition, creating two equal compartments of 2.0 ft³ each. Compartment A contains steam at 14.7 psia and 500°F, and compartment B contains steam at 160 psia and 600°F. The partition is removed, and the steam reaches a new equilibrium state. Determine the entropy production.

*8.29 A 10-hp compressor handles 5.0 lbm/min of air from 14.7 psia and 60°F to 90 psia and 330°F. The surroundings' temperature is 60°F. Determine (a) the entropy change of the air in the compressor; (b) the entropy change of the surroundings; (c) the entropy production.

*8.30 An adiabatic container has two compartments, one containing 1 lbm of oxygen at 80°F and 750 psia and the other containing 0.5 lbm of oxygen at 440°F and 1500 psia. The partition is removed. What is the change of entropy of the gas?

*8.31 Air enters an adiabatic nozzle at 45 psia and 720°R with an initial velocity of 150 ft/sec. The exit pressure is 15 psia, and the process is reversible. Determine the exit velocity using (a) constant specific heats; (b) the gas tables for air.

*8.32 Methane is compressed isothermally from 125°F and 15 psia to 30 psia. The surroundings are at 70°F. Is the process internally reversible? Is it externally and internally reversible?

*8.33 A piston-cylinder contains 0.4 lbm of steam at 200 psia and 550°F and expands adiabatically to 14.7 psia. What is the maximum work that the steam can produce in the expansion process?

*8.34 A piston-cylinder contains 2 lbm of ammonia, initially at 40% quality and a pressure of 60 psia and finally as a saturated liquid at 60°F. What is the change of entropy of the ammonia? Can this process occur adiabatically?

*8.35 A piston-cylinder contains 0.2 lbm of R 12 as a saturated vapor at −10°F. The refrigerant is compressed adiabatically until the final volume is 0.067 ft³. Is it possible for the discharge pressure to be 100 psia? 125 psia?

*8.36 During the heat-addition process of a Carnot engine, the cycle temperature is 1440°R, and the reservoir supplying the heat is constant at 1800°R. Fifteen hundred Btu of heat is transferred. Determine the entropy change of the system and of the reservoir, and the entropy production.

*8.37 A tank contains 3 lbm of air at 14.7 psia and 900°R. Heat is transferred from a constant-temperature heat reservoir at 1800°R until the air temperature is 1440°R. The system's boundary has a constant temperature of 1500°R during the heat-transfer process. Determine the system entropy production.

*8.38 A constant-volume system contains neon at 25 psia and 120°F. The system receives energy until the temperature rises to 330°F. Determine the entropy production if (a) the system receives adiabatic paddle work to raise the temperature; (b) the system receives heat from a constant-temperature heat reservoir at 440°F, and the system boundary is 440°F during the heat-addition process.

*8.39 An adiabatic cylinder contains a frictionless, thermally conducting piston initially held in place by a pin. One side of the piston contains 7.0 ft³ of air at 900°R and 50 psia, and the other side contains 7.0 ft³ of neon at 540°R and 75 psia. The pin is removed. Determine (a) the equilibrium temperature and pressure of the gases; (b) the entropy change of each gas.

*8.40 Air is flowing through an adiabatic, horizontal duct. Measurements at the A end indicate a temperature of 152°F, a pressure of 15.4 psia, and a velocity of 225 ft/sec. At the B end of the duct the temperature is 80°F, the pressure is 13.2 psia, and the velocity is 915 ft/sec. What is the flow direction, A to B or B to A?

*8.41 Four hundred lbm/min of helium at 75 psia and 80°F enters an insulated device where the work performed is zero. The fluid divides into two equal streams leaving the device, both at 15 psia and one at 185°F and the other at an unknown temperature. Neglecting changes in kinetic and potential energies, what is the exit temperature of the second stream? Is it possible for the device to operate?

*8.42 Steam flows steadily through a turbine from inlet conditions of 700 psia, 700°F, and negligible velocity to exit conditions of 1 atm, saturated vapor, of 350 ft/sec. Heat transfer, 10 Btu/lbm, from the turbine casing occurs at an average temperature of 200°F. Determine (a) the work per unit mass; (b) the rate of entropy production in the turbine per unit mass of steam flowing through the turbine.

*8.43 Consider the turbine in Problem *8.42; let the control also envelop part of the power plant, which is at a temperature of 95°F. What is the entropy production per unit mass of steam in the control volume under this circumstance?

*8.44 A compressor receives 420 ft³/min of air at 80°F and 15 psia and compresses it to 100 psia and 550°F. Heat loss per unit mass from the compressor surface at 200°F is 6 Btu/lbm. (a) Determine the power required, neglecting changes in kinetic and potential energies. (b) Determine the entropy production for the compressor.

*8.45 A 20-ft³ tank is initially evacuated. A valve connecting it to a very large supply of steam at 80 psia and 600°F is opened, and steam flows into the tank until the pressure is 80 psia. If the process is adiabatic, determine (a) the final steam temperature in the tank; (b) the entropy produced in the tank.

*8.46 A 10-ft³ tank is to be charged with air from a large supply that is at 120°F and 150 psia. The tank is charged adiabatically until the pressure in the tank equals that of the supply. Determine (a) the temperature of the air in the tank; (b) the entropy produced in the process.

COMPUTER PROBLEMS

C8.1 Using the ideal-gas relationships, develop a TK Solver or spreadsheet model to calculate the values of entropy function, relative pressure, and relative specific volume for air. Assume constant specific heat and that at 273°K the relative pressure is 1.0 and the entropy function is 2.42. Produce a table for temperatures between 300°K and 1500°K in increments of 50°K and compare your results with values in Table A.2. Explain any differences.

C8.2 Repeat Problem C8.1, but use the equation in Table 5.4 to account for the variation of specific heat with temperature. Compare your results with the values in Table A.2 and your results from Problem C8.1.

C8.3 Repeat Problem C8.2, but assume the entropy function at 273°K is 1.0. Calculate the final pressure of air expanded isentropically from 1500°K and 1000 kPa to 300°K, using the tables produced in this problem and in Problem C8.2. Explain.

C8.4 Use AIR. TK to compute the work produced by air expanding isentropically in a turbine from 600°C to 100 kPa for inlet pressures varying from 200 kPa to 2000 kPa in increments of 100 kPa.

C8.5 One lbm/sec of steam expands isentropically in a nozzle from 1000 psia and 1000°F to 50 psia. Modify STEAM.TK to compute the nozzle area as a function of pressure.

9

Availability Analysis

The second law of thermodynamics has been stated in several ways, the Kelvin-Planck statement, the entropic formulation, and the energy-quality view. In thermodynamic applications one of the most fruitful ways to see the effects of the second law is through availability analysis. In this chapter you will

- Develop a philosophic understanding of availability as applied to open and closed systems;
- Understand the implications of entropy increases and of decreases in work potential;
- Use the relatively new concept of second-law efficiency;
- Explore special-case applications of available energy analysis.

9.1 INTRODUCTION

As we indicated in Chapter 1 the world is running out of petroleum and mineral resources, meaning we must create more efficient ways of using energy. The first-law analyses you performed dealt with the equality of the quantity of energy transferred. They do not speak to the quality of this energy. Availability analysis allows us to develop an expression for the quality of energy, and through second-law analysis we can determine changes in this quality as an energy transformation occurs.

Availability relates to the usefulness of the energy to perform work for us, its *work potential,* called by some its potential for use. The greater this work potential is, the greater the quality or value of the energy. We soon realize that this potential is decreased by irreversibilities in the energy-transfer process. Unlike the first law, where total energy is conserved in energy transformations, net availability inevitably decreases in those transformations. Wise design seeks to minimize this decrease, preserving, not conserving, the energy's availability.

Let us consider a tank containing compressed air. The air's temperature is the same as the surroundings, it has no velocity relative to the surroundings, and it is not elevated above the ground; hence it has no gravitational potential energy relative to the surroundings. The only difference between the air in the tank and in the surroundings is a difference in pressure. It is possible to devise ways to use this greater pressure to produce work by expanding the system to the surroundings' pressure. The maximum work that the system can produce in reaching equilibrium with the surroundings is a measure of the system's availability. Work is a path function, depending on the initial and final states and the connecting process path. It is not a system property. The maximum system work relative to the surroundings is a property, however.

When determining the air's availability, the initial state is specified, hence it is not a variable; the process path must be a reversible path connecting the initial and final state, hence no irreversibilities are considered; and in the final state the system is in thermodynamic equilibrium with the surroundings, called the *dead* state, hence it is fixed.

The system is in the *dead state* when its temperature and pressure are identical to the surroundings; its velocity and elevation are zero relative to the surroundings. In addition, all other thermodynamic potentials must be the same as the surroundings, for example, the chemical potential, surface tension, magnetic potential, electrical potential. Were this not the case, it would be possible to extract work from the system using the potential difference between the system and surroundings.

For instance, if the temperature of the system is different from that of the surroundings, a heat engine can operate between the system and surroundings. Analogous devices can use other potential differences between the system and surroundings. If the system is initially at the dead state, no work can be produced. The atmosphere, a source of tremendous energy, is at the dead state; hence its work potential is zero.

Availability is defined as the maximum work a system can produce in going from an initial state to the dead state, following reversible processes. It represents the upper limit that any work-producing device can deliver. Of course, there will be a difference between the work a machine actually produces and the change of availability. The challenge for engineers is to reduce this difference. Availability depends on the system initial state and the dead state, so it is a system/surroundings property.

The term *availability* was developed in the United States, whereas the term *exergy* is used in Europe. In this text the terms are synonymous, though some authors may define them with slight differences.

9.2 AVAILABILITY ANALYSIS FOR CLOSED SYSTEMS

If the system is in some state other than the dead state, it will change spontaneously toward the dead state. In Figure 8.9 this dead state is indicated by the apex of the spheroid. The change by the system to the dead state from any other state will occur without work being supplied from an outside source. Hence, the work done by the system in changing to the dead state will never be less than zero. It will be zero when the system is initially at the dead state,

$$W_{\max} \geq 0 \qquad (9.1)$$

where $W_{\max}$ is the maximum useful work that can be produced by the system as it interacts with the surroundings in achieving the dead state. Availability of a system state is defined as the maximum work that can result from the interaction of the system and surroundings.

Now let us find an expression for the availability. The symbols p_0, T_0, E_0, V_0, and S_0 will represent the pressure, temperature, energy, volume, and entropy of the surroundings, which are in the dead state. The system will have properties p and T and values of E, V, and S for the energy, volume, and entropy.

When the system changes state, an infinitesimal amount of heat δQ may be added; if δQ is negative, heat flows from the system. A certain amount of work will be delivered, and the maximum amount of work, $\delta W_{\max}$, that can be delivered is that work done when the heat flow is reversible. For the heat transfer to be reversible, it is necessary to place a reversible engine between the system and the surroundings (Figure 9.1). Let the temperature of the system rejecting the heat be T and the temperature of the surroundings be T_0. The work,

$$\delta W_q = \delta Q \frac{T_0 - T}{T} \qquad (9.2)$$

found from the Carnot efficiency, is always positive. The sign of the heat flow must be considered.

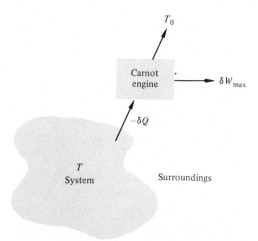

Figure 9.1 Development of availability.

Other forms of work may occur between the system and the surroundings. There may be a change in surface tension, electrical work, and so on. All these forms of work will be lumped in one term, δW. Consider, however, a change in the system volume, dV. This change of system volume is resisted by the pressure of the surroundings, p_0. The lumped work is resisted by the pressure-volume work, $p_0 \, dV$. The net work that can be delivered by the system is

$$\delta W - p_0 \, dV$$

The net work is the sum of all the individual forms of work:

$$\delta W_{\text{net}} = \delta W - p_0 \, dV + \frac{T_0 - T}{T} \, \delta Q$$

$$\delta W_{\text{net}} = \delta W - \delta Q - p_0 \, dV + T_0 \frac{\delta Q}{T}$$

The first law of thermodynamics for a closed system, $\delta Q = dE + \delta W$, yields the following, where the heat transfer is a reversible process:

$$\delta W_{\text{net}} = -dE - p_0 \, dV + T_0 \, dS \tag{9.3}$$

The term E is considered in the most general sense, including kinetic, potential, capillary energies, among others. Notice that if the function $dE + p_0 \, dV - T_0 \, dS$ decreases, the work increases. The maximum work that can be accomplished by any system at state 1 to state 0 can be represented by the decrease in this function:

$$W_{\text{max}} = -\int_1^0 (dE + p_0 \, dV - T_0 \, dS) \tag{9.4}$$

The function $E + p_0 V - T_0 \, dS$ is called the *closed-system availability*, $\mathscr{A}$. A similar function exists for open systems. Thus,

$$W_{\text{max}} = \mathscr{A}_1 - \mathscr{A}_0 \tag{9.5}$$

The availability, $\mathscr{A}$, is a property, but it is bound by the particular values of T_0 and p_0. However, the availability varies as a function of system state, and the change in availability of a system in going from state 1 to state 2 is

$$\mathscr{A}_2 - \mathscr{A}_1 = \int_1^2 d\mathscr{A} = \int_1^0 d\mathscr{A} + \int_0^2 d\mathscr{A} = \int_1^0 d\mathscr{A} - \int_2^0 d\mathscr{A}$$

$$\mathscr{A}_2 - \mathscr{A}_1 = \int_1^0 (dE + p_0 \, dV - T_0 \, dS) - \int_2^0 (dE + p_0 \, dV - T_0 \, dS)$$

$$\mathscr{A}_2 - \mathscr{A}_1 = (E_2 + p_0 V_2 - T_0 S_2) - (E_1 + p_0 V_1 - T_0 S_1) \tag{9.6}$$

As noted, the term E includes all energy terms the system mass possesses. In this text we have limited E to the traditional values, hence $E = U + \text{K.E.} + \text{P.E.}$ Equation (9.6) becomes

$$\mathscr{A}_2 - \mathscr{A}_1 = (U_2 - U_1) + [(\text{K.E.})_2 - (\text{K.E.})_1]$$
$$+ [(\text{P.E.})_2 - (\text{P.E.})_1] + p_0(V_2 - V_1) - T_0(S_2 - S_1) \qquad (9.7)$$

The concept of availability may be expressed in terms of another thermodynamic function, and since this function is tabulated, the availability may be calculated. The Gibbs function, or free energy, G, is defined as

$$G = H - TS$$
$$dG = dH - T\,dS - S\,dT$$
$$H = E + pV$$
$$dH = dE + p\,dV + V\,dp$$
$$dG = dE + p\,dV + V\,dp - T\,dS - S\,dT \qquad (9.8)$$

Conditions of constant temperature and pressure, where the temperature is T_0 and the pressure is p_0, yield for the Gibbs function

$$dG = dE + p_0\,dV - T_0\,dS \qquad (9.9)$$

Note that equation (9.9) is the same as equation (9.6) on the differential basis. The change in availability of a system therefore is equal to the change in the Gibbs function of the system at constant temperature and pressure:

$$d\mathscr{A} = (dG)_{T,p} \qquad (9.10)$$

This is particularly useful when evaluating the availability of systems in which chemical reactions occur. These reactions may often be considered at constant temperature and pressure.

The reversible work for a closed system changing from state 1 to state 2 is determined by finding the change in the availability:

$$W_{\text{rev}} = \mathscr{A}_1 - \mathscr{A}_2 \text{ kJ} \qquad (9.11)$$

Thus, the reversible work is associated with a decrease in the system's availability. Figure 9.2 illustrates this.

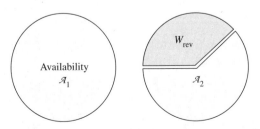

Figure 9.2 Illustration of the relationship between reversible work and availability change.

Figure 9.3 Relationship between reversible work, actual work, and irreversibility.

The actual work, W_{act}, that a system performs is less than the reversible work because of irreversibilities. The difference between the actual work and the reversible work is the irreversibility, I.

$$I = W_{rev} - W_{act} \tag{9.12}$$

Figure 9.3 illustrates the relationship between reversible work, actual work, and irreversibility.

The expression for the change of availability can be expanded to include all heat transfer in the change of entropy expression, and from this we can gain an insight into factors causing the availability change.

The change of entropy from equation (8.29) is

$$S_2 - S_1 = \sum_i \frac{Q_i}{T_i} - \Delta S_{prod}$$

If this is substituted into equation (9.6), the following results:

$$\mathcal{A}_2 - \mathcal{A}_1 = E_2 - E_1 + p_0(V_2 - V_1) - T_0 \sum_i \frac{Q_i}{T_i} - T_0 \Delta S_{prod}$$

From the first law,

$$E_2 - E_1 = \sum_i Q_i - W$$

the expression for the change of availability becomes

$$\underbrace{\mathcal{A}_2 - \mathcal{A}_1}_{\text{Availability change}} = \underbrace{\sum_i \left(1 - \frac{T_0}{T_i}\right) Q_i - [W - p_0(V_2 - V_1)]}_{\text{availability transfers}} - \underbrace{T_0 \Delta S_{prod}}_{\text{availability destruction}} \tag{9.13}$$

The first term on the right-hand side of equation (9.13) can be interpreted as the changes in availability associated with heat transfer into or out of the system. The second term on the right-hand side refers to work interactions between the system and the surroundings. The destruction of availability due to irreversibilities is shown by the third term on the right-hand side. This may also be interpreted as the loss of work potential, or irreversibility, I. Thus, the irreversibility expression given in terms of equation (9.12) may also be determined in terms of entropy production.

$$I = W_{rev} - W_{act} = T_0 \Delta S_{prod} \tag{9.14}$$

Sometimes it may be desirable to write the availability balance equation on a time basis. Taking the derivative of equation (9.13) with respect to time yields

$$\frac{d\mathcal{A}}{dt} = \sum_i \left(1 - \frac{T_0}{T_i}\right) \dot{Q}_i - \left[\dot{W} - p_0 \frac{dV}{dt}\right] - \dot{i} \tag{9.15}$$

When examining the entropic formulation of the second law, we noted that the entropy change of an isolated system is greater than or equal to zero. When examining the change of availability of an isolated system, we note that it is equal to the irreversibility,

$$\Delta \mathscr{A}_{isol} = -I_{isol} \qquad (9.16)$$

Thus, irreversibilities act to reduce the work potential, the change of availability, of an isolated system.

The availability balance affords us the opportunity when examining thermal systems to find ways to minimize the avoidable losses of availability. Looking at the terms that increase the entropy production unnecessarily is often the first step in establishing strategies in this regard.

Example 9.1

Three kg of water exist as a saturated vapor at a pressure of 2 MPa, a velocity of 50 m/s, and an elevation of 15 m. A change of state occurs such that the water is a saturated liquid at 500 kPa, the velocity is 10 m/s, and the elevation is 5 m. Determine the availabilities of the initial and final states and the change in availability. Assume $T_0 = 25°C$, $p_0 = 1$ atm, and $g = 9.8$ m/s².

Solution

Given: Water with initial and final states defined as well as the dead state conditions.

Find: The change in availability at the initial and final states and the net change.

Sketch and Given Data:

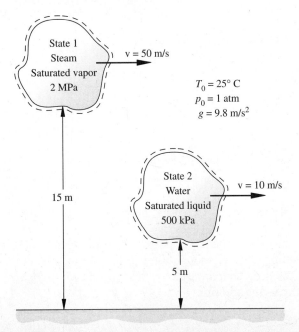

Figure 9.4

Assumptions:

1. The water is a closed system with equilibrium property values at the initial and final states.
2. $T_0 = 298°K$, $p_0 = 1$ atm, $g = 9.8$ m/s^2.

Analysis: The availability at any state is

$$\mathcal{A} = m\left[(u - u_0) + p_0(v - v_0) - T_0(s - s_0) + \frac{v^2}{2} + gz\right]$$

Determine the property values at the dead state by noting that at 25°C and 1 atm water is a subcooled liquid with properties essentially equal to those of a saturated liquid at 25°C. Thus, $u_0 = 103.95$ kJ/kg, $v_0 = 0.001\ 003$ m^3/kg, and $s_0 = 0.3626$ kJ/kg-K.

At state 1 the property values of water are $u_1 = 2600.34$ kJ/kg, $v_1 = 0.09963$ m^3/kg, and $s_1 = 6.3397$ kJ/kg-K. The availability at state 1 is

$$\mathcal{A}_1 = (3\text{ kg})\Big[(2600.34 - 103.95\text{ kJ/kg}) + (100\text{ N})(0.099\ 63 - 0.001\ 003\text{ m}^3/\text{kg})$$

$$- (298°\text{K})(6.3397 - 0.3626\text{ kJ/kg-K}) + \frac{50^2\text{ m}^2/\text{s}^2}{2(1000\text{ J/kJ})}$$

$$+ \frac{(9.8\text{ m/s}^2)(15\text{ m})}{(1000\text{ J/kJ})}\Big]$$

$$\mathcal{A}_1 = 2179.4\text{ kJ}$$

The property values at state 2 are $u_2 = 639.24$ kJ/kg, $v_2 = 0.001\ 093$ m^3/kg, and $s_2 = 1.8584$ kJ/kg-K. The availability at state 2 is

$$\mathcal{A}_2 = (3\text{ kg})\Big[(639.24 - 103.95\text{ kJ/kg}) + (100\text{ N})(0.001\ 093 - 0.001\ 003\text{ m}^3/\text{kg})$$

$$- (298°\text{K})(1.8584 - 0.3626\text{ kJ/kg/K}) + \frac{10^2\text{ m}^2/\text{s}^2}{2(1000\text{ J/kJ})}$$

$$+ \frac{(9.8\text{ m/s}^2)(15\text{ m})}{(1000\text{ J/kJ})}\Big]$$

$$\mathcal{A}_2 = 268.9\text{ kJ}$$

The change of availability between states 2 and 1 is

$$\mathcal{A}_2 - \mathcal{A}_1 = 268.9 - 2179.4 = -1910.5\text{ kJ}$$

Comments:

1. The availability of the system decreased, hence the negative sign on change of availability.
2. The availability at a given state is relative to the property values at the dead state.
3. When the availability change is sought (as is most often the situation), T_0 and p_0 are needed, but the dead state internal energy, specific volume, and entropy are not needed, as the terms add out.

Example 9.2
A quantity of air, 0.7 m³, is at 1500 kPa and 500°K with negligible velocity and at a zero datum. For $T_0 = 25°C$ and $p_0 = 100$ kPa, determine the availability of the air.

Solution

Given: Air at an initial temperature, pressure, and volume.

Find: The air's availability.

Sketch and Given Data:

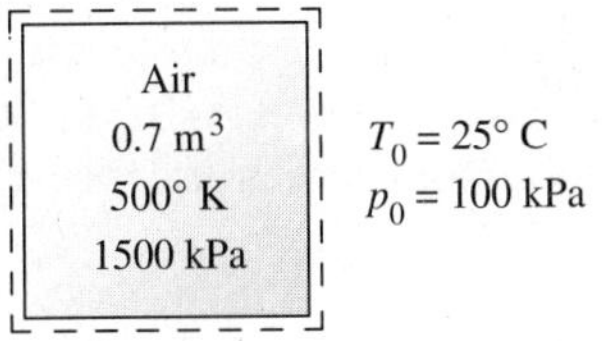

Figure 9.5

Assumptions:

1. Air is an ideal gas.
2. The air is a closed system.
3. The changes in kinetic and potential energies may be neglected.

Analysis: The availability for a closed system subject to the assumptions is

$$\mathcal{A}_1 = m[(u_1 - u_0) + p_0(v_1 - v_0) - T_0(s_1 - s_0)]$$

Employing the ideal-gas equations of state,

$$u_1 - u_0 = c_v(T_1 - T_0) = (0.7176 \text{ kJ/kg-K})(500 - 298°\text{K})$$

$$u_1 - u_0 = 144.95 \text{ kJ/kg}$$

$$v_1 = \frac{RT_1}{p_1} = \frac{(0.287 \text{ kJ/kg-K})(500°\text{K})}{(1500 \text{ kPa})} = 0.0956 \text{ m}^3/\text{kg}$$

$$v_0 = \frac{RT_0}{p_0} = \frac{(0.287 \text{ kJ/kg-K})(298°\text{K})}{(100 \text{ kPa})} = 0.8553 \text{ m}^3/\text{kg}$$

$$p_0(v_1 - v_0) = (100 \text{ kN/m}^2)(0.0956 - 0.8553 \text{ m}^3/\text{kg}) = -75.97 \text{ kJ/kg}$$

$$s_1 - s_0 = (1.0047 \text{ kJ/kg-K})\left(\ln \frac{500}{298}\right) - (0.287 \text{ kJ/kg-K})\left(\ln \frac{1500}{100}\right)$$

$$= -0.2573 \text{ kJ/kg-K}$$

$$T_0(s_1 - s_0) = (298°\text{K})(-0.2573 \text{ kJ/kg-K}) = -76.66 \text{ kJ/kg}$$

The mass is

$$m = \frac{p_1 V_1}{RT_1} = (1500 \text{ kN/m}^2)(0.7 \text{ m}^3)/[(0.287 \text{ kJ/kg-K})(500^\circ \text{K})]$$

$$m = 7.317 \text{ kg}$$

$$\mathscr{A}_1 = (7.317 \text{ kg})[+144.95 - 75.97 - (-76.66) \text{ kJ/kg}]$$

$$\mathscr{A}_1 = 1065.6 \text{ kJ}$$

Comments: The positive value of the availability indicates that the air has the potential to perform work.

Example 9.3

A 5-kg steel billet is heated to 1200°K and then quenched by placing it in 50 kg of water, initially at 20°C. The average specific heat of the steel is 0.5 kJ/kg-K. Consider the steel and water as forming an isolated system and determine the irreversibility for the process.

Solution

Given: Steel billet and quenching water at known initial conditions.

Find: The irreversibility.

Sketch and Given Data:

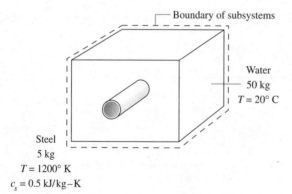

Steel
5 kg
$T = 1200^\circ$ K
$c_s = 0.5$ kJ/kg–K

Figure 9.6

Assumptions:

1. Steel and water form an isolated system.
2. No heat or work crosses the system boundary.
3. The changes in kinetic and potential energies are zero.
4. Water and steel are considered incompressible.

Analysis: The irreversibility of an isolated system is equal to minus the change of the system's availability (equation [9.16]). The change in the availability is the sum of the subsystems' availability changes,

$$\mathcal{A}_{isol} = [\Delta U + p_0\,\Delta V - T_0\,\Delta S]_{steel} + [\Delta U + p_0\,\Delta V - T_0\,\Delta S]_{water}$$

The internal energy changes are equal and opposite to one another; thus, they add out. The ΔV term is zero for each substance, as steel and water were assumed incompressible. The only term remaining is the entropy change term. Refer to Example 8.4 for its calculation.

The entropy production, the sum of the ΔS terms, is 4.11 kJ/K. Thus, the $\Delta\mathcal{A}$ of an isolated system becomes

$$\Delta\mathcal{A} = -(298°K)(4.11\text{ kJ/K}) = -1224.78\text{ kJ}$$

and the irreversibility is

$$I = 1224.78\text{ kJ}$$

Comments: The effect of the entropy production is to reduce the system's availability, its work potential. The energy of the system remains constant, however.   ▬

Example 9.4
A piston-cylinder contains helium at 200 psia, a volume of 2 ft³, and a temperature of 300°F. It receives heat at a temperature of 800°F from a solar furnace. The heat addition process in the piston-cylinder is constant-pressure and proceeds until the volume doubles. The dead state, and the conditions of the surroundings, is 77°F and 1 atm. Determine the availability change of the system.

Solution

Given: A system with known properties receives heat from a constant-temperature solar furnace.

Find: The change of system availability.

Sketch and Given Data:

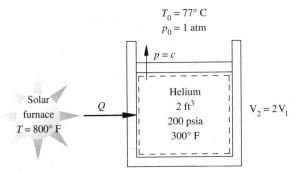

Figure 9.7

Assumptions:

1. The piston-cylinder is a closed system.
2. Helium may be treated as an ideal gas.

3. There is no heat transfer from the system to the surroundings.
4. The changes in kinetic and potential energies are zero.

Analysis: The change of system availability is given by equation (9.13):

$$\Delta \mathscr{A} = \left(1 - \frac{T_0}{T_i}\right) Q_i - [W - p_0(V_2 - V_1)] - T_0 \, \Delta S_{prod}$$

The helium properties may be found in Table A.1: $c_p = 1.241$ Btu/lbm-R, $R = 386.04$ ft-lbf/lbm-R.

The heat flow, Q_i, is not known at this point and must be found from a first-law analysis. The mass of helium within the cylinder is found from the ideal-gas equation of state.

$$m = \frac{p_1 V_1}{R T_1} = \frac{(200 \text{ lbf/in.}^2)(144 \text{ in.}^2/\text{ft}^2)(2 \text{ ft}^3)}{(386 \text{ ft-lbf/lbm-R})(760°\text{R})} = 0.196 \text{ lbm}$$

For a constant-pressure process, the heat transfer is equal to the change of system enthalpy. For an ideal gas, $\Delta H = mc_p(T_2 - T_1)$. Before this can be solved, T_2 must be determined. Again, for a constant-pressure process using an ideal gas, $T/V = C$ and thus, $T_2 = 2T_1 = 2(760) = 1520°\text{R}$.

$$Q = \Delta H = (0.196 \text{ lbm})(1.241 \text{ Btu/lbm-R})(1520 - 760°\text{R})$$

$$Q = 184.8 \text{ Btu}$$

The system work, the work the helium did in moving the piston, is

$$W = \int p \, dV = p(V_2 - V_1) = \frac{(200 \text{ lbf/in.}^2)(144 \text{ in.}^2/\text{ft}^2)(4 - 2 \text{ ft}^3)}{(778.16 \text{ ft-lbf/Btu})}$$

$$W = 74.02 \text{ Btu}$$

The work against the surroundings is

$$p_0(V_2 - V_1) = \frac{(14.7 \text{ lbf/in.}^2)(144 \text{ in.}^2/\text{ft}^2)(4 - 2 \text{ ft}^3)}{(778.16 \text{ ft-lbf/Btu})} = 5.44 \text{ Btu}$$

The change of entropy of the system is

$$\Delta S_{sys} = mc_p \ln [T_2/T_1] = (0.196 \text{ lbm})(1.241 \text{ Btu/lbm-R})(\ln 2)$$

$$\Delta S_{sys} = 0.1686 \text{ Btu/R}$$

The expression representing that portion of the heat added that can be used for work is

$$\left(1 - \frac{T_0}{T_i}\right) Q_i = \left(1 - \frac{537}{1260}\right)(184.8 \text{ Btu}) = 106.04 \text{ Btu}$$

The entropy change of the furnace as it transfers the heat at constant temperature is

$$\Delta S_{\text{furn}} = \frac{Q}{T} = \frac{-184.8 \text{ Btu}}{1260°\text{R}} = -0.1467 \text{ Btu/R}$$

The entropy production is found by adding this to the system entropy change, thus

$$\Delta S_{\text{prod}} = +0.1686 - 0.1467 = 0.0219 \text{ Btu/R}$$

$$T_0 \Delta S_{\text{prod}} = (537°\text{R})(0.0219 \text{ Btu/R}) = 11.76 \text{ Btu}$$

The change in system availability is found by substituting into equation (9.13).

$$\Delta \mathscr{A} = 106.04 - 74.02 + 5.44 - 11.76 = 25.7 \text{ Btu}$$

Comment: In this problem the value of 106.04 Btu represents the work that a heat engine could have performed if it received the heat. The piston expanded, producing a net work of 68.58 Btu (the algebraic sum of 74.02 and 5.44). In addition the entropy of the system increased, resulting in a net entropy production when added to the entropy decrease of the furnace. The 11.76 Btu represents the work potential irrevocably lost due to irreversibilities. ▪

9.3 FLOW AVAILABILITY

Let us consider a steady-state open system that performs work and exchanges heat with the surroundings. The first law for such a system would be

$$\dot{W} = \dot{m}(e_1 - e_2) + \dot{Q} \tag{9.17}$$

where $e = h + v^2/2 + gz$. The process for such a device is shown by the expansion 1 to 2 in Figure 9.8.

In addition in Figure 9.8, a reversible path connecting states 1 and 2 is shown, 1-a-b-2. The process 1-a is an isentropic expansion, a-b is an isothermal heat addition, and b-2 is an isentropic compression. The reversible work for process 1-2 may be found by summing the works for the reversible processes connecting states 1 and 2.

$$\dot{W}_{1a} = \dot{m}(e_1 - e_a)$$
$$\dot{Q}_{1a} = 0$$
$$\dot{W}_{ab} = \dot{m}(e_a - e_b) + \dot{Q}_{ab}$$
$$\dot{Q}_{ab} = \dot{m}T_0(s_2 - s_1)$$
$$\dot{W}_{b2} = \dot{m}(e_b - e_2)$$
$$\dot{Q}_{b2} = 0$$

The reversible power is

$$\dot{W}_{\text{rev}} = \dot{W}_{1a} + \dot{W}_{ab} + \dot{W}_{b2}$$
$$\dot{W}_{\text{rev}} = \dot{m}(e_1 - e_2) - \dot{m}T_0(s_1 - s_2) \tag{9.18}$$

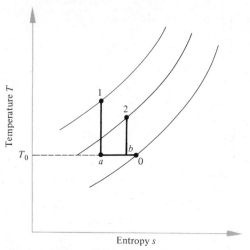

Figure 9.8 A T-s diagram illustrating reversible and irreversible processes between states 1 and 2.

Equation (9.18) represents the maximum power the device can produce when operating between fixed states 1 and 2.

We see in determining the reversible work for the process 1-2 that the reversible process never reached the surrounding pressure, p_0. A question arises as to what maximum amount of work, that is reversible work, could be obtained from a substance at state 1 if the final state were in equilibrium with the surroundings at p_0 and T_0 (i.e., state 0 in Figure 9.8). In this case the reversible work path is 1-a, a-0. The reversible work, referenced to T_0 and p_0, is called *flow availability, ψ*.

$$\dot{W}_{1\text{-}0} = \dot{W}_{1\text{-}a} + \dot{W}_{a\text{-}0}$$

$$\dot{W}_{a\text{-}0} = \dot{m}(e_a - e_0) + \dot{Q}_{a\text{-}0}$$

$$\dot{Q}_{a\text{-}0} = \dot{m}T_0(s_0 - s_1)$$

$$\dot{W}_{1\text{-}0} = \dot{m}(e_1 - e_0) + \dot{m}T_0(s_0 - s_1)$$

Including the kinetic and potential energies and writing the maximum work on a unit mass basis yields

$$\psi_1 = \left(h_1 + \frac{v_1^2}{2} + gz_1 \right) - (h_0 + gz_0) - T_0(s_1 - s_0) \tag{9.19}$$

Notice that the final velocity is zero because the system must be in equilibrium with the surroundings. Equation (9.18) may be written as

$$\dot{W}_{\text{rev}} = \dot{m}(\psi_1 - \psi_2) \tag{9.20}$$

The availability expression for situations involving fluid flow, equation (9.19), is also called exergy.

A flow availability balance may be determined as well. It is possible to derive such a balance as was done for the second law in Chapter 8, or we can opt to modify directly the rate equation, equation (9.15), for a closed system. Choosing the latter and noting that equation (9.15) has been modified to reflect the availability with the flow of material into and out of the control volume, we get

$$\underbrace{\frac{d\mathcal{A}_{cv}}{dt}}_{\substack{\text{Rate of} \\ \text{availability change}}} = \underbrace{\left(1 - \frac{T_0}{T_i}\right)\dot{Q}_i - \left(\dot{W}_{cv} - p_0 \frac{dV_{cv}}{dt}\right) + \sum_i \dot{m}_i\psi_i - \sum_e \dot{m}_e\psi_e}_{\text{rate of availability transfer}} - \underbrace{\dot{I}_{cv}}_{\substack{\text{rate of} \\ \text{availability destruction}}} \quad (9.21)$$

where $\dot{m}_i$ represents the mass flows into the control volume and $\dot{m}_e$ represents the mass flows out of the control volume. The irreversibility term, $\dot{I}_{cv}$, may be found by evaluating $T_0\,\Delta\dot{S}_{prod}$ from an availability balance.

Equation (9.21) is most often evaluated for steady-state, steady-flow situations with a single fluid. Thus, $d\mathcal{A}_{cv}/dt = 0$, $dV_{cv}/dt = 0$, $\dot{m}_i = \dot{m}_e = \dot{m}$. Equation (9.21) becomes

$$0 = \left(1 - \frac{T_0}{T_i}\right)\dot{Q}_i - \dot{W}_{cv} + \dot{m}(\psi_1 - \psi_2) - \dot{I}_{cv} \quad (9.22)$$

Example 9.5

A turbine receives 20 kg/s of air at $1200\,°K$ and 800 kPa with a velocity of 50 m/s. The air leaves at 100 kPa and $690\,°K$ with a velocity of 30 m/s. The heat loss from the turbine casing is 500 kW at $500\,°K$. Neglect changes in potential energy. Find the availability supplied, the work, and the irreversibility. $T_0 = 25\,°C$ and $p_0 = 100$ kPa.

Solution

Given: Air entering and leaving a control volume at known states with heat transfer and work from the control volume.

Find: The availability of the air entering the turbine, the portion converted to work, and the irreversibility.

Sketch and Given Data:

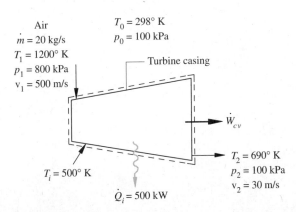

Figure 9.9

Assumptions:

1. Air is an ideal gas with constant specific heats.
2. The process is steady-state.
3. Heat transfer from the control volume occurs at a constant temperature.
4. The change in potential energy may be neglected.

Analysis: A glance at the steady-state availability rate, equation (9.22), indicates that the work from the control volume is needed. In addition, inspection of the term for specific availability change indicates that the entropy change must be determined. Employing the first law for an open system yields

$$\dot{Q} + \dot{m}(h + \text{k.e.})_{\text{in}} = \dot{W} + \dot{m}(h + \text{k.e.})_{\text{out}}$$

Substitute numerical values; use the ideal-gas equations of state for enthalpy.

$$\dot{W} = -500 \text{ kW} + (20 \text{ kg/s}) \Bigg[(1.0047 \text{ kJ/kg-K})$$

$$(1200 - 690°\text{K}) + \frac{(50^2 - 30^2 \text{m}^2/\text{s}^2)}{2(1000 \text{ J/kJ})} \Bigg]$$

$$\dot{W} = 9763.9 \text{ kW}$$

$$w = \dot{W}/\dot{m} = 488.2 \text{ kJ/kg}$$

The change of entropy needed in the availability expression is $(s_1 - s_2)$; note that this is the negative of the traditional difference.

$$s_1 - s_2 = c_p \ln [T_1/T_2] - R \ln [p_1/p_2]$$

$$s_1 - s_2 = (1.0047 \text{ kJ/kg-K}) \ln (1200/690)$$

$$- (0.287 \text{ kJ/kg-K}) \ln (800/100)$$

$$s_1 - s_2 = -0.0408 \text{ kJ/kg-K}$$

The specific flow availability is

$$\psi_1 - \psi_2 = (h_1 - h_2) - T_0(s_1 - s_2) + \frac{v_1^2 - v_2^2}{2}$$

and substituting yields

$$\psi_1 - \psi_2 = (1.0047 \text{ kJ/kg-K})(1200 - 690°\text{K})$$

$$- (298°\text{K})(-0.0408 \text{ kJ/kg-K}) + \frac{(50^2 - 30^2 \text{m}^2/\text{s}^2)}{2(1000 \text{ J/kJ})}$$

$$\psi_1 - \psi_2 = 524.5 \text{ kJ/kg}$$

Solve equation (9.22) for $\dot{I}_{cv}$.

$$\dot{I}_{cv} = \left(1 - \frac{T_0}{T_i}\right) \dot{Q}_i - \dot{W}_{cv} + \dot{m}(\psi_1 - \psi_2)$$

$$\dot{I}_{cv} = (1 - 298/500)(-500) - 9763.9 + 20(525.4) \text{ kW}$$

$$\dot{I}_{cv} = 542.1 \ kW$$

$$\frac{\dot{I}_{cv}}{\dot{m}} = \frac{542.1}{20} = 27.1 \ kJ/kg$$

An availability balance yields useful information as to where the availability entering the turbine went:

Net availability to turbine $(\psi_1 - \psi_2) = 525.4 \ kJ/kg$

Availability consumption

$$\text{Work} = 488.2 \ kJ/kg \ (92.9\%)$$
$$\text{Irreversibility} = 27.1 \ kJ/kg \ (5.2\%)$$
$$\text{Portion of heat transferred} = \frac{202}{20} = 10.1 \ kJ/kg \ (1.9\%)$$
$$\text{Total} = 525.4 \ kJ/kg \ (100\%)$$

Comments:

1. The irreversibility term may be found from equation (9.22) or from evaluating $T_0 \ \Delta S_{prod}$.
2. The availability balance is useful in finding design improvements for increasing the transfer of availability into work. ▬

Example 9.6

An adiabatic heat exchanger in a power plant receives 50,000 lbm/hr of steam with a quality of 95% and a pressure of 20 psia and condenses it to a saturated liquid at 20 psia. In the process the heat exchanger raises the temperature of 250,000 lbm/hr of water from 100°F to an unknown temperature. Determine the availability changes of the steam and water and the irreversibility for the heat exchanger. Neglect changes in kinetic and potential energies. $T_0 = 77°F$ and $p_0 = 1$ atm.

Solution

Given: An adiabatic heat exchanger condenses steam and heats water. Steam and water conditions are given.

Find: The changes in availability for the steam and water and the irreversibility for the heat exchanger.

Sketch and Given Data:

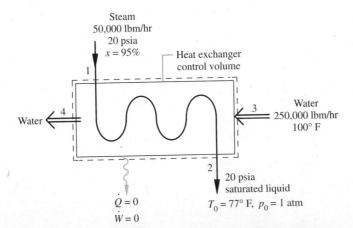

Figure 9.10

Assumptions:

1. The process is steady-state.
2. Heat flow to the surroundings and the work is zero.
3. The changes in kinetic and potential energies are zero.
4. The water does not vaporize; the pressure is sufficiently high to assure it remains a subcooled liquid.

Analysis: Before proceeding to the availability analysis, we must first determine the outlet condition of the water from a first-law analysis of the heat exchanger. Invoking assumptions 2 and 3, the first law becomes

$$\dot{m}_s h_{s_{in}} + \dot{m}_w h_{w_{in}} = \dot{m}_s h_{s_{out}} + \dot{m}_w h_{w_{out}}$$

The properties of steam and water may be found from the steam tables to be $h_{w_{in}} = 67.92$ Btu/lbm, $s_{w_{in}} = 0.12895$ Btu/lbm-R, $h_{s_{in}} = 1108.4$ Btu/lbm, $s_{s_{in}} = 1.6620$ Btu/lbm-R, $h_{s_{out}} = 196.4$ Btu/lbm, and $s_{s_{out}} = 0.3355$ Btu/lbm-R. The enthalpy of the water leaving the heat exchanger is

$$h_{w_{out}} = 67.92 \text{ Btu/lbm} + \frac{(50,000 \text{ lbmstm/hr})(1108.4 - 196.4 \text{ Btu/lbmstm})}{(250,000 \text{ lbmwtr/hr})}$$

$$h_{w_{out}} = 250.32 \text{ Btu/lbm}$$

and invoking assumption 4, $s_{w_{out}} = s_f = 0.4110$ Btu/lbm-R.

The change of availability of the steam is

$$(\psi_1 - \psi_2)_s = (h_1 - h_2) - T_0(s_1 - s_2)$$

$$(\psi_1 - \psi_2)_s = (1108.4 - 196.4 \text{ Btu/lbm}) - (537°R)(1.6620 - 0.3355 \text{ Btu/lbm-R})$$

$$(\psi_1 - \psi_2)_s = 199.7 \text{ Btu/lbm}$$

$$\dot{m}(\psi_1 - \psi_2)_s = (199.7)(50,000) = 9.985 \times 10^6 \text{ Btu/hr}$$

Write the availability change of the water as a positive quantity.

$$(\psi_4 - \psi_3)_w = (h_4 - h_3) - T_0(s_4 - s_3)$$

$$(\psi_4 - \psi_3)_w = (250.32 - 67.92 \text{ Btu/lbm}) - (537°R)$$

$$(0.4110 - 0.12895 \text{ Btu/lbm-R})$$

$$(\psi_4 - \psi_3) = 30.94 \text{ Btu/lbm}$$

$$\dot{m}(\psi_4 - \psi_3)_w = (30.94)(250,000) = 7.735 \times 10^6 \text{ Btu/hr}$$

The irreversibility may be found by finding the entropy production from equation (8.42), noting there is no heat transfer, and multiplying it by T_0.

$$0 = \dot{m}_s s_{s_{in}} + \dot{m}_w s_{w_{in}} - \dot{m}_s s_{s_{out}} - \dot{m}_w s_{w_{out}} + \Delta\dot{S}_{prod}$$

$$0 = (50,000 \text{ lbm/hr})(1.6620 - 0.3355 \text{ Btu/lbm})$$

$$+ (250,000 \text{ lbm/hr})(0.12895 - 0.4110 \text{ Btu/lbm}) + \Delta\dot{S}_{prod}$$

$$\Delta\dot{S}_{prod} = 4187.5 \text{ Btu/hr-R}$$

The irreversibility is then

$$\dot{i}_{cv} = T_0 \, \Delta\dot{S}_{prod} = (537°R)(4187.5 \text{ Btu/hr-R}) = 2.249 \times 10^6 \text{ Btu/hr}$$

Comment: An availability balance is illustrative: the steam entered with an availability of 9.985×10^6 Btu/hr; of that, 7.735×10^6 Btu/hr or 77.5% was transferred to the water; the irreversibility indicates that the balance was destroyed by irreversibilities in the transfer process.

9.4 SECOND-LAW EFFICIENCY

It is important to conserve energy and energy's quality, its work potential. Conserving energy is made possible by using less for a given task or improving the first-law efficiency for a process—for example, turning off extra lights or driving cars with higher gas mileage. Additionally, it is possible to conserve energy quality by matching source quality and system quality needs, improving the second-law efficiency of the process. This points in the direction of using energy with a high quality, or availability, in those situations requiring it but not in those that do not. This assures that people in the near future will have the energy resources necessary. We should examine the practice of burning oil, which has a high availability, for a situation like space heating, which requires heat with a low availability.

Processes

The second-law efficiency for a process may be defined as

$$\eta_2 = \frac{\text{the change of availability of the system}}{\text{the change of availability of the source}} \qquad (9.23)$$

We will consider three types of processes: power-producing, power-consuming, and heat-transfer. Figure 9.11 illustrates the steady-state expansion of a fluid through

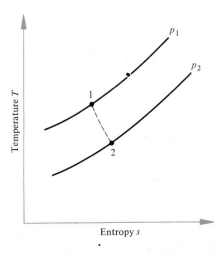

Figure 9.11 A *T-s* diagram showing adiabatic expansion in a turbine.

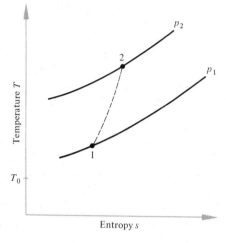

Figure 9.12 A T-s diagram for a pump or compressor.

an adiabatic turbine in an irreversible process. In this instance the change of the availability of the system is the actual work output of the turbine, w_{sys}. The actual work produced is less than the change of availability because irreversibilities have diminished it. Thus,

$$\Delta \psi_{\text{source}} = w_{\text{rev}}$$

and the second-law efficiency is

$$\eta_2 = \frac{w_{\text{sys}}}{w_{\text{rev}}} \tag{9.24}$$

A power-consuming device such as a pump or compressor uses work to increase the availability of the fluid passing through the device, shown in Figure 9.12. The increase in availability is manifested as an increase in pressure and/or temperature of the fluid. The change of availability of the system is equal to the reversible work; the change of availability of the source is equal to the work input from the source. Writing this so the availability change is positive yields

$$\eta_2 = \frac{\psi_2 - \psi_1}{w_{\text{source}}} \tag{9.25}$$

The last process we will consider is heating. Let us consider a solar collector, used to heat air or water, which in turn is used for home heating. The collector receives a total amount of energy Q_A from the sun at temperature T_A. Of this total energy some is lost to the surroundings Q_C and the remainder Q_B is transferred to the water at temperature T_B. Figure 9.13 illustrates this process. Note that $Q_A \neq Q_B$. The changes of the system and source availabilities are

$$\Delta \mathscr{A}_{\text{sys}} = Q_B \left(1 - \frac{T_0}{T_B} \right)$$

$$\Delta \mathscr{A}_{\text{source}} = Q_A \left(1 - \frac{T_0}{T_A} \right)$$

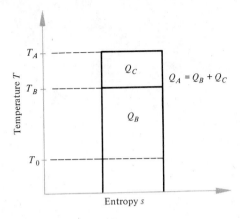

Figure 9.13 A T-s diagram for a nonadiabatic solar collector.

and

$$\eta_2 = \frac{1 - (T_0/T_B)}{1 - (T_0/T_A)}\, \eta_1 \tag{9.26}$$

where

$$\eta_1 = \frac{Q_B}{Q_A}$$

The first-law efficiency of the collector, η_1, is a measure of the adiabatic effectiveness of the collector.

For all heat exchangers the change of availability of the system in equation (9.23) is the availability change of the cold fluid, and the change of availability of the source in equation (9.23) corresponds to the availability change of the hot fluid. The adiabatic effectiveness is included in these terms and cannot be separated as in equation (9.26). Thus, the second-law efficiency of the heat exchanger in Example 9.6 is 77.5%.

Cycles

The second-law efficiency for a cycle is slightly different than for a process in that a cycle's purpose is to perform a given task, be it cooling or power production. The second-law efficiency for a cycle is defined as

$$\eta_2 = \frac{\text{minimum availability change necessary to perform a given task}}{\text{actual availability change required to perform a given task}} \tag{9.27}$$

We will analyze two cycles, a general heat engine cycle and a refrigeration cycle. Figure 9.14 illustrates a T-s diagram for a general heat engine cycle, 1234. In addition,

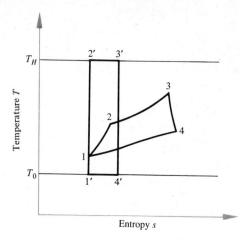

Figure 9.14 A T-s diagram for a general heat engine cycle.

a Carnot heat engine is illustrated by the cycle $1'2'3'4'$. The task of the heat engine is to perform work, w_{net}. In performing the work the cycle operates between temperature limits T_1 and T_3. The high- and low-temperature heat reservoirs are at T_H and T_0. The energy supplied by the high-temperature reservoir is Q_H; the availability change is

$$\Delta \mathscr{A}_{act} = \left(1 - \frac{T_0}{T_H}\right) Q_H$$

What is the minimum availability change required? This is the net work of a Carnot engine operating between T_H and T_0, as illustrated in Figure 9.14. Thus,

$$\Delta \mathscr{A}_{min} = W_{net}$$

The second-law efficiency is

$$\eta_2 = \frac{W_{net}}{(1 - T_0/T_H)Q_H} = \frac{W_{net}/Q_H}{1 - T_0/T_H}$$

$$\eta_2 = \eta_1/\eta_{Carnot} \tag{9.28}$$

If the actual cycle has many reversibilities, it will cause the first-law efficiency to be low. Consequently, the second-law efficiency will likewise be low. The minimum available energy required for the task is the net work of a Carnot engine operating between the reservoir temperature limits.

For a refrigeration system, illustrated in Figure 9.15, the given task is to provide cooling, Q_{4-1}. The actual refrigeration cycle is 1234. To produce this amount of

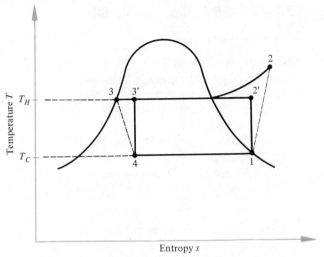

Figure 9.15 A T-s diagram for a refrigeration system.

cooling, work must be supplied; hence, the availability actually used is W_{net}. We must determine the minimum available energy required to provide cooling, $Q_{4\text{-}1}$. This is the cooling supplied by a reversed Carnot cycle, $1'2'3'4$. The minimum availability change is equal to the net work of the reversed Carnot cycle.

$$\Delta \mathscr{A}_{min} = \left(\frac{T_H}{T_C} - 1\right) Q_{4\text{-}1}$$

The second-law efficiency is

$$\eta_2 = \frac{(T_H/T_C - 1)Q_{4\text{-}1}}{W_{net}}$$

Let

$$\beta = COP$$

$$\eta_2 = \frac{\beta}{\beta_{Carnot}} \tag{9.29}$$

Second-law efficiencies may be used in energy-conservation programs, evaluating the available energy usage. The first-law efficiency evaluates the energy quantity utilization, whereas the second-law efficiency evaluates the energy quality utilization.

9.5 AVAILABLE ENERGY—A SPECIAL CASE OF AVAILABILITY

Practicing engineers often are concerned with steady-state, steady-flow devices that transfer heat. When we examine the change of availability for those devices, notably heat exchangers, it reduces to a simple expression. This expression is identical to the concept of available energy that has been used by engineers for a long time and is referenced in many application areas, notably in power plant design and analysis.

To determine the available energy of any heat source, it is possible to imagine the heat source's supplying energy to a reversible heat engine. For any engine, even the Carnot engine, only a portion of the heat received can be used for work. This is the available portion of the heat supplied. The rest of the heat is energy unavailable to do work. Thus,

Energy added = available energy + unavailable energy

$$Q \quad = \quad \text{A.E.} \quad + \quad \text{U.E.} \tag{9.30}$$

The Carnot engine produces the maximum possible work for a given heat input, so this will be equal to the available portion of the heat supplied. Let us consider that heat is added to a system over a temperature range from T_1 to T_2. This may be approximated by a sum of infinitesimal amounts of heat δQ, added at a temperature T. This infinitesimal amount of heat will be added to an imaginary differential Carnot engine. Figure 9.16 illustrates this. The total shaded area is the amount of heat, δQ. The Carnot engine must operate between two temperature limits. The lowest temperature at which it can reject heat is the temperature of the surroundings, T_0. If the temperature of heat rejection were below that of the surroundings, heat would have to be transferred from a cooler body to a warmer body—an impossibility.

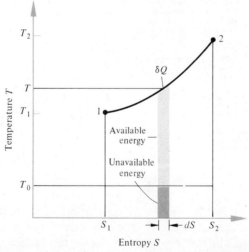

Figure 9.16 A T-s diagram illustrating the available and unavailable portions of a differential amount of heat.

The amount of work possible for the Carnot engine, δW_{max}, can be related to the heat added by the Carnot efficiency; hence, using equations (7.1) and (7.8) yields

$$\delta W_{max} = \delta Q \left(1 - \frac{T_0}{T}\right)$$

however,

$$\delta(\text{A.E.}) = \delta W_{max}$$

$$\delta(\text{A.E.}) = \delta Q \left(1 - \frac{T_0}{T}\right)$$

We now integrate from state 1 to state 2:

$$\int_1^2 \delta(\text{A.E.}) = \int_1^2 \delta Q - T_0 \int_1^2 \frac{\delta Q}{T}$$

$$(\text{A.E.})_{1\text{-}2} = Q_{1\text{-}2} - T_0(S_2 - S_1) \tag{9.31}$$

The unavailable energy from equation (9.30) is

$$(\text{U.E.})_{1\text{-}2} = + T_0(S_2 - S_1) \tag{9.32}$$

Equations (9.31) and (9.32) apply for a reversible or irreversible process between states 1 and 2.

A Special Case — Turbines

Practice in a field sometimes differs with classical presentations. Such is the case in turbine analysis. In practice a turbine is considered to be adiabatic, with an expansion as illustrated in Figure 9.17. In the ideal turbine the fluid expands isentropically from

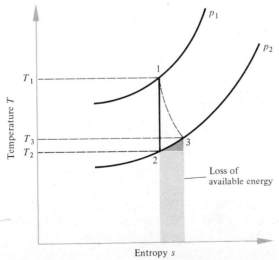

Figure 9.17 A T-s diagram for flow through a turbine.

inlet to exit, or state 1 to 2. The maximum work is the isentropic enthalpy drop from
inlet to exit. This applies to a turbine stage, as well as to the overall turbine. Thus,

$$(a.e.)_{1\text{-}2} = w_{max} = (h_1 - h_2)_{s=c} \text{ kJ/kg} \quad [\text{Btu/lbm}] \qquad (9.33)$$

However, in actual turbines irreversibilities do not permit isentropic expansion of the
fluid. The ideal exit pressure is p_2, and the actual fluid exit is at this pressure but at
some temperature greater than T_2. Since the fluid does not expand to the temperature
T_2, it does not utilize all its available energy; thus, there is a loss of available energy
utilization by the turbine. If we look at the difference between the availabilities at
states 2 and 3, we find

$$\psi_3 - \psi_2 = (h_3 - h_2) - T_0(s_3 - s_2)$$

which is the energy that could be used for work. Comparing this to equation (9.33),
we find that the loss of available energy due to irreversible expansion is

$$(a.e.)_{loss} = -T_0(s_3 - s_2)$$

The loss of available energy is represented by the rectangular lightly shaded area in
Figure 9.17. The darkened triangular area is the available energy of the exit fluid that
could drive a Carnot engine.

At this point let us reflect on the adiabatic turbine. The isentropic enthalpy drop
across the turbine is the available energy of the entering fluid. In an actual turbine,
irreversibilities decrease this amount by an amount $T_0(s_3 - s_2)$. The turbine can only
use an amount $(h_1 - h_3)$; the rest is lost from the turbine's point of view.

We have seen that the second-law efficiency compares the change of availability
across a turbine to the work delivered by the turbine. A more commonly encountered
turbine efficiency is known by several names—the *turbine internal efficiency*, the
turbine adiabatic efficiency, or the *turbine isentropic efficiency*. Regardless of the
name, this efficiency compares the work the fluid performed across the turbine,
the actual enthalpy change, to the isentropic enthalpy change or the available energy
change of the fluid across the turbine. Thus,

$$\eta_{isen} = \frac{\Delta h_{act}}{\Delta h_{isen}}$$

Referring to Figure 9.17, this becomes

$$\eta_{isen} = \frac{h_1 - h_3}{h_1 - h_2} \qquad (9.34)$$

Example 9.7
A constant-temperature system at 720°K receives 500 kJ of heat from a constant-
temperature heat reservoir at 835°K. The temperature of the surroundings, T_0, is
280°K. Determine the available portion of the heat given up by the reservoir and
received by the system, and the net loss of available energy.

Solution

Given: A given amount of heat added to a constant-temperature system from a constant-temperature reservoir.

Find: The available energy change of the heat reservoir and the net change of available energy.

Sketch and Given Data:

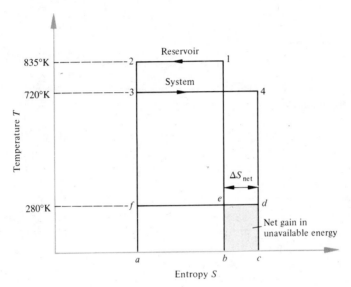

Figure 9.18

Assumptions: The rejection and addition processes are reversible and isothermal.

Analysis: The determination of the entropy change for an isothermal process is

$$\Delta S = \frac{Q}{T}$$

In this case both processes are isothermal, so the entropy change of the reservoir is

$$(S_2 - S_1) = \frac{-500 \text{ kJ}}{835°\text{K}} = -0.5988 \text{ kJ/K}$$

and of the system is

$$(S_4 - S_3) = \frac{500 \text{ kJ}}{720°\text{K}} = 0.6944 \text{ kJ/K}$$

The change of available energy of the reservoir is

$$(A.E.)_{1\text{-}2} = Q_{1\text{-}2} - T_0(S_2 - S_1) = (-500 \text{ kJ}) - (280°\text{K})(-0.5988 \text{ kJ/K})$$
$$(A.E.)_{1\text{-}2} = -332.34 \text{ kJ}$$

The change of available energy of the system is

$$(A.E.)_{3\text{-}4} = Q_{3\text{-}4} - T_0(S_4 - S_3) = (+500 \text{ kJ}) - (280°\text{K})(0.6944 \text{ kJ/K})$$
$$(A.E.)_{3\text{-}4} = +305.56 \text{ kJ}$$

The net change of available energy is

$$(A.E.)_{net} = (A.E.)_{1\text{-}2} + (A.E.)_{3\text{-}4} = -332.34 + 305.56 = -26.78 \text{ kJ}$$

The negative value indicates a loss of available energy due to irreversibilities in the heat-transfer process.

Comment: A T-S diagram is very useful in visualizing heat, available energy, and unavailable energy. The first law states that energy is conserved, so the areas 12*ab* and 34*ca* must be equal. We can determine graphically that $(S_4 - S_3)$ must be greater than $|S_2 - S_1|$ for the first law to be valid. Thus, there is a net increase in entropy due to the irreversible heat-transfer process. This is denoted as ΔS_{net}. If the heat transfer between the reservoir and the system were done reversibly and the system and reservoir temperatures were equal, there would be no net entropy increase. The available portion of the heat when it leaves the reservoir is denoted by area 12*fe*, and the available portion of the heat the system receives is denoted by area 34*df*. The unavailable energy of the heat leaving the reservoir is denoted by area *efab* and the unavailable portion of the heat entering the system is area *fdca*. The net gain in unavailable energy is denoted by area *edcb;* notice that this is equal in magnitude to the net loss of available energy. It is sometimes easier to find the net gain of unavailable energy $([U.E.]_{net} = T_0 \Delta S_{net})$ than the net loss of available energy.   ▬

Example 9.8
An adiabatic turbine receives 5 kg/s of a gas ($c_p = 1.09$ kJ/kg-K and $k = 1.3$) at 700 kPa and 1000°C and discharges it at 150 kPa and 665°C. Determine the second-law and isentropic efficiencies of the turbine. $T_0 = 25$°C and $p_0 = 1$ atm.

Solution

Given: The inlet and exit states of an adiabatic turbine as well as gas property information.

Find: The second-law and isentropic efficiencies.

Sketch and Given Data:

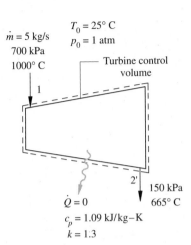

$T_0 = 25°$ C
$\dot{m} = 5$ kg/s
$p_0 = 1$ atm
700 kPa
1000° C

Turbine control
volume

1

2'   150 kPa
$\dot{Q} = 0$ 665° C
$c_p = 1.09$ kJ/kg–K
$k = 1.3$

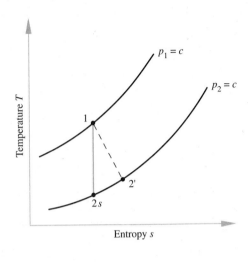

Figure 9.19

Assumptions:

1. Fluid is an ideal gas with constant specific heats.
2. The conditions are steady-state.
3. The heat loss is zero.
4. The changes in kinetic and potential energies may be neglected.

Analysis: From the information about the ideal gas, determine the gas constant.

$$k = c_p/c_v = 1.3 \quad \text{and} \quad c_p = 1.09 \quad \text{thus } c_v = 0.838 \text{ kJ/kg-K}$$
$$R = c_p - c_v \quad \text{thus } R = 0.252 \text{ kJ/kg-K}$$

The second-law efficiency for a turbine is the work that the turbine performs, or indicated work, in this situation determined by a first-law analysis, divided by the change in availability of the fluid in going from state 1 to state 2'.

$$\eta_2 = \frac{w_{\text{sys}}}{w_{\text{rev}}}$$

The system work is found from a first-law analysis of the turbine:

$$\dot{Q} + \dot{m}[h_1 + (\text{k.e.})_1 + (\text{p.e.})_1] = \dot{W} + \dot{m}[h_{2'} + (\text{k.e.})_{2'} + (\text{p.e.})_{2'}]$$

Invoking assumptions 2, 3, and 4 yields

$$w_{\text{sys}} = (h_1 - h_{2'}) = c_p(T_1 - T_{2'}) = (1.09 \text{ kJ/kg-K})(1273 - 938°\text{K})$$
$$w_{\text{sys}} = 365.2 \text{ kJ/kg}$$

The reversible work or change of availability is

$$(\psi_1 - \psi_{2'}) = (h_1 - h_{2'}) - T_0(s_1 - s_{2'})$$

$$(\psi_1 - \psi_{2'}) = 365.2 \text{ kJ/kg} - (298°K)$$

$$\left[(1.09 \text{ kJ/kg-K}) \left(\ln \left(\frac{1273}{938} \right) \right) - (0.252 \text{ kJ/kg-K}) \left(\ln \left(\frac{700}{150} \right) \right) \right]$$

$$(\psi_1 - \psi_{2'}) = 381.6 \text{ kJ/kg}$$

The second-law efficiency is

$$\eta_2 = \frac{365.2}{381.6} = 0.956 \quad \text{or } 95.6\%$$

The isentropic efficiency is found by dividing the actual enthalpy change across the turbine by the isentropic enthalpy drop across the turbine. The temperature for an isentropic expansion across the turbine is indicated by the state $2s$. Using the reversible adiabatic relationships yields

$$T_{2s} = 1273 \left(\frac{150}{700} \right)^{0.3/1.3} = 892°K$$

$$\eta_s = \frac{(h_1 - h_{2'})}{(h_1 - h_{2s})} = \frac{(T_1 - T_{2'})}{(T_1 - T_{2s})} = \frac{(1273 - 938)}{(1273 - 892)}$$

$$\eta_s = 0.879$$

Comment: The second-law efficiency is greater than the isentropic efficiency because only the availability that is destroyed is charged against the turbine. In the case of the isentropic efficiency, the turbine is charged for all the available energy whether or not it was used. ▪

CONCEPT QUESTIONS

1. What is the dead state?
2. What is the difference between maximum work and the change of availability between two states?
3. Describe the relationship between reversible work and changes in availability.
4. What is the difference between reversible work and actual work?
5. Do the surroundings affect a system's availability?
6. If the entropy production for a process is zero, are the irreversibilities necessarily zero?
7. What is irreversibility? What causes it to increase?
8. Describe availability destruction in terms of entropy change.
9. How do heat and work affect changes in system availability?
10. What is entropy production?
11. What does it mean that availability at a given state is positive?

12. Does the energy of a system change when the system's availability decreases? Why?

13. What is the purpose of second-law efficiencies?

14. Describe the difference between available energy and availability.

15. What is the difference between the isentropic and second-law turbine efficiencies?

16. Describe the change of available energy as applied to turbines. Contrast this to the change of availability across the turbine.

PROBLEMS (SI)

9.1 Determine the availability of a unit mass for each of the following, assuming the system is at rest, at zero elevation, and $I_0 = 27°C$ and $p_0 = 1$ atm.
  (a) Dry saturated steam at 5 MPa.
  (b) Refrigerant 12 at 1 MPa and 90°C.
  (c) Air at 500°K and 1000 kPa.
  (d) Water as a saturated liquid at 100°C.
  (e) Water as a saturated liquid at 10°C.

9.2 A compressed-air tank contains 0.5 m³ of air at 300°K and 1500 kPa. Determine the availability if $T_0 = 300°K$ and $p_0 = 100$ kPa.

9.3 A child's balloon, filled with helium, is released and at a given time is found to be 100 m above the ground, traveling at a velocity of 10 m/s relative to the ground. The pressure of the helium in the 3-liter balloon is atmospheric, and the temperature is 25°C. Determine the helium's availability, assuming $T_0 = 25°C$ and $p_0 = 1$ atm.

9.4 A hot-air balloon is 10 m in diameter and contains air at 600°K and 1 atm. It is floating at an elevation of 100 m with a velocity of 2 m/s. Determine the hot air's availability, assuming $T_0 = 300°K$ and $p_0 = 1$ atm.

9.5 The availability of a tank filled with air at 600°K and 500 kPa is 8000 kJ. Determine the tank's volume if $T_0 = 300°K$ and $p_0 = 100$ kPa.

9.6 Derive an expression for the availability of an ideal gas in terms of temperatures and pressures if the effects of kinetic and potential energies are neglected.

9.7 Derive an expression for the availability of an ideal gas in terms of temperatures and specific heat at constant pressure, given that its initial state is at temperature T and pressure p_0. Neglect the effects of kinetic and potential energies.

9.8 A 2-m³ tank contains steam at 5 MPA and 500°C. The tank is cooled until the pressure is reduced to 100 kPa. Determine the change in availability, assuming $T_0 = 300°K$ and $p_0 = 100$ kPa.

9.9 Air at initial conditions of 450°C and 300 kPa undergoes a process to a final state of 280°K and 80 kPa. $T_0 = 300°K$ and $p_0 = 100$ kPa. Determine the availability per unit mass at the initial state and at the final state.

9.10 Determine the change of availability of a unit mass of steam initially at 1000 kPa and 300°C and undergoing the following processes where the final volume is always 1.5 times the initial volume: (a) constant entropy; (b) constant temperature; (c) constant pressure. Assume $T_0 = 300°K$ and $p_0 = 100$ kPa.

9.11 A piston-cylinder contains steam at 500 kPa and 300°C and expands at constant pressure until the temperature is equal to that of the surroundings. The surroundings are at

$T_0 = 30°C$ and $p_0 = 100$ kPa. Find, per unit mass, (a) the heat; (b) the work; (c) the availability transfer with the heat and work.

9.12 A rigid insulated container holds a unit mass of air at 300°K and 100 kPa and receives 100 kJ of paddle work. $T_0 = 300°K$ and $p_0 = 100$ kPa. Considering the air as the system, determine its change of availability, the availability transfer of the work, and the irreversibility.

9.13 An adiabatic container has two compartments of equal volume, one containing 0.2 kg of helium at 750 kPa and 330°K and the other completely evacuated. A valve connecting the two compartments is opened, and the helium expands throughout both compartments. Determine the final temperature and pressure of the helium and the irreversibility, assuming $T_0 = 300°K$ and $p_0 = 100$ kPa.

9.14 Two solid blocks form an isolated system. One block has a mass of 5 kg, a temperature of 1000°K, and a specific heat of 0.6 kJ/kg-K. The second block has a mass of 7 kg, a temperature of 500°K, and a specific heat of 0.8 kJ/kg-K. $T_0 = 300°K$. Determine (a) the equilibrium temperature when the two blocks are brought together; (b) the irreversibility that occurs.

9.15 A rigid tank contains 1 kg air at 400°K and 100 kPa and receives heat from a constant-temperature reservoir at 800°K until the air temperature increases to 600°K. The tank surface temperature during the heat-addition process is 800°K. Assume $T_0 = 300°K$ and $p_0 = 100$ kPa. Determine the heat transfer, its availability transfer, and the irreversibility for the process.

9.16 An electric motor produces 3.0 kW of shaft power while using 15 A at 220 V. The motor's outer surface is 50°C. $T_0 = 300°K$. Determine the irreversibility rate and the heat's availability transfer rate in kW.

9.17 A cylindrical rod, insulated except for its ends, is connected to two constant-temperature reservoirs, one maintained at 800°K and the other at 400°K. The heat-transfer rate through the rod is 8 kW. $T_0 = 300°K$. Determine the rod's irreversibility rate.

9.18 A rigid tank contains a unit mass of air at 300°K and 100 kPa. The air's temperature is doubled in each of the following cases. Determine the irreversibility in each case. $T_0 = 300°K$ and $p_0 = 100$ kPa. (a) The tank is adiabatic and receives paddle work. (b) The tank receives heat from a constant-temperature reservoir at 800°K, and the tank's surface is 800°K while receiving the heat.

9.19 An electric kiln uses wire as a heating element. A steady-state condition occurs when the wire is at a temperature of 2000°K, the kiln walls are at a temperature of 750°K, and the electrical power through the wire is 10 kW. $T_0 = 300°K$. (a) Considering the wire as the system, determine its irreversibility rate. (b) Considering the space between the wire and the kiln walls as the system, determine its irreversibility rate.

9.20 Two kg of air at 400°K and 1000 kPa is contained in a piston-cylinder. The air expands at constant temperature until the pressure is 200 kPa, while receiving heat from a thermal reservoir at a temperature of 800°K. $T_0 = 300°K$ and $p_0 = 100$ kPa. Determine the heat transfer and work and the availability transfer associated with each, as well as the change in the air's availability.

9.21 Steam flows at a velocity of 300 m/s at a temperature of 500°C, a pressure of 1000 kPa, and an elevation of 100 m. Determine the specific flow availability if $T_0 = 300°K$, $p_0 = 100$ kPa, and $g = 9.8$ m/s².

9.22 Determine the specific availability and the specific flow availability in kJ/kg, where potential energy effects are neglected, $T_0 = 300°K$, and $p_0 = 100$ kPa: (a) steam at

5 MPa, 500°C, and 100 m/s; (b) nitrogen at 10 MPa, 700°C, and 300 m/s; (c) 12 at 900 kPa, 60°C, and 50 m/s.

9.23 A steam turbine receives steam at 50 m/s, 5000 kPa, and 400°C, and the steam exits as a saturated vapor at 150°C with a velocity of 100 m/s. Heat transfer from the turbine casing is 25 kJ/kg of steam, and the casing is at a temperature of 200°C. Determine per unit mass of steam flowing through the turbine and for $T_0 = 300°$K and $p_0 = 100$ kPa (a) the work done by the steam; (b) the heat's availability transfer; (c) the irreversibility.

9.24 An adiabatic steam turbine receives steam at 2.5 MPa and 500°C, discharges it at 50 kPa, and produces 820 kJ/kg of work. $T_0 = 300°$K and $p_0 = 100$ kPa. Determine the irreversibility rate per unit mass of steam.

9.25 Thirty kg/s of steam flows through a throttling valve where the inlet conditions to the valve are 5 MPa and 300°C and the exit condition is 500 kPa. $T_0 = 300°$K and $p_0 = 100$ kPa. Determine the specific flow availability change and the irreversibility rate across the valve.

9.26 Ten kg/s of air enters a turbine at 500 kPa and 700°K with a velocity of 100 m/s and leaves at 100 kPa and 460°K with a velocity of 50 m/s. Heat transfer occurs at a rate of 150 kW, and the average surface temperature is 600°K. $T_0 = 300°$K and $p_0 = 100$ kPa. Determine the change of availability of air across the turbine.

9.27 Two kg/s of air enters an adiabatic compressor at 300°K and 100 kPa and is compressed to 500 kPa and 520°K. $T_0 = 300°$K and $p_0 = 100$ kPa. Determine the power required and the change of availability of the air.

9.28 A refrigeration compressor receives 300 kg/h of R 12 as a saturated vapor at −10°C and compresses it adiabatically to 0.90 MPa and 55°C. $T_0 = 300°$K and $p_0 = 100$ kPa. Determine the power required, change in availability of the refrigerant, and irreversibility rate.

9.29 A compressor receives 0.2 m³/s of air at 27°C and 100 kPa and compresses it to 700 kPa and 290°C. Heat loss per unit mass from the compressor surface at 100°C is 20 kJ/kg. $T_0 = 300°$K and $p_0 = 100$ kPa. Determine (a) the change in the air's availability across the compressor; (b) the availability transfer rate of the heat.

9.30 An adiabatic steam turbine receives steam at 10 MPa and 500°C and expands it to 100 kPa. $T_0 = 300°$K and $p_0 = 100$ kPa. Determine the work produced per unit mass and the change in availability for turbine isentropic efficiencies of 100%, 80%, and 50%.

9.31 Hydrogen enters an adiabatic nozzle with a flow rate of 5 kg/s, a temperature of 500°C, a pressure of 800 kPa, and a velocity of 15 m/s. It exits at 150 kPa and 240°C. $T_0 = 300°$K and $p_0 = 100$ kPa. Determine the gas's exit velocity and the change in availability.

9.32 An air heater is located in a closed gas turbine cycle. The air enters the heater at 512°K and 690 kPa and is heated to 1067°K. The pressure drops 138 kPa across the heater. Determine the percentage of loss of availability due to this pressure drop.

9.33 An evaporator, a counterflow heat exchanger, in a refrigeration system receives 5 kg/s of R 12 at 30% quality and −10°C and evaporates it at constant pressure until it is a saturated vapor. Twenty kg/s of air enters the evaporator at 60°C and 1 atm and is cooled at constant pressure. $T_0 = 300°$K and $p_0 = 1$ atm. Determine the availability changes of both the air and the refrigerant, and the irreversibility rate.

9.34 Methane enters a compressor with a steady volumetric flow rate of 0.3 m³/s at 37°C and 200 kPa. It is compressed to 350 kPa isothermally. $T_0 = 300°$K and $p_0 = 100$ kPa. Determine the heat and work availability transfer rates and the irreversibility.

9.35 A manufacturing process requires 50 kg/s of water at 80°C and 1 atm. This can be obtained by mixing dry saturated steam at 1 atm with subcooled water at 20°C. Determine for $T_0 = 300°K$ and $p_0 = 1$ atm (a) the water and steam flow rates; (b) the irreversibility of the process.

9.36 A manufacturing process requires steam at 2500 kPa and 400°C. Steam is available at another location at 10 MPa and 500°C. A suggestion is made to throttle the steam to the desired pressure and then cool it to the desired temperature through heat transfer to the surroundings, which are at 33°C. Determine for $T_0 = 300°K$ and $p_0 = 100$ kPa the total availability destruction for the process per unit mass.

9.37 Referring to Problem 9.36, another engineer suggests expanding the steam through an adiabatic turbine to the desired pressure and then heating or cooling the steam to the desired temperature. Investigate the availability destruction in this scenario. What assumptions are necessary?

9.38 Ten kg/s of air enters an adiabatic turbine at 500°C and 600 kPa with a velocity of 50 m/s and exits at 500°K and 100 kPa with a velocity of 10 m/s. Determine for $T_0 = 300°K$ and $p_0 = 100$ kPa (a) the power produced; (b) the isentropic and second-law efficiencies.

9.39 Fifteen kg/s of air enters an adiabatic compressor at 120 kPa and 20°C with a velocity of 60 m/s and exits at 400 kPa and 160°C with a velocity of 90 m/s. $T_0 = 300°K$ and $p_0 = 100$ kPa. Determine the power required, the change of availability, and the second-law efficiency.

9.40 A compressor receives 5 kg/s of air at 100 kPa and 35°C and compresses it to 800 kPa and 175°C. Heat transfer from the compressor to the surroundings is 450 kW. Determine the compressor power and its second-law efficiency if $T_0 = 300°K$ and $p_0 = 100$ kPa.

9.41 Determine the available energy per unit mass of furnace gas, $c_p = 1.046$ kJ/kg-K, when it is cooled from 1260°K to 480°K at constant pressure. The surroundings are at 295°K.

9.42 Steam is contained in a constant-pressure closed system at 200 kPa and 200°C and is allowed to reach thermal equilibrium with the surroundings, which are at 26°C. Find the loss of available energy per kg of steam.

9.43 A tank contains dry saturated steam at 80°C. Heat is added until the pressure reaches 200 kPa. The lowest available temperature is 20°C. Determine the available portion of the heat added per kg.

9.44 If 2 kg of air at 21°C is heated at constant pressure until the absolute temperature doubles, determine for the process (a) the heat required; (b) the entropy change; (c) the unavailable energy; (d) the available energy if $T_0 = 10°C$.

9.45 A constant-volume container holds air at 102 kPa and 300°K. A paddle wheel does work on the air until the temperature is 422°K. The air is now cooled by the surroundings (at 289°K) to its original state. Determine (a) the paddle work required (adiabatic) per kg; (b) the available portion of the heat removed per kg.

9.46 Air is contained in a rigid tank at 138 kPa and 335°K. Heat is transferred to the air from a constant-temperature reservoir at 555°K until the pressure is 207 kPa. The temperature of the surroundings is 289°K. Determine per kg of air (a) the heat transferred; (b) the available portion of the heat transferred; (c) the loss of available energy due to irreversible heat transfer.

9.47 Air having a mass of 2.5 kg is cooled from 210 kPa and 205°C to 5°C at constant volume. All the heat is rejected to the surroundings at -4°C. Determine the available portion of the heat removed. Draw a T-s diagram and label the available and unavailable portions of the energy rejected.

9.48 A tank contains water vapor at a pressure of 75 kPa and a temperature of 100°C. Heat is added to the water vapor until the pressure is tripled. The lowest available temperature is 16°C. Find the available portion of the heat added per kg.

9.49 A boiler produces dry saturated steam at 3.5 MPa. The furnace gas enters the tube bank at a temperature of 1100°C, leaves at a temperature of 430°C, and has an average specific heat $c_p = 1.046$ kJ/kg-K over this temperature range. Neglecting heat losses from the boiler and for water entering the boiler as a saturated liquid with a flow rate of 12.6 kg/s, determine for $T_0 = 21$°C (a) the heat transfer; (b) the loss of available energy of the gas; (c) the gain of available energy of the steam; (d) the entropy production.

9.50 A steam turbine uses 12.6 kg/s of steam and exhausts to a condenser at 5 kPa and 90% quality. Cooling water enters the condenser at 21°C and leaves at the steam temperature. Determine (a) the heat rejected; (b) the net loss of available energy; (c) the entropy production.

9.51 Steam enters a turbine at 6.0 MPa and 500°C and exits at 10 kPa with a quality of 89%. The flow rate is 9.07 kg/s. Determine (a) the loss of available energy; (b) the available energy entering the turbine; (c) the available portion of the steam exiting the turbine. $T_0 = T_{sat}$ @ 10 kPa.

9.52 An air turbine receives air at 825 kPa and 815°C. The air leaves the turbine at 100 kPa and 390°C. The lowest available temperature is 300°K. Determine (a) the available energy entering the turbine; (b) the entropy increase across the turbine.

9.53 Air expands from 825 kPa and 500°K to 140 kPa and 500°K. Determine (a) the Gibbs function at the initial conditions; (b) the maximum work; (c) the entropy change.

9.54 A Carnot engine receives 2000 kJ of heat at 1100°C from a heat reservoir at 1400°C. Heat is rejected at 100°C to a reservoir at 50°C. The expansion and compression processes are isentropic. Determine (a) the entropy production for one cycle; (b) the second-law efficiencies of the heat-addition process and the cycle.

9.55 In a steam generator the steam-generating tubes receive heat from hot gases passing over the outside surface, evaporating water inside the tubes. The flue gas flow rate is 20 kg/s with an average specific heat of 1.04 kJ/kg-K. The gas temperature decreases from 650°C to 400°C while generating steam at 300°C. The water enters the tubes as a saturated liquid and leaves with a quality of 90%. $T_0 = 27$°C. Determine (a) the water flow rate; (b) the net change of available energy; (c) the second-law efficiency.

9.56 A Carnot engine receives and rejects heat with a 20°C temperature differential between it and the heat reservoirs. The expansion and compression processes have pressure ratios of 50. For 1 kg of air as the substance, cycle temperature limits of 1000°K and 300°K, $T_0 = 280$°K, determine the second-law efficiency.

9.57 A parabolic collector receives 1.1 kW of solar radiation at 200°C. This energy is used to evaporate R 12 from a subcooled liquid at 30°C and 3.34 MPa to a saturated vapor at 3.34 MPa. $T_0 = 30$°C. Determine (a) refrigerant flow rate; (b) the entropy production; (c) second-law efficiency.

9.58 A constant-pressure heat source of air decreases in temperature from 1000°K to 500°K when delivering 1000 kJ to a Carnot engine at a high temperature of 500°K and a low temperature of 300°K. $T_0 = 300$°K. Find the irreversibility per cycle.

9.59 A steam turbine receives 1 kg/s of steam at 2.0 MPa and 400°C and expands it adiabatically to saturated vapor at 100 kPa. Assume $T_0 = 25°C$ and $p_0 = 100$ kPa. Determine (a) the reversible work; (b) the available energy of the fluid entering.

9.60 A rigid insulated container holds 1 kg of carbon dioxide at 300°K and is dropped from a balloon 3.0 km above the earth's surface. Assume $T_0 = 300°K$ and $p_0 = 100$ kPa. Determine (a) the change of entropy of the CO_2 after the container hits the ground; (b) the change of availability of the CO_2.

PROBLEMS (ENGLISH UNITS)

*9.1 Determine the availability of a unit mass for each of the following, assuming the system is at rest, at zero elevation, and $T_0 = 77°F$ and $p_0 = 1$ atm.
(a) Dry saturated steam at 500 psia.
(b) Refrigerant 12 at 200 psia and 200°F.
(c) Air at 900°R and 150 psia.
(d) Water as a saturated liquid at 212°F.
(e) Water as a saturated liquid at 40°F.

*9.2 A compressed-air tank contains 15 ft³ of air at 77°F and 250 psia. Determine the availability if $T_0 = 77°F$ and $p_0 = 14.7$ psia.

*9.3 A child's balloon, filled with helium, is released and at a given time is found to be 300 ft above the ground, traveling at a velocity of 30 ft/sec relative to the ground. The pressure of the helium in the 1-ft³ balloon is atmospheric, and the temperature is 70°F. Determine the helium's availability, assuming $T_0 = 70°F$ and $p_0 = 1$ atm.

*9.4 A hot-air balloon is 30 ft in diameter and contains air at 620°F and 1 atm. It is floating at an elevation of 300 ft with a velocity of 5 ft/sec. Determine the hot air's availability, assuming $T_0 = 77°F$ and $p_0 = 1$ atm.

*9.5 The availability of a tank filled with air at 620°F and 75 psia is 8000 Btu. Determine the tank's volume if $T_0 = 77°F$ and $p_0 = 14.7$ psia.

*9.6 A 50-ft³ tank contains steam at 600 psia and 700°F. The tank is cooled until the pressure is reduced to 14.7 psia. Determine the change in availability, assuming $T_0 = 77°F$ and $p_0 = 14.7$ psia.

*9.7 Air at initial conditions of 840°F and 45 psia undergoes a process to a final state of 40°F and 12 psia. $T_0 = 77°F$ and $p_0 = 14.7$ psia. Determine the availability per unit mass at the initial state and at the final state.

*9.8 Determine the change of availability of a unit mass of steam initially at 300 psia and 500°F and undergoing the following processes where the final volume is always 1.5 times the initial volume: (a) constant entropy; (b) constant temperature; (c) constant pressure. Assume $T_0 = 77°F$ and $p_0 = 14.7$ psia.

*9.9 A piston-cylinder contains steam at 100 psia and 400°F cools at constant pressure until the temperature is equal to that of the surroundings. Find, per unit mass, the heat, the work, and the availability transfer with the heat and work. The surroundings are at $T_0 = 70°F$ and $p_0 = 14.7$ psia.

*9.10 A rigid insulated container holds a unit mass of air at 77°F and 14.7 psia and receives 77,800 ft-lbf of paddle work. $T_0 = 77°F$ and $p_0 = 14.7$ psia. Considering the air as the system, determine its change of availability, the availability transfer of the work, and the irreversibility.

*9.11 An adiabatic container has two compartments of equal volume, one containing 0.5 lbm of helium at 100 psia and 135°F and the other completely evacuated. A valve connecting the two compartments is opened, and the helium expands throughout both compartments. Determine the final temperature and pressure of the helium and the irreversibility, assuming $T_0 = 77°F$ and $p_0 = 14.7$ psia.

*9.12 Two solid blocks form an isolated system. One block has a mass of 5 lbm, a temperature of 1000°R, and a specific heat of 0.6 Btu/lbm-F. The second block has a mass of 7 lbm, a temperature of 500°R, and a specific heat of 0.8 Btu/lbm-F. $T_0 = 77°F$. Determine (a) the equilibrium temperature when the two blocks are brought together; (b) the irreversibility that occurs.

*9.13 A rigid tank contains 1 lbm air at 260°F and 15 psia and receives heat from a constant-temperature reservoir at 1000°F until the air temperature increases to 620°F. The tank surface temperature during the heat-addition process is 1000°F. Assume $T_0 = 77°F$ and $p_0 = 14.7$ psia. Determine the heat transfer, its availability transfer, and the irreversibility for the process.

*9.14 An electric motor produces 3.0 kW of shaft power while using 15 A at 220 V. The motor's outer surface is 110°F. $T_0 = 77°F$. Determine the irreversibility rate and the heat's availability transfer rate in kW.

*9.15 A cylindrical rod, insulated except for its ends, is connected to two constant-temperature reservoirs, one maintained at 800°F and the other at 400°F. The heat-transfer rate through the rod is 450 Btu/min. $T_0 = 77°F$. Determine the rod's irreversibility rate.

*9.16 A rigid tank contains a unit mass of air at 68°F and 14.7 psia. The air's absolute temperature is doubled in each of the following cases. Determine the irreversibility in each case. $T_0 = 77°F$ and $p_0 = 14.7$ psia. (a) The tank is adiabatic and receives paddle work. (b) The tank receives heat from a constant-temperature reservoir at 1400°R, and the tank's surface is 1400°R while receiving the heat.

*9.17 An electric kiln uses wire as a heating element. A steady-state condition occurs when the wire is at a temperature of 3000°R, the kiln walls are at a temperature of 1500°R, and the electrical power through the wire is 10 kW. $T_0 = 77°F$. (a) Considering the wire as the system, determine its irreversibility rate. (b) Considering the space between the wire and the kiln walls as the system, determine its irreversibility rate.

*9.18 Two lbm of air at 720°R and 150 psia is contained in a piston-cylinder. The air expands at constant temperature until the pressure is 30 psia, while receiving heat from a thermal reservoir at a temperature of 800°F. $T_0 = 77°F$ and $p_0 = 14.7$ psia. Determine the heat transfer and work and the availability transfer associated with each, as well as the change in the air's availability.

*9.19 Steam flows at a velocity of 1000 ft/sec at a temperature of 900°F, a pressure of 150 psia, and an elevation of 300 ft. Determine the specific flow availability if $T_0 = 77°F$, $p_0 = 14.7$ psia, and $g = 32.174$ ft/sec^2.

*9.20 Determine the specific availability and the specific flow availability in Btu/lbm for the following, where potential energy effects are neglected, $T_0 = 77°F$, and $p_0 = 14.7$ psia: (a) steam at 200 psia, 500°F, and 300 ft/sec; (b) nitrogen at 6000 psia, 1260°F, and 1000 ft/sec; (c) R 12 at 175 psia, 160°F, and 150 ft/sec.

*9.21 A steam turbine receives steam at 150 ft/sec, 800 psia, and 700°F, and the steam exits as a saturated vapor at 240°F with a velocity of 300 ft/sec. Heat transfer from the turbine casing is 10 Btu/lbm of steam, and the casing is at a temperature of 400°F. Determine per unit mass of steam flowing through the turbine and for $T_0 = 77°F$ and $p_0 = 14.7$ psia (a) the work done by the steam; (b) the heat's availability transfer; (c) the irreversibility.

*9.22 An adiabatic steam turbine receives steam at 600 psia and 600°F, discharges it at 8 psia, and produces 270 Btu/lbm of work. Determine the irreversibility rate per unit mass of steam. $T_0 = 77°F$ and $p_0 = 14.7$ psia.

*9.23 Sixty-five lbm/sec of steam flows through a throttling valve where the inlet conditions to the valve are 800 psia and 600°F and the exit condition is 100 psia. $T_0 = 77°F$ and $p_0 = 14.7$ psia. Determine the specific flow availability change and the irreversibility rate across the valve.

*9.24 Twenty-five lbm/sec of air enters a turbine at 75 psia and 800°F with a velocity of 300 ft/sec and leaves at 15 psia and 370°F with a velocity of 150 ft/sec. Heat transfer occurs at a rate of 9000 Btu/min, and the average surface temperature is 620°F. $T_0 = 77°F$ and $p_0 = 14.7$ psia. Determine the change of availability of air across the turbine.

*9.25 Five lbm/sec of air enters an adiabatic compressor at 77°F and 14.7 psia and is compressed to 75 psia and 475°F. $T_0 = 77°F$ and $p_0 = 14.7$ psia. Determine the power required and the change of availability of the air.

*9.26 A refrigeration compressor receives 11 lbm/min of R 12 as a saturated vapor at −10°F and compresses it adiabatically to 150 psia and 150°F. $T_0 = 77°F$ and $p_0 = 14.7$ psia. Determine the power required, the change in availability of the refrigerant, and the irreversibility rate.

*9.27 A compressor receives 425 ft³/min of air at 77°F and 14.7 psia and compresses it to 105 psia and 530°F. Heat loss per unit mass from the compressor surface at 200°F is 9 Btu/lbm. $T_0 = 77°F$ and $p_0 = 14.7$ psia. Determine (a) the change in the air's availability across the compressor; (b) the availability transfer rate of the heat.

*9.28 An adiabatic steam turbine receives steam at 1500 psia and 800°F and expands it to 14.7 psia. $T_0 = 77°F$ and $p_0 = 14.7$ psia. Determine the work produced per unit mass and the change in availability for turbine isentropic efficiencies of 100%, 80%, and 50%.

*9.29 Hydrogen enters an adiabatic nozzle with a flow rate of 15 lbm/sec, a temperature of 900°F, a pressure of 120 psia, and a velocity of 45 ft/sec. It exits at 20 psia and 375°F. $T_0 = 77°F$ and $p_0 = 14.7$ psia. Determine the gas's exit velocity and the change in availability.

*9.30 An air heater is located in a closed gas turbine cycle. The air enters the heater at 920°R and 100 psia and is heated to 1920°R. The pressure drops 20 psia across the heater. Determine the percentage of loss of availability due to this pressure drop. $T_0 = 537°R$.

*9.31 An evaporator, a counterflow heat exchanger, in a refrigeration system receives 15 lbm/sec of R 12 at 30% quality and −10°F and evaporates it at constant pressure until it is a saturated vapor. Forty-five lbm/sec of air enters the evaporator at 110°F and 1 atm and is cooled at constant pressure. $T_0 = 77°F$ and $p_0 = 1$ atm. Determine the availability changes of both the air and the refrigerant, and the irreversibility rate.

*9.32 Methane enters a compressor with a steady volumetric flow rate of 650 ft³/min at 100°F and 30 psia. It is compressed to 45 psia isothermally. $T_0 = 77°F$ and $p_0 = 14.7$ psia. Determine the heat and work availability transfer rates and the irreversibility.

*9.33 A manufacturing process requires 800 gpm of water at 175°F and 1 atm. This can be obtained by mixing dry saturated steam at 1 atm with subcooled water at 65°F. Determine for $T_0 = 77°F$ and $p_0 = 1$ atm (a) the water and steam flow rates; (b) the irreversibility of the process.

*9.34 A manufacturing process requires steam at 400 psia and 700°F. Steam is available at another location at 1500 psia and 900°F. A suggestion is made to throttle the steam to the desired pressure and then cool it to the desired temperature through heat transfer to the surroundings, which are at 100°F. Determine for $T_0 = 77°F$ and $p_0 = 14.7$ psia the total availability destruction for the process per unit mass.

*9.35 Referring to Problem *9.34, another engineer suggests expanding the steam through an adiabatic turbine to the desired pressure and then heating or cooling the steam to the desired temperature. Investigate the availability destruction in this scenario. What assumptions are necessary?

*9.36 Twenty-five lbm/sec of air enters an adiabatic turbine at 900°F and 90 psia with a velocity of 150 ft/sec and exits at 440°F and 15 psia with a velocity of 30 ft/sec. Determine for $T_0 = 77°F$ and $p_0 = 14.7$ psia (a) the power produced; (b) the isentropic and second-law efficiencies.

*9.37 Thirty-five lbm/sec of air enters an adiabatic compressor at 17.4 psia and 68°F with a velocity of 180 ft/sec and exits at 58 psia and 320°F with a velocity of 250 ft/sec. $T_0 = 77°F$ and $p_0 = 14.7$ psia. Determine the power required, the change of availability, and the second-law efficiency.

*9.38 A compressor receives 11 lbm/sec of air at 14.7 psia and 100°F and compresses it to 120 psia and 350°F. Heat transfer from the compressor to the surroundings is 450 kW. Determine the compressor power and its second-law efficiency if $T_0 = 77°F$ and $p_0 = 14.7$ psia.

*9.39 A 10-ft³ tank contains dry saturated steam at 15 psia. Heat is added until the pressure reaches 30 psia. For $T_0 = 40°F$, determine the available portion of the heat added.

*9.40 A boiler produces dry saturated steam at 500 psia. The furnace gas enters the tube bank at a temperature of 2000°F, leaves at a temperature of 800°F, and has an average specific heat $c_p = 0.25$ Btu/lbm-R over this temperature range. Neglecting heat losses from the boiler and for the water entering the boiler as a saturated liquid with a flow rate of 1×10^5 lbm/hr, determine for $T_0 = 70°F$ (a) the heat transfer; (b) the loss of available energy of the gas; (c) the gain of available energy of the steam; (d) the entropy production.

*9.41 A tank contains water vapor at a pressure of 13 psia and a temperature of 210°F. Heat is added to the water vapor until the pressure is tripled. The lowest available temperature is 60°F. Find the available portion of the heat added.

*9.42 Five lbm of air is cooled from 30 psia and 400°F to 40°F at constant volume. All the heat is rejected to the surroundings at 25°F. Determine the available portion of the heat removed. Draw a T-s diagram and label the available and unavailable portions of the energy rejected.

*9.43 Steam enters a turbine at 900 psia and 900°F and exits at 2 psia with a quality of 89%. The flow rate is 20 lbm/sec. Determine (a) the loss of available energy; (b) the available energy entering the turbine; (c) the available portion of the steam exiting the turbine; (d) the reversible work; (e) the second-law efficiency. $T_0 = 537°R$.

*9.44 An air turbine receives air at 120 psia and 1500°F. The air leaves the turbine at 15 psia and 730°F. $T_0 = 530°R$. Determine (a) the available energy entering the turbine; (b) the entropy change across the turbine; (c) the reversible work; (d) η_2.

COMPUTER PROBLEMS

C9.1 Air is compressed isentropically in a compressor from atmospheric conditions of 100 kPa and 300°K to various discharge pressures. Develop a computer model for the compressor to determine the change of availability of the air for discharge pressures of 500 kPa, 1 MPa, 5 MPa, and 10 MPa. You may use the TK Solver model in AIR.TK.

C9.2 A turbine with an isentropic efficiency of 92% adiabatically expands steam, exhausting it to a condenser where the pressure is 1 psia. Develop a model for the turbine to compute the change of availability and available energy across the turbine for the following steam inlet conditions: 1000°F and 100 psia, 300 psia, 500 psia, 1000 psia, 1500 psia, and 2000 psia; 1000 psia and 500°F, 600°F, 700°F, 800°F, 900°F, and 1000°F. Assume $T_0 = 70°F$. You may use the TK Solver model STMCYCLE.TK.

C9.3 A refrigeration compressor adiabatically compresses saturated R 12 vapor from an inlet pressure of 300 kPa to a variety of discharge pressures. Develop a compressor model with second-law efficiencies of 80%, 85%, and 90% and determine the change of availability for discharge pressures from 600 kPa to 1200 kPa in increments of 100 kPa. Also, compute the availability change for a discharge pressure of 850 kPa and inlet pressures ranging from 150 kPa to 400 kPa in 50-kPa increments. You may use the compressor model from R12CYCLE.TK.

10

Thermodynamic Relationships

The problems in previous chapters were based on the assumption that the property values for the various substances were available. The specific heats, the tables of properties, and so on, all have to be determined from experimental data. The question arises as to how, from an experiment, the necessary properties can be determined. Experimenters cannot blindly set up an experiment and hope to achieve meaningful results. They must know what to vary in the experiment and what to hold constant. They are further limited in that only four properties are measurable: temperature, pressure, volume, and mass. All remaining properties must be calculated from these four or from changes these properties undergo under certain conditions.

In order to achieve the desirable result of tests that are useful, experimenters must be theoreticians as well. They must develop a mathematical model for thermodynamic properties and use this in determining the functional relationship between the four properties. This is the purpose of the chapter: to show you the method used and to illustrate how properties may be determined. One may have a good model of a substance, but it is only a mathematical model, not the substance itself. This is true of all sciences that predict system behaviors.

In this chapter you will

- Physically and mathematically interpret differentials;
- Apply mathematical analysis to equations of state;
- Learn of Maxwell's relations and their usefulness;

- Develop a deeper understanding of variations in enthalpy, internal energy, and specific heats;
- Discover how tables of properties are created.

10.1 INTERPRETING DIFFERENTIALS AND PARTIAL DERIVATIVES

To develop a model we must first understand differential equations and partial derivatives. Consider $Z = f(X, Y)$; then

$$dZ = \left(\frac{\partial Z}{\partial X}\right)_Y dX + \left(\frac{\partial Z}{\partial Y}\right)_X dY \qquad (10.1)$$

where dZ is the total differential; $(\partial Z/\partial X)_Y$ tells us how Z varies with respect to X at constant Y through a differential distance dX. The second term, $(\partial Z/\partial Y)_X$, shows how Z varies with respect to Y at constant X through a differential distance dY. The sum of both terms tells how dZ varies overall. If Z is assumed to be a continuous function with continuous derivatives, then

$$\frac{\partial^2 Z}{\partial X \partial Y} = \frac{\partial^2 Z}{\partial Y \partial X} \qquad (10.2)$$

Equation (10.2) is an important relationship and is valid for exact differentials, differentials that are point, not path, functions. This is true of properties, which depend only on the system state, but it is not true of the inexact differentials, heat and work.

To illustrate the inexact differential, consider the differential of mechanical work, $\delta W = p \, dV$. Let us assume momentarily that work is an exact differential. Then

$$W = f(p, V)$$

$$dW = \left(\frac{\partial W}{\partial p}\right)_V dp + \left(\frac{\partial W}{\partial V}\right)_p dV$$

The first term on the right-hand side is zero because no work is accomplished at constant volume: this tells us, then, that the only work is work at constant pressure. This, obviously, is not true, and the hypothesis is invalid.

Physical Interpretation of Partial Derivatives

Before we continue with the mathematical development of our thermodynamic model, let us first consider the physical significance of the partial derivative and how these derivatives may be used in developing a thermodynamic surface. The importance of understanding the difference between independent and dependent variables is a key to the understanding of the mathematical model. Any two variables are independent if one may be held constant while the other is varied. The changing of one variable will not change the other variable, thus demonstrating their independ-

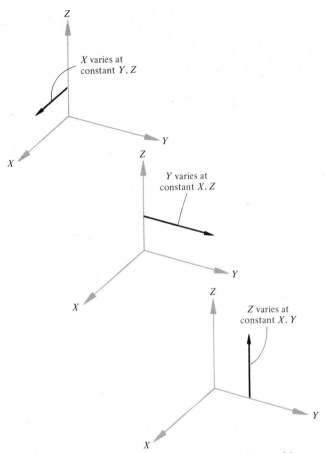

Figure 10.1 A set of three independent variables, with one varying while two remain constant.

ence from one another. Figure 10.1 shows a set of three independent variables, each of which may be varied while the other two remain constant.

When X, Y, and Z are used to describe some surface, a functional relationship is developed between the variables, and the concept of dependence and independence appears. Let these variables describe a sphere of radius R; then

$$X^2 + Y^2 + Z^2 = R^2 \tag{10.3}$$

An octant of the sphere is shown in Figure 10.2, and the lines on the surface demonstrate how two variables change while the third is held constant.

On the surface only two variables may be independent; the third is dependent. For any two points X_1 and Y_1 within the domain of $X^2 + Y^2 \le R^2$, there is only a single value of Z, call it Z_1, such that the point at (X_1, Y_1, Z_1) will be on the surface described by equation 10.3. At any other value the point, or state, will not be on the

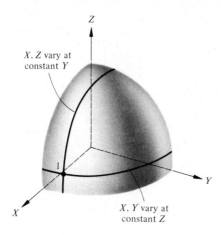

Figure 10.2 The variation of two variables with the third dependent, keeping the line on the surface.

surface. This is shown mathematically by solving equation (10.3) for any variable, in this case Z:

$$Z = (R^2 - X^2 - Y^2)^{1/2} \qquad (10.4)$$

Since R is a constant, this may be written as

$$Z = f(X, Y) \qquad (10.5)$$

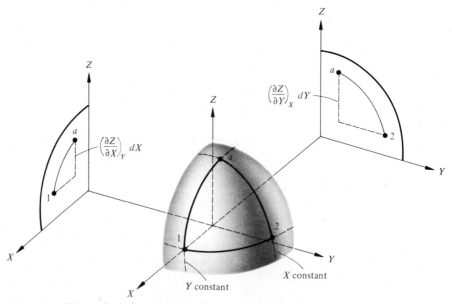

Figure 10.3 Illustration of a true projection of curves $a1$ and $a2$ in the XZ and ZY planes, respectively.

Hence, we get equation (10.1),

$$dZ = \left(\frac{\partial Z}{\partial X}\right)_Y dX + \left(\frac{\partial Z}{\partial Y}\right)_X dY$$

The partial derivatives are graphically shown in Figure 10.3.

The equilibrium surfaces used in the previous chapters are thermodynamic surfaces. We will be able to develop the mathematical tools used in making these surfaces. The steam tables and other tables of properties rely on these techniques in their development.

10.2 AN IMPORTANT RELATIONSHIP

Another mathematical relationship between derivatives may be developed as follows. Let us consider three variables, X, Y, and Z, and suppose that

$$Z = f(X, Y)$$

then

$$dZ = \left(\frac{\partial Z}{\partial X}\right)_Y dX + \left(\frac{\partial Z}{\partial Y}\right)_X dY \qquad (10.6a)$$

but the relationship may also be written as

$$Y = f(X, Z)$$

$$dY = \left(\frac{\partial Y}{\partial X}\right)_Z dX + \left(\frac{\partial Y}{\partial Z}\right)_X dZ \qquad (10.6b)$$

Solving for dZ,

$$dZ = \frac{dY - (\partial Y/\partial X)_Z dX}{(\partial Y/\partial Z)_X}$$

and substituting in equation (10.6a) yields

$$dY - \left(\frac{\partial Y}{\partial X}\right)_Z dX = \left(\frac{\partial Y}{\partial Z}\right)_X \left(\frac{\partial Z}{\partial X}\right)_Y dX + \left(\frac{\partial Y}{\partial Z}\right)_X \left(\frac{\partial Z}{\partial Y}\right)_X dY$$

and, rearranging,

$$\left[1 - \left(\frac{\partial Y}{\partial Z}\right)_X \left(\frac{\partial Z}{\partial Y}\right)_X\right] dY = \left[\left(\frac{\partial Y}{\partial X}\right)_Z + \left(\frac{\partial Y}{\partial Z}\right)_X \left(\frac{\partial Z}{\partial X}\right)_Y\right] dx \qquad (10.7)$$

Since X and Y are independent variables, any values may be assigned to them and equation (10.7) will be valid. Let $X = $ const; then $dX = 0$,

$$\left(\frac{\partial Y}{\partial Z}\right)_X \left(\frac{\partial Z}{\partial Y}\right)_X = 1 \qquad (10.8)$$

or

$$\left(\frac{\partial Z}{\partial Y}\right)_X = \frac{1}{(\partial Y / \partial Z)_X} = \left(\frac{\partial Z}{\partial Y}\right)_X \tag{10.9}$$

Equation (10.9) proves that partial derivatives may be inverted. If Y is held constant in equation (10.7), then

$$\left(\frac{\partial Y}{\partial X}\right)_Z + \left(\frac{\partial Y}{\partial Z}\right)_X \left(\frac{\partial Z}{\partial X}\right)_Y = 0 \tag{10.10}$$

If we multiply equation (10.10) by $(\partial X / \partial Y)_Z$,

$$\left(\frac{\partial X}{\partial Y}\right)_Z \left(\frac{\partial Y}{\partial X}\right)_Z + \left(\frac{\partial X}{\partial Y}\right)_Z \left(\frac{\partial Y}{\partial Z}\right)_X \left(\frac{\partial Z}{\partial X}\right)_Y = 0$$

but the first term is equal to one, by equation (10.8), yielding

$$\left(\frac{\partial Y}{\partial Z}\right)_X \left(\frac{\partial Z}{\partial X}\right)_Y \left(\frac{\partial X}{\partial Y}\right)_Z = -1 \tag{10.11}$$

Equation (10.11) is very important and will be used later.

10.3 APPLICATION OF MATHEMATICAL METHODS TO THERMODYNAMIC RELATIONS

The first law for a closed system on a unit mass basis is

$$\delta q = du + p\, dv \tag{10.12}$$

If the process is reversible,

$$\delta q = T\, ds$$

hence

$$du = T\, ds - p\, dv \tag{10.13}$$

Also,

$$dh = du + p\, dv + v\, dp$$

$$dh = T\, ds + v\, dp \tag{10.14}$$

Two other thermodynamic relationships that are frequently used are the Helmholtz function, a, and the Gibbs function, g. Written on a unit mass basis,

$$a = u - Ts \tag{10.15}$$

$$g = h - Ts \tag{10.16}$$

Taking the derivative of equation (10.15) yields

$$da = du - T\, ds - s\, dT$$

but, from equation (10.13),

$$du = T \, ds - p \, dv$$

hence

$$da = -p \, dv - s \, dT \tag{10.17}$$

And taking the derivative of equation (10.16) yields

$$dg = dh - T \, ds - s \, dT$$

but

$$dh - T \, ds = v \, dp$$

hence

$$dg = v \, dp - s \, dT \tag{10.18}$$

Equations (10.13), (10.14), (10.17), and (10.18) are of the same general form. In each case, we are dealing with a thermodynamic property expressed as a function of other thermodynamic properties. Under the constraint of the state postulate with a single quasi-static work mode, any property can be expressed as the function of two other properties, provided these properties are independent. For example, considering equation (10.13), we can write

$$u = f(s, v)$$

then

$$du = \left(\frac{\partial u}{\partial s}\right)_v ds + \left(\frac{\partial u}{\partial v}\right)_s dv \tag{10.19}$$

Comparing equation (10.19) and (10.13) yields

$$T = \left(\frac{\partial u}{\partial s}\right)_v \qquad -p = \left(\frac{\partial u}{\partial v}\right)_s \tag{10.20}$$

Relationships between the other differentials may be developed from equations (10.14), (10.17), and (10.18) in a similar manner. They are summarized below:

$$
\begin{aligned}
\left(\frac{\partial u}{\partial s}\right)_v &= T & \left(\frac{\partial h}{\partial s}\right)_p &= T \\[2mm]
\left(\frac{\partial a}{\partial v}\right)_T &= -p & \left(\frac{\partial u}{\partial v}\right)_s &= -p \\[2mm]
\left(\frac{\partial g}{\partial p}\right)_T &= v & \left(\frac{\partial h}{\partial p}\right)_s &= v \\[2mm]
\left(\frac{\partial a}{\partial T}\right)_v &= -s & \left(\frac{\partial g}{\partial T}\right)_p &= -s
\end{aligned}
\tag{10.21}
$$

10.4 MAXWELL'S RELATIONS

Another group of relations, called *Maxwell's relations,* may be developed in the following manner. Differentiating equation (10.20),

$$\left(\frac{\partial T}{\partial v}\right)_s = \frac{\partial^2 u}{\partial v \, \partial s} \qquad -\left(\frac{\partial p}{\partial s}\right) = \frac{\partial^2 u}{\partial s \, \partial v}$$

but, from equation (10.2),

$$\frac{\partial^2 u}{\partial v \, \partial s} = \frac{\partial^2 u}{\partial s \, \partial v}$$

hence

$$\left(\frac{\partial T}{\partial v}\right)_s = -\left(\frac{\partial p}{\partial s}\right)_v \tag{10.22}$$

Equation (10.22) is one Maxwell relation; the others may be derived in a similar manner, and the following will result:

$$\left(\frac{\partial T}{\partial p}\right)_s = \left(\frac{\partial v}{\partial s}\right)_p$$

$$\left(\frac{\partial p}{\partial T}\right)_v = \left(\frac{\partial s}{\partial v}\right)_T \tag{10.23}$$

$$\left(\frac{\partial v}{\partial T}\right)_p = -\left(\frac{\partial s}{\partial p}\right)_T$$

It should be recognized that this is not a complete listing, but a function of equations (10.13), (10.14), (10.17), and (10.18). If more terms, such as surface work, were added to the first-law equation, there would be more Maxwell relations. However, the ones listed in equation (10.23) will be sufficient for the applications in this text. Should others be needed, they may readily be developed.

10.5 SPECIFIC HEATS, ENTHALPY, AND INTERNAL ENERGY

The specific heat at constant volume was developed by considering a constant-volume process; hence the first law is written as $\delta q = du = c_v \, dT$, and c_v is defined as

$$c_v \equiv \left(\frac{\partial u}{\partial T}\right)_v = \left(\frac{\partial u}{\partial s}\right)_v \left(\frac{\partial s}{\partial T}\right)_v = T\left(\frac{\partial s}{\partial T}\right)_v \tag{10.24}$$

In a similar manner for a constant-pressure process, $\delta q = dh = c_p\, dT$, and c_p is defined as

$$c_p \equiv \left(\frac{\partial h}{\partial T}\right)_p = \left(\frac{\partial h}{\partial s}\right)_p \left(\frac{\partial s}{\partial T}\right)_p = T\left(\frac{\partial s}{\partial T}\right)_p \qquad (10.25)$$

It is possible to develop an expression for enthalpy and internal energy in terms of the specific heats and temperature and pressure. This is the intent of the chapter, to use the measurable properties to obtain nonmeasurable properties.

Since u can be expressed as $u = f(T, v)$,

$$du = \left(\frac{\partial u}{\partial T}\right)_v dT + \left(\frac{\partial u}{\partial v}\right)_T dv \qquad (10.26)$$

$$T\, ds = du + p\, dv$$

For $T = C$, it follows that

$$T\left(\frac{\partial s}{\partial v}\right)_T = \left(\frac{\partial u}{\partial v}\right)_T + p$$

$$\left(\frac{\partial u}{\partial v}\right)_T = T\left(\frac{\partial s}{\partial v}\right)_T - p \qquad (10.27)$$

By use of a Maxwell relation, equation (10.27) becomes

$$\left(\frac{\partial u}{\partial v}\right)_T = T\left(\frac{\partial p}{\partial T}\right)_v - p \qquad (10.28)$$

and equation (10.26) becomes

$$du = c_v\, dT + \left[T\left(\frac{\partial p}{\partial T}\right)_v - p\right] dv \qquad (10.29)$$

Notice that the change of internal energy may be calculated in terms of measurable properties.

Since h can be expressed as $h = f(T, p)$,

$$dh = \left(\frac{\partial h}{\partial T}\right)_p dT + \left(\frac{\partial h}{\partial p}\right)_T dp \qquad (10.30)$$

and

$$T\, ds = dh - v\, dp \qquad (10.31)$$

Hence

$$\left(\frac{\partial h}{\partial p}\right)_T = T\left(\frac{\partial s}{\partial p}\right)_T + v$$

If we use a Maxwell relation,

$$\left(\frac{\partial h}{\partial p}\right)_T = -T\left(\frac{\partial v}{\partial T}\right)_p + v \qquad (10.32)$$

Substituting equation (10.32) into equation (10.30) yields

$$dh = c_p \, dT + \left[v - T \left(\frac{\partial v}{\partial T} \right)_p \right] dp \qquad (10.33)$$

The functional variables were chosen in each case so that the first term would be the specific heat. All that is needed to integrate equation (10.33) is a correlation between T, p, and v.

Example 10.1

Assume that air has an equation of state $p(v - b) = RT$. Assuming the specific heats are known, determine an expression for the change in the specific internal energy.

Solution

Given: Gas equation of state.

Find: Expression for the change in the gas's specific internal energy.

Analysis: The general expression for the change of specific internal energy for a simple compressible substance is given by equation (10.29),

$$du = c_v \, dT + \left[T \left(\frac{\partial p}{\partial T} \right)_v - p \right] dv$$

Solve the equation of state for p/T:

$$\frac{p}{T} = \frac{R}{v - b}$$

Take the derivative of the equation with respect to temperature, holding specific volume constant.

$$\frac{T(\partial p / \partial T)_v - p(1)}{T^2} = 0$$

Solving for $T(\partial p / \partial T)$ yields

$$T(\partial p / \partial T)_v = p$$

and substituting into equation (10.29) yields

$$du = c_v \, dT + [p - p] \, dv = c_v \, dT$$

Comment: The gas equation of state in this example is that of an ideal gas, the Clausius equation of state. For more complicated equations of state, this simplification will not occur. ◼

Entropy and Specific Heat Relationship

The equation for the change of entropy may be found in the following manner. The results lead to a very important thermodynamic relationship between the specific heat at constant volume and constant pressure.

Since s can be expressed as $s = f(T, p)$,

$$ds = \left(\frac{\partial s}{\partial T}\right)_p dT + \left(\frac{\partial s}{\partial p}\right)_T dp \qquad (10.34)$$

Using equation (10.25) this becomes

$$ds = c_p \frac{dT}{T} + \left(\frac{\partial s}{\partial p}\right)_T dp$$

Using a Maxwell relation from equation (10.23),

$$ds = c_p \frac{dT}{T} - \left(\frac{\partial v}{\partial T}\right)_p dp \qquad (10.35)$$

We now let

$$s = f(T, v)$$

$$ds = \left(\frac{\partial s}{\partial T}\right)_v dT + \left(\frac{\partial s}{\partial v}\right)_T dv$$

$$ds = c_v \frac{dT}{T} + \left(\frac{\partial s}{\partial v}\right)_T dv$$

$$ds = c_v \frac{dT}{T} + \left(\frac{\partial p}{\partial T}\right)_v dv \qquad (10.36)$$

Equations (10.35) and (10.36) both express the change of entropy, but they use different specific heats. Again, an equation of state is needed to determine the partial derivatives.

Specific Heat Difference

Another relationship denoting the difference in specific heats may be found in the following manner. Let us equate equations (10.35) and (10.36):

$$c_p \frac{dT}{T} - \left(\frac{\partial v}{\partial T}\right)_p dp = c_v \frac{dT}{T} + \left(\frac{\partial p}{\partial T}\right)_v dv$$

$$dT = \frac{T(\partial v/\partial T)_p \, dp + T(\partial p/\partial T)_v \, dv}{c_p - c_v} \qquad (10.37)$$

Since $T = T(p, v)$,

$$dT = \left(\frac{\partial T}{\partial p}\right)_v dp + \left(\frac{\partial T}{\partial v}\right)_p dv \qquad (10.38)$$

We equate the coefficients of dp or dv in equations (10.37) and (10.38). The following is found by equating the coefficients of dp:

$$(c_p - c_v)\left(\frac{\partial T}{\partial p}\right)_v = T\left(\frac{\partial v}{\partial T}\right)_p$$

$$c_p - c_v = T\left(\frac{\partial v}{\partial T}\right)_p \left(\frac{\partial p}{\partial T}\right)_v \qquad (10.39)$$

From equation (10.11), we may write for p, v, and T

$$\left(\frac{\partial v}{\partial T}\right)_p \left(\frac{\partial T}{\partial p}\right)_v \left(\frac{\partial p}{\partial v}\right)_T = -1$$

from which

$$\left(\frac{\partial p}{\partial T}\right)_v = -\left(\frac{\partial v}{\partial T}\right)_p \left(\frac{\partial p}{\partial v}\right)_T \qquad (10.40)$$

Substituting equation (10.40) into equation (10.39),

$$c_p - c_v = -T\left(\frac{\partial v}{\partial T}\right)_p \left(\frac{\partial v}{\partial T}\right)_p \left(\frac{\partial p}{\partial v}\right)_T$$

$$c_p - c_v = -T\left(\frac{\partial v}{\partial T}\right)_p^2 \left(\frac{\partial p}{\partial v}\right)_T \qquad (10.41)$$

Three conclusions follow from applying experimental observations to equation (10.41).

1. Since $(\partial v/\partial T)_p$ is very small for liquids and solids, the difference between c_p and c_v is essentially zero; only one specific heat for a liquid or solid is usually tabulated with designation of constant pressure or constant volume.
2. As T approaches zero, c_p approaches c_v.
3. For all known substances $(\partial p/\partial v)_T < 0$, and $(\partial v/\partial T)_p^2 > 0$ all the time; hence $c_p \geq c_v$.

Incompressible Substances

Very often it is possible to consider solids and liquids as incompressible substances. When this is the case, simplifications in the expressions for internal energy, enthalpy, and entropy are possible. For an incompressible substance the total and partial derivatives of specific volume are zero. This allows great simplification in some property equations of state. Thus, the difference between specific heats (equation [10.41]) is zero and

$$c_p = c_v = C$$

Furthermore, the expression for the change of internal energy (equation [10.29]) becomes

$$du = cdT$$

$$u_2 - u_1 = \int_1^2 cdT$$

The expression for the change of enthalpy (equation [10.33]) becomes

$$dh = cdT + vdp$$

$$h_2 - h_1 = u_2 - u_1 + v(p_2 - p_1)$$

and the expression for the change of entropy (equation [10.36]) becomes

$$ds = c\frac{dT}{T}$$

$$s_2 - s_1 = \int_1^2 c\frac{dT}{T}$$

The specific heats of most solids and liquids vary linearly with temperature, so the previous equations may be readily evaluated.

10.6 CLAPEYRON EQUATION

It is possible to predict values of enthalpy for changes in phase of a substance in thermodynamic equilibrium. By equilibrium we mean that the system is in mechanical equilibrium (both phases are at the same pressure) and in thermal equilibrium (both phases are at the same temperature). There is a direct correlation between the enthalpy change in vaporization and the volume change in vaporization.

The relationship that we want is $(\partial h/\partial v)_p$, but enthalpy cannot be calculated, so

$$\left(\frac{\partial h}{\partial v}\right)_p = \left(\frac{\partial h}{\partial s}\right)_p \left(\frac{\partial s}{\partial v}\right)_p$$

and from equation (10.21)

$$\left(\frac{\partial h}{\partial s}\right)_p = T$$

Hence

$$\left(\frac{\partial h}{\partial v}\right)_p = T\left(\frac{\partial s}{\partial v}\right)_p \tag{10.42}$$

Fortunately, in the saturated region a constant-pressure process is also a constant-temperature process, so

$$\left(\frac{\partial s}{\partial v}\right)_p = \left(\frac{\partial s}{\partial v}\right)_T = \left(\frac{\partial p}{\partial T}\right)_v \tag{10.43}$$

from equation (10.23).

Since the pressure is a function only of temperature when two phases are present (see Figure 4.5), the partial derivative may be written as a total derivative in this saturated region. Equation (10.42) becomes

$$\left(\frac{\partial h}{\partial v}\right)_p = T\left(\frac{dp}{dT}\right) \tag{10.44}$$

Denoting the change of a property from a saturated liquid to a saturated vapor phase as *fg*, equation (10.44) becomes

$$\frac{h_{fg}}{v_{fg}} = T\left(\frac{dp}{dT}\right)$$

or

$$h_{fg} = v_{fg} T\left(\frac{dp}{dT}\right) \tag{10.45}$$

Equation (10.45), which is the Clapeyron equation, enables us to determine the change in enthalpy during a phase change by measuring the volume change, the temperature, and the slope of the vapor pressure curve. If the assumption is made that the specific volume of the vapor is much larger than that of the liquid and if we employ the ideal-gas law to determine the specific volume of the vapor, the enthalpy of vaporization may be readily calculated. Let us assume that

$$v_g \gg v_f$$

and

$$v_g = \frac{RT}{p}$$

then

$$h_{fg} = \frac{RT^2}{p}\frac{dp}{dT} \tag{10.46}$$

or

$$\frac{dp}{p} = \frac{h_{fg}}{R}\frac{dT}{T^2} \tag{10.47}$$

If equation (10.47) is integrated, the following results:

$$\ln p = -h_{fg}/(RT) + \text{constant of integration}$$

or

$$\ln p = A + B/T \tag{10.48}$$

where A and B are constants. Equation (10.48) is commonly used to represent the relationship between saturated temperature and pressure for a substance. It suggests that plotting $\ln p$ versus $1/T$ will give a straight line.

Example 10.2

Determine the pressure of saturated steam at 40°C if at 35°C the pressure is 5.628 kPa, the enthalpy of vaporization is 2418.6 kJ/kg, and the specific volume is 25.22 m³/kg.

Solution

Find: The pressure of saturated steam, knowing the temperature and other properties at a different state.

Given: Steam properties at an adjacent state.

Assumptions:

1. The enthalpy of vaporization is essentially constant between the two states.
2. The specific volume of the liquid is much less than that of the vapor and may be neglected.
3. The ideal-gas law may be employed to determine the specific volume of the vapor.

Analysis: The Clapeyron equation provides us with the means to determine the pressure. Integrate equation (10.47) between two states, yielding

$$\ln\left(\frac{p_2}{p_1}\right) = \frac{h_{fg}}{R}\left(\frac{1}{T_1} - \frac{1}{T_2}\right)$$

The value of R may be looked up in the ideal-gas table or calculated from the given data. The later method is preferable, as R will vary because of non-ideal-gas effects.

$$R = \frac{v_1 p_1}{T_1} = \frac{(25.22 \text{ m}^3/\text{kg})(5.628 \text{ kN}/\text{m}^2)}{(308°\text{K})} = 0.46 \text{ kJ}/\text{kg-K}$$

Substituting into the previous equation yields

$$\ln\left(\frac{p_2}{p_1}\right) = \frac{2418.6 \text{ kJ}/\text{kg}}{0.46 \text{ kJ}/\text{kg-K}}\left(\frac{1}{308°\text{K}} - \frac{1}{313°\text{K}}\right)$$

$$\ln\left(\frac{p_2}{p_1}\right) = 0.2727$$

and solving for the pressure p_2 yields

$$p_2 = 7.392 \text{ kPa}$$

This corresponds very well with the saturated pressure at 40°C, found in the steam table to be 7.389 kPa.

Comment: The Clapeyron equation is a very powerful tool in determining properties of saturated pure substances. Its usefulness decreases at states near the critical state, as the assumption concerning the differences in specific volume is no longer valid.

Example 10.3

Use equation (10.48) to develop an equation to predict the pressure of saturated steam at a given temperature between 100°C and 200°C. Use TK Solver to calculate the values of A and B. Compare the results for 150°C with the steam tables.

Solution

Given: Steam pressure and temperature at 100°C and 200°C from steam tables.

Find: The values of the constants A and B in equation (10.48) for saturated steam between 100°C and 200°C.

Assumptions:

1. The enthalpy of vaporization is essentially constant.
2. The specific volume of the liquid is much less than that of the vapor.
3. The ideal-gas law may be employed to determine the specific volume of the vapor.

Analysis: Using TK Solver, enter equation (10.48) twice in the Rule Sheet, once for each of the two data points. Enter the saturated temperatures and pressures in the Rule Sheet, guess values for A and B, and solve.

```
============================ VARIABLE SHEET ================
St Input ——— Name ——— Output ——— Unit ——— Comment ———

             A          17.539583
             B          -4821.657
     101.3   p1                       kPa
     373.15  T1                       degK
     1554.7  p2                       kPa
     473.15  T2                       degK




=========================== RULE SHEET ====================
S Rule ————————————————————————————————————————————————————
  LN(p1)  = A + B/T1
  LN(p2)  = A + B/T2
```

With the values for A and B determined, enter 150°C (423.15°K) and solve for the saturated pressure.

```
================ VARIABLE SHEET ================
St Input ——— Name —— Output —— Unit —— Comment ———

   17.539583    A
   -4821.657    B
                p      466.33706    kPa
   423.15       T                   degK




================= RULE SHEET =================
S Rule ———
  LN(p)  = A + B/T
```

This compares well with the value in the steam tables at 150°C of 476.3 kPa.

Comments:

1. The equation predicted the saturation pressure at 150°C with an accuracy of about 2%.
2. Attempting to use the equation over a wider range would result in greater errors, since the assumptions made would be less valid. ▪

10.7 IMPORTANT PHYSICAL COEFFICIENTS

Several other important coefficients are used to determine actual properties of a substance. From knowledge of the coefficients, property changes may be determined. Let us consider that $V = f(T, p)$; then

$$dV = \left(\frac{\partial V}{\partial T}\right)_p dT + \left(\frac{\partial V}{\partial p}\right)_T dp$$

We divide each partial derivative by V, which results in the following coefficients:

$$\alpha \equiv \frac{1}{V}\left(\frac{\partial V}{\partial T}\right)_p \tag{10.49}$$

and

$$\beta_T \equiv -\frac{1}{V}\left(\frac{\partial V}{\partial p}\right)_T \qquad (10.50)$$

where α is the coefficient of *thermal expansion* and β_T is the coefficient of *isothermal compressibility*. The negative sign is included in equation (10.50) because the volume decreases with increasing pressure, and vice versa, so the partial derivative is always negative and the minus sign allows β_T always to be positive. The reciprocal of the isothermal compressibility is the *isothermal bulk modulus*, B_T;

$$B_T \equiv -V\left(\frac{\partial p}{\partial V}\right)_T \qquad (10.51)$$

The difference between the specific heats may be expressed in terms of these coefficients. By equation (10.41),

$$c_p - c_v = -T\left(\frac{\partial v}{\partial T}\right)_p^2 \left(\frac{\partial p}{\partial v}\right)_T$$

but

$$\frac{B_T}{v} = -\left(\frac{\partial p}{\partial v}\right)_T$$

and

$$\alpha^2 v^2 = \left(\frac{\partial v}{\partial T}\right)_p^2$$

Hence,

$$c_p - c_v = \frac{T B_T \alpha^2 v^2}{v}$$

$$c_p - c_v = \frac{T \alpha^2 v}{\beta_T} \qquad (10.52)$$

The change in entropy at constant temperature may also be found when entropy is considered as a function of temperature and pressure. Then

$$ds = \left(\frac{\partial s}{\partial p}\right)_T dp \quad \text{for } T = c$$

but from the Maxwell relations,

$$\left(\frac{\partial s}{\partial p}\right)_T = -\left(\frac{\partial v}{\partial T}\right)_p = -v\alpha \qquad (10.53)$$

and

$$(ds)_T = -v\alpha(dp)_T$$

$$(s_2 - s_1)_T = -v\alpha(p_2 - p_1)_T \qquad (10.54)$$

Thus, by knowing tabulated coefficients we can calculate the property change of certain systems. For instance, the *Handbook of Chemistry and Physics* lists these and other coefficients for a great variety of substances.

The homework problems at the end of the chapter illustrate some typical cases where the coefficients may be calculated if an equation of state is known, as in the case of gases. The ideal-gas law will yield much simplified results, but these results can be helpful in confirming your understanding of the meaning of the various coefficients.

Joule-Thomson Coefficient

A throttling process produces no change in enthalpy; hence for an ideal gas the temperature remains constant. For a real gas, however, the throttling process will cause the temperature to increase or decrease. The Joule-Thomson coefficient, μ, relates this change and is defined as

$$\mu \equiv \left(\frac{\partial T}{\partial p}\right)_h \qquad (10.55)$$

A positive value of μ indicates that the temperature decreased as the pressure decreased; a cooling effect is thus observed. This is true for almost all gases at ordinary pressures and temperatures. The exceptions are hydrogen (H_2), neon, and helium, which have a temperature increase with a pressure decrease, hence $\mu < 0$. Even for these gases there is a temperature above which the Joule-Thomson coefficient changes from negative to positive. At this *inversion temperature,* $\mu = 0$.

Let us analyze the Joule-Thomson coefficient to gain insight into why the coefficient changes sign.

$$dh = T\,ds + v\,dp$$

$$s = f(T, p)$$

$$ds = \left(\frac{\partial s}{\partial T}\right)_p dT + \left(\frac{\partial s}{\partial p}\right)_T dp$$

$$dh = T\left(\frac{\partial s}{\partial T}\right)_p dT + T\left(\frac{\partial s}{\partial p}\right)_T dp + v\,dp$$

For constant T, $dT = 0$.

$$dh = \left[T\left(\frac{\partial s}{\partial p}\right)_T + v\right] dp$$

or

$$\left(\frac{\partial h}{\partial p}\right)_T = T\left(\frac{\partial s}{\partial p}\right)_T + v \tag{10.56}$$

From equation (10.11) for $h = f(T, p)$

$$\left(\frac{\partial T}{\partial p}\right)_h \left(\frac{\partial p}{\partial h}\right)_T \left(\frac{\partial h}{\partial T}\right)_p = -1$$

$$\left(\frac{\partial T}{\partial p}\right)_h = -\frac{1}{(\partial h/\partial T)_p}\left(\frac{\partial h}{\partial p}\right)_T \tag{10.57}$$

and

$$\left(\frac{\partial h}{\partial T}\right)_p \equiv c_p \tag{10.58}$$

Therefore, we substitute equations (10.32) and (10.58) into equation (10.57):

$$\mu = -\frac{1}{c_p}\left(\frac{\partial h}{\partial p}\right)_T \tag{10.59}$$

$$\mu = -\frac{1}{c_p}\left[-T\left(\frac{\partial v}{\partial T}\right)_p + v\right] \tag{10.60}$$

For an ideal gas, $pv = RT$.

$$\left(\frac{\partial v}{\partial T}\right)_p = \frac{R}{P} = \frac{v}{T}$$

and substituting into equation (10.60) yields

$$\mu = 0 \tag{10.61}$$

For an ideal gas, $\mu = 0$, but that is not a necessary and sufficient condition that the gas is ideal. Remember, at the inversion temperature, $\mu = 0$ also.

To illustrate why the Joule-Thomson coefficient changes sign, we substitute $h = u + pv$ into equation (10.59):

$$\mu = +\frac{1}{c_p}\left[-\left(\frac{\partial u}{\partial p}\right)_T - \left(\frac{\partial(pv)}{\partial p}\right)_T\right] \tag{10.62}$$

The first term in the brackets denotes the deviation from Joule's law, which states that the internal energy is a function only of temperature. On expansion, there is an increase in the molecular potential energy, and hence $(\partial u/\partial p)_T$ is negative. This results in a positive μ and a temperature decrease. The second term in the brackets indicates the deviation from Boyle's law (that v varies inversely with p) for a real gas. For most gases at low temperatures and pressures, $(\partial(pv)/\partial p)_T$ is negative; however, it changes sign at higher temperatures and pressures.

Figure 10.4 illustrates the plot of μ on the T-p and T-s diagrams. Lines of constant enthalpy are also noted.

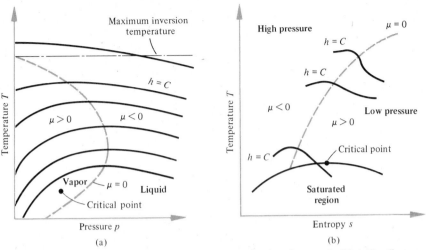

Figure 10.4 (a) A T-p diagram showing the locus of points for $\mu = 0$. (b) A T-s diagram showing the locus of points for $\mu = 0$.

10.8 REDUCED COORDINATES FOR VAN DER WAALS EQUATION OF STATE

In Chapter 5 we used a generalized compressibility chart for gases to determine the deviation from the ideal-gas equation of state. Because $Z = pv/RT = 1$ for an ideal gas, use of the generalized compressibility chart enabled us to use one chart for all gases in Chapter 5.

Why is this so? The van der Waals equation of state will demonstrate that the use of reduced coordinates leads to an equation of state in terms of these coordinates. The constants for the various gases are eliminated.

The van der Waals equation of state is

$$p = \frac{RT}{v - b} - \frac{a}{v^2} \tag{10.63}$$

Figure 10.5 shows a p-v diagram with a plot of isotherms for a typical pure substance. The critical isotherm is of importance. Examination of the experimental data has shown that at least the first two derivatives are zero at the critical point:

$$\left(\frac{\partial p}{\partial v}\right)_T = 0 \tag{10.64}$$

and

$$\left(\frac{\partial^2 p}{\partial v^2}\right)_T = 0 \tag{10.65}$$

If equations (10.64) and (10.65) are applied to equation (10.63), we may determine

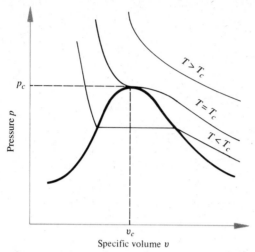

Figure 10.5 A p-v diagram for a typical pure substance.

the critical properties from the van der Waals equation:

$$\left(\frac{\partial p}{\partial v}\right)_T = -\frac{RT_c}{(v_c - b)^2} + \frac{2a}{v_c^3} = 0 \tag{10.66}$$

$$RT_c = \frac{2a}{v_c^3}(v_c - b)^2 \tag{10.67}$$

$$\left(\frac{\partial^2 p}{\partial v^2}\right)_T = +\frac{2RT_c}{(v_c - b)^3} - \frac{6a}{v_c^4} = 0$$

$$RT_c = \frac{3a}{v_c^4}(v_c - b)^3 \tag{10.68}$$

Solving equations (10.67) and (10.68) for v_c yields

$$v_c = 3b \tag{10.69}$$

and substituting this in equation (10.67) results in

$$T_c = \frac{8a}{27Rb} \tag{10.70}$$

Substitution into equation (10.63) yields for the critical pressure

$$p_c = \frac{a}{27b^2} \tag{10.71}$$

Using the critical properties, we rewrite the van der Waals equation of state with reduced coordinates:

$$p = \frac{p_r a}{27b^2} \qquad v = 3bv_r \qquad T = \frac{8aT_r}{27Rb}$$

Substituting these into equation (10.63) yields

$$p_r = \frac{8T_r}{3v_r - 1} - \frac{3}{v_r^2} \qquad\qquad (10.72)$$

Equation (10.72) illustrates that an equation of state may be expressed in terms of the reduced coordinates and be applicable for any gas. However, no substance obeys van der Waals equation of state exactly, since it is an approximation of actual gas behavior.

CONCEPT QUESTIONS

1. What are the differences between partial and ordinary differentials?
2. Given that $u = u(x, y)$, under what conditions is the partial derivative, $(\partial u/\partial y)_x$, equal to du/dy?
3. For $u = u(x, y)$, is the partial derivative, $(\partial u/\partial y)_x$, still a function of x?
4. A function of one variable may be represented as $f(x)$ and its derivative as df/dx. Can the derivative be evaluated by finding dx/df and taking the inverse?
5. Why is the Clapeyron equation useful in thermodynamics?
6. What assumptions, if any, are used in the Clapeyron equation?
7. Describe the Joule-Thomson coefficient.
8. Describe the Joule-Thomson maximum inversion temperature.
9. In a throttling process the pressure always decreases; does the temperature also always decrease?

PROBLEMS (SI)

10.1 Derive an expression for the change of internal energy of a gas using the van der Waals equation of state.

10.2 Derive an expression for the change of enthalpy of a gas using (a) the ideal-gas equation of state; (b) the van der Waals equation of state.

10.3 Show that equation (10.11) is valid when the variables are p, v, and T and are related by the ideal-gas equation of state.

10.4 Prove that heat is an inexact differential $[Q(T, s)]$.

10.5 Using the Maxwell relation in equations (10.22) and (10.11), develop the three remaining relations given in equation (10.23).

10.6 The following information is obtained from the steam tables of the vaporization process: 1378 kPa at 467.6°K, 1413 kPa at 468.8°K, and 1448 kPa at 469.9°K. At 1413 kPa, $v_{fg} = 0.1383$ m³/kg. Determine the latent heat of vaporization at 1413 kPa.

10.7 Use the Clapeyron relation to determine the change of enthalpy of (a) steam at 2000 kPa; (b) ammonia at 40°C; (c) R 12 at 15°C.

10.8 Determine the difference between c_p and c_v for a gas that obeys (a) the ideal-gas equation of state; (b) the van der Waals equation of state.

10.9 Derive an expression for the change of entropy of a gas that obeys the van der Waals equation of state.

10.10 The Berthelot equation of state is $(p + a/Tv^2)(v - b) = RT$, where a and b are constants. Show that, at the critical isotherm, (a) $a = (9/8)Rv_cT_c^2$; (b) $b = v_c/3$.

10.11 The Dieterici equation of state is $p(v - b)e^{a/RTv} = RT$, where a and b are constants. Show that, at the critical isotherm, (a) $a = 2RT_cv_c$; (b) $= v_c/2$.

10.12 Calculate the coefficient of thermal expansion and the coefficient of isothermal compressibility for a gas that obeys (a) the ideal-gas equation of state; (b) the van der Waals equation of state.

10.13 Compute the coefficient of thermal expansion for methane at $32°C$ and 1400 kPa using (a) the ideal-gas equation of state; (b) the van der Waals equation of state.

10.14 Derive an expression for the change of enthalpy and entropy at constant temperature, using the van der Waals equation of state.

10.15 Two kg of oxygen is expanded isothermally from $380°K$ and 700 kPa while obeying the van der Waals equation of state. Determine the changes in internal energy, enthalpy, and entropy for the process.

10.16 Determine the Joule-Thomson coefficient for a gas with the following equation of state: (a) $p(v - b) = RT$; (b) $(p + a/v^2)(v - b) = RT$.

10.17 Determine the average Joule-Thomson coefficient for steam that is throttled from 1.1 MPa and $280°C$ to 140 kPa.

10.18 Determine the average Joule-Thomson coefficient for ammonia that is throttled from 2.0 MPa and $140°C$ to 500 kPa.

10.19 Show that the Joule-Thomson coefficient equals $\mu = (v/c_p)(T\alpha - 1)$.

10.20 One kg of methane occupies 0.0094 m³ when the pressure is 10 MPa. Determine the gas's temperature using the van der Waals equation of state and compare the results with those from the ideal-gas law and the generalized compressibility chart.

10.21 The specific volume of steam at $350°C$ is 2.5 m³/kg. Determine the pressure using (a) the ideal-gas equation of state; (b) the van der Waals equation of state; (c) the Redlich-Kwong equation of state. (d) Compare the results with that found from the steam tables.

10.22 The specific volume of R 12 at $90°C$ is 0.020 m³/kg. Determine the pressure using (a) the ideal-gas equation of state; (b) the van der Waals equation of state; (c) the Redlich-Kwong equation of state. (d) Compare the results with that found from the refrigerant tables.

10.23 A 0.2-m³ tank contains steam at $725°C$ and 1000 kPa. Determine the mass in the tank using (a) the ideal-gas law; (b) the van der Waals equation of state; (c) the generalized compressibility chart.

10.24 A tank contains 1.5 kg of oxygen at 5000 kPa and $190°K$. The temperature of the oxygen is lowered, and the pressure becomes 4000 kPa. Determine the tank's volume and the final temperature using (a) the ideal-gas equation of state; (b) the compressibility chart; (c) the Redlich-Kwong equation of state.

10.25 Two kg of air occupies a volume of 0.05 m³ at a temperature of $318°C$. The air expands isothermally until the pressure is 1390 kPa. Using the van der Waals equation of state determine (a) the initial pressure; (b) the final volume; (c) the work.

10.26 Evaluate dv from the ideal-gas law. Show that equation (10.2) is satisfied.

10.27 One of the following expressions for the change of pressure will yield an equation of state. Determine the equation of state.

$$dp = \frac{2(v-b)}{RT} dv + \frac{(v-b)^2}{RT^2} dT$$

$$dp = \frac{-RT}{(v-b)^2} dv + \frac{R}{(v-b)} dT$$

10.28 Given that $x = x(y, v)$, $y = y(z, v)$, and $z = z(x, v)$, show that

$$\left(\frac{\partial x}{\partial y}\right)_v \left(\frac{\partial y}{\partial z}\right)_v \left(\frac{\partial z}{\partial x}\right)_v = 1$$

10.29 The maximum density for liquid water at atmospheric pressure occurs at a temperature of $4°C$. What can you determine about $(\partial s/\partial p)_T$ at temperatures of $3°C$, $4°C$, and $5°C$?

10.30 Using the van der Waals and then the Redlich-Kwong equations of state, develop expressions for $[s(v_2, T) - s(v_1, T)]$ and $[h(v_2, T) - h(v_1, T)]$.

10.31 A gas's pvT behavior can be described by the compressibility factor $Z = 1 + Bp/RT$, where B is a function of temperature. Derive expressions for $[h(p_2, T) - h(p_1, T)]$ and $[s(p_2, T) - s(p_1, T)]$.

10.32 A gas's pvT behavior can be described by the compressibility factor $Z = 1 + Bv^{-1} + Cv^{-2}$, where B and C are functions of temperature. Derive an expression for $[s(v_2, T) - s(v_1, T)]$.

10.33 A gas's pvT behavior at certain states can create a compressibility factor of $Z = 1 - ApT^{-4}$, where A is a constant. Derive an expression for the difference in specific heats, $c_p - c_v$.

10.34 For solids it is often assumed that $c_p = c_v$. For copper at $230°C$, the density is 8930 kg/m^3, $\alpha = 54.1 \times 10^{-6}$ K^{-1} and $\beta_T = 0.838 \times 10^{-11}$ m^2/N. Determine the percentage of error in making the assumption.

10.35 Determine the maximum Joule-Thomson inversion temperature, expressed in terms of the critical temperature, using the van der Waals and Redlich-Kwong equations of state.

PROBLEMS (English Units)

*10.1 The equation of state for a gas is $p(v - b) = RT$, where b is an experimental constant. Find the expression for the coefficient of isothermal compressibility.

*10.2 Five lbm of ammonia are compressed isothermally from 100 psia to 200 psia at a temperature of $220°F$. Use the van der Waals equation of state to determine the changes of internal energy, enthalpy, and entropy.

*10.3 Steam at 6000 psia and $1100°F$ is throttled adiabatically to atmospheric pressure. Determine the Joule-Thomson coefficient and the final temperature.

*10.4 Evaluate both terms of the Maxwell relation in equation (10.22) independently for steam at 600 psia and $550°F$.

*10.5 Determine the constant-volume and constant-pressure specific heats of steam at 350 psia and $500°F$ by means of equations (10.24) and (10.25).

*10.6 Determine the difference between the specific heats of constant pressure and volume from equation (10.41) for steam at 350 psia and 500°F.

*10.7 Two lbm of methane occupies 0.30 ft³ when the pressure is 1450 psia. Determine the gas's temperature using the van der Waals equation of state, and compare the results with those from the ideal-gas law and the generalized compressibility chart.

*10.8 The specific volume of steam at 620°F is 1.50 ft³/lbm. Determine the pressure using (a) the ideal-gas equation of state; (b) the van der Waals equation of state; (c) the Redlich-Kwong equation of state. (d) Compare the results with that found from the steam tables.

*10.9 The specific volume of R 12 at 190°F is 0.275 ft³/lbm. Determine the pressure using (a) the ideal-gas equation of state; (b) the van der Waals equation of state; (c) the Redlich-Kwong equation of state. (d) Compare the results with that found from the refrigerant tables.

*10.10 A 7.0-ft³ tank contains steam at 1300°F and 150 psia. Determine the mass in the tank using (a) the ideal-gas law; (b) the van der Waals equation of state; (c) the generalized compressibility chart.

*10.11 A tank contains 3.0 lbm of oxygen at 725 psia and 340°R. The temperature of the oxygen is lowered, and the pressure becomes 580 psia. Determine the tank's volume and the final temperature, using (a) the ideal-gas equation of state; (b) the compressibility chart; (c) the Redlich-Kwong equation of state.

*10.12 A volume of 1.75 ft³ holds 4.5 lbm of air at a temperature of 600°F. The air expands isothermally until the pressure is 200 psia. Using the van der Waals equation of state, determine (a) the initial pressure; (b) the final volume; (c) the work.

COMPUTER PROBLEMS

C10.1 For steam from 25°C to 350°C, plot the inverse of the absolute temperature versus the natural log of the absolute pressure ($1/T$ versus $\ln p$) using at least 20 data points.

C10.2 Using TK Solver, determine the values of A and B for equation (10.48) for saturated steam between 25°C and 350°C. Calculate the errors at 100°C, 200°C, and 300°C.

C10.3 With the addition of a third parameter to equation (10.48), the Antoine equation results: $\ln p = A + B/(T - C)$. Use TK Solver to calculate the values of A, B, and C for steam from 25°C to 350°C and compare the results with those in Problem C10.2.

C10.4 Steam at 300°C is throttled to 100 kPa from 2000 kPa, 1000 kPa, and 500 kPa. Determine (a) the final temperature of the steam; (b) the average Joule-Thomson coefficient.

11

Nonreacting Ideal-Gas and Gas-Vapor Mixtures

In the previous chapters we often assumed a substance was a pure substance or acted like one. Very often that is not the case; for example, the products of combustion from an automobile are a mixture of gases and water vapor. And atmospheric air is a mixture of several gases, including water vapor, all of which are included in more complete analyses. In this chapter you will

- Determine what is an ideal-gas mixture;
- Learn to convert a mass basis to a mole basis;
- Find mixture properties knowing the constituent properties;
- Establish the limits for modeling air–water vapor mixtures as ideal-gas mixtures;
- Understand relative and absolute humidity;
- Learn about wet-bulb and dry-bulb temperatures;
- Determine when the dew point occurs.

11.1 IDEAL-GAS MIXTURES

The gases in a gas mixture are called "components of the mixture." A given component i will have a mass m_i, and the total mixture will have a mass m, where

$$m = \sum_i m_i \tag{11.1}$$

The mass fraction, x_i, is defined as

$$x_i = \frac{m_i}{m} \tag{11.2}$$

The component has n_i moles in a total mixture of n moles, where $n = \sum n_i$; we can define the mole fraction, y_i, as

$$y_i = \frac{n_i}{n} \tag{11.3}$$

Let us consider that we have the components of a gas mixture separated and existing at the same temperature, T, and pressure, p. The equation of state may be written for each of these gases as

$$pV_i = n_i \overline{R} T \tag{11.4}$$

The properties for the mixture of the gases would have a volume V; the properties of the mixture are denoted by the absence of a subscript. For the mixture, the ideal-gas equation of state is

$$pV = n \overline{R} T \tag{11.5}$$

and we find the volume fraction of the mixture by dividing equation (11.4) by equation (11.5):

$$\frac{V_i}{V} = \frac{n_i}{n} = y_i \tag{11.6}$$

Thus, the volume fraction of a gas mixture is equal to the mole fraction of the mixture.

Amagat's Law

Amagat's law of additive volumes is as follows:

> The total volume of a mixture of gases is equal to the sum of the volumes that would be occupied by each component at the mixture temperature, T, and pressure, p.

This applies rigorously to ideal gases:

$$V = \sum_i V_i \tag{11.7}$$

This is illustrated in Figure 11.1 for three components, 1, 2, 3, of an ideal-gas mixture.

Another approach in analyzing gas mixtures applies when the components occupy the total mixture volume, V, at the same temperature, T. Figure 11.2 illustrates

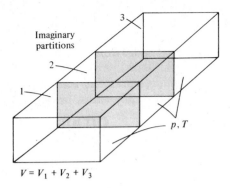

Imaginary
partitions

p, T

$V = V_1 + V_2 + V_3$

Figure 11.1 Visualization of Amagat's law
of additive volumes.

this case. The ideal-gas equation of state may be written for each component and for
the total mixture:

$$p_i V = n_i \overline{R} T \qquad (11.8a)$$

and

$$p V = n \overline{R} T \qquad (11.8b)$$

where equation (11.8a) is the ideal-gas equation of state for the ith component and
equation (11.8b) is the ideal-gas equation of state for the total mixture. Let us divide

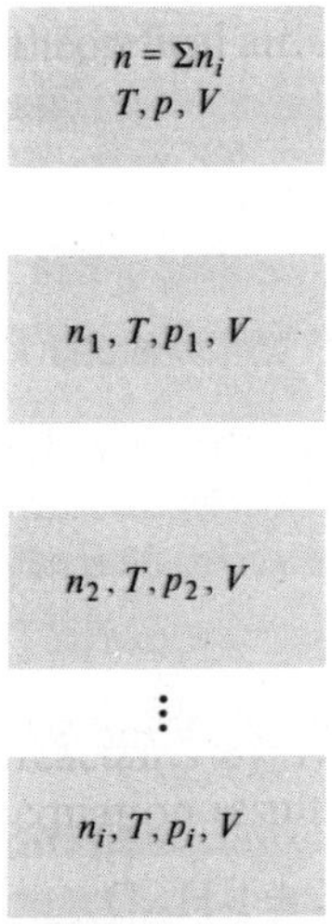

$n = \Sigma n_i$
T, p, V

n_1, T, p_1, V

n_2, T, p_2, V

$\vdots$

n_i, T, p_i, V

Figure 11.2 Visualization of Dalton's law of
partial pressures.

equation (11.8a) by equation (11.8b):

$$\frac{p_i}{p} = \frac{n_i}{n} = y_i \tag{11.9}$$

Thus, the ratio of the partial pressure, the pressure of the ith component occupying the mixture volume at the same temperature and volume, to the total pressure, is equal to the mole fraction.

Dalton's Law

Dalton's law of partial pressure states:

> The total mixture pressure, p, is the sum of the pressures that each gas would exert if it were to occupy the vessel alone at volume V and temperature T.

This is rigorously true only for ideal gases. Equation (11.9) leads us to the same conclusion:

$$p_i = p\frac{n_i}{n}$$

$$\sum_i p_i = p\sum_i \frac{n_i}{n} = p \tag{11.10}$$

$$p = \sum_i p_i$$

Equation (11.10) is the algebraic statement of Dalton's law of partial pressures.

Mixture Properties

The total mixture properties, such as internal energy, enthalpy, and entropy, may be determined by adding the properties of the components at the mixture conditions. The internal energy and enthalpy are functions of temperature only; hence

$$U = n\bar{u} = \sum_i n_i\bar{u}_i$$
$$H = n\bar{h} = \sum_i n_i\bar{h}_i \tag{11.11}$$

where $\bar{u}_i$ and $\bar{h}_i$ are the internal energy and enthalpy of the ith component on the mole basis, energy per unit mole, at the mixture temperature.

The entropy of a component is a function of the temperature and pressure. The entropy of a mixture is the sum of the component entropies,

$$S = n\bar{s} = \sum_i n_i\bar{s}_i \tag{11.12}$$

where $\bar{s}_i$ is the entropy per mole of the ith component at the mixture temperature, T, and its partial pressure, p_i.

The specific heats of a mixture may be found by expanding the equations for enthalpy and internal energy of a mixture:

$$c_p = \sum_i x_i c_{pi} \tag{11.13}$$

$$c_v = \sum_i x_i c_{vi} \tag{11.14}$$

The key to analyzing gas mixtures and changing from a mass to a mole analysis, and vice versa, is a unit balance. The following examples illustrate several ways of finding mixture properties.

Example 11.1

The products of combustion from a diesel engine have the following molal analysis: $CO_2 = 10.2\%$, $CO = 0.4\%$, $H_2O = 14.3\%$, $O_2 = 1.9\%$, and $N_2 = 73.2\%$. Determine the molecular weight of the products and the mass fraction of each component.

Solution

Given: A gas mixture with molal analysis given.

Find: The gas mixture molecular weight and the mass fraction for each component.

Assumption: Each component and the mixture behaves like an ideal gas.

Analysis: The mixture molecular weight is found from

$$M = \sum_i y_i M_i \text{ kg mixture/kgmol mixture}$$

Substituting into this yields

$$M = (0.102)(44.01) + (0.004)(28.01) + (0.143)(18.016)$$

$$+ (0.019)(32.0) + (0.732)(28.016) = 28.29 \text{ kg mixture/kgmol mixture}$$

The mass fractions, x_i's, may be found from the relationship

$$x_i = (y_i)\left(\frac{1}{M}\right)(M_i)$$

$$\frac{(\text{kg})_i}{(\text{kg})_{\text{mix}}} = \frac{(\text{kgmol})_i}{(\text{kgmol})_{\text{mix}}} \cdot \frac{(\text{kgmol})_{\text{mix}}}{(\text{kg})_{\text{mix}}} \cdot \frac{\text{kg}_i}{(\text{kgmol})_i}$$

Thus,

$$x_{CO_2} = (0.102)(44.01)/(28.29) = 0.159 \text{ kg } CO_2/\text{kg mixture}$$

$$x_{CO} = (0.004)(28.01)/(28.29) = 0.004 \text{ kg } CO/\text{kg mixture}$$

$$x_{H_2O} = (0.143)(18.016)/(28.29) = 0.091 \text{ kg } H_2O/\text{kg mixture}$$

$$x_{O_2} = (0.019)(32.0)/(28.29) = 0.021 \text{ kg } O_2/\text{kg mixture}$$

$$x_{N_2} = (0.732)(28.016)/(28.29) = 0.725 \text{ kg } N_2/\text{kg mixture}$$

$$\text{total} = 1.000$$

Comment: When there is a significant difference between the individual and the mixture molecular weights, there will be a significant difference between the mass and mole fractions.

Example 11.2

A mixture of 0.4 lbm of helium and 0.2 lbm of oxygen is compressed polytropically from 14.7 psia and 60°F to 60 psia according to $n = 1.4$. Determine the final temperature, the heat, and the work for the process.

Solution

Given: A gas mixture is compressed polytropically between two states.

Find: The final temperature, heat, and work for the process.

Sketch and Given Data:

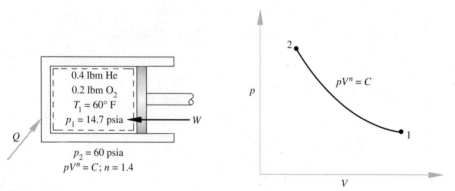

Figure 11.3

Assumptions:

1. The individual gases and the gas mixture behave like ideal gases.
2. The changes in kinetic and potential energies can be neglected.

Analysis: The system is a closed system, and the heat and work expressions for polytropic processes, developed in Chapter 6, may be employed. First, find the final temperature.

$$T_2 = T_1 \left(\frac{p_2}{p_1} \right)^{(n-1)/n} = (520°\text{R}) \left(\frac{60}{14.7} \right)^{0.4/1.4} = 777.2°\text{R}$$

The expression for work for a polytropic process may be written

$$W = \frac{mR(T_2 - T_1)}{1 - n}$$

In this case R is the mixture gas constant and must be determined.

$$R = \sum_i x_i R_i$$

where

$$x_{He} = \frac{0.4}{0.6} = 0.667 \quad \text{and} \quad x_{O_2} = \frac{0.2}{0.6} = 0.333$$

Thus,

$$R = \left(0.667 \ \frac{\text{lbm}_{He}}{\text{lbm mix}}\right)\left(386.04 \ \frac{\text{ft-lbf}}{\text{lbm-R}}\right) + \left(0.333 \ \frac{\text{lbm}_{O_2}}{\text{lbm mix}}\right)\left(48.29 \ \frac{\text{ft-lbf}}{\text{lbm-R}}\right)$$

$$= 273.6 \ \text{ft-lbf/lbm-R}$$

$$R = 0.3516 \ \text{Btu/lbm-R}$$

The work becomes

$$W = \frac{(0.6 \ \text{lbm})(0.3516 \ \text{Btu/lbm-R})(777.2 - 520°\text{R})}{-0.4} = -135.6 \ \text{Btu}$$

The heat is found by solving the first-law equation for a closed system, which requires that the change in internal energy for the mixture be evaluated.

$$\Delta U = m_{He} c_{p_{He}}(T_2 - T_1) + m_{O_2} c_{p_{O_2}}(T_2 - T_1)$$

$$\Delta U = (0.4 \ \text{lbm})(1.241 \ \text{Btu/lbm-R})(777.2 - 520°\text{R})$$

$$+ (0.2 \ \text{lbm})(0.2194 \ \text{Btu/lbm-R})(777.2 - 520°\text{R})$$

$$\Delta U = 139.0 \ \text{Btu}$$

From the first law,

$$Q = \Delta U + W = 139 - 135.6 = 3.4 \ \text{Btu}$$

Comment: Any equation developed for an ideal gas can be used for ideal-gas mixtures. The mixture properties need to be evaluated.

Entropy of Mixing

When two gases are mixed together, there will be an increase in entropy. This is called the *entropy of mixing,* illustrated in Figure 11.4. There are n_a moles of the first component of a gas mixture and n_b moles of the second component; they exist at the same temperature and pressure. The partition separating the components is removed, and the gases mix adiabatically. For constant temperature, the change of entropy is

$$S_2 - S_1 = -n\bar{R} \ln\left(\frac{p_2}{p_1}\right)$$

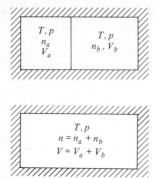

Figure 11.4 Illustration of entropy of mixing.

and for the a component, the final pressure is p_a, its partial pressure, and the initial pressure is p. Therefore,

$$(S_2 - S_1)_a = -n_a \overline{R} \ln \left(\frac{p_a}{p} \right) = -n_a \overline{R} \ln (y_a)$$

$$(S_2 - S_1)_b = -n_b \overline{R} \ln \left(\frac{p_b}{p} \right) = -n_b \overline{R} \ln (y_b)$$

and the entropy change of the mixture is the sum of the component entropy changes:

$$S_2 - S_1 = -n_a \overline{R} \ln (y_a) - n_b \overline{R} \ln (y_b) \tag{11.15}$$

If the components a and b are the same gas, the entropy change in mixing is zero. This is because we cannot distinguish between the gases to determine the partial pressures, and the final pressure therefore is also the partial pressure of the gas. In the case discussed for distinguishable components, we may generalize equation (11.15) for any number of components at the same temperature and pressure:

$$S_2 - S_1 = -\overline{R} \sum_i n_i \ln (y_i) \tag{11.16}$$

We see that there will be an entropy increase, because the sign of the natural logarithm of a fraction is negative.

Let us consider the second law for an isolated system, $\Delta S \geq 0$. The entropy is greater than zero due to irreversibilities that occur within the system. The mixing process is a totally irreversible process, and we should expect the entropy to increase as we demonstrated it does.

The total entropy change for a system may include the entropy change of mixing as well as other entropy changes due to irreversibilities and heat transfer.

Example 11.3

An adiabatic tank has two compartments joined by a valve. One compartment contains 0.7 kg of carbon dioxide at 500°K and 6000 kPa, and the other contains 0.3 kg of nitrogen at 600 kPa and 300°K. The valve is opened and the gases mix. Determine the final mixture temperature and pressure and the entropy produced.

Solution

Given: An adiabatic tank with two separated gases at known initial states are mixed.

Find: The final temperature and pressure and the entropy produced by the irreversible mixing of the gases.

Sketch and Given Data:

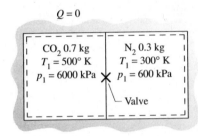

Figure 11.5

Assumptions:

1. The individual gases and the gas mixture behave like ideal gases.
2. The total system is the sum of the two subsystems.
3. Heat and work are zero.
4. The changes in kinetic and potential energies may be neglected.

Analysis: The first law for a closed system, $Q = \Delta U + W$, subject to assumption 3 yields

$$U_2 - U_1 = 0$$

The internal energy change of the mixture is the sum of the internal energy changes of each constituent gas; thus,

$$m_{CO_2} c_{v_{CO_2}}(T_2 - T_1) + m_{N_2} c_{v_{N_2}}(T_2 - T_1) = 0$$

$$(0.7 \text{ kg})(0.6552 \text{ kJ/kg-K})(T_2 - 500°K)$$

$$+ (0.3 \text{ kg})(0.7431 \text{ kJ/kg-K})(T_2 - 300°K) = 0$$

$$T_2 = 434.6°K$$

The mixture gas constant may be determined from the ideal-gas law where

$$R = \sum_i x_i R_i$$

$$R = (0.7)(0.1889) + (0.3)(0.2968) = 0.2213 \text{ kJ/kg-K}$$

The total volume for the system is the sum of the two initial constituent volumes.

$$V_{1N_2} = \frac{mRT_1}{p_1} = \frac{(0.3 \text{ kg})(0.2968 \text{ kJ/kg-K})(300°\text{K})}{600 \text{ kN/m}^2} = 0.04452 \text{ m}^3$$

$$V_{1CO_2} = \frac{mRT_1}{p_1} = \frac{(0.7 \text{ kg})(0.1889 \text{ kJ/kg-K})(500°\text{K})}{6000 \text{ kN/m}^2} = 0.01102 \text{ m}^3$$

$$V_{total} = 0.05572 \text{ m}^3$$

Solving the mixture ideal-gas law for pressure yields

$$p_2 = \frac{mRT_2}{V_2} = \frac{(1.0 \text{ kg})(0.2213 \text{ kJ/kg-K})(434.6°\text{K})}{0.05572 \text{ m}^3} = 1726 \text{ kPa}$$

The entropy produced from the irreversible mixing is found from the second law for closed systems,

$$S_2 - S_1 = \int_1^2 \frac{\overset{0}{\cancel{\delta Q}}}{T} + \Delta S_{prod}$$

and

$$S_2 - S_1 = \sum_i \Delta S_i$$

The expression for entropy change of an ideal gas is

$$(S_2 - S_1) = mc_p \ln (T_2/T_1) - mR \ln (p_2/p_1)$$

The partial pressures of the gases at state 2 are

$$(p_2)_{N_2} = \frac{mRT_2}{V_2}$$

$$(p_2)_{N_2} = \frac{(0.3 \text{ kg})(0.2968 \text{ kJ/kg-K})(434.6°\text{K})}{(0.055\ 72 \text{ m}^3)} = 694.5 \text{ kPa}$$

$$(p_2)_{CO_2} = 1726 - 694.5 = 1031.5 \text{ kPa}$$

$$(S_2 - S_1)_{CO_2} = (0.7 \text{ kg})(0.844 \text{ kJ/kg-K})[\ln (434.6/500)]$$

$$- (0.7 \text{ kg})(0.1889 \text{ kJ/kg-K})[\ln (1031.5/6000)]$$

$$= +0.150\ 00 \text{ kJ/K}$$

$$(S_2 - S_1)_{N_2} = (0.3 \text{ kg})(1.0399 \text{ kJ/kg-K})[\ln (434.6/300)]$$

$$- (0.3 \text{ kg})(0.2968 \text{ kJ/kg-K})[\ln (694.5/600)]$$

$$= +0.1025 \text{ kJ/K}$$

The total entropy produced is

$$\Delta S_{prod} = 0.1500 + 0.1025 = +0.2525 \text{ kJ/K}$$

Comment: The entropy production term increases because of mixing of gases that are initially at different temperatures, that are initially at different pressures, and that are distinguishable from one another.

Example 11.4

Two gaseous streams, 50 m³/min of air at 40°C and 1 atm and 20 m³/min of helium at 100°C and 1 atm, are adiabatically mixed to form a mixture at 1 atm. Determine the mass and mole fractions of the mixture, the mixture temperature, and the rate of entropy production.

Solution

Given: Two gaseous streams at known conditions mix adiabatically at constant pressure.

Find: The mixture mass and mole fractions, the mixture temperature, and the entropy production.

Sketch and Given Data:

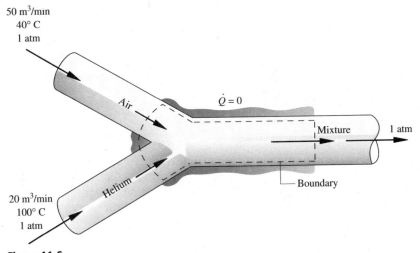

Figure 11.6

Assumptions:

1. The conditions are steady-state.
2. The changes in kinetic and potential energies may be neglected.
3. The components and the mixture are ideal gases.
4. Air may be treated as a single component.
5. The heat and work transfer are zero.

Analysis: The mass flow rates may be determined from the ideal-gas law.

$$\dot{m}_a = \frac{p\dot{V}}{RT} = \frac{(101.3 \text{ kN/m}^2)(50 \text{ m}^3/\text{min})}{(0.287 \text{ kJ/kg-K})(313^\circ\text{K})(60 \text{ s/min})} = 0.9397 \text{ kg/s}$$

$$\dot{m}_{\text{He}} = \frac{p\dot{V}}{RT} = \frac{(101.3 \text{ kN/m}^2)(20 \text{ m}^3/\text{min})}{(2.077 \text{ kJ/kg-K})(373^\circ\text{K})(60 \text{ s/min})} = 0.0436 \text{ kg/s}$$

$$\dot{m}_{\text{total}} = 0.9397 + 0.0436 = 0.9833 \text{ kg/s}$$

The mass fractions are

$$x_a = \frac{0.9397}{0.9833} = 0.955$$

$$x_{\text{He}} = \frac{0.0436}{0.9833} = 0.045$$

$$\text{total} = 1.000$$

Convert the mass flow rate to mole flow rate and find the mole fractions.

$$\dot{n}_a = (0.9397 \text{ kg/s})(1/28.97 \text{ kg/kgmol}) = 0.0324 \text{ kgmol/s}$$

$$\dot{n}_{\text{He}} = (0.0436 \text{ kg/s})(1/4.003 \text{ kg/kgmol}) = 0.0109 \text{ kgmol/s}$$

$$\dot{n}_{\text{total}} = 0.0433 \text{ kgmol/s}$$

$$y_a = \frac{0.0324}{0.0433} = 0.748$$

$$y_{\text{He}} = \frac{0.0109}{0.0433} = 0.252$$

For steady state and steady flow, the first law for an open system is

$$\dot{Q} + \dot{m}_a(h + \text{k.e.} + \text{p.e.})_a + \dot{m}_{\text{He}}(h + \text{k.e.} + \text{p.e.})_{\text{He}} = \dot{W} + \dot{m}_{\text{mix}}(h + \text{k.e.} + \text{p.e.})_{\text{mix}}$$

Applying the assumptions and dividing by the mixture mass flow rate yields

$$x_a h_a + x_{\text{He}} h_{\text{He}} = h_m$$

Using the ideal-gas equation of state for enthalpy,

$$x_a c_{p_a} T_a + x_{\text{He}} c_{p_{\text{He}}} T_{\text{He}} = c_{p_m} T_m$$

Noting that

$$c_{p_m} = \sum c_{p_i} x_i$$

$$c_{p_m} = (0.955 \text{ kg}_{\text{air}}/\text{kg}_{\text{mix}})(1.0047 \text{ kJ/kg}_{\text{air}}\text{-K})$$

$$+ (0.045 \text{ kg}_{\text{He}}/\text{kg}_{\text{mix}})(5.1954 \text{ kJ/kg}_{\text{He}}\text{-K}) = 1.1933 \text{ kJ/kg-K}$$

and substituting in the first-law expression,

$$(0.955)(1.0047 \text{ kJ/kg-K})(313°K)$$

$$+ (0.045)(5.1954 \text{ kJ/kg-K})(373°K) = (1.1933 \text{ kJ/kg-K})T_m$$

yields

$$T_m = 324.8°K = 51.8°C$$

The entropy production is brought about by irreversible mixing of distinguishable steams, a change in component temperature, and a change in component partial pressure. Equation (8.43a) may be used for each component, yielding

$$\dot{m}_a(s_2 - s_1)_a + \dot{m}_{He}(s_2 - s_1)_{He} = \Delta\dot{S}_{prod}$$

Since the entropy change is a function of temperature and pressure, the partial pressure of each component must be found at the final mixture state.

$$p_a = y_a p_{total} = (0.748)(101.3) = 75.8 \text{ kPa}$$

$$p_{He} = y_{He} p_{total} = (0.252)(101.3) = 25.5 \text{ kPa}$$

The entropy change for each component is

$$(s_2 - s_1) = c_p \ln(T_2/T_1) - R \ln(p_2/p_1)$$

and applying this to each component yields

$$(s_2 - s_1)_a = (1.0047 \text{ kJ/kg-K})[\ln(324.8/313)]$$

$$- (0.287)[\ln(75.8/101.3)] = 0.1204 \text{ kJ/kg-K}$$

$$\dot{m}_a(s_2 - s_1)_a = (0.9397 \text{ kg/s})(0.1204 \text{ kJ/kg-K}) = 0.1131 \text{ kW/K}$$

$$(s_2 - s_1)_{He} = (5.1954 \text{ kJ/kg-K})[\ln(324.6/373)]$$

$$- (2.077 \text{ kJ/kg-K})[\ln(25.5/101.3)] = 2.1429 \text{ kJ/kg-K}$$

$$\dot{m}_{He}(s_2 - s_1) = (0.0436 \text{ kg/s})(2/1429 \text{ kJ/kg-K}) = 0.0934 \text{ kW/K}$$

$$\Delta\dot{S}_{prod} = 0.1131 + 0.0934 = 0.2065 \text{ kW/K}$$

Comment: It is usually easier to solve problems on the mass basis than on the mole basis. In solving the first-law equation we assumed that the enthalpy of all components could be expressed as $c_p T$. This presumes that the enthalpy is zero at absolute zero, which is accurate in this instance but is not always true for all substances, as we will see in the next chapter.

11.2 GAS-VAPOR MIXTURES

The practicing engineer frequently encounters mixtures of vapors and gases. The products of combustion contain water vapor and gas oxides; the carburetor of an automobile has a mixture of gasoline vapor and air. The most common mixture is that of water vapor and air. This is important in heating and cooling problems.

The analysis of gas-vapor mixtures may be performed quite easily and accurately if the following assumptions are made: (1) the solid or liquid phase contains no dissolved gases; (2) the gaseous phase can be treated as an ideal-gas mixture; and (3) the equilibrium between the condensed phase and vapor phase is independent of the gaseous mixture.

Relative Humidity, Specific Humidity, and Dew Point

There are several frequently used terms that are often misunderstood. Let us define them before proceeding with the analysis.

Consider the cylinder in Figure 11.7(a) and let the substance inside the container be a gas-vapor mixture, where the vapor is superheated at its partial pressure. Water vapor in atmospheric air exists in such a condition. Heat is transferred to the surroundings at constant pressure, as indicated by the line 1-2 in Figure 11.7(b). At point 2, the vapor is a saturated vapor for that pressure; this is the *dew point*. If further heat is transferred, some liquid condensation will occur and the partial pressure will be reduced to state 3, with the saturated liquid at state 4.

When the vapor in the mixture is at state 2, a saturated vapor for its partial pressure, the mixture is called a *saturated mixture;* if the mixture is air, it is called *saturated air.* This is a misnomer, since it is the water vapor in the air, not the air, that is saturated.

The relative humidity, ϕ, of a mixture is the ratio of the mass of vapor in a unit volume to the mass of vapor that the volume could hold if the vapor were saturated at the mixture temperature. The vapor can be considered an ideal gas, and the properties with the subscript g are the saturated vapor properties:

$$\phi = \frac{m_v}{m_g} = \frac{p_v V / RT}{p_g V / RT} = \frac{p_v}{p_g} \qquad (11.17)$$

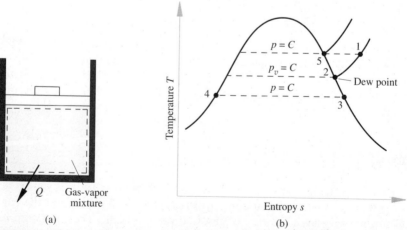

Figure 11.7 (a) A constant-pressure process with heat transfer. (b) A *T-s* diagram illustrating the dew point.

and from Figure 11.7(b),

$$\phi = \frac{p_1}{p_5}$$

The humidity ratio of an air–water vapor mixture, or specific humidity, ω, is defined as the ratio of the mass of water vapor in a given volume of mixture to the mass of air in the same volume. Let m_v be the mass of water vapor and m_a be the mass of air (without vapor) present in the mixture. Thus,

$$\omega = \frac{m_v}{m_a} \qquad (11.18)$$

If the ideal-gas law is used,

$$\omega = \frac{R_a p_v}{R_v p_a}$$

where

$$R_a = 0.287 \text{ kJ/kg-K}$$

$$R_v = 0.4615 \text{ kJ/kg-K}$$

$$\omega = 0.622 \frac{p_v}{p_a} \qquad (11.19)$$

and using the definition of relative humidity, equation (11.19) relates the two humidity ratios:

$$\phi = \frac{\omega p_a}{0.622 p_g} \qquad (11.20)$$

We can use the ideal-gas law to describe behavior up to 65°C. Above this temperature the saturation of water in air is high, and nonideal behavior of the vapor-gas mixture creates too great a discrepancy.

Example 11.5

A tank, $6 \times 4 \times 4$ m, contains an air–water vapor mixture at 38°C. The atmospheric pressure is 101 kPa, and the relative humidity is 70%. The temperature in the tank is lowered to 10°C at constant volume. Determine the humidity ratio (initial and final), the dew point, the mass of air and mass of water vapor (initial and final), and the heat transferred.

Solution

Given: A tank at initial conditions is cooled to a final temperature at constant volume.

Find: The heat transferred and the initial and final values of the humidity ratio, the mass of water vapor, the dew point, and the mass of air.

Sketch and Given Data:

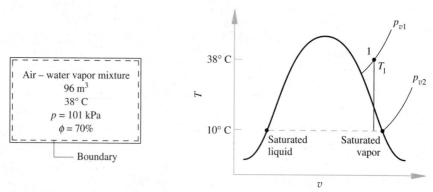

```
┌ ─ ─ ─ ─ ─ ─ ─ ─ ─ ─ ─ ─ ─ ┐
│  Air – water vapor mixture  │
│          96 m³              │
│          38° C              │
│        p = 101 kPa          │
│        φ = 70%              │
└ ─ ─ ─ ─ ─ ─ ─ ─ ─ ─ ─ ─ ─ ┘
        └── Boundary
```

Figure 11.8

Assumptions:

1. The air and water vapor may be considered ideal gases.
2. The changes in kinetic and potential energies may be neglected.
3. When a liquid phase is present, the vapor phase exists as a saturated vapor, and the liquid is a saturated liquid at the system temperature.

Analysis: The total volume is 96 m³. From the steam tables at 38°C,

$$p_g = 6.687 \text{ kPa}$$

$$p_v = (0.70)(6.687) = 4.681 \text{ kPa}$$

The dew point is T_{sat} at $p = 4.681$ kPa:

$$T_{dew} = 31.5°C$$

$$p_a = p - p_v = 101 - 4.7 = 96.3 \text{ kPa}$$

Solving equation (11.20) for ω,

$$\omega_1 = 0.622 \frac{p_g \phi}{p_a} = \frac{\left(0.622 \dfrac{\text{kg vapor}}{\text{kg air}}\right)(6.687 \text{ kPa})(0.7)}{(96.3 \text{ kPa})} = 0.032 \text{ kg vapor/kg air}$$

The mass of air is found from the ideal-gas law:

$$m_a = \frac{p_a V}{R_a T} = \frac{(96.3 \text{ kN/m}^2)(96 \text{ m}^3)}{(0.287 \text{ kJ/kg-K})(311°K)} = 103.6 \text{ kg}$$

Since $\omega = m_v/m_a$,

$$m_v = \omega m_a$$

$$m_v = (0.0302 \text{ kg vapor/kg air})(103.6 \text{ kg air}) = 3.13 \text{ kg vapor}$$

At 10°C the mixture is saturated because the temperature is less than the dew point temperature of 31.5°C.

$$p_{v2} = p_{g2} = 1.2287 \text{ kPa}$$

The air pressure decreases also as the temperature is lowered at constant volume. The measure may be found from the ideal-gas law.

$$p_{a2} = \frac{m_a R_a T}{V} = \frac{(103.6 \text{ kg})(0.287 \text{ kJ/kg-K})(283°\text{K})}{(96 \text{ m}^3)} = 87.6 \text{ kPa}$$

The total pressure is $87.6 + 1.2 = 88.8$ kPa at state 2. The mass of water condensed may be found by determining the humidity ratio at state 2.

$$\omega_2 = \frac{0.622 \, p_{v2}}{p_{a2}} = \frac{\left(0.622 \, \dfrac{\text{kg vapor}}{\text{kg air}}\right)(1.2 \text{ kPa})}{(87.6 \text{ kPa})} = 0.0085 \text{ kg vapor/kg air}$$

Hence,

$$m_1 = m_a(\omega_1 - \omega_2)$$

$$m_1 = (103.6 \text{ kg air})(0.0302 - 0.0085 \text{ kg vapor/kg air}) = 2.25 \text{ kg water}$$

The mass of vapor at state 2 is found by

$$m_{v2} = 3.13 - 2.25 = 0.88 \text{ kg vapor}$$

The first law for a closed system, noting the assumptions, is

$$Q = \Delta U + W$$

The work is zero for constant-volume processes; hence

$$Q = U_2 - U_1$$

where the total internal energy at any state is the sum of the component internal energies at that state.

$$U_1 = m_a u_{a1} + m_{v1} u_{v1}$$

$$U_2 = m_a u_{a2} + m_{v2} u_{v2} + m_1 u_{12}$$

Subtracting these equations yields

$$U_2 - U_1 = m_a(u_{a2} - u_{a1}) + m_{v2} u_{v2} + m_1 u_{12} - m_{v1} u_{v1}$$

We can evaluate the air property change using tables or the ideal-gas equation of state for internal energy. Use the saturated steam tables to evaluate the internal energy of

the vapor and liquid at the appropriate state. The initial vapor state is slightly superheated but that effect may be neglected, and the internal energy is assumed to be a
saturated vapor at the mixture temperature.

$$U_2 - U_1 = (103.6 \text{ kg air})(0.7176 \text{ kJ/kg-K})(283 - 311°\text{K})$$
$$+ (0.88 \text{ kg vapor})(2389.3 \text{ kJ/kg}) + (2.25 \text{ kg liquid})(41.4 \text{ kJ/kg})$$
$$- (3.13 \text{ kg})(2427.7 \text{ kJ/kg})$$
$$U_2 - U_1 = -2081.6 + 2102.6 + 93.2 - 7598.7 = -7484.5 \text{ kJ}$$
$$Q = \Delta U = -7484.5 \text{ kJ}$$

Comments:

1. The temperature datum for the internal energy for air (when u is zero) is different
   than the steam datum. Thus, the internal energy terms must be separated by
   substance and state.
2. Very often the mass of liquid at the final state is ignored in energy calculations;
   note its magnitude compared to the other terms.
3. Notice that the latent heat caused by the condensation of water vapor accounts for
   a large portion of the heat transferred.

Example 11.6

Five hundred ft^3/min of moist air at 40°F and 80% relative humidity enters a duct
where heating occurs at a constant pressure of 14.7 psia. The air exits at 90°F.
Determine the heat flux and the relative humidity of the exit air.

Solution

Given: The volume flow rate of air at a known state entering a constant-pressure
heating duct and the air exit temperature.

Find: The heat flux and the relative humidity of the air leaving.

Sketch and Given Data:

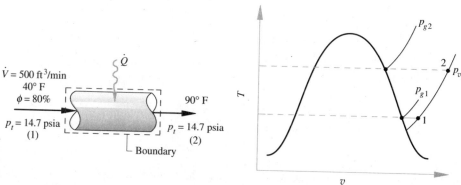

Figure 11.9

Assumptions:

1. The air, the water vapor, and the mixture behave like ideal gases.
2. The process is steady-state.
3. The changes in kinetic and potential energies may be neglected.
4. The work across the control volume is zero.

Analysis: In this situation the mass flow rate of the air and water vapor must be determined before undertaking the first-law analysis. Use the definition of the relative humidity to determine the partial pressure of the water vapor and then find the air's partial pressure. p_{g1} at $40°F = 0.1218$ psia.

$$P_{v1} = \phi p_{g1} = (0.8)(0.1218) = 0.097 \text{ psia}$$

$$p_a = p_{total} - p_{v1} = 14.7 - 0.097 = 14.6 \text{ psia}$$

$$\dot{m}_a = \frac{p_a \dot{V}}{R_a T} = \frac{(14.6 \text{ lbf/in.}^2)(144 \text{ in.}^2/\text{ft}^2)(500 \text{ ft}^3/\text{min})}{(53.34 \text{ ft-lbf/lbm-R})(500°R)}$$

$$\dot{m}_a = 39.41 \text{ lbm/min}$$

The water vapor flow rate is found by determining the humidity ratio and multiplying this by the air mass flow rate. Alternatively, one could solve the ideal-gas law directly for the water vapor flow, but this methodology is not used in air conditioning practice.

$$\omega_1 = 0.622 \frac{p_v}{p_a} = \frac{(0.622)(0.097 \text{ psia})}{(14.6 \text{ psia})} = 0.00413 \text{ lbm vapor/lbm air}$$

At state 2 the partial pressure of the water vapor is the same as it was at state 1, as is the partial pressure of air, for there was no mass addition or deletion. However, p_{g2} is different, and hence the relative humidity is different than it was at state 1. p_{g2} at $90°F = 0.70024$ psia, and $p_{v2} = 0.097$ psia.

$$\phi_2 = \frac{(0.097 \text{ psia})}{(0.70024 \text{ psia})} = 0.138 \quad \text{or } 13.8\%$$

The first law for open systems with steady-state conditions is

$$\dot{Q} + \dot{m}_a(h + \text{k.e.} + \text{p.e.})_{a1} + \dot{m}_v(h + \text{k.e.} + \text{p.e.})_{v1}$$
$$= \dot{W} + \dot{m}_a(h + \text{k.e.} + \text{p.e.})_{a2} + \dot{m}_v(h + \text{k.e.} + \text{p.e.})_{v2}$$

Applying the assumptions and combining terms yields

$$\dot{Q} = \dot{m}_a(h_{2a} - h_{1a}) + \dot{m}_v(h_{2v} - h_{1v})$$

Using the equation of state for enthalpy of an ideal gas and that the enthalpy of the water vapor is essentially equal to the saturated vapor enthalpy at the mixture temperature yields

$$h_{v2} = h_g \text{ at } 90°F = 1100.8 \text{ Btu/lbm}$$

$$h_{v1} = h_g \text{ at } 40°F = 1078.9 \text{ Btu/lbm}$$

$$\dot{Q} = (39.41 \text{ lbm/min})(0.24 \text{ Btu/lbm-R})(550 - 500°R)$$

$$+ (0.163 \text{ lbm/min})(1100.8 - 1078.9 \text{ Btu/lbm})$$

$$\dot{Q} = 472.9 + 3.6 = 476.5 \text{ Btu/min}$$

Comments:

1. The relative humidity decreases because the partial pressure to which the air could be saturated with water vapor increases as the temperature increases.
2. The water vapor contribution to the heat flux is small unless a phase change occurs.
3. The substances need to be separated when finding the enthalpy differences, as different reference datums are used in defining when the value of enthalpy is zero.

Adiabatic Saturation

The adiabatic saturation process is an important process in the study of air–water vapor mixtures. In this process, as the name implies, the mixture is saturated with water vapor adiabatically. To visualize this, consider the schematic diagram in Figure 11.10(a). An unsaturated ($\phi < 100\%$) air–water vapor mixture enters an insulated duct. In the bottom of the duct lies water, which evaporates and becomes water vapor in the mixture. The heat from the vaporization must come from the enthalpy of the incoming mixture. Since enthalpy is a function of temperature, the temperature of the mixture decreases. If the mixture and liquid water are in contact for long enough, the mixture will leave as a saturated mixture at the adiabatic saturated temperature. For this to be a steady-state problem, the changes in kinetic and potential energies must be zero. The water must be supplied at the adiabatic saturation temperature, or a correction must be made for heating the liquid to the temperature. Since this is an open system, the first law is

$$\dot{m}_a h_{a_1} + \dot{m}_{v_1} h_{v_1} + \dot{m}_{l_1} h_{l_1} = \dot{m}_a h_{a_2} + \dot{m}_{v_2} h_{v_2}$$

$$h_{a_1} + \omega_1 h_{v_1} + (\omega_2 - \omega_1) h_{l_1} = h_{a_2} + \omega_2 h_{v_2}$$

$$\omega_1(h_{v_1} - h_{l_1}) = h_{a_2} - h_{a_1} + \omega_2 h_{fg_2} \tag{11.21}$$

$$\omega_1(h_{v_1} - h_{l_1}) = c_{pa}(T_2 - T_1) + \omega_2(h_{fg_2})$$

The adiabatic saturation temperature, T_2, is a function, then, of the inlet temperature, pressure, relative humidity, and exit pressure. Figure 11.10(b) illustrates the adiabatic saturation temperature on a T-s diagram.

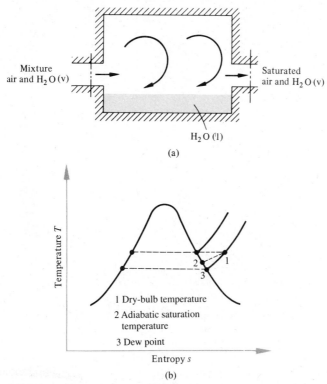

Figure 11.10 (a) An adiabatic saturation unit for steady flow. (b) A *T-s* diagram illustrating the adiabatic saturation temperature.

11.3 PSYCHROMETER

To determine the relative humidity of an air–water vapor mixture, a psychrometer uses two thermometers. A cotton wick saturated with water covers the bulb of one. The dry-bulb thermometer measures the temperature of the air–water vapor flow. If the air–water vapor flow is not saturated, water will evaporate from the wick on the wet bulb. The energy for this evaporation comes, in part, from the internal energy of the mercury in the thermometer, which then decreases, causing a drop in the temperature of the wet-bulb thermometer. Eventually a steady state is reached in which the change of temperature of the air, water vapor, and thermometer is zero with respect to time. This requires that the relative velocity between the air mixture and the wet-bulb thermometer be greater than 3.5 m/s, minimizing the effect of radiative heat transfer and making convective heat transfer a predominant mode.

For air at atmospheric pressure, there is very little difference between the wet-bulb temperature and the adiabatic saturation temperature. This is not necessarily true at other pressures or for mixtures other than the air–water vapor mixture.

Several items affect the wet-bulb temperature reading: conduction along the

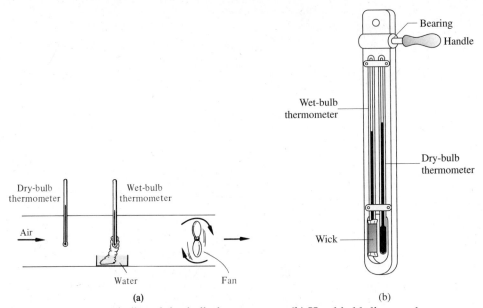

Figure 11.11 (a) Wet-bulb and dry-bulb thermometers. (b) Hand-held sling psychrometer.

thermometer stem, radiant heat transfer from the surroundings to the wet bulb, and a boundary layer between the wet bulb and the air. The adiabatic saturation temperature is an equivalent temperature and not affected by these items. The reason the wet- and dry-bulb temperatures determine the state of the mixture is not readily apparent. Two independent properties are necessary for the determination of the state. The dry-bulb temperature is one such property. By knowing the wet-bulb temperature we may determine the vapor pressure for the mixture, which combined with the static pressure, typically atmospheric, defines the second property. Figure 11.11(a) illustrates a psychrometer. In a sling psychrometer, Figure 11.11(b), the two bulbs are in a casing and are whirled around to achieve the necessary relative velocity.

The TK Solver model PSYCHRO.TK permits the convenient determination of the properties of air–water vapor mixtures. PSYCHRO2.TK analyzes processes of such mixtures. With the input of any two independent properties (dry-bulb temperature, wet-bulb temperature, dew point, relative humidity, humidity ratio), all the other properties can be determined. When analyzing processes using PSYCHRO2.TK, the change in dry-bulb temperature, humidity ratio, or enthalpy can be used as an input for the second point.

Example 11.7
Repeat Example 11.6 using PSYCHRO2.TK to compute the relative humidity of the exit air.

Solution

Given: The dry-bulb temperature and relative humidity of moist air entering a constant-pressure heating duct and the exit dry-bulb temperature.

Find: The relative humidity of the exit air.

Sketch and Given Data: See Figure 11.9.

Assumptions: Same as for Example 11.6.

Analysis: For the heating process given, there will be no change in humidity ratio between inlet and exit. Entering the given temperatures and relative humidity, and zero for the change in humidity ratio, PSYCHRO2.TK calculates the exit relative humidity.

```
================= VARIABLE SHEET =================
St Input — Name — Output —— Unit —— Comment ——————————

                                    ***Psychrometric Process***

                                        INLET CONDITIONS
   40      DB1                  degF   Dry-Bulb Temperature
           WB1    37.523        degF   Wet-Bulb Temperature
           DP1    34.327        degF   Dew Point Temperature
   80      RH1                  %      Relative Humidity
           W1     .0041951      lb/lb  Humidity Ratio
           h1     14.126        BTU/lb Total Enthalpy
           V1     12.688        ft3/lb Specific Volume

                                        OUTLET CONDITIONS
   90      DB2                  degF   Dry-Bulb Temperature
           WB2    60.366        degF   Wet-Bulb Temperature
           DP2    34.327        degF   Dew Point Temperature
           RH2    13.978        %      Relative Humidity
           W2     .0041951      lb/lb  Humidity Ratio
           h2     26.22         BTU/lb Total Enthalpy
           V2     13.956        ft3/lb Specific Volume

                                        PROCESS VARIABLES
14.696     pb                   psia   Barometric Pressure
           delDB  50            F      Change in Dry-Bulb Temperature
   0       delW                 lb/lb  Change in Humidity Ratio
           delh   12.094        BTU/lb Change in Enthalpy
```

Comments:

1. The exit relative humidity agrees closely with that calculated in Example 11.6.
2. The change in enthalpy computed by PSYCHRO2.TK can be used to calculate the heat flux of the process.

CONCEPT QUESTIONS

1. Describe mass fractions and mole fractions.

2. Is it possible for the mass and mole fractions of a mixture of CO and N_2 to be the same? Why?

3. For an ideal-gas mixture the sum of the mass fractions is unity. Is this also true for non-ideal-gas mixtures? Why?

4. Is a mixture of ideal gases also an ideal gas? Why?

5. Describe Dalton's law of partial pressure.

6. Describe Amagat's law of additive volumes.

7. Consider a gas mixture with several components. Which component will have the higher partial pressure, the one with the largest number of moles or with the greatest mass?

8. A tank contains two ideal gases. The tank is heated, and the temperature and pressure of the tank increase. Do the partial pressures of each component increase and remain in the same proportion?

9. Explain how the internal energy change of a mixture is determined.

10. Explain how entropy change of a mixture is determined for adiabatic conditions and for nonadiabatic conditions.

11. What is the difference between dry air and atmospheric air?

12. Under what circumstances can the water vapor in air be treated as an ideal gas?

13. Describe vapor pressure.

14. What is the difference between relative humidity and specific humidity?

15. Explain wet- and dry-bulb temperatures.

16. What is the adiabatic saturation temperature?

17. Explain the dew point temperature.

18. In the summer a glass of cold water will frequently have condensation on the outer surface. Why?

19. At what condition are the wet- and dry-bulb temperatures the same?

PROBLEMS (SI)

11.1 A gaseous mixture has the following volumetric analysis: O_2, 30%; CO_2, 40%; N_2, 30%. Determine (a) the analysis on a mass basis; (b) the partial pressure of each component if the total pressure is 100 kPa and the temperature is 32°C; (c) the molecular weight of the mixture.

11.2 A gaseous mixture has the following analysis on a mass basis: CO_2, 30%; SO_2, 30%; He, 20%; N_2, 20%. For a total pressure and temperature of 101 kPa and 300°K, determine (a) the volumetric (molal) analysis; (b) the component partial pressures; (c) the mixture gas constant; (d) the mixture specific heats.

11.3 A cubical tank, 1 m on a side, contains a mixture of 1.8 kg of nitrogen and 2.8 kg of an unknown gas. The mixture pressure and temperature are 290 kPa and 340°K. Determine (a) the molecular weight and gas constant of the unknown gas; (b) the volumetric analysis.

11.4 A mixture of ideal gases at 30°C and 200 kPa is composed of 0.20 kg CO_2, 0.75 kg N_2, and 0.05 kg He. Determine the mixture volume.

11.5 Equal masses of hydrogen and oxygen are mixed. The mixture is maintained at 150 kPa and 25°C. For each component determine the volumetric analysis and its partial pressure.

11.6 A 3-m^3 drum contains a mixture at 101 kPa and 35°C of 60% methane and 40% oxygen on a volumetric basis. Determine the amount of oxygen that must be added at 35°C to change the volumetric analysis to 50% for each component. Determine also the new mixture pressure.

11.7 A gaseous mixture of propane, nitrogen, and hydrogen has partial pressures of 83 kPa, 102 kPa, and 55 kPa, respectively. Determine (a) the volumetric analysis; (b) the gravimetric (mass) analysis.

11.8 A mixture of gases contains 20% N_2, 40% O_2, and 40% CO_2 on a mass basis. The mixture pressure and temperature are 150 kPa and 300°K. (a) Consider the mixture to be heated in a 20-m^3 tank to 600°K; find the heat required. (b) Consider the mixture to be flowing steadily at 1 kg/s through a heat exchanger until the temperature is doubled; find the heat rate required.

11.9 A rigid insulated tank, as shown in Figure 11.4, is divided into two sections by a membrane. One side contains 0.5 kg of nitrogen at 200 kPa and 320°K, and the other side contains 1.0 kg of helium at 300 kPa and 400°K. The membrane is removed. Determine (a) the mixture temperature and pressure; (b) the change of entropy for the system; (c) the change of internal energy of the system.

11.10 A rigid insulated tank, as shown in Figure 11.4, contains 0.28 m^3 of nitrogen and 0.14 m^3 of hydrogen. The pressure and temperature of each gas is 210 kPa and 93°C. The membrane separating the gases is removed. Determine the entropy of mixing.

11.11 Ethylene is stored in a 5.6-liter spherical vessel at 260°C and 2750 kPa. To protect against explosion, the vessel is enclosed in another spherical vessel with a volume of 56 liters and filled with nitrogen at 260°C and 10.1 MPa. The entire assembly is maintained at 260°C in a furnace. The inner vessel ruptures. Determine (a) the final pressure; (b) the entropy change.

11.12 An air–water vapor mixture at 101 kPa, 35°C, and 80% relative humidity. Determine (a) the dew point; (b) the humidity ratio; (c) the partial pressure of air; (d) the mass fraction of water vapor.

11.13 An air–water vapor mixture at 138 kPa, 43°C, and 50% relative humidity is contained in a 1.4 m^3 tank. The tank is cooled to 21°C. Determine (a) the mass of water condensed; (b) the partial pressure of water vapor initially; (c) the final mixture pressure; (d) the heat transferred.

11.14 Given for an air–water vapor mixture that $T_{mix} = 60°C$, $p_{mix} = 300$ kPa, and $\phi = 50.1\%$, find (a) the dew point; (b) the humidity ratio.

11.15 Given for an air–water vapor mixture that $T_{mix} = 70°C$ and $p_{mix} = 200$ kPa and $p_a = 180$ kPa, find (a) the dew point; (b) humidity ratio; (c) relative humidity.

11.16 The molal analysis of a gas mixture at 300°K and 100 kPa is 65% N_2, 25% CO_2, and 10% O_2. Determine (a) the mass fraction of each component; (b) its partial pressure; (c) the volume occupied by 20 kg of the mixture.

11.17 A gas mixture has components with the following mass fractions: 50% CO_2, 20% CO, and 30% He. The mixture temperature and pressure are 50°C and 150 kPa. Determine

(a) the mole fractions; (b) the partial pressure of each component; (c) the mixture gas constant.

11.18 A 50-liter tank contains 0.6 kg of nitrogen and 0.6 kg of carbon dioxide at 300°K. Heat is transferred until the temperature rises to 400°K. Determine the entropy increase of the gas.

11.19 An adiabatic compressor receives 2 kg/s of a gas mixture and compresses it from 280°K and 100 kPa to 450°K and 500 kPa. The mixture's mass fractions are 50% N_2, 30% CO_2, and 20% O_2. Determine (a) the power required; (b) the entropy production.

11.20 A mixture of 0.5 kgmol of nitrogen and 0.4 kgmol of ammonia is compressed isothermally from 500°K and 100 kPa to 400 kPa. Determine (a) the work required; (b) the heat transferred; (c) the entropy production.

11.21 A turbine receives 1.5 kg/s of a gas mixture at 800 kPa and 1000°K and expands it to a pressure of 100 kPa isentropically. The mixture molal analysis is 60% N_2, 20% CO_2, and 20% water vapor. Determine (a) the exit temperature; (b) the power developed.

11.22 A mixture containing 50% He and 50% N_2 on a mass basis enters a nozzle at 450°K and 500 kPa with a velocity of 50 m/s. It expands isentropically through the nozzle. The exit velocity from the nozzle is 300 m/s. Determine the exit temperature and pressure.

11.23 A gas mixture with a molal analysis of 60% N_2 and 40% O_2 enters an adiabatic compressor at 1.5 kg/s, 100 kPa, and 290°K. The discharge pressure is 500 kPa, and the discharge temperature is 500°K. Determine (a) the second-law efficiency; (b) the power; (c) the entropy production.

11.24 Two m³ of gas 1 at 300°K and 100 kPa mixes adiabatically with 6 m³ of gas 2 at 300°K and 250 kPa. Determine (a) the final pressure of the mixture; (b) the entropy change of each gas.

11.25 A quantity of neon, 0.5 kg at 20°C and 100 kPa, is contained in an adiabatic tank. Another adiabatic tank contains 0.7 kg of nitrogen at 390°K and 500 kPa. A valve connecting the tanks is opened, and the gases achieve equilibrium. Determine (a) each tank's volume; (b) the final mixture pressure; (c) the entropy production.

11.26 The tanks in Problem 11.25 are nonadiabatic, and the final mixture temperature is 330°K. Determine (a) the final pressure; (b) the heat transfer; (c) the entropy change of each component.

11.27 Two kg of nitrogen at 67°C and 100 kPa mix adiabatically with (a) six kilograms of oxygen, (b) six kilograms of nitrogen, both of which are at the same initial conditions as the 2 kg of nitrogen. Determine the entropy production.

11.28 An adiabatic tank has two compartments, one containing 2 kgmol of carbon dioxide at 300°K and 200 kPa and the other containing 2 kgmol of nitrogen at 420°K and 500 kPa. The gases mix, and 900 kJ of energy is added by a resistance heater. Determine (a) the mixture final temperature and pressure; (b) the change in availability; (c) the irreversibility of $T_0 = 300°K$.

11.29 Two kg/s of helium flows steadily into an adiabatic mixing chamber at 87°C and 400 kPa and mixes with nitrogen entering at 287°C and 400 kPa. The mixture leaves at 350 kPa with a molal analysis of 50% He and 50% N_2. Determine (a) the temperature of the mixture leaving the chamber; (b) the rate of entropy production.

11.30 An industrial process requires a mixture of nitrogen and helium at 120 kPa. The mixture is created by adiabatically combining 3 kg/s of nitrogen at 160 kPa and 310°K and 1.5 kg/s of helium at 175 kPa and 415°K. The mixture leaves the chamber at 150 kPa and is throttled to 120 kPa. Determine (a) the temperature of the mixture exiting the chamber and exiting the throttling valve; (b) the entropy production across the chamber and across the throttling valve.

11.31 The temperature of the inside surface of a room's exterior wall is 15°C, while the temperature of the air in the room is 23°C. What is the maximum relative humidity the air in the room can have before condensation occurs?

11.32 Cold water flows through a pipe in the basement of a home at a temperature of 12°C. The air in the basement has a temperature of 22°C. What is the maximum relative humidity the air can have before condensation occurs?

11.33 A 100-m³ tank contains atmospheric air at 27°C with a humidity ratio of 0.008 kg vapor/kg air. What mass of water must be removed to lower the relative humidity to 25%?

11.34 A room contains atmospheric air at 25°C with a humidity ratio of 0.01 kg vapor/kg air. What is the relative humidity and the dew point?

11.35 A volume, 0.6 m³, of air at 100 kPa and 27°C with a relative humidity of 50% is compressed isothermally until water condenses. At what pressure does the condensation first occur?

11.36 A mixture of nitrogen and water at 65°C and 100 kPa with a molal analysis of 85% N_2 and 15% water vapor is cooled at constant pressure. At what temperature does the water vapor first condense?

11.37 A piston-cylinder with an initial volume of 0.75 m³ contains air at 45°C, 120 kPa, and 60% relative humidity. The system is cooled at constant total pressure until the air temperature is 30°C. What are the system work and heat transfer?

11.38 A tank contains air initially at 90°C, 300 kPa, and 40% relative humidity. Heat transfer occurs until the air is 25°C. Determine (a) the heat transferred; (b) the temperature at which condensation occurs.

11.39 A 5-m³ tank contains air at 400°K, 500 kPa, and 10% relative humidity. The air is cooled until the temperature is 300°K. Determine (a) the final pressure; (b) the heat transferred; (c) the change of entropy.

11.40 Three kg/s of air at 320°K, 300 kPa, and relative humidity of 35% expands isentropically to an exit pressure. Determine the lowest exit pressure possible without condensation occurring.

11.41 A flow of air, 2.6 kg/s, at 100 kPa, 350°K, and 30% relative humidity enters a heat exchanger and is mixed with another stream of air with a flow rate of 2.0 kg/s, a pressure of 100 kPa, a temperature of 300°K, and a relative humidity of 40%. Determine the temperature of the exiting mixture.

PROBLEMS (English Units)

*11.1 A 10-ft³ tank contains a gas mixture at 100 psia and 100°F. The composition is 40% O_2 and 60% CH_4 on a mass basis. It is desired to have a mixture at 50% O_2 and 50% CH_4 at the same temperature and pressure. How much mixture must be removed and how much oxygen added to achieve this?

*11.2 A gas mixture has the following volumetric analysis: 20% N_2, 30% CO_2, 25% He, and 25% CH_4. Determine (a) the mass fractions; (b) R_m; (c) c_{pm}.

*11.3 Two fluid streams mix adiabatically, 5000 ft^3/min of CO_2 at 50 psia and 150°F in one stream, and 3000 ft^3/min of CH_4 at 20 psia and 70°F in the other. Determine (a) the mixture temperature (b) y_i, x_i; (c) R_m, M_m; (d) the entropy production if the discharge pressure is 20 psia.

*11.4 Referring to Figure 11.4, let gas a be 1 lbm of argon at 100°F and 20 psia and gas b be 2 lbm of helium at 100°F and 20 psia. The membrane is removed; find the entropy production.

*11.5 Referring to Figure 11.4, let gas a be 5 ft^3 of air at 200 psia and 200°F and gas b be 3 lbm of helium at 100 psia and 100°F. Determine (a) the final mixture temperature and pressure; (b) the entropy production.

*11.6 The molal analysis of a gas mixture at 500°R and 15 psia is 65% N_2, 25% CO_2, and 10% O_2. Determine (a) the mass fraction of each component; (b) its partial pressure; (c) the volume occupied by 20 lbm of the mixture.

*11.7 A gas mixture has components with the following mass fractions: 50% CO_2, 20% CO, and 30% He. The mixture temperature and pressure are 100°F and 20 psia. Determine (a) the mole fractions; (b) the partial pressure of each component; (c) the mixture gas constant.

*11.8 A 15-gal tank contains 1.3 lbm of nitrogen and 1.3 lbm of carbon dioxide at 77°F. Heat is transferred until the temperature rises to 250°F. Determine the entropy increase of the gas.

*11.9 An adiabatic compressor receives 250 lbm/min of a gas mixture and compresses it from 40°F and 14.7 psia to 350°F and 73.5 psia. The mixture's mass fractions are 50% N_2, 30% CO_2, and 20% O_2. Determine (a) the power required; (b) the entropy production.

*11.10 A mixture of 0.5 pmol of nitrogen and 0.4 pmol of ammonia is compressed isothermally from 800°R and 14.6 psia to 60 psia. Determine (a) the work required; (b) the heat transferred; (c) the entropy production.

*11.11 A turbine receives 4.5 lbm/s of a gas mixture at 120 psia and 1000°R and expands it to a pressure of 15 psia isentropically. The mixture molal analysis is 60% N_2, 20% CO_2, and 20% water vapor. Determine (a) the exit temperature; (b) the power developed.

*11.12 A mixture containing 50% He and 50% N_2 on a mass basis enters a nozzle at 350°F and 75 psia with a velocity of 150 ft/sec. It expands isentropically through the nozzle. The exit velocity from the nozzle is 900 ft/sec. Determine the exit temperature and pressure.

*11.13 A gas mixture with a molal analysis of 60% N_2 and 40% O_2 enters an adiabatic compressor at 1.5 lbm/sec, 14.7 psia, and 62°F. The discharge pressure is 74 psia, and the discharge temperature is 440°F. Determine (a) the second-law efficiency; (b) the power; (c) the entropy production.

*11.14 Two ft^3 of gas 1 at 600°R and 15 psia mixes adiabatically with 6 ft^3 of gas 2 at 600°R and 37.5 psia. Determine (a) the final pressure of the mixture; (b) the entropy change of each gas.

*11.15 A quantity of neon, 1.2 lbm at 68°F and 1 atm, is contained in an adiabatic tank. Another adiabatic tank contains 1.5 lbm of nitrogen at 240°F and 5 atm. A valve connecting the tanks is opened, and the gases achieve equilibrium. Determine (a) each tank's volume; (b) the final mixture pressure; (c) the entropy production.

*11.16 The tanks in Problem *11.15 are nonadiabatic, and the final mixture temperature is 150°F. Determine (a) the final pressure; (b) the heat transfer, (c) the entropy change of each component.

*11.17 Two lbm of nitrogen at 150°F and 14.7 psia mixes adiabatically with (a) 6 lbm of oxygen, (b) 6 lbm of nitrogen, both of which are at the same initial conditions as the 2 lbm of nitrogen. Determine the entropy production.

*11.18 An adiabatic tank has two compartments, one containing 2 pmol of carbon dioxide at 77°F and 30 psia and the other containing 2 pmol of nitrogen at 300°F and 75 psia. The gases mix, and 900 Btu of energy is added by a resistance heater. Determine (a) the mixture final temperature and pressure; (b) the change in availability; (c) the irreversibility if $T_0 = 77°F$.

*11.19 Two lbm/sec of helium flow steadily into an adiabatic mixing chamber at 190°F and 60 psia and mix with nitrogen entering at 550°F and 60 psia. The mixture leaves at 55 psia with a molal analysis of 50% He and 50% N_2. Determine (a) the temperature of the mixture leaving the chamber; (b) the rate of entropy production.

*11.20 An industrial process requires a mixture of nitrogen and helium at 18 psia. The mixture is created by adiabatically combining 3 lbm/sec of nitrogen at 23 psia and 100°F and 1.5 lbm/sec of helium at 26 psia and 290°F. The mixture leaves the chamber at 22 psia and is throttled to 18 psia. Determine (a) the temperature of the mixture exiting the chamber and exiting the throttling valve; (b) the entropy production across the chamber and across the throttling valve.

*11.21 The temperature of the inside surface of a room's exterior wall is 60°F, while the temperature of the air in the room is 73°F. What is the maximum relative humidity the air in the room can have before condensation occurs?

*11.22 Cold water flows through a pipe in the basement of a home at a temperature of 54°F. The air in the basement has a temperature of 72°F. What is the maximum relative humidity the air can have before condensation occurs?

*11.23 A 100-ft³ tank contains atmospheric air at 80°F with a humidity ratio of 0.008 lbm vapor/lbm air. What mass of water must be removed to lower the relative humidity to 25%?

*11.24 A room contains atmospheric air at 77°F with a humidity ratio of 0.01 lbm vapor/lbm air. What is the relative humidity and the dew point?

*11.25 Three ft³ of air at 14.7 psia and 80°F with a relative humidity of 50% are compressed isothermally until water condenses. At what pressure does the condensation first occur?

*11.26 A mixture of nitrogen and water at 148°F and 1 atm with a molal analysis of 85% nitrogen and 15% water vapor is cooled at constant pressure. At what temperature does the water vapor first condense?

*11.27 A piston-cylinder with an initial volume of 0.75 ft³ contains air at 113°F, 17.6 psia, and 60% relative humidity. The system is cooled at constant total pressure until the air temperature is 86°F. What are the system work and heat transfer?

*11.28 A 10 ft^3 tank contains air initially at 195°F, 44 psia, and 40% relative humidity. Heat transfer occurs until the air is 77°F. Determine (a) the heat transferred; (b) the temperature at which condensation occurs.

*11.29 A 5-ft^3 tank contains air at 260°F, 75 psia, and 10% relative humidity. The air is cooled until the temperature is 80°F. Determine (a) the final pressure; (b) the heat transferred; (c) the change of entropy.

*11.30 Three lbm/sec of air at 120°F, 35 psia, and 35% relative humidity expands isentropically to an exit pressure. Determine the lowest exit pressure possible without condensation occurring.

*11.31 A flow of air, 2.6 lbm/sec, at 1 atm, 170°F, and 30% relative humidity enters a heat exchanger and is mixed with another stream of air with a flow rate of 2.0 lbm/sec, a pressure of 1 atm, a temperature of 80°F, and a relative humidity of 40%. Determine the temperature of the exiting mixture.

COMPUTER PROBLEMS

C11.1 The readings from a sling psychrometer are 90°F dry-bulb temperature and 70°F wet-bulb temperature. Use PSYCHRO.TK to determine the relative humidity.

C11.2 Using PSYCHRO.TK, compute the relative humidity, humidity ratio, and enthalpy of air with a constant wet-bulb temperature of 65°F for dry-bulb temperatures between 65°F and 90°F.

C11.3 Using PSYCHRO2.TK, compute the adiabatic saturation temperature of air at a dry-bulb temperature of 90°F and 40% relative humidity.

12

Reactive Systems

Very often the first thought that occurs when the word "combustion" is mentioned is the burning or oxidation of hydrocarbon fuels such as gasoline, wood and coal. The change of chemical energy into thermal energy is fundamental to most power-producing devices, such as automotive engines and gas turbines. This chapter will discuss hydrocarbon fuels primarily; however, the fundamentals developed are also applicable to other chemical reactions, such as the oxidation of food in living organisms. In this chapter you will

- Develop a greater understanding of hydrocarbon fuels and their oxidation;
- Analyze the combustion process;
- Determine the energy released in combustion reactions;
- Calculate the maximum temperature a combustion process can have and learn of ways to control the temperature;
- Calculate the effects of dissociation on high-temperature combustion;
- Analyze fuel cells from a first-law and second-law view;
- Further your understanding of acid rain and global warming.

12.1 HYDROCARBON FUELS

One of the basic constituents in the combustion reaction is the fuel; we will be dealing with hydrocarbon fuels—solid, liquid, or vapor. The most important form of solid hydrocarbon is coal, which is mined in several grades, ranging from anthracite (hard) to bituminous (soft). Coal is a mixture of carbon, hydrogen, oxygen, nitrogen, sulfur, water, and a noncombustible solid material, ash.

The liquid hydrocarbons, such as gasoline, kerosene, and fuel oil, are obtained by the distillation of petroleum. Their advantages over solid fuel are cleanliness and ease of handling and storage. The chemical form of the liquid hydrocarbons is C_xH_y, the values of subscripts x and y depending on the hydrocarbon family. Any fuel, such as gasoline, is actually a mixture of many hydrocarbons. Except in sophisticated analysis, the predominant hydrocarbon is assumed to be the only one present, or an average of the several constituent hydrocarbons is taken. Alcohols are sometimes used as fuels in internal-combustion engines. In alcohol, one of the hydrogen atoms in the hydrocarbon is replaced by the OH radical. The resulting hydrocarbon is a *carbohydrate* and is written C_xH_zOH. In this text, unless specifically noted, we will consider gasoline to be a single hydrocarbon, C_8H_{18} (octane), and diesel fuel to be a single hydrocarbon, $C_{12}H_{26}$ (dodecane).

Gaseous hydrocarbon fuels also are a mixture of various constituent hydrocarbons. They have nearly complete combustion and are very clean. The products of their combustion do not have sulfur components, which have adverse environmental effects. There are great differences between natural gas, hydrocarbon fuel found underground, and manufactured gas. These differences lie in the proportion of the constituents found in each, as illustrated in Table 12.1.

12.2 COMBUSTION PROCESS

The combustion process is the oxidation of the fuel constituents, and we may write an equation describing the reaction. During the combustion process the total mass remains the same, so in balancing reaction equations we are applying the law of conservation of mass. Actually there is a slight reduction in mass from Einstein's equation $E = mc^2$, but this is extremely small, on the order of 10^{-10} kg/kg fuel, and can be neglected.

TABLE 12.1 VOLUMETRIC ANALYSIS OF SOME FUEL-GAS MIXTURES (NUMBERS ARE PERCENTAGES)

	CO	H_2	CH_4	C_2H_6	C_4H_8	O_2	CO_2	N_2
Coal gas	9	53.6	25	—	3	0.4	3	6
Producer gas	29	12	2.6	0.4	—	—	4	52
Blast furnace gas	27	2	—	—	—	—	11	60
Natural gas I	1	—	93	3	—	—	—	3
Natural gas II	—	—	80	18	—	—	—	2
Natural gas III	—	1	93	3.5	—	—	2	0.5

We will first consider the complete oxidation of carbon and, in so doing, define the terms commonly used in combustion-reaction analysis.

$$\text{Reactants}\quad\text{Products}$$

$$C + O_2 \rightarrow CO_2 \tag{12.1}$$

In this reaction, carbon and oxygen are the initial substances, that is, the *reactants*. They undergo a chemical reaction, yielding carbon dioxide, the final substance, or *product*. Furthermore, we see that

$$1 \text{ mol C} + 1 \text{ mol O}_2 \rightarrow 1 \text{ mol CO}_2$$

and since there are 12 kg/kgmol for carbon, 32 kg/kgmol for oxygen, and 44 kg/kgmol for carbon dioxide,

$$12 \text{ kg C} + 32 \text{ kg O}_2 \rightarrow 44 \text{ kg CO}_2$$

Similarly, the hydrogen in the fuel reacts with oxygen to form water

$$H_2 + 0.5 \, O_2 \rightarrow H_2O$$

$$1 \text{ mol H}_2 + 0.5 \text{ mol O}_2 \rightarrow 1 \text{ mol H}_2O$$

$$2 \text{ kg H}_2 + 16 \text{ kg O}_2 \rightarrow 18 \text{ kg H}_2O$$

When a hydrocarbon fuel is completely oxidized, the resulting products are primarily carbon dioxide and water. Let us consider methane as the fuel:

$$CH_4 + 2O_2 \rightarrow CO_2 + 2H_2O \tag{12.2}$$

The water may exist as a solid (s), liquid (l), or vapor (v), depending on the final product temperature and pressure. This is an important consideration when we perform energy balances later in the chapter. In the oxidation process, many reactions occur before the final products in equation (12.2) are formed. These intermediate reactions constitute an important area of investigation, but we will be concerned with only the initial and final products.

Combustion with Air

Most combustion processes depend on air, not pure oxygen. Air contains many constituents, particularly oxygen, nitrogen, argon, and other vapors and inert gases. Its volumetric, or molal, composition is approximately 21% oxygen, 78% nitrogen, and 1% argon. Since neither nitrogen nor argon enters into the chemical reaction, we will assume that the volumetric air proportions are 21% oxygen and 79% nitrogen, and that for 100 mol of air, there are 21 mol of oxygen and 79 mol of nitrogen, or

$$\frac{79}{21} = 3.76 \text{ mol N}_2/\text{mol O}_2$$

To account for the argon, which we include as nitrogen, we use 28.16 as the equivalent molecular weight of nitrogen. This is the molecular weight of what is called *atmospheric nitrogen*. Pure nitrogen has a molecular weight of 28.016. In analyses in which great accuracy is desired, this distinction should be made. However, we will consider the nitrogen in the air to be pure.

12.3 THEORETICAL AIR

The combustion of methane in air is

$$CH_4 + 2O_2 + 2(3.76)N_2 \rightarrow CO_2 + 2H_2O + 7.52N_2 \qquad (12.3)$$

The nitrogen does not enter into the reaction, but it must be accounted for. There is 3.76 mol nitrogen per mole of oxygen, and since 2 mol of oxygen is necessary for the oxidation of methane, (2)(3.76) mol of nitrogen is present. In equation (12.3) only the minimum amount of oxygen necessary to complete the reaction is included in the equation. The minimum amount of air required to oxidize the reactants is the *theoretical air.* When combustion is achieved with theoretical air, no oxygen is found in the products. In practice, this is not possible. Additional oxygen is required to achieve complete combustion of the reactants. The *excess air* is needed because the fuel is of finite size, and each droplet must be surrounded by more than the necessary number of oxygen molecules to assure oxidation of all the hydrocarbon molecules. This excess air is usually expressed as a percentage of the theoretical air. Thus, if 25% more air than is theoretically required is used, this is expressed as 125% theoretical air, or 25% excess air. There is 1.25 times as much air actually supplied than is theoretically required. The combustion of methane in 125% theoretical air is thus

$$CH_4 + (1.25)(2)O_2 + (1.25)(2)(3.76)N_2 \rightarrow$$
$$CO_2 + 2H_2O + 9.4N_2 + 0.5O_2 \qquad (12.4)$$

Equation (12.4) is balanced by first balancing the oxidation equation for theoretical air and then multiplying the theoretical air by 1.25 to account for the 125% theoretical air and adding a term to the products for the excess oxygen. The amount of nitrogen and oxygen appearing in the products is determined by a mass balance on each term.

If the excess air is insufficient to provide complete combustion, not all the carbon will be oxidized to carbon dioxide; some will be oxidized to carbon monoxide. When there is considerably less theoretical air, unburned hydrocarbons will be present in the products. This is the soot or black smoke that sometimes pours from chimneys and smokestacks when one or more of the following three conditions for complete combustion have not been met:

1. The air-fuel mixture must be at the ignition temperature.
2. There must be sufficient oxygen to assure complete combustion.
3. The oxygen must be in intimate contact with the fuel.

Smoky products of combustion during start-up operations usually result from a failure to satisfy requirements 1 and 3. To balance the combustion equation when incomplete combustion occurs, we need information about the products. For instance, assume that, with theoretical air, the oxidation of carbon is 90% complete in the combustion of methane; then

$$CH_4 + 2O_2 + 2(3.76)N_2 \rightarrow$$
$$0.9CO_2 + 0.1CO + 2H_2O + 7.2N_2 + 0.05O_2 \quad (12.5)$$

12.4 AIR/FUEL RATIO

Two important terms in the combustion process give us the proportion of air to fuel; these are the *air/fuel ratio*, $r_{a/f}$, and its reciprocal, the *fuel/air ratio*, $r_{f/a}$. Both may be expressed in terms of the moles or the mass of the fuel and air present.

Example 12.1
A fuel oil is burned with 50% excess air, and the combustion characteristics of the fuel oil are similar to $C_{12}H_{26}$. Determine the air/fuel ratio on a mass and mole basis and the volumetric (molal) analysis of the products of combustion.

Solution

Given: A specified fuel oil and the excess air to oxidize it completely.

Find: The air/fuel ratios and the volumetric product analysis.

Assumptions:

1. The combustion of the fuel is complete; no CO is formed.
2. The molal ratio of nitrogen to oxygen is 3.76.
3. The products behave like an ideal gas.

Analysis: First the reaction equation must be balanced for 50% excess air, or 150% theoretical air. Apply the conservation of mass to each reactant for 100% theoretical air.

$$C_{12}H_{26} + aO_2 + 3.76aN_2 \rightarrow bCO_2 + cH_2O + dN_2$$

C balance $12 = b$

H$_2$ balance $26 = 2c$ $c = 13$

O$_2$ balance $a = b + c/2 = 12 + 6.5 = 18.5$

N$_2$ balance $3.76a = d = (3.76)(18.5) = 69.56$

This balances the equation for 100% theoretical air. Now multiply the air terms of the reactants by 1.5 and add an excess oxygen term to the products. The following equation results:

$$C_{12}H_{26} + 27.75O_2 + 104.34N_2 \rightarrow 12CO_2 + 13H_2O + 104.34N_2 + 9.25O_2$$

The air/fuel ratio on the mole basis is found by dividing the moles of air by the moles of fuel in the reaction equation.

$$r_{a/f} = \frac{(27.75 + 104.34 \text{ mol air})}{(1 \text{ mol fuel})} = 132.1 \text{ mol air/mol fuel}$$

The ratio on the mass basis is found by multiplying the moles by their appropriate molecular weight. The molecular weight for $C_{12}H_{26}$ may be found in Appendix Table C.1.

$$r_{a/f} = \frac{(132.1 \text{ mol air})(28.97 \text{ kg/kgmol air})}{(1 \text{ mol fuel})(170.328 \text{ kg/kgmol fuel})} = 22.46 \text{ kg air/kg fuel}$$

In determining the volumetric analysis of the products, find first the total number of moles of product and then each component's fraction of the total.

The total moles of product is

$$12 + 13 + 104.34 + 9.25 = 138.59 \text{ mol}$$

The molal analysis is then

$$CO_2 = 12/138.59 = 0.0866$$

$$H_2O = 13/138.59 = 0.0938$$

$$N_2 = 104.34/138.59 = 0.7529$$

$$O_2 = 9.25/138.59 = 0.0667$$

Comment: When balancing equations with excess air, start with the 100% theoretical air equation and modify it for excess air. ▬

Example 12.2
Determine the dew point of the products at 1 atm for the liquid fuel oxidized in Example 12.1 if the air supplied is dry at 28°C, and if the air supplied is at 28°C with 50% relative humidity.

Solution

Given: The balanced reaction for a hydrocarbon fuel and its total pressure and reactant air condition.

Find: The dew point of the products.

Assumptions:

1. The products behave like an ideal gas.
2. The molal ratio of nitrogen to oxygen is 3.76.

Analysis: The balanced reaction equation from Example 12.1 is

$$C_{12}H_{26} + 27.75O_2 + 104.34N_2 \rightarrow 12CO_2 + 13H_2O + 104.34N_2 + 9.25O_2$$

The mole fraction of water in the products is 0.0938. From Dalton's law of partial pressures, the vapor pressure of water contributing to the total pressure is

$$(0.0938)(101.3 \text{ kPa}) = 9.5 \text{ kPa}$$

The saturation temperature corresponding to this pressure is $45°C$, which is also the dew point temperature of the products when dry air is used in the combustion process.

When atmospheric air is used with water vapor present, the reactants side carries a water vapor term, which is additive to the water vapor formed from the combustion reaction. First we need to find the water vapor present in air at $28°C$ with 50% relative humidity.

Using the terminology of Chapter 11, $p_g = 3.8194$ kPa and the vapor pressure of water, p_v, is $(0.5)(3.8194) = 1.9097$ kPa. The rest of the reactant pressure is supplied by the air. The liquid fuel does not contribute to the total pressure. Thus, p_a is

$$p_a = 101.3 - 1.9 = 99.4 \text{ kPa}$$

The humidity ratio is

$$\omega = 0.622 \frac{p_v}{p_a} = \frac{(0.622)(1.9)}{(99.4)} = 0.0118 \text{ kg vapor/kg air}$$

This needs to be converted to an equivalent value on the mole basis, ω'.

$$\omega' = (0.0118 \text{ kg vapor/kg air}) \left(\frac{28.97 \text{ kg air}}{1 \text{ kgmol air}} \right) \left(\frac{1 \text{ kgmol vapor}}{18.016 \text{ kg vapor}} \right)$$

$$\omega' = 0.019 \text{ mol vapor/mol air}$$

In this problem 132.1 mol of air oxidizes 1 mol of fuel, so

$$n_{H_2O} = (0.019 \text{ mol vapor/mol air})(132.1 \text{ mol air}) = 2.51 \text{ mol vapor}$$

This is additive to the total moles of vapor in the product and to the total number of moles of gas in the product. Thus, there are 15.51 mol of water vapor in 141.1 mol of product. The partial pressure of the water vapor is

$$p_v = (101.3)(15.51/141.1) = 11.1 \text{ kPa}$$

The saturation temperature, or dew point, associated with this pressure is $48°C$.

Comments:

1. Most often dry air is used in reaction equations. The gain in accuracy of using moist air is offset by the increased complexity of the calculations.
2. The dew point is often sought in combustion reactions because when the temperature of the products falls below this value, water precipitates. Should precipitation occur, the liquid water will contain dissolved gases, forming a corrosive substance. To prevent this, the temperature of the products of combustion is kept well above the dew point in smokestacks and exhaust piping.
3. In steam generators burning fuel oil, atomizing steam is often used to atomize fuel droplets. A typical value is 0.03–0.05 kg steam/kg fuel. A similar analysis would be made to find the moles of water used and their effect in raising the dew point of the products.

Coal has played an important role as an energy source in the past and will become increasingly important as oil and natural gas become more expensive. Coal has a variable chemical composition, since it originally was vegetation. It is essentially a hydrocarbon fuel, with impurities such as entrapped water and ash that represent the inorganic matter that does not oxidize. The ultimate analysis of coal includes all the constituent mass fractions. The proximate analysis, also on a mass basis, is often used to define the percentage of carbon in the coal; that is,

$$\% \text{ C} = 100\% - \% \text{ ash} - \% \text{ moisture} - \% \text{ volatiles}$$

The volatiles are those compounds that evaporate at a low temperature when coal is heated. The following example illustrates these concepts.

Example 12.3

An ultimate analysis of coal yields the following composition: 74% C, 5% H_2, 6% O_2, 1% S, 1.2% N_2, 3.8% H_2O, and 9% ash. Determine the theoretical air/fuel ratio.

Solution

Given: Ultimate analysis of coal.

Find: Theoretical air/fuel ratio.

Assumptions:

1. The molal ratio of nitrogen to oxygen is 3.76.
2. The products behave like an ideal gas.

Analysis: Since ash does not enter into the oxidation of coal, determine the mole fractions of the coal's constituents on an ashless basis. First divide the mass fractions by 0.91, then divide by the molecular mass. Finally, divide by the total moles to determine the mole fraction.

	x_i	M_i	x_i/M_i	y_i
C	0.813	12.0	0.0677	0.674
H_2	0.055	2.0	0.0275	0.274
O_2	0.066	32.0	0.0021	0.021
S	0.011	32.0	0.0003	0.003
N_2	0.013	28.0	0.0005	0.005
H_2O	0.042	18.0	0.0023	0.023
	1.000		0.1004	1.000

The combustion equation for 1 mol of ashless fuel is

$$\overbrace{0.674C + 0.274H_2 + 0.021O_2 + 0.003S + 0.005N_2 + 0.023H_2O}^{\text{Fuel}}$$

$$+ \overbrace{aO_2 + bN_2}^{\text{Air}} \rightarrow cCO_2 + dH_2O + eSO_2 + fN_2$$

Perform a mass balance on each term of the reactants.

$$\text{C balance} \quad c = 0.674$$

$$\text{H}_2 \text{ balance} \quad 2d = 2(0.274 + 0.023) \quad d = 0.297$$

$$\text{S balance} \quad e = 0.003$$

$$\text{O}_2 \text{ balance} \quad 0.021 + 0.0115 + a = 0.674 + 0.148 + 0.003 \quad a = 0.792$$

$$\text{N}_2 \text{ balance} \quad (0.005) + (3.76)(0.792) = f \quad f = 2.983$$

The air/fuel ratio on the mole basis is

$$r_{alf} = \frac{(4.76)(0.792) \text{ mol air}}{(1 \text{ mol fuel})} = 3.77 \text{ kgmol air/kgmol fuel}$$

The air/fuel ratio on the mass basis requires that we determine the molecular weight of the fuel. This must be found on an ashless basis, the one represented in the combustion reaction, from $M = \sum_i y_i M_i$. Thus,

$$M_{coal} = (0.674)(12) + (0.274)(2.016) + (0.021)(32.0)$$

$$+ (0.003)(32) + (0.005)(28.016) + (0.023)(18.016)$$

$$M_{coal} = 9.96 \text{ kg/kgmol}$$

$$r_{alf} = \frac{(3.77 \text{ kgmol air})(28.97 \text{ kg/kgmol air})}{(1 \text{ mol fuel})(9.96 \text{ kg/kgmol fuel})}$$

$$r_{alf} = 10.96 \text{ kg air/kg fuel}$$

Comment: The fuel must be represented by an ashless percentage to correctly balance the reaction equation.

Acid Rain and the Greenhouse Effect

Residual fuel and coal often have constituents like sulfur that when burned cause problems. Sulfur forms sulfur dioxide and sulfur trioxide in the combustion process, which when combined with water form sulfuric acid. This is one of the primary constituents of acid rain, as is nitric oxide, formed in high-temperature combustion processes involving air. The high temperature causes diatomic nitrogen to dissociate into monatomic nitrogen, which is chemically active and combines with oxygen to form nitric oxide. When nitric oxide dissolves in water, nitric acid is formed.

The carbon dioxide formed in the combustion process is increasing the greenhouse effect of the earth's atmosphere and accelerating the warming of the earth. Carbon dioxide concentrations in the atmosphere are 25% higher than preindustrial levels. Carbon dioxide is not alone in creating the greenhouse effect, but it is a primary contributor. The average global temperatures are 0.6°C higher than 100 years ago, and they are rising. Computer models predict an average temperature increase of between 2.5 and 4.5°C by the end of the next century. These temperature rises will change the world climate. Growing seasons in the Northern Hemisphere may increase by several weeks, but this is offset by drought conditions in many of the currently temperate regions. Melting of polar ice caps will cause sea levels to rise, flooding coastal areas.

The greenhouse effect is the reason the temperature of an automobile's interior with the windows closed is warmer than the outside air on a sunny day. When its wavelength is short, sunlight passes through glass, which is viewed as being transparent to the radiation. The sun's radiation hits a surface inside the car, is absorbed, and is reradiated. The wavelength of the reradiation is longer, and the glass is not transparent to this radiation, but opaque and traps the energy within the car's interior. This is also how a garden greenhouse works. Certain gases, carbon dioxide for instance, act like glass in solar radiation, forming a "greenhouse" around the earth.

We can calculate the magnitude of these terms quite easily. The gravimetric analysis of the fuel tells us the percentage of the various constituents in each unit mass of fuel. For instance, fuel oil may contain carbon, hydrogen, sulfur, water, and ash. Let's say that the fuel contains 1% sulfur and 90% carbon and that the cleaning processes on the stack remove 98% of the sulfur dioxide from the combustion gases. Calculate the kilograms of sulfur and carbon dioxide that are released daily from a power plant that burns 3 000 000 kg of fuel daily.

Initially let us determine the total mass of sulfur that will be emitted from the stack. The daily mass of fuel burned is

$$\dot{m}_f = 3\ 000\ 000\ \text{kg/day}$$

$$\dot{m}_S = (0.01)(3\ 000\ 000) = 30\ 000\ \text{kg/day}$$

However, 98% of this is contained, so only 2% of this value leaves the stack.

$$(\dot{m}_S)_{\text{emitted}} = 600\ \text{kg/day}$$

The sulfur reacts with oxygen to produce sulfur dioxide according to the following reaction equation:

$$S + O_2 \rightarrow SO_2$$

$$32\ \text{kg S} + 32\ \text{kg O}_2 = 64\ \text{kg SO}_2$$

Thus, for every 32 kg of sulfur oxidized (burned), 64 kg of sulfur dioxide is formed. Let's just consider the sulfur that escapes.

$$\frac{64\ \text{kg SO}_2}{32\ \text{kg S}} \times 600\ \text{kg S/day} = 1200\ \text{kg SO}_2\ \text{released/day}$$

This amounts to 1.32 tons per day from this one plant. Imagine the amount that could be released if scrubbers were not installed. Not all industries nor all countries use exhaust gas scrubbers.

The tons of carbon dioxide produced may be found in a similar fashion.

$$\dot{m}_C = (0.9)(3\ 000\ 000) = 2\ 700\ 000 \text{ kg/day}$$

$$C + O_2 \rightarrow CO_2$$

Thus, there are $44/12 = 3.667$ kg of carbon dioxide produced for each kilogram of carbon completely oxidized.

$$\dot{m}_{CO_2} = (3.667 \text{ kg } CO_2/\text{kg C})(2\ 700\ 000 \text{ kg C}) = 9\ 900\ 000 \text{ kg } CO_2/\text{day}$$

This is equal to 10,890 tons per day—from one power plant.

12.5 PRODUCTS OF COMBUSTION

In power plants and other facilities using large amounts of fuel, it is important that the burning be as efficient as possible. Tiny increases in efficiency (even a fraction of 1%) can save thousands of dollars. One important factor affecting the efficiency is the amount of excess air. If not enough air is used, combustion will be incomplete, and not all the chemical energy of the fuel will be used. If too much air is used, the heat released by combustion is wasted in heating this excess air. The object is to oxidize the fuel completely with the smallest amount of air. This will yield the greatest release of energy per mass of air. How may this be determined?

Orsat Analysis

An analysis of the products of combustion tells us how much of each product was formed. The Orsat apparatus performs this task by measuring carbon dioxide, carbon monoxide, and oxygen volumetrically. The combustion gas is passed through various chemicals, which absorb carbon dioxide, carbon monoxide, and oxygen. The volumetric decrease of the combustion gas is noted at each step, and the decrease in volume divided by the initial volume gives the percentage of each combustion product. The remaining volume is assumed to be nitrogen. The Orsat gives the volumetric proportions on a dry basis; the amount of water vapor cannot be determined, since it is condensed as a liquid in the sampling and measuring processes. Since the analysis usually occurs at room temperature and pressure, which is below the dew point of most hydrocarbon products of combustion, the error involved is quite small.

The Orsat analysis cannot measure partially combusted hydrocarbons, nor can it measure carbon. These quantities are important in dealing with internal-combustion engines, where the combustion is not complete and we want to analyze reactions occurring before the final products are formed. There also may be nitric oxides formed at high temperatures. Measuring devices, such as gas chromatographs can determine these compounds, but these problems are beyond the scope of this book. They involve reaction times and partial reactions; we will consider only complete combustion and no time-dependent reactions at all.

The traditional Orsat apparatus can take only periodic measurements. In order to provide power plant operators with continuous information about the combustion process, various electronic instruments have been developed. The most popular of these are direct-insertion (in situ) oxygen analyzers using a zirconium oxide (ZrO_2) sensor cell. The sensor cell is installed directly in the stack gas flow. The gases of combustion pass across one side of the cell, and the other side is exposed to atmospheric air. A heater assembly maintains the cell at a constant temperature of about 815°C. At this temperature, oxygen ions pass through the ZrO_2 cell, generating a voltage proportional to the logarithm of the ratio of the oxygen partial pressures. In many modern boiler installations, the output from the analyzer is used by the combustion control system to automatically regulate the air/fuel ratio.

The principles used in an Orsat analysis will be demonstrated in the examples. We can determine the air/fuel ratio as well as balance the reaction equation.

Example 12.4

Fuel oil, $C_{12}H_{26}$, is burned in air at atmospheric pressure. The Orsat analysis for the products of combustion yields 13.1% CO_2, 2.0% O_2, 0.2% CO, and 84.7% N_2. Determine the mass air/fuel ratio, the percentage of theoretical air, and the combustion equation.

Solution

Given: The Orsat analysis of the products of a known fuel.

Find: The mass air/fuel ratio, the percentage of theoretical air, and the balanced combustion equation.

Assumptions:

1. The products behave individually and collectively like an ideal gas.
2. The molal ratio of nitrogen to oxygen in air is 3.76.

Analysis: Write the combustion equation for 100 mol of dry products, as this allows us to write the percentages as the number of moles. The unknown coefficients on the reactant side of the combustion equation may be determined by applying the conservation of mass to each reactant.

$$aC_{12}H_{26} + bO_2 + cN_2 \rightarrow 13.1CO_2 + 0.2CO + 2.0O_2 + 84.7N_2 + dH_2O$$

$$\text{C balance} \quad 12a = 13.1 + 0.2 \quad \therefore a = \frac{13.3}{12}$$

$$\text{N}_2 \text{ balance} \quad c = 84.7$$

$$\text{N}_2/\text{O}_2 \text{ ratio} \quad \frac{c}{b} = 3.76 \quad \therefore b = 22.52$$

$$\text{H}_2 \text{ balance} \quad 2d = \left(\frac{13.3}{12}\right)(26) \quad \therefore d = 14.41$$

Divide the equation by a to determine the combustion equation for 1 mol fuel.

$$C_{12}H_{26} + 20.3O_2 + 76.4N_2 \rightarrow 11.8CO_2 + 0.18CO + 1.8O_2 + 76.4N_2 + 13H_2O$$

The air/fuel ratio is

$$r_{a/f} = \frac{(20.3 + 76.4 \text{ mol air})(28.97 \text{ kg/kgmol})}{(1 \text{ mol fuel})(170.328 \text{ kg/kgmol})} = 16.45 \text{ kg air/kg fuel}$$

The balanced reaction equation for 100% theoretical air is

$$C_{12}H_{26} + 18.5O_2 + 18.5(3.76)N_2 \rightarrow 12CO_2 + 13H_2O + 69.56N_2$$

and the air/fuel ratio is

$$r_{a/f} = \frac{(18.5 + 69.56 \text{ mol air})(28.97 \text{ kg/kgmol})}{(1 \text{ mol fuel})(170.328 \text{ kg/kgmol})} = 14.98 \text{ kg air/kg fuel}$$

The percentage of theoretical air is

$$\text{Theoretical air} = \frac{16.45}{14.98} \times 100 = 109.8\%$$

Comment: It is crucial to include water on the products side of the combustion equation, as the Orsat and other analyses do not account for this.   ▪

Example 12.5
An unknown hydrocarbon fuel, burned in air, has the following Orsat analysis: 12.5% CO_2, 0.3% CO, 3.1% O_2, and 84.1% N_2. Determine the mass air/fuel ratio, the fuel composition on a mass basis, and the percentage of theoretical air.

Solution

Given: The Orsat analysis of an unknown fuel burned in air.

Find: The mass composition of the fuel, the mass air/fuel ratio, and the percentage of theoretical air.

Assumptions:

1. The products behave individually and collectively like an ideal gas.
2. The molal ratio of nitrogen to oxygen in air is 3.76.

Analysis: Write the reaction equation for 100 mol of dry products and perform a mass balance for each of the unknown quantities.

$$C_aH_b + cO_2 + dN_2 \rightarrow 12.5CO_2 + 0.3CO + 3.1O_2 + 84.1N_2 + eH_2O$$

$$\text{C balance} \quad a = 12.5 + 0.3 = 12.8$$

$$\text{N}_2 \text{ balance} \quad d = 84.1 \qquad \frac{d}{c} = 3.76 \qquad c = 22.36$$

$$\text{O}_2 \text{ balance} \quad 22.36 = 12.5 + \frac{0.3}{2} + 3.1 + \frac{e}{2} \qquad e = 13.2$$

$$\text{H}_2 \text{ balance} \quad b = 2e = 26.4$$

The balanced reaction equation is

$$C_{12.8}H_{26.4} + 22.35O_2 + 84.1N_2 \rightarrow$$

$$12.5CO_2 + 0.3CO + 3.1O_2 + 84.1N_2 + 13.2H_2O$$

The molecular weight of the fuel is

$$M_f = (12.01 \text{ kg/kgmol C})(12.8 \text{ mol C}) + (1.008 \text{ kg/kgmol H})(26.4 \text{ mol H})$$

$$M_f = 180.34 \text{ kg/kgmol}$$

The air/fuel ratio on a mass basis is

$$r_{a/f} = \frac{(22.36 + 84.1 \text{ mol air})(28.97 \text{ kg/kgmol})}{(1 \text{ mol fuel})(180.34 \text{ kg/kgmol})} = 17.1 \text{ kg air/kg fuel}$$

The reaction equation balanced for 100% theoretical air is

$$C_{12.8}H_{26.4} + 19.4O_2 + 73N_2 \rightarrow 12.8CO_2 + 13.2H_2O + 73N_2$$

and the air/fuel ratio is

$$r_{a/f} = \frac{(19.4 + 73 \text{ mol air})(28.97 \text{ kg/kgmol})}{(1 \text{ mol fuel})(180.34 \text{ kg/kgmol})} = 14.84 \text{ kg air/kg fuel}$$

The percentage of theoretical air is

$$100 \times \frac{17.1}{14.8} = 115.5\%$$

The fuel composition on a mass basis is

$$C = \frac{(12.8 \text{ mol C})(12.01 \text{ kg/kgmol})}{(1 \text{ mol fuel})(180.34 \text{ kg/kgmol})} = 0.852 \quad \text{or } 85.2\%$$

$$H = \frac{(26.4 \text{ mol H})(1.008 \text{ kg/kgmol})}{(1 \text{ mol fuel})(180.34 \text{ kg/kgmol})} = 0.148 \quad \text{or } 14.8\%$$

Comment: It is not necessary to know the fuel type to establish the combustion equation. The products of combustion would indicate the other combustible materials in the fuel but would not indicate compounds such as water, nitrogen, and ash.

The TK Solver model ORSAT.TK can be used to analyze the products of a combustion process. The back-solving capabilities of TK Solver permit the same model to determine the stack gas analysis, given the fuel analysis and excess air, or the fuel composition and excess air, given the Orsat stack gas analysis.

Example 12.6
Repeat Example 12.5 using ORSAT.TK.

Solution

Given: The Orsat analysis of an unknown fuel burned in air.

Find: The mass composition of the fuel, the mass air/fuel ratio, and the percentage of excess air.

Assumptions:

1. The products behave individually and collectively like an ideal gas.
2. The molal ratio of nitrogen to oxygen in air is 3.76.

Analysis: Load ORSAT.TK into TK Solver and input the data into the Variable Sheet. Note that zero is entered for the sulfur, nitrogen, oxygen, and ash content of the fuel. The solved Variable Sheet is thus

```
=================== VARIABLE SHEET ===================
St Input —— Name —— Output —— Unit —— Comment ————————

                                    ***ORSAT ANALYSIS***

                              Fuel Ultimate Analysis
             C        .85135      kg/kg      Carbon
             H        .14865      kg/kg      Hydrogen
    0        S                    kg/kg      Sulfur
    0        N                    kg/kg      Nitrogen
    0        O                    kg/kg      Oxygen
    0        Ash                  kg/kg      Ash

                              Orsat Analysis—Dry Basis
   12.5      CO2%                 %          Carbon Dioxide
   3.1       O2%                  %          Oxygen
   .3        CO%                  %          Carbon Monoxide

                              Air—Fuel Ratio
             ThRaf    14.799      kg/kg      Stoichiometric Air—Fuel Ratio
             ExAir%   15.162      %          Excess Air
             Raf      17.043      kg/kg      Actual Air—Fuel Ratio
             ThmolO2  .10775                 Theoretical Moles Oxygen

             molCO2   .070887
             molH2    .073733
             molSO2   0
             molN2    .46658
             molO2    .016337
             molTot   .55381
```

Comment: ORSAT.TK permits convenient analysis of the products of a combustion process.

12.6 ENTHALPY OF FORMATION

In the previous chapters the working substance has always been homogeneous and never changed its chemical composition during any process. Tables were developed describing the properties of these substances — for example, the steam tables — and in these tables there was always an arbitrary reference base. The enthalpy of saturated liquid water at 0°C is zero. This is an arbitrary base, but it does not matter since we deal with changes in the enthalpy as with changes in any other property. In chemical reactions, however, the substance in the system changes during the course of the process, so the capricious use of arbitrary standards for each substance would make the energy analysis of the process impossible.

To overcome this difficulty, the enthalpy of all elements is assumed to be zero at an arbitrary reference state of 25°C (77°F) and 1 atm pressure. The enthalpy of formation of a compound is its enthalpy at this temperature and pressure.

Consider a steady-state combustion process in which 1 mol of carbon and 1 mol of oxygen at the reference state of 25°C and 1 atm pressure combine to produce 1 mol of carbon dioxide. Heat is transferred, so the carbon dioxide finally exists at the reference state. Figure 12.1 demonstrates this process.

The reaction equation is

$$C + O_2 \rightarrow CO_2$$

Let H_R be the total enthalpy of all the reactants and H_P be the total enthalpy of all the products per mole of fuel. The first law for this process is

$$Q + H_R = H_P \tag{12.6}$$

or

$$Q + \sum_R n_i \bar{h}_i = \sum_P n_j \bar{h}_j \tag{12.7}$$

where the summations are over all the reactants and all the products, and $Q = \dot{Q}/\dot{n}_f$.

Since the enthalpy of all the reactants is zero (they are all elements), we find that

$$Q = H_P = -393\ 757\ \text{kJ} \tag{12.8}$$

where the heat transferred has been carefully measured and found to be $-393\ 757$ kJ. The sign is negative because heat is flowing from the control volume, opposite to the assigned direction in equation (12.6). The enthalpy of carbon dioxide at the reference state, 25°C and 1 atm pressure, is the *enthalpy of formation* and is

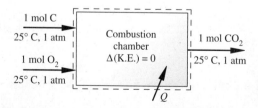

1 mol C
25° C, 1 atm

Combustion
chamber
Δ(K.E.) = 0

1 mol CO$_2$
25° C, 1 atm

1 mol O$_2$
25° C, 1 atm

Q

Figure 12.1 A steady-state combustion process with heat transfer.

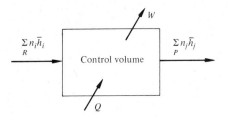

Figure 12.2 A steady-flow system where a chemical reaction transfers heat and produces work.

designated by the symbol $\bar{h}_f^\circ$. Therefore,

$$(\bar{h}_f^\circ)_{CO_2} = -393\ 757 \text{ kJ/kgmol} \tag{12.9}$$

The negative sign for the enthalpy of formation is due to the reaction's being exothermic; heat is released from the combustion of carbon and oxygen. Since the energy must be conserved in the reaction and 393 757 kJ left the control volume, the energy of the carbon dioxide must be less than the energy of the reactants by an amount equal to the heat transferred. The enthalpy of the reactants is zero at 25°C and 1 atm; therefore the enthalpy of carbon dioxide at 25°C and 1 atm must be negative.

Table C.1 lists the enthalpies of formation for several substances.

12.7 FIRST-LAW ANALYSIS FOR STEADY-STATE REACTING SYSTEMS

In general, the steady-state combustion process will transfer heat and produce work, as illustrated in Figure 12.2. An energy balance would yield

$$Q + \sum_R n_i \bar{h}_i = W + \sum_P n_j \bar{h}_j \tag{12.10}$$

where $W = \dot{W}/\dot{n}_f$. The enthalpies of formation may be readily used in this analysis, since they are all relative to the same reference base. Usually the analysis is done per mole of fuel, thus the units are kilojoules per kilogram-mole. The following examples illustrate important concepts that may be used in combustion energy analysis.

Example 12.7
Propane, C_3H_8, undergoes a steady-state, steady-flow reaction with atmospheric air. Determine the heat transfer per mole of fuel entering the combustion chamber. The reactants and products are at 25°C and 1 atm pressure.

Solution

Given: A known fuel completely oxidizing in a combustion chamber at a steady state.

Find: The heat transfer per mole of fuel.

Sketch and Given Data:

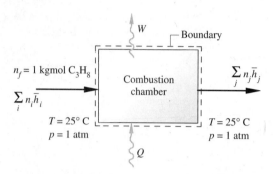

Figure 12.3

Assumptions:

1. The water in the products is a liquid, as the temperature is below the dew point.
2. The molal ratio of nitrogen to oxygen in air is 3.76.
3. No work is done.
4. The changes in kinetic and potential energies may be neglected.

Analysis: The balanced reaction equation for 100% theoretical air is

$$C_3H_8 + 5O_2 + 5(3.76)N_2 \rightarrow 3CO_2 + 4H_2O(l) + 18.8N_2$$

The first-law equation with assumptions 3 and 4 invoked yields

$$Q + \sum_R n_i \bar{h}_i = \sum_P n_j \bar{h}_j$$

Since the enthalpy of all the elements at 25°C and 1 atm is zero, the summation terms contain only the enthalpies of formation for the compounds, which are found in Table C.1.

$$\sum_R n_i \bar{h}_i = (\bar{h}_f^\circ)_{C_3H_8} = -103\,909 \text{ kJ/kgmol}$$

$$\sum_P n_j \bar{h}_j = 3(\bar{h}_f^\circ)_{CO_2} + 4(\bar{h}_f^\circ)_{H_2O(l)} = -2\,325\,311 \text{ kJ/kgmol}$$

$$Q = -2\,221\,402 \text{ kJ/kgmol fuel}$$

Comment: In most cases, neither the reactants nor the products are at the reference condition of 25°C and 1 atm pressure. In these cases we must account for the property change between the reference state and the actual state. Table C.2 tabulates the change in enthalpy between the reference state and the actual state, $(\bar{h}^\circ - \bar{h}_{298}^\circ)$ in kilojoules per kilogram-moles. The temperature of the reference state in absolute notation is 298°K. The superscript ° denotes that the pressure is 1 atm. If these tabulated values are not available, we have to use the ideal-gas law, $\bar{h} = \bar{c}_p T$, determining the specific heat by the best means available. For example, if the specific heat varied as a function of temperature and this functional relationship were known, this would be used. If tables of property values existed, such as the gas tables or steam tables, then these, too, could be used.

Most problems in this text are at 1 atm pressure. This is the standard pressure and the one found in most combustion reactions. When it is necessary to include the pressure effects, as when calculating entropy changes, the reactants and products must be expanded to the reference state and the change in doing so included in the calculation. This is a very small effect and need not be included in most engineering applications.

In general, the first law for a steady-state, steady-flow reaction may be written as

$$\dot{Q} + \sum_R \dot{n}_i[\bar{h}_f^\circ + (\bar{h}^\circ - \bar{h}_{298}^\circ)]_i = \dot{W} + \sum_P \dot{n}_j[\bar{h}_f^\circ + (\bar{h}^\circ - \bar{h}_{298}^\circ)]_j \quad (12.11)$$

where changes in kinetic and potential energies are considered negligible. This will always be assumed for problems in this text.

Example 12.8

A diesel engine uses dodecane, $C_{12}H_{26}(v)$, for fuel. The fuel and air enter the engine at 25°C. The products of combustion leave at 600°K, and 200% theoretical air is used. The heat loss from the engine is measured at 232 000 kJ/kgmol fuel. Determine the work for a fuel flow rate of 1 kgmol/h.

Solution

Given: A diesel engine receives fuel and air at 25°C and a known ratio of fuel to air. Complete combustion occurs, with the products' temperature known as well as the heat transfer from the engine.

Find: The power for a known fuel flow rate.

Sketch and Given Data:

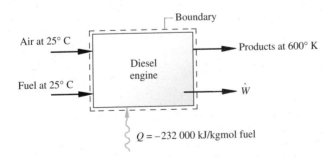

Figure 12.4

Assumptions:

1. The products behave like ideal gases.
2. The molal ratio of nitrogen to oxygen in air is 3.76.
3. The changes in kinetic and potential energies may be neglected.

Analysis: The balanced reaction equation for 200% theoretical air is

$$C_{12}H_{26}(v) + 2(18.5)O_2 + 2(18.5)(3.76)N_2 \rightarrow$$

$$12CO_2 + 13H_2O + 18.5O_2 + 139.12N_2$$

Apply the first law per mole of fuel

$$Q + \sum_R n_i[\bar{h}_f^\circ + (\bar{h}^\circ - \bar{h}_{298}^\circ)]_i = W + \sum_P n_j[\bar{h}_f^\circ + (\bar{h}^\circ - \bar{h}_{298}^\circ)]_j$$

The heat transfer per mole of fuel is $Q = -232\ 000$ kJ/kgmol fuel. The reactant and product energy terms may now be evaluated:

$$\sum_R n_i[\bar{h}_f^\circ + (\bar{h}^\circ - \bar{h}_{298}^\circ)]_i = 1(\bar{h}_f^\circ)_{C_{12}H_{26(v)}}$$

$$= -290\ 971 \text{ kJ/kgmol}$$

$$\sum_P n_i[\bar{h}_f^\circ + (\bar{h}^\circ - \bar{h}_{298}^\circ)]_i = 12(-393\ 757 + 12\ 916)_{CO_2}$$

$$+ 13(-241\ 971 + 10\ 498)_{H_2O}$$

$$+ 139.12(8891)_{N_2} + 18.5(9247)_{O_2}$$

$$= -6\ 171\ 255 \text{ kJ/kgmol fuel}$$

Substitute these terms into the first-law equation.

$$Q + H_R = W + H_P$$

$$-232\ 000 \text{ kJ/kgmol fuel} - 290\ 971 \text{ kJ/kgmol fuel}$$

$$= W - 6\ 171\ 255 \text{ kJ/kgmol fuel}$$

$$W = 5\ 648\ 284 \text{ kJ/kgmol fuel}$$

Multiply the work per mole of fuel by the fuel flow rate.

$$\dot{W} = \dot{n}_f W = \frac{(1 \text{ kgmol/h})(5\ 648\ 284 \text{ kJ/kgmol})}{(3600 \text{ s/h})} = 1568.9 \text{ kW}$$

Comment: The first law is usually solved per unit mole of fuel, as the reaction equation is balanced for 1 mol of fuel. The rate of energy change is calculated by multiplying the energy change per mole by the molar fuel flow rate.   ▬

Example 12.9

A gas turbine generating unit produces 600 kW and uses $C_8H_{18}(l)$ as a fuel at 77°F; 400% theoretical air is used, and the air enters the unit at 100°F. The products of combustion leave at 800°F. The heat transfer to the surroundings is 50,000 Btu/hr. Determine the fuel flow rate in pounds mass per hour for complete combustion.

Solution

Given: A gas turbine unit with known fuel and air temperatures entering and products' temperature leaving. In addition the heat transfer and the power produced are specified.

Find: The fuel flow rate.

Sketch and Given Data:

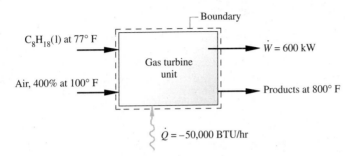

Figure 12.5

Assumptions:

1. The products behave like ideal gases.
2. The molal ratio of nitrogen to oxygen in air is 3.76.
3. The changes in kinetic and potential energies may be neglected.

Analysis: The balanced reaction equation for 400% theoretical air is

$$C_8H_{18}(l) + 4(12.5)O_2 + 4(12.5)(3.76)N_2 \rightarrow 8CO_2 + 9H_2O + 37.5O_2 + 188N_2$$

Determine the values of H_R and H_P for the 1 mol of fuel oxidized in the combustion equation.

$$H_R = \sum_R n_i[\bar{h}_f^\circ + (\bar{h}^\circ - \bar{h}_{537}^\circ)]_i = 1(-107,532)_f + 50(165)_{O_2} + 188(162)_{N_2}$$

$$H_R = -68,826 \text{ Btu/pmol fuel}$$

$$H_P = \sum_P n_j[\bar{h}_f^\circ + (\bar{h}^\circ - \bar{h}_{537}^\circ)]_j$$

$$H_P = 8(-169,297 + 7641)_{CO_2} + 9(-104,036 + 6102)_{H_2O}$$
$$+ 37.5(5378)_{O_2} + 188(5135)_{N_2}$$

$$H_P = -1,007,599 \text{ Btu/pmol fuel}$$

$$H_P - H_R = -938,773 \text{ Btu/pmol fuel}$$

The first law is

$$\dot{Q} + \dot{n}_f H_R = \dot{W} + \dot{n}_f H_P$$

$$\dot{Q} - \dot{W} = \dot{n}_f(H_P - H_R)$$

$$600 \text{ kW} = 2,047,680 \text{ Btu/hr}$$

Substituting into the first-law equation,

$$-50,000 \text{ Btu/hr} - 2,047,680 \text{ Btu/hr} = \dot{n}_f(-938,773 \text{ Btu/pmol fuel})$$

$$\dot{n}_f = 2.235 \text{ pmol/hr}$$

The molecular weight of the fuel is 114.23 lbm/pmol; hence the mass flow rate is

$$\dot{m}_f = (2.235 \text{ pmol/hr})(114.23 \text{ lbm/pmol}) = 255.3 \text{ lbm/hr}$$

Comment: Care must be exercised that all units in the first-law equation are the same. Do not mix terms that have units of energy per mole of fuel with those that have energy per unit time.

Example 12.10

A mixture of methane and oxygen, in the proper ratio for complete combustion and at 25°C and 1 atm, reacts in a constant-volume calorimeter bomb. Heat is transferred until the products of combustion are at 400°K. Determine the heat transfer per mole of methane and the final pressure.

Solution

Given: Known reactants oxidize in a constant-volume container. The products are cooled to 400°K.

Find: The heat transferred per mole of fuel and the final pressure.

Sketch and Given Data:

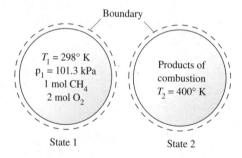

| State 1 | State 2 | **Figure 12.6** |

Assumptions:

1. The reactants and products behave like ideal gases.
2. The water in the products is a vapor.
3. No work is done.
4. The changes in kinetic and potential energies may be neglected.

Analysis: The combustion equation is

$$CH_4 + 2O_2 \rightarrow CO_2 + 2H_2O$$

The first-law equation for constant volume, invoking assumptions 3 and 4, is

$$Q = U_2 - U_1 = \sum_P n_j \bar{u}_j - \sum_R n_i \bar{u}_i$$

where the internal energy may be calculated from the ideal-gas law as

$$\bar{u} = \bar{h} - \bar{R}T$$

Solve for the reactant and product internal energies.

$$\sum_R n_i \bar{u}_i = \sum_R n_i [\bar{h}_f^\circ + (\bar{h}^\circ - \bar{h}_{298}^\circ) - \bar{R}T]_i$$

$$U_R = (1 \text{ mol})[-74\,917 \text{ kJ/kgmol fuel} + 0$$

$$- (8.3143 \text{ kJ/kgmol-K})(298°K)]_f$$

$$+ (2 \text{ mol})[0 + 0 - (8.3143 \text{ kJ/kgmol-K})(298°K)]_{O_2}$$

$$U_R = -82\,350 \text{ kJ/kgmol fuel}$$

$$\sum_P n_j \bar{u}_j = U_P = \sum_P n_j [\bar{h}_f^\circ + (\bar{h}^\circ - \bar{h}_{298}^\circ) - \bar{R}T]_j$$

$$U_P = (1 \text{ mol})[-393\,757 \text{ kJ/kgmol fuel}$$

$$+ 4008 \text{ kJ/kgmol fuel} - (8.3143 \text{ kJ/kgmol-K})(400°K)]_{CO_2}$$

$$+ (2 \text{ mol})[-241\,971 \text{ kJ/kgmol fuel}$$

$$+ 3452 \text{ kJ/kgmol fuel} - (8.3143 \text{ kJ/kgmol-K})(400°K)]_{H_2O}$$

$$U_P = -876\,764 \text{ kJ/kgmol fuel}$$

The heat transfer is

$$Q = U_2 - U_1 = U_P - U_R = -794\,414 \text{ kJ/kgmol fuel}$$

The pressure may be found from the ideal-gas law applied to the reactants and products.

$$p_1 V_1 = n_R \bar{R} T_1$$

$$p_2 V_2 = n_P \bar{R} T_2$$

Divide one equation into the other and solve for p_2.

$$p_2 = p_1 \frac{n_P T_2}{n_R T_1} = \frac{(101.3 \text{ kPa})(3 \text{ mol})(400°K)}{(3 \text{ mol})(298°K)} = 136 \text{ kPa}$$

Comments:

1. The expression for the internal energy can be used only if the substance is an ideal gas.
2. Using the molal ideal-gas law with the universal gas constant allows us to model reactants and products with the same equation. Care must be taken to use the appropriate number of moles for the reactant side and the product side. It is unusual that the numbers of moles are equal.

12.8 ADIABATIC FLAME TEMPERATURE

So far we have assumed that heat transfer and work occur during a combustion process. If no work, no heat transfer, and no change in kinetic or potential energy should occur, all the thermal energy would go into raising the temperature of the products of combustion. When the combustion is complete under these circumstances, the maximum amount of chemical energy has been converted into thermal energy and the temperature of the products is at its maximum. This temperature is called the *adiabatic flame temperature.*

Should the combustion be incomplete or excess air be used, the temperature of the mixture would be less than the maximum adiabatic flame temperature. Excess air is used in engine design to keep the temperature within metallurgical limits. If combustion is incomplete, not all the chemical energy is converted into thermal energy; hence the temperature will be lower than the maximum that is possible. When excess air is used, the thermal energy must be used to raise the temperature of a greater mass; hence the temperature rise for the fixed amount of thermal energy will not be as great as the maximum. A third factor that reduces temperature is dissociation of the combustion products. The dissociation reaction is endothermic; it uses some of the available thermal energy to proceed. We will discuss this reaction later.

The following example calculates the adiabatic flame temperature. It is a trial-and-error process.

Example 12.11

Gaseous propane is burned with 100% and 400% theoretical air at 25°C. Determine the adiabatic flame temperature in each case.

Solution

Given: The complete oxidation of propane with specified amounts of theoretical air at 25°C.

Find: The adiabatic flame temperature for each case.

Assumptions:

1. The products and reactants behave like ideal gases.
2. No work is done.
3. The heat transfer is zero.
4. The changes in kinetic and potential energies may be neglected.
5. The molal ratio of nitrogen to oxygen in air is 3.76.

Analysis: The reaction equation balanced for 100% theoretical air is

$$C_3H_8 + 5O_2 + 5(3.76)N_2 \rightarrow 3CO_2 + 4H_2O + 18.8N_2$$

Applying the first-law equation with assumptions 2, 3, and 4 yields

$$H_R = H_P$$

$$H_R = \sum_R n_i[\bar{h}_f^\circ + (\bar{h}^\circ - \bar{h}_{298}^\circ)]_i$$

$$H_R = 1(-103\ 909)_f = -103\ 909 \text{ kJ/kgmol fuel}$$

$$H_P = \sum_P n_j[\bar{h}_f^\circ + (\bar{h}^\circ - \bar{h}_{298}^\circ)]_j$$

$$H_P = 3[-393\ 757 \text{ kJ/kgmol fuel} + (\bar{h}^\circ - \bar{h}_{298}^\circ)]_{CO_2} + 4[-241\ 971 \text{ kJ/kgmol fuel}$$
$$+ (\bar{h}^\circ - \bar{h}_{298}^\circ)]_{H_2O} + 18.8[(\bar{h}^\circ - \bar{h}_{298}^\circ)]_{N_2}$$

At this point the temperature of the products must be guessed at. Assume that $T_P = 2500°K$. Then

$$H_P = +10\ 845 \text{ kJ/kgmol fuel}$$

and

$$H_P - H_R = +114\ 754 \text{ kJ/kgmol fuel}$$

Try another temperature, 2300°K, for the products.

$$H_P = -206\ 068 \text{ kJ/kgmol fuel}$$

and

$$H_P - H_R = -102\ 159 \text{ kJ/kgmol fuel}$$

Figure 12.7 shows a plot of $(H_P - H_R)$ versus T. Where the line connecting the two points intersects the $(H_P - H_R) = 0$ axis is the linear approximation of the temperature that will satisfy the first-law equation. The temperature that will balance the equation is $T = 2394°K$, the adiabatic flame temperature for 100% theoretical air.

Now balance the combustion equation for 400% theoretical air.

$$C_3H_8 + (4)(5)O_2 + (4)(5)(3.76)N_2 \rightarrow 3CO_2 + 4H_2O + 15O_2 + 75.2N_2$$

The first law with assumptions 2, 3, and 4 yields

$$H_P = H_R$$

and

$$H_R = 1(-103\ 909)_f = -103\ 909 \text{ kJ/kgmol fuel}$$

Assume that $T_P = 1000°K$; hence

$$H_P = +10\ 612 \text{ kJ/kgmol fuel}$$

and

$$H_P - H_R = +114\ 521 \text{ kJ/kgmol fuel}$$

Assume that $T_P = 900°K$; hence

$$H_P = -317\ 290 \text{ kJ/kgmol fuel}$$

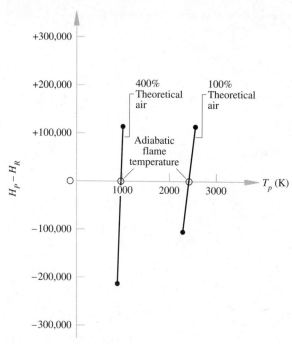

Figure 12.7 Graph for Example 12.11.

and

$$H_P - H_R = -213\ 381 \text{ kJ/kgmol fuel}$$

From the plot in Figure 12.7, the value for the adiabatic flame temperature is $942°K$ for 400% theoretical air.

Comments:

1. Using excess air brings about a significant temperature decrease. This can be an important engineering tool in designing optimum energy transfer in equipment such as steam generators and automotive engines.
2. The assumption of linear extrapolation between temperatures, used in Figure 12.7, is reasonable if the temperature changes are not large.

12.9 ENTHALPY OF COMBUSTION, HEATING VALUE

The enthalpy of combustion, h_{RP}, is the difference between the enthalpies of the products and reactants at the same temperature, T, and pressure. Thus,

$$\bar{h}_{RP} = H_P - H_R$$
$$\bar{h}_{RP} = \sum_P n_j[\bar{h}_f° + (\bar{h}_T° - \bar{h}_{298}°)]_j - \sum_R n_i[\bar{h}_f° + (\bar{h}_T° - \bar{h}_{298}°)]_i \qquad (12.12)$$

The enthalpy of combustion usually is expressed in the units of kilojoules per kilogram of fuel. This is also called the *heating value* of the fuel at constant pressure, because the heat is transferred at constant pressure in an open system and is equal to the same enthalpy difference. The enthalpy of combustion of various fuels is given in Table C.3, where the values are given at the reference state.

The internal energy of combustion, $\bar{u}_{RP}$, is the difference between the internal energies of the products and reactants and may be written as

$$\bar{u}_{RP} = U_P - U_R$$

$$\bar{u}_{RP} = \sum_P n_j[\bar{h}_f^\circ + (\bar{h}_T^\circ - \bar{h}_{298}^\circ) - \bar{R}T]_j$$

$$- \sum_R n_i[\bar{h}_f^\circ + (\bar{h}_T^\circ - \bar{h}_{298}^\circ) - \bar{R}T]_i \qquad (12.13)$$

where all the reactants and products have been considered as gases. Should this not be the case, then $p\bar{v}$ must be used in place of $\bar{R}T$ in determining the internal energy. Since this is equal to the heat transferred at constant volume, the constant-volume heating value of the fuel is also equal to $\bar{u}_{RP}$. Note that equation (12.13) may be written more compactly as

$$\bar{u}_{RP} = \bar{h}_{RP} - \bar{R}T \left(\sum_P n_j - \sum_R n_i \right) \qquad (12.14)$$

We have seen that it is important to know whether the water in the products of combustion is in the liquid or vapor phase. The *higher heating value* of a fuel indicates that the water in the products of combustion is liquid; thus the latent heat of water is included in determining the heat transferred. The *lower heating value* means that the water in the products of combustion exists as a vapor; thus, the latent heat will not be included in determining $\bar{h}_{RP}$, and it will nave a lower numerical value than the higher heating value.

Example 12.12

Calculate the enthalpy of combustion—the heating value—of liquid propane at 25°C when the water in the products is a liquid and is a vapor. The enthalpy of evaporation is 370 kJ/kg.

Solution

Given: The combustion of liquid propane at 25°C with its enthalpy of evaporation known.

Find: The higher and lower values of the enthalpy of combustion, or the heating value, of the propane.

Assumptions:

1. Combustion is complete, and both reactants and products are at the same temperature.
2. The ideal-gas law applies to gaseous reactants and products.

Analysis: Write the balanced reaction equation with only oxygen. Nitrogen is not included, as the temperature of products and reactants is the same and nitrogen is not dissociated. The enthalpy of the nitrogen, an element, is zero at 25°C.

$$C_3H_8(l) + 5O_2 \rightarrow 3CO_2 + 4H_2O(l)$$

From the definition of heating value,

$$\bar{h}_{RP} = (H_P - H_R)$$

The enthalpy of the reactants is

$$H_R = \sum_i n_i (\bar{h}_f^\circ)_i$$

$$H_R = (\bar{h}_f^\circ)_{C_3H_8(l)} = (\bar{h}_f^\circ)_{C_3H_8(v)} - Mh_{fg}$$

$$H_R = -103\ 909 - 44.1(370) = -120\ 226 \text{ kJ/kgmol fuel}$$

For the product, enthalpy is

$$H_P = \sum_j n_j (\bar{h}_f^\circ)_j$$

$$H_P = 3(\bar{h}_f^\circ)_{CO_2} + 4(\bar{h}_f^\circ)_{H_2O(l)}$$

$$H_P = 3(-393\ 757) + 4(-286\ 010) = -2\ 325\ 311 \text{ kJ/kgmol fuel}$$

The higher heating value, where the water is a liquid, is

$$\bar{h}_{RP} = \frac{(-2\ 325\ 311 + 120\ 226 \text{ kJ/kgmol fuel})}{(44.1 \text{ kg/kgmol})} = -50\ 002 \text{ kJ/kg}$$

For the case in which the water in the products remains a vapor, the enthalpy of the reactants is the same, but the enthalpy of the products is

$$H_P = \sum_j n_j (\bar{h}_f^\circ)_j$$

$$H_P = 3(\bar{h}_f^\circ)_{CO_2} + 4(\bar{h}_f^\circ)_{H_2O(v)}$$

$$H_P = 3(-393\ 757) + 4(-241\ 971)$$

$$H_P = -2\ 149\ 155 \text{ kJ/kgmol}$$

The lower heating value of propane is

$$\bar{h}_{RP} = \frac{(-2\ 149\ 155 + 120\ 226 \text{ kJ/kgmol fuel})}{(44.1 \text{ kg/kgmol})} = -46\ 007 \text{ kJ/kg}$$

Comment: Because the heating value is defined as the enthalpy difference between reactants and products at the same temperature, elements that have the same coefficient on each side of the reaction equation do not enter into the energy calculation.

∎

Example 12.13

Calculate the enthalpy of combustion of ethane, C_2H_6, at 500°K. Assume the average value of the specific heat at constant pressure for ethane to be 2.22 kJ/kg-K.

Solution

Given: The combustion of ethane at 500°K and its specific heat at constant pressure.

Find: The heating value, the enthalpy of combustion, of ethane at 500°K.

Assumptions:

1. Ethane is an ideal gas, as are the reactants and products.
2. Only oxygen is needed to determine the combustion reaction, as nitrogen would have the same value on both sides of the reaction equation.

Analysis: The balanced reaction equation is

$$C_2H_6 + 3.5O_2 \rightarrow 2CO_2 + 3H_2O(v)$$

The definition of the heating value is $\bar{h}_{RP} = H_P - H_R$ at the same temperature. The enthalpy of the products is

$$H_P = \sum_j n_j[\bar{h}_f^\circ + (\bar{h}_T^\circ - \bar{h}_{298}^\circ)]_j$$

$$H_P = 2(-393\ 757 + 8314) + 3(-241\ 971 + 6920) = -1\ 476\ 039 \text{ kJ/kgmol fuel}$$

In calculating the enthalpy of the reactants, the enthalpy of the change of ethane from 298°K is $\bar{c}_p(T - 298)$, found from the ideal-gas equation of state for enthalpy change. The reactant enthalpy is

$$H_R = \sum_i n_i[\bar{h}_f^\circ + (\bar{h}_T^\circ - \bar{h}_{298}^\circ)]_i$$

$$H_R = 1[-84\ 718 \text{ kJ/kgmol} + (2.22 \text{ kJ/kg-K})$$

$$(30.07 \text{ kg/kgmol})(500 - 298°\text{K})] + 3.5(6088 \text{ kJ/kgmol})$$

$$H_R = -49\ 925 \text{ kJ/kgmol fuel}$$

The enthalpy of combustion of ethane is

$$\bar{h}_{RP} = \frac{(-1\ 476\ 039 + 49\ 925 \text{ kJ/kgmol fuel})}{(30.07 \text{ kg/kgmol})} = -47\ 426 \text{ kJ/kg}$$

Comment: When determining the heating value of a substance at other than 25°C, the enthalpy change of that substance to the desired temperature, 500°K in this example, must be included. This often involves finding the specific heat of a substance as illustrated. ■

12.10 SECOND-LAW ANALYSIS

The combustion process and most chemical reactions proceed very rapidly. As we have seen previously, processes that proceed rapidly do so irreversibly. The entropy change for such a reaction indicates how irreversibly the reaction occurred, and, knowing this, we can determine the available portion of the thermal energy released by the combustion process.

As with enthalpy, we must be able to refer all the substance entropies to a common reference state or base. In this manner entropies of different substances may be added and subtracted, since they are common to the same base. The third law of thermodynamics states that the entropy of all pure substances is zero at zero degrees absolute. This gives us a reference base, and entropy measured from this base is called *absolute entropy.* In Tables C.1 and C.2 the values of absolute entropy at 1 atm pressure, $\bar{s}°$, are given for various temperatures for several substances. In these tables and by customary practice, 1 atm is essentially 100 kPa.

To account for the reaction's not occurring at standard conditions, the absolute entropy at some state 2 is

$$\bar{s}_2 = \bar{s}° + \Delta \bar{s}_{0\text{-}2}$$

where $\Delta \bar{s}_{0\text{-}2}$ is the difference between the absolute entropy at reference conditions and the absolute entropy at state 2. For an ideal gas,

$$(\bar{s}_2 - \bar{s}_0) = \bar{\phi}_2 - \bar{\phi}_0 - \bar{R} \ln (p_2/p_0)$$

If the pressure is expressed in atmospheres, further simplifications are possible. At reference conditions $p_0 = 1$ atm; thus,

$$(\bar{s}_2 - \bar{s}_0) = \bar{\phi}_2 - \bar{\phi}_0 - \bar{R} \ln (p_2)$$

Note that $s_0 = \bar{\phi}_0$, which allows us to write

$$\bar{s}_2 = \bar{\phi}_2 - \bar{R} \ln (p_2)$$

or in general

$$\bar{s} = \bar{\phi} - \bar{R} \ln (p) \tag{12.15}$$

If values of ϕ are not available in the gas tables, then $\int c_p \, dT/T$ must be evaluated.

In Chapter 9 we studied available energy and introduced the concept of availability, $\mathcal{A}$. We mentioned that availability was particularly important when dealing with chemical reactions. Furthermore, we showed that the change of availability was equal to the maximum work possible for a process between two states, and that if this process took place at constant temperature and pressure, the change in availability was equal to the change in the Gibbs function,

$$d\mathcal{A} = (dG)_{T_0, p_0} \tag{12.16}$$

If we integrate this expression between initial and final conditions, then

$$W_{\text{max}} = \mathcal{A}_1 - \mathcal{A}_2 = \sum_R n_i \bar{g}_i - \sum_P n_j \bar{g}_j \tag{12.17}$$

where the maximum work is written as a positive quantity. The availability decreases between the initial and final conditions.

Since $g = h - Ts$ and $T = T_0$, then for the reactive systems under consideration

$$g = h - T_0 s$$

Just as enthalpy of formation was necessary when analyzing compounds in first-law analysis of reactive systems, so is the Gibbs function of formation, $\bar{g}_f^\circ$, necessary in second-law analysis. The Gibbs function of formation is given in Table C.1 for several compounds at reference conditions of 25°C and 1 atm pressure. The Gibbs function of all elements is zero at reference conditions. The following example illustrates this.

Example 12.14
Determine the Gibbs function of $H_2O(l)$ at 25°C and 1 atm.

Solution

Given: The oxidation of hydrogen with oxygen to form liquid water at 25°C and 1 atm.

Find: The Gibbs function of liquid water.

Assumptions:

1. The gases in the combustion equation behave like ideal gases.
2. All the water formed is a liquid.

Analysis: The balanced reaction equation is

$$2H_2 + O_2 \rightarrow 2H_2O(l)$$

The difference is the Gibbs function for the reactants and products:

$$G_P - G_R = (H_P - H_R) - T_0(S_P - S_R)$$

$$G_P - G_R = \sum_P n_j (\bar{h}_f^\circ)_j - \sum_R n_i (\bar{h}_f^\circ)_i$$

$$- T_0 \left[\sum_P n_j (\bar{s}_{298}^\circ)_j - \sum_R n_i (\bar{s}_{298}^\circ)_i \right]$$

Since the enthalpy and the enthalpy of formation of elements are zero at 25°C and

1 atm, the difference in the Gibbs function is

$$G_P - G_R = 2(\bar{h}_f^\circ)_{H_2O(l)} - 298[2(\bar{s}_{298}^\circ)_{H_2O} - 2(\bar{s}^\circ)_{H_2} - (\bar{s}^\circ)_{O_2}]$$

$$G_P - G_R = (2 \text{ moles})(-286\ 010 \text{ kJ/kgmol})$$

$$- (298°K)[(2 \text{ moles})(69.980 \text{ kJ/kgmol-K})$$

$$- (2 \text{ moles})(130.684 \text{ kJ/kgmol-K})$$

$$- (1 \text{ mole})(205.142 \text{ kJ/kgmol-K})$$

$$G_P - G_R = -474\ 708 \text{ kJ/kgmol}$$

The Gibbs function of formation of all elements is zero, hence $G_R = 0$ and the difference is the Gibbs function of products. $G_P = -474\ 708$ for 2 mol H_2O.

$$G_P = 2(\bar{g}_f^\circ)_{H_2O(l)} \qquad \therefore (\bar{g}_f^\circ)_{H_2O(l)} = -237\ 354 \text{ kJ/kgmol} \qquad \blacksquare$$

Example 12.15

Propane at 25°C and 1 atm is burned with 400% theoretical air at 25°C and 1 atm. The reaction takes place adiabatically, and all the products leave at 1 atm. The temperature of the surroundings is 25°C. Compute the entropy change and the maximum work and compare these to values from an isothermal reaction.

Solution

Given: The adiabatic combustion of propane with 400% theoretical air with initial temperature and pressure given as well as the final pressure.

Find: The entropy change and the maximum work for adiabatic and isothermal combustion.

Sketch and Given Data:

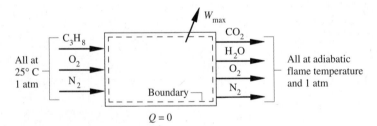

$Q = 0$ Figure 12.8

Assumptions:

1. The combustion is complete.
2. The molal ratio of nitrogen to oxygen in air is 3.76.
3. The reactants and products behave like ideal gases.
4. The changes in kinetic and potential energies may be neglected.

Analysis: The balanced combustion equation is

$$C_3H_8 + 4(5)O_2 + (4)(5)(3.76)N_2 \rightarrow 3CO_2 + 4H_2O + 15O_2 + 75.2N_2$$

If the reaction is adiabatic, the temperature of products is equal to the adiabatic flame temperature. This was calculated in Example 12.11 to be 942°K. The entropy of products and reactants may now be calculated.

$$S_R = \sum_R n_i(\bar{s}_i^\circ)_{298} = (\bar{s}_{C_3H_8}^\circ + 20\bar{s}_{O_2}^\circ + 75.2\bar{s}_{N_2}^\circ)_{298}$$

$$S_R = 1(270.09 \text{ kJ/kgmol-K}) + 20(205.142 \text{ kJ/kgmol-K})$$

$$+ 75.2(191.611 \text{ kJ/kgmol-K})$$

$$S_R = 18\ 782 \text{ kJ/kgmol fuel-K}$$

$$S_P = \sum_P n_j(\bar{s}_j^\circ)_{942} = (3\bar{s}_{CO_2}^\circ + 4\bar{s}_{H_2O}^\circ + 15\bar{s}_{O_2}^\circ + 75.2\bar{s}_{N_2}^\circ)_{942}$$

$$S_P = 3(266.04 \text{ kJ/kgmol-K}) + 4(230.22 \text{ kJ/kgmol-K})$$

$$+ 15(241.47 \text{ kJ/kgmol-K}) + 75.2(226.19 \text{ kJ/kgmol-K})$$

$$S_P = 22\ 351 \text{ kJ/kgmol fuel-K}$$

The irreversibility, I, is

$$I = T_0 (S_P - S_R)$$

$$= (298°K)(22\ 351 - 18\ 782 \text{ kJ/kgmol fuel-K})$$

$$I = 1\ 063\ 562 \text{ kJ/kgmol fuel}$$

The maximum work for this case may be found by finding the change in the availability of the reactants and products. The Gibbs function is tabulated only for reference conditions, so it cannot be used in this situation.

$$W_{max} = \mathscr{A}_1 - \mathscr{A}_2 = G_R - G_P = (H_R - T_0S_R) - (H_P - T_0S_P)$$

Since the temperature is the adiabatic flame temperature, $H_R = H_P$. The maximum work reduces to

$$W_{max} = T_0(S_P - S_R) = 1\ 063\ 562 \text{ kJ/kgmol fuel}$$

$$W_{max} = \frac{(1\ 063\ 562 \text{ kJ/kgmol fuel})}{(44.099 \text{ kg/kgmol fuel})}$$

$$W_{max} = 24\ 117 \text{ kJ/kg fuel}$$

This represents the maximum possible work an engine could perform if the products of combustion were at 942°K. For an isothermal reaction, we may find the maximum work by finding the change in the Gibbs function between the reactants and products at 25°C and 1 atm pressure.

$$W_{max} = G_R - G_P = \sum_R n_i \bar{g}_i^\circ - \sum_P n_j \bar{g}_j^\circ$$

$$G_R = 1(-23\ 502) + 0 + 0 = -23\ 502 \text{ kJ/kgmol fuel}$$

$$G_P = 3(-394\ 631) + 4(-237\ 327) + 0 + 0$$

$$= -2\ 133\ 201 \text{ kJ/kgmol fuel}$$

$$W_{max} = 47\ 840 \text{ kJ/kg fuel}$$

Comment: The isothermal maximum work represents the ideal energy available from the combustion reaction to do work. If we find the actual work that an engine does using the combustion process as the energy source (such as an internal combustion engine) and compare the two, we can determine the efficiency of the engine. Actually, the heating value of the fuel, not the change in the Gibbs function, is used in current engineering practice.

Example 12.15 assumes all the products left at 1 atm pressure. Actually, the total pressure is constant and is the sum of the partial pressures of all the constituents. The following example illustrates this more realistic situation.

Example 12.16
Propane reacts with 400% theoretical air in an isothermal reaction where the initial and final conditions are 25°C and 1 atm pressure. The surroundings temperature is also 25°C. Determine the maximum work.

Solution

Given: Propane oxidizing with 400% theoretical air in an isothermal reaction at 25°C and 1 atm pressure.

Find: The maximum work.

Sketch and Given Data:

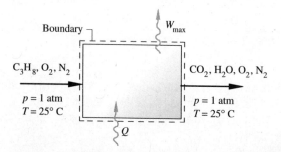

Figure 12.9

Assumptions:

1. Reactants and products behave like ideal gases.
2. The molal ratio of nitrogen to oxygen in air is 3.76.
3. The changes in kinetic and potential energies may be neglected.

Analysis: The combustion equation is

$$C_3H_8 + 20O_2 + 75.2N_2 \rightarrow 3CO_2 + 4H_2O + 15O_2 + 75.2N_2$$

The values of entropy must be calculated for the partial pressure at which they exist.

$$\bar{s}_i^\circ = \bar{s}^\circ - \bar{R} \ln \left(\frac{p_i}{p_0} \right)$$

The reaction is isothermal, so the temperature term in the entropy equation adds out. The partial pressure of the *i*th component is

$$p_i = y_i p_0$$

where y_i is the mole fraction of the *i*th component. Thus,

$$\bar{s}_i^\circ = \bar{s}^\circ + \bar{R} \ln \left(\frac{1}{y_i} \right)$$

Note that $1/y_i = n/n_i$ and $\bar{R} = 8.3143$ kJ/kgmol-K. For the reactants,

	n	$1/y_i$	$\bar{R} \ln (1/y_i)$	$\bar{s}^\circ$	$\bar{s}_i$
C_3H_8	1	96.2	37.967	270.065	308.032
O_2	20	4.81	13.059	205.142	218.201
N_2	75.2	1.279	2.046	191.611	193.657
	96.2				

For the products,

	n	$1/y_i$	$\bar{R} \ln (1/y_i)$	$\bar{s}^\circ$	$\bar{s}_i$
CO_2	3	32.4	28.918	213.795	242.713
H_2O	4	24.3	26.526	188.833	215.359
O_2	15	6.48	15.537	205.142	220.679
N_2	75.2	1.292	2.130	191.611	193.741
	97.2				

The maximum work is equal to the change in the Gibbs function or expressed in terms of enthalpy and entropy as

$$W_{max} = H_R - H_P - T_0(S_R - S_P)$$

$$H_R = \sum_i n_i (\bar{h}_f^{\circ} + \Delta h)_i$$

$$H_R = 1(-103\ 909\ \text{kJ/kgmol}) + 0 = -103\ 909\ \text{kJ/kgmol fuel}$$

$$H_P = \sum_j n_j(\bar{h}_f^{\circ} + \Delta h)_j$$

$$H_P = 3(-393\ 757\ \text{kJ/kgmol})_{CO_2}$$

$$+ 4(-241\ 971\ \text{kJ/kgmol})_{H_2O} + 0 + 0$$

$$= -2\ 149\ 155\ \text{kJ/kgmol fuel}$$

$$H_R - H_P = +2\ 045\ 246\ \text{kJ/kgmol fuel}$$

$$S_R = \sum_i n_i \bar{s}_i$$

$$S_R = 1(308.032\ \text{kJ/kgmol-K})$$

$$+ 20(218.201\ \text{kJ/kgmol-K}) + 75.2(193.657\ \text{kJ/kgmol-K})$$

$$= 19\ 235.1\ \text{kJ/kgmol fuel-K}$$

$$S_P = \sum_j n_j \bar{s}_j$$

$$S_P = 3(242.713\ \text{kJ/kgmol-K})$$

$$+ 4(215.359\ \text{kJ/kgmol-K}) + 15(220.679\ \text{kJ/kgmol-K})$$

$$+ 75.2(193.741\ \text{kJ/kgmol-K})$$

$$S_P = 19\ 469.1\ \text{kJ/kgmol fuel-K}$$

$$T_0(S_R - S_P) = (298°\text{K})(19235.1 - 19469.1\ \text{kJ/kgmol fuel-K})$$

$$= -69\ 732\ \text{kJ/kgmol fuel}$$

$$W_{max} = 2\ 045\ 246 + 69\ 732 = 2\ 114\ 978\ \text{kJ/kgmol fuel}$$

$$W_{max} = \frac{(2\ 114\ 978\ \text{kJ/kgmol fuel})}{(44.099\ \text{kg/kgmol fuel})} = 47960\ \text{kJ/kg fuel}$$

Comment: The value for the work including the pressure is essentially the same as when the pressure effect was neglected in the previous example. Thus, unless the desired accuracy of the problem demands it, we will not include the pressure effect. This sort of effect would be included in precise scientific measurements and calculations.

12.11 CHEMICAL EQUILIBRIUM AND DISSOCIATION

Thus far in considering chemical reactions we have considered complete reactions, but this is not the whole picture. In considering thermodynamic equilibrium, we concentrated on thermal and mechanical equilibrium. Likewise, in studying chemical reactions, we must study chemical equilibrium. The following discussion is by no means as extensive as would be found in a textbook on chemical engineering thermodynamics.

As a reaction proceeds, some of the products dissociate into the original reactants; when chemical equilibrium is reached, the reaction is proceeding in both directions, so there is no net change in either the reactants or the products. This reaction may be written as

$$v_a A + v_b B + \cdots \rightleftharpoons v_x X + v_y Y + \cdots \tag{12.18}$$

where v_i is the stoichiometric coefficient for the balanced reaction equation, A and B are the reactants, and X and Y are the products. The difference between n_i, the moles of i, and v_i, its stoichiometric coefficient, is that n_i refers to the moles available, whereas v_i denotes the equation requirements.

How can we determine when chemical equilibrium has been achieved? When a system is in equilibrium, it can no longer produce work. The Gibbs function can be used to determine the maximum work of a system at constant temperature and pressure. Thus, by determining the Gibbs function of the reactants and of the products, we can determine whether or not a chemical reaction can occur. When the Gibbs functions of the reactants and products are equal, no work may be done and chemical equilibrium is achieved, or

$$(dG)_{T,p} = 0 \tag{12.19}$$

for chemical equilibrium.

It is possible to show that the change in the Gibbs function at constant temperature and pressure is the criterion for chemical equilibrium without use of the maximum-work concept.

Heretofore the Gibbs function has been written for a single species; however, for systems of many species we may write that

$$G = G(T,p,n_1,n_2, \ldots , n_i) \tag{12.20}$$

where n_i represents the moles of species i at any instant. The change in the total Gibbs function is

$$dG = \left(\frac{\partial G}{\partial p}\right)_{T,n_i} dp + \left(\frac{\partial G}{\partial T}\right)_{p,n_i} dT + \sum_i \left(\frac{\partial G}{\partial n_i}\right)_{T,p,n_j} dn_i \tag{12.21}$$

The partial derivative in the last term in equation (12.21) evaluates the change in G with respect to n_i with T, p, and all other species, n_j, held constant. The partial derivatives in equation (12.21) are defined as

$$\left(\frac{\partial G}{\partial p}\right)_{T,n_i} = V \qquad \left(\frac{\partial G}{\partial T}\right)_{p,n_i} = -S \qquad \left(\frac{\partial G}{\partial n_i}\right) = \mu_i \tag{12.22}$$

In equation (12.22), μ_i is the chemical potential of the ith component. The chemical potential of species i is the driving force in changing the moles of species i.

Let us apply the change of the Gibbs function to the reaction in equation (12.18), occurring at constant temperature and pressure:

$$(dG)_{T,p} = \sum_i \mu_i \, dn_i$$

$$(dG)_{T,p} = \mu_a \, dn_a + \mu_b \, dn_b + \mu_x \, dn_x + \mu_y \, dn_y \qquad (12.23)$$

The changes in mole numbers are dependent on one another, since the products come from the reactants. Assume a forward reaction—the left-hand side is being depleted—and let us further assume an infinitesimal change, so a proportionality constant may be used to describe the change in the mole numbers. Thus,

$$dn_a = -kv_a \qquad dn_b = -kv_b \qquad dn_x = +kv_x \qquad dn_y = +kv_y \qquad (12.24)$$

The negative and positive values of k reflect the depletion and gain of species. Equation (12.23) becomes

$$(dG)_{T,p} = (-\mu_a v_a - \mu_b v_b + \mu_x v_x + \mu_y v_y)k \qquad (12.25)$$

Since we are interested in chemical equilibrium, assume that the small change occurs at the point of equilibrium where $(dG)_{T,p} = 0$. Thus,

$$\mu_a v_a + \mu_b v_b = \mu_x v_x + \mu_y v_y \qquad (12.26)$$

for chemical equilibrium.

The concept of work may be related to equation (12.23) by recalling equation (3.21), where work is defined as an intensive property acting on the change of its related extensive property.

Consider a system containing methane and oxygen, existing at the same temperature and pressure as the surroundings. The system is in mechanical and thermal equilibrium. Is it in chemical equilibrium? No, because if we computed the Gibbs function for the reaction, we would find that $G_R > G_P$, so a reaction could occur. Nothing will happen, however, until a spark is introduced into the mixture. Then a reaction will occur until $G_R = G_P$. Thus, we find that the Gibbs function has two purposes:

1. It tells us whether or not a reaction can occur. If a system is in chemical equilibrium, $(dG)_{T,p} = 0$, no reaction can occur.
2. It is the thermodynamic potential that causes the constant-temperature and constant-pressure reaction to occur. It does not tell us how fast the reaction will occur.

Consider the reaction

$$n_a A + n_b B \rightleftharpoons n_x X + n_y Y \qquad (12.27)$$

and let the reactants A and B flow slowly into a box so their relative amounts remain constant. A reaction takes place, and products X and Y are formed. We will prescribe that the amounts of the products formed are proportional to the amounts of A and B, and that the ratio of the moles of the products to the moles of the reactants is a constant. For this to occur, products X and Y must be removed from the box in such a

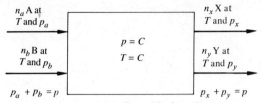

Figure 12.10 Van't Hoff equilibrium box.

manner that the mass entering the box equals the mass leaving. The reactants and products leave at the same temperature and pressure, and the reaction also occurs at this temperature. This is called a *Van't Hoff equilibrium box.* Figure 12.10 illustrates this process.

Since the Gibbs function is used in determining chemical equilibrium, let us find the Gibbs function for each of the substances in the box.

$$dG = V\,dp - S\,dT$$

For $T = C$,

$$dG = V\,dp$$

For an ideal gas, $V = n\overline{R}T/p$

$$dG = n\overline{R}T\,\frac{dp}{p}$$

Integrating,

$$(G_2 - G_1) = n\overline{R}T \ln\left(\frac{p_2}{p_1}\right) \tag{12.28}$$

We then apply equation (12.28) to the conditions in the box. We let $p_2 = p_i$, the partial pressure of the given component in the box. Thus, for component A

$$(G_{A_2} - G_{A_1})_T = n_a \overline{R}T \ln\left(\frac{p_a}{p}\right) \tag{12.29a}$$

where p is the total pressure.

Similarly, expressions may be written for the other substances. For substance B,

$$(G_{B_2} - G_{B_1})_T = n_b \overline{R}T \ln\left(\frac{p_b}{p}\right) \tag{12.29b}$$

For substance X,

$$(G_{X_2} - G_{X_1})_T = n_x \overline{R}T \ln\left(\frac{p}{p_x}\right) \tag{12.29c}$$

For substance Y,

$$(G_{Y_2} - G_{Y_1})_T = n_y \overline{R}T \ln\left(\frac{p}{p_y}\right) \tag{12.29d}$$

Adding equation (12.29a) to (12.29d) yields

$$(G_{A_2} + G_{B_2})_T - (G_{A_1} + G_{B_1})_T + (G_{X_2} + G_{Y_2})_T - (G_{X_1} + G_{Y_1})_T$$

$$= n_a \overline{R} T \ln \left(\frac{p_a}{p}\right) + n_b \overline{R} T \ln \left(\frac{p_b}{p}\right) + n_x \overline{R} T \ln \left(\frac{p}{p_x}\right) \qquad (12.30)$$

$$+ n_y \overline{R} T \ln \left(\frac{p}{p_y}\right)$$

Since equilibrium conditions exist within the box, the sum of the Gibbs function of the reactants must equal the sum of the Gibbs functions of the products in the box. Hence,

$$(G_{A_2} + G_{B_2})_T - (G_{X_1} + G_{Y_1})_T = 0 \qquad (12.31)$$

Combining equations (12.30) and (12.31),

$$(G_{X_2} + G_{Y_2})_T - (G_{A_1} + G_{B_1})_T$$

$$= \overline{R} T \ln \left(\frac{p_a}{p}\right)^{n_a} + \overline{R} T \ln \left(\frac{p_b}{p}\right)^{n_b} + \overline{R} T \ln \left(\frac{p}{p_x}\right)^{n_x} + \overline{R} T \ln \left(\frac{p}{p_y}\right)^{n_y}$$

We let the total pressure, p, equal 1 atm and express the partial pressures in atmospheres. Simplifications result if the pressures are so expressed.

$$(G_P - G_R)_T = \overline{R} T \ln (p_a)^{n_a} + \overline{R} T \ln (p_b)^{n_b} - \overline{R} T \ln (p_x)^{n_x} - \overline{R} T \ln (p_y)^{n_y}$$

$$(G_P - G_R)_T = \overline{R} T \ln \left(\frac{p_a^{n_a} p_b^{n_b}}{p_x^{n_x} p_y^{n_y}}\right) = - \overline{R} T \ln \left(\frac{p_x^{n_x} p_y^{n_y}}{p_a^{n_a} p_b^{n_b}}\right) \qquad (12.32)$$

The equilibrium constant, K_p, is defined as

$$K_p \equiv \frac{p_x^{n_x} p_y^{n_y}}{p_a^{n_a} p_b^{n_b}} \qquad (12.33)$$

Thus, equation (12.32) becomes

$$(G_p - G_R)_T + \overline{R} T \ln (K_p) = 0$$

K_p is a constant only for a given temperature for an ideal gas. Values for the natural log of the equilibrium constant K_p are given in Table C.4.

When the temperature of a gas is increased, there is a tendency for the gas to dissociate. So at any temperature there is an equilibrium mixture of the gas and its dissociated products. The dissociation process is endothermic and, as such, tends to reduce the total energy of a system to a minimum value for a given temperature. The effect of dissociation reduces the adiabatic flame temperature, as some of the chemical energy is used in the dissociation process instead of raising the thermal energy. The following examples illustrate the effect of dissociation.

Example 12.17
Determine the percentage of dissociation of carbon dioxide into carbon monoxide and oxygen at 3800°K and 1 atm pressure.

Solution

Given: The dissociation reaction of carbon dioxide into carbon monoxide and oxygen at a given temperature and pressure.

Find: The percentage of dissociation.

Assumption: Reactants and products behave like ideal gases.

Analysis: From Table C.4 the $\ln (K_p) = 1.170$ for the reaction $CO + \frac{1}{2}O_2 \rightarrow CO_2$. Let x denote the dissociation of 1 mol of carbon dioxide. The reaction equation is

$$x CO + \frac{x}{2} O_2 \rightleftharpoons (1 - x) CO_2$$

The total number of moles present at equilibrium is $(2 + x)/2$. The partial pressure for each component, p_i, may be expressed in terms of the mole fraction and the total pressure, p.

$$p_{CO} = \frac{x}{(2 + x)/2} p = \frac{2x}{2 + x} p$$

$$p_{O_2} = \frac{x}{2 + x} p$$

$$p_{CO_2} = \frac{2(1 - x)}{2 + x} p$$

Since the total pressure is 1 atm, we may solve equation (12.33) for the equilibrium constant, K_p.

$$K_p = \frac{p_{CO_2}^1}{(p_{CO}^1)(p_{O_2}^{0.5})} = \frac{(2 + x)^{1/2}(1 - x)}{x^{3/2}}$$

Solve this equation for x by a trial-and-error process, knowing that $\ln (K_p) = 1.170$.

$x = 0.3$	$K_p = 6.46$	$\ln (K_p) = 1.865$
$x = 0.4$	$K_p = 3.673$	$\ln (K_p) = 1.301$
$x = 0.425$	$K_p = 3.231$	$\ln (K_p) = 1.173$

Thus 42.5% of the carbon dioxide has dissociated at 3800°K.

Comment: The dissociation reaction uses some of the chemical energy released by combustion and thus reduces the adiabatic flame temperature. In determining the adiabatic flame temperature, we must find the amount of dissociation before calculating the enthalpy.

Example 12.18

Hydrogen is burned steadily with 100% theoretical air at 1 atm pressure and standard temperature. Determine the adiabatic flame temperature and the maximum temperature, taking the dissociation of H_2O into H_2 and O_2 into account.

Solution

Given: The steady-state oxidation of hydrogen with air for adiabatic conditions with known initial conditions.

Find: The maximum temperature possible with and without dissociation occurring.

Assumptions:

1. The reactants and products behave like ideal gases.
2. The work and heat are zero.
3. The changes in kinetic and potential energies may be neglected.

Analysis: The combustion equation for the oxidation of hydrogen is

$$H_2 + 0.5O_2 + 1.88N_2 \rightarrow H_2O + 1.88N_2$$

The adiabatic flame temperature occurs when $H_R = H_P$. In this case, the reactant enthalpy is zero, as the fuel is hydrogen, an element. Following the techniques described in Example 12.11, determine the adiabatic flame temperature to be 2525°K.

The combustion equation accounting for dissociation of water in the products of combustion is

$$H_2 + 0.5O_2 + 1.88N_2 \rightarrow (1 - x)H_2O + xH_2 + \frac{x}{2}O_2 + 1.88N_2$$

where x is the fraction of water that dissociates.

1. First-law analysis:

$$H_R = H_P$$
$$H_R = 0$$

2. Guess a temperature: look up $\ln (K_p)$.
3. Guess a value for x: calculate K_p—the correct value of x will have the same value of $\ln (K_p)$ as in step 2.

4. Using the value of x found in step 3, calculate H_p for first-law analysis.
5. Repeat steps 2–4 until $H_P = H_R$.

Guess $T = 2200$ K:

$$\ln (K_p) = -6.774$$

$$K_p = \frac{(p_{O_2})^{1/2}(p_{H_2})^1}{(p_{H_2O})^1}$$

Moles of products at equilibrium $= 2.88 + 0.5x$. Find the partial pressures.

$$p_{H_2O} = \left[\frac{(1-x)}{(2.88 + 0.5x)}\right](p)$$

$$p_{H_2} = \left[\frac{x}{(2.88 + 0.5x)}\right](p)$$

$$p_{O_2} = \left[\frac{0.5x}{(2.88 + 0.5x)}\right](p)$$

$$p = 1 \text{ atm}$$

$$K_p = \frac{0.7071x^{3/2}}{(1-x)(2.88 + 0.5x)^{1/2}}$$

x	K_p	$\ln (K_p)$
0.1	0.0145	−4.23
0.05	0.00488	−5.32
0.02	0.00120	−6.72

Assume $x = 0.02$ for the initial calculation; the products side of the combustion equation is

$$0.98H_2O + 0.02H_2 + 0.01O_2 + 1.88N_2$$

Guess $T = 2400°$K:

$$\ln (K_p) = -5.625$$

A value of $x = 0.04$ balances $\ln (K_p)$ and yields $H_P = -5302$ kJ/kgmol fuel. Plot $(H_P - H_R)$ versus T; the line intersects $(H_P - H_R) = 0$ at $T = 2433°$K. The value of $x = 0.045$ essentially balances the first-law equation.

Thus, 4.5% of the water dissociated, lowering the maximum temperature by 92°K.

Comment: In hydrocarbon combustion the dissociation of the hydrocarbon should be considered, as well as the reaction of nitrogen and oxygen to form nitric oxide and the dissociation of other products. All of these reactions are endothermic, tending to reduce the products' temperature. ▬

In Example 12.18 only one dissociation reaction is considered. A number of other possible reactions could be also be considered. One such reaction is the dissociation of water into hydrogen and hydroxyl. As additional reactions are considered, additional unknowns are added, and the problem solution becomes more complicated. Solving a problem with more than one dissociation equation without a computer is a daunting proposition. The TK Solver model COMBUST.TK solves combustion process problems involving up to three dissociation reactions: water into hydrogen and oxygen, water into hydrogen and hydroxyl, and carbon dioxide into carbon monoxide and oxygen. COMBUST.TK contains data on ideal-gas enthalpies from Table C.2 and equilibrium constants from Table C.4.

Example 12.19

Repeat Example 12.18 using COMBUST.TK and assuming the reaction products are only oxygen, nitrogen, water, hydrogen, and hydroxyl.

Solution

Given: The steady-state oxidation of hydrogen with air for adiabatic conditions with known initial conditions.

Find: The maximum temperature possible with dissociation of water taken into account.

Assumptions:

1. The reactants and products behave like ideal gases.
2. The work and heat are zero.
3. The changes in kinetic and potential energies may be neglected.
4. Two dissociation reactions will be considered,

$$H_2O \rightarrow H_2 + 0.5O_2 \quad \text{and} \quad H_2O \rightarrow 0.5H_2 + OH$$

Analysis: Load COMBUST.TK into TK Solver. In the Variable Sheet, enter the enthalpy of formation of the fuel, the temperature of the reactants, the pressure of the combustion process, the percentage of theoretical oxygen, the theoretical moles of oxygen required per mole of fuel, and the theoretical moles of carbon dioxide and water produced per mole of fuel. Note that the last three values come from the balanced reaction equation without dissociation. The solved Variable Sheet should look like this:

```
================================ VARIABLE SHEET ================================
St Input — Name — Output — Unit — Comment ──────────────────────────

                                           ***COMBUSTION ANALYSIS***

   0      hfuel                kJ/kgmol  Fuel Enthalpy of Formation
 298      Tr            degK              Temperature of Reactants
          Tp          2403.56  degK      Temperature of Products
   1      mThH2O               mol/mol   Theoretical Moles H2O/Moles Fuel
   0      mThCO2               mol/mol   Theoretical Moles CO2/Moles Fuel
  .5      mThO2                mol/mol   Theoretical Moles O2/Moles Fuel
 100      %ThO2                %         Percentage of Theoretical Oxygen
          Hr            0      kJ/kgmol  Enthalpy of Reactants
          Hp            0      kJ/kgmol  Enthalpy of Products

          x            .0170274          Unknown in Balanced Equation
          y            .0225089          Unknown in Balanced Equation
          z             0                Unknown in Balanced Equation

                                         Equilibrium Constants
          KpH2O2      .00367477            For H2O <==> H2 + 0.5 O2
          KpH2OH      .00297794            For H2O <==> 0.5 H2 + OH
          KpCOO2      .0213755             For CO2 <==> CO + 0.5 O2

                                         Pressures
   1      P                     atm        Combustion Process
          pH2O        .324396   atm        Partial - H2O
          pCO2         0        atm        Partial - CO2
          pO2         .0058548  atm        Partial - O2
          pH2         .0155794  atm        Partial - H2
          pOH         .00773957 atm        Partial - OH
          pCO          0        atm        Partial - CO

                                         Moles of Constituents/Mole of Fuel
          mH2O        .943436   mol/mol    Water
          mCO2         0        mol/mol    Carbon Dioxide
          mN2         1.88      mol/mol    Nitrogen
          mO2         .0170274  mol/mol    Oxygen
          mH2         .0453093  mol/mol    Hydrogen
          mOH         .0225089  mol/mol    Hydroxyl
          mCO          0        mol/mol    Carbon Monoxide
          Mp          2.90828   mol/mol    Total Products
```

Comment: Considering the dissociation of water into hydrogen and hydroxyl as well as hydrogen and oxygen results in an adiabatic flame temperature of 2403.56°K, about 30°K lower than that calculated in Example 12.18. The additional dissociation of water results in further lowering of the products' temperature.

12.12 STEAM GENERATOR EFFICIENCY

The first-law efficiency of any device is output divided by input. This is no exception for steam generators. The only question to be answered is what we define as input and output.

$$\eta_{\text{stm gen}} = \frac{\text{output}}{\text{input}} = \frac{\text{energy to water/steam across steam generator}}{\text{energy supplied by fuel}} \tag{12.34}$$

The energy supplied to the water is the change of enthalpy of the water from the feedwater entering to the superheated steam leaving. All flows must be considered if some of the steam is extracted before entering the superheater. The energy supplied by the fuel is the fuel flow rate times the fuel's higher heating value.

A power plant's efficiency is also denoted by its specific fuel consumption and heat rate. The specific fuel consumption, sfc, is

$$\text{sfc} = \frac{\dot{m}_{\text{fuel}}}{\dot{W}_{\text{net}}} = \frac{\text{kg}}{\text{kW-s}} \left[\frac{\text{lbm}}{\text{hp-hr}} \right] \tag{12.35}$$

The difficulty with using equation (12.35) is that it does not account for the fuel's heating value. In comparing plant and/or steam generator designs, we should know the heating value of the fuel. The heat rate eliminates this problem, and modern convention recommends its use over specific fuel consumption; that is,

$$\text{Heat rate} = \frac{(\dot{m}_{\text{fuel}})(h_{RP})}{\dot{W}_{\text{net}}} \frac{\text{kW}}{\text{kW}} \left[\frac{\text{Btu}}{\text{hp-hr}} \right] \tag{12.36}$$

12.13 FUEL CELLS

A fuel cell transforms chemical energy into electrical energy through a series of catalyst-aided oxygen-reduction reactions. Unlike many energy conversion systems such as the steam power plant and the internal-combustion engine, the fuel cell generates electricity in a continuous and direct process. In this manner, the excessive losses that occur in multistep energy conversion systems may be avoided using a single-step process. Thus, fuel cells tend to have higher theoretical energy conversion efficiencies.

The fuel cell is composed of an electrolyte sandwiched between electrodes known as the anode and cathode, as shown in Figure 12.11. Unlike the conventional battery, however, the fuel cell consumes externally supplied fuels. The fuel (such as hydrogen) and the oxidizer (normally oxygen in the form of air) are supplied to the anodic and cathodic sides of the cell, respectively. The electrochemical reaction between these two results in the transfer to electrons and the production of a voltage between the two electrodes. When a load is connected in series to these electrodes, a current, calculable by Ohm's law, will result. The theoretical voltages that arise from these fuel cell reactions normally range between 1.0 and 1.3 V dc at 1–2 kW/m² of electrode.

As an example, let us analyze the most common fuel cell, the hydrogen-oxygen

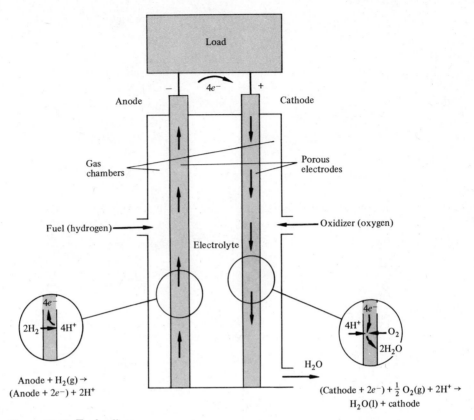

Figure 12.11 Fuel cell.

fuel cell. Fuel as hydrogen enters the fuel cell and loses an electron, giving the anodic plate a negative charge. The half-cell reaction for the anode is

$$\text{Anode} + H_2(g) \rightarrow (\text{anode} + 2e^-) + 2H^+$$

At the cathode, the oxygen molecules supplied by the air pick up electrons, giving that plate a positive charge. The half-cell reaction for the cathode is

$$(\text{Cathode} + 2e^-) + \tfrac{1}{2}O_2(g) + 2H^+ \rightarrow H_2O(l) + \text{cathode}$$

When a load is connected in series with the electrodes, a current develops as a function of the induced voltage. The excess electrons in the anode travel through the load to the cathode providing power. The H^+ ions formed at the anode according to the half-cell reaction migrate through the electrolyte solution to the cathode. The oxygen then combines with the H^+ ions to form water.

Thermal efficiency is typically defined to measure the performance of a heat engine. Because the performance of a fuel cell cannot be defined by heat and work transfers, a "fuel cell efficiency" can be defined as

$$\eta_{fc} = \frac{\text{maximum work}}{\text{change of overall enthalpy}} = \frac{\Delta G}{\Delta H} \tag{12.37}$$

where ΔG equals the change in the Gibbs function and ΔH the change in enthalpy. It is important not to confuse this with thermal efficiency for heat engines, as the two cannot be directly compared. The Gibbs function is defined as

$$G = H - TS$$

Therefore, at standard temperature and pressure (25°C, 1 atm),

$$\Delta G^\circ = G_p - G_R = (H_p - H_R) - T_0(S_p - S_R)$$

$$\Delta G^\circ = \sum_p n_j(\bar{h})_j - \sum_R n_i(\bar{h})_i - T_0 \left[\sum_p n_j \bar{s}_j^\circ - \sum_R n_i \bar{s}_i^\circ \right]$$

The values of h and s for substances are found in Tables C.1 and C.2. To simplify calculations, the hydrogen-oxygen fuel cell reaction may be written as

$$H_2(g) + \tfrac{1}{2}O_2(g) \rightarrow H_2O(l)$$

Calculate the change in the Gibbs function at standard temperature and pressure, substituting the tabulated values. The Gibbs function of an element is zero at standard conditions; thus, G_R is zero and

$$\Delta G^\circ = G_p = (\bar{g}_f^\circ)_{H_2O(l)} = -237\ 327\ \text{kJ/kgmol}$$

The change in enthalpy for the overall fuel cell reaction is

$$\Delta H = H_p = (h_f^\circ)_{H_2O(l)} = -286\ 010\ \text{kJ/kgmol}$$

The ideal hydrogen-oxygen fuel cell efficiency is

$$\eta_{fc} = 0.83 \quad \text{or } 83\%$$

The fuel cell voltage is found from

$$\Delta G = -n\mathscr{F}V \tag{12.38}$$

where ΔG is the change in Gibbs function per mole of fuel, n is the number of electrons transferred per molecule of fuel, $\mathscr{F}$ is Faraday's constant, 96 500 kJ/V—kgmol, and V is the fuel cell voltage.

TABLE 12.2   THEORETICAL FUEL CELL PERFORMANCE

Fuel cell reaction	$-\Delta H$ (kJ/kgmol)	$-\Delta G$ (kJ/kgmol)	η_{fc}	V(V)
$H_2(g) + \frac{1}{2}O_2(g) \rightarrow H_2O(l)$				
At 298°K	286 010	237 327	0.830	1.229
At 1000°K	—	—	—	—
$H_2(g) + \frac{1}{2}O_2(g) \rightarrow H_2O(g)$				
At 298°K	241 971	228 729	0.945	1.184
At 1000°K	247 856	175 940	0.777	0.998
$CO(g) + \frac{1}{2}O_2(g) \rightarrow CO_2(g)$				
At 298°K	283 161	257 405	0.909	1.333
At 1000°K	282 796	195 797	0.692	1.014
C (graphite) $+ O_2(g) \rightarrow CO_2(g)$				
At 298°K	393 757	394 631	1.002	1.002
At 1000°K	394 828	396 042	1.003	1.026
$C_3H_8(g) + 5O_2(g) \rightarrow 3CO_2(g) + 4H_2O(g)$				
At 298°K	2 044 884	2 075 023	1.015	1.075
At 1000°K	2 046 558	2 149 953	1.051	1.124

Hence

$$V = \frac{\Delta G}{n\mathcal{F}} = \frac{+237\ 327}{(2)(96\ 500)} = 1.229 \text{ V}$$

A number of other fuels may be used in fuel cells, some of which are listed in Table 12.2. Each of these fuels reacts and may be analyzed in a similar fashion. The fuel cell reaction for carbon monoxide fuel is

$$CO(g) + \frac{1}{2}O_2(g) \rightarrow CO_2(g)$$

To calculate

$$\Delta G^\circ = (H_P - H_R) - T_0(S_P - S_R)$$

H_R does not equal zero because $(\bar{h}_f^\circ)_{CO(g)}$ must be included in it. Similarly, G_R does not equal zero. Substituting, we find that

$$\Delta G^\circ = G_P - G_R$$

$$\Delta G^\circ = -393\ 757 - (-110\ 596)$$

$$- 298[213.795 - 197.653 - \tfrac{1}{2}(205.142)] = -257\ 405 \text{ kJ/kgmol}$$

The change of enthalpy for the fuel cell reaction is

$$\Delta H = H_P - H_R$$

$$\Delta H = -393\ 757 - (-110\ 596) = -283\ 161\ \text{kJ/kgmol}$$

The fuel cell efficiency is

$$\eta_{fc} = \frac{\Delta G}{\Delta H} = \frac{-257\ 405}{-283\ 161} = 0.909 \quad \text{or } 90.9\%$$

and the fuel cell voltage is

$$V = \Delta G/n\mathscr{F} = \frac{+257\ 405}{(2)(96\ 500)} = 1.333\ \text{V}$$

The fuel cell efficiency and voltage may be calculated at other than standard conditions. Consider the carbon monoxide fuel cell at 1000°K and 1 atm pressure.

$$\Delta G = \left\{ \sum_P n_j[\bar{h}_f^\circ + (\bar{h} - \bar{h}_{298})]_j - \sum_R n_i[\bar{h}_f^\circ + (\bar{h} - \bar{h}_{298})] \right\}$$

$$- T\left[\sum_P n_j(\bar{s}^\circ)_j - \sum_R n_j(\bar{s}^\circ)_i \right]$$

$$\Delta G = [(-393\ 757 + 33\ 405) - (-110\ 596 + 21\ 686\ \text{kJ/kgmol})]$$

$$- 0.5(22\ 707\ \text{kJ/kgmol}) - 1000[269.325 - 234.531$$

$$- 0.5(243.585\ \text{kJ/kgmol})]$$

$$\Delta G = -195\ 797\ \text{kJ/kgmol}$$

The change in the fuel cell enthalpy is $\Delta H = -282\ 796$ kJ/kgmol. The fuel cell efficiency is

$$\eta_{fc} = \frac{\Delta G}{\Delta H} = \frac{-195\ 797}{-282\ 796} = 0.692 \quad \text{or } 69.2\%$$

and the fuel cell voltage is

$$V = \frac{\Delta G}{n\mathscr{F}} = \frac{+195\ 797}{(2)(96\ 500)} = 1.014\ \text{V}$$

As may be seen in Table 12.2, some ideal fuel cell efficiencies may exceed unity. These cases correspond to processes in which the fuel cell absorbs heat from the surroundings.

In actual fuel cell operation, losses occurring during electron transfer at the electrode, during mass transport through concentration gradients, and during electron transport through the electrolyte result in energy losses that may exceed 25% of the ideal energy generated.

CONCEPT QUESTIONS

1. Does the presence of nitrogen in air affect the combustion process? How?
2. What effect does the water vapor in air have on the combustion process?
3. What is the dew point temperature? How is this determined?
4. Describe the air/fuel ratio.
5. Is the air/fuel ratio on the mass and mole basis the same?
6. Can water be used as a fuel in a combustion process?
7. What are differences between systems where reactions occur and those where none occurs?
8. Will mixing fuel with oxygen be sufficient to start a combustion process?
9. What is "excess air"?
10. Is there a difference between complete combustion and theoretical combustion?
11. Describe the use of an Orsat analyzer. What is meant by "dry basis"?
12. What is the heating value of a fuel?
13. What causes the numerical differences between higher and lower heating values of fuels?
14. What is the difference between the enthalpy of combustion and the enthalpy of formation?
15. What is an endothermic reaction? An exothermic reaction?
16. What is the adiabatic flame temperature?
17. How are the entropy values of gases at other than 1 atm pressure determined?
18. What is the value in determining the Gibbs function of formation of a compound?
19. What characterizes chemical equilibrium?
20. Will a fuel burn more completely at $2000°K$ or at $3000°K$? Why?
21. What is the energy conversion process in a fuel cell?
22. Why is the ideal efficiency of some fuel cells greater than one?

PROBLEMS (SI)

12.1 A fuel mixture of 50% C_7H_{16} and 50% C_8H_{18} is oxidized with 20% excess air. Determine (a) the mass of air required for 50 kg of fuel; (b) the volumetric analysis of products of combustion.

12.2 A gas turbine power plant receives an unknown type of hydrocarbon fuel. Some of the fuel is burned with air, yielding the following Orsat analysis of the products of combustion: 10.5% CO_2, 5.3% O_2, and 84.2% N_2. Determine (a) the percentage, by mass, of carbon and hydrogen in the fuel; (b) the percentage of theoretical air.

12.3 What mass of liquid oxygen is required to completely burn 1000 kg of liquid butane, C_4H_{10}, on a rocket ship?

12.4 One kg of sugar, $C_{12}H_{22}O_{11}$, is completely oxidized with theoretical air. Determine (a) the volumetric product analysis; (b) the mass of air required at standard temperature and pressure.

12.5   With 110% theoretical air, 1 kgmol of methane is completely oxidized. The products of combustion are cooled and completely dried at atmospheric pressure. Determine (a) the partial pressure of oxygen in the products; (b) the mass in kg of water removed.

12.6   An unknown hydrocarbon had the following Orsat analysis when burned with air: 11.94% CO_2, 2.26% O_2, 0.41% CO, and 85.39% N_2. Determine (a) the air/fuel ratio on a mass basis; (b) the percentage of carbon and hydrogen in the fuel on a mass basis.

12.7   Write the combustion equation for gaseous dodecane and theoretical air. Determine (a) the fuel/air ratio on the mass basis; (b) the fuel/air ratio on the mole basis; (c) the mass of fuel/mass of water formed; (d) the molecular weight of the reactants; (e) the molecular weight of the products; (f) the ratio of moles of reactants to moles of products.

12.8   A coal sample has the following ultimate analysis on a dry basis: 81% C, 2.5% H_2, 0.6% S, 3.0% O_2, 1.0% N_2, and 11.9% ash. Determine the reaction equation for 100% theoretical air.

12.9   The ultimate analysis of a coal sample is 77% C, 3.5% H_2, 1.8% N_2, 4.5% O_2, 0.7% S, 6.5% ash, and 6.0% H_2O. Determine the reaction equation for 120% theoretical air.

12.10 How many m^3 of air at 20°C will be required to completely oxidize 1 m^3 of gas with the following volumetric analysis? The pressure is atmospheric. 46% CH_4, 38% CO, 5% O_2, and 11% N_2.

12.11 Given the following ultimate analysis on the dry basis of a coal sample, 80% C, 4.5% H, 4.5% O_2, 1.0% N_2, 1.0% S, and 9.0% ash, and that the heats of combustion of carbon, hydrogen, and sulfur are 33 700, 141 875, and 9300 kJ/kg, respectively, determine the higher heating value of the coal in kJ/kg.

12.12 A gaseous mixture containing 60% CH_4, 30% C_2H_6, and 10% CO is burned with 12% excess air. The combustion air is supplied to the furnace by a forced draft fan that increases the pressure by 76 mm of water. Determine for complete combustion (a) the molal air/fuel ratio; (b) the fan power when 0.8 m^3/s of fuel is burned at standard temperature and pressure.

12.13 An adiabatic container has a mixture of oxygen and carbon monoxide in it. Determine whether there is sufficient oxygen for complete combustion if the mixture is 33% O_2 and 67% CO on (a) a mole basis; (b) a mass basis.

12.14 A sample of coal has an ultimate analysis of 78% C, 3% S, 15% ash, and 2% H_2 and is completely oxidized with 120% theoretical air. Determine the amount of sulfur dioxide produced in kg/kg coal burned.

12.15 A coal sample has an ultimate analysis of 80.0% C, 4.0% O_2, 4.5% H_2, 1.7% N_2, 1.5% S, and 8.3% ash. Determine the mass air/fuel ratio for complete combustion with 100% theoretical air.

12.16 Garbage, or municipal waste, has an ultimate analysis of 80.5% C, 5.0% H_2, 1.6% S, 1.5% N_2, and 5.5% O_2, with the balance ash. Determine the balanced reaction equation and the mass air/fuel ratio.

12.17 Determine the heating value at 25°C and 1 atm of the municipal waste described in Problem 12.17.

12.18 Hydrazine is burned with 30% excess air. Determine the mass air/fuel ratio and the water vapor condensed if the products are cooled to 20°C in units of mol/mol fuel.

12.19 A fuel with a mass analysis of 83% C, 12% H_2, and 5% O_2 is burned with 100% theoretical air. Determine the molal analysis of the products with and without water vapor considered.

12.20 Ethylene burns with 30% excess air at 30°C and 50% relative humidity. What is the dew point of the products?

12.21 A residual fuel with a mass analysis of 90% C, 8% H_2, and 2% S is burned with air at 40°C and 50% relative humidity. In addition steam atomization is used, requiring 0.05 kg steam/kg fuel. Determine the dew point of the products.

12.22 The volumetric analysis of the products of combustion on a dry basis for the oxidation of octane in air is 9.19% CO_2, 0.24% CO, 7.48% O_2, and 83.09% N_2. Determine the percentage of theoretical air used in the combustion process.

12.23 Propane burns with 90% theoretical air to form only carbon dioxide, carbon monoxide, water and nitrogen. Determine the balanced reaction equation.

12.24 The volumetric analysis of a natural gas is 40% C_3H_8, 40% C_2H_6, and 30% CH_4. It burns with air yielding the following dry molal product analysis: 11.4% CO_2, 1.2% CO, 1.7% O_2, and 85.7% N_2. Determine the mass air/fuel ratio.

12.25 A fuel C_xH_y burns with air. The products have the following molal analysis on a dry basis: 11% CO_2, 0.5% CO, 2% CH_4, 1.5% H_2, 6% O_2, and 79% N_2. Determine (a) the percentage of excess air; (b) the fuel composition.

12.26 The volumetric analysis of the dry products of combustion of a hydrocarbon fuel on a dry basis is 11% CO_2, 1% CO, 3% O_2, and 85% N_2. Determine (a) the air/fuel ratio; (b) the percentage on a mass basis of carbon and hydrogen in the fuel.

12.27 Octane is burned with 150% theoretical air. The air is at 25°C and has 50% relative humidity. Determine (a) the balanced reaction equation; (b) the dew point of the products; (c) the dew point of the products if dry air were used.

12.28 A diesel engine uses $C_{12}H_{26}$ in a ratio of 1 to 30 (fuel to air by mass). The engine exhausts at 327°C into a heating system where the room temperature is 25°C. Determine the percentage of fuel available for heating. Assume complete combustion of the fuel with air and fuel entering at 25°C.

12.29 An internal-combustion engine uses liquid octane for fuel and 150% theoretical air at 25°C and 100 kPa. The products of combustion leave the engine at 260°C. The heat loss is equal to 20% of the work. Determine (a) the work/kgmol; (b) the dew point; (c) the kg/s of fuel required to produce 400 kW.

12.30 The coal in Problem 12.15 is used in a power plant with an overall efficiency of 40%. The net power produced is 1000 MW. Determine the fuel's heating value and the tons of coal required per day.

12.31 How many tons of carbon dioxide and sulfur dioxide are produced daily by the power plant in Problem 12.30?

12.32 Five kg/s of ethane gas enters a furnace at 25°C and 1 atm pressure and burns with 100% theoretical air at the same temperature and pressure. The products leave at 227°C. Determine the rate of heat transfer to the surroundings.

12.33 Five m^3/s of methane gas enters a furnace at 25°C and 1 atm and burns with 110% theoretical air at the same temperature and pressure. The products leave at 500°K. Determine (a) the air's volumetric flow rate; (b) the heat transfer to the surroundings.

12.34 Propane gas at 25°C and 1 atm pressure enters a furnace and burns with 120% theoretical air at the same temperature and pressure. On a mole basis, 94% of the carbon in the fuel is completely oxidized to carbon dioxide, the remainder to carbon monoxide.

The heat transfer from the furnace is measured to be 1400 MJ/kgmol fuel. Determine the temperature of the products leaving the furnace.

12.35 A tank contains a gaseous mixture of oxygen and ethene at 25°C and 1 atm. The molal analysis of the reactants is 25% ethene and 75% oxygen. The combustion is complete, and the products are cooled to 600°K. Determine the heat transfer from the tank per kgmol of ethene.

12.36 A power plant operates with an overall efficiency of 40%. The plant uses methane as the fuel and air, both at 25°C and 1 atm. The products of combustion of the steam generator leave at 400°K. Determine the mass flow rate of methane per 1000 kW of power produced.

12.37 A tank contains 1 kgmol of butane and 200% theoretical air at 25°C and 1 atm. Combustion occurs, and heat is transferred from the tank until the products' temperature is 800°K. Determine (a) the heat transfer from the tank; (b) the final pressure of the products in the tank.

12.38 Determine the higher and lower heating values of gaseous propane at 25°C and 1 atm in kJ/kg.

12.39 Determine the higher heating value of hydrogen at 25°C and 1 atm in kJ/kg.

12.40 Determine the adiabatic flame temperature of butane with 100% theoretical air if all reactants are at 25°C and 1 atm.

12.41 Determine the adiabatic flame temperature of butane with 100% oxygen if all reactants are at 25°C and 1 atm.

12.42 Carbon monoxide at 25°C and 1 atm enters a combustion chamber and burns with 100% theoretical air. Determine the adiabatic flame temperature.

12.43 Liquid octane at 25°C and 1 atm steadily enters an adiabatic combustion chamber and burns with air at 500°K and 1 atm. The products leave at 1300°K. Determine the percentage of excess air supplied.

12.44 A mixture of methane and 150% theoretical air is contained in an adiabatic tank at 25°C and 1 atm pressure. Combustion occurs. Determine the temperature and pressure of the products.

12.45 Equal moles of hydrogen and carbon monoxide are mixed with theoretical air in an insulated rigid vessel at standard temperature and pressure. The mixture is ignited by a spark. Complete oxidation occurs. Determine (a) the maximum temperature; (b) the maximum pressure.

12.46 A steam turboelectric generator plant has the following test data:

Orsat analysis of stack gas	12.1% CO_2, 3.71% O_2
Stack gas temperature	232°C
Feedwater temperature	288°C
Steam leaves at	16 MPa, 570°C
Steam flow rate	453 kg/s
Fuel oil flow rate	23.3 kg/s
Fuel	$C_{12}H_{26}$
Air and fuel enter at	25°C

Determine (a) the excess air in kg/s; (b) the steam generator thermal efficiency.

12.47 A furnace burns natural gas that has the following volumetric analysis: 90% CH_4, 7% C_2H_6, and 3% C_3H_8. The gas flow is 0.02 m³/s, and 25% excess air is required for complete combustion. The natural gas and air enter at 25°C and 1 atm pressure. The exhaust gas has a temperature of 1060°C. Determine (a) the volumetric analysis of products of combustion; (b) the dew point of the products; (c) the thermal energy used in the furnace; (d) the exit gas velocity (the stack has a 1-m diameter).

12.48 A coal-fired steam boiler had the following test data: coal consumption 1.26 kg/s, and coal heating value 27 900 kJ/kg. The Orsat analysis of products is 11% CO_2, 8% O_2, 1% CO, and 80% N_2. The stack temperature is 227°C, and the ambient temperature is 25°C. The refuse is removed from the ash pit at 0.095 kg/s. The coal initially contains 5% ash and 80% C. Determine (a) the thermal energy loss to the products of combustion; (b) the heat loss in kW due to incomplete combustion; (c) the loss of energy release due to unburned carbon in the refuse.

12.49 A coal-fired steam generating plant was operated for a year with an average flue gas analysis of 13% CO_2, 0% CO, 6.25% O_2, and 10% combustible matter to the ash pit. An attempt to improve efficiency was made, and the second-year average was 15% CO_2, 0.1% CO, and 3.9% O_2, and 16% combustible matter to the ash pit. Coal with a 7% ash content and a heating value of 33 000 kJ/kg, dry, was used. At the end of the second year it was found that the efficiency had remained the same, but the cost of operation had increased. Why?

12.50 A test of an oil-fired steam generator indicated that 12.77 kg of water evaporated per kg of oil burned. The boiler pressure was 1.4 MPa, the superheat temperature was 245°C, the feedwater temperature was 34°C, and the boiler efficiency was 82.8%. Determine the fuel's heating value.

12.51 A steam generator generates 401 430 kg of steam in a 4-h period. The steam pressure is 2750 kPa at 370°C. The temperature of the water supplied to the steam generator is 138°C. If the steam generator efficiency is 82.5% and the coal has a heating value of 32 200 kJ/kg, find the average amount of coal burned per h.

12.52 A steam generator uses coal containing 75% C and 8% ash and is completely burned, leaving no unburned residue. The air enters the furnace at 32°C and 80% relative humidity. The stack temperature is 360°C. The average flue gas analysis is 12.6% CO_2, 6.2% O_2, and 1% CO. Determine (a) the percentage of excess air; (b) the complete analysis of the fuel (include percentage of hydrogen); (c) the m³ of flue gas per kg of coal burned; (d) the m³ of air per kg of coal burned.

12.53 Compute the adiabatic flame temperature of gaseous methane, ethane, and octane for steady combustion in 100% theoretical air. Compare the resultant temperatures.

12.54 Octane is burned with 200% theoretical air in a steady-flow process. The total pressure is 1 atm, and the reactants enter at standard temperature. The products leave at the adiabatic flame temperature. Determine (a) the entropy increase during combustion; (b) the maximum work; (c) the maximum work for isothermal combustion.

12.55 Methane reacts with 200% theoretical air at standard conditions. The surrounding temperature is also at 25°C. Calculate the maximum work for an isothermal reaction, considering the partial pressure of the reactants and products.

12.56 Calculate the equilibrium constant for the complete oxidation of methane. The total pressure is 1 atm.

12.57 Calculate the percentage of dissociation of oxygen, $O_2 \rightarrow 2O$, at 4000°K and 1 atm pressure.

12.58 Calculate the percentage of dissociation of oxygen, $O_2 \rightarrow 2O$, at 4000°K and 3 atm pressure.

12.59 Methane is steadily burned with 100% theoretical air at standard conditions. Determine the maximum temperature, considering the dissociation of carbon dioxide.

12.60 Determine the Gibbs function at 25°C and 1 atm for the reaction $C + O_2 \rightarrow CO_2$ in units of kJ/kgmol C. Use (a) enthalpy and entropy data; (b) Gibbs function of formation data.

12.61 Methane enters an adiabatic reactor at 25°C and 1 atm and reacts with air also entering at 25°C and 1 atm. The products leave at 1 atm. Determine the entropy production in kJ/K per kgmol of methane entering.

12.62 A mixture of gaseous propane and 150% theoretical air enter a furnace at 25°C and 1 atm. Complete combustion occurs, and the products exit at 1000°K and 1 atm. The furnace is water-cooled, with water entering as a saturated liquid at 200 kPa and leaving as a saturated vapor at the same pressure. Determine (a) the mass flow rate of water per kgmol of fuel; (b) the rate of entropy production per kgmol of fuel; (c) the irreversibility rate per kgmol of fuel if $T_0 = 25$°C.

12.63 Oxygen and hydrogen at 25°C and 1 atm enter a fuel cell steadily, and liquid water exits at 25°C and 1 atm. Determine the maximum theoretical work the fuel cell can develop per kgmol of hydrogen.

12.64 In a coal gasification process carbon reacts with water vapor to form carbon monoxide and hydrogen when heat is applied. The figure illustrates the process and the temperatures of the reactants and products. Determine per kgmol of carbon (a) the electrical energy required in kW; (b) the maximum work; (c) the irreversibility rate.

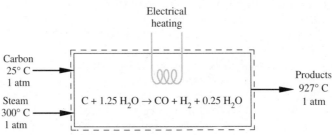

Problem 12.64

12.65 A small internal-combustion engine uses liquid octane as the fuel, which enters at 25°C and 1 atm, as does the air. The mass flow rate of the fuel is 0.86 kg/h, the engine develops 1.5 kW, and the products leave at 400°C. The dry molal analysis of the products is 11.4% CO_2, 2.9% CO, 1.6% O_2, and 84.1% N_2. Determine the heat transfer from the engine in kW.

12.66 A furnace operates steadily and uses liquid dodecane as a fuel that enters at 25°C and 1 atm as does the 20% excess air used in the combustion process. Heat transfer occurs to the furnace walls, and the products exit the furnace at 150°C and 1 atm. Determine per kgmol of fuel (a) the change of availability across the furnace; (b) the availability of the products relative to T_0; (c) the irreversibility rate.

12.67 Determine the higher and lower heating values of coal at 25°C and 1 atm given the following mass analysis: 49.8% C, 19.4% ash, 14.1% H_2O, 6.8% O_2, 6.4% S, and 3.5% H_2.

12.68 Calculate the equilibrium constant, expressed as $\log_{10} K$, for the reaction $CO_2 \leftrightharpoons CO + 0.5O_2$ at $1000°K$ and $4000°K$, and compare the values in Table C.4.

12.69 One kgmol of carbon dioxide dissociates into a mixture of carbon dioxide, carbon monoxide, and oxygen at $2800°K$. Determine the equilibrium composition if the mixture pressure is (a) 1 atm; (b) 20 atm.

12.70 One kgmol of water dissociates into a mixture of water, hydrogen, and oxygen. Determine the equilibrium composition at $3400°K$ if the mixture pressure is (a) 1 atm; (b) 20 atm.

12.71 In an internal-combustion engine the local flame temperature in the combustion process reaches $2800°K$. Determine the composition of the dissociation reaction $0.5N_2 + 0.5O_2 \rightarrow NO$ at a pressure of 1 atm.

12.72 Determine the ideal-cell voltage and efficiency for a hydrogen-oxygen fuel cell at $1000°K$ and 1 atm.

12.73 Determine the ideal-cell voltage and efficiency for a methane-oxygen fuel cell at $298°K$ and $1000°K$ and 1 atm.

12.74 For a hydrogen-oxygen fuel cell, compute the flow rates necessary for the cell to produce 25 kW at $298°K$ and 1 atm.

12.75 For the methane-oxygen fuel cell operating at $1000°K$, determine the fuel flow rate to produce 50 kW.

PROBLEMS (ENGLISH UNITS)

*12.1 Write the reaction equation for hydrogen with 120% theoretical air. Determine the mass of hydrogen required if 2000 lbm of air is available.

*12.2 One lbm of carbon is oxidized with one-half the carbon forming carbon dioxide and the other half forming carbon monoxide. Determine (a) the mass of air required; (b) the volumetric analysis of the products.

*12.3 A hydrocarbon fuel C_nH_{xn} requires equal masses of oxygen for the carbon and hydrogen. Determine x and n.

*12.4 Hydrogen peroxide, H_2O_2, is used to oxidize nonane, C_9H_{20}. The products are carbon dioxide and water. (a) Write the reaction equation. (b) The products are at 1 atm; find the partial pressure of water and the dew point.

*12.5 The dry volumetric analysis of the products of the combustion of a hydrocarbon fuel is 13.6% CO_2, 0.8% CO, 0.4% CH_4, 0.4% O_2, and 84.8% N_2. Determine (a) the reaction equation and find the x and n of the fuel, C_nH_{xn}; (b) find the percentage of excess or deficient air.

*12.6 An adiabatic container has a mixture of oxygen and carbon monoxide in it. Determine whether there is sufficient oxygen for complete combustion if the mixture is 33% O_2 and 67% CO on (a) a mole basis; (b) a mass basis.

*12.7 A sample of coal has an ultimate analysis of 78% C, 3% S, 15% ash, and 2% H_2 and is completely oxidized with 120% theoretical air. Determine the amount of sulfur dioxide produced in lbm/lbm coal burned.

*12.8 A coal sample has an ultimate analysis of 80.0% C, 4.0% O_2, 4.5% H_2, 1.7% N_2, 1.5% S, and 8.3% ash. Determine the mass air/fuel ratio for complete combustion with 100% theoretical air.

*12.9   Garbage, or municipal waste, has an ultimate analysis of 80.5% C, 5.0% H_2, 1.6% S, 1.5% N_2, and 5.5% O_2, with the balance ash. Determine the balanced reaction equation and the mass air/fuel ratio.

*12.10 Hydrazine is burned with 30% excess air. Determine the mass air/fuel ratio and the water vapor condensed if the products are cooled to 60°F in units of mol/mol fuel.

*12.11 A fuel with a mass analysis of 83% C, 12% H_2, and 5% O_2, is burned with 100% theoretical air. Determine the molal analysis of the products with and without water vapor considered.

*12.12 Acetylene burns with 30% excess air at 90°F and 50% relative humidity. What is the dew point of the products?

*12.13 A residual fuel with a mass analysis of 90% C, 8% H_2, and 2% S is burned with air at 100°F and 50% relative humidity. In addition steam atomization is used, requiring 0.05 lbm steam/lbm fuel. Determine the dew point of the products.

*12.14 The volumetric analysis of the products of combustion on a dry basis for the oxidation of octane in air is 9.19% CO_2, 0.24% CO, 7.48% O_2, and 83.09% N_2. Determine the percentage of theoretical air used in the combustion process.

*12.15 Propane burns with 90% theoretical air to form only carbon dioxide, carbon monoxide, water, and nitrogen. Determine the balanced reaction equation.

*12.16 The volumetric analysis of a natural gas is 40% C_3H_8, 40% C_2H_6, 30% CII_4. It burns with air, yielding the following dry molal product analysis: 11.4% CO_2, 1.2% CO, 1.7% O_2, and 85.7% N_2. Determine the mass air/fuel ratio.

*12.17 A fuel C_xH_y burns with air. The products have the following molal analysis on a dry basis: 11% CO_2, 0.5% CO, 2% CH_4, 1.5% H_2, 6% O_2, and 79% N_2. Determine (a) the percentage of excess air; (b) the fuel composition.

*12.18 The volumetric analysis of the dry products of combustion of a hydrocarbon fuel on a dry basis is 11% CO_2, 1% CO, 3% O_2, and 85% N_2. Determine (a) the air/fuel ratio; (b) the percentage on a mass basis of carbon and hydrogen in the fuel.

*12.19 The coal in Problem *12.8 is used in a power plant with an overall efficiency of 40%. The net power produced is 1000 MW. Determine the fuel's heating value and the tons of coal required per day.

*12.20 How many tons of carbon dioxide and sulfur dioxide are produced daily by the power plant in Problem *12.19?

*12.21 Five lbm/sec of ethane gas enters a furnace at 77°F and 1 atm pressure and burns with 100% theoretical air at the same temperature and pressure. The products leave at 440°F. Determine the rate of heat transfer to the surroundings.

*12.22 Five ft³/sec of methane gas enters a furnace at 77°F and 1 atm and burns with 110% theoretical air at the same temperature and pressure. The products leave at 900°R. Determine (a) the air's volumetric flow rate; (b) the heat transfer to the surroundings.

*12.23 Propane gas at 77°F and 1 atm pressure enters a furnace and burns with 120% theoretical air at the same temperature and pressure. On a mole basis 94% of the carbon in the fuel is completely oxidized to carbon dioxide, the remainder to carbon monoxide. The heat transfer from the furnace is measured to be 844,000 Btu/pmol fuel. Determine the temperature of the products leaving the furnace.

*12.24 A tank contains a gaseous mixture of oxygen and ethene at 77°F and 1 atm. The molal analysis of the reactants is 25% ethene and 75% oxygen. The combustion is complete, and the products are cooled to 620°F. Determine the heat transfer from the tank per pmol of ethene.

*12.25 A power plant operates with an overall efficiency of 40%. The plant uses methane as the fuel and air, both at 77°F and 1 atm. The products of combustion of the steam generator leave at 260°F. Determine the mass flow rate of methane per 1000 kW of power produced.

*12.26 A tank contains 1 pmol of butane and 200% theoretical air at 77°F and 1 atm. Combustion occurs, and heat is transferred from the tank until the products' temperature is 980°F. Determine (a) the heat transfer from the tank; (b) the final pressure of the products in the tank.

*12.27 Determine the higher and lower heating values of gaseous propane at 77°F and 1 atm in Btu/lbm.

*12.28 Determine the higher heating value of hydrogen at 77°F and 1 atm in Btu/lbm.

*12.29 Determine the heating value at 77°F and 1 atm of the municipal waste described in Problem *12.9.

*12.30 A mixture of gaseous octane at 1080°R specific heat with constant pressure of 0.395 Btu/lbm-R reacts with 150% theoretical air also at 1080°R. Determine the adiabatic flame temperature.

*12.31 A natural gas has the following volumetric analysis: 22.6% C_2H_6 and 77.4% CH_4. Determine the higher and lower heating values of the gas.

*12.32 Benzene $C_6H_6(g)$ is oxidized with 200% theoretical air. The reactants are at 1 atm and 537°R. Determine the maximum work possible.

*12.33 The exhaust from an automobile engine is the same as in Problem *12.5. The engine consumes 15 lbm/hr of fuel and is located in a garage with dimensions 10 × 30 × 50 ft. If a concentration of 1 part carbon monoxide to 100,000 parts air is hazardous to human life and the surroundings remain at 77°F and 1 atm, what is the maximum time the engine can safely run?

*12.34 Determine the adiabatic flame temperature of butane with 100% theoretical air if all reactants are at 77°F and 1 atm.

*12.35 Determine the adiabatic flame temperature of butane with 100% oxygen if all reactants are at 77°F and 1 atm.

*12.36 Carbon monoxide at 77°F and 1 atm enters a combustion chamber and burns with 100% theoretical air. Determine the adiabatic flame temperature.

*12.37 Liquid octane at 77°F and 1 atm steadily enters an adiabatic combustion chamber and burns with air at 440°F and 1 atm. The products leave at 1880°F. Determine the percentage of excess air supplied.

*12.38 A mixture of methane and 150% theoretical air is contained in an adiabatic tank at 77°F and 1 atm pressure. Combustion occurs. Determine the products' temperature and pressure.

*12.39 Determine the Gibbs function at 77°F and 1 atm for the reaction $C + O_2 \rightarrow CO_2$ in units of Btu/pmol C. Use (a) enthalpy and entropy data; (b) Gibbs function of formation data.

*12.40 Methane enters an adiabatic reactor at 77°F and 1 atm and reacts with air also entering at 77°F and 1 atm. The products leave at 1 atm. Determine the entropy production in Btu/R per pmol of methane entering.

*12.41 A mixture of gaseous propane and 150% theoretical air enters a furnace at 77°F and 1 atm. Complete combustion occurs, and the products exit at 1340°F and 1 atm. The furnace is water-cooled, with water entering as a saturated liquid at 30 psia and leaving as a saturated vapor at the same pressure. Determine (a) the mass flow rate of water per pmol of fuel; (b) the rate of entropy production per pmol of fuel; (c) the irreversibility rate per pmol of fuel if $T_0 = 77°F$.

*12.42 Oxygen and hydrogen at 77°F and 1 atm enter a fuel cell steadily, and liquid water exits at 77°F and 1 atm. Determine the maximum theoretical work the fuel cell can develop per pmol of hydrogen.

*12.43 A small internal-combustion engine uses liquid octane as the fuel, which enters at 77°F and 1 atm, as does the air. The mass flow rate of the fuel is 1.9 lbm/hr, the engine develops 2 hp, and the products leave at 750°F. The dry molal analysis of the products is 11.4% CO_2, 2.9% CO, 1.6% O_2, and 84.1% N_2. Determine the heat transfer from the engine in Btu/min.

*12.44 A furnace operates steadily and uses liquid dodecane as a fuel, which enters at 77°F and 1 atm as does the 20% excess air used in the combustion process. Heat transfer occurs to the furnace walls, and the products exit the furnace at 300°F and 1 atm. Determine per pmol of fuel (a) the change of availability across the furnace; (b) the availability of the products relative to T_0; (c) the irreversibility rate.

*12.45 Determine the higher and lower heating values of coal at 77°F and 1 atm, given the following mass analysis; 49.8% C, 19.4% ash, 14.1% H_2O, 6.8% O_2, 6.4% S, and 3.5% H_2.

*12.46 Calculate the equilibrium constant, expressed as $\log_{10} K$, for the reaction $CO_2 \rightleftharpoons CO + 0.5O_2$ at 1800°R and 7200°R, and compare the values in Table C.4.

*12.47 One pmol of carbon dioxide dissociates into a mixture of carbon dioxide, carbon monoxide, and oxygen at 5040°R. Determine the equilibrium composition if the mixture pressure is (a) 1 atm; (b) 20 atm.

*12.48 One pmol of water dissociates into a mixture of water, hydrogen, and oxygen. Determine the equilibrium composition at 6120°R if the mixture pressure is (a) 1 atm; (b) 20 atm.

*12.49 In an internal-combustion engine the local flame temperature in the combustion process reaches 5040°R. Determine the composition of the dissociation reaction $0.5N_2 + 0.5O_2 \rightleftharpoons NO$ at a pressure of 1 atm.

COMPUTER PROBLEMS

C12.1 Methane is being burned in air. The excess air is varied from 0 to 100% in steps. Use ORSAT.TK (or develop a spreadsheet template or computer program) to calculate the percentage of carbon dioxide and oxygen in the combustion products. Plot the results versus percentage of excess air.

C12.2 Repeat Problem C12.1 for the combustion of fuel oil, $C_{12}H_{26}$, in air. Plot the results and compare them with those from Problem C12.1.

C12.3 Use COMBUST.TK to calculate the adiabatic flame temperature of hydrogen being burned in air at 1 atm. Assume the reaction goes to completion, that is, no dissociation. Vary the excess air from 0 to 100% in steps and plot the results.

C12.4 Repeat Problem C12.3 for the complete combustion of liquid $C_{12}H_{26}$ in air.

C12.5 Use COMBUST.TK to calculate the adiabatic flame temperature of hydrogen being burned in air at 1 atm. Assume that the only products are water, oxygen, nitrogen, hydrogen, and hydroxyl. Vary the excess air from 0 to 100% in steps and plot the results. Compare your results with those from Problem C12.3.

C12.6 Repeat Problem C12.5 for the combustion of liquid $C_{12}H_{26}$. Assume that the only products are water, carbon dioxide, oxygen, nitrogen, hydrogen, hydroxyl, and carbon monoxide. Compare your results with those from Problem C12.4.

C12.7 Carbon is burned in 100% theoretical oxygen to produce carbon dioxide. Calculate the percentage of dissociation of the carbon dioxide into carbon monoxide and oxygen for product temperatures between 2000°K and 4000°K and product pressures of (a) 1 atm; (b) 5 atm; (c) 20 atm. Plot the results.

C12.8 Use COMBUST.TK to investigate the effect of changing the pressure on the combustion of liquid $C_{12}H_{26}$ in 100% theoretical air. Vary the pressure from 1 to 40 atm and calculate the adiabatic flame temperature and the moles of oxygen, hydrogen, hydroxyl, and carbon monoxide in the products. Comment on the results.

13

Internal-Combustion Engines

We all know that one of the most important power-producing devices is the internal-combustion engine, of which the automobile engine is one example. Why is this called an internal-combustion engine? Because the heat addition, the combustion process, occurs within the engine. The chemical energy of the fuel is converted to thermal energy (heat), which in turn is converted into mechanical energy, or work. The work per unit time is the power propelling the car. The fuel used in internal-combustion engines is a hydrocarbon mixture, such as gasoline, diesel fuel, alcohol, or gas. In this chapter we will investigate

- Air-standard cycle models of internal-combustion engines;
- The effect of compression ratio on engine performance;
- The differences between Diesel and automotive internal-combustion engines;
- Factors that improve engine performance;
- The performance of actual engine cycles;
- Efficiencies used in actual engine testing.

13.1 INTRODUCTION

Attempts have been made since the mid-1800s to develop an internal-combustion engine. Attempts were made to operate an engine on gunpowder. The exploding gunpowder would force a piston upward. The downward motion of the piston would engage a rachet and turn a shaft, which was connected to a load. This engine was not a success, and other avenues were explored. In 1862 Beau de Rochas developed the theoretical steps necessary to have an efficient engine. However, it remained for Nikolaus A. Otto, who developed the theory independently, to construct an operable engine. The first successful internal-combustion engine was built by Otto in 1876.

Another example of the internal-combustion engine is the Diesel engine, developed by Rudolf Diesel, who wished to operate an engine on powdered coal. It exploded. Later designs, in which the engine operated on liquid fuel, were successful.

Figure 13.1 is a schematic diagram of a piston-cylinder internal-combustion engine. The engine cylinder is described as having a *bore* (its diameter) and a *stroke* (the maximum length the piston moves in one direction). When the piston is at the topmost extreme, it is said to be at *top dead center.* When it is at the bottommost extreme, it is said to be at *bottom dead center.* The volume swept by the piston in moving between top dead center and bottom dead center is the *displacement volume.* The sum of the cylinder displacement volumes is the *engine displacement.* In automobiles an engine might be referred to as having a 2.5-liter displacement or a 153-in.[3] displacement.

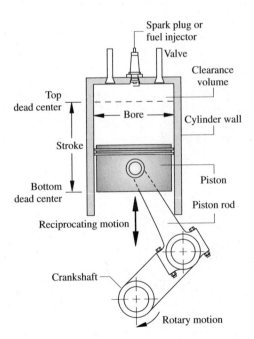

Figure 13.1 Nomenclature for a reciprocating internal-combustion engine.

The volume between the engine head and the piston when the piston is at top dead center is the *clearance volume.* The *compression ratio, r,* is defined as being the cylinder volume at bottom dead center divided by the cylinder volume at top dead center.

13.2 AIR-STANDARD CYCLES

Both Otto's and Diesel's engines operate on open cycles. The products of combustion leaving the engine cannot be continuously reused in a closed system. Fresh air must be drawn in.

Many spark-ignition and compression-ignition engines operate on a four-stroke cycle, illustrated in Figure 13.2. Near the top of the compression stroke the spark plug fires, and combustion of the air/fuel mixture begins. The combustion of the fuel continues as the piston moves downward on the power stroke, with the pressure ideally remaining constant, but actually increasing, then decreasing once combustion ceases. The four strokes in the cycle are

1. Intake stroke: The intake valve is open; the exhaust valve is initially open, then it closes and the piston moves down, bringing fresh air/fuel mixture into the cylinder.
2. Compression stroke: Both intake and exhaust valves are closed, and the air/fuel mixture is compressed by the upward piston movement.
3. Power stroke: Both intake and exhaust valves are closed; spark ignition and combustion occur, with the resultant pressure increase forcing the piston downward.
4. Exhaust stroke: The exhaust valve is open, the intake valve is closed, and the upward movement of the piston forces the products of combustion (exhaust) from the engine.

A compression-ignition, or Diesel, engine has no spark plug; a fuel injector injects the fuel at high pressure into the cylinder. Only air is compressed to a high temperature and pressure, high enough to ignite the fuel when it is injected. The temperature and pressure at the end of the compression stroke are higher than in a spark-ignition engine.

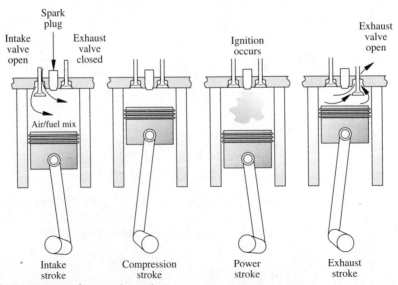

Figure 13.2 A four-stroke cycle.

It is theoretically convenient to have these engines operating on a thermodynamic cycle. The theoretical engines are called *air-standard* engines—air is the working fluid in the engine. Instead of fuels being burned, heat is added from an external source; and instead of exhausting the products of combustion, a heat sink is used to remove heat from the air and return the air to its original state.

Air-Standard Otto Cycle

The air-standard Otto cycle has the following processes:

1. Starting with the piston at bottom dead center, compression proceeds isentropically from state 1 to state 2.
2. Heat is added at constant volume from state 2 to state 3.
3. Expansion occurs isentropically from state 3 to state 4.
4. Heat is rejected at constant volume from state 4 to state 1.

Figure 13.3 illustrates the T-S and p-V diagrams for the cycle. The mass of air remains constant throughout this thermodynamic cycle. This is different from an actual internal-combustion engine cycle, which will be discussed later in the chapter.

Let us calculate the thermal efficiency, η_{th}, for the Otto cycle. The thermal efficiency is defined as the work produced (the desired effect) divided by the heat added (what it costs to achieve that effect).

$$\eta_{th} = \frac{W_{net}}{Q_{in}} = \frac{\Sigma A}{Q_{in}} \tag{13.1}$$

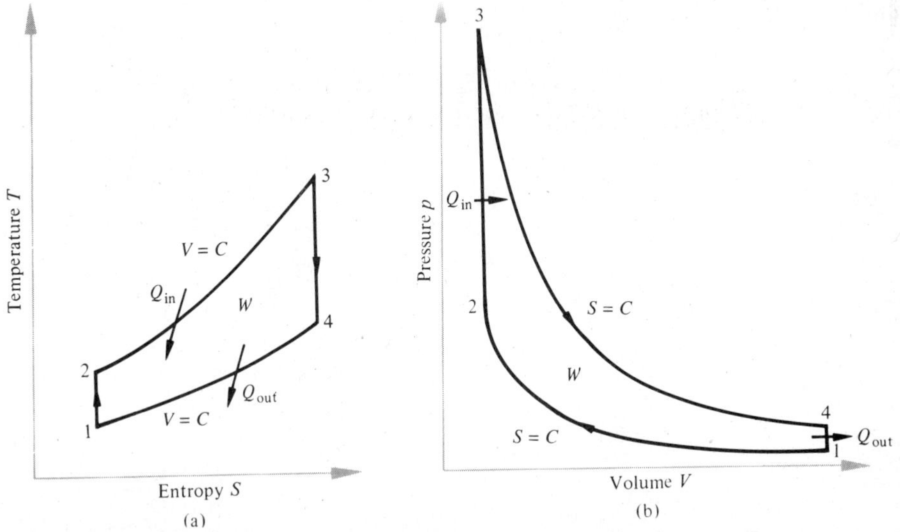

Figure 13.3 (a) The T-S diagram for the air-standard Otto cycle. (b) The p-V diagram for the air-standard Otto cycle.

The heat is added at constant volume from state 2 to state 3. Since the system is closed, because the mass is constant, the first law tells us that

$$Q_{2\text{-}3} = U_3 - U_2 = mc_v(T_3 - T_2) \tag{13.2}$$

Heat is rejected at constant volume from state 4 to state 1. The first law again states

$$Q_{4\text{-}1} = U_1 - U_4 = -mc_v(T_4 - T_1) \tag{13.3}$$

Substituting equations (13.2) and (13.3) into equation (13.1) yields

$$\eta_{\text{th}} = 1 - \frac{T_4 - T_1}{T_3 - T_2} \tag{13.4}$$

The compression ratio, r, is defined as

$$r = \frac{\text{volume at bottom dead center}}{\text{volume at top dead center}} \tag{13.5}$$

and for the Otto cycle this becomes

$$r = \frac{V_1}{V_2} = \frac{V_4}{V_3} \tag{13.6}$$

Also, the temperature is related by an isentropic process between state 1 and state 2 (equation 6.17b), so

$$\frac{T_2}{T_1} = (r)^{k-1} \tag{13.7}$$

and, similarly,

$$\frac{T_3}{T_4} = (r)^{k-1}$$

so

$$\frac{T_3}{T_4} = \frac{T_2}{T_1} \tag{13.8}$$

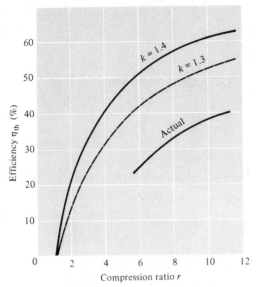

Figure 13.4 The thermal efficiency of the air-standard and actual engines as a function of compression ratio.

Equation (13.4) may be further simplified by eliminating T_3 and T_4, yielding

$$\eta_{th} = 1 - \frac{1}{(r)^{k-1}} \tag{13.9}$$

Thus, the thermal efficiency of the Otto cycle is a function of the compression ratio only. As the compression ratio increases, the efficiency increases. In an actual engine the compression is limited by the temperature at state 2. If this temperature is too great, the gasoline-air mixture will ignite spontaneously and at the incorrect time.

The efficiency of the Otto cycle is much greater than that of an actual engine. Figure 13.4 shows the graph of efficiency versus compression ratio for two values of k and indicates where the actual engine efficiency may lie. The value of the thermal efficiency for $k = 1.4$ is the *cold-air standard efficiency*, and for $k = 1.3$, the *hot-air standard efficiency*. As we will see, the temperature of the air throughout most of the cycle is quite high, $1000-2000°K$, and the value of k at these temperatures is less than at lower temperature levels. When we analyze the engine as an open system considering the combustion process, the results will be more accurate. Finally, the effects of dissociation and nonideal considerations must be considered before the engine may be correctly modeled.

Example 13.1

An engine operates on the air-standard Otto cycle. The conditions at the start of compression are $27°C$ and 100 kPa. The heat added is 1840 kJ/kg. The compression ratio is 8. Determine the temperature and pressure at the end of each process in the cycle, the thermal efficiency, and the mean effective pressure.

Solution

Given: An engine operating on the air-standard Otto cycle with known compression ratio, initial conditions, and heat added.

Find: The cycle state points, the thermal efficiency, and the mean effective pressure.

Sketch and Given Data:

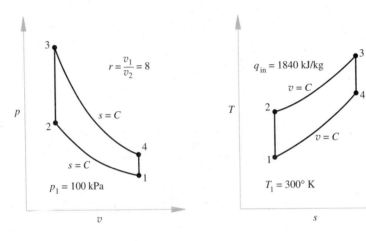

Figure 13.5

Assumptions:

1. The air in the piston-cylinder is a closed system.
2. The air is an ideal gas.
3. The changes in kinetic and potential energies may be neglected.

Analysis: The temperature and pressure at state 1 are given: $T_1 = 300°$K, $p_1 =$ 100 kPa. Follow the processes around the cycle. The process from state 1 to state 2 is isentropic, so the reversible adiabatic relationships for an ideal gas may be used.

$$pV^k = C \quad \text{and} \quad V_1/V_2 = r = 8$$

$$\frac{T_2}{T_1} = \left(\frac{V_1}{V_2}\right)^{k-1}$$

$$T_2 = (300°\text{K})(8)^{0.4} = 689.2°\text{K}$$

$$\frac{p_2}{p_1} = \left(\frac{V_1}{V_2}\right)^{k}$$

$$p_2 = (100 \text{ kPa})(8)^{1.4} = 1837.9 \text{ kPa}$$

The process from state 2 to state 3 is constant volume, and we have noted that the heat transfer is equal to the change of internal energy from first-law analysis.

$$q = u_3 - u_2 = c_v (T_3 - T_2)$$
$$1840 \text{ kJ/kg} = (0.7176 \text{ kJ/kg-K})(T_3 - 689.2°\text{K})$$
$$T_3 = 3253.5°\text{K}$$
$$p_3 = p_2 \left(\frac{T_3}{T_2}\right) = (1837.9 \text{ kPa}) \left(\frac{3253.5}{689.2}\right) = 8676.1 \text{ kPa}$$

The process from state 3 to state 4 is isentropic; hence the reversible adiabatic ideal-gas relationships may be used.

$$pV^k = C \quad \text{and} \quad V_3/V_4 = 1/r$$
$$\frac{T_4}{T_3} = \left(\frac{V_3}{V_4}\right)^{k-1}$$
$$T_4 = (3253.5°\text{K}) \left(\frac{1}{8}\right)^{0.4} = 1416.2°\text{K}$$
$$\frac{p_4}{p_3} = \left(\frac{V_3}{V_4}\right)^{k}$$
$$p_4 = (8676.1 \text{ kPa}) \left(\frac{1}{8}\right)^{1.4} = 472.1 \text{ kPa}$$

The thermal efficiency may be found from equation (13.9) for an ideal gas with constant specific heats as in this model:

$$\eta_{\text{th}} = 1 - \frac{1}{(r)^{k-1}} = 1 - \frac{1}{(8)^{0.4}} = 0.565 \quad \text{or } 56.5\%$$

The mean effective pressure is defined as

$$p_m = \frac{w_{\text{net}}}{\text{displacement volume}} = \frac{w_{\text{net}}}{v_1 - v_2}$$

The specific volumes are found from the ideal-gas equation of state.

$$v_1 = \frac{RT_1}{p_1} = \frac{(0.287 \text{ kJ/kg-K})(300°\text{K})}{(100 \text{ kN/m}^2)} = 0.861 \text{ m}^3/\text{kg} \qquad v_4 = v_1$$

$$v_2 = \frac{v_1}{r} = (0.861/8) = 0.1076 \text{ m}^3/\text{kg} \qquad v_3 = v_2$$

$$w_{\text{net}} = (\eta_{\text{th}})(q_{\text{in}})$$
$$w_{\text{net}} = (1840 \text{ kJ/kg})(0.565) = 1039.6 \text{ kJ/kg}$$

Hence the mean effective pressure is

$$p_m = \frac{(1039.6 \text{ kJ/kg})}{(0.861 - 0.1076 \text{ m}^3/\text{kg})} = 1379.9 \text{ kPa}$$

Comments:

1. The thermal efficiency of the Otto cycle is high, in part because the maximum temperatures and pressures are also very high in this air-standard cycle. This is because the specific heat variation with temperature is neglected, there is no dissociation, and heat is added at constant volume. In an actual engine none of the preceding is true. The advantages of using the air-standard model lie with its simplicity and its giving a sense of direction to changes that affect an actual engine. We can determine that increasing the compression ratio improves the thermal efficiency, without running experiments to verify this.

2. The mean effective pressure is a useful measure in comparing engines operating on different cycles—the greater the mean effective pressure, the smaller the engine may be for a given work output. For instance, the mean effective pressure in a Carnot cycle is quite low, about one-tenth the value in this example. ▬

Sometimes the clearance volume in the piston-cylinder of a reciprocating engine is expressed as a percentage of the total displacement volume; it is called the percentage of clearance, c, and is defined as

$$c = \frac{\text{clearance volume}}{\text{displacement volume}} = \frac{V_2}{V_1 - V_2}$$

Example 13.2

An engine operating on the air-standard Otto cycle has a 15% clearance volume and a total displacement volume of 1 ft³ and operates at 2500 rpm. The heat added is 700 Btu/lbm. Determine the maximum temperature and pressure, the thermal efficiency, the power, and the available portion of the heat rejected. The inlet conditions are 77°F and 14.5 psia, and $T_0 = 77°F$. Use air tables, not constant–specific heat relationships, when analyzing the cycle.

Solution

Given: An air-standard Otto cycle with the air characterized by the air tables. The engine displacement, revolutions per minute, heat added, and inlet conditions are known.

Find: The peak temperature and pressure, the mean effective pressure, the power produced, and the available portion of the heat rejected.

Sketch and Given Data:

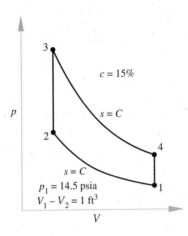
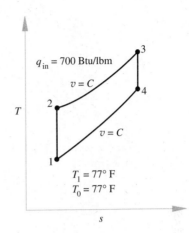

Figure 13.6

Assumptions:

1. The air in the piston-cylinder is an ideal gas with variable specific heats as characterized by air property data.
2. The changes in kinetic and potential energies may be neglected.
3. The air in the piston-cylinder is a closed system.

Analysis: The properties for air are found in Table A.13. The definition of percentage of clearance allows us to define the following volumes:

$$V_2 = V_3 = cV_{1\text{-}2} = 0.15 \text{ ft}^3$$
$$V_4 = V_1 = V_{1\text{-}2} + V_2 = 1.15 \text{ ft}^3$$

From inspection of the p-v and T-s diagrams, note that the maximum temperature and pressure occur at state 3. Determine the properties at each of the cycle state points as follows. From Table A.13, $u_1 = 91.53$ Btu/lbm, $p_{r_1} = 1.3593$, and $v_{r_1} = 146.34$. Process 1-2 is reversible adiabatic; hence

$$\frac{v_{r_2}}{v_{r_1}} = \frac{v_2}{v_1} \qquad v_{r_2} = 146.34 \left(\frac{0.15}{1.15}\right) = 19.08$$

From Table A.13, $p_{r_2} = 23.06$, $T_2 = 1187°$R, and $u_2 = 206.63$ Btu/lbm.

$$\frac{p_2}{p_1} = \frac{p_{r_2}}{p_{r_1}} \qquad p_2 = (14.5 \text{ psia}) \left(\frac{23.06}{1.3593}\right) = 245.98 \text{ psia}$$

The process 2-3 is constant volume, and heat is added; hence

$$q = u_3 - u_2$$

$$700 \text{ Btu/lbm} = u_3 - 206.63 \text{ Btu/lbm}$$

$$u_3 = 906.63 \text{ Btu/lbm}$$

Interpolating for this value of internal energy, find the following properties: $p_{r_3} = 4967$, $v_{r_3} = 0.3273$, and $T_3 = T_{max} = 4395°\text{R}$.

$$p_3 = p_2 \left(\frac{T_3}{T_2}\right) = (245.98 \text{ psia}) \left(\frac{4395}{1187}\right) = 910.76 \text{ psia} = p_{max}$$

The process 3-4 is reversible adiabatic, and the relative pressure and specific volume relationships may be used.

$$\frac{v_{r_4}}{v_{r_3}} = \frac{v_4}{v_3} = \frac{v_1}{v_2} \qquad v_{r_4} = 0.3273 \left(\frac{1.15}{0.15}\right) = 2.5093$$

$$u_4 = 446.8 \text{ Btu/lbm} \qquad p_{r_4} = 350.5$$

$$p_4 = p_3 \left(\frac{p_{r_4}}{p_{r_3}}\right) = (910.76 \text{ psia}) \left(\frac{350.5}{4967}\right) = 64.27 \text{ psia}$$

The process 4-1 is constant volume with heat rejection occurring; hence

$$q = u_1 - u_4$$

$$q = 91.53 - 446.8 = -355.27 \text{ Btu/lbm}$$

The net work for the cycle may be represented by the algebraic sum of the heats into and out of the cycle.

$$w_{net} = \sum q = 700 - 355.27 = 344.73 \text{ Btu/lbm}$$

and the thermal efficiency is

$$\eta_{th} = \frac{w_{net}}{q_{in}} = \frac{344.73}{700} = 0.49 \quad \text{or } 49\%$$

Note that the thermal efficiency cannot be evaluated by equation (13.9) because the assumption of constant specific heat is no longer invoked when using the air tables.

The power is found by determining the work per cycle for the mass in the piston-cylinder and multiplying by the number of cycles that occur per unit time.

$$\dot{W} = mw_{net}n$$

The mass is found from the ideal-gas equation of state.

$$m = \frac{p_1 V_1}{RT_1} = \frac{(14.5 \text{ lbf/in.}^2)(144 \text{ in.}^2/\text{ft}^2)(1.15 \text{ ft}^3)}{(53.34 \text{ ft-lbf/lbm-R})(537°\text{R})} = 0.0838 \text{ lbm}$$

$$n = 2500 \text{ cycles/min}$$

and the power is

$$\dot{W} = (0.0838 \text{ lbm/cycle})(2500 \text{ cycle/sec})(344.73 \text{ Btu/lbm})$$

$$\dot{W} = 72{,}220.9 \text{ Btu/min} = 1703.3 \text{ hp}$$

The available portion of the heat out is

$$(\text{a.e.})_{4\text{-}1} = q_{4\text{-}1} - T_0(s_1 - s_4)$$

The entropy change for air with variable specific heats using the air tables from state 1 to state 4 is

$$s_1 - s_4 = \phi_1 - \phi_4 - R \ln\left(\frac{p_1}{p_4}\right)$$

$$s_1 - s_4 = (0.59945 - 0.98005 \text{ Btu/lbm-R})$$

$$- \frac{(53.34 \text{ ft-lbf/lbm-R})}{(778.169 \text{ ft-lbf/Btu})} \ln\left(\frac{14.5}{64.27}\right)$$

$$s_1 - s_4 = -0.2785 \text{ Btu/lbm-R}$$

$$(\text{a.e.})_{4\text{-}1} = -355.27 \text{ Btu/lbm} - (537°\text{R})(-0.2785 \text{ Btu/lbm-R})$$

$$(\text{a.e.})_{4\text{-}1} = -205.72 \text{ Btu/lbm}$$

$$\text{a.e. \% of } q_{\text{out}} = \frac{(-205.72 \text{ Btu/lbm})(100)}{(-355.27 \text{ Btu/lbm})} = 57.9\%$$

The high temperature at state 4 indicates that a substantial amount of available energy is not being used.

Comments:

1. The air tables compensate for air's variable specific heat but treat air as an ideal gas in terms of its p, v, and T relationships.
2. In an air-standard cycle, one complete cycle occurs each revolution. This is not the case for an actual engine.

Air-Standard Diesel Cycle

The Diesel cycle, developed by Rudolf Diesel, is characterized by constant-pressure heat addition, constant-volume heat rejection, and isentropic compression and expansion processes. This engine is a compression-ignition type; the air is compressed to a high temperature, fuel is injected into the air, ignition is due to the high air temperature, and combustion occurs at constant pressure. The piston expands isentropically to bottom dead center, where heat is rejected at constant volume. Figure 13.7 illustrates the p-V and T-S diagrams for the cycle. In the air standard, cycle heat, not fuel, is added.

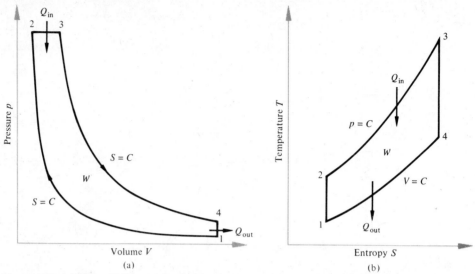

Figure 13.7 (a) The p-V diagram for the air-standard Diesel cycle. (b) The T-S diagram for the air-standard diesel cycle.

The processes in the air-standard Diesel cycle are

1. Starting with the piston at bottom dead center, compression occurs isentropically from state 1 to state 2.
2. Heat is added at constant pressure from state 2 to state 3.
3. Expansion occurs isentropically from state 3 to state 4.
4. Heat rejection occurs at constant volume from state 4 to state 1.

The thermodynamic cycle is complete. Let us calculate the thermal efficiency for the Diesel cycle. To accomplish this, we must find the expressions for the heat supplied and heat rejected and substitute these into equation (13.1). The heat supplied is at constant pressure, and for a closed system the first law tells us that

$$Q_{2\text{-}3} = H_3 - H_2 = mc_p(T_3 - T_2) \tag{13.10}$$

The heat rejection occurs at constant volume, and the first law for a closed system tells us that

$$Q_{4\text{-}1} = U_1 - U_4 = -mc_v(T_4 - T_1) \tag{13.11}$$

Substituting in equation (13.1) yields

$$\eta_{\text{th}} = 1 - \frac{1}{k}\left(\frac{T_4 - T_1}{T_3 - T_2}\right) \tag{13.12}$$

Because the cycle is not symmetrical, the expression for the thermal efficiency cannot be further reduced as can that for the Otto cycle. We note that $V_1/V_2 \neq V_4/V_3$.

The thermal efficiency of the Diesel cycle is slightly less than that of the Otto cycle for the same heat addition. This is because part of the expansion process is occurring while heat is being added in the Diesel cycle, whereas the expansion of air in the Otto cycle occurs after all the heat is added. An important factor favoring diesels, however is that the compression ratio can be much greater in the Diesel cycle than in the Otto cycle because only air, not an air-fuel mixture, is being compressed. Thus, for actual engines, the Diesel-cycle engine is more efficient than the Otto-cycle engine because the possible compression ratio is greater. The maximum allowable temperature after combustion is about the same for both engines, so the advantage of high peak temperatures in the Otto cycle does not accrue to the actual engine.

Example 13.3

An engine operates on the air-standard Diesel cycle. The conditions at the start of compression are 27°C and 100 kPa. The heat supplied is 1840 kJ/kg, and the compression ratio is 16. Determine the maximum temperature and pressure, the thermal efficiency, and the mean effective pressure.

Solution

Given: An engine operating on the air-standard Diesel cycle with known initial temperature and pressure, compression ratio, and heat added.

Find: The cycle maximum temperature and pressure, the thermal efficiency, and the mean effective pressure.

Sketch and Given Data:

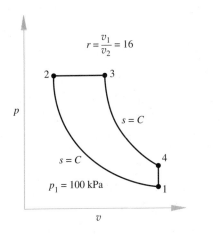
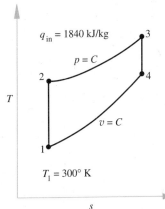

Figure 13.8

Assumptions:

1. The air in the piston-cylinder is a closed system.
2. The air in the system is an ideal gas with constant specific heats.
3. The changes in kinetic and potential energies may be neglected.

Analysis: Use the ideal-gas relationships and determine the cycle state points. At state 1, $T_1 = 300°K$, $p_1 = 100$ kPa, and

$$v_1 = \frac{RT}{p_1} = \frac{(0.287 \text{ kJ/kg-K})(300°K)}{(100 \text{ kN/m}^2)} = 0.861 \text{ m}^3/\text{kg}$$

The process from state 1 to state 2 is isentropic; hence,

$$pv^k = C \qquad r = \frac{v_1}{v_2} = 16 \qquad v_2 = \frac{0.861}{16} = 0.0538 \text{ m}^3/\text{kg}$$

$$\frac{T_2}{T_1} = \left(\frac{v_1}{v_2}\right)^{k-1} \qquad T_2 = (300°K)(16)^{0.4} = 909.4°K$$

$$\frac{p_2}{p_1} = \left(\frac{v_1}{v_2}\right)^{k} \qquad p_2 = (100 \text{ kPa})(16)^{1.4} = 4850.3 \text{ kPa}$$

The process from state 2 to state 3 is constant-pressure with heat addition occurring; hence

$$q = (h_3 - h_2) = c_p(T_3 - T_2)$$

$$1840 \text{ kJ/kg} = (1.0047 \text{ kJ/kg-K})(T_3 - 909.4°K)$$

$$T_3 = T_{\text{max}} = 2740.8°K$$

$$p_{\text{max}} = p_3 = p_2 = 4850.3 \text{ kPa}$$

$$v_3 = \frac{RT_{\text{max}}}{p} = \frac{(0.287 \text{ kJ/kg-K})(2740.8°K)}{(4850.3 \text{ kN/m}^2)} = 0.1622 \text{ m}^3/\text{kg}$$

The process from state 3 to state 4 is isentropic, and $v_4 = v_1$; hence

$$pv^k = C$$

$$\frac{T_4}{T_3} = \left(\frac{v_3}{v_4}\right)^{k-1}$$

$$T_4 = (2740.8°K)\left(\frac{0.1622}{0.861}\right)^{0.4} = 1405.7°K$$

$$\frac{p_4}{p_3} = \left(\frac{v_3}{v_4}\right)^{k}$$

$$p_4 = (4850.3 \text{ kPa})\left(\frac{0.1622}{0.861}\right)^{1.4} = 468.6 \text{ kPa}$$

The process from state 4 to state 1 is constant-volume with heat rejection occurring; the first law yields

$$q = (u_1 - u_4) = c_v(T_1 - T_4)$$

$$q_{\text{out}} = (0.7176 \text{ kJ/kg-K})(300 - 1405.7°K) = -793.4 \text{ kJ/kg}$$

$$w_{net} = \sum q = (1840 - 793.4 \text{ kJ/kg}) = 1046.6 \text{ kJ/kg}$$

$$\eta_{th} = \frac{w_{net}}{q_{in}} = \frac{1046.6}{1840} = 0.569 \quad \text{or } 56.9\%$$

The mean effective pressure is

$$p_m = \frac{w_{net}}{v_1 - v_2} = \frac{(1046.6 \text{ kJ/kg})}{(0.861 - 0.0538 \text{ m}^3/\text{kg})} = 1296.6 \text{ kPa}$$

Comments:

1. Because the piston moves downward as heat is added, the peak temperature and pressure are less than in the Otto-cycle engine.
2. The thermal efficiency could have been calculated using equation (13.12), but in this case the additional data are needed for other parts of the solution process. In general, it is wiser to use the efficiency definition in terms of heat and work, not of temperatures, as the former uses no assumptions and the latter has restrictive requirements. ▪

Cutoff Ratio

Another ratio is used in describing Diesel engine performance, the *cutoff ratio*, r_c. It is defined as

$$r_c = \frac{\text{volume at end of heat addition}}{\text{volume at start of heat addition}} = \frac{V_3}{V_2} \tag{13.13}$$

The cutoff percentage, R_c, is defined as

$$R_c = \frac{V_3 - V_2}{V_1 - V_2} \times 100 \tag{13.14}$$

In an actual engine the cutoff ratio refers to the volume at the start of injection compared to that at the end of fuel injection. It is also sometimes defined as the volume change during injection (heat addition) divided by the displacement volume. Using equation (13.13) the thermal efficiency for an air-standard Diesel engine, equation (13.12), may be reduced to

$$\eta_{th} = 1 - \frac{1}{(r)^{k-1}} \left[\frac{r_c^k - 1}{k(r_c - 1)} \right] \tag{13.15}$$

Figure 13.9 illustrates the effect of cutoff and compression ratios on the Diesel-cycle efficiency. As the cutoff ratio increases, there is more power. This means there is a longer period of heat addition or, in an actual engine, a longer period of fuel injection, and hence more energy input. However, there is a limit to r_c. If the cutoff percentage is more than 10% of the stroke, smoking tends to occur in an actual engine because there is not sufficient time for the combustion process to be completed before

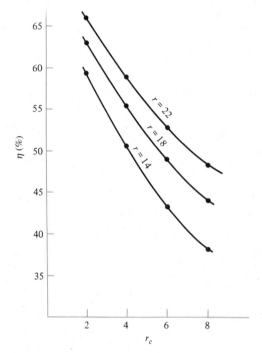

Figure 13.9 Cycle efficiency as a function of cutoff and compression ratios.

the exhaust valve opens. Thus, the products of combustion leave the cylinder before all the chemical energy has been converted into thermal energy, and unburned fuel is present. There is a lower limit on r_c; there must be some heat added, or fuel consumption, to overcome the friction of the various moving parts, as well as the piston and cylinder, and the friction associated with bearings in the engine.

Air-Standard Stirling and Ericsson Cycles

The Stirling and Ericsson cycles are two less well-known thermodynamic cycles. A regenerator is the key element in both cycles; heat must be stored in the regenerator during one part of the cycle and reused in another part. The development of such a heat exchanger has proved to be a very limiting factor in the development of these cycles into operating machinery.

The Stirling cycle, illustrated in Figure 13.10, is characterized by the following processes: From state 1 to state 2, heat is added at constant temperature, causing the volume to increase. The air is then forced through a regenerator from state 2 to state 3 at constant volume. In the regenerator the air is progressively cooled, and the thermal energy, transferred from the air, is stored in the regenerator. From state 3 to state 4, heat is rejected at constant temperature, and the volume decreases. The air is now forced back into the regenerator from state 4 to state 1. The entering air is cool, and heat is ideally transferred from the regenerator to the air. All the heat stored is returned to the air, and the air exits the regenerator at state 1. Thus, in the T-S

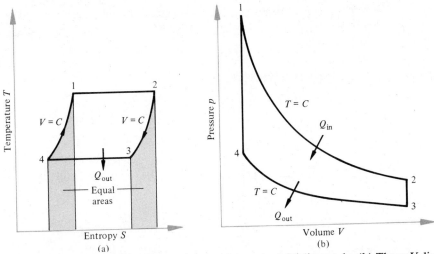

Figure 13.10 (a) The T-S diagram for the air-standard Stirling cycle. (b) The p-V diagram for the air-standard Stirling cycle.

diagram, the areas denoting the heat transferred are equal. Obviously this can happen only in an ideal regenerator; otherwise a temperature difference must exist between two systems for there to be heat transfer. Two problems with the Stirling hot-air engine are the regenerator design and constant-volume regeneration.

To overcome the constant-volume regeneration, Ericsson developed a constant-pressure regenerative cycle, illustrated in Figure 13.11.

Why were these cycles devised? The thermal efficiencies of both cycles match that of the Carnot cycle. The work per cycle of the Ericsson cycle lies between that of

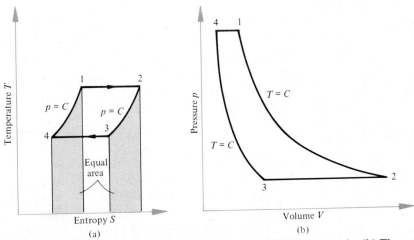

Figure 13.11 (a) The T-S diagram for the air-standard Ericsson cycle. (b) The p-V diagram for the air-standard Ericsson cycle.

the Stirling and Carnot cycles. This is significant when applied to practical engines, where the work per cycle is very important: the greater the work per cycle, the smaller the engine.

Stirling-Cycle Thermal Efficiency

Let us calculate the thermal efficiency of the Stirling cycle. The same method may be used for the Ericsson cycle, yielding the identical thermal efficiency.

$$\eta_{th} = \frac{W_{net}}{Q_{in}} = \frac{\Sigma Q}{Q_{in}}$$

The first law for a closed system is

$$Q = \Delta U + W$$

$$\Delta U = 0 \quad \text{for } T = C$$

$$Q_{in} = W = mRT_1 \ln\left(\frac{V_2}{V_1}\right)$$

Similarly,

$$Q_{out} = -mRT_3 \ln\left(\frac{V_3}{V_4}\right)$$

However, $V_2 = V_3$ and $V_4 = V_1$, so

$$\eta_{th} = \frac{T_1 - T_3}{T_1} = 1 - \frac{T_L}{T_H} \tag{13.16}$$

where T_L is the low temperature and T_H is the high temperature. The thermal efficiency is the same as in the Carnot cycle. This is not so surprising when we realize the heat is transferred by the same process in both engines.

The Stirling engine is gaining in interest today in both power-producing and power-consuming cycles. The reversed Stirling cycle is used for gas liquefaction and cryogenic work because advances in heat transfer make better regenerator designs possible. Because of the high thermal efficiency potential, actual engine designs have been developed, but they do not adhere to constant-volume regeneration. The major problem for the engine is that the air is heated from an external source, with resulting inefficiencies. The engine has also been used in solar-power systems, with the solar energy acting as the heat source. This chapter, however, is concerned more with the internal-combustion engine, particularly the Diesel and Otto cycles.

Air-Standard Dual Cycle

Neither the air-standard Otto cycle nor the air-standard Diesel cycle approximates the cycles of actual engines. An air-standard approximation, the dual cycle, was developed to compensate for nonideal behavior in both engine types. In this cycle, heat is added at constant volume and at constant pressure. Figure 13.12 illustrates the p-V and T-S diagrams. The heat addition simulates the behavior of either engine, as both engines experience pressure and volume changes during the combustion process. The equations for work, thermal efficiency, and heat supplied and rejected may be calculated in a similar manner to that for the Otto and Diesel cycles.

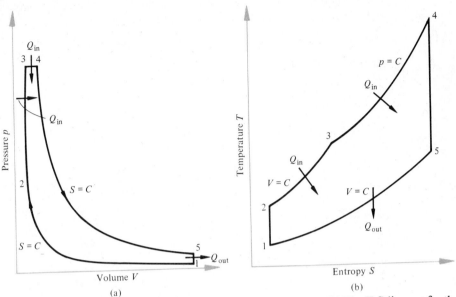

Figure 13.12 (a) The p-V diagram for the air-standard dual cycle. (b) The T-S diagram for the air-standard dual cycle.

Analyzing Engine Cycles Using TK Solver

In Chapter 5 the TK Solver model AIR.TK was introduced to analyze processes involving air. AIRCYCLE.TK is an extension of that model to four-point cycles with air as the working fluid. As does AIR.TK, AIRCYCLE.TK accounts for the variation of specific heat with temperature and uses the Redlich-Kwong, rather than the ideal-gas, equation of state.

AIRCYCLE.TK can be used to analyze Otto, Diesel, Stirling, Ericsson, dual, and similar engine cycles. Because the model includes real-gas properties, it produces more accurate results than does air-standard cycle analysis. AIRCYCLE.TK does, however, require the iterative solution of a number of simultaneous equations. Convergence can be a problem if the initial guesses used in the solution are not well selected. OTTO.TK and DIESEL.TK are slightly modified versions of AIRCYCLE.TK with the appropriate cycle rules added and suitable variable guesses entered. These models can be used to investigate Otto and Diesel cycles and illustrate how to modify AIRCYCLE.TK to model other cycles.

Example 13.4
Use DIESEL.TK to compute the maximum pressure and temperature, thermal efficiency, and mean effective pressure of a Diesel cycle with the same conditions as in Example 13.3.

Solution

Given: An engine operating on the Diesel cycle with known initial temperature and pressure, compression ratio, and heat added.

Find: The maximum pressure and temperature, thermal efficiency, and mean effective pressure.

Sketch and Given Data: See Figure 13.8 on page 445.

Assumptions:

1. The air in the piston-cylinder is a closed system.
2. The changes in kinetic and potential energies may be neglected.

Analysis: Enter the input data into the DIESEL.TK Variable Sheet.

VARIABLE SHEET

St	Input	Name	Output	Unit	Comment
					Diesel-Cycle Analysis
					POINT 1
	100	p1		kPa	Pressure (kPa, MPa, psia)
	300	T1		degK	Temperature (degK, degC, degR, degF)
		v1	.86073	m3/kg	Specific Volume (m3/kg, ft3/lbm)
					POINT 2
		p2	4662.4	kPa	Pressure (kPa, MPa, psia)
		T2	863.63	degK	Temperature (degK, degC, degR, degF)
		v2	.053795	m3/kg	Specific Volume (m3/kg, ft3/lbm)
					POINT 3
		p3	4662.4	kPa	Pressure (kPa, MPa, psia)
		T3	2382.1	degK	Temperature (degK, degC, degR, degF)
		v3	.14747	m3/kg	Specific Volume (m3/kg, ft3/lbm)
					POINT 4
		p4	466.58	kPa	Pressure (kPa, MPa, psia)
		T4	1397.9	degK	Temperature (degK, degC, degR, degF)
		v4	.86073	m3/kg	Specific Volume (m3/kg,ft3/lbm)
					CHANGE (POINT 2 - POINT 1)
		DELh21	592	kJ/kg	Enthalpy (kJ/kg, BTU/lbm)
		DELu21	430.21	kJ/kg	Internal Energy (kJ/kg, BTU/lbm)
	0	DELs21		kJ/kg-K	Entropy (kJ/kg-K, B/lbm-R)
					CHANGE (POINT 3 - POINT 2)
	1840	DELh32		kJ/kg	Enthalpy (kJ/kg, BTU/lbm)
		DELu32	1404.1	kJ/kg	Internal Energy (kJ/kg, BTU/lbm)
		DELs32	1.2206	kJ/kg-K	Entropy (kJ/kg-K, B/lbm-R)
					CHANGE (POINT 4 - POINT 3)
		DELh43	-1219.9	kJ/kg	Enthalpy (kJ/kg, BTU/lbm)
		DELu43	-937.36	kJ/kg	Internal Energy (kJ/kg, BTU/lbm)
	0	DELs43		kJ/kg-K	Entropy (KJ/kg-K, B/lbm-R)
					CHANGE (POINT 1 - POINT 4)
		DELh14	-1212.1	kJ/kg	Enthalpy (kJ/kg, BTU/lbm)
		DELu14	-897	kJ/kg	Internal Energy (kJ/kg, BTU/lbm)
		DELs14	-1.2206	kJ/kg-K	Entropy (kJ/kg-K, B/lbm-R)
					CYCLE PARAMETERS
		Eff	51.25	%	Thermal Efficiency
		Pm	1168.6	kPa	Mean Effective Pressure
	16	CR			Compression Ratio

Comment: The maximum pressure and temperature, thermal efficiency, and mean effective pressure computed are all lower than those calculated in Example 13.3 with an air-standard analysis, due to the real-gas effects included in the TK Solver model.

13.3 ACTUAL DIESEL AND OTTO CYCLES

One of the more difficult concepts in engine analysis is that of the actual cycle. The thermodynamic cycle, used for analyzing the air-standard cycle, is not an actual engine cycle. The mass of air and products of combustion are actually continually undergoing change. As we know, it is impossible to recombust air, so a fresh air supply must continually be drawn into the engine and the products of combustion removed.

Four-Stroke Cycle

Let us consider the four-stroke mechanical cycle in Figure 13.2. In the spark-ignition engine, the air/fuel mixture is compressed, the spark plug discharges, and a spark ignites the fuel mixture. The combustion process is very rapid and occurs over a small volume change in the cylinder; thus, the ideal process is a constant-volume heat addition. (If the engine is compression-ignition — a Diesel cycle, for example — then at the top of the compression stroke a selected amount of fuel is injected into the cylinder and ignites because of the high air temperature. The combustion of the fuel continues as the piston expands on the power stroke, and the pressure remains essentially constant during the combustion process.) Because there is only one power stroke for every two revolutions, it is necessary to know the operating cycle when calculating engine power.

Since not all the products of combustion are removed from the cylinder on the exhaust stroke, there is a dilution of the incoming air charge by the remaining products. The greater the clearance volume, the greater the dilution.

Actual p-V Diagram

The *p-V* diagram for a spark-ignition engine is illustrated in Figure 13.13, with the lines for the Otto cycle superimposed. The compression process from state 1 to state 2 is not adiabatic, so the actual pressure is less than the ideal. Ignition occurs before top dead center, allowing time for the combustion to develop. The combustion process does not occur at constant volume. There is energy loss to the cylinder walls and piston head, and the piston is moving downward; thus, the peak pressure is less than that in the Otto cycle. The expansion process is nonadiabatic, hence the lower pressures from state 3 to state 4. We notice that the exhaust valve opening occurs before bottom dead center, and this reduces the power produced (it reduces area I). Why is this done? Area II represents the work the engine must do in pushing out the exhaust products, state 4 to state 5, and pulling in the fresh air charge, state 5 to state 1. A trade-off must be made that will allow the net area, I – II, to be a maximum. If the

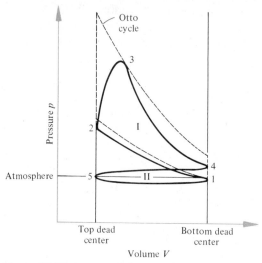

Figure 13.13 An actual four-stroke spark-ignition engine p-V diagram with the Otto cycle superimposed.

exhaust valve opens too early, then the area I reduction cannot compensate for the resulting decrease in area II. For this condition the exhaust pressure at state 4 is lower, and the line from 4 to 5 is more horizontal. If the exhaust valve is opened too late, the work to remove the exhaust gases is greater because of a higher pressure at state 4. This is indicated by an increase in area II. Each engine is unique and is a function of the operating conditions as well as the design. In the design we would want the resistance in the intake and exhaust systems to be as small as possible. The p-V diagram for a compression-ignition engine is very similar to that shown in Figure 13.13.

Two-Stroke Cycle

Another cycle that the spark-ignition and compression-ignition engines use is the two-stroke cycle, in which all four events occur in two strokes, or one revolution of the engine. The two-stroke cycle is found on compression-ignition engines, in large and small power ranges, and on small spark-ignition engines such as those in lawn mowers, chain saws, and motorcycles. Figure 13.14 illustrates the p-V diagram for a two-stroke cycle diesel engine that is not supercharged. To assist the exhaust process in the two-stroke cycle engine, the engines are equipped with scavenging blowers, which raise the inlet air pressure to 13–35 kPa above atmospheric pressure. Thus, the intake air pushes the exhaust gas out. This is not the same as supercharging, which raises the inlet pressure much higher. The work required in operating the scavenging air pump or compressor is charged against the engine, so the net work of the two-stroke cycle engine is reduced by this amount. In addition to this loss of work, the combustion process often does not go as far toward completion as it does in the

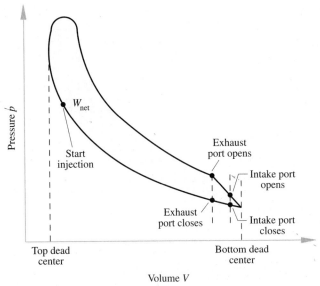

Figure 13.14 A p-V diagram for the two-stroke Diesel cycle (nonsupercharged).

four-stroke cycle engine, so that the work of a two-stroke cycle engine is not twice, but only about one and a half times, that of a four-stroke cycle engine.

13.4 CYCLE COMPARISONS

Of all heat engines, the internal-combustion engine has the highest thermal efficiency. This is because the maximum temperature in the cylinder may reach as high as 2400°C during the combustion process. The metal parts of the engine do not come in contact with this temperature, since it occurs only in the gas mixture for a small portion of the cycle. The economic trade-offs between the Otto, or spark-ignition engine, and the Diesel, or compression-ignition engine, are such that at high speeds, 4000–6000 rev/min, and reasonably low power, 150–225 kW, the spark-ignition engine is more advantageous as well as lighter. In the middle range (several hundred kilowatts) the Diesel and Otto engines overlap, and at higher powers the Diesel engine dominates. Diesel engines are commonly found in trucks, buses, and auxiliary or emergency power generators, and as the main propulsion engine on ships. A typical marine Diesel engine size is around 20 000 kW, with a bore and stroke of 0.8 × 2.5 m, and these engines may have ratings up to 40 000 kW. The Diesel engine uses a less expensive fuel than the Otto engine, but both engines require precise timing for the combustion process. The combustion process is not continuous and must occur cyclically and be completed in times of 10^{-3} s.

In the spark-ignition engine the fuel mixture is ignited by an electric spark from the spark plug. The temperature of the air-fuel mixture in the vicinity of the spark plug is raised above the ignition temperature, and combustion occurs. When the air-fuel mixture is correct, the flame of this localized mixture spreads throughout the cylinder, burning the mixture. The fuel/air ratio must be $0.055 \leq r_{fla} \leq 0.10$ or combustion will not occur. The propagation of the flame throughout the cylinder occurs very quickly, but the crank travel may be 20° to 30°.

13.5 ACTUAL CYCLE ANALYSIS

In considering an internal-combustion engine as an open system, which indeed it is, greater accuracy may be achieved in the determination of engine work and efficiency. Let us consider an internal-combustion engine with p-V and T-s diagrams as illustrated in Figure 13.15. Let there be no pressure drop or temperature increase in the air as it enters the engine at state 1. The total mass in the engine at state 1 comprises the fresh air charge drawn in, the fuel drawn in, and the unpurged products left in the clearance volume at T_5. Let z represent the fraction of unpurged products, or

$$z = \frac{\text{mass of unpurged products}}{\text{total mass of reactants}}$$

Thus, the internal energy of the reactants, u_{r_1}, at state 1 is

$$u_{r_1} = (1 - z)u_{a_1} + zu_{z_1} + r_{f/r}u_{RP} + r_{f/r}u_{f_1} \qquad \frac{\text{energy}}{\text{mass reactant}}$$

The internal energy of the unpurged products, u_{z_1}, is evaluated at p_1 and T_5. The internal energy of the incoming air is u_{a_1}. Note that $r_{f/r}$ is the fuel/reactant ratio, not the fuel/air ratio. The term u_{f_1} is the change of internal energy of the fuel above the reference datum at which u_{RP} is evaluated. The compression process from state 1 to

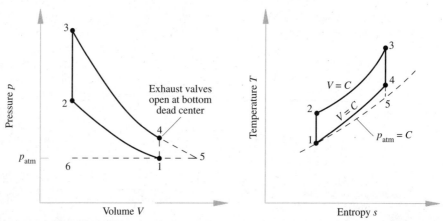

Figure 13.15 (a) The p-V diagram for an open Otto-cycle engine. (b) The T-s diagram for the same engine.

state 2 is reversible adiabatic, so

$$u_{r_2} = (1 - z)u_{a_2} + zu_{z_2} + r_{f/r}u_{RP} + r_{f/r}u_{f_2}$$

and the work is

$$w_{1\text{-}2} = u_{r_1} - u_{r_2}$$

The combustion process is also adiabatic.

$$u_{p_3} - u_{r_2} = r_{f/r}u_{RP}$$

The expansion process is reversible adiabatic to state 4, bottom dead center. At this point the exhaust valves open, and no further work is done. The exhaust flows out, and the unpurged products remain at T_5 and p_{atm}.

The work from state 3 to state 4 is

$$w_{3\text{-}4} = w_{p_3} - w_{p_4}$$

The net work is the sum of the compressive and expansive work terms,

$$w_{net} = w_{1\text{-}2} + w_{3\text{-}4}$$

and the efficiency is

$$\eta_{th} = \frac{w_{net}}{r_{f/r}h_{RP}}$$

where the entire enthalpy of formation is chargeable against the engine, although we did not use it in the combustion process.

Example 13.5

An engine operates on the open Otto cycle and has a compression ratio of 7.5. The air/fuel ratio is 30 kg air/kg fuel, and the engine uses a gaseous octane with a lower heating value of 44 232 kJ/kg. The air enters the engine at 27°C and 100 kPa. Determine the work per kilogram of air and the thermal efficiency of the engine.

Solution

Given: An engine operating on the open Otto cycle with known compression ratio, inlet conditions, and fuel type.

Find: The engine's thermal efficiency and the work produced per kilogram of air entering.

Sketch and Given Data:

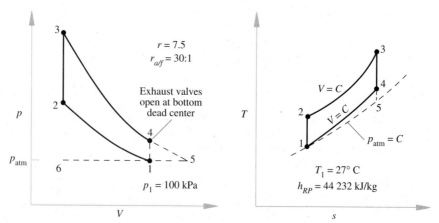

Figure 13.16

Assumptions:

1. The value of k for the reactants and products is 1.3 to compensate for the substances' not being air and the higher average cycle temperature.
2. Neglect the unpurged fraction of products and the variation of u_{RP} with temperature.
3. The gases behave like ideal gases.
4. The changes in kinetic and potential energies may be neglected.

Analysis: Determine the internal energies, pressures, and temperatures around the cycle. At state 1 the internal energy is

$$u_{r_1} = u_{a_1} + r_{f/a}u_{RP} \qquad T_1 = 300°\text{K} \qquad r_{f/a} = 0.0333 \text{ kg fuel/kg air}$$

The process from state 1 to state 2 is reversible adiabatic; hence

$$T_2 = T_1(r)^{k-1} = (300°\text{K})(7.5)^{0.3} = 549°\text{K}$$

$$p_2 = p_1(r)^k = (100 \text{ kPa})(7.5)^{1.3} = 1372.7 \text{ kPa}$$

$$u_{r_2} = u_{a_2} + r_{f/a}u_{RP}$$

The work of compression from state 1 to state 2 is

$$w_{1\text{-}2} = (1 + r_{f/a})(u_{r_1} - u_{r_2})$$

When the terms are subtracted, $r_{f/a}u_{RP}$ adds out and

$$w_{1\text{-}2} = (1 + r_{f/a})(u_{a_1} - u_{a_2})$$
$$w_{1\text{-}2} = (1.0333 \text{ kg})(0.7176 \text{ kJ/kg-K})(300 - 549°\text{K})$$
$$= -184.6 \text{ kJ/kg air}$$

The process from state 2 to state 3 is constant-volume. For the open Otto cycle, combustion is a constant-volume process; the energy released is u_{RP}, where $u_{RP} = h_{RP} - RT$. The temperature is taken at the reference datum, 25°C, for u_{RP} because we do not have a method to calculate the variation of h_{RP} with temperature. This results in an error of about 1%, which is less than the error involved in assuming an isentropic compression and expansion.

$$u_{RP} = 44\ 232 \text{ kJ/kg} - \frac{(8.3143 \text{ kJ/kg-K})(298°\text{K})}{114.23}$$
$$= 44\ 210.0 \text{ kJ/kg}$$

$$(1 + r_{f/a})(u_{p_3} - (1 + r_{f/a})u_{r_2} = r_{f/a}u_{RP}$$
$$(1.0333)u_{p_3} = (1.0333 \text{ kg})(0.7176 \text{ kJ/kg-K})(549°\text{K})$$
$$+ (0.0333 \text{ kg fuel/kg air})(44210 \text{ kJ/kg fuel})$$
$$u_{p_3} = 1819.4 \text{ kJ/kg air} = (1 + r_{f/a})c_v T$$
$$1819.4 \text{ kJ/kg air} = (1.0333 \text{ kg air})(0.7176 \text{ kJ/kg-K})(T_3)$$
$$T_3 = 2453.7°\text{K}$$

The pressure is found from ideal-gas relationships at constant volume.

$$p_3 = p_2\left(\frac{T_3}{T_2}\right) = (1372.7 \text{ kPa})\left(\frac{2453.7}{549}\right) = 6135.1 \text{ kPa}$$

The process from state 3 to state 4 is reversible adiabatic; hence

$$T_4 = T_3\left(\frac{v_3}{v_4}\right)^{k-1} = (2453.7°\text{K})\left(\frac{1}{7.5}\right)^{0.3} = 1340.6°\text{K}$$

In addition the products expand to atmospheric pressure when the exhaust valve opens, and state 5 temperature may be determined from reversible adiabatic relationships as

$$T_5 = T_3 \left(\frac{p_5}{p_3}\right)^{(k-1)/k} = (2453.7°K)\left(\frac{100}{6135.1}\right)^{0.3/1.3} = 948.9°K$$

The work produced during the expansion process is

$$w_{3\text{-}4} = (1 + r_{fla})(u_{p_3} - u_{p_4})$$
$$w_{3\text{-}4} = (1.0333 \text{ kg})(0.7176 \text{ kJ/kg-K})(2453.7 - 1340.6°K)$$
$$w_{3\text{-}4} = 825.4 \text{ kJ/kg air}$$
$$w_{net} = 825.4 - 184.6 = 640.8 \text{ kJ/kg air}$$

and the thermal efficiency is

$$\eta_{th} = \frac{w_{net}}{r_{fla}h_{RP}} = \frac{(640.8 \text{ kJ/kg air})}{(0.0333 \text{ kg fuel/kg air})(44\,232 \text{ kJ/kg fuel})} = 0.435$$

Comments:

1. In an actual engine the cylinder walls, and in larger engines the cylinder heads and pistons, are cooled by water. This means that the assumption of isentropic compression and expansion is not valid, and more information must be known about the heat removed from the engine. If this quantity is given, we may assume the exhaust gas of the adiabatic engine is cooled by this amount of heat to give a better approximation of the actual engine exit temperature. The heat released to the cooling water, as a percentage of total energy, varies with engine load and type, but an indicative value for large Diesel engines is 25%.
2. The entire heating value is charged against the engine in the efficiency expression, even though it was not used in the combustion process.
3. The pressure at state 4 may be calculated for an isentropic expansion. This is pressure in the cylinder when the exhaust valves open. In this example the pressure is 447 kPa.

Example 13.6
A Diesel engine operates on the open Diesel cycle and receives air at 77°F and 14.5 psia and fuel, dodecane, at the same temperature. The air/fuel ratio is 30 lbm air/lbm fuel, the compression ratio is 15, and the unpurged products are 10% of the inlet air charge. The lower heating value of dodecane is 18,964 Btu/lbm. Determine the engine's net work per unit mass of mixture and its thermal efficiency.

Solution

Given: A Diesel engine operating on the open Diesel cycle with inlet air and fuel conditions specified as well as the compression ratio.

Find: The engine's thermal efficiency and net work per unit mass of mixture.

Sketch and Given Data:

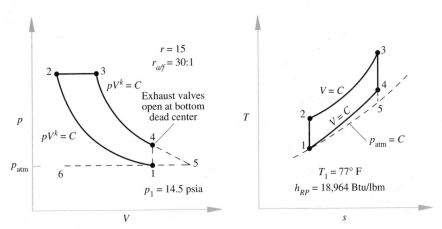

Figure 13.17

Assumptions:

1. The products have the following gas properties: $c_{vp} = 0.180$ Btu/lbm-R, $c_{pp} = 0.249$ Btu/lbm-R, and $k = 1.383$.
2. The gases behave like ideal gases.
3. The changes in kinetic and potential energies may be neglected.

Analysis: For a Diesel engine, the process of heat addition is at constant pressure so the equations for determining the expansive and compressive works are different than those of the Otto cycle. Using the same terminology as before,

$$u_{r_1} = (1 - z)u_{a_1} + zu_{x_1}$$

The internal energy at the end of compression does not include a fuel term, since the fuel is injected after the compression process; thus,

$$u_{r2} = (1 - z)u_{az} + zu_{x2}$$

The combustion process is adiabatic.

$$h_{r2} + r_{f/r}h_{f/2} + r_{f/2}h_{RP} = (1 + r_{f/r})h_{p3} \qquad \text{(a)}$$

where h_{f2} represents the enthalpy of the fuel above the reference datum at which h_{RP} is evaluated.

The work during the expansion processes is

$$w_{2\text{-}3\text{-}4} = p_2(v_3 - v_2) + (u_{p3} - u_{p4})$$

and the net work is

$$w_{net} = w_{2\text{-}3\text{-}4} - w_{1\text{-}2} = w_{2\text{-}3\text{-}4} - (u_{r2} - u_{r1})$$

$$w_{net} = h_{p3} - h_{r2} + u_{r1} - (1 + r_{flr})u_{p4}$$

$$w_{net} = r_{flr}h_{RP} + u_{r1} - (1 + r_{flr})u_{p4} \qquad \text{(b)}$$

We must assume an exhaust temperature for the unpurged products at state 5 to calculate the temperature of the reactants at state 1. Let us use a value of $1380°$R. We will see how this was determined at the end of the problem solution.

$$m_r c_{vr} T_r = m_a c_{va} T_a + m_p c_{vp} T_p$$

$$(1.0 \text{ lbm})(0.1723 \text{ Btu/lbm-R})(T_r) = (0.9 \text{ lbm})(0.1714 \text{ Btu/lbm-R})(537°\text{R})$$

$$+ (0.1 \text{ lbm})(0.188 \text{ Btu/lbm-R})(1380°\text{R})$$

$$T_1 = T_r = 624.9°\text{R} = 164.9°\text{F}$$

The specific heat for the reactants is

$$c_{pr} = (0.9)(0.24 \text{ Btu/lbm-R}) + (0.1)(0.249 \text{ Btu/lbm-R})$$

$$c_{pr} = 0.241 \text{ Btu/lbm-R}$$

and similarly, $k = 1.398$. Process 1-2 is reversible adiabatic.

$$T_2 = T_1 \left(\frac{v_1}{v_2}\right)^{k-1} = (624.9°\text{R})(15)^{0.398}$$

$$T_2 = 1836.2°\text{R}$$

$$p_2 = p_1 \left(\frac{v_1}{v_2}\right)^{k} = (14.7 \text{ psia})(15)^{1.398} = 647.8 \text{ psia}$$

Next determine r_{flr}:

$$r_{flr} = \left(\frac{\text{lbm fuel}}{\text{lbm air}}\right)\left(\frac{\text{lbm air}}{\text{lbm reactants}}\right)$$

$$r_{flr} = (0.0333)(0.9) = 0.02997 = 0.03 \text{ lbm fuel/lbm reactants}$$

Substitute into equation (a) to find T_{p3}, where $h_f = 0$ because the fuel is injected at $77°$F where $h = c_p T$ for a gas.

$$(c_{pr})(T_2) + r_{flr}h_{RP} = (1 + r_{flr})(c_{pp})(T_{p3})$$

Substituting yields

$$(0.241 \text{ Btu/lbm-R})(1836.2°\text{R}) + (0.03 \text{ lbm fuel/lbm reactants})$$

$$\times (18,964 \text{ Btu/lbm fuel})$$

$$= (1.03 \text{ lbm products})(0.249 \text{ Btu/lbm products})(T_{p3})$$

$$T_{p3} = 3943.7°\text{R}$$

The cutoff ratio is defined as

$$r_c = \frac{V_3}{V_2} = \frac{T_3}{T_2}$$

In addition the expansion ratio, r_e, is defined as

$$r_e = \frac{V_4}{V_3}$$

$$V_1 = V_4$$

$$r_e = \frac{V_1}{V_3} = \left(\frac{V_1}{V_2}\right)\left(\frac{V_2}{V_3}\right) = \frac{r}{r_c}$$

This ratio assists us in determining the remaining cycle temperatures.

$$r_c = \frac{3943.7}{1836.2} = 2.147$$

$$r_e = \frac{15}{2.147} = 6.986$$

$$T_4 = T_3 \left(\frac{v_3}{v_4}\right)^{k-1} = (3943.7°R)\left(\frac{1}{6.986}\right)^{0.383} = 1873.1°R$$

$$T_5 = T_3 \left(\frac{p_5}{p_3}\right)^{(k-1)/k} = (3943.7°R)\left(\frac{14.7}{647.8}\right)^{0.383/1.383} = 1382°R$$

From equation (b) the net work may be determined.

$$w_{net} = (0.03 \text{ lbm fuel/lbm reactants})(18{,}964 \text{ Btu/lbm fuel})$$

$$+ (0.1723 \text{ Btu/lbm-R})(624.9°R)$$

$$- (1.03 \text{ lbm products})(0.18 \text{ Btu/lbm-R})(1873.1°R)$$

$$w_{net} = 329.3 \text{ Btu/lbm}$$

and the thermal efficiency is

$$\eta_{th} = \frac{(329.3 \text{ Btu/lbm})}{(0.03 \text{ lbm fuel/lbm reactants})(18{,}964 \text{ Btu/lbm fuel})} = 0.579 \quad \text{or } 57.9\%$$

Comments:

1. Normally, the problem requires that one iterate on $T_5 = T_p$, which determines the value of u_{r1}, until convergence occurs. In this case a fortuitous guess for T_5 allowed us not to do this.

2. The analysis may be improved by using a value for k, as in Example 13.5, which accounts for some nonadiabatic effects, specific heat variation with temperature, and irreversibilities.

3. Actual analysis of engines must include dissociation effects and variation of specific heat with temperature in addition to heat loss from the cylinder. The heat loss reduces the temperatures in the cylinder and hence the work done by the gas.

13.6 ENGINE PERFORMANCE ANALYSIS

An increase in thermal efficiency of a spark-ignition engine occurs if the combustion time is decreased — a greater temperature can be achieved before expansion and more work accomplished for the same energy supplied. One method is to increase the flame velocity. As the flame velocity increases, however, the engine begins to run roughly, because of unbalanced pressure waves moving across the piston top. The rapidly burning fuel in one region causes a localized pressure increase. The unbalanced pressure causes a pressure wave to move across the cylinder trying to achieve equilibrium. If localized self-ignition of the mixture occurs, severe pressure rises will be created. The pressure waves produced may be supersonic and move across the cylinder very quickly, producing a knocking noise. The combined effects of autoignition and the noise are called *detonation.* This obviously is not good for the engine surfaces. Pitting may occur on the metal surfaces, and cracks may develop from high-speed, cyclic knocking. The formation and movement of shock waves is an irreversible process. We know that energy thus spent cannot be used for work and has a deleterious effect on engine efficiency. There are fuel additives that resist detonation. The octane rating of fuel is a measure of its resistance to detonation. The greater the octane number, the more resistant the fuel is to detonation.

In compression-ignition engines a spark plug is not needed to ignite the fuel; the temperature of the compressed air is sufficiently high to accomplish this. Fuel is sprayed into the cylinder at the proper instant, and the hot air causes the fuel oil droplets to vaporize. Then, as ignition temperature is reached, combustion occurs. The delay between the injection and the combustion is called *ignition delay.* The fuel droplets are a cool liquid when they enter and must be broken into a fine spray so vaporization and combustion can occur. As in the spark-ignition engine, a discontinuous combustion process can cause knocking. One means of preventing autoignition is use of a fuel with a low ignition temperature. The fuel will ignite as it is injected into the cylinder, and the flame will sustain the combustion as more fuel is injected. The ignitability of the fuel is rated by its cetane number — the greater the ignitability, the higher the rating. Considerable excess air must also be provided in the diesel engine to ensure combustion of the liquid fuel droplets. If the fuel is heated to a high temperature in the absence of oxygen, cracking occurs, and carbon is formed. This is manifested as a sooty exhaust.

At reduced power the compression-ignition engine is more efficient than the spark-ignition engine. This is due, in part, to the combustion process. At a reduced load less fuel is needed. In the spark-ignition engine the air flow must be restricted to maintain the correct fuel/air ratio. The throttling loss is irreversible, and the loss of available energy reduces the thermal efficiency of the engine. In the compression-ignition engine, on the other hand, the fuel and air systems are separate, and the fuel can be decreased independently of the air with no resulting throttling losses. The fuel/air ratio may be smaller than 0.01, and, as the ratio decreases, the engine efficiency approaches that of the air-standard Diesel engine.

13.7 WANKEL ENGINE

The Wankel engine operates by transferring rotary motion from the engine's rotary piston to the car's drive shaft. In a piston-cylinder engine the vertical motion of the piston is transferred to the rotary motion of the crankshaft by the connecting rod. The Wankel engine is an internal-combustion engine and must have the same four processes: intake, compression, power, and exhaust. Figure 13.18 illustrates the continuous cycle. On the three sides of the rotary piston different processes are continuously occurring. There are few moving parts in the engine; there are, for instance, ports instead of exhaust and inlet valves. An air and fuel mixture is continuously drawn in from state 1 to state 4. At state 4, the inlet port is closed by one of the lobes. Compression continues from state 5 to state 8. At state 9, the mixture ignites, and the power from the burning mixture, demonstrated by increased pressure and temperature, forces the rotor lobe from state 10 to state 12. The exhaust port is uncovered and the burned mixture exhausts from state 13 to state 18. The process now repeats itself. Notice that each revolution has three power strokes. Therefore, the power may be produced in a more compact volume than in a piston-cylinder engine.

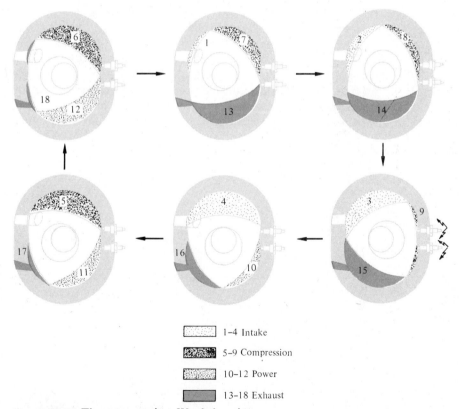

1–4 Intake

5–9 Compression

10–12 Power

13–18 Exhaust

Figure 13.18 The processes in a Wankel engine.

One of the potential problems with the engine is in properly sealing the space between the lobes and the rotor housing. Also, the combustion process is similar to the two-stroke internal-combustion engine, and thus the exhaust contains comparatively large amounts of unburned hydrocarbons. This results in a decrease in overall efficiency compared with the piston-cylinder engine. However, the high power produced per unit engine volume makes the Wankel engine attractive for certain applications.

13.8 ENGINE EFFICIENCIES

Similar to a gas compressor, the function of the intake stroke of the engine is to draw air into the cylinder. The term *volumetric efficiency* describes how efficiently air is drawn into the engine. The ideal volumetric efficiency, η_v, is the ratio of the volume (mass) of air actually drawn in divided by the maximum possible amount of air that could be drawn in, the displacement volume (mass). Either mass or volume may be used in defining volumetric efficiency, though volume is used more frequently.

$$\eta_v = \frac{\text{actual volume}}{\text{displacement volume}} \qquad (13.17)$$

Figure 13.19 illustrates the intake stroke of a four-stroke cycle internal-combustion engine. Notice that at top dead center some unburned gas remains and expands as the piston moves downward, until the pressure in the cylinder is less than the intake manifold pressure and air flows into the cylinder. Thus, the volume of the new air charge for each intake stroke is $V_1 - V_4$, while the piston displacement, V_{PD}, is

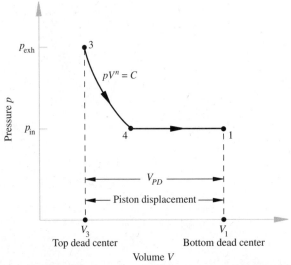

Figure 13.19 Intake stroke of a four-stroke cycle internal-combustion engine.

$V_1 - V_3$. The volumetric efficiency is

$$\eta_v = \frac{V_1 - V_4}{V_{PD}} \tag{13.18}$$

From the definition of percentage of clearance, $V_3 = cV_{PD}$, noting that $V_1 = V_{PD} + cV_{PD}$. Substituting into equation (13.18) yields

$$\eta_v = 1 + c - c(p_3/p_1)^{1/n} \tag{13.19}$$

In analyzing equation (13.19) we should note that the volumetric efficiency decreases as the clearance increases and as the pressure at top dead center increases. An increase of either effect will cause the mass of air entering to be less because of a greater mass of trapped air at top dead center.

In an actual engine, additional effects cause the volumetric efficiency to decrease. The air in the intake manifold must be at a higher pressure than that in the cylinder in order for the air to flow. There are frictional effects to overcome in flowing around intake valves, as well as flow irreversibilities in the cylinder itself. In addition, the cylinder walls are hot, which raises the temperature of the incoming air. These combined effects serve to reduce the mass, hence the volume, of surrounding air that can be drawn into the engine. To account for these effects, the volumetric efficiency is reduced because the pressure inside the cylinder at state 1, the beginning of compression, is less than the pressure of the surrounding air; and the heating-effect term is the ratio of the temperature of the surrounding air, T_0, to the temperature of the gas at state 1. Thus,

$$\eta_{v(act)} = \eta_v(T_0/T_1) \tag{13.20}$$

There are basic measurements and indications of an engine's performance. From the measurements we get an indication of how well the engine is producing power. As a fluid passes through an engine, it performs work on the engine; for example, it moves a piston or turns a turbine wheel. This is *indicated work,* or W_i, the work the fluid indicates it has accomplished. This is usually expressed in power, the time rate of doing work, as the indicated power $\dot{W}_i$, expressed in kilowatts. Not all the work produced by the fluid is delivered to the engine shaft, however; some of it must overcome the friction in the engine, and we have to consider the *pumping losses* in the exhaust and intake processes. This is the *frictional work,* W_f and the power is the *friction power,* $\dot{W}_f$. The remaining portion of the indicated work can be used to drive an automobile, propel a ship, or accomplish whatever purpose the engine may have. The work is called the *shaft* or *brake work,* W_b, and the associated power is the *brake power,* $\dot{W}_b$.

Indicated work − friction work = brake work

$$W_i \quad - \quad W_f \quad = \quad W_b$$

Several efficiencies may be determined for an engine: two engine efficiencies are the brake engine efficiency, η_b; and the indicated engine efficiency, η_i. All efficiencies are the desired effect divided by the expense of producing the effect.

$$\eta_{engine} = \frac{\text{actual work of a system}}{\text{work of corresponding ideal system}} \tag{13.21}$$

Thus,

$$\eta_b = \frac{\text{brake work}}{\text{theoretical work}} = \frac{W_b}{W} \tag{13.22}$$

$$\eta_i = \frac{\text{indicated work}}{\text{theoretical work}} = \frac{W_i}{W} \tag{13.23}$$

The mechanical efficiency of an engine, η_m, is an indication of how well the engine could convert the indicated work into brake work, or

$$\eta_m = \frac{\text{brake work}}{\text{indicated work}} = \frac{W_b}{W_i} \tag{13.24}$$

There are thermal efficiencies for the engine also. We know that thermal efficiencies are

$$\eta_{\text{th}} = \frac{\text{work output}}{\text{energy input}} \tag{13.25}$$

and since we are discussing two work terms, there are two thermal efficiencies. The brake thermal efficiency, η_{tb}, is

$$\eta_{tb} = \frac{\text{brake work}}{\text{energy in}} = \frac{\dot{W}_b}{\dot{m}_f h_{RP}} \tag{13.26}$$

and the indicated thermal efficiency, η_{ti}, is

$$\eta_{ti} = \frac{\text{indicated work}}{\text{energy in}} = \frac{\dot{W}_i}{\dot{m}_f h_{RP}} \tag{13.27}$$

The higher heating value of the fuel is typically used. There are many ways to relate the various efficiencies by algebraic manipulation, which is best left for the individual problem.

13.9 POWER MEASUREMENT

We have discussed the various work forms and the engine efficiencies. How are these work terms measured experimentally? The brake work may be measured by determining the horsepower generated by the rotating engine shaft. For low-speed, low-horsepower engines, this may be accomplished with the use of a prony brake (Figure 13.20). For higher powers and speeds, a hydraulic or electric dynamometer is used, and the output is typically measured electrically. Let us consider the prony brake, as this is the simplest case.

The brake work is dissipated as friction on a water-cooled friction band, similar to a brake shoe on an automobile. Work is defined as a force acting through a

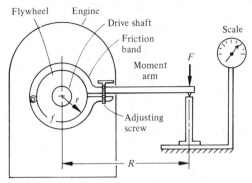

Figure 13.20 A schematic diagram of a prony brake.

distance. For one revolution of the engine, a point on the perimeter of the flywheel will move a distance $2\pi r$. A uniform friction force f acts against the flywheel throughout the revolution; thus, the work dissipated as friction during the revolution is

$$\text{Work during one revolution} = \text{force} \times \text{distance} = 2\pi rf$$

The turning moment, rf, produced by the drive shaft is balanced by an equal and opposite moment, which is the product of the length of the moment arm, R, and the force on the scale, F. Thus

$$rf = RF$$

$$\text{Work during one revolution} = 2\pi RF$$

If the number of revolutions per unit time is known, then the time rate of doing work, or power, may be calculated.

$$\text{Power} = 2\pi RFn$$

where n is the revolutions per minute. To determine the brake power in watts,

$$\dot{W}_b = (2\pi R \text{ m/rev})(F \text{ N})(n \text{ rev/s})$$

$$\dot{W}_b = 2\pi RFn \text{ w} \tag{13.28a}$$

To determine the brake power in horsepower,

$$\text{bhp} = \frac{(2\pi R \text{ ft/rev})(F \text{ lbf})(n \text{ rev/min})}{33{,}000 \text{ ft-lbf/min-hp}}$$

$$\text{bhp} = \frac{2\pi RFn}{33{,}000} \tag{13.28b}$$

The product of the moment arm, R, and the force, F, is called the *torque*, τ, of the engine. The torque is the turning moment exerted by a tangential force acting at a distance from the axis of rotation. Thus, equations (13.28a) and (13.28b) could be written as

$$\dot{W}_b = 2\pi\tau n \text{ W} \tag{13.29a}$$

$$\text{bhp} = \frac{2\pi\tau n}{33,000} \tag{13.29b}$$

Since the torque of an engine is frequently referred to and is sometimes confused with power, the following example should help keep the two distinct. Torque is the capacity to do work, whereas power is the rate at which work may be done. Let us consider a truck pulling a load. The torque will determine whether or not the truck is able to pull the load; the power will determine how fast the load may be pulled.

If a test is being run on the prony brake, the brake arm will exert an initial force, or *tare*. This must be subtracted from the scale reading to find the force exerted by the engine.

The indicated work is found by determining the *p-V* diagram of the engine. An indicator card of the cycle is taken either mechanically or with an oscilloscope. The work is the net area enclosed by the cycle processes. This area divided by the length of stroke will give the mean effective pressure. It may be helpful to review the section in Chapter 7 in which the concept of mean effective pressure was developed. The *indicated mean effective pressure, p_{mi},* is determined by knowing the indicated work.

$$p_{mi} = \frac{W_i}{\text{piston displacement}} \tag{13.30}$$

The brake mean effective pressure, p_{mb}, is

$$p_{mb} = \frac{W_b}{\text{piston displacement}} \tag{13.31}$$

The power may be determined in terms of the mean effective pressure.

$$\text{Piston displacement} = LA$$

where L is the length of the stroke and A is the area of the piston.

$$p_{mi}LA = \frac{\text{indicated work}}{\text{revolutions}}$$

$$p_{mi}LAn = \text{indicated power}$$

$$\dot{W}_i = p_{mi}LAn \text{ W} \tag{13.32a}$$

$$\text{ihp} = \frac{p_{mi}LAn}{33,000} \text{ hp} \tag{13.32b}$$

$$\dot{W}_b = p_{mb}LAn \text{ W} \tag{13.33a}$$

$$\text{bhp} = \frac{p_{mb}LAn}{33,000} \text{ hp} \tag{13.33b}$$

The important concept to remember about mean effective pressure is that it is an artificial pressure based on the work of the cycle.

Example 13.7

A spark-ignition engine produces 224 kW while using 0.0169 kg/s of fuel. The fuel has a higher heating value of 44 186 kJ/kg, and the engine has a compression ratio of 8. The friction power is found to be 22.4 kW. Determine η_{tb}, η_{ti}, η_m, η_b, and η_i.

Solution

Given: A spark-ignition engine with the power it produces, the frictional power, the fuel consumption, and the fuel type.

Find: Various engine efficiencies.

Sketch and Given Data:

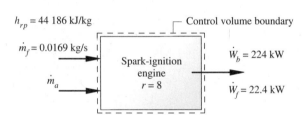

$h_{rp} = 44\ 186$ kJ/kg

$\dot{m}_f = 0.0169$ kg/s

Control volume boundary

Spark-ignition engine
$r = 8$

$\dot{W}_b = 224$ kW

$\dot{W}_f = 22.4$ kW

$\dot{m}_a$

Figure 13.21

Assumption: The theoretical cycle for the spark-ignition engine is the air-standard Otto cycle.

Analysis: Determine the thermal efficiency to be

$$\eta_{tb} = \frac{\dot{W}_b}{\dot{m}_f h_{RP}} = \frac{(224\ \text{kW})}{(0.0169\ \text{kg/s})(44\ 186\ \text{kJ/kg})} = 0.30$$

The indicated power is

$$\dot{W}_i = \dot{W}_b + \dot{W}_f = 246.4\ \text{kW}$$

The indicated thermal efficiency is

$$\eta_{ti} = \frac{\dot{W}_i}{\dot{m}_f h_{RP}} = \frac{(246.4\ \text{kW})}{(0.0169\ \text{kg/s})(44\ 186\ \text{kJ/kg})} = 0.33$$

The mechanical efficiency is

$$\eta_m = \frac{\dot{W}_b}{\dot{W}_i} = \frac{224}{246.4} = 0.909$$

The brake engine efficiency requires that we know the theoretical power produced.

From the assumption the theoretical cycle is the air-standard Otto cycle. Hence,

$$\eta_{\text{Otto}} = 1 - \frac{1}{(r)^{k-1}} = 1 - 0.435 = 0.565$$

$$\eta_{\text{Otto}} = \frac{\dot{W}}{\dot{Q}_{\text{in}}} = 0.565 = \frac{\dot{W}}{(0.0169 \text{ kg/s})(44 \ 186 \text{ kJ/kg})}$$

$$\dot{W} = 421.9 \text{ kW}$$

The brake engine efficiency is defined as

$$\eta_b = \frac{\dot{W}_b}{\dot{W}} = \frac{224}{421.9} = 0.53$$

and the indicated engine efficiency is

$$\eta_i = \frac{\dot{W}_i}{\dot{W}} = \frac{246.4}{421.9} = 0.58$$

Comment: Engine efficiency is a function of how the ideal work is calculated. Frequently, the air-standard cycle is used for calculating the theoretical work. ▬

Example 13.8
A six-cylinder automotive engine with a bore and stroke of $3\frac{1}{2} \times 3\frac{1}{2}$ in. has a fuel consumption of 0.50 lbm/bhp-hr at 3000 rpm; bhp = 115; ihp = 140. The thermal efficiency of the ideal cycle is 47%, and the fuel has a heating value of 19,000 Btu/lbm. Compute the mechanical efficiency and the brake and indicated thermal efficiencies; the brake and indicated engine efficiencies; and the brake and indicated mean effective pressures.

Solution

Given: An engine with its fuel consumption, size, power ratings, thermal efficiency, and fuel type.

Find: Various engine efficiencies and mean effective pressures.

Sketch and Given Data:

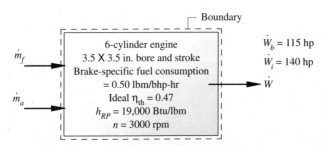

Figure 13.22

Analysis: The mechanical efficiency is

$$\eta_m = \frac{bhp}{ihp} = \frac{115}{140} = 0.821 \quad \text{or } 82.1\%$$

The fuel flow rate is

$$\dot{m}_f = (0.5 \text{ lbm/bhp-hr})(115 \text{ bhp}) = 57.5 \text{ lbm/hr}$$

and the brake thermal efficiency is

$$\eta_{tb} = \frac{bhp}{\dot{m}_f h_{RP}} = \frac{(115 \text{ hp})(2545 \text{ Btu/hr-hp})}{(57.5 \text{ lbm/hr})(19{,}000 \text{ Btu/lbm})} = 0.267 \quad \text{or } 26.7\%$$

The indicated thermal efficiency is

$$\eta_{ti} = \frac{(140 \text{ hp})(2545 \text{ Btu/hr-hp})}{(57.5 \text{ lbm/hr})(19{,}000 \text{ Btu/lbm})} = 0.326 \quad \text{or } 32.6\%$$

The brake engine efficiency is

$$\eta_b = \frac{bhp}{\dot{W}}$$

where the theoretical power is determined from the thermal efficiency for an ideal cycle.

$$\dot{W} = 0.47 \, (\dot{Q}_{in}) = \frac{(0.47)(57.5 \text{ lbm/hr})(19{,}000 \text{ Btu/lbm})}{(2545 \text{ Btu/hr-hp})} = 201.7 \text{ hp}$$

Thus,

$$\eta_b = \frac{115}{201.7} = 0.57 \quad \text{or } 57\%$$

The indicated engine efficiency is

$$\eta_i = \frac{ihp}{\dot{W}} = \frac{140}{201.7} = 0.694 \quad \text{or } 69.4\%$$

The mean effective pressures are found from equations (13.30) and (13.31). The piston displacement per unit time is

$$\dot{V}_{PD} = \frac{6\pi d^2 Ln}{4}$$

$$\dot{V}_{PD} = \frac{\pi}{4} \times 6 \times \left(\frac{3.5}{12}\text{ ft}\right)^2 \times \left(\frac{3.5}{12}\text{ ft}\right) \times 3000\text{ rpm}$$

$$= 350.8\text{ ft}^3/\text{min}$$

$$p_{mi} = \frac{(140\text{ hp})(33\ 000\text{ ft-lbf/min-hp})}{(350.8\text{ ft}^3/\text{min})(144\text{ in.}^2/\text{ft}^2)} = 91.5\text{ lbf/in.}^2$$

and

$$p_{mb} = \frac{(115\text{ hp})(33\ 000\text{ ft-lbf/min-hp})}{(350.8\text{ ft}^3/\text{min})(144\text{ in.}^2/\text{ft}^2)} = 75.1\text{ lbf/in.}^2$$

Comment: The mean effective pressure is an artificial pressure based on the particular type of work produced. It is important when comparing engine specifications, such as mean effective pressure, that the comparisons be based on the same power produced.

CONCEPT QUESTIONS

1. What are the assumptions implicit in the air-standard cycles?
2. What are the differences between the air-standard Otto and Diesel cycles?
3. What are the processes that characterize the air-standard Otto cycle?
4. What are the processes that characterize the air-standard Diesel cycle?
5. Will the efficiency of an Otto cycle be greater than that of a Carnot cycle, both operating between the same temperature limits?
6. How does the efficiency of an air-standard Otto cycle vary with compression ratio and gas specific heat ratio?
7. Which engine, spark-ignition or compression-ignition, operates at a higher compression ratio? Why?
8. What is the purpose of the air-standard dual cycle?
9. What is the purpose of regenerative heating in the Stirling and Ericsson cycles?
10. What is compression ratio, cutoff ratio, and expansion ratio? Describe them in terms of piston movement.
11. What is the difference between compression-ignition and spark-ignition engines?
12. How are combustion and exhaust processes modeled in the air-standard cycles?
13. What are the differences between two- and four-stroke cycle engines?
14. Can the mean effective pressure of a cycle be less than atmospheric pressure? Why?
15. How are volumetric efficiency, clearance volume, and percentage of clearance related?

16. Describe the four-stroke cycle internal-combustion engine, noting differences between compression-ignition and spark-ignition engines.

17. What is the effect of unpurged products in the clearance volume on engine performance?

18. At part-load operation, which engine is more efficient, the compression-ignition or spark-ignition? Why?

19. What is mechanical efficiency?

20. What is the difference between indicated and brake mean effective pressure?

21. What is the difference between torque and power?

22. How does increasing the compression ratio affect the volumetric efficiency?

PROBLEMS (SI)

13.1 An air-standard Otto cycle has a compression ratio of 8.0 and has air conditions at the beginning of compression of 100 kPa and 25°C. The heat added is 1400 kJ/kg. Determine (a) the four cycle state points; (b) the thermal efficiency; (c) the mean effective pressure.

13.2 An engine operates on an air-standard Otto cycle with a compression ratio of 9 to 1. The pressure and temperature at the end of the compression stroke are 800 kPa and 700°C. Determine the net cycle work per kg if the pressure at the end of heat addition is 3.0 MPa.

13.3 An engine operates on the air-standard Otto cycle. The pressure and temperature at the beginning of isentropic compression are 120 kPa and 35°C. The peak pressure and temperature are 4.8 MPa and 2500°C. Determine (a) the net cycle work in kJ/kg; (b) the cycle efficiency.

13.4 An air-standard Otto cycle uses 0.1 kg of air and has a 17% clearance. The intake conditions are 98 kPa and 37°C, and the energy release during combustion is 1600 kJ/kg. Using the air tables, determine (a) the compression ratio; (b) the pressure and temperature at the four cycle state points; (c) the displacement volume; (d) the thermal efficiency; (e) the work; (f) the cycle second-law efficiency.

13.5 An air-standard Otto cycle has the following cycle states, where state 1 is at the beginning of the isentropic compression: $p_1 = 101$ kPa, $T_1 = 333°$K, $V_1 = 0.28$ m^3, $T_3 = 2000°$K, $r = 5$. Determine (a) the remaining cycle state points; (b) the thermal efficiency; (c) the heat added; (d) the heat rejected; (e) if $T_1 = T_0$, the available portion of the heat rejected.

13.6 A four-cylinder spark-ignition engine with a compression ratio of 8 has pistons with a bore of 9 cm and a stroke of 10 cm. The air pressure at the beginning of compression is 98 kPa, and the temperature is 37°C. The engine may be modeled by the air-standard Otto cycle. The maximum cycle temperature is 1700°K. If the engine produces 75 kW of power, determine (a) the heat supplied per cylinder; (b) the thermal efficiency; (c) the rpm required.

13.7 An air-standard Otto cycle has a compression ratio of 7.5. The maximum and minimum cycle temperatures are 1600° and 300°K, and the minimum pressure is 100 kPa. Determine (a) the cycle efficiency; (b) the change of entropy during heat addition; (c) the change of availability per unit mass during the expansion process.

13.8 A four-cylinder engine with a 9.5-cm bore and an 8.75-cm stroke has a 10% clearance. The engine rotates at 2500 rpm. The conditions at the beginning of compression are 17°C and 98 kPa. The maximum cycle temperature is 2900°K. The engine may be assumed to operate on an air-standard Otto cycle. Determine the cycle work and the power produced by the engine.

13.9 An air-standard Diesel cycle receives 28.5 kJ/cycle of heat while operating at 300 rpm. At the beginning of compression, $p_1 = 100$ kPa, $T_1 = 305°$K, and $V_1 = 0.0425$ m^3. At the beginning of heat addition, the pressure is 3450 kPa. Determine (a) p, V, and T at each cycle state point; (b) the work; (c) the power; (d) the mean effective pressure.

13.10 An engine operates on the air-standard Diesel cycle with a compression ratio of 18. The pressure and temperature at the beginning of compression are 120 kPa and 43°C. The maximum temperature is 1992°K, and the heat added is 1274 kJ/kg. Determine (a) the maximum pressure; (b) the temperature at the beginning of heat addition.

13.11 An air-standard Diesel cycle has a compression ratio of 20 and a cutoff ratio of 3. Inlet pressure and temperature are 100 kPa and 27°C. Determine (a) the heat added per kg; (b) the net work per kg.

13.12 An air-standard Diesel-cycle engine operates as follows: at the end of expansion the pressure is 240 kPa and the temperature is 550°C; at the end of compression the pressure is 4.2 MPa and the temperature is 700°C. Determine (a) the compression ratio; (b) the cutoff ratio; (c) the heat added per kg of air; (d) the cycle efficiency; (e) the cycle second-law efficiency.

13.13 In an air-standard Diesel cycle, the compression ratio is 17. The cutoff ratio, the ratio of the volume after heat addition to that before heat addition (V_3/V_2), is 2.5 : 1. The air conditions at the beginning of compression are 101 kPa and 300°K. Determine (a) the thermal efficiency; (b) the heat added per kg of air; (c) the mean effective pressure.

13.14 In an air-standard Diesel cycle, the air is compressed isentropically from 26°C and 105 kPa to 3.7 MPa. The entropy change during heat rejection is -0.6939 kJ/kg-K. Determine (a) the heat added per kg of air; (b) the thermal efficiency; (c) the maximum temperature; (d) the temperature at the start of heat rejection.

13.15 A four-cylinder compression-ignition engine with a compression ratio of 18 has pistons with a bore of 9 cm and a stroke of 10 cm. The air pressure at the beginning of compression is 98 kPa, and the temperature is 37°C. The engine may be modeled by the air-standard Diesel cycle. The maximum cycle temperature is 1700°K. If the engine produces 75 kW of power, determine (a) the heat supplied per cylinder; (b) the thermal efficiency; (c) the rpm required.

13.16 An air-standard Diesel cycle has a compression ratio of 14. The maximum and minimum cycle temperatures are 1600° and 300°K, and the minimum pressure is 100 kPa. Determine (a) the cycle efficiency; (b) the change of entropy during heat addition; (c) the change of availability per unit mass during the expansion process.

13.17 A four-cylinder engine with a 9.5-cm bore and an 8.75-cm stroke has a 7% clearance. The engine rotates at 2500 rpm. The conditions at the beginning of compression are 17°C and 98 kPa. The maximum cycle temperature is 2900°K. The engine may be assumed to operate on an air-standard Diesel cycle. Determine the cycle work and the power produced by the engine.

13.18 A Diesel engine operates on the air-standard cycle and has a displacement of 3.5 liters. The engine's compression ratio is 19, and the cutoff ratio is 2.4. The conditions at the beginning of compression are 37°C, 98 kPa, and 3.7 liters. If the engine operates at

1500 rpm, determine (a) the expansion ratio; (b) the cycle work; (c) the power developed.

13.19 The compression ratio of an air-standard dual cycle is 12, and at the beginning of compression the pressure is 100 kPa, the volume is 1.2 liters, and the temperature is 37°C. During the heat-addition processes, 0.4 kJ is transferred at constant volume and 1.0 kJ at constant pressure. Determine (a) the cycle thermal efficiency; (b) the pressure at the beginning of heat rejection.

13.20 A dual-cycle engine has a compression ratio of 14. The air state at the beginning of compression is 100 kPa and 300°K. Fifteen hundred kJ/kg of heat is added during the heat-addition processes, with one-third at constant volume and two-thirds at constant pressure. (a) Find the change of availability during the heat-addition processes. (b) Compare this with the change of availability during heat addition of an air-standard Diesel cycle with the same cycle conditions.

13.21 In an air-standard dual cycle, the isentropic compression starts at 100 kPa and 300°K. The compression ratio is 13, the maximum temperature is 2750°K, and the maximum pressure is 6894 kPa. Determine (a) the cycle work per kg; (b) the heat added per kg; (c) the mean effective pressure.

13.22 An engine operates on an air-standard Stirling cycle with regeneration. At the beginning of isothermal compression the air pressure is 150 kPa. At the end of isothermal compression the pressure is 300 kPa. The peak cycle pressure and temperature are 1.5 MPa and 850°C. Determine (a) the heat added per kg; (b) the heat rejected per kg; (c) the thermal efficiency; (d) the cycle second-law efficiency.

13.23 An engine operates on the air-standard Stirling cycle. The pressure and temperature at the beginning of isothermal compression are 3.5 MPa and 150°C. The engine thermal efficiency is 40%. Determine the heat transferred to the regenerator per kg.

13.24 An engine operates on an air-standard Stirling cycle. The pressure and temperature at the beginning of isothermal compression are 700 kPa and 100°C. The engine has a compression ratio of 3 and a mean effective pressure of 1.0 MPa. Determine the heat transferred to the regenerator per kg.

13.25 An Ericsson cycle uses helium as the working fluid. The isothermal compression process begins at 300°K and 120 kPa, and 175 kJ/kg of heat is rejected. Heat addition occurs at 1100°K. Determine (a) the cycle maximum pressure; (b) the net work produced per unit mass; (c) the thermal efficiency.

13.26 An automobile engine, operating on the open Otto cycle, receives air at 100 kPa and 27°C and fuel (octane) at the same temperature. The engine has a compression ratio of 8. The engine has an air/fuel ratio of 25. The heat loss from the engine is equal to 30% of the work produced. Determine (a) the maximum temperature and pressure; (b) the work produced per cycle; (c) the heat loss; (d) the available portion of the products if $T_0 = 300°K$.

13.27 A test is run on a high-speed four-cycle spark-ignition engine operating at 5000 rpm. Air enters at 100 kPa and 27°C, and fuel, with a heating value of 43 000 kJ/kg, enters at the same temperature. The maximum temperature is 2100°K. Cooling water enters the engine at 21°C and leaves at 43°C with a flow of 27 kg water/kg fuel. Use the open Otto cycle for determining necessary state points. For a six-cylinder engine with a bore and stroke of 9 × 9 cm and a compression ratio of 7.5, determine (a) the fuel/air ratio; (b) the fuel flow rate; (c) the power produced; (d) the water flow rate; (e) the exhaust temperature.

13.28 A stationary internal-combustion engine operates on the Otto cycle and develops 335 kW at 2200 rpm with full throttle at standard atmospheric pressure and temperature. The mechanical efficiency is 80%, and the brake specific fuel consumption is 7.6×10^{-5} kg/s-kW. The engine is moved to a higher elevation, and tests indicate that only 261 kW will be produced at full throttle. The thermal efficiency is assumed to be constant. Determine (a) the percentage of change in the volumetric efficiency at the new elevation; (b) the elevation; (c) the fuel tank required for 2 days' fuel, if the fuel density is 672 kg/m³.

13.29 A six-cylinder spark-ignition engine has a bore and stroke of 10.9×10.5 cm. The engine requires 0.0035 kg/s of $C_8H_{18}(l)$ when operating at half-load with a speed of 3000 rpm. The reduction of engine speed to axle speed is 3.78:1. The tires have an effective radius of 35.5 cm. (a) Determine the car speed in km/h and the fuel consumption in km/liter (the specific gravity may be assumed to be 0.85). (b) The air/fuel ratio on the mass basis is 15.3:1; the products of combustion leave the engine at 900°K, with air and fuel inlet temperature of 25°C. Determine the percentage of the heat release lost to the products of combustion.

13.30 An automotive engine is rated at 150 kW at 4500 rpm. The fuel has a heating value of 46 500 kJ/kg. The overall thermal efficiency is 27%. Find the brake-specific fuel consumption.

13.31 In an air-standard Ericsson cycle the maximum pressure is 4.1 MPa and the minimum pressure is 210 kPa. The heat supplied is 581 kJ/kg, and the minimum temperature is 21°C. Determine (a) the cycle work; (b) the heat rejected; (c) the heat stored in the regenerator; (d) the entropy change during heat addition.

13.32 An engine operating on the air-standard Stirling cycle is examined and found to have the following conditions: $p_1 = 725$ kPa, $T_1 = 590°K$, and $V_1 = 0.0567$ m³ at the beginning of isothermal expansion; $V_2/V_1 = 1.5$; $T_3 = 300°K$. For one cycle, determine (a) the work; (b) the thermal efficiency; (c) the mean effective pressure; (d) the heat rejected; (e) the heat added.

13.33 An eight-cylinder Diesel engine with a bore and stroke of 10×10 cm operates at 2000 rpm. Dodecane ($C_{12}H_{26}[l]$) fuel is used with 80% excess air. The air enters the engine at 100 kPa and 37°C and is compressed to 3.0 MPa. The heat loss from the engine is one-third of the work produced. Use the open-system diesel cycle to calculate state points. Determine (a) the compression ratio; (b) the fuel consumption; (c) the thermal efficiency; (d) the power produced; (e) the engine-cooling water required if the water enters at 21°C and leaves at 49°C.

13.34 A Diesel engine with a compression ratio of 14.5 starts the compression stroke with air at 101 kPa and 312°K. Fuel with a heating value of 43 260 kJ/kg is used with a ratio of 0.0333 kg fuel/kg air. The airflow to the engine is measured to be 0.10 m³/s. Use the ideal-gas laws for air and determine (a) the maximum temperature; (b) the mean effective pressure; (c) the maximum power.

13.35 A four-cylinder, four-cycle Diesel engine has a bore and stroke of 30×53 cm and operates at 250 rpm. The fuel is natural gas with a heating value of 37 200 kJ/m³, measured at 25°C and 1 atm pressure. The theoretical air/fuel ratio is 10.3 m³ air/m³ gas. At full load, the engine requires 25% excess air. The indicated thermal efficiency is 25%, and the mechanical efficiency is 75%. Determine (a) the full-load power; (b) the full-load fuel consumption in m³/s; (c) the brake mean effective pressure at full load.

13.36 A four-stroke cycle Diesel engine with a compression ratio of 16 drives a 500-kW generator at 1200 rpm. The generator efficiency is 90%. At this condition the engine's air/fuel ratio is 20, and the brake-specific fuel consumption is 6.75×10^{-5} kg/s-kW. The inlet air conditions are 90 kPa and 35°C. The engine volumetric efficiency is 85%. The stroke is 1.2 times the bore. Determine for a six-cylinder engine (a) the bore and stroke per cylinder; (b) the clearance volume per cylinder; (c) the total fuel consumption in kg/s; (d) the thermal efficiency if the fuel is $C_{12}H_{26}$.

13.37 A Diesel engine develops 750 kW at 200 rpm when the ambient pressure is 100 kPa and the temperature is 17°C. The air/fuel ratio is 23 kg air/kg fuel, and 7.6×10^{-5} kg/s of fuel is consumed per brake kW developed. Determine for $h_{RP} = 43\ 200$ kJ/kg (a) the thermal efficiency; (b) the fuel consumption for 52 kW if the thermal efficiency is constant; (c) the second-law cycle efficiency.

13.38 A prony brake is used to analyze the performance of a small, double-acting, reciprocating steam engine. The engine operates at 200 rpm and has an indicated mean effective pressure of 516 kPa, a stroke of 17.7 cm, a bore of 15 cm, a moment arm of 58 cm, and a scale reading of 667 N. The tare reading is 44 N. Determine (a) the brake mean effective pressure; (b) the indicated power; (c) the friction power; (d) the mechanical efficiency.

13.39 An engine has 14 cylinders with a 13.6-cm bore and a 15.2-cm stroke and develops 2850 kW at 250 rpm. The clearance volume of each cylinder is 380 cm³. Determine (a) the compression ratio; (b) the brake mean effective power.

13.40 A test on a one-cylinder Otto cycle engine yields the following data: 950 N-m torque; 758 kPa mean effective pressure; 28-cm bore; 30.5-cm stroke; 300 rpm; and 0.003-kg/s fuel consumption with a heating value of 41 860 kJ/kg. Determine (a) the engine thermal efficiency; (b) the engine mechanical efficiency; (c) the fuel cost per h if fuel costs 50 cents per liter. The specific gravity is 0.82.

13.41 Calculate the bore and stroke of a six-cylinder engine that delivers 22.4 kW at 1800 rpm with a ratio of bore to stroke of 0.71. Assume the mean effective pressure in the cylinder is 620 kPa and the mechanical efficiency is 85%.

13.42 A spark-ignition internal-combustion engine must be designed that will have six cylinders and a four-stroke cycle and produce 261 N-m of torque at 3000 rpm with an indicated mean effective pressure of 813 kPa. The mechanical efficiency is 78%, the brake thermal efficiency is 23%, the clearance is 15.5%, and the ratio of bore to stroke is 1.0. At the beginning of compression $p_1 = 99$ kPa, $T_1 = 77$°C, and $k = 1.31$. Determine (a) the brake hp; (b) the compression ratio; (c) the bore and stroke; (d) fuel consumption in kg/h if the lower heating value is 44 000 kJ/kg.

13.43 A six-cylinder, 70×90–cm, two-stroke cycle Diesel engine develops 2240 kW at 128 rpm. The cutoff ratio is 2.45, the expansion ratio is 5.3, $k = 1.33$, $p_1 = 97$ kPa, and $T_1 = 55$°C. The indicated mean effective pressure is 620 kPa. During the 30-min test 286 kg of fuel with a lower heating value of 42 566 kJ/kg was consumed. Determine (a) the indicated power; (b) the mechanical efficiency; (c) the brake engine efficiency.

13.44 A six-cylinder Diesel engine with a bore and stroke of 44×63 cm operates at 225 rpm and produces 560 kW brake power. The fuel consumption is 136 kg/h. The engine's mechanical efficiency is 85%, and the ideal cycle efficiency is 52.2%. Determine (a) the indicated power; (b) the indicated mean effective pressure; (c) the brake engine efficiency; (d) the brake mean effective pressure.

13.45 A six-cylinder four-stroke cycle spark-ignition engine with a compression ratio of 9.5 must be designed to produce 67.1 kW with a torque of 194 N-m. At these conditions the mechanical efficiency is 78%, and the brake mean effective pressure is 550 kPa. For the air-standard cycle, $p_1 = 101$ kPa, $T_1 = 308°K$, and $k = 1.32$. The fuel flow rate is 0.353 kg/h-kW, where the power is indicated power and the fuel has a lower heating value of 43 970 kJ/kg. The ratio of the piston bore to stroke is 1.1. Determine (a) the bore and stroke; (b) the indicated thermal efficiency; (c) the brake engine efficiency.

PROBLEMS (English Units)

*13.1   An air-standard Otto cycle has an initial temperature of 100°F, a pressure of 14.7 psia, and a compression pressure $p_2 = 356$ psia. The pressure at the end of heat addition is 1100 psia. Determine (a) the compression ratio; (b) the thermal efficiency; (c) the percentage of clearance; (d) the maximum temperature.

*13.2   An air-standard Otto cycle has 15% clearance and starts compression at 14.7 psia and 100°F. The heat added is 525 Btu/lbm. Determine (a) the compression ratio; (b) the pressures and temperatures around the cycle at states 1 to 4; (c) the thermal efficiency.

*13.3   A four-cylinder spark-ignition engine with a compression ratio of 8 has pistons with a bore of 3.54 in. and a stroke of 3.94 in. The air pressure at the beginning of compression is 14.6 psia, and the temperature is 100°F. The engine may be modeled by the air-standard Otto cycle. The maximum cycle temperature is 3060°R. If the engine produces 100 hp, determine (a) the heat supplied per cylinder; (b) the thermal efficiency; (c) the rpm required.

*13.4   An air-standard Otto cycle has a compression ratio of 7.5. The maximum and minimum cycle temperatures are 2940° and 540°R, and the minimum pressure is 14.7 psia. Determine (a) the cycle efficiency; (b) the change of entropy during heat addition; (c) the change of availability per unit mass during the expansion process.

*13.5   A four-cylinder engine with a 3.75-in. bore and a 3.4-in. stroke has a 10% clearance. The engine rotates at 2500 rpm. The conditions at the beginning of compression are 65°F and 14.5 psia. The maximum cycle temperature is 5220°R. The engine may be assumed to operate on an air-standard Otto cycle. Determine the cycle work and the power produced by the engine.

*13.6   A four-cylinder compression-ignition engine with a compression ratio of 18 has pistons with a bore of 3.5 in. and a stroke of 3.9 in. The air pressure at the beginning of compression is 14.6 psia, and the temperature is 100°F. The engine may be modeled by the air-standard Diesel cycle. The maximum cycle temperature is 3060°R. If the engine produces 100 hp, determine (a) the heat supplied per cylinder; (b) the thermal efficiency; (c) the rpm required.

*13.7   An air-standard Diesel cycle has a compression ratio of 14. The maximum and minimum cycle temperatures are 2940° and 540°R, and the minimum pressure is 14.7 psia. Determine (a) the cycle efficiency; (b) the change of entropy during heat addition; (c) the change of availability per unit mass during the expansion process.

*13.8   An air-standard Diesel-cycle engine operates on 1 ft³ of air at 14.5 psia and 140°F. The compression ratio is 14, and the cutoff is 6.2% of the stroke. Determine (a) the temperatures and pressures around the cycle; (b) the net work; (c) the heat added; (d) the efficiency.

*13.9   A one-cylinder Diesel engine operates on the air-standard cycle and receives 27 Btu/rev. The inlet pressure is 14.7 psia, the inlet temperature is 90°F, and the volume at bottom dead center is 1.5 ft³. At the end of compression, $p_2 = 500$ psia. Determine (a) the cycle state points; (b) the power if the engine runs at 300 rpm; (c) the mean effective pressure.

*13.10 A four-cylinder engine with a 3.75-in. bore and a 3.4-in. stroke has a 7% clearance. The engine rotates at 2500 rpm. The conditions at the beginning of compression are 65°F and 14.6 psia. The maximum cycle temperature is 5220°R. The engine may be assumed to operate on an air-standard Diesel cycle. Determine the cycle work and the power produced by the engine.

*13.11 A Diesel engine operates on the air-standard cycle and has a displacement of 210 in.^{3}. The engine's compression ratio is 19, and the cutoff ratio is 2.4. The conditions at the beginning of compression are 100°F, 14.6 psia, and 221 in.^{3}. If the engine operates at 1500 rpm, determine (a) the expansion ratio; (b) the cycle work; (c) the power developed.

*13.12 An air-standard dual cycle is characterized by the following: $p_1 = 14.1$ psia, $T_1 = 80°F$, $p_3 = 470$ psia, and $r = 9$; heat addition at $p = C$ is 100 Btu/lbm. Determine (a) the cycle state points; (b) the thermal efficiency; (c) the mean effective pressure.

*13.13 The compression ratio of an air-standard dual cycle is 12, and the pressure at the beginning of compression is 14.7 psia, the volume is 75 in.^{3}, and the temperature is 100°F. During the heat-addition processes, 0.4 Btu is transferred at constant volume and 1.0 Btu at constant pressure. Determine (a) the cycle thermal efficiency; (b) the pressure at the beginning of heat rejection.

*13.14 A dual-cycle engine has a compression ratio of 14. The air state at the beginning of compression is 14.7 psia and 77°F. Six hundred forty-five Btu/lbm of heat is added during the heat-addition processes, with one-third at constant volume and two-thirds at constant pressure. Find the change of availability during the heat-addition processes. Compare this with the change of availability during heat addition of an air-standard Diesel cycle with the same cycle conditions.

*13.15 An Ericsson cycle uses helium as the working fluid. The isothermal compression process begins at 540°R and 17.6 psia, and 75 Btu/lbm of heat is rejected. Heat addition occurs at 1980°R. Determine (a) the cycle maximum pressure; (b) the net work produced per unit mass; (c) the thermal efficiency.

*13.16 A six-cylinder, four-stroke cycle, spark-ignition engine has a compression ratio of 9.5 and must produce 90 bhp with 415 ft-lbf of torque. The mechanical efficiency is 78%, and brake mean effective pressure is 80 psia. For the air-standard cycle $p_1 = 14.7$ psia, $T_1 = 100°F$, and $k = 1.32$. The indicated specific fuel consumption is 0.58 lbm/ihp-hr with a lower heating value of 18,900 Btu/lbm. Determine (a) the bore and stroke if their ratio is 1.1; (b) the brake engine efficiency.

*13.17 An adiabatic four-stroke cycle, six-cylinder Diesel engine has a bore of 12 in. and a stroke of 15 in. and operates at 500 rpm, receiving air at 90°F and 14.5 psia. The compression ratio is 18, and the dodecane fuel is injected at 100°F with a ratio of 0.0444 lbm fuel/lbm air. Assume the products of combustion have properties as in Example 13.6, but let $k = 1.3$ for products and reactants. Determine (a) the percentage of unpurged products; (b) the thermal efficiency; (c) the power produced.

*13.18 Consider a Diesel engine as an open system where it operates at 300 rpm, receiving 6 lbm/min of air at 90°F and 14.7 psia and exhausting at 800°F and 14.7 psia. Fuel is consumed at the rate of 16 lbm/hr and has a heating value of 18,500 Btu/lbm. The exhaust has properties per Example 13.6. Determine (a) the power if the heat and power are equal; (b) the efficiency.

*13.19 A spark-ignition internal-combustion engine needs to be designed that will have six cylinders and a four-stroke cycle and produce 193 ft-lbf of torque at 3000 rpm with an indicated mean effective pressure of 118 psia. The mechanical efficiency is 78%, the brake thermal efficiency is 23%, the clearance is 15.5%, and the ratio of bore to stroke is 1.0. At the beginning of compression $p_1 = 14.5$ psia, $T_1 = 170°F$, and $k = 1.31$. Determine (a) the bhp; (b) the compression ratio; (c) the bore and stroke; (d) the fuel consumption in lbm/hr if the lower heating value is 18,900 Btu/lbm.

*13.20 A six-cylinder, 28×36–in., two-stroke cycle Diesel engine develops 3000 bhp at 128 rpm. The cutoff ratio is 2.45, the expansion ratio is 5.3, $k = 1.33$, $p_1 = 14.1$ psia, and $T_1 = 130°F$. The indicated mean effective pressure is 90 psia. During the 1-hr test 1260 lbm of fuel with a lower heating value of 18,300 Btu/lbm was consumed. Determine (a) the indicated power; (b) the mechanical efficiency; (c) the brake engine efficiency.

*13.21 A six-cylinder Diesel engine with a bore and stroke of 17.5×25 in. operates at 225 rpm and produces 750 bhp. The fuel consumption is 300 lbm/hr. The engine's mechanical efficiency is 85%, and the ideal cycle efficiency is 52.2%. Determine (a) the indicated power; (b) the indicated mean effective pressure; (c) the brake engine efficiency; (d) the brake mean effective pressure.

COMPUTER PROBLEMS

C13.1 Develop a computer program, spreadsheet template, or TK Solver model to compute the thermal efficiency of an air-standard Otto cycle. Compute the thermal efficiency of the cycle for compression ratios between 6 and 11 for specific heat ratios of 1.3, 1.35, and 1.4, and plot the results.

C13.2 Develop a computer program, spreadsheet template, or TK Solver model to compute the thermal efficiency of an air-standard Diesel cycle. For a specific heat ratio of 1.4, compute the thermal efficiency of the cycle for compression ratios between 12 and 22 and for cutoff ratios of 2, 4, 6, and 8, and plot the results. Repeat for a specific heat ratio of 1.3.

C13.3 Develop a computer program, spreadsheet template, or TK Solver model to analyze an open Diesel cycle similar to the cycle in Example 13.6. Investigate the effects of (a) using higher values of specific heat to account for the variation with temperature; (b) different air/fuel ratios.

C13.4 Using the TK Solver model OTTO.TK, compute the thermal efficiency of an ideal Otto cycle with compression ratios between 6 and 11. Plot the results and compare them to those for the air-standard Otto cycle from Problem C13.1.

C13.5 Using the TK Solver model DIESEL.TK, compute the thermal efficiency of an ideal diesel cycle with compression ratios between 12 and 22, and for heat inputs 500 kJ/kg, 1500 kJ/kg, and 3000 kJ/kg. Plot the results and compare them to those for the air-standard Diesel cycle from Problem C13.2.

C13.6 Modify OTTO.TK or DIESEL.TK to permit the analysis of an ideal dual cycle. Compute the thermal efficiency for compression ratios of 8 and 16 and a heat input of 1500 kJ/kg. Vary the percentage of the heat input during the constant-pressure process from 25% to 75%.

14

Gas Turbines

Gas turbines are used in a variety of ways to produce power for industrial and societal needs. One of the most apparent uses of gas turbines is in jet aircraft engines. Other gas turbines designs provide electrical power for utilities, and still others are used as main and auxiliary propulsion in ships and offshore oil platforms. An important factor in the selection of a gas turbine power plant is that it is very compact and lightweight for the power produced. In this chapter we will

- Consider fundamental gas turbine cycle parameters;
- Develop the air-standard Brayton-cycle model;
- Investigate the effect of turbine, compressor, and combustion efficiency on cycle performance;
- Model actual open-system gas turbines;
- Discover methods to improve cycle thermal efficiency;
- Analyze jet engines for use on aircraft.

14.1 FUNDAMENTAL GAS TURBINE CYCLE

For the gas turbine to produce any work, the hot gases must expand from a high pressure to a low pressure. Therefore, the gases must first be compressed. If after the compression the fluid were expanded through the turbine, the power produced would

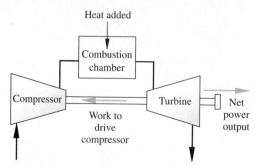

Figure 14.1 A simple open gas turbine cycle.

equal that used by the compressor, provided that both the turbine and compressor functioned ideally. If heat were added to the fluid before it reached the turbine, raising its temperature, then an increase in the power output could be achieved. Figure 14.1 illustrates this. If more and more thermal energy could be added to the fluid, more and more power output could be produced. Unfortunately this cannot occur; the turbine blades have a metallurgical thermal limit. If the gases continuously enter at a temperature higher than this, the combined thermal and material stresses in the blade will cause it to fail. Typically, inlet temperatures of 1300°K may be found on industrial turbines, and inlet temperatures on experimental models are up to 1500°K. Note that the turbine and compressor rotors are mechanically coupled, which allows a portion of the turbine work to drive the compressor, with the balance being the net work produced.

14.2 CYCLE ANALYSIS

The gas turbine cycle may be either closed or open. The open cycle is more common; atmospheric air is continuously drawn into the compressor, heat is added to the air by the combustion of fuel, and the fluid expands through the turbine and exhausts to the atmosphere. This is illustrated in Figure 14.1. In the closed cycle, the heat must be added to the fluid in a heat exchanger from an external source, such as a nuclear power plant, and the fluid must be cooled after it leaves the turbine and before it enters the compressor. Figure 14.2 illustrates this.

The air-standard Brayton cycle is the ideal closed-system gas turbine cycle. It is characterized by constant-pressure heat addition and heat rejection and isentropic compression and expansion processes. Air is the working fluid and may be considered an ideal gas. Figure 14.2 illustrates the schematic for this cycle, and Figure 14.3 illustrates the p-V and T-S diagrams for the cycle.

The thermal efficiency, η_{th}, of the Brayton cycle may be found as follows:

$$\eta_{th} = \frac{W_{net}}{Q_{in}} = \frac{Q_{in} - Q_{out}}{Q_{in}} = 1 - \frac{Q_{out}}{Q_{in}} = 1 - \frac{(h_4 - h_1)}{(h_3 - h_2)}$$

$$\eta_{th} = 1 - \frac{c_p(T_4 - T_1)}{c_p(T_3 - T_2)} = 1 - \frac{T_4 - T_1}{T_3 - T_2} \tag{14.1}$$

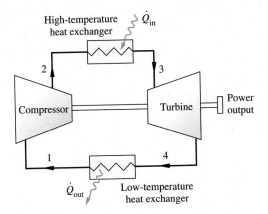

High-temperature
heat exchanger

Compressor

Turbine

Power
output

$\dot{Q}_{in}$

$\dot{Q}_{out}$

Low-temperature
heat exchanger

Figure 14.2 A simple closed gas turbine cycle.

The pressure ratio, r_p, is defined as

$$r_p = \frac{p_2}{p_1} \tag{14.2}$$

and from isentropic expansion and compression processes, we find that

$$\frac{T_2}{T_1} = (r_p)^{(k-1)/k}$$

$$\frac{T_4}{T_3} = \left(\frac{1}{r_p}\right)^{(k-1)/k}$$

Therefore,

$$\frac{T_2}{T_1} = \frac{T_3}{T_4} \tag{14.3}$$

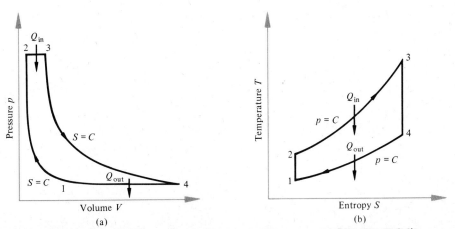

Figure 14.3 (a) The p-V diagram for the air-standard Brayton cycle. (b) The T-S diagram for the air-standard Brayton cycle.

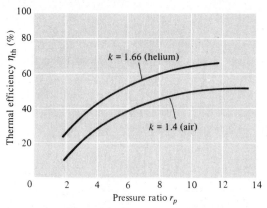

Figure 14.4 The variation of thermal efficiency with pressure ratio for the air-standard Brayton cycle.

We eliminate T_4 from equation (14.1) by means of equation (14.3):

$$\eta_{th} = 1 - \frac{1}{(r_p)^{(k-1)/k}} \tag{14.4}$$

Thus, for the Brayton cycle the thermal efficiency is a function of the pressure ratio, r_p. Figure 14.4 illustrates this for two values of k, one of air, $k = 1.4$, and the other of helium, $k = 1.66$.

The maximum temperature does have an effect on the optimum performance. The previous derivation does not account for a fixed temperature T_3. If T_3 and T_1 are fixed, there will be an optimum pressure ratio to produce a maximum amount of work W_{net}. The variable temperature is T_2, the temperature of the fluid leaving the compressor.

$$W_{net} = mc_p(T_3 - T_4) - mc_p(T_2 - T_1) \tag{14.5}$$

but

$$T_4 = \frac{T_3 T_1}{T_2}$$

$$W_{net} = mc_p\left(T_3 - \frac{T_3 T_1}{T_2} - T_2 + T_1\right)$$

For W_{net} to be a maximum, $dW_{net}/dT_2 = 0$, yielding

$$T_2 = \sqrt{T_3 T_1}$$

Also,

$$r_p = \frac{p_2}{p_1} = \left(\frac{T_2}{T_1}\right)^{k/(k-1)} = \left(\frac{T_3}{T_1}\right)^{k/2(k-1)}$$

$$(r_p)^{(k-1)/k} = \frac{T_2}{T_1} = \frac{T_3}{T_4} \qquad \therefore (r_p)^{2(k-1)/k} = \frac{T_3}{T_1}$$

hence

$$T_2 = T_4 \tag{14.6}$$

Thus, when equation (14.6) is substituted in equation (14.5), an expression for the optimum net work in terms of the maximum temperature and the inlet temperature is obtained. The optimum net work, $W_{net} = mc_p(T_3 + T_1 - 2\sqrt{T_3 T_1})$, occurs when the air temperature at the compressor exit is equal to the air temperature at the turbine exit. For industrial use $(T_3/T_1) = 3.5 \rightarrow 4$, whereas for aircraft engines $(T_3/T_1) = 5.0 \rightarrow 5.5$, with the use of air-cooled blades. The pressure ratio may be calculated as in the previous derivation, if the optimum temperature is known.

14.3 EFFICIENCIES

One of the greatest problems gas turbine manufacturers had to overcome was poor compressor efficiencies. During the 1930s and 1940s, compressor efficiencies of 60% were common. Turbine efficiencies were greater, but the combination meant that virtually all the turbine work went to drive the compressor. Only by the gas dynamic analysis of the fluid passing through compressor and turbine blading could design changes be made—changes that now result in compressor efficiencies between 75 and 90% and turbine efficiencies commonly a few percent better. Thus, for a compressor with an efficiency of 85% for a given power and pressure ratio, the turbine would usually have an efficiency of about 87%.

The compressor may use more than 50% of the total work the turbine produces. The compressor efficiency, η_c, and the turbine efficiency, η_t, are defined as

$$\eta_c = \frac{W_{ideal}}{W_{act}} = \frac{(h_2 - h_1)_s}{h_{2'} - h_1} \tag{14.7}$$

$$\eta_t = \frac{W_{act}}{W_{ideal}} = \frac{h_3 - h_{4'}}{(h_3 - h_4)_s} \tag{14.8}$$

Figure 14.5 illustrates the states on an h-s diagram.

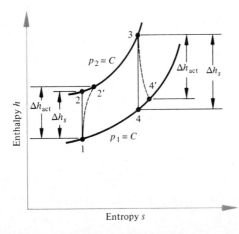

Figure 14.5 The h-s diagram showing actual and ideal compressor and turbine states.

The following examples illustrate the effects of component efficiencies on the overall cycle efficiency.

Example 14.1

An air-standard Brayton cycle has air enter the compressor at 27°C and 100 kPa. The pressure ratio is 10, and the maximum allowable temperature in the cycle is 1350°K. Determine the pressure and temperature at each state in the cycle, and the compressor work, the turbine work, and the cycle efficiency per kilogram of air.

Solution

Given: The initial temperature and pressure, the maximum temperature, and the pressure ratio for an air-standard Brayton cycle.

Find: The cycle state points, the various work terms, and the cycle efficiency.

Sketch and Given Data:

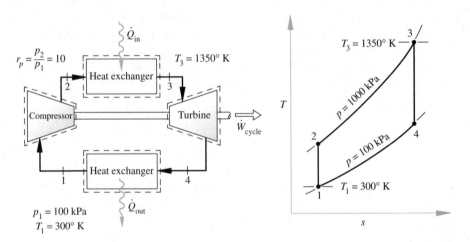

Figure 14.6

Assumptions:

1. Each component is analyzed as a steady-state open system.
2. The processes are for the air-standard Brayton cycle.
3. Air behaves like an ideal gas.
4. The changes in kinetic and potential energies may be neglected.

Analysis: At state 1 the pressure and temperature are $p_1 = 100$ kPa and $T_1 = 300°$K. Knowing the pressure ratio and the process, isentropic from state 1 to state 2, allows us to find state 2.

$$r_p = \frac{p_2}{p_1} = 10$$

$$p_2 = 1000 \text{ kPa}$$

$$T_2 = T_1(r_p)^{(k-1)/k} = (300°\text{K})(10)^{0.286} = 579.6°\text{K}$$

The process from state 2 to state 3 is constant-pressure, and the maximum temperature occurs at state 3; hence,

$$p_3 = p_2 = 1000 \text{ kPa} \qquad T_3 = T_{max} = 1350°\text{K}$$

The process from state 3 to state 4 is isentropic, and $p_4 = p_1 = 100$ kPa.

$$T_4 = T_3 \left(\frac{p_4}{p_3}\right)^{(k-1)/k} = (1350)(0.10)^{0.286} = 698.8°\text{K}$$

The first-law analysis for steady state for the turbine, compressor, and high-temperature heat exchanger yields

$$w_c = -(h_2 - h_1) = -c_p(T_2 - T_1)$$

$$w_c = -(1.0047 \text{ kJ/kg-K})(579.6 - 300°\text{K}) = -280.9 \text{ kJ/kg}$$

$$w_t = (h_3 - h_4) = c_p(T_3 - T_4)$$

$$w_t = (1.0047 \text{ kJ/kg-K})(1350 - 698.8°\text{K}) = 654.3 \text{ kJ/kg}$$

$$w_{net} = \sum w = 373.4 \text{ kJ/kg}$$

$$q_{in} = (h_3 - h_2) = c_p(T_3 - T_2)$$

$$q_{in} = (1.0047 \text{ kJ/kg-K})(1350 - 579.6°\text{K}) = 774.0 \text{ kJ/kg}$$

$$\eta_{th} = \frac{w_{net}}{q_{in}} = 0.482 \quad \text{or } 48.2\%$$

Check:

$$\eta_{th} = 1 - \frac{1}{(r_p)^{(k-1)/k}} = 0.482$$

Comment: Notice that 43% of the turbine work is used to drive the compressor. The remainder is available as net work. ▬

Example 14.2

In an air-standard Brayton cycle, air enters the compressor at 27°C and 100 kPa. The pressure ratio is 10, and the maximum allowable temperature is 1350°K. The compressor and turbine efficiencies are 85%, and pressure drops 27 kPa between the compressor discharge and the turbine inlet. Determine the pressure and temperature at each state in the cycle, the compressor and turbine work, and the cycle thermal efficiency. Compare the results with those in Example 14.1.

Solution

Given: The same air-standard cycle as in Example 14.1 but including efficiency terms and a pressure drop.

Find: The cycle state points, the turbine and compressor work, and the cycle thermal efficiency. Compare the results to those of Example 14.1.

Sketch and Given Data:

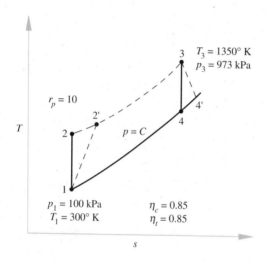

Figure 14.7

Assumptions:

1. Each component is analyzed as a steady-state open system.
2. The processes are for the air-standard Brayton cycle, modified for the effect of component efficiency and pressure drop as appropriate.
3. Air behaves like an ideal gas.
4. The changes in kinetic and potential energies may be neglected.

Analysis: The temperature and pressure at state 1 are $T_1 = 300°$K and $p_1 = 100$ kPa. From Example 14.1 we know the isentropic value of T_2 and can compute the actual value using the compressor efficiency term. The pressure actually and ideally at state 2 is 1000 kPa.

$$\eta_c = 0.85 = \frac{579.6 - 300°\text{K}}{T_{2'} - 300°\text{K}}$$

$$T_{2'} = 628.9°\text{K}$$

$$p_{2'} = p_2 = 1000 \text{ kPa}$$

During the process from state 2 to state 3, pressure drops 27 kPa. The maximum temperature occurs at state 3; hence $T_3 = T_{max} = 1350°$K, and $p_3 = 973$ kPa. The ideal process from state 3 to state 4 is reversible adiabatic; hence,

$$T_4 = T_3 \left(\frac{p_4}{p_3}\right)^{(k-1)/k} = 1350 \left(\frac{100}{973}\right)^{0.286} = 704.3°\text{K}$$

The turbine efficiency term allows us to calculate the actual exit temperature from the turbine.

$$\eta_t = 0.85 = \frac{1350°\text{K} - T_{4'}}{1350 - 704.1°\text{K}}$$

$$T_{4'} = 801.0°\text{K}$$

Notice that state 4 from Example 14.1 could not be used as the pressure at state 3 is different in this problem because of the pressure drop that occurred from state 2 to state 3.

A first-law analysis for each component yields

$$w_c = -(h_{2'} - h_1) = -c_p(T_{2'} - T_1)$$

$$w_c = -(1.0047 \text{ kJ/kg-K})(628.9 - 300°\text{K}) = -330.4 \text{ kJ/kg}$$

$$w_t = (h_3 - h_{4'}) = c_p(T_3 - T_{4'})$$

$$w_t = (1.0047 \text{ kJ/kg-K})(1350 - 801°\text{K}) = 551.6 \text{ kJ/kg}$$

$$q_{in} = (h_3 - h_{2'}) = c_p(T_3 - T_{2'})$$

$$q_{in} = (1.0047 \text{ kJ/kg-K})(1350 - 628.9°\text{K}) = 724.5 \text{ kJ/kg}$$

$$w_{net} = \sum w = 221.2 \text{ kJ/kg}$$

The overall efficiency is

$$\eta_{th} = \frac{w_{net}}{q_{in}} = 0.305 \quad \text{or } 30.5\%$$

The change in component energy output or input is

$$\text{Change in } w_c = \frac{(49.5)(100)}{280.9} = 17.6\% \text{ increase}$$

$$\text{Change in } w_t = \frac{(102.7)(100)}{654.3} = 15.7\% \text{ decrease}$$

$$\text{Change in } q_{in} = \frac{(49.5)(100)}{774.0} = 6.4\% \text{ decrease}$$

$$\text{Change in } \eta_{th} = \frac{(0.177)(100)}{0.482} = 36.7\% \text{ decrease}$$

Comments:

1. Notice that increasing the compressor work required and decreasing the turbine work produced dramatically decreases the overall cycle efficiency.
2. The heat added decreases slightly because the air reaching the heat exchanger is at a higher temperature because of compressor irreversibilities, and thus less heat is required to reach the maximum temperature of 1350°K.
3. Equation (14.4) cannot be used to determine the thermal efficiency for the cycle, as the processes are not ideal.　　　　　　　　　　　　　　　　　▬

In the last chapter, we used the TK Solver model AIRCYCLE.TK to analyze Otto and diesel cycles. AIRCYCLE.TK can also be used to analyze Brayton cycles.

Example 14.3
Solve Example 14.2 using AIRCYCLE.TK and compare the results.

Solution

Given:　A simple Brayton cycle operating in the same conditions as the air-standard cycle in Example 14.2.

Find:　The cycle state points, the turbine and compressor work, and the cycle thermal efficiency.

Sketch and Given Data:　See Figure 14.7 on page 490.

Assumptions:

1. Each component is analyzed as a steady-state open system.
2. The changes in kinetic and potential energies may be neglected.

Analysis:　Load AIRCYCLE.TK into TK Solver. In the Rule Sheet, enter the following:

$$\text{Eth} = (-\text{DELh43} - \text{DELh21})/\text{DELh32}$$

to calculate the thermal efficiency. Enter the given pressure and temperature data, enter zero for the change in entropy for the compression and expansion processes (DELs21 and DELs43) in the Variable Sheet, and solve. To account for the compressor and turbine efficiencies, delete the entries for DELs21 and DELs43, and enter input values for DELh21 and DELh43 as the values calculated previously and divided and multiplied by .85, respectively. Solve the model again. The solved Variable Sheet should look like this:

```
============================== VARIABLE SHEET ==============================
St  Input —— Name —— Output — Unit —— Comment ————————————————

                                        ***Four-Point Air Cycle***

                                        POINT 1
      100     p1                  kPa   Pressure (kPa, MPa, psia)
      27      T1                  degC  Temperature (degK, degC, degR, degF)
              v1       .86116     m3/kg Specific Volume (m3/kg, ft3/lbm)

                                        POINT 2
      1000    p2                  kPa   Pressure (kPa, MPa, psia)
              T2       617.72     degK  Temperature (degK, degC, degR, degF)
              v2       .17776     m3/kg Specific Volume (m3/kg, ft3/lbm)

                                        POINT 3
      973     p3                  kPa   Pressure (kPa, MPa, psia)
      1350    T3                  degK  Temperature (degK, degC, degR, degF)
              v3       .399       m3/kg Specific Volume (m3/kg, ft3/lbm)

                                        POINT 4
      100     p4                  kPa   Pressure (kPa, MPa, psia)
              T4       855.91     degK  Temperature (degK, degC, degR, degF)
              v4       2.4574     m3/kg Specific Volume (m3/kg,ft3/lbm)

                                        CHANGE (POINT 2 - POINT 1)
      325.24  DELh21              kJ/kg Enthalpy (kJ/kg, BTU/lbm)
              DELu21   234.08     kJ/kg Internal Energy (kJ/kg, BTU/lbm)
              DELs21   .081074    kJ/kg-K Entropy (kJ/kg-K, B/lbm-R)

                                        CHANGE (POINT 3 - POINT 2)
              DELh32   829.44     kJ/kg Enthalpy (kJ/kg, BTU/lbm)
              DELu32   619.24     kJ/kg Internal Energy (kJ/kg, BTU/lbm)
              DELs32   .88673     kJ/kg-K Entropy (kJ/kg-K, B/lbm-R)

                                        CHANGE (POINT 4 - POINT 3)
     -571.42  DELh43              kJ/kg Enthalpy (kJ/kg, BTU/lbm)
              DELu43  -429.59     kJ/kg Internal Energy (kJ/kg, BTU/lbm)
              DELs43   .1245      kJ/kg-K Entropy (KJ/kg-K, B/lbm-R)

                                        CHANGE (POINT 1 - POINT 4)
              DELh14  -583.26     kJ/kg Enthalpy (kJ/kg, BTU/lbm)
              DELu14  -423.74     kJ/kg Internal Energy (kJ/kg, BTU/lbm)
              DELs14  -1.0923     kJ/kg-K Entropy (kJ/kg-K, B/lbm-R)

              Eth      .2968            Thermal Efficiency
```

Comments:

1. Because of the variable specific heat in AIRCYCLE.TK, the calculated temperatures and energies are slightly different than those for the air-standard analysis. The thermal efficiency calculated is slightly lower.
2. Instead of the two-step approach taken to solve the problem with AIRCYCLE.TK, the model could have been modified to include the equations for the compressor and turbine efficiencies, thus permitting a single-step solution. ▪

14.4 OPEN-CYCLE ANALYSIS

In the open gas turbine cycle, the air entering the compressor is increased in pressure. It then enters the combustion chamber, where fuel is added, and the combustion process raises the temperature of the products of combustion, which enter the turbine. The products of combustion leave the turbine and exhaust to the atmosphere. Figure 14.8 illustrates this cycle schematically, and Figure 14.9 shows the *T-s* diagram for the cycle. Note that air is the working fluid up to the combustion chamber and that a different fluid, the products of combustion, is the working substance after the combustion chamber. In analyzing this power cycle, we should use the properties of each fluid. In dealing with air, we may use the air tables, or if none is available, the ideal-gas law. In dealing with the products of combustion, we may use the 400% theoretical air tables or the ideal-gas law. The specific heats of the air and the products of combustion will be slightly different, with the specific heat of the products of combustion being greater than that of air.

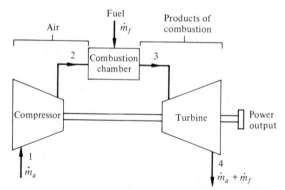

Figure 14.8 Schematic diagram of an open gas turbine cycle.

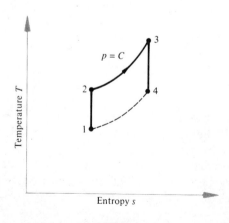

Figure 14.9 The *T-s* diagram for the gas turbine in Figure 14.8.

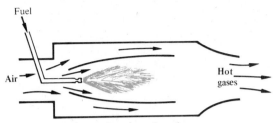

Figure 14.10 A combustion chamber.

Let us look at the combustion chamber to find out why 400% theoretical air is a good assumption when evaluating the products of combustion. Figure 14.10 illustrates the combustion chamber, or combustor, for a gas turbine. Metallurgical limitations of the turbine section put a limit on the temperature of the gases leaving the combustion chamber. We noted when we studied adiabatic flame temperatures in Chapter 12 that combustion temperatures may easily exceed this limit. To prevent the gas temperature from exceeding the allowable exit temperature from the combustion chamber, excess air is used. For a gas turbine, the gases typically have properties similar to those of 400% theoretical air. Not all the excess air is used in the combustion process, as that would cool the air-fuel mixture and cause incomplete combustion. Some of the air passes around the sides of the combustion chamber, cooling the metal walls and mixing with the products of combustion in the latter part of the combustion chamber.

The following example illustrates the analysis of an open-system gas turbine unit where the dissimilarity in the working substances and the combustion chamber analysis are included.

Example 14.4

A gas turbine unit produces 600 kW while operating under the following conditions: the inlet air pressure and temperature are 100 kPa and 300°K; the pressure ratio is 10; the fuel is similar to $C_{12}H_{26}$ and has a ratio of 0.015 kg fuel/kg air; the products of combustion are similar to 400% theoretical air. Calculate the air flow rate, the total turbine work, the compressor work, and the thermal efficiency.

Solution

Given: The power produced from a gas turbine unit, the air inlet conditions, the fuel/air ratio, the fuel type, and the pressure ratio.

Find: Find the air flow rate required, the total turbine and compressor powers, and the thermal efficiency.

Sketch and Given Data:

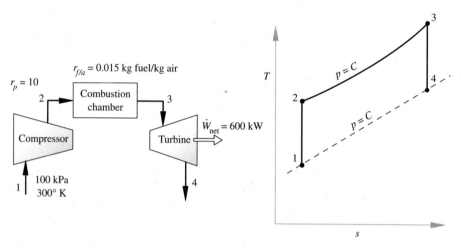

Figure 14.11

Assumptions:

1. Each component is analyzed as a steady-state open system.
2. The processes are for the ideal, open gas turbine cycle.
3. Air and products of combustion behave like ideal gases with variable specific heats.
4. The changes in kinetic and potential energies may be neglected.
5. The compressor, combustion chamber, and turbine are adiabatic.

Analysis: In this problem the air has specific heats that vary with temperature, so Table A.2 is used. At state 1 $h_1 = 300.19$ kJ/kg, and $p_{r_1} = 1.3860$. The first law for an open system applied to the compressor and invoking assumptions 4 and 5 yields

$$\dot{m}_a h_1 = \dot{m}_a h_2 + \dot{W}_c$$
$$\dot{W}_c = -\dot{m}_a(h_2 - h_1)$$

Since no internal efficiencies are given, $h_2 = (h_2)_s$. For an isentropic process,

$$p_{r_2} = p_{r_1}(r_p) = (1.3860)(10) = 13.86$$
$$h_2 = 579.8 \text{ kJ/kg}$$

The adiabatic combustion chamber is the next element in the cycle. It too is an open system, and so the first law applied to it is

$$\dot{m}_a h_2 + \dot{m}_f h_{RP} = (\dot{m}_a + \dot{m}_f)h_3$$

where $\bar{h}_{RP}$ is the enthalpy of combustion. From Table C.3, $h_{RP} = 44\ 102$ kJ/kg.

$$h_2 + r_{f/a}h_{RP} = (1 + r_{f/a})h_3$$

$$(579.8\ \text{kJ/kg}) + (0.015\ \text{kg fuel/kg air})(44\ 102\ \text{kJ/kg fuel})$$

$$= (1.015\ \text{kg products/kg air})h_3$$

$$h_3 = 1241.3\ \text{kJ/kg}$$

The substance leaving the combustion chamber is 400% theoretical air, found in Table A.3. Thus, $p_{r_3} = 220.1$.

The flow through the turbine may be considered isentropic, so

$$p_{r_4} = p_{r_3}\left(\frac{1}{r_p}\right) = 22.01$$

$$h_4 = 662.5\ \text{kJ/kg}$$

Applying the first law for open systems to the turbine while invoking assumptions 4 and 5 yields

$$\dot{m}_a(1 + r_{f/a})h_3 = \dot{W}_c + \dot{W}_{\text{net}} + \dot{m}_a(1 + r_{f/a})h_4$$

$$\dot{W}_c = -\dot{m}_a(h_2 - h_1)$$

$$\dot{W}_{\text{net}} = 600\ \text{kW}$$

Solve for $\dot{m}_a$.

$$(\dot{m}_a\ \text{kg/s})(1.015\ \text{kg products/kg air})(1241.3 - 662.5\ \text{kJ/kg}) = 600\ \text{kW} -$$

$$(\dot{m}_a\ \text{kg/s})(579.8 - 300.19\ \text{kJ/kg})$$

$$\dot{m}_a = 1.949\ \text{kg/s}$$

The total turbine power produced is

$$\dot{W}_t = \dot{m}_a(1 + r_{f/a})(h_3 - h_4)$$

$$\dot{W}_t = (1.949\ \text{kg air/s})(1.015\ \text{kg products/kg air})(1241.3 - 662.5\ \text{kJ/kg products})$$

$$\dot{W}_t = 1145\ \text{kW}$$

and the total compressor power is

$$\dot{W}_c = -\dot{m}_a(h_2 - h_1)$$

$$\dot{W}_c = -(1.949\ \text{kg air/s})(579.8 - 300.19\ \text{kJ/kg air}) = -545.0\ \text{kW}$$

The unit's thermal efficiency is

$$\eta_{th} = \frac{\dot{W}_{net}}{\dot{Q}_{in}} = \frac{600 \text{ kW}}{(1.949 \text{ kg air/s})(0.015 \text{ kg fuel/kg air})(44\ 102 \text{ kJ/kg fuel})}$$

$$\eta_{th} = 0.465$$

Comment: The fuel flow rate may be calculated by multiplying the fuel/air ratio by the air flow rate. The size of inlet ducts and fuel tanks can be determined from information provided by first-law analysis.

14.5 COMBUSTION EFFICIENCY

The combustion process may not proceed to completion, so not all the chemical energy is converted into thermal energy. The temperature of the products of combustion will not be as high as ideally expected. To account for this, we introduce a combustion efficiency η_{cc}, which tells us the percentage of energy released from the fuel. Therefore, the thermal energy generated in a combustion chamber per kilogram of air is $r_{f/a}\eta_{cc}h_{RP}$, not $r_{f/a}h_{RP}$.

14.6 REGENERATION

The thermal efficiency of the typical gas turbine unit is quite low. We observe that the exhaust temperature of the turbine is quite high, indicating that a large portion of available energy is not being used. This high-temperature energy could be used to preheat the combustion air before it enters the combustor. This increases the overall efficiency by decreasing the fuel, hence heat, added. Theoretically, the net work from the turbine does not change. In an actual unit, the net work will decrease because of pressure effects, which will be discussed later. The optimum place for preheating is after the air has been compressed; Figure 14.12(a) illustrates a schematic for the system and Figure 14.12(b) illustrates a typical diagram for the unit. Figure 14.13 shows the T-s diagram for the regenerative cycle. If the heat exchanger were 100% effective, the temperature at state x would be equal to the temperature at state 4. However, the heat exchanger is not 100% effective. To denote how effectively the heat exchanger operates, we define a coefficient called the *regenerator effectiveness*, $\mathscr{E}_{reg}$.

$$\mathscr{E}_{reg} = \frac{\text{actual heat transferred}}{\text{maximum possible heat transferred}}$$

$$\mathscr{E}_{reg} = \frac{h_x - h_2}{(1 + r_{f/a})(h_4 - h_c)} \tag{14.9}$$

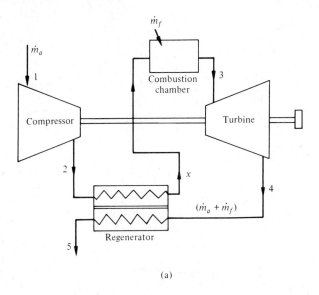

(a)

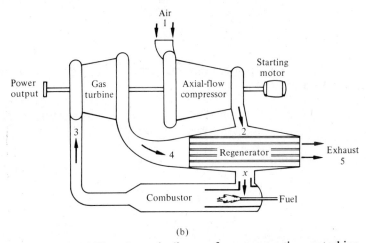

(b)

Figure 14.12 (a) The schematic diagram for a regenerative gas turbine cycle. (b) An open-cycle gas turbine with regenerative heating.

where h_c is the enthalpy of the products at temperature T_2. The actual heat transferred is equal to the heat picked up by the air. The maximum possible heat transfer would occur when the products of combustion are cooled down to the temperature of the air entering the regenerator.

The following examples illustrate the effect of a regenerator on the gas turbine cycle.

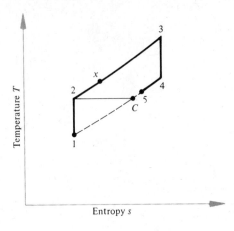

Figure 14.13 The *T-s* diagram for the regenerative gas turbine cycle.

Example 14.5

A gas turbine unit receives air at 100 kPa and 300°K and compresses it adiabatically to 620 kPa with a compressor that has an isentropic efficiency of 88%. The fuel has a heating value of 44 186 kJ/kg, and the fuel/air ratio is 0.017 kg fuel/kg air. The turbine internal efficiency is 90%. Determine per unit mass the turbine work, the compressor work, and the thermal efficiency.

Solution

Given: A steady-state open system gas turbine with known fuel/air ratio, inlet conditions, and turbine and compressor efficiencies.

Find: The turbine and compressor work per unit mass and the unit's overall efficiency.

Sketch and Given Data:

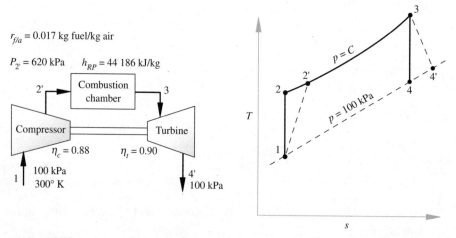

Figure 14.14

Assumptions:

1. Each component is analyzed as a steady-state open system.
2. Air and products of combustion behave like ideal gases with variable specific heats.
3. The changes in kinetic and potential energies may be neglected.
4. The compressor, combustion chamber, and turbine are adiabatic.
5. The combustion process is constant-pressure.

Analysis: Determine the cycle state points and solve for the work terms and the efficiency. Knowing the temperature at state 1 yields from Table A.2 $h_1 = 300.19$ kJ/kg air, and $p_{r_1} = 1.3860$. The isentropic process from state 1 to state 2 is

$$p_{r_2} = 1.3860 \left(\frac{620}{100} \right) = 8.593$$

$$h_2 = 506.04 \text{ kJ/kg}$$

Using the compressor internal efficiency allows calculation of state 2':

$$\eta_c = 0.88 = \frac{h_2 - h_1}{h_{2'} - h_1} = \frac{506.04 - 300.19}{h_{2'} - 300.19} \text{ kJ/kg}$$

$$h_{2'} = 534.11 \text{ kJ/kg air}$$

The process from state 2' to state 3 is constant-pressure; hence the first-law equation yields

$$h_{2'} + r_{fla}h_{RP} = (1 + r_{fla})h_3$$

$$(534.11 \text{ kJ/kg air}) + (0.017 \text{ kg fuel/kg air})(44\ 186 \text{ kJ/kg fuel})$$

$$= (1.017 \text{ kg products/kg air})h_3$$

$$h_3 = 1263.8 \text{ kJ/kg}$$

From Table A.3 $T_3 = 1166°$K, and $p_{r_3} = 236.0$. The ideal expansion process through the turbine from state 3 to state 4 is isentropic; hence

$$p_{r_4} = (236) \left(\frac{100}{620} \right) = 38.06$$

$$h_4 = 771.3 \text{ kJ/kg}$$

The actual state 4′ is found from the turbine internal efficiency:

$$\eta_t = 0.90 = \frac{h_3 - h_{4'}}{h_3 - h_4} = \frac{1263.8 - h_{4'}}{1263.8 - 771.3} \text{ kJ/kg}$$

$$h_{4'} = 820.6 \text{ kJ/kg}$$

The first-law analysis for the turbine is

$$w_t = (1 + r_{f/a})(h_3 - h_{4'})$$

$$w_t = (1.017 \text{ kJ products/kg air})(1263.8 - 820.6 \text{ kJ/kg products}) = 450.7 \text{ kJ/kg air}$$

The first-law analysis for the compressor is

$$w_c = -(h_{2'} - h_1) = -(534.11 - 300.19 \text{ kJ/kg air}) = -233.92 \text{ kJ/kg air}$$

The net work is the algebraic sum of the compressor and turbine works.

$$w_{net} = w_t + w_c = 216.78 \text{ kJ/kg air}$$

The energy added per unit mass of air is

$$q_{in} = r_{f/a}h_{RP} = (0.017 \text{ kg fuel/kg air})(44\ 186 \text{ kJ/kg fuel}) = 751.2 \text{ kJ/kg air}$$

and the thermal efficiency is

$$\eta_{th} = \frac{w_{net}}{q_{in}} = \frac{216.78}{751.2} = 0.288 \quad \text{or } 28.8\%$$

Comment: The turbine and compressor efficiencies dramatically reduce the overall thermal efficiency of the gas turbine unit. ▬

Example 14.6

A regenerator with an effectiveness of 60% is added to the gas turbine unit of Example 14.5. There are pressure drops across the regenerator. There is a 20-kPa drop on the air side and a 3.4-kPa drop on the exhaust side of the regenerator. Determine the new fuel/air ratio to achieve the same value of T_3. In addition determine the new net work and cycle efficiency.

Solution

Given: The gas turbine unit from Example 14.5 now includes a regenerator of known effectiveness with known pressure drops.

Find: The new fuel/air ratio, net work, and unit efficiency.

Sketch and Given Data:

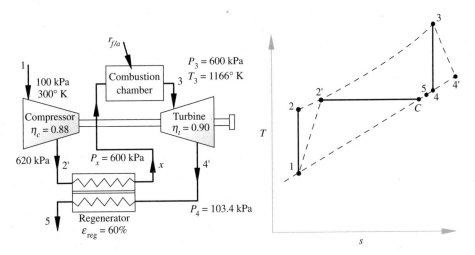

Figure 14.15

Assumptions:

1. Each component is analyzed as a steady-state open system.
2. Air and products of combustion behave like ideal gases with variable specific heats.
3. The changes in kinetic and potential energies may be neglected.
4. The compressor, combustion chamber, regenerator, and turbine are adiabatic.
5. The combustion process is constant-pressure.

Analysis: Certain property values from Example 14.5 remain the same: $h_1 = 300.19$ kJ/kg air, $h_{2'} = 534.11$ kJ/kg air, and $w_c = -233.92$ kJ/kg air. The effect of the regenerator pressure drop will cause the properties at state 3 and state 4 to change: $p_3 = 620 - 20 = 600$ kPa; $T_3 = 1166°$K; $h_3 = 1263.8$ kJ/kg products; $p_{r_3} = 236.0$. The pressure at state 4 is 103.4 kPa. The ideal process from state 3 to state 4 is

$$p_{r_4} = (236.0)\left(\frac{103.4}{600}\right) = 40.67$$

$$h_4 = 785.5 \text{ kJ/kg products}$$

The turbine efficiency allows the determination of state 4′ to be

$$\eta_t = 0.90 = \frac{h_3 - h_{4'}}{h_3 - h_4} = \frac{1263.8 - h_{4'}}{1263.8 - 785.5}$$

$$h_{4'} = 833.3 \text{ kJ/kg products}$$

From the definition for regenerator effectiveness,

$$0.60 = \frac{h_x - h_{2'}}{(1 + r_{f/a})(h_{4'} - h_c)}$$

where $h_c = 541$ kJ/kg at $T_c = T_{2'} = 530.1°$K; and from the first-law analysis of the combustion chamber,

$$h_x + r_{f/a}h_{RP} = (1 + r_{f/a})h_3$$

the two unknowns $r_{f/a}$ and h_x may be determined. Solving these equations yields

$$0.60 = \frac{h_x - 534.11 \text{ kJ/kg air}}{(1 + r_{f/a})(833.3 - 541 \text{ kJ/kg products})}$$

$$h_x + (r_{f/a})(44\ 186 \text{ kJ/kg fuel}) = (1 + r_{f/a})(1263.8 \text{ kJ/kg products})$$

$$r_{f/a} = 0.01286 \text{ kg fuel/kg air}$$

The first-law analysis of the turbine yields

$$w_t = (1 + r_{f/a})(h_3 - h_{4'})$$
$$w_t = (1.01286 \text{ kg products/kg air})(1263.8 - 833.3 \text{ kJ/kg products})$$
$$= 436.0 \text{ kJ/kg air}$$

and the net work is

$$w_{\text{net}} = w_t + w_c = 202.08 \text{ kJ/kg air}$$

The energy added in the combustion chamber is

$$q_{\text{in}} = r_{f/a}h_{RP} = (0.01286 \text{ kg fuel/kg air})(44\ 186 \text{ kJ/kg fuel}) = 568.2 \text{ kJ/kg air}$$

and the new thermal efficiency is

$$\eta_{\text{th}} = \frac{w_{\text{net}}}{q_{\text{in}}} = \frac{202.08}{568.2} = 0.355 \quad \text{or } 35.5\%$$

Comment: The overall effect of the regenerator is to improve the cycle thermal efficiency significantly. The turbine work decreased due to pressure drops within the regenerator. The loss of work was more than offset by a greater decrease in the fuel added to reach the maximum temperature, yielding the overall efficiency improvement.

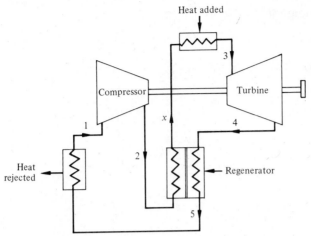

Figure 14.16 The closed regenerative Brayton cycle.

Closed Regenerative Cycle

The closed air-standard Brayton cycle may also use a regenerator. The same losses may occur, but the closed system allows us the opportunity to investigate analytically the effect the regenerator has on the cycle efficiency. Figure 14.16 illustrates the schematic diagram, and Figure 14.17 illustrates the T-s diagram for the cycle.

The thermal efficiency may be written as

$$\eta_{\text{th(reg)}} = \frac{W_{\text{net}}}{Q_{\text{in}}} = 1 - \frac{Q_{\text{out}}}{Q_{\text{in}}} = 1 - \frac{T_5 - T_1}{T_3 - T_x} \tag{14.10}$$

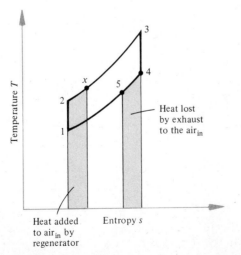

Figure 14.17 The T-s diagram for the closed regenerative Brayton cycle.

For an ideal counterflow heat exchanger,

$$T_x = T_4 \quad \text{and} \quad T_5 = T_2$$

and the thermal efficiency is

$$\eta_{th(reg)} = 1 - \frac{T_2 - T_1}{T_3 - T_4} \tag{14.11}$$

From equation (14.6),

$$\frac{T_2}{T_1} = \frac{T_3}{T_4} = (r_p)^{(k-1)/k}$$

so equation (14.11) becomes

$$\eta_{th(reg)} = 1 - \left(\frac{T_1}{T_3}\right)(r_p)^{(k-1)/k} \tag{14.12}$$

Thus, in the ideal cycle with regeneration, the overall thermal efficiency is a function of the pressure ratio and the ratio of minimum to maximum temperature occurring in the cycle. A plot of equations (14.12) and (14.4) is given in Figure 14.18, where $T_3/T_1 = t$.

In actual cycles the temperature and pressure ratios are important. The closed-system analysis does permit us to find that adding a regenerator may or may not increase the cycle efficiency. When pressure drops due to the regenerator are considered, the desirability of adding a regenerator must be reevaluated. The ultimate analysis is an economic one—will adding the generator save money, balancing the increased capital cost and maintenance against the fuel savings? All engineering analyses must eventually undergo the same economic projection: Is it worthwhile? As

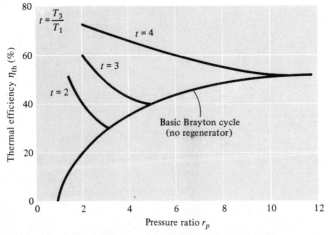

Figure 14.18 The thermal efficiency of the air-standard regenerative Brayton cycle for different values of t and r_p.

fuel costs increase, the desirability of efficiency-producing additions to the basic cycle increases. Thus, what has been accepted practice may no longer be valid when the economic constraints change.

14.7 REHEATING AND INTERCOOLING

Other methods for improving the cycle efficiency are available. Although they may not all be useful on small gas turbine units, large units usually profit by incorporating them. Improvement in the cycle efficiency can be made by reheating the fluid after it has performed some work. In the gas turbine unit this is accomplished by passing the products of combustion through another combustion chamber. Since there is more than sufficient air for combustion, we can inject some more fuel. The reheated products of combustion return to the turbine. Figure 14.19(a) illustrates the sche-

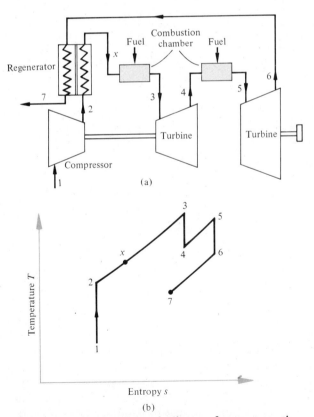

Figure 14.19 (a) The schematic diagram for a regenerative-reheat gas turbine cycle. (b) The *T-s* diagram for the regenerative-reheat gas turbine cycle.

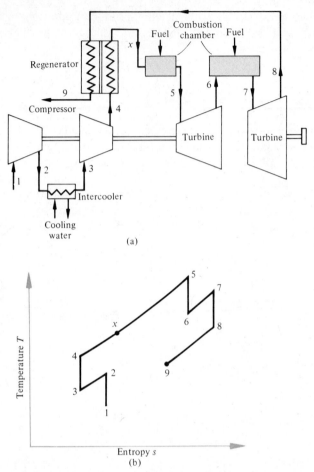

Figure 14.20 (a) The schematic diagram for a regenerative-reheat gas turbine cycle with intercooling. (b) The *T-s* diagram for a regenerative-reheat gas turbine cycle with intercooling.

matic for a reheat-regenerative gas turbine cycle, and Figure 14.19(b) illustrates the *T-s* diagram. The products of combustion reentering the turbine at state 5 are usually at the same temperature as those entering the turbine at state 3.

Another way to improve the cycle efficiency is to reduce the work of the compressor. This may be accomplished by compressing in stages and, with an intercooler, cooling the air as it passes from one stage to another. The following example illustrates the combined effects of reheating and intercooling on a gas turbine unit. Figure 14.20(a) illustrates the schematic arrangement of the equipment, and Figure 14.20(b) shows the *T-s* diagram.

Example 14.7

A gas turbine unit operates on a regenerative-reheat cycle with compressor interstage cooling. Air enters the compressor at 100 kPa and 290°K and is compressed to 410 kPa; it is cooled at constant pressure until the temperature drops by 13°C and is finally compressed to 750 kPa. The regenerator has an effectiveness of 70%. The products of combustion enter the turbine at 1350°K and expand to 410 kPa, where they are reheated to 1350°K. The exhaust pressure is 100 kPa. The fuel has a heating value of 44 186 kJ/kg. Determine the fuel/air ratio in each combustion chamber, the compressor work, the total turbine work, and the overall thermal efficiency where all expansion and compressor processes are isentropic and no pressure drops occur in the regenerator.

Solution

Given: A gas turbine unit with regenerative heating, reheat, and intercooling. All necessary pressures and temperatures are included that describe the cycle, as well as regenerator effectiveness.

Find: Determine the cycle efficiency, which requires determining the initial component works and heat transfer.

Sketch and Given Data:

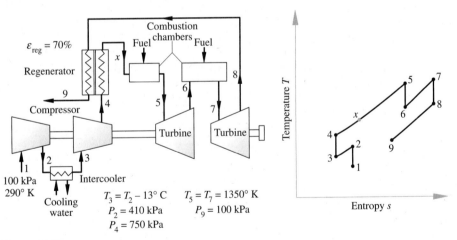

Figure 14.21

Assumptions:

1. Each component is analyzed as a steady-state open system.
2. Air and products of combustion behave like ideal gases with variable specific heats.
3. The changes in kinetic and potential energies may be neglected.

4. The compressor, combustion chamber, regenerator, intercooler, and turbine are adiabatic.
5. The combustion process is constant-pressure.

Analysis: This cycle is complex, so we proceed in an orderly fashion around the cycle, determining the cycle state points as possible. Then we use first-law analysis to determine the work and heat transfers that occur. At state 1, $h_1 = 290.17$ kJ/kg, $p_1 = 100$ kPa, $T_1 = 290°$K, and $p_{r_1} = 1.2311$.

The process from state 1 to state 2 is isentropic.

$$p_{r_2} = (1.2311) \left(\frac{410}{100}\right) = 5.048$$

$$h_2 = 434.68 \text{ kJ/kg} \qquad T_2 = 433°\text{K}$$

The process from state 2 to state 3 is constant-pressure: $T_3 = 420°$K, $h_3 = 421.26$ kJ/kg, and $p_{r_3} = 4.522$.

The process from state 3 to state 4 is isentropic.

$$p_{r_4} = (4.522) \left(\frac{750}{410}\right) = 8.272$$

$$T_4 = 498°\text{K} \qquad h_4 = 500.58 \text{ kJ/kg}$$

$$T_c = 498°\text{K}$$

$$h_c = 507.1 \text{ kJ/kg} \quad (400\% \text{ theoretical air})$$

State 5 is given; hence $T_5 = 1350°$K, and $h_5 = 1487.8$ kJ/kg (400% theoretical air).

The process from state 5 to state 6 is isentropic; hence

$$p_{r_6} = (438) \left(\frac{410}{750}\right) = 239.44$$

$$h_6 = 1268.5 \text{ kJ/kg} \qquad T_6 = 1170°\text{K}$$

State 7 is specified; hence $T_7 = 1350°$K, $h_7 = 1487.8$ kJ/kg, and $p_{r_7} = 438.0$.
The process from state 7 to state 8 is isentropic.

$$p_{r_8} = (438) \left(\frac{100}{410}\right) = 106.83$$

$$T_8 = 963°\text{K} \qquad h_8 = 1022.8 \text{ kJ/kg}$$

Find h_x, r_{f/a_1}, and r_{f/a_2} by first-law analysis:

$$(1 + r_{f/a_1})h_6 + r_{f/a_2}h_{RP} = (1 + r_{f/a_1} + r_{f/a_2})h_7$$

$$h_x + r_{f/a_1}h_{RP} = (1 + r_{f/a_1})h_5$$

and by the definition of regenerator effectiveness,

$$0.70 = \frac{h_x - h_4}{(1 + r_{f/a_1} + r_{f/a_2})(h_g - h_c)}$$

Solve these equations simultaneously for h_x, r_{f/a_1}, and r_{f/a_2} where r_{f/a_1} is the fuel/air ratio in the first combustion chamber and r_{f/a_2} is the fuel/air ratio in the second combustion chamber.

```
====================== VARIABLE SHEET ======================
St Input ——— Name ——— Output ——— Unit ——— Comment ———
              rfa1      .01449979  kg/kg
              rfa2      .00521052  kg/kg
              hx        868.68522  kJ/kg



======================= RULE SHEET =========================
S Rule ————
   (1+rfa1)*1268.5+rfa2*44186=(1+rfa1+rfa2)*1487.8
   hx+rfa1*44186=(1+rfa1)*1487.8
   0.70=(hx-500.58)/((1+rfa1+rfa2)*(1022.8-507.1))
```

The first-law analysis of the compressor yields

$$w_c = -(h_2 - h_1) - (h_4 - h_3)$$

$$w_c = -[(434.68 - 290.17) + (500.58 - 421.26)\text{kJ/kg air}]$$

$$w_c = -223.8 \text{ kJ/kg air}$$

The first-law analysis of the turbine yields

$$w_t = (1 + r_{f/a_1})(h_5 - h_6) + (1 + r_{f/a_1} + r_{f/a_2})(h_7 - h_8)$$

$$w_t = (1.0145 \text{ kg products/kg air})(1487.8 - 1268.5 \text{ kJ/kg products})$$

$$+ (1.01971 \text{ kg products/kg air})(1487.8 - 1022.8 \text{ kJ/kg products})$$

$$w_t = 696.6 \text{ kJ/kg air}$$

The net work is

$$w_{net} = w_t + w_c = 472.8 \text{ kJ/kg air}$$

The total energy added is

$$q_{in} = (r_{f/a_1} + r_{f/a_2})h_{RP}$$

$$q_{in} = (0.01971 \text{ kg fuel/kg air})(44\ 186 \text{ kJ/kg fuel}) = 870.9 \text{ kJ/kg air}$$

and the overall thermal efficiency is

$$\eta_{th} = \frac{w_{net}}{q_{in}} = \frac{472.8}{870.9} = 0.543 \quad \text{or } 54.3\%$$

Comment: The reheating and intercooling improve the effectiveness of the regenerator. The temperatures of the fluids are further apart when they enter the regenerator, so more heat may be transferred for a given tube surface or regenerator size. The intercooling may be accomplished with either air or water. The cost of intercooling is quite low, so it offers a substantial advantage in improving overall efficiency. ▬

14.8 AIRCRAFT GAS TURBINES

The aircraft industry has pioneered many of the advances made in gas turbine design. The space and weight limitations of the airplane engine make the gas turbine an ideal unit because it can provide large amounts of power in a small volume with a very low weight. The stationary gas turbine and the jet engine differ greatly in the manner by which work is produced. The jet engine gas turbine provides work to drive the compressor (and, on a turboprop, the propeller); the exhaust gases from the turbine are expanded through a nozzle, increasing the velocity. The change in velocity of the gas entering and leaving the gas turbine unit causes a force to be exerted on the aircraft. Figure 14.22(a) illustrates the gas turbine unit schematically, and Figure

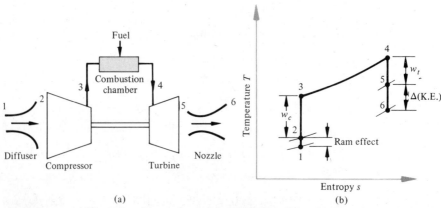

Figure 14.22 (a) The gas turbine unit used for aircraft engines. (b) The corresponding *T-s* diagram.

14.22(b) illustrates the *T-s* diagram. Located before the compressor is the *diffuser,* a nozzle that decreases velocity and increases pressure. This is called the *ram effect.* Figure 14.22(b) is drawn for isentropic expansion and compression processes. Since no actual process is isentropic, we deal in terms of efficiencies for each component —diffuser, compressor, turbine, and nozzle.

Energy Analysis of Jet Engines

Since the unit produces no net work, as we have used the term in previous sections (all the work of the turbine drives the compressor), let us find out how the aircraft moves. The force that moves the airplane is caused by the gas change of momentum through the engine. The analysis may be made thermodynamically, on an energy basis, or on the impulse-momentum principle. We will use both approaches, the thermodynamic method first.

Consider a turbojet engine driving a plane at a velocity v_p. The relative velocities of the gas entering and leaving the engine depend on where the observer is located. The observer on the plane finds that the air enters the engine at state i. He makes an energy balance on the engine, Figure 14.23, and finds that, from the first law of thermodynamics for an open system,

$$h_i + (\text{k.e.})_i + r_{f/a}h_{RP} = (1 + r_{f/a})h_e + (1 + r_{f/a})(\text{k.e.})_e \qquad (14.13)$$

$$h_i - (1 + r_{f/a})h_e = (1 + r_{f/a})(\text{k.e.})_e - (\text{k.e.})_i - r_{f/a}h_{RP}$$

where

$$(\text{k.e.})_i = \frac{v_i^2}{2} \quad \text{and} \quad v_i = v_p$$

$$(\text{k.e.})_e = \frac{v_e^2}{2}$$

where v_e is the relative exit velocity.

The observer on the ground sees the same engine and makes an energy analysis based upon Figure 14.24. In this case energy is added to the system as above, but work is also observed, as a force is moving through a distance. The fuel enters with kinetic

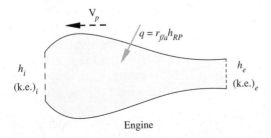

Figure 14.23 The energy flows for a turbojet engine with the observer on the plane.

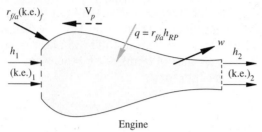

Figure 14.24 The energy flows for a turbojet engine with the observer on the ground.

energy, since it is located on the plane moving at velocity v_p. Thus, an energy balance
yields

$$h_1 - (1 + r_{f/a})h_2 = (1 + r_{f/a})(k.e.)_2 - (k.e.)_1 - r_{f/a}h_{RP} - r_{f/a}(k.e.)_f + w \qquad (14.14)$$

where $(k.e.)_2 = v_2^2/2$ and $v_2 = v_e - v_p$, and where $-v_p$ is the speed of the plane in the
opposite direction to gas flow.

$$(k.e.)_1 = \frac{v_1^2}{2} \quad \text{but } v_1 = 0 \text{ (still air)}$$

$$(k.e.)_f = \frac{v_p^2}{2}$$

The temperature is not a function of where the observer is located, so

$$h_i = h_1 \quad \text{and} \quad h_e = h_2 \qquad (14.15)$$

Thus, equations (14.13) and (14.14) are equal, and the right-hand sides may be
equated.

$$w = (1 + r_{f/a})[(k.e.)_e - (k.e.)_2] + r_{f/a}(k.e.)_f - (k.e.)_i \qquad (14.16)$$

Substituting for the kinetic energy yields

$$w = v_e v_p \left[1 + r_{f/a} - \frac{v_p}{v_e} \right] \text{ J/kg air} \qquad (14.17)$$

To further visualize the work term, think of the fluid as it passes through the
engine. In passing through the compressor, it experiences work input. In passing
through the turbine, the fluid performs work to drive the compressor and auxiliaries.
Because of friction, the fluid performs additional work on the engine. The turbine
and compressor work terms are essentially equal, and the friction work appears as
heat. The work term in equation (14.17) represents the atmospheric drag on the
control volume, moving through a distance, which will be equal to the propulsive
work.

The same result may be obtained by conventional impulse-momentum analyses of the engine. The propulsive force in a jet engine is brought about by a continuous change in the momentum of the gas as it passes through the engine. From basic mechanics, $F + d(mv)/d\tau$, where (mv) is the momentum. Consider the engine to be moving at velocity v_p (observer on the plane), and the momentum entering the engine is $\dot{m}_a v_i = \dot{m}_a v_p$, where $\dot{m}_a$ is the air flow rate. An amount of fuel $\dot{m}r_{f/a}$ is added to the engine, and the momentum leaving the engine is $\dot{m}_a(1 + r_{f/a}) v_e$ where v_e is the exit velocity relative to the engine. The force generated by the change in momentum is

$$F = \dot{m}_a v_e \left[1 + r_{f/a} - \frac{v_p}{v_e} \right] \tag{14.18}$$

where the force generated per kilogram of air, denoted in equation (14.18), is called the *specific thrust*.

Work is a force through a distance, and the plane travels a distance v_p during 1 s; thus

$$\dot{W} = F v_p = \dot{m}_a v_e v_p \left[1 + r_{f/a} - \frac{v_p}{v_e} \right] \text{W} \tag{14.19}$$

There is no difference between equations (14.19) and (14.17), other than that equation (14.19) is derived for an air flow rate of $\dot{m}_a$. This is what we would have expected, as the engine will not perform more or less work as a function of the analysis.

Maximum Work and Propulsive Efficiency

There are two more terms to define: the *maximum work* developed as a function of plane speed and the *propulsive efficiency*. The maximum work developed may be found by finding the derivative of work with respect to plane velocity from equation (14.19).

$$\frac{d\dot{W}}{dv_p} = 0 = \left(1 + r_{f/a} - \frac{2v_p}{v_e} \right)$$
$$v_p = \frac{v_e(1 + r_{f/a})}{2} \tag{14.20}$$

Thus, the maximum work will be developed when the plane speed is about one-half the gas exit velocity.

The propulsive efficiency, η_p, is defined as the ratio of the work of the propulsive force divided by the sum of this same work plus the unused kinetic energy of the engine relative to the ground:

$$\eta_p = \frac{w}{w + (k.e.)_{unused}}$$

$$(k.e.)_{unused} = \frac{(1 + r_{f/a})(v_e - v_p)^2}{2}$$

$$w = v_e v_p \left[1 + r_{f/a} - \frac{v_p}{v_e} \right] \qquad (14.21)$$

After considerable algebraic operations and with the assumption that $r_{f/a}$ is much less than 1, so it may be neglected, we finally determine that the propulsive efficiency is

$$\eta_p = \frac{2}{1 + (v_e/v_p)} \qquad (14.22)$$

This is often called the *Froude efficiency.* From an analysis of equations (14.22) and (14.18), we find that

When $v_p = 0$, the specific thrust is a maximum, but $\eta_p = 0$. This occurs at takeoff, where maximum thrust is needed.

When $v_p = v_e$, $\eta_p = 100\%$, but the specific thrust is zero. Thus, v_e should be greater than v_p but not substantially greater.

There is a great deal more information on jet engines and gas turbine units in general. We have only scratched the surface, presenting concepts from which more detailed analyses are developed. A thorough understanding of gas dynamics is very important for advanced work in this area. The following example illustrates some of the concepts developed in this section.

Example 14.8

A jet engine in which all expansions and compressions are isentropic is operating at an altitude where the entrance pressure is 83 kPa and the entrance temperature is 230°K. The aircraft is moving at a velocity of 305 m/s. Fuel is added so the maximum temperature is 1400°K and the compressor discharge pressure is 480 kPa. Neglect the mass of the fuel, assuming air as the substance throughout. Determine the pressure to which the turbine expands, the exit velocity of the combustion gas from the engine, the specific thrust, the work, the propulsive efficiency, and the thermal efficiency.

Solution

Given: An aircraft's jet engine with the inlet air conditions, the maximum temperature, and the compressor discharge pressure.

Find: The turbine exit pressure, gas velocity exiting the engine, the engine's work, the specific thrust, the propulsive efficiency, and the overall thermal efficiency.

Sketch and Given Data:

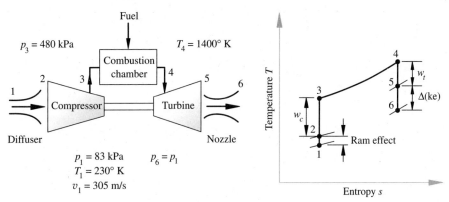

Figure 14.25

Assumptions:

1. Each component is analyzed as a steady-state open system.
2. Air and products of combustion behave like ideal gases with variable specific heats.
3. The mass of fuel may be ignored in determining the compressor and turbine work.
4. The compressor, combustion chamber, and turbine are adiabatic.
5. The combustion process is constant-pressure.

Analysis: Determine the enthalpies located around the gas turbine cycle, using Table A.2.

$$h_1 = 230.01 \qquad p_{r_1} = 0.5477$$

$$h_2 = h_1 + \frac{v_1^2}{(2)(1000)} = 230.01 \text{ kJ/kg} + \frac{(305 \text{ m/s})^2}{(2)(1000 \text{ J/kJ})}$$

$$h_2 = 276.52 \text{ kJ/kg}$$

From the air tables $p_{r_2} = 1.0420$.

The pressure at state 2 is determined from the isentropic compression from state 1 to state 2.

$$p_2 = p_1 \left(\frac{p_{r_2}}{p_{r_1}}\right) = (83 \text{ kPa})\left(\frac{1.0420}{0.5477}\right) = 157.9 \text{ kPa}$$

Pressure at state 3 is given as $p_3 = 480$ kPa. The process from state 2 to state 3 is isentropic.

$$p_{r_3} = p_{r_2} \left(\frac{p_3}{p_2} \right) = (1.0420) \left(\frac{480}{157.9} \right) = 3.168$$

$$h_3 = 380.48 \text{ kJ/kg}$$

From a first-law analysis of the compressor,

$$w_c = -(h_3 - h_2) = -(380.48 - 276.52 \text{ kJ/kg}) = -103.96 \text{ kJ/kg}$$

The combustion temperature is given as $1400°\text{K}$; hence $h_4 = 1515.41$ kJ/kg and $p_{r_4} = 450.5$.

Since all the turbine work provides for the compressor work,

$$w_t = -w_c = (h_4 - h_5) = +103.96 \text{ kJ/kg} = (1515.41 - h_5 \text{ kJ/kg})$$

$$h_5 = 1411.45 \text{ kJ/kg}$$

$$p_{r_5} = 344.9$$

The process from state 4 to state 5 is isentropic; hence the turbine discharge pressure is

$$p_5 = p_4 \left(\frac{p_{r_5}}{p_{r_4}} \right) = (480 \text{ kPa}) \left(\frac{344.9}{450.5} \right) = 367.5 \text{ kPa}$$

$$p_6 = 83 \text{ kPa}$$

The process from state 5 to state 6 is isentropic.

$$p_{r_6} = p_{r_5} \left(\frac{p_6}{p_5} \right) = (344.9) \left(\frac{83}{367.5} \right) = 77.9$$

$$h_6 = 941.8 \text{ kJ/kg}$$

From a first-law analysis of the engine, equation (14.13) yields

$$h_1 + (\text{k.e.})_1 + q_{in} = h_6 + (\text{k.e.})_6$$

$$230.01 + \frac{(305 \text{ m/s})^2}{2(1000 \text{ J/kJ})} + (1515.41 - 380.48 \text{ kJ/kg})$$

$$= (941.8 \text{ kJ/kg}) + \frac{(v_6 \text{ m/s}^2)}{2(1000 \text{ J/kJ})}$$

$$v_6 = v_e = 969.2 \text{ m/s}$$

From equation (14.18), the specific thrust is

$$\frac{F}{\dot{m}_a} = v_e \left[1 - \frac{v_p}{v_e} \right] = 969.2 \left[1 - \frac{305}{969.2} \right] = 664.2 \text{ kN/kg/s}$$

From equation (14.19) the work is

$$w = v_e v_p \left[1 - \frac{v_p}{v_e} \right] = \frac{(305 \text{ m/s})(969.2 \text{ m/s})}{(1000 \text{ J/kJ})} \left[1 - \frac{305}{969.2} \right] = 202.58 \text{ kJ/kg}$$

From equation (14.22) the propulsive efficiency is

$$\eta_p = \frac{2}{1 + (v_e/v_p)} = \frac{2}{1 + 969.2/305} = 0.48$$

The overall thermal efficiency is the net work produced, the work available to move the aircraft, divided by the energy supplied,

$$\eta_{th} = \frac{w}{q_{in}} = \frac{w}{h_4 - h_3} = \frac{202.58}{(1515.41 - 380.48)} = 0.18 \quad \text{or } 18\%$$

Comments: Including the mass of fuel in the previous expressions adds complication at the expense of clarity. In addition there are pressure drops from the compressor discharge to the turbine inlet and efficiencies associated with all expansion and compression processes.

CONCEPT QUESTIONS

1. What are the processes that characterize the air-standard Brayton cycle?
2. What is the effect of raising the pressure ratio in an air-standard Brayton cycle? Is this effect valid if the maximum cycle temperature is fixed? Why?
3. When analyzing a Brayton cycle, a closed system, are the components considered open or closed systems? Why?
4. In a gas turbine is all the turbine work available as net work? Why?
5. How are irreversibilities in fluid flow through the compressor and turbine accounted for?
6. Why do turbine and compressor efficiencies have such a significant effect on the overall cycle performance?
7. Why does a pressure drop from the compressor discharge to the turbine inlet decrease the turbine work if the temperature of the gas entering the turbine is constant?
8. Why does an air-standard Brayton-cycle analysis predict a higher cycle thermal efficiency than does one using the air tables or AIRCYCLE.TK?
9. Why does a regenerator improve the efficiency of a Brayton cycle?
10. Explain regenerator effectiveness.
11. What would be the temperature of the products exiting an ideal regenerator?
12. A regenerator is added to a closed Brayton cycle, all processes remaining ideal. Under what circumstances will there be no improvement in cycle thermal efficiency?

13. Explain reheating and intercooling in an open gas turbine cycle.

14. How does the jet engine propulsion system differ from the open-cycle gas turbine?

15. What are the purposes of the diffuser and nozzle in a turbojet engine?

16. What is the ram effect?

17. What is the specific thrust? How is this related to power?

18. Explain propulsive efficiency.

PROBLEMS (SI)

When performing an open-system analysis of a gas turbine in the following problems, use the 400% theoretical air tables, unless otherwise instructed, for the products of combustion.

14.1 An air-standard Brayton cycle has a pressure ratio of 8. The air properties at the start of compression are 100 kPa and 25°C. The maximum allowable temperature is 1100°C. Determine (a) the thermal efficiency; (b) the net work; (c) the heat added.

14.2 A Brayton cycle uses helium as the working substance. The pressure ratio is 4, but the low pressure is 500 kPa. The temperature at the start of compression is 300°K. The heat added is 600 kJ/kg. Determine (a) the cycle work per kg; (b) the maximum temperature; (c) the displacement volume per kg.

14.3 The same as Problem 14.2 except that the low pressure is 105 kPa. Compare the results of parts (a) and (c).

14.4 An air-standard Brayton cycle has a net power output of 100 kW. The working substance is air, entering the compressor at 30°C, leaving the high-temperature heat exchanger at 750°C, and leaving the turbine at 300°C. Determine (a) the compressor pressure ratio; (b) the compressor work in kJ/kg; (c) the mass flow rate of air; (d) the thermal efficiency; (e) the second-law efficiency.

14.5 A furnace needs hot pressurized gas at 200 kPa. This gas is to be provided by the exhaust from a gas turbine operating on the Brayton cycle. The turbine will produce no power beyond that required by the compressor. The compressor inlet conditions are 100 kPa and 290°K. The turbine inlet temperature is 815°C. Determine the compressor pressure ratio.

14.6 In an air-standard Brayton cycle the compressor inlet conditions are 100 kPa and 280°K. The turbine inlet conditions are 1000 kPa and 1167°K. The turbine produces 11 190 kW. Determine (a) the air flow rate; (b) the compressor power; (c) the cycle efficiency; (d) the pressure ratio for maximum work; (e) the second-law efficiency.

14.7 A 75-kW air-standard Brayton cycle is designed for maximum work. The compressor inlet conditions are 100 kPa and 27°C, and the pressure ratio is 5.5. Determine (a) the turbine inlet temperature; (b) the cycle efficiency; (c) the air flow rate.

14.8 The same as Problem 14.1, except that the compressor and turbine efficiencies are equal to 90%.

14.9 A Brayton cycle uses argon as the working substance. At the beginning of compression, the temperature is 335°K and the pressure is 480 kPa. The compression process is adiabatic with discharge conditions of 645°K and 1930 kPa. The argon is heated and enters the turbine at 1390°K and 1930 kPa and expands adiabatically to 890°K and

480 kPa. Determine (a) the compressor efficiency; (b) the turbine efficiency; (c) the thermal efficiency.

14.10 A gas turbine turboelectric plant is to provide 30 000 kW to the electric generator. The turbine unit receives air at 100 kPa and 295°K. The maximum temperature is 1200°K, and the maximum pressure is 400 kPa. Neglecting the mass of the fuel, determine (a) the overall thermal efficiency; (b) energy added in the combustion chamber; (c) compressor power; (d) total turbine power; (e) volumetric flow rate of air at inlet conditions in m^3/s.

14.11 For an air-standard Brayton cycle the air flow rate entering the compressor is 10 m^3/s at 300°K and 100 kPa. The pressure ratio is 12, and the cycle maximum temperature is 1100°K. Determine (a) the net power developed; (b) the cycle thermal efficiency.

14.12 An air-standard Brayton cycle is characterized by minimum and maximum temperatures of 300°K and 1500°K. The compressor inlet pressure is 100 kPa. Determine (a) the pressure ratio that yields the optimum thermal efficiency; (b) the value of the thermal efficiency.

14.13 An air-standard Brayton cycle has compressor inlet conditions of 310°K and 98 kPa and turbine inlet conditions of 882 kPa and 1200°K. The heat transferred to the air in the high-temperature heat exchanger is 25 MW. Determine the net power produced, assuming (a) constant specific heats; (b) variable specific heats.

14.14 An air-standard Brayton cycle has compressor inlet conditions of 300°K and 400 kPa. The pressure ratio is 12, and the cycle maximum temperature is 1300°K. Determine the cycle efficiency if the substance is (a) air; (b) helium.

14.15 A gas turbine receives 4.72 m^3/s of air at 27°C and 100 kPa. The compression is adiabatic with discharge at 517 kPa and 220°C. Dodecane is used with a ratio of 0.015 kg fuel/kg air. The turbine exit pressure and temperature are 105 kPa and 454°C. Determine (a) the thermal efficiency; (b) the turbine efficiency; (c) the compressor efficiency; (d) the maximum temperature; (e) the total turbine power; (f) the compressor power.

14.16 A gas turbine unit is to drive a natural gas compressor located on a pipeline. The natural gas (assume it is methane) has inlet conditions to the compressor of 140 kPa and 295°K and an exit pressure of 690 kPa. The isentropic compressor efficiency is 85%, and the gas flow is 9.5 m^3/s at inlet conditions. The gas turbine unit receives air at 101 kPa and 300°K; the isentropic compressor discharges it at 550 kPa. The fuel/air ratio is 0.0165 kg fuel/kg air, and $h_{RP} = 44\ 000$ kJ/kg. The products of combustion expand isentropically through the turbine to 101 kPa. Determine (a) the gas turbine unit thermal efficiency; (b) the power required to drive the natural gas compressor; (c) the air flow rate required to drive the compressor.

14.17 A gas turbine unit has compressor inlet conditions of 100 kPa and 310°K. The compressor discharge pressure is 700 kPa, and the temperature is 565°K. Fuel enters the combustion chamber and raises the air temperature to 1200°K. The turbine discharge temperature is 710°K, and the pressure is 100 kPa. Determine (a) the compressor and turbine adiabatic efficiencies; (b) the cycle thermal efficiency.

14.18 The turbine in Problem 14.17 has a gas flow rate entering it of 10 m^3/s. Determine (a) the net power produced; (b) the fuel consumption in kg/min.

14.19 A gas turbine unit receives 4.72 m^3/s of air at 27°C and 100 kPa and compresses it isentropically to 450 kPa (gage). In the combustion chamber, fuel with a heating value of 43 200 kJ/kg is added so the maximum temperature is 1250°K. The turbine exhausts to atmospheric pressure. Determine (a) the fuel flow rate; (b) the unit thermal efficiency; (c) the turbine exit temperature; (d) the available portion of the energy of the products of combustion leaving the turbine, if $T_0 = 25°C$.

14.20 An air turbine operates between pressures of 410 kPa and 100 kPa and receives 0.45 kg/s of air at 650°C. (a) For an ideal turbine, what is the power developed? (b) When operating under these conditions an actual turbine develops 111.9 kW and has a discharge temperature of 150°C. The turbine blades are water-cooled, with water entering at 10°C and leaving at 38°C. Determine the water flow rate in kg/s.

14.21 Air enters the combustion chamber of a gas turbine unit at 550 kPa, 227°C, and 43 m/s. The products of combustion leave the combustor at 517 kPa, 1004°C, and 140 m/s. Liquid fuel enters with a heating value of 43 000 kJ/kg. The combustor efficiency is 95%. Determine the fuel/air ratio.

14.22 A regenerator in a gas turbine unit receives air from the compressor at 400 kPa and 455°K. The air leaves at 395 kPa and 77°K. The products of combustion enter at 105 kPa and 823°K and leave at 102 kPa and 611°K. The specific heat of air is 1.0047 kJ/kg-K and of the products of combustion is 1.044 kJ/kg-K. The air flow rate is 22.89 kg/s, and the fuel flow rate is 0.231 kg/s. Determine (a) the fuel/air ratio; (b) the regenerator effectiveness; (c) the net entropy change across the regenerator.

14.23 A gas turbine unit is equipped with a regenerator. The compression process is isentropic with air entering the compressor at 290°K and 100 kPa. The pressure ratio is 8. The maximum allowable temperature entering the turbine is 1400°K. The turbine exhausts to the regenerator, with an effectiveness of 60% at 100 kPa. The expansion process through the turbine is isentropic. Determine, for $h_{RP} = 43\,000$ kJ/kg, (a) the fuel/air ratio; (b) the temperature entering the combustor; (c) the thermal efficiency; (d) the temperature of the products of combustion leaving the regenerator; (e) the thermal efficiency if the regenerator had not been installed.

14.24 The same as Problem 14.23, except that the compressor efficiency is 87%, the turbine efficiency is 90%, and there is a 40-kPa pressure drop between the compressor and turbine and a 2.7-kPa pressure drop between the turbine exhaust and atmosphere.

14.25 A regenerative gas turbine unit has two turbines; the first, located immediately following the combustion chamber, drives the compressor, and its discharge enters a second turbine that drives a generator. In addition, a regenerator receives the exhaust from the second turbine and the discharge from the compressor. Each turbine has an isentropic efficiency of 87%, and the compressor has an isentropic efficiency of 85%. The effectiveness of the regenerator is 80%. The turbine inlet temperature is 1300°K, and the fuel burned is dodecane. The electric power generated is 20 MW. The compressor inlet conditions are 300°K and 100 kPa, and the compressor pressure ratio is 10. Determine (a) the volume flow rate of air at compressor inlet conditions; (b) the fuel flow rate in kg/min; (c) the thermal efficiency; (d) the temperature of the products leaving the regenerator.

14.26 Rework Problem 14.25 with a compressor discharge temperature of 600°K, the temperature of the air leaving the regenerator at 700°K, and a ratio of 0.019 kg fuel/kg air. The net power, turbine efficiencies, pressure ratio, and compressor inlet conditions remain the same. Determine (a) the regenerator effectiveness; (b) the thermal efficiency; (c) the compressor isentropic efficiency.

14.27 Air enters a gas turbine unit's compressor at 100 kPa and 300°K and is compressed in two stages to 1200 kPa. Intercooling occurs between the stages at 400 kPa, and the temperature entering the second stage is 330°K. The turbine inlet conditions are 1150 kPa and 1200°K. The fuel used in the combustion chamber has a heating value of 44 000 kJ/kg. Determine (a) the thermal efficiency of the unit; (b) the fuel/air ratio.

14.28 The turbine in Problem 14.27 is replaced with a two-stage reheat turbine, where reheating occurs at a pressure of 400 kPa and raises the temperature entering the second stage by 150°K. Determine (a) the fuel/air ratio for the second combustion chamber; (b) the unit's thermal efficiency.

14.29 In designing a gas turbine for maximum efficiency, a decision is made to use intercooling of the compressor. The air is delivered from 100 kPa and 290°K to a final discharge pressure of 950 kPa. There are two stages of compression, with intercooling at the optimum interstage pressure. The intercooling cools the air temperature to 25°C of the inlet temperature. The regenerator has an effectiveness of 65%, and the maximum allowable turbine inlet temperature is 1350°K. All expansion and compression processes are isentropic. Determine, for $h_{RP} = 43\,000$ kJ/kg, (a) the thermal efficiency; (b) the fuel/air ratio; (c) the turbine work per kg; (d) the compressor work per kg; (e) the heat removed in the intercooler; (f) the available energy of the products of combustion leaving the regenerator; (g) the thermal efficiency with no intercooling.

14.30 The same as Problem 14.29 except that (a) there is a 7.0-kPa drop in the intercooler; (b) there is a 3.4-kPa drop between the turbine exhaust and atmosphere; (c) there is a 34-kPa drop between the compressor and the turbine; (d) the turbine and compressor efficiencies are 90%.

14.31 A closed gas turbine unit uses neon as the working fluid. There are two stages of compression, with ideal intercooling, from 480 kPa and 290°K to 2.75 MPa. A regenerator has an effectiveness of 60% and delivers the neon to the high-temperature heat exchanger, where it is heated to 890°K. There are two turbine stages, the first discharging at 1170 kPa. Following the first turbine stage, a heat exchanger raises the neon temperature to 890°K. The second turbine exhausts at 480 kPa to the regenerator. All expansion and compression processes are isentropic. Determine (a) the compressor work per kg; (b) the turbine work per kg; (c) the heat added per kg; (d) the overall thermal efficiency; (e) the power output for a neon flow rate of 45 kg/s; (f) the available portion of the heat rejected in the low-temperature heat exchanger if $T_0 = 275$°K.

14.32 The same as Problem 14.31 except that (a) the compressor and turbine efficiencies are 90%; (b) there is a pressure drop of 1.5% through each heat exchanger.

14.33 A jet plane is traveling at 0.309 km/s and has an engine that develops a thrust of 13 344 N. The gas exiting the engine has a relative velocity of 340 m/s, and the fuel/air ratio is 0.02 kg fuel/kg air. Determine (a) the air flow rate; (b) the propulsive efficiency; (c) the fuel flow rate.

14.34 A jet aircraft is flying with a velocity of 965 km/h at an altitude of 12.2 km, where the pressure is 20 kPa and the temperature is 220°K. The air enters an ideal diffuser and leaves the combustor at 1350°K and 100 kPa. The fuel has a heating value of 43 000 kJ/kg. All expansion and compression processes are isentropic. Determine (a) the compressor work per kg; (b) the fuel/air ratio; (c) the pressure entering the nozzle; (d) the specific thrust; (e) the propulsive efficiency; (f) the thermal efficiency; (g) the total thrust for an airflow of 32 kg/s.

14.35 A turbojet aircraft has a velocity of 280 m/s and flies at an altitude of 6 km. The air conditions at 6 km are 48 kPa and 260°K. The compressor pressure ratio is 14, and the turbine inlet temperature is 1300°K. Determine for the ideal cycle (a) the pressure at the turbine exit; (b) the exhaust gas velocity; (c) the propulsive efficiency.

14.36 A turbojet aircraft has a velocity of 325 m/s at an altitude of 10 km. The ambient air conditions are 30 kPa and 240°K. The compressor pressure ratio is 12, and the turbine inlet temperature is 1300°K. The mass flow rate of air entering the compressor is 40 kg/s, and the fuel has a heating value of 43 000 kJ/kg. Determine for ideal gas with constant specific heats and treating the products as air (a) the exhaust gas velocity; (b) the total propulsive power; (c) the fuel consumption.

14.37 Calculate Problem 14.36 using variable specific heats for the gases and with turbine and compressor isentropic efficiencies of 85%.

14.38 A turbojet-powered aircraft is held stationary on the ground by its brakes. The engine's compressor has a pressure ratio of 12, the ambient air is at 310°K and 100 kPa, and the air flow rate entering the engine is 6 kg/s. The fuel has a heating value of 43 000 kJ/kg and a flow rate of 0.12 kg/s. Assume all the processes to be ideal. Determine the force necessary to hold the plane stationary.

14.39 An aircraft driven by a turbojet engine is flying at an altitude of 10 km where the air temperature is 230°K and the pressure is 23 kPa. The compressor pressure ratio is 12, and the turbine inlet temperature is 1250°K. The turbine and compressor isentropic efficiencies are 87% and 85%, respectively. Determine (a) the velocity at the nozzle exit; (b) the pressure entering the nozzle.

14.40 The engine in Problem 14.39 has an afterburner inserted between the turbine and nozzle, which raises the temperature of the gas entering the nozzle to 1100°K. Determine the velocity at the nozzle exit.

14.41 In the turboprop engine shown, the turbine's rotor is connected to the compressor and to the propeller. The engine is ideally designed such that turbine work is divided equally between the compressor and the propeller. The gas from the turbine discharges to a nozzle for additional thrust. Consider such an engine where the air enters the diffuser at 200 m/s, 40 kPa, 240°K, and a flow rate of 40 kg/s. The compressor pressure ratio is 11, and the turbine inlet temperature is 1200°K. The fuel used has a heating value of 43 000 kJ/kg. All processes are ideal. Determine (a) the fuel consumption; (b) the power delivered to the propeller and to the compressor; (c) the velocity from the nozzle.

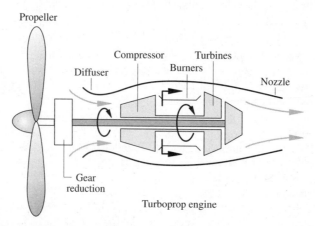

Problem 14.41

14.42 There is 0.567 m³/min of air available at 6.2 mPa and 135°C. Air is needed at 750 kPa and between 15°C and 38°C. A proposal suggests using an air turbine to expand the air and perform useful work. (a) If the turbine internal efficiency is 80%, what is the exhaust temperature? (b) What heat must be added in part (a) in a heat exchanger to raise the temperature of the air to 15°C? (c) An alternative method is to throttle the air before it enters the turbine. What should be the inlet pressure of the turbine? (d) What power is produced in (a) and (c)?

PROBLEMS (English Units)

*14.1   An air-standard Brayton cycle has temperature limits of 100°F and 1200°F and $p_1 = 15$ psia. Determine (a) the pressure ratio for maximum work; (b) the thermal efficiency.

*14.2   A manufacturing facility needs hot gas at a pressure of 30 psia for its operation. The gas is to be the exhaust from a gas turbine whose only function is to drive the compressor. The compressor receives air at 15 psia and 60°F, and the turbine inlet temperature is 1540°F. Determine the pressure ratio of the compressor if the compressor and turbine efficiencies are (a) 100%; (b) 85%.

*14.3   For an air-standard Brayton cycle the air flow rate entering the compressor is 350 ft³/sec at 77°F and 1 atm. The pressure ratio is 12, and the cycle maximum temperature is 2100°R. Determine (a) the net power developed; (b) the cycle thermal efficiency.

*14.4   An air-standard Brayton cycle is characterized by minimum and maximum temperatures of 540°R and 2700°R. The compressor inlet pressure is 1 atm. Determine (a) the pressure ratio that yields the optimum thermal efficiency; (b) the value of the thermal efficiency.

*14.5   An air-standard Brayton cycle has compressor inlet conditions of 100°F and 14.5 psia and turbine inlet conditions of 130 psia and 2160°R. The heat transferred to the air in the high-temperature heat exchanger is 24 000 Btu/sec. Determine the net power produced, assuming (a) constant specific heats; (b) variable specific heats.

*14.6   An air-standard Brayton cycle has compressor inlet conditions of 77°F and 60 psia. The pressure ratio is 12, and the cycle maximum temperature is 2300°R. Determine the cycle efficiency if the substance is (a) air; (b) helium.

*14.7   A gas turbine unit has compressor inlet conditions of 1 atm and 100°F. The compressor discharge pressure is 103 psia, and the temperature is 557°F. Fuel enters the combustion chamber and raises the air temperature to 1700°F. The turbine discharge temperature is 818°F, and the pressure is 1 atm. Determine (a) the compressor and turbine adiabatic efficiencies; (b) the cycle thermal efficiency.

*14.8   The turbine in Problem *14.7 has a gas flow rate entering it of 350 ft³/sec. Determine (a) the net power produced; (b) the fuel consumption in kg/min if the heating value is 18,500 Btu/lbm.

*14.9   A gas turbine unit receives 10,000 ft³/min of air at 77°F and 14.6 psia and compresses it isentropically to 65 psia. In the combustion chamber, fuel with a heating value of 18,600 Btu/lbm is added so the maximum temperature is 1800°F. The turbine exhausts to atmospheric pressure. Determine (a) the fuel flow rate; (b) the unit thermal efficiency; (c) the turbine exit temperature; (d) the availability of the products of combustion leaving the turbine, if $T_0 = 77°F$.

*14.10   A gas turbine receives 10,000 ft³/min of air at 77°F and 14.6 psia. The compression is adiabatic with discharge at 75 psia and 430°F. Dodecane is used with a ratio of 0.015 lbm fuel/lbm air. The turbine exit pressure and temperature are 15.2 psia and

850°F. Determine (a) the thermal efficiency; (b) the turbine efficiency; (c) the compressor efficiency; (d) the maximum temperature; (e) the total turbine power; (f) the compressor power.

*14.11 A gas turbine unit is equipped with a regenerator. The compression process is isentropic with air entering the compressor at 530°R and 14.6 psia. The pressure ratio is 8. The maximum allowable temperature entering the turbine is 2500°R. The turbine exhausts to the generator, with an effectiveness of 60%, at 14.6 psia. The expansion process through the turbine is isentropic. Determine for $h_{RP} = 18,500$ Btu/lbm (a) the fuel/air ratio; (b) the temperature entering the combustor; (c) the thermal efficiency; (d) the temperature of the products of combustion leaving the regenerator; (e) the thermal efficiency if the regenerator had not been installed.

*14.12 The same as Problem *14.11, except the compressor efficiency is 87%, the turbine efficiency is 90%, and there is a 6-psia pressure drop between the compressor and turbine and a 0.4-psia pressure drop between the turbine exhaust and atmosphere.

*14.13 A regenerative gas turbine unit has two turbines; the first, located immediately following the combustion chamber, drives the compressor, and its discharge enters a second turbine that drives a generator. In addition, a regenerator receives the exhaust from the second turbine and the discharge from the compressor. Each turbine has an isentropic efficiency of 87%, and the compressor has an isentropic efficiency of 85%. The effectiveness of the regenerator is 80%. The turbine inlet temperature is 1880°F, and the fuel burned is dodecane. The electric power generated is 20 MW. The compressor inlet conditions are 77°F and 1 atm, and the compressor pressure ratio is 10. Determine (a) the volume flow rate of air at compressor inlet conditions; (b) the fuel flow rate in lbm/min; (c) the thermal efficiency; (d) the temperature of the products leaving the regenerator.

*14.14 Consider Problem *14.13 where the compressor discharge temperature is 620°F, the temperature of the air leaving the regenerator is 800°F, and the fuel/air ratio is 0.019 lbm fuel/lbm air. The net power, turbine efficiencies, pressure ratio, and compressor inlet conditions remain the same. Determine (a) the regenerator effectiveness; (b) the thermal efficiency; (c) the compressor isentropic efficiency.

*14.15 Air enters a gas turbine unit's compressor at 14.7 psia and 77°F and is compressed in two stages to 175 psia. Intercooling occurs between the stages at 55 psia, and the temperature entering the second stage is 135°F. The turbine inlet condition is 170 psia and 1700°F. The fuel used in the combustion chamber has a heating value of 18,500 Btu/lbm. Determine (a) the thermal efficiency of the unit; (b) the fuel/air ratio.

*14.16 The turbine in Problem *14.15 is replaced with a two-stage reheat turbine, where reheating occurs at a pressure of 60 psia and raises the temperature entering the second stage by 270°F. Determine (a) the fuel/air ratio for the second combustion chamber; (b) the unit's thermal efficiency.

*14.17 A turbojet aircraft has a velocity of 920 ft/sec and flies at an altitude of 20,000 ft. The air conditions at 20,000 ft are 7 psia and 5°F. The compressor's pressure ratio is 14, and the turbine inlet temperature is 2340°R. Determine for the ideal cycle (a) the pressure at the turbine exit; (b) the exhaust gas velocity; (c) the propulsive efficiency.

*14.18 A turbojet aircraft has a velocity of 1066 ft/sec at an altitude of 33,000 ft. The ambient air conditions are 4.4 psia and −28°F. The compressor pressure ratio is 12, and the turbine inlet temperature is 2340°R. The mass flow rate of air entering the compressor is 88 lbm/sec, and the fuel has a heating value of 18,500 Btu/lbm. Determine for ideal gas with constant specific heats and treating the products as air (a) the exhaust gas velocity; (b) the total propulsive power; (c) the fuel consumption.

*14.19 Calculate Problem *14.18 using variable specific heats for the gases and with turbine and compressor isentropic efficiencies of 85%.

*14.20 A turbojet-powered aircraft is held stationary on the ground by its brakes. The engine's compressor has a pressure ratio of 12, the ambient air is at 100°F and 14.7 psia, and the air flow rate entering the engine is 13.2 lbm/sec. The fuel has a heating value of 18,500 Btu/lbm and a flow rate of 0.264 lbm/sec. Assume all the processes to be ideal. Determine the force necessary to hold the plane stationary.

*14.21 An aircraft driven by a turbojet engine is flying at an altitude of 33,000 ft where the air temperature is −46°F and the pressure is 3.4 kPa. The compressor pressure ratio is 12, and the turbine inlet temperature is 2250°R. The turbine and compressor isentropic efficiencies are 87% and 85%, respectively. Determine (a) the velocity at the nozzle exit; (b) the pressure entering the nozzle.

*14.22 The engine in Problem *14.21 has an afterburner inserted between the turbine and nozzle, which raises the temperature of the gas entering the nozzle to 1980°R. Determine the velocity at the nozzle exit.

*14.23 In a turboprop engine, the turbine's rotor is connected to the compressor and to the propeller. The engine is ideally designed such that turbine work is divided equally between the compressor and the propeller. The gas from the turbine discharges to a nozzle for additional thrust. (See the figure for Problem 14.41.) Consider such an engine where the air enters the diffuser at 640 ft/sec, 5.9 psia, −28°F, and a flow rate of 88 lbm/sec. The compressor pressure ratio is 11, and the turbine inlet temperature is 2160°R. The fuel used has a heating value of 18,500 Btu/lbm. All processes are ideal. Determine (a) the fuel consumption; (b) the power delivered to the propeller and to the compressor; (c) the velocity from the nozzle.

COMPUTER PROBLEMS

C14.1 Develop a computer program, spreadsheet template, or TK Solver model to calculate the thermal efficiency of an air-standard Brayton cycle using equation (14.4). Calculate the cycle thermal efficiency for pressure ratios between 3 and 15 and for specific heat ratios of 1.3, 1.4, 1.5, 1.6, and 1.7. Plot the results.

C14.2 Develop a computer program, spreadsheet template, or TK Solver model to calculate the thermal efficiency of an air-standard regenerative Brayton cycle using equation (14.12). For a specific heat ratio of 1.4, calculate the cycle thermal efficiency for pressure ratios between 3 and 15 and for temperature ratios of 3, 4, and 5. Plot the results with those for the simple air-standard cycle.

C14.3 Use AIRCYCLE.TK to calculate the thermal efficiency of a simple Brayton cycle with pressure ratios between 3 and 15. Assume compressor inlet conditions of 100 kPa and 300°K and a maximum cycle temperature of 1300°K. Compare your results with those for the air-standard cycle in Problem C14.1.

C14.4 Develop a computer program or spreadsheet template or modify AIRCYCLE.TK to calculate the thermal efficiency of a Brayton cycle with variable compressor and turbine efficiencies. Calculate the thermal efficiency for pressure ratios between 4 and 20 and for compressor and turbine efficiencies of 100, 90, 80, and 70%. Plot the results.

C14.5 Use the model developed for Problem C14.4 to determine the effect of varying the maximum cycle temperature (T3). Calculate the thermal efficiency of a simple Brayton cycle with a pressure ratio of 10, compressor and turbine efficiencies of 100%, and maximum cycle temperatures between 1000°K and 1400°K. Repeat for component efficiencies of 85%.

C14.6 Develop a computer program or spreadsheet template or modify AIRCYCLE.TK to analyze a regenerative Brayton cycle. For a cycle with compressor and turbine efficiencies of 85%, a maximum cycle temperature of 1300°K, and a regenerator effectiveness of 100%, calculate the thermal efficiency for pressure ratios of 4 to 20. Repeat for a regenerator effectiveness of 70%.

C14.7 Develop a computer program or spreadsheet template or modify AIRCYCLE.TK to analyze a Brayton cycle with intercooling and regeneration. Use the model to determine the optimum pressure to intercool the compressor. For a cycle with conditions similar to Problem C14.6, calculate the cycle thermal efficiency for a range of intercooling pressures. Plot the results.

15

Vapor Power Systems

In Chapters 13 and 14 we analyzed gas power cycles, particularly internal-combustion engines and gas turbines. In this chapter we will look at vapor power systems, systems that generate power using a pure substance as the working fluid. As is true for all power systems, there is heat input, heat rejection, and net work produced. In the case of vapor power systems, the heat may come from combustion of hydrocarbons (coal, oil, gas, wood, garbage) or solar, geothermal, or nuclear energy. The thermodynamic cycle that receives the energy is essentially the same—it does not matter what provides the heat input, only that heat is available.

In vapor power cycles the fluid is alternately vaporized and condensed, flowing through the cycle continuously. The most common working substance used in vapor power cycles is water. Steam is widely used for running engines (turbines), and in the historical as well as the thermodynamic sense, the steam engine was the motive force of the industrial revolution.

Steam-powered electric generating plants with the variety of energy sources mentioned above produce the vast majority of electric power in the world. Along with internal-combustion engines, gas turbines, and hydroelectric plants, virtually all the power produced comes from one of these sources. An important addition to these traditional plants is cogeneration. This will be discussed and described in this chapter. In cogeneration, the waste heat from the primary power-producing process is used as the heat input to another cycle, using another fluid. The recovered waste heat may be used for other heating requirements or for producing additional power. In this chapter we will investigate

- The fundamental vapor power cycle, the Rankine cycle;
- The effects of regeneration on the Rankine cycle's performance and efficiency;
- The effects of reheat on the Rankine cycle's performance and efficiency;
- Supercritical and binary vapor cycles;
- Cogenerative power plants using internal-combustion engines and gas turbines;
- Geothermal energy;
- Steam turbine performance curves;
- Actual heat balance considerations.

15.1 VAPOR POWER PLANTS

Figure 15.1 illustrates a typical hydrocarbon-fuel power plant. The boundary indicates the thermodynamic system. In this is a steam generator, or boiler, where heat is transferred from the hot combustion gases to the water, which changes phase to steam. This energy conversion from a source—the combustion gases, nuclear energy, solar-collector energy, geothermal energy—to the water is what distinguishes these power plants from one another. The steam enters the turbine, where thermal

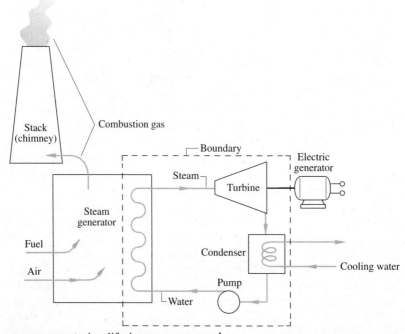

Figure 15.1 A simplified vapor power cycle.

energy is converted to mechanical energy. Nozzles in the turbine increase the steam's velocity, and the steam's kinetic energy is converted to the rotative motion of the turbine rotor. The rotor, through reduction gears, turns an electric generator. The steam exits the turbine at low pressures and temperatures and is condensed in the condenser. The liquid water is pumped from the condenser back into the steam generator, and the cycle repeats.

One design consideration for power plants is the source of cooling water to the condenser. The source may be a river, lake, or ocean, or perhaps a cooling tower is used and the cooling water is recycled. Cooling towers are examined in Chapter 16. The cooling process may lead to environmental problems. The warm cooling water may cause localized changes in the water ecology, such as an abrupt thermal gradient that negatively impacts aquatic life. When evaporative cooling towers are used, the vapor produced from them can be a pollutant as well.

The combustion gases must be processed before discharge to the atmosphere. Certain devices and processes remove particulate matter and gases, such as sulfur dioxide, that contribute to acid rain. Baghouses and electrostatic precipitators remove ash and particulate matter from the combustion gases. Flue gas desulfurization systems, "scrubbers," remove sulfur dioxide. Both wet and dry scrubbers rely on passing the gas through a chemical reagent, such as lime. A chemical reaction occurs, forming a new chemical, calcium sulfate or sulfite. The new compound is collected and removed as a solid or liquid, and the gas leaves with greatly reduced sulfur dioxide.

15.2 THE CARNOT CYCLE

The Carnot cycle is the most efficient heat cycle operating between two temperature limits. A T-s diagram (Figure 15.2) illustrates a steam Carnot cycle. Saturated water at state 1 is evaporated, at constant temperature and pressure, to state 2, where it is a

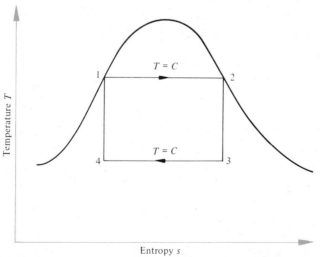

Figure 15.2 A T-s diagram for a Carnot cycle using steam as the working substance.

saturated vapor. Further heating at constant temperature would necessitate a pressure drop. The steam enters the engine at state 2 and expands isentropically, doing work, until state 3 is reached. The steam-water mixture partially condenses at constant temperature and pressure until state 4 is reached. At state 4 a compressor isentropically compresses this vapor-liquid mixture to state 1. Some of the work developed in going from state 2 to state 3 is returned in the compression process.

One of the difficulties that arise in this situation is that the turbine will have to handle low-quality steam. Steam with a quality less than 85 to 90% has too much moisture, and the liquid impingement and erosion of the blading is detrimental. It is also difficult to design a device to compress the liquid-vapor mixture, not to mention to control a partial condensation process.

15.3 THE IDEAL RANKINE CYCLE

The *Rankine cycle* overcomes many of the operational difficulties encountered with the Carnot cycle when the working fluid is a vapor. In this cycle the heating and cooling processes occur at constant pressure. Figure 15.3 illustrates the Rankine cycle on a *T-s* diagram and the equipment used in the cycle.

Following the cycle from state 1, we see that the water enters the steam generator as a subcooled liquid at pressure p. The energy supplied in the steam generator raises the state of the water from that of a subcooled liquid to that of a saturated liquid and, further, to that of a saturated vapor at state 2. The vapor leaves the steam generator at state 2 and enters a steam turbine, where it expands isentropically to state 3. It enters the condenser at this point and is condensed at constant pressure from state 3 to state 4. At state 4 the water is a saturated liquid at the pressure in the condenser. The liquid cannot enter the steam generator, which is at a higher pressure, until its pressure is raised to that of the steam generator. A pump performs this very easily, in contrast to the compressor in the Carnot vapor cycle, and raises the pressure of the liquid to p,

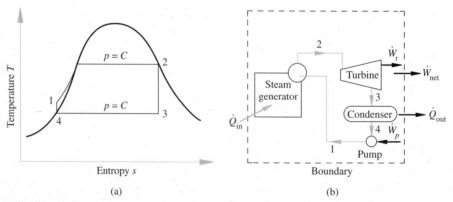

Figure 15.3 (a) A *T-s* diagram for the Rankine cycle. (b) A schematic diagram for the Rankine cycle.

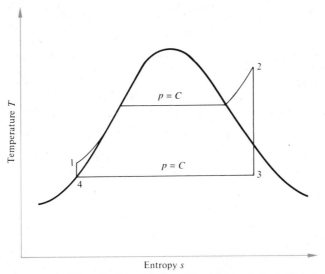

Figure 15.4 The Rankine cycle with superheated steam at state 2.

the steam generator pressure. The liquid is now a subcooled liquid at state 1, and the cycle is complete. Problems would still exist if the steam entered the turbine as a saturated vapor, in that the moisture content of the steam would be too high as it passed through the turbine, resulting in liquid impingement and erosion of the blading.

Since the Rankine cycle is characterized by constant-pressure heating, there is no reason to stop heating the steam when it reaches the saturated vapor state. The customary practice is to superheat the steam to a much higher temperature. Figure 15.4 illustrates how the superheating shifts the isentropic expansion process to the right, thus preventing high moisture content of the steam when it exits the turbine. A typical value for the temperature of the steam at state 2 is 500–600°C. Metallurgical limitations prevent higher values. The pressure is not limited, and a wide range of pressure will be found.

Energy Analysis of the Ideal Rankine Cycle

The first law for a steady-state open system is

$$\dot{Q} + \dot{m}(h + \text{k.e.} + \text{p.e.})_{\text{in}} = \dot{W} + \dot{m}(h + \text{k.e.} + \text{p.e.})_{\text{out}}$$

Apply this equation to each of the components in the ideal Rankine cycle.

The steam's expansion process in the ideal turbine is isentropic, and the changes in the kinetic and potential energies are zero for the ideal cycle; hence the turbine work is

$$\dot{W}_t = \dot{m}(h_2 - h_3)_s \tag{15.1}$$

The turbine work is often written on a unit mass basis, found by dividing equation (15.1) by the mass flow rate.

$$w_t = (h_2 - h_3)_s \tag{15.2}$$

In the condenser the steam vapor condenses at constant temperature and pressure, where the final state leaving the condenser is a saturated liquid. The changes in the kinetic and potential energies of steam are zero, and no work is done. Hence the heat leaving the steam is

$$\dot{Q}_{out} = \dot{m}(h_4 - h_3) \tag{15.3}$$

which on a unit mass basis is

$$q_{out} = (h_4 - h_3) \tag{15.4}$$

The condensed steam enters a pump where the water is isentropically compressed to the boiler pressure. Again, the changes in the kinetic and potential energies are zero. The first-law equation reduces to

$$\dot{W}_p = -\dot{m}(h_1 - h_4)_s \tag{15.5}$$

On a unit mass basis this is

$$w_p = -(h_1 - h_4)_s \tag{15.6}$$

The change of enthalpy for isentropic processes is $dh = v\,dp$, and equation (15.5) becomes

$$\dot{W}_p = -\dot{m}\int_4^1 v\,dp \tag{15.7}$$

If water is considered to be incompressible for the pressure range across the pump, the specific volume is constant, and the pump work becomes

$$\dot{W}_p = -\dot{m}v_4(p_1 - p_4) \tag{15.8}$$

In the steam generator, heat is added to the water at constant pressure, the changes of kinetic and potential energies are zero, the work is zero, and the first-law equation reduces to

$$\dot{Q}_{in} = \dot{m}(h_2 - h_1) \tag{15.9}$$

and dividing by the mass flow rate yields the heat added per unit mass.

$$q_{in} = (h_2 - h_1) \tag{15.10}$$

The thermal efficiency of the cycle is the net work divided by the heat added.

$$w_{net} = w_t + w_p \tag{15.11}$$

$$\eta_{th} = w_{net}/q_{in} \tag{15.12}$$

This may also be written just in terms of heat, as the algebraic sum of the cycle heat flows is equal to the net work for ideal and actual cycles, $w_{net} = q_{in} + q_{out}$, where q_{out} is negative.

Example 15.1

A steam power plant operates on the ideal Rankine cycle. The steam enters the turbine at 7.0 MPa and 550°C. It discharges to the condenser at 20 kPa. Determine the cycle thermal efficiency.

Solution

Given: A power plant operating on the ideal Rankine cycle with specified temperature and pressures.

Find: The cycle thermal efficiency.

Sketch and Given Data:

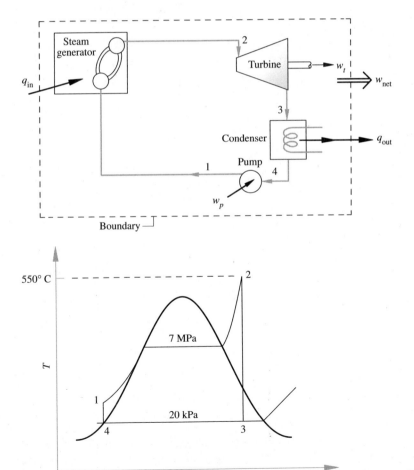

Figure 15.5

Assumptions:

1. Each component within the cycle may be treated as a steady-state open system.
2. The processes are reversible, constant pressure in the heat exchangers and isentropic in the pump and turbine.
3. The water leaves the condenser as a saturated liquid.
4. The changes in kinetic and potential energies may be neglected.
5. Liquid water may be considered incompressible.

Analysis: In solving Rankine-cycle problems, first find all the cycle enthalpy and other relevant property values. When these are known, proceed with the first-law analysis.

At the turbine inlet, state 2, look in the superheated steam tables to find $h_2 = 3530.9$ kJ/kg and $s_2 = 6.9460$ kJ/kg-K. The process from state 2 to state 3 is isentropic, hence $s_3 = s_2$.

$$s_3 = s_f + x_3 s_{fg}$$

$$6.9460 \text{ kJ/kg-K} = 0.8311 \text{ kJ/kg-K} + x_3(7.0762 \text{ kJ/kg-K})$$

$$x_3 = 0.8641$$

$$h_3 = h_f + x_3 h_{fg}$$

$$h_3 = 251.96 \text{ kJ/kg} + (0.8641)(2357.9 \text{ kJ/kg}) = 2289.5 \text{ kJ/kg}$$

State 4 is a saturated liquid at 20 kPa, so $h_4 = 251.96$ kJ/kg and $v_3 = 0.001\,017$ m³/kg. The enthalpy at state 1 may be found by

$$h_1 = h_4 + v_4(p_1 - p_4)$$

$$h_1 = 251.96 \text{ kJ/kg} + (0.001\,017 \text{ m}^3/\text{kg})(7000 - 20 \text{ kN/m}^2) = 259.05 \text{ kJ/kg}$$

At this point we may undertake the first-law analysis. Find the energy terms per unit mass.

$$w_t = h_2 - h_3 = 3530.9 - 2289.5 = 1241.4 \text{ kJ/kg}$$

$$w_p = h_4 - h_1 = 251.96 - 259.05 = -7.1 \text{ kJ/kg}$$

$$w_{net} = w_t + w_p = 1241.4 - 7.1 = 1234.3 \text{ kJ/kg}$$

$$q_{in} = h_2 - h_1 = 3530.9 - 259.05 = 3271.8 \text{ kJ/kg}$$

$$\eta_{th} = w_{net}/q_{in} = 1234.3/3271.8 = 0.377 \quad \text{or } 37.7\%$$

Comments:

1. The quality of the steam leaving the turbine is low enough for moisture droplets to form, causing erosion of the turbine blading.
2. The heat rejected in the condenser, 2037.5 kJ/kg, represents 62.2% of the heat added in the steam generator. This is a large fraction. In addition the temperature of this rejected heat is lower than the saturated steam temperature of 60°C, so its usefulness from an availability viewpoint is limited.

Discussion of Actual Cycle Components

In a turbine thermal energy is converted to mechanical energy. When we analyze the terms in the first-law equation, we find for actual turbines that heat loss is quite small, as the turbines are insulated; hence adiabatic conditions are reasonable. Furthermore, the height of turbine inlet to outlet is also quite small, less than 3 m, so the change of potential energy across a turbine is negligible. The kinetic energy of the steam entering a turbine is reasonably low, so the entrance kinetic energy term is essentially zero when compared to the other energy terms. The exit kinetic energy term is much higher than the inlet term and cannot be neglected. This is not because of irreversibilities in flow process, but because the steam must have sufficient velocity to exit the rapidly spinning (3000–10 000 rpm) turbine blades. The kinetic energy leaving a well-designed turbine may be 200 m/s.

In the condenser heat is rejected to cooling water as the steam in the cycle condenses. Physically, the steam passes over the outside of tubes in a shell-and-tube heat exchanger while cooling water passes through the tubes. This accounts for a substantive decrease in the cycle efficiency because the latent heat of the steam is transferred to the cooling water. A substantial portion of the energy that was supplied in the steam generator to convert the water to steam is now lost from the system. An energy balance on the condenser will yield an expression for this loss. Figure 15.6 illustrates the condenser, considered as the system, and the various energy forms that should be considered. The potential energy has been neglected in this case, as the effect of elevation is small. The condenser is adiabatic, so there is no heat flow to the surroundings, and the difference in velocities of the cooling water in and out is very small, hence the change in kinetic energy of the cooling water is essentially zero. Making an energy balance across the condenser yields

$$\dot{m}_{sys}(h_3 - h_4) + \dot{m}_{sys}[(\text{k.e.})_3 - (\text{k.e.})_4] = \dot{m}_{H_2O}(h_{out} - h_{in}) \qquad (15.13)$$

The exit steam velocity from the turbine to the condenser may be high, and so the entering kinetic energy should be included in the energy balance. The conservation of mass equation for the system will give us an estimate of the difference between the inlet and exit velocities of the steam and saturated liquid:

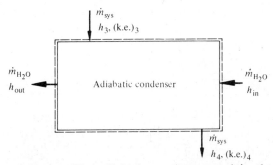

Figure 15.6 The adiabatic condenser used in the Rankine cycle, with mass and energy flows noted.

$$v_4 = v_3 \left(\frac{A_3}{A_4}\right)\left(\frac{v_4}{v_3}\right) \tag{15.14}$$

Picking typical values for the specific volumes at $35\,°C$ (approximately $100\,°F$) shows that $v_4 = 0.001\ 006$ and $v_3 = 25.22$. Allowing an area ratio of $A_3/A_4 = 100$, equation (15.14) becomes

$$v_4 = 0.004 v_3$$

We may conclude that the exit velocity of the saturated liquid from the condenser is very small compared with the inlet velocity of the steam, and thus the exit kinetic energy leaving is also very small compared to the inlet kinetic energy. With these considerations in mind, equation (15.13) becomes

$$\dot{m}_{sys}[(h_3 - h_4) + (\text{k.e.})_3] = \dot{m}_{H_2O}(h_{out} - h_{in}) \tag{15.15}$$

where the term on the right-hand side denotes the energy rejected from the system to the surroundings, the cooling water. This energy, about 2100 to $2300\ kJ/kg$ (system), may cause thermal pollution if the power plant is not correctly situated to minimize thermal-energy impact.

For the pump, adiabatic compression is a reasonable assumption. The pump casing is insulated, and very little heat is lost. Furthermore, the change in elevation is slight, less than a meter for most pumps, so the change in potential energy is essentially zero. The change in kinetic energy is essentially zero, as the pipe diameter does not decrease in size in most installations, and the water is viewed as incompressible. Should greater accuracy be desired for the variation of specific volume with pressure, a linear variation is a good choice. In this text and in most overall cycle analysis problems, however, the assumption of incompressibility is reasonable.

Steam generators, sometimes referred to as boilers, create steam from water. There are many design considerations for the steam generator, but basically the water enters a steam drum at a temperature less than saturated. The water passes from this large drum into small tubes where it is heated to its saturation point with some vaporization occurring. Hot gases used for heating the water may be formed by burning fuel or may be the high-temperature exhaust from a diesel or gas turbine engine. This mixture of steam and liquid water reenters the steam drum, where the steam rises to the top and the liquid is recirculated. The steam, now a saturated vapor, passes through superheater tubes where its temperature is increased by further heat transfer from the hot gases. The superheated steam leaves through insulated piping to the steam turbine.

15.4 FACTORS CONTRIBUTING TO CYCLE IRREVERSIBILITIES AND LOSSES

Irreversibilities are associated with each of the cycle components and the piping connecting the components to a system. Fluid friction and heat loss to the surroundings are the most common causes of irreversibility.

In the turbine the ideal expansion process is isentropic, but as the steam flows through the turbine blading, fluid friction occurs, increasing the steam's entropy so it is greater at exit than at inlet conditions. This is accounted for by the turbine internal, or isentropic, efficiency. This value is determined experimentally by the manufacturer. Once it is known, the value may be used in computing the actual work the steam indicates it did. The turbine isentropic efficiency is defined as

$$\eta_t = \frac{(h_2 - h_3)_{\text{act}}}{(h_2 - h_3)_s} \tag{15.16}$$

Fluid flow irreversibilities significantly reduce the turbine work, hence the net work from the cycle. Heat loss from the turbine is negligible. The turbine isentropic efficiency is a determination of how well the available energy is used. The denominator, $(h_2 - h_3)_s$, is the available energy of the steam; the numerator, $(h_2 - h_3)_{\text{act}}$, is the actual amount of available energy that is used as work. Once the isentropic efficiency of the turbine is known, the actual enthalpy of the steam at the turbine exit may be calculated.

In the pump, frictional effects mean more work is required than the ideal to raise the water's pressure to a higher value. Frictional effects in fluid flow through the pump and between the pump's impeller and the water increase the entropy. Again, the pump may be considered adiabatic, as heat loss is negligible. The pump isentropic efficiency allows us to determine the actual state of the water leaving the pump and to determine the entropy increase across the pump. The isentropic efficiency is

$$\eta_p = \frac{\dot{m}(h_1 - h_4)_s}{\dot{W}_p} = \frac{(h_1 - h_4)_s}{(h_1 - h_4)_{\text{act}}} \tag{15.17}$$

The internal efficiency of the pump is the ratio of the increase in the available energy of the fluid passing through the pump divided by the available energy (the work) necessary to achieve that effect. It may also be viewed as the cost, the actual pump work, in moving the water to a higher energy state as indicated by the discharge enthalpy. In Rankine cycles the pump work is much smaller than the turbine work, so the net effect of pump inefficiency on cycle efficiency is small.

In actual condensers and steam generators there are also pressure drops. There is a pressure gradient across the condenser. The discharge pressure from the pump must be greater than the boiler pressure to overcome the pressure drops across valves and piping to the boiler. A pressure drop within the steam generator is caused by fluid friction. These are discussed in greater detail in the section covering actual heat balance practice.

The irreversibilities examined so far are within the Rankine cycle. In addition, there are irreversibilities external to the cycle but still within the components analyzed. For instance, irreversibilities are associated with the heat transfer from the hot gases to the water/steam, as well as with the combustion process itself. Similarly, the condenser and the cooling water have associated irreversibilities.

Example 15.2

A 100-MW steam power plant operates on the Rankine cycle but with turbine and pump efficiencies of 85%. The steam enters the turbine at 7.0 MPa and 550°C. It discharges to the condenser at 20 kPa. Determine the cycle thermal efficiency, the steam flow rate, and the cooling-water flow rate in the condenser if cooling water enters at 20°C and leaves at 35°C.

Solution

Given: A power plant operating on the Rankine cycle with specified temperature, pressures, and pump and turbine isentropic efficiencies.

Find: The cycle thermal efficiency, the steam flow rate, and the cooling-water flow rate.

Sketch and Given Data:

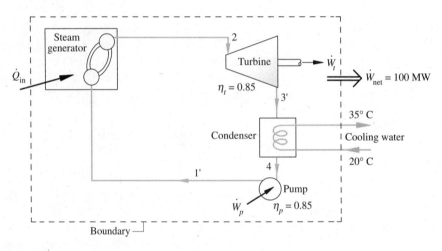

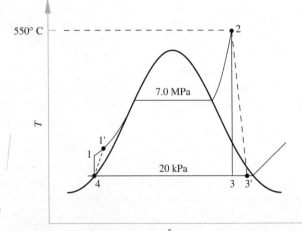

Figure 15.7

Assumptions:

1. Each component within the cycle may be treated as a steady-state open system.
2. The processes are reversible and constant-pressure in the heat exchangers and irreversible and adiabatic in the pump and turbine.
3. The water leaves the condenser as a saturated liquid.
4. The changes in kinetic and potential energies may be neglected.
5. Liquid water may be considered incompressible.

Analysis: Determine the enthalpies around the cycle. Use the values from Example 15.1 as appropriate and then use the turbine and pump isentropic efficiencies to determine the actual state values: $h_2 = 3530.9$ kJ/kg, $h_{3s} = 2289.5$ kJ/kg, $h_4 = 251.96$ kJ/kg, and $h_1 = 259.05$ kJ/kg.

$$\eta_t = 0.85 = \frac{h_2 - h_{3'}}{(h_2 - h_3)_s} = \frac{3530.9 - h_{3'}}{3530.9 - 2289.5}$$

$$h_{3'} = 2475.7 \text{ kJ/kg}$$

$$\eta_p = 0.85 = \frac{h_1 - h_4}{h_{1'} - h_4} = \frac{259.05 - 251.96}{h_{1'} - 251.96}$$

$$h_{1'} = 260.3 \text{ kJ/kg}$$

The various work and heat terms per unit mass are found from the first-law analysis.

$$w_t = h_2 - h_{3'} = 3530.9 - 2475.7 = 1055.2 \text{ kJ/kg}$$

$$w_p = -(h_{1'} - h_4) = -(260.3 - 251.96) = -8.3 \text{ kJ/kg}$$

$$w_{net} = w_t + w_p = 1055.2 - 8.3 = 1046.9 \text{ kJ/kg}$$

$$q_{in} = h_2 - h_1 = 3530.9 - 251.96 = 3270.6 \text{ kJ/kg}$$

The cycle thermal efficiency is

$$\eta_{th} = \frac{w_{net}}{q_{in}} = \frac{1046.9}{3270.6} = 0.320 \quad \text{or } 32.0\%$$

The flow rate of steam through the Rankine cycle is found by dividing the total net power by the net work per unit mass.

$$\dot{m}_p = \frac{\dot{W}_{net}}{w_{net}} = \frac{100\ 000 \text{ kJ/s}}{1046.9 \text{ kJ/kg}} = 95.5 \text{ kg/s}$$

Performing a first-law analysis on the condenser (equation [15.15]) yields

$$\dot{m}_{cw} = \frac{\dot{m}_p(h_{3'} - h_4)}{(h_{out} - h_{in})}$$

$$h_{in} = h_f \text{ at } 20°C = 82.9 \text{ kJ/kg}$$

$$h_{out} = h_f \text{ at } 35°C = 146.2 \text{ kJ/kg}$$

$$\dot{m}_{cw} = \frac{(95.5 \text{ kg steam/s})(2475.7 - 251.96 \text{ kJ/kg steam})}{(146.2 - 82.9 \text{ kJ/kg water})}$$

$$\dot{m}_{cw} = 3355 \text{ kg water/s}$$

Comments:

1. The effect of pump and turbine efficiencies caused the overall cycle thermal efficiency to decrease from 37.7% to 32.0%, a significant decline. Note that the decrease in the turbine work was the primary cause of this decline. The quality of the exit steam rose from 86.4% to 94.3%.
2. Note the tremendous mass flow rate of water needed for cooling, approximately 53,180 gal/min.

15.5 IMPROVING RANKINE CYCLE EFFICIENCY

Steam power plants produce most of the electrical power in the world, consuming tremendous amounts of fuel in the process. Hence the attention focused on improving the cycle's efficiency: gains of a fraction of 1% yield large savings in fuel consumption and resource preservation.

The heat added, heat rejected, and net work per unit mass for the ideal Rankine cycle are illustrated in Figure 15.8. The heat added and rejected may be represented as areas on a *T-s* diagram, and the net work is the algebraic sum of these areas. To improve the cycle efficiency, the net work must increase, the heat added to the cycle must decrease, or both. We may visualize that the actual area under the curve from state 1 to state 2 could be replaced by an area of equal size, but rectangular. The ordinate would be the average temperature of heat addition, $(T_{in})_{avg}$, as illustrated in Figure 15.9. The cycle efficiency may be written as

$$\eta_{th} = 1 - \frac{q_{out}}{q_{in}} = \frac{w_{net}}{q_{in}} = 1 - \frac{(T_{out})_{avg}}{(T_{in})_{avg}} \tag{15.18}$$

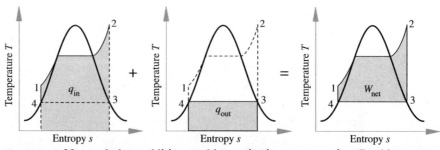

Figure 15.8 Net work, heat addition, and heat rejection processes in a Rankine cycle.

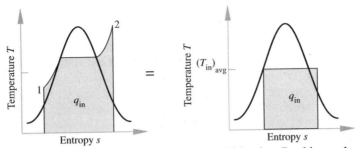

Figure 15.9 Average temperature of heat addition in a Rankine cycle.

The cycle efficiency may be increased by raising the average temperature of heat addition or by lowering the average temperature of heat rejection. Relating a cycle to absolute temperature is possible only for reversible cycles; thus equation (15.18) is a useful approximation when considering actual cycles.

Lowering Condenser Pressure

Steam entering the condenser in a Rankine cycle is almost always in the mixture region; thus, lowering the pressure lowers the saturation temperature of the steam and hence the cycle's average temperature of heat rejection. The only limitation here is the surroundings' temperature. The condenser in the limit can approach the surroundings' temperature and for actual heat exchangers must be greater than that value. Consider a condenser that is cooled by river water at 20°C. A 10°C temperature difference, required for effective heat transfer, results in a condenser saturation temperature of 30°C, or 4.25 kPa. It cannot be lowered below this without an excessively large condenser and very high cooling-water flow. In addition, steam turbines are designed to operate with a certain discharge pressure; reducing the pressure below this point causes the turbine efficiency to decrease as the steam overexpands in the last turbine stage, increasing irreversibility.

Increasing Steam Generator Pressure

The steam generator pressure may be raised, increasing the average temperature of heat addition. The major portion of the heat transfer occurs during the phase transition (boiling), so raising the boiling temperature by raising the temperature improves the thermal efficiency. Raising only the pressure has liabilities. When the average temperature of heat addition is increased, state 2 at the turbine inlet is shifted to the position designated 2' in Figure 15.10. During isentropic turbine expansion from state 2' to state 3', liquid droplets will begin to form earlier in the expansion as compared to the process between state 2 and state 3. This will lead to turbine deterioration as mentioned earlier. Also, while there is an overall net work increase created by raising the pressure, Figure 15.10 illustrates that it is not as significant as one might assume.

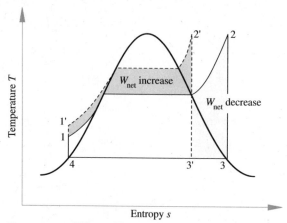

Figure 15.10 Effect of increasing steam generator pressure in a Rankine cycle.

If the condenser pressure is 10 kPa and the superheated steam temperature entering the turbine is 550°C, increasing the boiler pressure from 3.0 MPa to 8.0 MPa to 16.0 MPa, increases the efficiency from 36.4% to 40.2% to 42.4%. The net work changes from 1228 kJ/kg to 1334 kJ/kg to 1372 kJ/kg, respectively.

The pressure may be raised enough to exceed the critical pressure, creating a *supercritical Rankine cycle.* No boiling occurs; rather, the water gradually changes from dense, low-temperature liquid to low-density, high-temperature vapor, as illustrated in Figure 15.11.

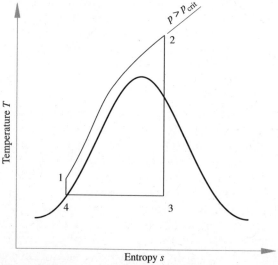

Figure 15.11 A supercritical Rankine cycle.

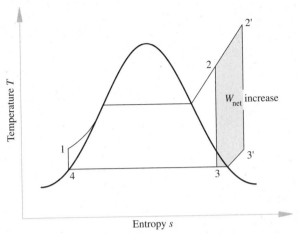

Figure 15.12 Effect of increasing superheated steam temperature in a Rankine cycle.

Raising the Superheat Temperature

Raising the superheat temperature raises the average temperature of heat addition. Figure 15.12 illustrates that it also increases the net work performed in the cycle and shifts the turbine discharge state further toward the superheat region. Increases in superheat temperature, combined with increasing steam generator pressure, provide a complementary combination of effects. The pressure increase moves the turbine exit state toward greater moisture, which is compensated by heat moving it toward the superheat region. Some modern power plants operate at pressures greater than the critical pressure, 30 MPa, and temperatures of 600°C. Often power plants operate below critical pressure to reduce the complexity of the plant and improve reliability.

15.6 THE IDEAL REHEAT RANKINE CYCLE

In an extension of the concept of superheating, the steam may be reheated after it has partially expanded. This raises the average temperature of heat addition by adding the heat in stages. A significant portion of the work is accomplished by the steam when the pressure is such that the steam is saturated or nearly saturated. This is the correct place for the vapor to be resuperheated. The steam reenters the turbine and expands to condenser pressure. Figure 15.13 illustrates the schematic and *T-s* diagrams for the reheat cycle. The steam expands through the turbine until state 3 is reached, then is removed and reheated at constant pressure to state 4. The temperature of the steam at state 4 is usually the same as state 2. The steam reenters the turbine at state 4 and expands to the condenser pressure at state 5. The steam is condensed and pumped back to the steam generator, completing the cycle.

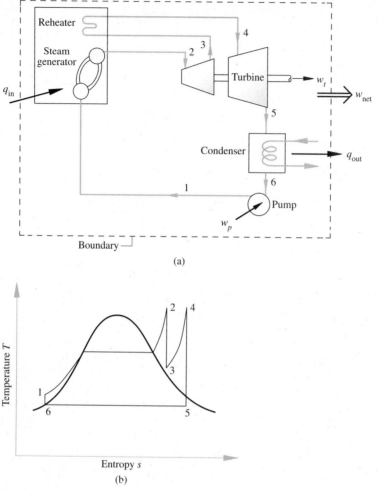

Figure 15.13 (a) A schematic diagram illustrating one stage of reheating. (b) The T-s diagram for one stage of reheating.

To calculate the thermal efficiency for the reheat cycle, the work and heat-added terms must be found. An energy balance on the turbine yields

$$\dot{W}_t = \dot{m}[(h_2 - h_3) + (h_4 - h_5)] \tag{15.19}$$

Equation (15.19) does not need an isentropic expansion but is valid for any turbine efficiency. The pump work is as previously calculated,

$$\dot{W}_p = \dot{m} \int_6^1 v \, dp$$

The energy added in the cycle occurs in the steam generator and may be found by an energy balance on the generator:

$$\dot{Q}_{in} = \dot{m}[(h_2 - h_1) + (h_4 - h_3)] \tag{15.20}$$

The thermal efficiency is then

$$\eta_{th} = \frac{[(h_2 - h_3) + (h_4 - h_5)] - \int_6^1 v\, dp}{(h_2 - h_1) + (h_4 - h_3)} \tag{15.21}$$

There is no theoretical limit to the number of reheat stages, but practically only one or two are used. Otherwise, the steam exiting the turbine tends to be superheated, and all the available energy cannot be extracted from it before it reaches condenser pressure.

Example 15.3

An ideal reheat Rankine cycle is characterized by a steam generator pressure of 10 MPa, superheat temperature of both the reheated steam and inlet turbine steam of 550°C, reheat pressure of 1000 kPa, and a condenser pressure of 10 kPa. Determine the cycle efficiency and compare it to the cycle efficiency of an ideal Rankine cycle without reheat.

Solution

Given: An ideal reheat Rankine cycle with required temperatures and pressures.

Find: The cycle efficiencies of the reheat cycle and a Rankine cycle with the same cycle minimum and maximum states.

Sketch and Given Data:

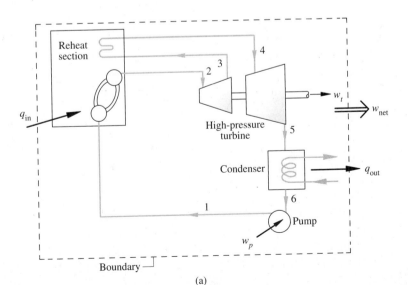

(a)

Figure 15.14

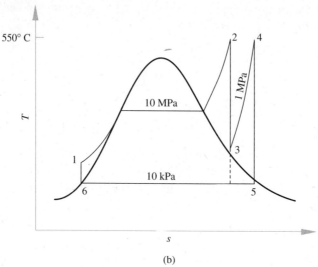

Figure 15.14 (*Continued*)

Assumptions:

1. Each component within the cycle may be treated as a steady-state open system.
2. The processes are reversible and constant-pressure in the heat exchangers and reversible and adiabatic in the pump and turbine.
3. The water leaves the condenser as a saturated liquid.
4. The changes in kinetic and potential energies may be neglected.
5. Liquid water may be considered incompressible.

Analysis: Perform the first-law analysis on the ideal Rankine cycle and determine its efficiency. The turbine inlet pressure is 10 MPa, and the temperature is 550°C; $h_2 = 3501.9$ kJ/kg and $s_2 = 6.7549$ kJ/kg-K. The process from state 2 to state 3 is isentropic; hence $s_3 = s_2$.

$$s_3 = s_f + x_3 s_{fg} = 6.7549 \text{ kJ/kg-K} = 0.6471 \text{ kJ/kg-K} + x_3(7.5028 \text{ kJ/kg-K})$$

$$x_3 = 0.814$$

$$h_3 = h_f + x_3 h_{fg}$$

$$h_3 = 191.8 \text{ kJ/kg} + (0.814)(2393.0 \text{ kJ/kg}) = 2139.9 \text{ kJ/kg}$$

State 4 is a saturated liquid at 10 kPa, with $h_4 = 191.8$ kJ/kg and $v_3 = 0.001\ 010$ m³/kg. The enthalpy at state 1 may be found by

$$h_1 = h_4 + v_4(p_1 - p_4)$$

$$h_1 = 191.8 \text{ kJ/kg} + (0.001\ 010 \text{ m}^3/\text{kg})(10\ 000 - 10 \text{ kN/m}^2) = 201.9 \text{ kJ/kg}$$

At this point we may undertake the first-law analysis. Find the energy terms per unit mass.

$$w_t = h_2 - h_3 = 3501.9 - 2139.9 = 1362.0 \text{ kJ/kg}$$

$$w_p = h_4 - h_1 = 191.8 - 201.9 = -10.1 \text{ kJ/kg}$$

$$w_{net} = w_t + w_p = 1362.0 - 10.1 = 1351.9 \text{ kJ/kg}$$

$$q_{in} = h_2 - h_1 = 3501.9 - 201.9 = 3300.0 \text{ kJ/kg}$$

$$\eta_{th} = w_{net}/q_{in} = 1351.9/3300.0 = 0.410 \quad \text{or } 41.0\%$$

For the reheat cycle the property values for states 1, 2, and 6 are known. At state 3′ the entropy is equal to s_2, and the pressure is 1000 kPa. From the superheated steam tables $h_{3'} = 2856.9$ kJ/kg. At state 4 the temperature and pressure are known, 1000 kPa and 550°C; hence $h_4 = 3586.4$ kJ/kg and $s_4 = 7.8942$ kJ/kg-K.

The process from state 4 to state 5 is isentropic; hence $s_5 = s_4$.

$$s_5 = s_f + x_5 s_{fg}$$

$$7.8942 \text{ kJ/kg-K} = 0.6471 \text{ kJ/kg-K} + x_5 \, (7.5028 \text{ kJ/kg-K})$$

$$x_5 = 0.966$$

$$h_5 = h_f + x_5 h_{fg} = 191.8 \text{ kJ/kg} + 0.966(2393 \text{ kJ/kg}) = 2503.2 \text{ kJ/kg}$$

The turbine work is

$$w_t = (h_2 - h_{3'}) + (h_4 - h_5)$$

$$w_t = (3501.9 - 2856.9) + (3586.4 - 2503.2) = 1728.2 \text{ kJ/kg}$$

The pump work is the same as in the Rankine cycle, $w_p = -10.1$ kJ/kg. The net work is

$$w_{net} = w_t + w_p = 1728.2 - 10.1 = 1718.1 \text{ kJ/kg}$$

The heat added is

$$q_{in} = (h_2 - h_1) + (h_4 - h_{3'})$$

$$q_{in} = (3501.9 - 201.9) + (3586.4 - 2856.9) = 4029.5 \text{ kJ/kg}$$

and the thermal efficiency is

$$\eta_{th} = \frac{w_{net}}{q_{in}} = \frac{1718.1}{4029.5} = 0.426 \quad \text{or } 42.6\%$$

Comment: Reheating caused an improvement in cycle efficiency, raising it from 41% to 42.6%. In power plants the efficiency gains and the resulting decrease in operating costs for fuel are weighed against increased capital cost and complexity of the system. On large systems the additional initial costs are usually worthwhile. On smaller systems this may not be the case.

15.7 THE IDEAL REGENERATIVE RANKINE CYCLE

Thus far we have learned how power is produced in a multiphase system and have seen some of the performance factors that must be considered, namely, superheating and internal efficiencies. Let us consider ways in which the thermal efficiency of the cycle may be improved.

We might say that if there were a way to preheat the water before it entered the steam generator, not as much energy would be added in the boiler. If some of the steam were taken from the turbine before it reached the condenser and were used to heat the feedwater, two purposes would be accomplished: the preheating would occur with no extra energy input, and the latent heat of vaporization would not be lost from the system in the condenser. A steam cycle using this type of feedwater heating is a *regenerative cycle*. Figure 15.15 illustrates such a cycle and its *T-s* diagram. Note that the only thermal-energy flow across the system boundary occurs in the steam generator and in the condenser. The feedwater heater is within the system boundary, so its effect is on only the heat added and the turbine work.

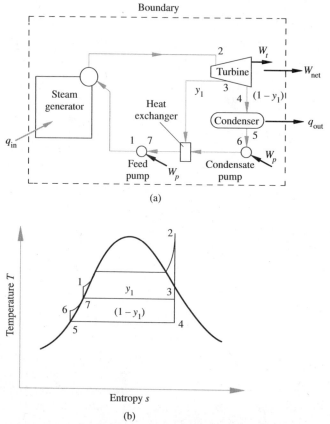

Figure 15.15 (a) The schematic diagram for a regenerative cycle with one stage of regenerative heating. (b) The *T-s* diagram for the regenerative cycle with one stage of feed heating.

Direct-Contact, Open Heat Exchanger

Some assumptions are implicit in the previous development for regenerative heaters. The first is that all the water leaving the feed heater leaves at the saturated-liquid enthalpy for the pressure of the entering steam, and the heater is a direct-contact or open type. In this type of heat exchanger, the incoming steam and water mix within the shell of the heater.

When calculating power plant problems, it is easiest to perform the operation per unit mass and, at the conclusion of the problem, incorporate the actual mass flow rates. In Figure 15.15 the symbol y_1 is the fraction of the total mass in the system that leaves the turbine to go to the feedwater heater.

$$y_1 = \frac{\text{mass of steam to feed heater}}{\text{total system mass}} \qquad (15.22)$$

How may this value be calculated? We perform an energy balance on the feedwater heater, subject to steady flow, with no changes in kinetic or potential energy, which yields

$$y_1 h_3 + (1 - y_1)h_6 = h_7 \qquad (15.23)$$

$$y_1 = \frac{h_7 - h_6}{h_3 - h_6}$$

where y_1 is the fraction of the total mass that is withdrawn for regenerative heating. When the turbine work and pump work are calculated, attention must be paid to the fraction of fluid passing through each device. The turbine work per kilogram in the system is

$$w_t = (h_2 - h_3) + (1 - y_1)(h_3 - h_4) \text{ kJ/kg [Btu/lbm]} \qquad (15.24)$$

The pump work per kilogram in the system will have to be modified by the fraction of the total mass that passes through it. There are two pumps, a condensate pump and a feed pump. Sometimes a booster pump is located before the feed pump, but typically two pumps characterize the basis cycle.

$$w_{p1} = (h_6 - h_5)(1 - y_1) \qquad w_{p2} = h_1 - h_7 \qquad (15.25)$$

$$w_p = w_{p1} + w_{p2}$$

This provides us with the tools necessary to calculate the cycle efficiency, but where is the heater situated?

Optimum Heater Location

Let us consider two cases, one with a heater using steam before it reaches the turbine, the other using steam after it has left the turbine. Neither case improves the cycle efficiency. In the latter case the temperature of the exhaust steam is that of the liquid leaving the condenser, and no heat transfer can occur. In the former case the heat exchanger acts only as a medium between the steam generator and the feedwater, and no gain in efficiency results.

Somewhere between the turbine inlet and exit lies the optimum location for a feed heater. The determination of the location is a trial-and-error procedure. For one feedwater heater, the optimum point is where the feedwater exit temperature from the heater is halfway between the saturated-steam temperature in the boiler and the condenser temperature. This is one-half the saturated temperature change for the cycle. When more regenerative heat exchangers are used, the temperature change between each pair of feedwater heaters should be equal. Thus, if two regenerative heaters are used, ideally one-third of the saturated temperature change occurs between the condenser and first heater, one-third of the saturated temperature change occurs between the two heaters, and one-third of the saturated temperature change occurs between the second heat exchanger and the boiler saturated temperature.

Example 15.4
An ideal regenerative Rankine cycle has a steam generator outlet pressure of 10 MPa and temperature of 550°C. The condenser pressure is 10 kPa, and extraction for one stage of regenerative heating occurs at 1000 kPa. Determine the cycle efficiency and compare it to the ideal Rankine cycle's efficiency.

Solution

Given: An ideal regenerative Rankine cycle with its necessary pressures and temperatures.

Find: The cycle thermal efficiency compared to that of an ideal Rankine cycle.

Sketch and Given Data:

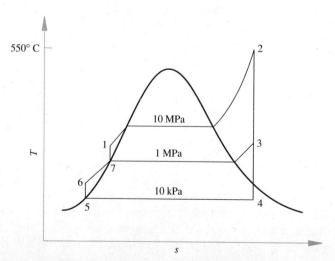

Figure 15.16

Assumptions:

1. Each component within the cycle may be treated as a steady-state open system.
2. The processes are reversible and constant-pressure in the heat exchangers and reversible and adiabatic in the pump and turbine.
3. The water leaves the condenser and heater as a saturated liquid.
4. The changes in kinetic and potential energies may be neglected.
5. Liquid water may be considered incompressible.

Analysis: Determine the cycle enthalpies first. Then perform an energy analysis for each component. The enthalpies for most of the states were found in Example 15.3: $h_2 = 3501.9$ kJ/kg, $h_3 = 2856.9$ kJ/kg, $h_4 = 2139.9$ kJ/kg, and $h_5 = 191.8$ kJ/kg.

$$h_6 = h_5 + v_5(p_7 - p_5)$$

$$h_6 = 191.8 \text{ kJ/kg} + (0.001\ 010 \text{ m}^3\text{/kg})(1000 - 10 \text{ kN/m}^2) = 192.8 \text{ kJ/kg}$$

$$h_7 = h_f \text{ at } 1000 \text{ kPa} = 762.5 \text{ kJ/kg}$$

$$h_1 = h_7 + v_7(p_1 - p_7)$$

$$h_1 = 762.5 \text{ kJ/kg} + (0.001\ 127 \text{ m}^3\text{/kg})(10\ 000 - 1000 \text{ kN/m}^2) = 772.6 \text{ kJ/kg}$$

Perform a first-law analysis of the heater, yielding the mass fraction y_1, per equation (15.23).

$$y_1 = \frac{h_7 - h_6}{h_3 - h_6} = \frac{772.6 - 192.8}{2856.9 - 192.8} = 0.2176$$

$$1 - y_1 = 0.7824$$

The turbine work is

$$w_t = (h_2 - h_3) + (1 - y_1)(h_3 - h_4)$$

$$w_t = (3501.9 - 2856.9) + (0.7824)(2856.9 - 2139.9) = 1206.0 \text{ kJ/kg}$$

The total pump work is

$$w_{p_1} = -(1 - y_1)(h_6 - h_5) = -(0.7824)(192.8 - 191.8) = -0.78 \text{ kJ/kg}$$

$$w_{p_2} = -(h_1 - h_7) = -(772.6 - 762.5) = -10.1 \text{ kJ/kg}$$

$$w_p = w_{p_1} + w_{p_2} = -10.88 \text{ kJ/kg}$$

The net work is

$$w_{\text{net}} = w_t + w_p = 1206.0 - 10.9 = 1195.1 \text{ kJ/kg}$$

The heat addition is

$$q_{\text{in}} = h_2 - h_1 = 3501.9 - 772.6 = 2729.3 \text{ kJ/kg}$$

The thermal efficiency is

$$\eta_{\text{th}} = \frac{w_{\text{net}}}{q_{\text{in}}} = \frac{1195.1}{2729.3} = 0.438 \quad \text{or } 43.8\%$$

Comparing this cycle to the ideal Rankine cycle operating with the same limits in Example 15.3, the net work decreases by 11.6%, the heat added decreases by 17.3%, and the efficiency increases by 6.8%.

Comment: In a regenerative Rankine cycle the work and heat per unit mass are less than in a Rankine cycle. The efficiency improvement through regenerative heating causes the heat term to decrease more than the net work term, yielding an improvement in cycle efficiency. In some installations, the valve allowing steam to flow to the regenerative heaters is closed, permitting more net work to be performed in times of peak demand. This does not typically occur in stationary power plants, but may be found aboard naval vessels and other power plants often subject to variable loads. Of course, the efficiency suffers dramatically.

Closed Feedwater Heaters

Another type of heat exchanger used in power plants is a shell-and-tube heat exchanger, or a closed heat exchanger. The steam and water are separated, with steam typically being on the shell side and water in the tubes. The assumption for the cycle remains that the liquid leaving the heat exchanger—condensed steam and water—has an enthalpy equal to the saturated-liquid enthalpy of the condensed steam. Figure 15.17 illustrates this type of heat exchanger. It is possible to have the saturated liquid pumped into the discharge line from the heat exchanger. This entails extra expense, however, a pump and motor that require maintenance and repair. Most commonly, the outlet is drained back to the shell side of a heat exchanger that operates at a lower pressure. A device called a *trap* allows the liquid water but not the steam to leave the heat exchanger. The trap is commonly an orifice type that allows some fluid flow at

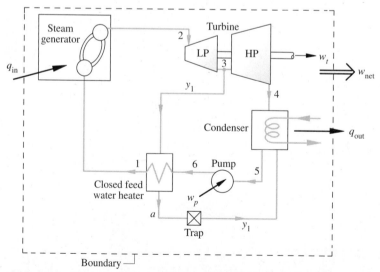

Figure 15.17 The use of traps with closed feedwater heaters.

all times and opens and closes to permit liquid flow. When the trap flows to a lower-pressure heat exchanger, the system is called a *cascade drain system*.

Actual systems use a combination of open and closed heat exchangers. The open heat exchanger has the additional benefit of deaerating the water, removing much of the dissolved oxygen. Water with dissolved oxygen is extremely corrosive at the high temperatures of a steam generator, causing pitting of metal surfaces that leads to failure of tubes and other components.

Example 15.5

Perform the same analysis as in Example 15.4, but let the heat exchanger be a closed heater with the drain returning to the condenser.

Solution

Given: An ideal regenerative Rankine cycle with a closed feedwater heater and with its necessary pressures and temperatures.

Find: The cycle thermal efficiency compared to that of an ideal Rankine cycle.

Sketch and Given Data:

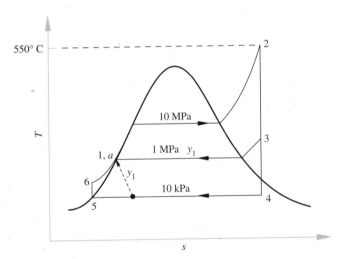

Figure 15.18

Assumptions:

1. Each component within the cycle may be treated as a steady-state open system.
2. The processes are reversible, constant-pressure in the heat exchangers and adiabatic in the pump and turbine.
3. The water leaves the condenser and heater as a saturated liquid.
4. The changes in kinetic and potential energies may be neglected.
5. Liquid water may be considered incompressible.

Analysis: Most of the enthalpies are available from Example 15.4 as follows: $h_2 = 3501.9$ kJ/kg, $h_3 = 2856.9$ kJ/kg, $h_4 = 2139.9$ kJ/kg, and $h_5 = 191.8$ kJ/kg.

$$h_6 = h_5 + v_5(p_6 - p_5)$$

$$h_6 = 191.8 + (0.001\ 010\ \text{m}^3/\text{kg})(10\ 000 - 10\ \text{kN/m}^2) = 201.9\ \text{kJ/kg}$$

$$h_1 = h_a = h_f \text{ at } 1000\ \text{kPa} = 762.5\ \text{kJ/kg}$$

Performing a first-law analysis on the heater allows the determination of y_1.

$$y_1 h_3 + h_6 = h_1 + y_1 h_a$$

$$y_1(h_3 - h_a) = h_1 - h_6$$

$$y_1 = \frac{h_1 - h_6}{h_3 - h_a} = \frac{(762.5 - 201.9)}{(2856.9 - 762.5)}$$

$$y_1 = 0.2677 \qquad (1 - y_1) = 0.7323$$

The turbine work is

$$w_t = (h_2 - h_3) + (1 - y_1)(h_3 - h_4)$$

$$w_t = (3501.9 - 2856.9) + (0.7323)(2856.9 - 2139.9) = 1170.0\ \text{kJ/kg}$$

The pump work is

$$w_p = -(h_6 - h_5) = -(201.9 - 191.8) = -10.1\ \text{kJ/kg}$$

The net work is

$$w_{\text{net}} = w_t + w_p = 1170.0 - 10.1 = 1159.9\ \text{kJ/kg}$$

The heat added is

$$q_{\text{in}} = h_2 - h_1 = 3501.9 - 762.5 = 2739.4\ \text{kJ/kg}$$

The thermal efficiency is

$$\eta_{\text{th}} = \frac{w_{\text{net}}}{q_{\text{in}}} = \frac{1159.9}{2739.4} = 0.423 \quad \text{or } 42.3\%$$

Compared to the Rankine cycle in Example 15.3 the net work decreases by 14.2%, the heat added decreases by 17.0%, and the efficiency increases by 3.2%.

Comments:

1. The efficiency is superior to that of a simple Rankine cycle, but the cascading of the drain caused some heat to be removed in the condenser that did not occur in direct-contact heaters.
2. Most cycles contain a combination of closed and direct-contact heat exchangers. This example exaggerates the effect of extra heat loss in the condenser, as there are no intermediate heat exchangers between the closed heater and the condenser.
3. Notice that only one pump is required in this system, reducing the cycle machinery complexity.

Multiple Regenerative Heaters

As we have seen, the overall efficiency of the power plant is improved when a feedwater heater is used. If one feedwater heater is good, maybe two are better, and so on. The efficiency does improve as more feedwater heaters are used, but the gain in efficiency is offset by the increase in capital cost and maintenance. Usually small plants, such as those on ships, have two regenerative heaters, and large stationary plants have six.

Let us consider a power plant with three feedwater heaters. We will develop expressions to calculate the thermal efficiency for this power plant. Figure 15.19(a) is a schematic drawing of the plant, and Figure 15.19(b) is a T-s diagram of the steam paths. The symbols y_1, y_2, and y_3 are the fractions of steam withdrawn at each feedwater heating state. The assumption is that all liquid leaves the feedwater heater as a saturated liquid at the bleed steam pressure.

The turbine work per kilogram in the system is

$$w_t = (h_2 - h_3) + (1 - y_1)(h_3 - h_4) + (1 - y_1 - y_2)(h_4 - h_5)$$
$$+ (1 - y_1 - y_2 - y_3)(h_5 - h_6) \tag{15.26}$$

The system pump work is found by adding the condensate and feed pump works:

$$w_{p_1} = -(1 - y_1 - y_2)(h_8 - h_7)$$
$$w_{p_2} = -(h_{11} - h_{10}) \tag{15.27}$$
$$w_p = w_{p_1} + w_{p_2}$$

The heat supplied per unit mass is

$$q_{in} = h_2 - h_1 \tag{15.28}$$

and the overall thermal efficiency, equation (15.12), is

$$\eta_{th} = \frac{w_{net}}{q_{in}}$$

To solve for the thermal efficiency, we must find the mass flow to each heater by performing an energy balance on each heater. The easiest method of solution for hand calculations is to take each heater in order and solve for the mass flow, knowing the mass flow from the previous heater. We make an energy balance on the high-pressure heater subject to its being adiabatic; the changes in kinetic and potential energies are zero, and no work is done. Thus,

$$\begin{matrix} \text{Energy} & \text{Energy} \\ \text{Into} & = \text{Out of} \\ \text{Heater} & \text{Heater} \end{matrix}$$

$$y_1 h_3 + (1)h_{11} = y_1 h_a + (1)h_1$$

$$y_1 = \frac{h_1 - h_{11}}{h_3 - h_a} \tag{15.29}$$

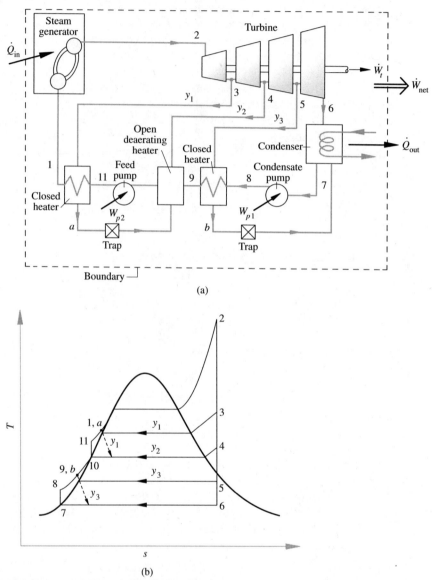

Figure 15.19 (a) The schematic diagram for a three-stage, multiple-regenerative Rankine cycle. (b) The T-s diagram for the cycle.

An energy balance on the deaerating heater, subject to the same constraints as above, yields

$$\text{Energy Into Heater} = \text{Energy Out of Heater}$$

$$y_2h_4 + y_1h_a + (1 - y_1 - y_2)h_9 = (1)h_{10}$$

$$y_2 = \frac{(h_{10} - h_9) - y_1(h_a - h_9)}{(h_4 - h_9)} \quad (15.30)$$

Since y_1 is known from equation (15.29), y_2 may be calculated. We perform an energy balance on the low-pressure heater as follows:

$$\text{Energy Into Heater} = \text{Energy Out of Heater}$$

$$y_3h_5 + (1 - y_1 - y_2)h_8 = y_3h_b + (1 - y_1 - y_2)h_9$$

$$y_3 = (1 - y_1 - y_2)\frac{(h_9 - h_8)}{(h_5 - h_6)} \quad (15.31)$$

Thus, all the fractional flow rates may be calculated and the turbine work and pump work calculated from equations (15.26) and (15.27). We may ask why the equations for the mass fraction balance are valid physically. What is happening in the heater to allow us to use enthalpies of the fluids entering and leaving? The entering liquid is heated by the condensing steam until it reaches the same temperature as the steam. A steam trap, similar to those on steam radiators, prevents the steam from leaving the heat exchanger until it is a liquid. Thus, all the liquid leaving the heat exchanger is at the same temperature. This heat exchanger is an ideal one; actually the water leaves somewhat cooler than the condensed steam temperature.

15.8 REHEAT–REGENERATIVE CYCLE

You might deduce that if the reheat and the regenerative cycles were combined, the overall efficiency would be further improved. The reheat–regenerative cycle usually has one reheat cycle and two or more stages of regenerative heating. The power plant has to be sufficiently large so the increased cost for the reheat piping and maintenance will be paid for by the improved thermal efficiency. Figure 15.20 illustrates a reheat–regenerative cycle with one stage of reheat and two stages of regenerative heating. The first reheat stage will occur at the same pressure as the first regenerative feedwater heater station. The extraction pressure is determined by optimization of the plant overall efficiency as a function of the extraction pressure. Results indicate this pressure is 16–22% of the turbine inlet pressure for a one-stage reheat–regenerative

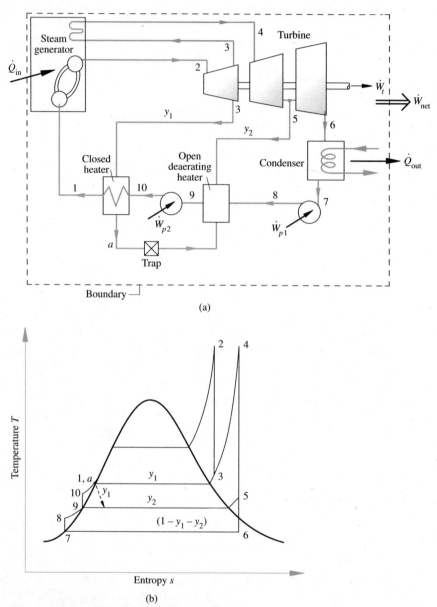

(a)

(b)

Figure 15.20 (a) The schematic diagram for a two-stage regenerative, one-stage reheat Rankine cycle. (b) The T-s diagram for the cycle.

cycle. The turbine work is

$$w_t = (h_2 - h_3) + (1 - y_1)(h_4 - h_5) + (1 - y_1 - y_2)(h_5 - h_6) \qquad (15.32)$$

The heat added to the steam is

$$q_{in} = (h_2 - h_1) + (1 - y_1)(h_4 - h_3) \qquad (15.33)$$

Before calculating the thermal efficiency, we perform an energy balance on the feedwater heater to obtain the fraction of steam withdrawn at each state:

$$y_1 = \frac{h_1 - h_{10}}{h_3 - h_a} \qquad (15.34)$$

$$y_2 = \frac{(h_9 - h_8) - y_1(h_a - h_8)}{(h_5 - h_8)} \qquad (15.35)$$

The pump work is

$$w_{p_1} = -(1 - y_1 - y_2)(h_8 - h_7)$$

$$w_{p_2} = -1\,(h_{10} - h_9)$$

$$w_p = w_{p_1} + w_{p_2} \qquad (15.36)$$

The thermal efficiency for the cycle may now be calculated as the heat added, and the work produced can be calculated.

Example 15.6

A steam power plant operates on a reheat–regenerative cycle with two stages of steam extraction for regenerative heating and one extraction for reheating. The turbine receives steam at 1500 psia and 1000°F; the steam expands isentropically to 300 psia, where a fraction is extracted for feedwater heating and the remainder is reheated at constant pressure until the temperature is 900°F. The steam reenters the turbine and expands to 50 psia, where a fraction is withdrawn for feedwater heating; the remainder expands isentropically in the turbine to 1 psia. The power plant produces 500 MW of power. The open deaerating heater is at 50 psia; the other heater is closed, and the pumps are located after the condenser and after the deaerating heater. Determine the cycle efficiency, the mass flow of steam leaving the steam generator and entering the high pressure turbine, the mass flow of steam entering the deaerating heater, and the heat rejected in the condenser.

Solution

Given: A two-stage-regenerative, one-stage-reheat cycle with specified temperatures and pressures.

Find: The cycle efficiency, the steam flow rates entering the turbine and the deaerating heater, and the total heat rejected in the condenser.

Sketch and Given Data: The plant schematic is illustrated in Figure 15.20(a).

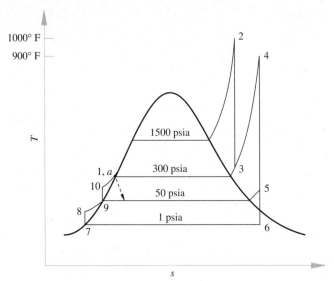

Figure 15.21

Assumptions:

1. Each component within the cycle may be treated as a steady-state open system.
2. The processes are reversible, constant-pressure in the heat exchangers and adiabatic in the pumps and turbine.
3. The water leaves the condenser and heaters as a saturated liquid.
4. The changes in kinetic and potential energies may be neglected.
5. Liquid water may be considered incompressible.

Analysis: Calculate the enthalpies around the cycle and the mass fractions before proceeding with the first-law analysis. $h_2 = 1490.9$ Btu/lbm and $s_2 = 1.6001$ Btu/lbm-R. Process 2–3 is isentropic, $s_3 = s_2$, $p_3 = 300$ psia, and $h_3 = 1287.2$ Btu/lbm. Process 3–4 is constant-pressure, $T_4 = 900°F$, $h_4 = 1473.5$ Btu/lbm, and $s_4 = 1.7582$ Btu/lbm-R. Process 4–5 is isentropic, $s_5 = s_4$, $p_5 = 50$ psia, and $h_5 = 1255.5$ Btu/lbm (steam is still superheated at this state). Process 5–6 is isentropic, $s_6 = s_5 = p_4$, and $p_6 = 1$ psia.

Determine the quality from the entropy term and solve for the enthalpy.

$h_6 = h_f + x_6 h_{fg} = 69.58 + (0.881)(1036.2) = 982.5$ Btu/lbm

$h_7 = h_f$ at 1 psia $= 69.58$ Btu/lbm

$h_8 = h_7 + \dfrac{v_7(p_8 - p_7)}{g}$

$= \dfrac{(69.58 \text{ Btu/lbm}) + (0.016137 \text{ ft}^3/\text{lbm})(50 - 1 \text{ lbf/in.}^2)(144 \text{ in.}^2/\text{ft}^2)}{778.16 \text{ (ft-lbf/Btu)}}$

$h_8 = 69.73$ Btu/lbm

$h_9 = h_f$ at 50 psia = 250.09 Btu/lbm

$$h_{10} = h_9 + \frac{v_9(p_{10} - p_9)}{\mathcal{J}}$$

$$= (250.09 \text{ Btu/lbm}) + \frac{(0.017273 \text{ ft}^3/\text{lbm})(1500 - 50 \text{ lbf/in.}^2)(144 \text{ in.}^2/\text{ft}^2)}{778.16 \text{ (ft-lbf/Btu)}}$$

$h_1 = h_a = h_f$ at 300 psia = 394.15 Btu/lbm

The mass fractions are found from equations (15.34) and (15.35).

$$y_1 = \frac{h_1 - h_{10}}{h_3 - h_a} = \frac{394.15 - 254.7}{1287.2 - 394.15} = 0.1562 \qquad 1 - y_1 = 0.8438$$

$$y_2 = \frac{(h_9 - h_8) - y_1(h_a - h_8)}{(h_5 - h_8)}$$

$$= \frac{(250.09 - 69.73) - (0.1562)(394.15 - 69.73)}{(1255.5 - 69.73)}$$

$$y_2 = 0.1094 \qquad 1 - y_1 - y_2 = 0.7344$$

Perform a first-law analysis for each component on a unit mass basis. The turbine work is

$$w_t = (h_2 - h_3) + (1 - y_1)(h_4 - h_5) + (1 - y_1 - y_2)(h_5 - h_6)$$

$$w_t = (1490.9 - 1287.2 \text{ Btu/lbm}) + (0.8438)(1473.5 - 1255.5 \text{ Btu/lbm})$$

$$+ (0.7344)(1255.5 - 982.5 \text{ Btu/lbm})$$

$$w_t = 588.14 \text{ Btu/lbm}$$

The pump work is

$$w_{p_1} = -(1 - y_1 - y_2)(h_8 - h_7) = -(0.7344)(69.73 - 69.58) = -0.11 \text{ Btu/lbm}$$

$$w_{p_2} = -1(h_{10} - h_9) = -(254.7 - 250.09) = -4.61 \text{ Btu/lbm}$$

$$w_p = w_{p_1} + w_{p_2} = -4.72 \text{ Btu/lbm}$$

The net work is

$$w_{\text{net}} = w_t + w_p = 588.14 - 4.72 = 583.42 \text{ Btu/lbm}$$

The heat added is

$$q_{\text{in}} = (h_2 - h_1) + (1 - y_1)(h_4 - h_3)$$

$$q_{\text{in}} = (1490.9 - 394.15) + (0.8438)(1473.5 - 1287.2) = 1253.95 \text{ Btu/lbm}$$

The thermal efficiency is

$$\eta_{\text{th}} = \frac{w_{\text{net}}}{q_{\text{in}}} = \frac{583.42}{1253.95} = 0.465 \quad \text{or } 46.5\%$$

The mass flow rate leaving the steam generator and entering the high-pressure turbine is found by dividing the total net power produced by the work produced per pound mass.

$$\dot{m}_{stm} = \frac{\dot{W}_{net}}{w_{net}} = \frac{(500\ 000\ \text{kW})(56.87\ \text{Btu/min-kW})}{(583.42\ \text{Btu/lbm})(60\ \text{sec/min})}$$

$$\dot{m}_{stm} = 812.3\ \text{lbm/sec}$$

The mass flow rate of steam entering the deaerating heater is found by multiplying the total system mass flow rate by the mass fraction that leaves to enter the heater.

$$\dot{m}_{dh} = y_2 \dot{m}_{stm} = (0.1094)(812.3\ \text{lbm/sec}) = 88.86\ \text{lbm/sec}$$

The heat rejected in the condenser may be found in two ways. First, perform a first-law analysis on the condenser, yielding

$$\dot{Q}_{out} = (1 - y_1 - y_2)(\dot{m}_s)(h_7 - h_6)$$

$$\dot{Q}_{out} = (0.7344)(812.3\ \text{lbm/sec})(69.58 - 982.5\ \text{Btu/lbm})$$

$$\dot{Q}_{out} = -544,605\ \text{Btu/sec} = -574.6\ \text{MW}$$

For all heat cycles, the sum of the heat flows is equal to the net work. Find the heat supplied by dividing the net power by the cycle efficiency or performing a first-law analysis on the steam generator.

$$\dot{Q}_{in} = \frac{\dot{W}_{net}}{\eta_{th}} = \frac{500}{0.465} = 1075.2\ \text{MW}$$

$$\dot{Q}_{out} = -\dot{Q}_{in} + \dot{W}_{net} = -1075.2 + 500 = -575.2\ \text{MW}$$

which is within round-off error of the $\dot{Q}_{out}$ found from the first-law analysis of the condenser.

Comments:

1. The mass fractions, while determined in terms of enthalpy differences, are physically that portion of the total inlet flow to the turbine that is extracted for regenerative heating. Thus, knowing the mass fraction and the total flow allows us to determine the fractional flow rate.
2. An ideal simple Rankine cycle, operating between the same upper and lower cycle limits, has a thermal efficiency of 41.8%. Thus, the combination of reheating and regeneration adds significantly to the overall thermal efficiency.
3. Notice that the effect of reheating moves the turbine end state, state 6, nearer the saturated vapor line. In this instance the exit quality was 88.1% for the reheat-regenerative cycle and 79.5% for the ideal cycle and for an ideal regenerative cycle.

In Chapter 4, TK Solver models SATSTM.TK and SHTSTM.TK, used to determine steam properties, were introduced. STMCYCLE.TK is based on the steam property functions from these two models and analyzes an ideal simple Rankine cycle. Note that STMCYCLE.TK is based on the turbine inlet steam being superheated and the outlet steam being a mixture. The model would need modification for other steam conditions. The rules in the model can easily be modified to include the effects of component efficiencies, reheat, and regenerative heating.

Another model included on the disk is STEAM.TK. When provided with inputs of two independent properties, this model determines if the steam is superheated or a mixture, and then determines the other properties. In many situations, we know the condition of the steam. In those cases, it is better to use SATSTM.TK or SHTSTM.TK to determine the steam properties. In other situations, such as determining the enthalpy of steam being extracted from a turbine, the steam condition may not be obvious. In those situations, STEAM.TK will be a useful and convenient tool. STEAM.TK is sensitive to the guesses used to begin the iteration. It may be necessary to experiment if the default values don't converge.

15.9 BINARY VAPOR CYCLES

We have seen that the more reversibly processes occur within a cycle, the greater will be the cycle efficiency. In a reversible process the heat should be transferred between two systems at the same temperature, or as near to the same temperature as possible. The temperature inside a furnace is around 1370°C. If heat at this temperature is used to evaporate water at 7.0 MPa, which has a saturated temperature of 286°C, considerable irreversibilities occur in the heat-transfer process.

A method that was explored in the binary vapor cycle is the use of a fluid other than water, one that evaporates at a temperature near the metallurgical limit. If this fluid could expand to the normal condenser temperature, around 32°C, the efficiency of the cycle would be improved. One fluid that meets the upper temperature limit is mercury. However, at 32°C the vapor pressure of mercury is on the order of 0.07 Pa. It is not possible to operate at this vacuum in power plants because of air-leakage complications. At a pressure of 7.0 kPa the temperature of mercury is 237°C. If steam is generated by the latent heat released by the condensing mercury, it could expand to normal condenser pressures. Figure 15.22 shows such a power plant schematically and its T-s diagram. The latent heat of mercury at 7.0 kPa is about 295 kJ/kg, and of water at 230°C is 1813.8 kJ/kg. Thus, there must be 6.1 kg Hg/kg water.

There are disadvantages to the binary cycle. Mercury is highly toxic; there are also heat-transfer problems and related equipment complications. The use of mercury in power plants operating on a binary system has not gone beyond a very limited effort made in the mid-1950s. If a fluid that does not pose the difficulties of mercury should be developed, the binary vapor power cycle would deserve attention. Very high efficiencies are possible, and the capital cost of such plants may be offset by savings due to reduced use of high-cost fuels.

15.10 BOTTOMING CYCLES AND COGENERATION

Thus far we have considered heat power cycles whose purpose is to convert thermal energy into work. In the process, heat near the temperature of the surroundings is rejected to the surroundings. Often this heat is useful as a heat source for another

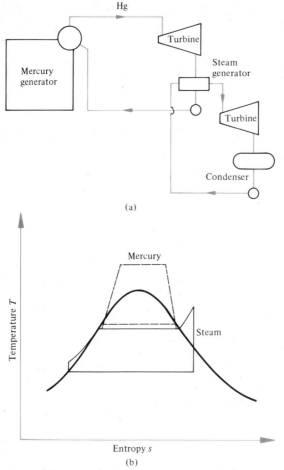

Figure 15.22 (a) A combined mercury-steam cycle. (b) The *T-s* diagram for the binary mercury-steam cycle with superimposed diagrams for each substance.

fluid, running on its own power cycle, or for process heating in an industry. This type of cycle is a *bottoming cycle* and is similar to the binary vapor cycle except that the steam condenser is also the bottoming fluid vapor generator, as illustrated in Figure 15.23, where the bottoming fluid is R 12.

For the bottoming cycle illustrated, the total net power produced is the sum of the net powers from the steam and R 12 cycles. The heat supplied from outside the system still goes only into the steam generator. The overall cycle efficiency is

$$\eta_{th} = \frac{\dot{W}_{\text{net stm}} + \dot{W}_{\text{net R12}}}{\dot{Q}_{\text{in}}} \qquad (15.37)$$

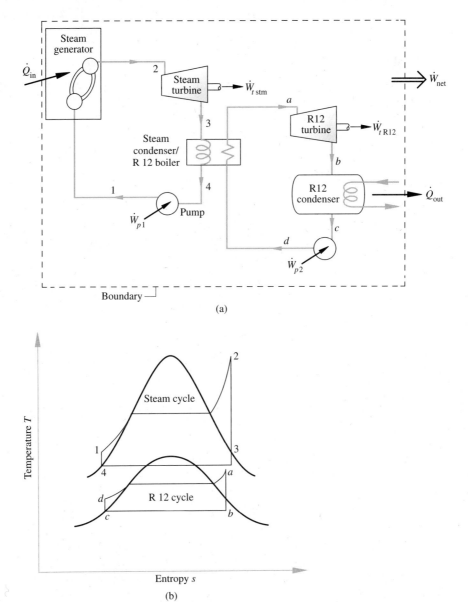

Figure 15.23 (a) The schematic diagram for an R 12 bottoming cycle. (b) the *T-s* diagram for the R 12 bottoming cycle with superimposed diagrams for each substance.

Example 15.7

A bottoming cycle is added to the Rankine cycle in Example 15.1. The cycle uses R 12, which leaves the condenser/R 12 generator at 1 MPa and 50°C. The R 12 condenser operates at 500 kPa. Determine the new cycle efficiency.

Solution

Given: A steam Rankine cycle and the R 12 states that define the bottoming cycle.

Find: The cycle efficiency for this configuration.

Sketch and Given Data:

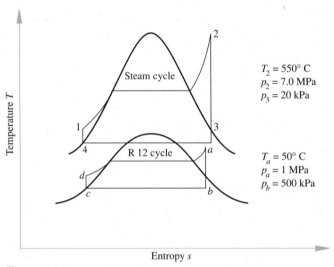

$T_2 = 550° \text{ C}$
$p_2 = 7.0 \text{ MPa}$
$p_3 = 20 \text{ kPa}$

$T_a = 50° \text{ C}$
$p_a = 1 \text{ MPa}$
$p_b = 500 \text{ kPa}$

Figure 15.24

Assumptions:

1. Each component within the cycle may be treated as a steady-state open system.
2. The processes are reversible, constant-pressure in the heat exchangers and adiabatic in the pumps and turbine.
3. The liquids leave the condenser and heaters as saturated liquids.
4. The changes in kinetic and potential energies may be neglected.
5. Liquid water and R 12 may be considered incompressible.

Analysis: The values of enthalpy and energy terms will be taken from Example 15.1 as needed. Determine the enthalpy values for the R 12 in the cycle as follows: $h_a = 210.16 \text{ kJ/kg}$ and $s_a = 0.7021 \text{ kJ/kg-K}$. The process from state a to state b is isentropic, $s_b = s_a$, $p_b = 500 \text{ kPa}$, and $h_b = 197.6 \text{ kJ/kg}$.

$$h_c = h_f \text{ at } 500 \text{ kPa} = 50.6 \text{ kJ/kg}$$

$$h_d = h_c + v_c(p_a - p_c)$$

$$h_d = 50.6 + (0.000\ 744)(1000 - 500) = 51.0\ \text{kJ/kg}$$

$$w_p = -(h_d - h_c) = -(51 - 50.6) = -0.4\ \text{kJ/kg}$$

$$w_t = h_a - h_b = 210.16 - 197.6 = 12.6\ \text{kJ/kg}$$

$$w_{net} = w_t + w_p = +12.2\ \text{kJ/kg R 12}$$

Find the ratio of mass of R 12 needed to the mass of steam condensed from a first-law analysis of the condenser/boiler.

$$\dot{m}_{stm}\,(h_3 - h_4) = \dot{m}_{R12}(h_a - h_d)$$

$$r = \frac{\dot{m}_{R12}}{\dot{m}_{stm}} = \frac{h_3 - h_4}{h_a - h_d}$$

$$r = \frac{(2289.5 - 251.96\ \text{kJ/kg steam})}{(210.16 - 51.0\ \text{kJ/kg R 12})} = 12.8\ \text{kg R 12/kg steam}$$

From Example 15.1 the net work is

$$(w_{net})_{stm} = 1234.3\ \text{kJ/kg steam}$$

Determine the net work of the R 12 cycle per unit mass of steam.

$$r(w_{net})_{R12} = (12.8\ \text{kg R 12/kg steam})(12.2\ \text{kJ/kg R 12})$$

$$r(w_{net})_{stm} = 156.16\ \text{kJ/kg steam}$$

The total net work is

$$w_{net} = 1234.3 + 156.16 = 1390.46\ \text{kJ/kg steam}$$

The new efficiency is

$$\eta_{th} = \frac{w_{net}}{q_{in}} = \frac{1390.46}{3271.8} = 0.425 \quad \text{or } 42.5\%$$

where the value of the heat supplied is found from Example 15.1.

Comments:

1. This is a significant increase in overall efficiency for the plant. Several premises, mostly economic, need to be examined when actual plants are operated. For instance, is the additional capital cost for the bottoming cycle worthwhile in terms of savings in fuel dollars?
2. The Rankine-cycle plant in Example 15.1 operates at a fairly high condenser temperature. As this temperature decreases, the advantage of the bottoming cycle decreases. Again, the bottoming cycle presumes that cooling is available at its condensing temperature of 15.5°C. ▪

   Figure 15.25 illustrates a simplified industrial process plant where a boiler generates steam that is used in a manufacturing process. There is no condenser, and the condensate returns are pumped back into the boiler. In this instance 150 kW of process heat is needed. As noted in the discussion of binary vapor cycles, the temper-

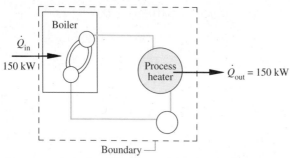

Figure 15.25 A simplified process heating facility.

ature of the combustion gases in the boiler is quite high — high availability — while the application of process heating has a low availability requirement. Thus, the second-law efficiency for the process would necessarily be low.

One way to raise the second-law efficiency is to have the steam from the boiler discharge to a turbine and the turbine exhaust to the process heater, as illustrated in Figure 15.26. There is no condenser, and the returns are pumped back to the boiler, completing the cycle. In this situation the turbine produces 25 kW, and steam flow is sufficient to provide 150 kW of heat to the process heater. The total heat supplied in the boiler increases to 175 kW.

The efficiency of a cogeneration process can be rated by using the *utilization factor, Y_{cg}*. Again, it follows the concept of output divided by input, where in this case the output is both the power generated and the heat delivered to the process heater.

$$Y_{cg} = \frac{\text{net power out} + \text{process heat out}}{\text{total heat input}} = \frac{\dot{W}_{net} + \dot{Q}_{out}}{\dot{Q}_{in}} \qquad (15.38)$$

For an ideal cogeneration plant the utilization factor is unity. In actual plants when

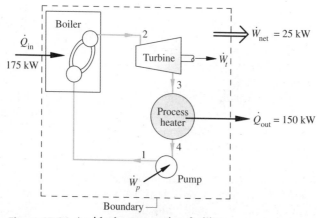

Figure 15.26 An ideal cogeneration facility.

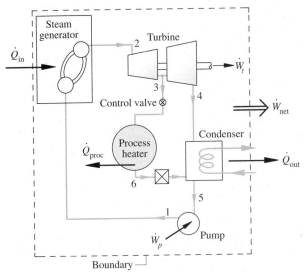

Figure 15.27 A cogeneration system using partially expanded steam.

boiler efficiency and nonadiabatic effects of heat distribution are included, the utilization factor decreases to about 70–75%.

Figure 15.27 illustrates a cogeneration facility where a fraction of the steam that has been partially expanded in a turbine is used for process heating. In the situation illustrated a control valve regulates the amount of steam the process receives. In actual systems, the process heater might receive the turbine bleed steam, as it is called, and, if that is not sufficient, additional steam directly from the steam generator. The definition of the utilization factor is the same.

Example 15.8

A cogenerative heating process, as shown in Figure 15.27, uses steam that has partially expanded through a steam turbine operating on the Rankine cycle for its heat source. The steam generator outlet conditions are 10 MPa and 500°C, and the steam flow rate is 6.0 kg/s. The extraction for process heating occurs at 500 kPa, and condensation occurs at 10 kPa. The process heater uses 3.0 kg/s of steam, and the return from the process heater is at the condenser saturated temperature. Determine the cycle efficiency and utilization factor for situations with and without the process heater.

Solution

Given: A cogenerative heating system where a defined portion of the turbine steam is used for process heating. The cycle state points and mass flow rates are given.

Find: The utilization factor with and without flow to the heater and the thermal efficiency for the same conditions.

Sketch and Given Data:

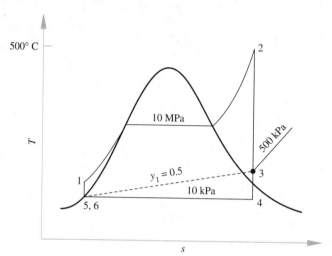

Figure 15.28

Assumptions:

1. Each component within the cycle may be treated as a steady-state open system.
2. The processes are reversible, constant-pressure in the heat exchangers and adiabatic in the pumps and turbine.
3. The liquid leaves the condenser and heaters as a saturated liquid at specified temperatures or pressures.
4. The changes in kinetic and potential energies may be neglected.
5. Liquid water may be considered incompressible.

Analysis: Determine the enthalpies around the cycle following the numbering sequence in Figure 15.27. Case 1, process heating: $h_2 = 3376.5$ kJ/kg, $s_2 = 6.5982$ kJ/kg-K; $p_3 = 500$ kPa, $s_3 = s_2$, $h_3 = 2654.0$ kJ/kg; $p_4 = 10$ kPa, $s_4 = s_3 = s_2$, $h_4 = 2089.9$ kJ/kg; $h_6 = h_5 = 191.8$ kJ/kg.

$$h_1 = h_5 + v_5(p_1 - p_5)$$

$$h_1 = 191.8 + (0.0010)(10\,000 - 10) = 201.9 \text{ kJ/kg}$$

Determine the fraction of the total flow used for process heating. This allows us to calculate the efficiency and utilization factor using unit masses.

$$y_1 = \frac{\dot{m}_{\text{proc}}}{\dot{m}_{\text{total}}} = \frac{3.0}{6.0} = 0.5$$

The heat added in the cycle is

$$q_{\text{in}} = h_2 - h_1$$

$$q_{\text{in}} = 3376.5 - 201.9 = 3174.6 \text{ kJ/kg}$$

The turbine, pump, and net works are

$$w_t = (h_2 - h_3) + (1 - y_1)(h_3 - h_4)$$

$$w_t = (3376.5 - 2654.0) + (0.5)(2654.0 - 2089.9) = 1004.6 \text{ kJ/kg}$$

$$w_p = -(h_1 - h_5) = -(201.9 - 191.8) = -10.1 \text{ kJ/kg}$$

$$w_{net} = w_t + w_p = 994.5 \text{ kJ/kg}$$

The process heat is

$$q_{proc} = y_1(h_3 - h_6)$$

$$q_{proc} = (0.5)(2654.0 - 191.8) = 1231.1 \text{ kJ/kg}$$

The utilization factor is

$$Y_{cg} = \frac{w_{net} + q_{proc}}{q_{in}}$$

$$Y_{cg} = \frac{994.5 + 1231.1}{3174.6} = 0.701 \quad \text{or } 70.1\%$$

and the thermal efficiency is

$$\eta_{th} = \frac{w_{net}}{q_{in}} = \frac{994.5}{3174.6} = 0.313 \quad \text{or } 31.3\%$$

Case 2, no process heat: In this situation all the steam expands through the turbine in an ideal Rankine cycle. The enthalpies from case 1 may be used. In this cycle the turbine inlet is state 2, turbine exit state 4, condenser outlet state 5, and pump outlet state 1. There is no extraction, so states 3 and 6 are not applicable. The turbine, pump, and net works are

$$w_t = h_2 - h_4 = 3376.5 - 2089.9 = 1286.6 \text{ kJ/kg}$$

$$w_p = -h_1 - h_5 = -(201.9 - 191.8) = -10.1 \text{ kJ/kg}$$

$$w_{net} = w_t + w_p = 1286.6 - 10.1 = 1276.5 \text{ kJ/kg}$$

The heat added is

$$q_{in} = h_2 - h_1 = (3376.5 - 201.9) = 3174.6 \text{ kJ/kg}$$

The thermal efficiency is

$$\eta_{th} = \frac{w_{net}}{q_{in}} = \frac{1276.5}{3174.6} = 0.402 \quad \text{or } 40.2\%$$

The utilization factor is

$$Y_{cg} = \frac{w_{net} + q_{proc}}{q_{in}} = \frac{1276.5 + 0}{3174.6} = 0.402 \quad \text{or } 40.2\%$$

Comments:

1. The utilization factor is the same as the cycle thermal efficiency when no process heating occurs.
2. When process heating occurs using partially expanded steam, the cycle efficiency is no longer relevant, and the utilization factor is needed to describe cycle performance. ▬

15.11 COMBINED GAS-VAPOR POWER CYCLES

In the combined cycle the waste heat or exhaust from a gas cycle, such as a gas turbine or diesel engine, is used as a heat source to generate steam. The steam is then used in a Rankine cycle, which in essence is a bottoming cycle to the gas power cycle.

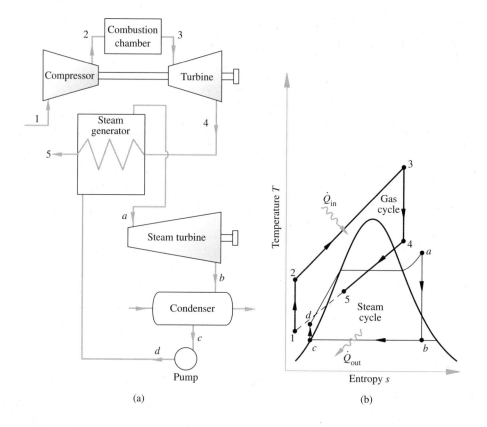

(a)

(b)

Figure 15.29 (a) A combined gas turbine–steam power cycle. (b) The superimposed *T-s* diagrams for the cycle.

The limitation on the temperature of gas entering a gas turbine is about 1260°C (2300°F), whereas in a steam power plant the maximum steam inlet temperature is about 600°C (1100°F). The gas turbine blades are cooled and sometimes coated with ceramics, permitting the higher operating temperature vis-à-vis the steam turbine. The potential high efficiency of the gas turbine unit, because of its high operating temperature and high average temperature of heat addition, is negated by the high exhaust temperatures. If the high-temperature exhaust is used as a heat source to generate steam in a waste heat boiler, a significant portion of the exhaust's availability can be recovered.

Figure 15.29 illustrates a simplified combined gas turbine–steam turbine plant. The steam cycle shown is an ideal simple Rankine cycle. Many actual plants include steam generators that produce steam at two or three different pressures. This increases the utilization of the thermal energy in the exhaust gases. The lower-pressure steam is used for regenerative heating and can also be reintroduced to the turbine at a later stage, increasing the power developed. Some larger plants include reheating to further increase energy utilization.

Gas turbines are not the only cycle that can be combined with the Rankine cycle. Large diesel engines have significant exhaust gas flows and high exhaust temperatures. The availability of this energy can be partially conserved by adding a waste heat boiler and operating a Rankine cycle. In some situations the steam from the boiler can be used for process heating requirements, perhaps as a heat source for freshwater evaporators, air conditioning systems, or heating systems. In these instances, the utilization factor would be a wiser way to judge the plant's operating effectiveness.

Example 15.9

A combined gas turbine–steam turbine power plant produces 50,000 hp net power. The exhaust from the gas turbine leaves the steam generator at 300°F. Steam leaves the boiler at 900 psia and 760°F. Air enters the gas turbine unit at 520°R and 14.6 psia and leaves the compressor at 146 psia. The gas turbine inlet temperature is 2500°R. Both the Rankine and the open gas turbine cycles are ideal. Calculate the flow rates required in each cycle, the overall efficiency, and the availability utilization of the gas turbine exhaust. The gas turbine inlet conditions are the surroundings temperature and pressure. The steam power plant has a condenser pressure of 2 psia. Assume air is the working fluid in the gas turbine cycle, so that the mass of fuel may be neglected.

Solution

Given: A combined gas turbine–steam turbine power plant with specifications of the gas turbine and the Rankine cycle as well as the total net power output.

Find: The steam and gas flow rates in each cycle, the overall efficiency, and the availability utilization of the gas turbine exhaust.

Sketch and Given Data:

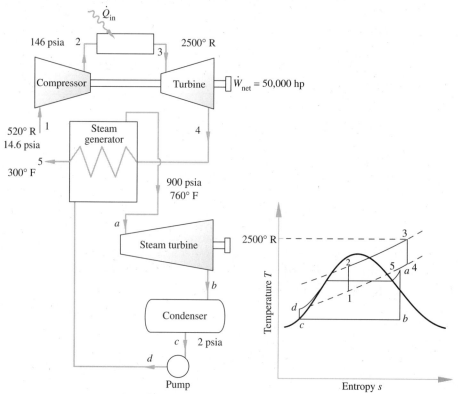

Figure 15.30

Assumptions:

1. Each component within the cycle may be treated as a steady-state open system.
2. The processes are reversible, constant-pressure in the heat exchangers and isentropic in the pump, turbines, and compressor.
3. The water leaves the condenser as a saturated liquid.
4. The changes in kinetic and potential energies may be neglected.
5. Liquid water may be considered incompressible.

Analysis: Determine the cycle state points and enthalpy values for the gas turbine and Rankine cycles. In this case AIRCYCLE.TK and STMCYCLE.TK were used to determine the cycle state points. The annotated Variable Sheets are shown on the facing page. To determine the ratio of steam flow to airflow, perform a first-law analysis on the steam generator. The outlet gas temperature ($300°$F) is specified, as is the inlet temperature ($T_4 = 1408.7°$R). Using AIR.TK, we find the change in enthalpy and the change in entropy for the air across the steam generator to be

```
══════════════════════════════ VARIABLE SHEET ═══════════════════════════════
St Input ── Name ── Output ── Unit ──── Comment ──────────────────────────────

                                         ***Four-Point Air Cycle***

                                            POINT 1
    14.6    P1                   psia       Pressure (kPa, MPa, psia)
    520     T1                   degR       Temperature (degK, degC, degR, degF)
            v1      13.188       ft3/lbm    Specific Volume (m3/kg, ft3/lbm)

                                            POINT 2
    146     P2                   psia       Pressure (kPa, MPa, psia)
            T2      990.15       degR       Temperature (degK, degC, degR, degF)
            v2      2.5185       ft3/lbm    Specific Volume (m3/kg, ft3/lbm)

                                            POINT 3
    146     P3                   psia       Pressure (kPa, MPa, psia)
    2500    T3                   degR       Temperature (degK, degC, degR, degF)
            v3      6.3559       ft3/lbm    Specific Volume (m3/kg, ft3/lbm)

                                            POINT 4
    14.6    P4                   psia       Pressure (kPa, MPa, psia)
            T4      1408.7       degR       Temperature (degK, degC, degR, degF)
            v4      35.757       ft3/lbm    Specific Volume (m3/kg,ft3/lbm)

            Qin     407.04       BTU/lbm
            Wnet    186.17       BTU/lbm
```

```
══════════════════════════════ VARIABLE SHEET ═══════════════════════════════
St Input ── Name ── Output ──── Unit ──── Comment ────────────────────────────
                                          ***Steam Cycle***

            Wturb   463.6        BTU/lbm   Turbine Work
            Wpump   2.6967       BTU/lbm   Pump Work
            Wnet    460.91       BTU/lbm   Net Work
            Qin     1273.5       BTU/lbm   Heat In
            Eth     36.194       %         Thermal Efficiency

                                    Point 1 - Feed Pump Discharge
            P1      900          psia      Pressure (kPa, MPa, psia)
            T1      128.72       degF      Temperature (degK, degC, degR, degF)
            v1      .016229      ft3/lbm   Specific Volume (m3/kg, ft3/lbm)
            h1      96.836       BTU/lbm   Enthalpy (kJ/kg, BTU/lbm)
            s1      .17464       B/lbm-R   Entropy (kJ/kg-K, B/lbm-R)

                                    Point 2 - Superheater Outlet
    900     P2                   psia      Pressure (KPa, MPa, psia)
    760     T2                   degF      Temperature (degK, degC, degR, degF)
            v2      .73847       ft3/lbm   Specific Volume (m3/kg, ft3/lbm)
            h2      1370.3       BTU/lbm   Enthalpy (kJ/kg, BTU/lbm)
            s2      1.562        B/lbm-R   Entropy (kJ/kg-K, B/lbm-R)

                                    Point 3 - Turbine Exhaust
            P3      2            psia      Pressure (kPa, MPa, psia)
            T3      126.02       degF      Temperature (degK, degC, degR, degF)
            v3      138.16       ft3/lbm   Specific Volume (m3/kg, ft3/lbm)
            h3      906.69       BTU/lbm   Enthalpy (kJ/kg, BTU/lbm)
            s3      1.562        B/lbm-R   Entropy (kJ/kg-K, B/lbm-R)
            x       .79508                 Quality

                                    Point 4 - Condenser Outlet
            P4      2            psia      Pressure (kPa, MPa, psia)
    126.02  T4                   degF      Temperature (degK, degC, degR, degF)
            v4      .016229      ft3/lbm   Specific Volume (m3/kg, ft3/lbm)
            h4      94.14        BTU/lbm   Enthalpy (kJ/kg, BTU/lbm)
            s4      .17464       B/lbm-R   Entropy (kJ/kg-K, B/lbm-R)
```

$$h_4 - h_5 = 162.9 \text{ Btu/lbm air}$$

$$s_4 - s_5 = 0.1535 \text{ Btu/lbm-R}$$

The change of steam enthalpy across the generator is

$$h_a - h_d = 1370.3 - 96.9 = 1273.4 \text{ Btu/lbm}$$

The first-law equation subject to the problem assumptions is

$$\dot{m}_{air}(h_4 - h_5) = \dot{m}_{stm}(h_a - h_d)$$

$$\frac{\dot{m}_{stm}}{\dot{m}_{air}} = \frac{h_4 - h_5}{h_a - h_d}$$

$$\frac{\dot{m}_{stm}}{\dot{m}_{air}} = \frac{162.9 \text{ Btu/lbm air}}{1273.4 \text{ Btu/lbm steam}} = 0.1279 \text{ lbm steam/lbm air}$$

The net total power produced is 50,000 hp. The mass flow rate of air may be determined as follows:

$$\dot{W}_{net} = \dot{m}_{air}\left[w_{net\ air} + \left(\frac{\dot{m}_{stm}}{\dot{m}_{air}}\right) w_{net\ stm} \right]$$

$$\dot{m}_{air} = \frac{(50{,}000 \text{ hp})(42.4 \text{ Btu/min-hp})}{(186.2 \text{ Btu/lbm air} + (0.1279 \text{ lbm steam/lbm air})(460.9 \text{ Btu/lbm steam})}$$

$$\dot{m}_{air} = 8648 \text{ lbm air/min}$$

The steam flow rate is

$$\dot{m}_{stm} = (0.1279) \text{ lbm steam/lbm air})(8648 \text{ lbm air/min}) = 1106 \text{ lbm steam/min}$$

The combined cycle thermal efficiency is

$$\eta_{th} = \frac{w_{net}}{q_{in}} = \frac{186.2 + (0.1279)(460.9)}{407.0} = \frac{245.1}{407.1} = 0.602 \quad \text{or } 60.2\%$$

The availability of exhaust gas is

$$(\varphi_4 - \varphi_5) = (h_4 - h_5) - T_0(s_4 - s_5)$$

From AIR.TK

$$h_4 - h_5 = 162.9 \text{ Btu/lbm air}$$

$$s_4 - s_5 = 0.1535 \text{ Btu/lbm air} - R$$

$$\varphi_4 - \varphi_5 = 162.9 - (520)(0.1535) = 83.08 \text{ Btu/lbm air}$$

The work produced by the steam cycle per pound mass of air is

$$w_{net/stm} = (0.1279 \text{ lbm steam/lbm air})(460.9 \text{ Btu/lbm steam}) = 58.9 \text{ Btu/lbm air}$$

Thus, the percentage of availability converted to work by the Rankine cycle is

$$(\Delta\varphi)_{used} = \frac{58.9}{83.08} = 0.709 \quad \text{or } 70.9\%$$

Comments:

1. The availability utilization of the exhaust is quite high, indicative of the high thermal efficiency.
2. The efficiency of the combined unit is higher than the efficiency of either unit by itself.

15.12 STEAM TURBINE REHEAT FACTOR AND CONDITION CURVE

In the preceding sections we analyzed various vapor power cycles. These cycles use a turbine to produce power, and this section analyzes, in more detail, the energy flow through the turbine. Although we use a steam turbine as the model, the analysis is valid for other turbines, such as a gas turbine. In the case of the gas turbine, the fluid is a mixture of air and oxidized hydrocarbons.

Let us consider an impulse turbine operating in the Rankine cycle. Of the total energy reaching the turbine, only ΔH_s is available for work. The term ΔH_s is the isentropic enthalpy drop between the inlet and exit pressures. Not all the available energy can be used; the following losses can and do occur:

1. Leakage of steam at shaft packings and between turbine stages;
2. Radiation losses to surroundings;
3. Kinetic energy loss to the condenser;
4. Reheating of the fluid caused by irreversibilities in fluid flow;
5. Fluid friction losses on the turbine rotor and blades.

Let us consider the turbine as the sum of individual stages. The steam enters the first stage and expands to pressure p_1; the available energy at this pressure is $(\Delta h_s)_1$. Of this available energy, a portion e_1 is turned into mechanical work. The remainder is reheat, Rh_1, rejected to the next stage. This reheat may be broken into two parts: q_{r1}, due to irreversibilities in the fluid flow, friction in the nozzles and blading, and leakage; and $(k.e.)_1$, the exit kinetic energy from the first stage. Thus,

$$(\Delta h_s)_1 - e_1 = Rh_1 = q_{r1} + (k.e.)_1 \tag{15.39}$$

The stage efficiency, η_{st}, indicates how well the available energy is converted into mechanical work.

$$(\eta_{st})_1 = \frac{e_1}{(\Delta h_s)_1} \tag{15.40}$$

The reheat from the first stage is passed on to the second stage, where $(\Delta h_s)_2$ is available energy, and e_2 is used as mechanical work. Eventually the reheat from the last stage is passed on to the condenser. The sum of the individual drops in enthalpy will be greater than ΔH_s because one stage's reheat is added to energy entering the successive stage. Thus,

$$\sum_i (\Delta h_s)_i = R_f \Delta H_s \tag{15.41}$$

where R_f is the reheat factor, a constant greater than 1 and usually $1.0 \le R_f \le 1.065$. If the stage efficiencies are the same for each stage (which, in general, they are not), the total mechanical work, E, may be expressed as

$$E = \sum_i e_i = \sum_i \eta_{st}(\Delta h_s)_i = \eta_{st} R_f \Delta H_s \qquad (15.42)$$

The stage efficiency will decrease when moisture is present in the steam. The following example illustrates several of these turbine concepts.

Example 15.10

Steam enters a six-stage impulse turbine at 3.5 MPa and 450°C. The stage efficiency is 80% for all stages, and the isentropic enthalpy drop per stage is 180 kJ/kg. Calculate the reheat factor and the end point. Denote the pressure in each stage.

Solution

Given: A six-stage impulse turbine and its stage efficiency and per-stage isentropic enthalpy drop.

Find: The reheat factor, end point, and pressure in each stage.

Sketch and Given Data:

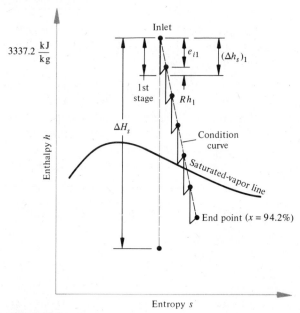

Figure 15.31

Assumptions:

1. Steam flows steadily through the turbine.
2. The turbine is adiabatic.

Analysis: The entering enthalpy, h_0, is 3337.2 kJ/kg, and the pressure is 3.5 MPa. In the first stage, 180 kJ/kg of energy is available, so proceed vertically (isentropically) down the *h-s* diagram until an enthalpy value of 3157 kJ/kg (3337.2 − 180) is reached. The pressure is 2.0 MPa. This would be the enthalpy of the steam entering the next stage if all available energy were used. However, only 80% is used; the remainder, 36 kJ/kg, is reheat passed on to the next stage. The reheat is added at constant pressure ($p = 2.0$ MPa) until a value of 3193 kJ/kg (3157 + 36) is reached. This is the condition of the steam entering the second stage.

The process repeats itself, with the following values denoting the pressure and enthalpy entering each stage:

Stage	Enthalpy (kJ/kg)	Pressure (kPa)
1	3337.2	3500
2	3193	2000
3	3049	1000
4	2905	460
5	2761	180
6	2617	62
Condenser	2473	18

The reheat factor, R_f, is defined as

$$R_f = \sum_i \frac{(\Delta h_s)_i}{\Delta H_s}$$

$$\Delta H_s = 3337.2 - 2294.6 = 1042.6 \text{ kJ/kg}$$

$$R_f = \frac{(6)(180)}{1042.6} = 1.036$$

The end point denotes the condition of the steam as it enters the condenser. From the *h-s* diagram, we find that the steam has a quality of 94.2% and a pressure of 18 kPa. The condition curve is a line joining the steam states entering the various stages. ◼

15.13 GEOTHERMAL ENERGY

Thermal energy stored below the earth's surface is called *geothermal energy*. Volcanic activity within the last 3 million years has brought molten rock, magma, to within 8 to 16 km of the earth's surface. Fractures in nearby surface rocks contain water, which, when heated by the cooling magma, increases in pressure. At times this heated water erupts as geysers or as hot springs. In other cases drilling may be necessary to bring it to the surface. The temperature of the water deposits ranges from 15° to 300°C, with dissolved solids ranging from 0.1 to 25%.

Figure 15.32 illustrates a model of a geothermal system. Surface runoff at state 1 percolates through the permeable rock. The water is heated by the convecting magma and tries to move upward through the rock of low permeability. Typically, the water cannot reach the surface, and a well is drilled, which allows the heated water at state 2 to flow to the surface. As the water rises, the pressure drops and flashing occurs (state 3), and a steam-water mixture exits the well at state 4.

The flashed steam system, illustrated schematically in Figure 15.33(a), is used in several countries. The steam-water mixture enters a separator, where the steam flashes and enters a turbine. The brine is discarded. This may not be easy, since the brine contains minerals and salts harmful to the surface ecology. The steam is condensed, states 5 and 6. The condensed steam may be recombined with the hot brine and injected into the ground, or it may be stored in a pond and used as cooling water. Figure 15.33(b) is the *T-s* diagram. The process 1-2-*a* is a throttling process.

There are plans to develop binary cycles, where the hot water is pumped under pressure from the bottom of the well through a heat exchanger. The cooling fluid in the heat exchanger will be a refrigerant, such as R 12, which will operate on a Rankine cycle.

The major areas of difficulty in using geothermal energy are (1) the relatively low pressures and temperatures of the wellhead water (about 700 kPa); (2) the extremely corrosive nature of the water; (3) the geologically unstable area where plants must be located. No new technology is needed to build the plants, however, and the projected cost per kilowatt is significantly less than in coal and nuclear plants. As sites for geothermal plants are discovered, more will certainly be built.

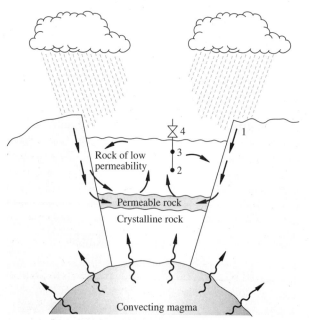

Figure 15.32 Hot-water geothermal system.

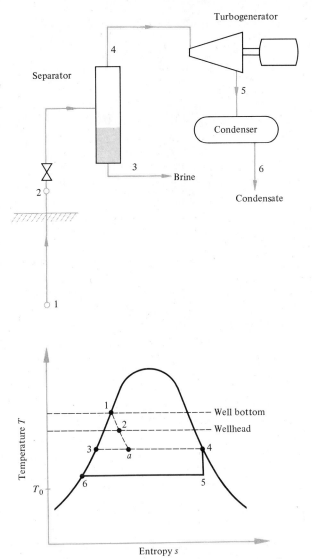

Figure 15.33 (a) Flashed steam system. (b) The T-s diagram for the system.

15.14 SECOND-LAW ANALYSIS OF VAPOR POWER CYCLES

Ideal Rankine cycles, be they simple, reheat, or regenerative, are internally reversible but may have external irreversibilities, such as those caused by heat transfer with a finite temperature difference. The cycles are internally reversible in that cycle processes are reversible, such as isentropic expansion, constant-pressure heat addition, and rejection. By analyzing the entire cycle, we can gain insight as to where the greatest irreversibilities occur.

The expression for irreversibility for steady-state open systems with multiple fluids that exchange heat with a constant-temperature heat source, or sink, is derived from Equation (8.41) by T_0. The result is

$$I = T_0 \Delta \dot{S}_{prod} = T_0 \left[\sum_i (\dot{m}_i s_i)_{out} - \sum_j (\dot{m}_j s_j)_{in} + \sum_i \frac{\dot{Q}_i}{T_i} \right] \qquad (15.43)$$

The irreversibilities due to heat transfer are the primary causes for the cycle irreversibility, particularly that caused by the difference between the combustion gas temperature in the furnace and the steam temperature in the boiler. The irreversibility in the condenser because of temperature differences between the cooling water and the condensing steam is much smaller.

Example 15.11
Consider the Rankine cycle in Example 15.1 and let the steam flow rate be 5 kg/s. The steam receives heat from the combustion gases in the furnace of the steam generator; assume the combustion gases change temperature from 2000°K to 425°K in flowing through the steam generator. The cooling water enters the condenser at 20°C and leaves at 40°C. T_0 is also 20°C. Determine the cycle irreversibility.

Solution

Given: An ideal Rankine cycle with combustion gas and cooling-water temperature changes.

Find: The cycle irreversibility.

Sketch and Given Data:

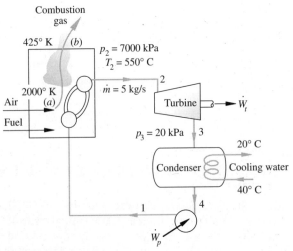

Figure 15.34

Assumptions:

1. The Rankine cycle has internally reversible processes.
2. The heat transfer in the steam generator and condenser can be considered adiabatic; no heat is lost to the surroundings.
3. The flow rate within the cycle is steady.
4. The combustion gas has properties similar to air.

Analysis: The values for enthalpy and entropy will be taken from Example 15.1. Proceed around the cycle, evaluating the irreversibility for each process from equation (15.43) and then summing the processes for the cycle's irreversibility.

Isentropic expansion of steam through the turbine from state 2 to state 3 is

$$\dot{I}_{2\text{-}3} = T_0(0) = 0$$

In this instance no heat is transferred to the surroundings; hence the process is externally and internally reversible, as the entropy does not change.

The compression process in the pump is isentropic; hence the process is internally and externally reversible; the irreversibility for this process, $\dot{I}_{4\text{-}1}$, is zero.

In the condensation of steam in the condenser from state 3 to state 4 heat transfers to the surroundings and there is a temperature difference between the system and the surroundings; hence the process is externally irreversible. The cooling-water flow rate through the condenser is found from a first-law analysis of the condenser.

The enthalpy and entropy terms for the cooling water may be determined by assuming they are equal to the saturated liquid values at that temperature: $h_i = 82.9$ kJ/kg, $s_i = 0.2914$ kJ/kg-K, $h_o = 167.3$ kJ/kg, and $s_o = 0.5697$ kJ/kg-K.

$$\dot{m}_s(h_3 - h_4) = \dot{m}_{cw}(h_o - h_i)$$

$$\dot{m}_{cw} = (5 \text{ kg steam/s}) \frac{(2289.5 - 251.96 \text{ kJ/kg steam})}{(167.3 - 82.9 \text{ kJ/kg water})} = 120.7 \text{ kg/s}$$

The entropy values for the Rankine cycle are $s_3 = 6.9460$ kJ/kg-K and $s_4 = 0.8311$ kJ/kg-K. The irreversibility is

$$\dot{I}_{34} = T_0 \left[\sum_i (\dot{m}_i s_i)_{\text{out}} - \sum_j (\dot{m}_j s_j)_{\text{in}} \right]$$

$$\dot{I}_{34} = (293 \text{ K})[(120.7 \text{ kg water/s})(0.5697 - 0.2914 \text{ kJ/kg water-K})]$$

$$+ (5 \text{ kg steam/s})(0.8311 - 6.9460 \text{ kJ/kg steam-K})]$$

$$\dot{I}_{34} = 883.8 \text{ kW}$$

Heating of the steam in the steam generator from state 1 to state 2 occurs at constant pressure; hence the process is internally reversible, but there is a temperature difference between the water in the steam generator and the combustion gas in the furnace, so there are external irreversibilities associated with the process.

The value of entropy of the water entering the boiler corresponds to the saturated-liquid value of entropy for h_1. Thus, $s_1 = 0.8521$ kJ/kg-K and $s_2 = 6.9460$ kJ/kg-K.

The flow rate of the gas is found from a first-law analysis of the steam generator. AIR.TK was used to find the enthalpy change and entropy change of gas, invoking assumption 4.

$$h_b - h_a = -1824.8 \text{ kJ/kg gas}$$

$$s_b - s_a = -1.7448 \text{ kJ/kg gas-K}$$

$$\dot{m}_g(h_a - h_b) = \dot{m}_s(h_2 - h_1)$$

$$\dot{m}_g = (5 \text{ kg steam/s}) \frac{(3530.9 - 259.05 \text{ kJ/kg steam})}{(1824.8 \text{ kJ/kg gas})}$$

$$\dot{m}_g = 8.96 \text{ kg/s}$$

The irreversibility is

$$\dot{I}_{12} = T_0 \left[\sum_i (\dot{m}_i s_i)_{\text{in}} - \sum_j (\dot{m}_j s_j)_{\text{out}} \right]$$

$$\dot{I}_{12} = (293°\text{K}) \, [(8.96 \text{ kg gas/s})(-1.7448 \text{ kJ/kg gas-K})$$

$$+ \, (5 \text{ kg steam/s})(6.9460 - 0.8521 \text{ kJ/kg steam-K})]$$

$$\dot{I}_{12} = 4347 \text{ kW}$$

The cycle irreversibility is

$$\dot{I}_{\text{total}} = 0 + 0 + 883.8 + 4347 = 5230.8 \text{ kW}$$

Comment: Over 80% of the irreversibility in this example occurs in the steam generator, not the condenser. This is because of the high temperature differences between the combustion gas and the steam. In the condenser the temperature difference between the water and condensing steam is not great; hence the irreversibility is lower.

15.15 ACTUAL HEAT BALANCE CONSIDERATIONS

The cycles that have been analyzed are essentially ideal, since considerations of pressure drop in the piping, steam leakage, and auxiliary steam consumption have not been considered. In this section we will review these and other impacts on actual heat balances.

Many large power plants have six or more feedwater heaters, so a heat balance becomes quite complex. There is only one deaerating heater, with the others being of shell-and-tube construction. In the deaerating heater, the steam mixes directly with the feedwater, and the resulting liquid is saturated at the bleed steam pressure. In the shell-and-tube heaters, the terminal temperature difference (TTD), that is, the difference between the feedwater leaving the heat exchanger and the saturated steam temperature, is typically 0°F when the entering steam is superheated and 10°F when the entering steam is saturated or hot water is used as the heat source. The latter situation occurs in drain coolers, where low-pressure heater drains heat the feedwater. Of course, for an open deaerating heater, the TTD is 0°F.

Pressure drops occur not only in the feedwater piping but also in steam piping. It is common to assume a 2½% pressure drop from the boiler to the turbine inlet and a 5°F temperature drop. For turbine bleed steam, assume a 7% pressure drop.

In calculating an actual heat balance, steam loss to leakage and other causes must be considered. If heavy oil is burned in the boiler, steam atomization inevitably will

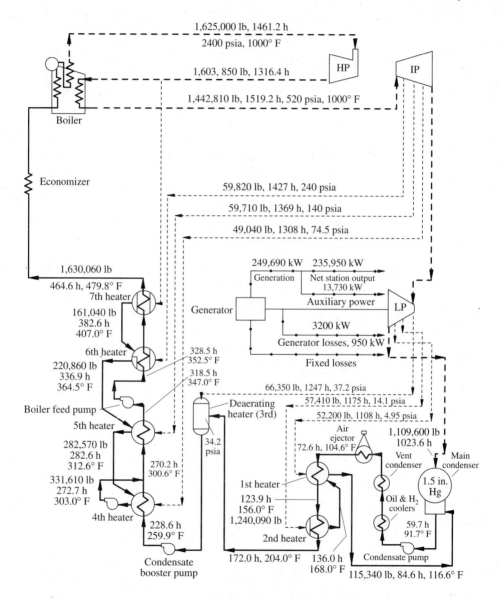

Figure 15.35 The heat balance for a 236-MW reheat–regenerative power plant. The *heavy dashed lines* represent the main flow of steam through the cycle and the *thin dashed lines* the flow of extraction steam. *Numbers* followed by *lb* represent the steam flow rate in lbm/hr, and *numbers* followed by *h* represent the enthalpy of the extracted steam in Btu/lbm.

be used, another loss from the system. Steam is used to atomize the fuel droplets, increasing the surface area per unit mass of fuel and improving the combustion process since air mixes better with fuel. Periodically the steam generator tubing surfaces must be blown clean. Soot blowers use steam to perform this task, another loss. Boilers must be "blown down" periodically to remove impurities.

All the pumps in the system require motors; the boiler fans that provide air also require large motors. Air ejectors or vacuum pumps must be run to maintain the condenser vacuum. These, too, must be included in actual heat balance. The steam generators are not 100% efficient in the conversion of chemical energy (i.e., the fuel's heating value) into thermal energy (i.e., the energy gained by the steam). The effi- ciency is around 90% for large steam generators and is defined as

$$\eta_{sg} = \frac{\dot{m}_s(h_2 - h_1)}{\dot{m}_{fuel}\, h_{RP}}$$

where h_{RP} is the heating value per unit mass of fuel. For a residual oil this value is approximately 18,500 Btu/lbm. For our purposes, consider the pump efficiency to be approximately 80% and the turbine internal efficiency to be 85%. Additionally, there are mechanical losses within the turbine and losses in the electric generator when converting the mechanical energy from the turbine into electrical energy.

Figure 15.35 illustrates an actual heat balance for a fossil-fuel reheat-regenerative power plant. The cycle has one stage of reheat, one open deaerating heater, and six closed regenerative heaters. A condensate booster pump is used in series with the feed pump so the entire pressure rise does not occur in one pump. Also, the pressure within the fourth and fifth heaters is much less, reducing the material strength the heat exchanger must have.

CONCEPT QUESTIONS

1. Why is high moisture content in steam detrimental to steam turbines?
2. Why is the Carnot cycle not used as the ideal model for steam power plants?
3. What are the processes that make up the ideal Rankine cycle?
4. What is the relationship between steam moisture content and quality?
5. Does the steam pressure increase in the steam generator as it is heated? Does the tempera- ture increase? When?
6. Explain the effect of raising the steam generator pressure in an ideal Rankine cycle on turbine work, heat added, and thermal efficiency.
7. Explain the effect of lowering the condenser pressure in an ideal Rankine cycle on turbine work, heat added, and thermal efficiency.
8. Explain the effect of increasing the superheated temperature entering the turbine in an ideal Rankine cycle on turbine work, heat added, thermal efficiency, and steam quality exiting the turbine.
9. Discuss the differences between ideal and actual steam vapor power cycles.
10. The design pressure of a steam condenser is 5 kPa. Cooling water is available at 20°C. Can the design pressure be obtained? Why?

11. What is the effect of reheating steam on the turbine work, the heat added, the heat rejected, the thermal efficiency, and the steam quality leaving the turbine?

12. What are the conditions that would cause reheating to decrease cycle efficiency?

13. Consider a one-stage regenerative Rankine cycle. What is the effect of regeneration on turbine work, heat in, heat rejected, and steam quality exiting the turbine?

14. Explain the differences between open and closed feedwater heaters.

15. Explain, using the concept of average temperatures of heat addition and heat rejection, the benefits of reheating, regenerative heating, raising boiler pressure, lowering condenser pressure, and superheating.

16. What is the difference between the utilization factor and thermal efficiency? Can they ever be the same?

17. Are regeneration and cogeneration the same effect? Why?

18. What is a bottoming cycle?

19. What is a binary vapor cycle?

20. Why is steam not used as the only fluid in a binary vapor cycle? In a bottoming cycle?

21. In combined gas turbine–steam cycles, where is the energy supplied?

22. The combined gas turbine–steam cycle is more efficient than either cycle by itself. Why?

23. What is the source of the greatest irreversibility in the Rankine cycle? In the Rankine reheat-regenerative cycle?

24. Discuss the combined gas turbine–vapor cycle in terms of the gas's availability entering the steam generator.

25. What is meant by a supercritical vapor power cycle? What are its advantages and disadvantages?

26. What is the steam turbine reheat factor?

27. What does a turbine's stage efficiency indicate?

28. Explain what a condition curve describes.

29. Why are not all geographic locations possibilities for a geothermal power plant?

30. Discuss the flashed steam geothermal power cycle.

31. What is terminal temperature difference (TTD) as applied to heat exchangers?

32. What does steam generator efficiency indicate?

33. What is a cascade heater drain system?

PROBLEMS (SI)

15.1 A Carnot cycle uses steam as the working substance and operates between pressures of 7.0 MPa and 7 kPa. Determine (a) the thermal efficiency; (b) the turbine work per kg; (c) the compressor work per kg.

15.2 The maximum steam temperature is 560°C and the lowest cycle temperature is 30°C. Compare the thermal efficiencies of Rankine and Carnot cycles operating with a maximum pressure of (a) 3.5 MPa; (b) 7.0 MPa.

15.3 In an ideal Rankine cycle steam enters the turbine at 8 MPa and 500°C. The condenser pressure is 7.5 kPa. The net power produced by the cycle is 100 MW. Determine (a) the heat transfer to the water in the steam generator; (b) the cycle thermal efficiency; (c) the mass flow rate of cooling water required if it enters the condenser at 20°C and leaves at 35°C.

15.4 Recalculate Problem 15.3 with turbine and pump efficiencies of 85% and 80%, respectively.

15.5 A Rankine cycle produces 100 MW of power with a condenser pressure of 7.5 kPa and an inlet turbine temperature of 500°C. Determine the cycle thermal efficiency and the steam flow rate required for an inlet turbine pressure of (a) 17.5 MPa; (b) 1750 kPa.

15.6 A steam power plant operates on the Rankine cycle, with steam leaving the boiler at 6.0 MPa and 500°C. The plant supervisor wants to know the effect of back pressure (condensing pressure) on the cycle efficiency. (a) Determine and graph the cycle efficiency for back pressures of 7, 15, 30, 45, and 70 kPa. (b) Discuss the results in terms of available energy.

15.7 A Rankine cycle is characterized by turbine inlet conditions of 10 MPa and 500°C. The condenser pressure is 5 kPa. The heat transfer to the steam in the boiler occurs at the rate of 7.0×10^6 kW. The cooling water in the condenser increases in temperature from 20° to 30°C. Determine (a) the net power produced; (b) the cooling-water flow rate in kg/s; (c) the cycle thermal efficiency.

15.8 Recalculate Problem 15.7 for boiler pressures of 30 000 kPa and 1000 kPa, all other conditions being the same. Compare the results to the original problem.

15.9 Concentrating solar collectors are used to provide the heat source for a Rankine cycle using water as the working substance. The design specifications require a power output of 10 MW. Commercially available collectors allow steam to be generated at 2500 kPa and 300°C. The cycle low pressure is assumed to be 7.5 kPa. Determine (a) the cycle thermal efficiency; (b) the steam flow rate.

15.10 A Rankine cycle produces 10 kW of net power, operates with R 12 as the working fluid, and uses solar collectors as the source for heat addition. The collectors receive solar radiation at a rate of 0.4 kW/m² of collector surface area. The turbine receives saturated vapor at 2000 kPa, and the condenser operates at 650 kPa. Determine the minimum area of collector surface required.

15.11 It is possible to construct a Rankine-cycle power plant using warm water near the ocean's surface as a heat source and cold water from the ocean's depth as a heat sink. For a particular plant operating with this configuration, the working substance is ammonia, the surface water is 27°C, and the cold water is 6°C. The ammonia is a saturated vapor entering the turbine and is 2°C less than the seawater temperature entering the turbine and 2°C more than the seawater temperature leaving the turbine. The cycle is to produce 50 MW of power. Determine (a) the cycle thermal efficiency; (b) the cycle efficiency when considering the power requirements of pumps to move the seawater, which amount to 15 MW.

15.12 An adiabatic steam turbine receives steam at 15 MPa and expands it to a condenser pressure of 7.5 kPa. The turbine isentropic efficiency is 85%. Determine the turbine inlet steam temperature such that the steam quality leaving the turbine is 92%.

15.13 A Rankine-cycle power plant produces 100 MW of power and is characterized by a steam generator outlet condition of 10 MPa and 500°C and a condenser pressure of 7.5 kPa. The condensate leaving the condenser is subcooled by 3.3°C. Because of frictional and nonadiabatic effects in the piping leading to the turbine, the turbine inlet conditions are 9.75 MPa and 475°C. The pump discharge pressure is 10.5 MPa, and the turbine and pump internal efficiencies are 85%. Determine (a) the cycle thermal efficiency; (b) the steam flow rate required; (c) the heat transfer from the steam pipe connecting the boiler and turbine.

15.14 The steam turbine in a Rankine cycle receives steam at 5 MPa and 450°C and expands it to 10 kPa and a quality of 95%. Heat loss from the turbine amounts to 100 kJ/kg of steam passing through the turbine. Determine (a) the turbine isentropic efficiency; (b) the irreversibility per unit mass if the turbine casing temperature is 200°C and $T_0 = 300°$K.

15.15 A Rankine-cycle power plant is to be designed with a maximum turbine inlet temperature of 525°C and a minimum condenser pressure of 5 kPa. The manufacturer guarantees a turbine isentropic efficiency of 85% and a pump efficiency of 80%. The manufacturer will make this guarantee for the turbine only if the exit steam condition from the turbine is 92% or greater. Determine the steam generator pressure that allows this.

15.16 A steam turbine in a Rankine cycle receives steam at 6.0 MPa and 480°C and exhausts at 15 kPa. At reduced loads, the governor valve throttles the steam entering the turbine to 2.2 MPa. Calculate the loss of availability per kg entering the steam turbine. ($T_0 = 300°$K)

15.17 Find the decrease in cycle efficiency for Problem 15.16.

15.18 A new power plant is to operate on the Rankine cycle. The following conditions have been established: maximum steam pressure of 9.1 MPa, maximum steam temperature of 460°C, maximum moisture in the exhaust steam of 15%, maximum condenser vacuum of 722 mm Hg (vacuum), and a turbine efficiency of 88%. Determine the optimum inlet temperature and pressure.

15.19 A Rankine cycle operates with the stage of reheat occurring at 20% of the boiler pressure. The boiler produces steam at 15 MPa and 500°C and reheats it to 450°C. Determine the cycle efficiency if the condenser pressure is 10 kPa.

15.20 A supercritical power plant generates steam at 25 MPa and 580°C. The condenser pressure is 7.0 kPa. Determine (a) the exit quality of the steam if it expands through a turbine in this power plant; (b) the cycle efficiency.

15.21 The supercritical power plant in Problem 15.20 has a reheat stage added at 3.5 MPa with a reheat temperature of 540°C. All other conditions are the same. Determine (a) the quality of steam entering the condenser; (b) the cycle efficiency.

15.22 A small chemical plant uses steam in its production area. A decision was made to use a reheat cycle with steam entering the turbine at 8.5 MPa and 480°C, being reheated to 440°C at 1.2 MPa, and condensing at 7 kPa. Determine (a) the net work per kg; (b) what percentage of the total heat supplied is constituted by the reheat; (c) the cycle efficiency; (d) the T-s diagram.

15.23 In a reheat Rankine cycle, steam first enters the turbine at 10 MPa and 600°C. It expands to an unknown pressure and is extracted and reheated to 500°C. Steam reenters the turbine and expands to 7.5 kPa with a quality of 92%. Determine (a) the reheat pressure; (b) the cycle thermal efficiency.

15.24 A reheat Rankine cycle is characterized by turbine inlet conditions of 10 MPa and 500°C. The steam expands until it is a saturated vapor and is reheated to 450°C. It reenters the turbine and expands to a condenser pressure of 5 kPa. Heat is transferred to the steam in the boiler at 7.0×10^6 kW. The cooling water in the condenser increases in temperature from 20° to 30°C. Determine (a) the net power produced; (b) the cooling-water flow rate in kg/s; (c) the heat transfer to the steam in the reheat process; (d) the cycle thermal efficiency.

15.25 A supercritical reheat Rankine cycle has three stages of reheat. The steam state entering the high-pressure turbine is 30 MPa and 550°C. The steam expands to 5 MPa and is reheated to 450°C, reenters and expands to 1000 kPa, and is reheated to 400°C. It reenters the turbine and exhausts at 7.5 kPa. Determine (a) the quality or degrees of superheat of the steam entering the condenser; (b) the net work; (c) the cycle thermal efficiency.

15.26 In the conceptual design stages of a power plant, consideration is given to a steam generator operating at 30 MPa and a maximum temperature of 550°C. The condenser pressure is 5 kPa. Should reheat be used in the cycle, and if so, how many stages of reheating would be needed? The steam leaving the turbine should not be superheated.

15.27 A regenerative Rankine cycle operates with one closed feedwater heater. The condensate from the heater passes through a steam trap and enters the condenser. The turbine inlet steam conditions are 10 MPa, 550°C, and 50 kg/s. The steam expands isentropically to 750 kPa, where extraction occurs for feedwater heating. The remaining steam expands to 7.5 kPa. Determine (a) the cycle thermal efficiency; (b) the mass flow rate of steam to the open feedwater heater; (c) the net power produced.

15.28 Recalculate Problem 15.27, this time including a turbine isentropic efficiency of 85% and a pump efficiency of 80%.

15.29 In some cycles drain pumps receive the condensate from the shell side of the closed feedwater heater and pump it back into the outlet piping from the tube side of the heater. Consider the following situation: Extraction steam enters the shell side of a heater at 1500 kPa and 300°C and condenses to a saturated liquid. The feedwater enters the tube side at 10 MPa and 100°C and leaves at 10 MPa and 250°C. A drain pump with an isentropic efficiency of 80% returns the condensate to the discharge line from the heater. Determine the temperature of the water in the discharge line after it receives the drain pump return.

15.30 A Rankine cycle has two stages of regeneration with a boiler pressure and temperature of 10 MPa and 500°C and a condenser pressure of 15 kPa. Determine (a) the optimum extraction pressures; (b) the cycle thermal efficiency.

15.31 A two-stage regenerative-cycle steam power plant has been selected to power a ship. The requirements are that 37 000 kW be provided. The gears between the turbine (high speed) and shaft (low speed) are 98% efficient. The maximum possible steam pressure and temperature leaving the boiler are 6.0 MPa and 500°C. The condenser has a design vacuum of 722 mm Hg (vacuum). The extraction points are at 150 kPa and 1.4 MPa. Determine (a) the T-s diagram; (b) the cycle efficiency; (c) the minimum steam flow rate; (d) the fractions of steam removed for feed heating; (e) the quality of steam entering the condenser.

15.32 The same as Problem 15.31 except the turbine and pump have internal efficiencies of 90%.

15.33 A steam power plant produces 1000 MW of electricity while operating on a three-stage regenerative cycle. The steam enters the turbine at 14 MPa and 580°C. Extractions for heating occur at 2.5 MPa, 700 kPa, and 150 kPa. The turbine exhausts at 15 kPa and has an internal efficiency of 92%. Determine (a) the T-s diagram; (b) the mass flow rate; (c) the heat supplied; (d) the fuel flow rate, if the energy release is 35 000 kJ/kg fuel; (e) the mass fractions y_1, y_2, y_3; (f) the cycle efficiency.

15.34 A 60 000-kW turbogenerator receives steam at 7.0 MPa and 500°C. There are two steam-extraction stages at 2.0 MPa and 200 kPa. The remainder of the steam at

2.0 MPa is reheated to 480°C. The turbine exhausts at 36 mm Hg (absolute). Determine (a) the *T-s* diagram; (b) the mass fractions extracted; (c) the cycle efficiency; (d) the mass flow rate of steam entering the turbine.

15.35 A project engineer wants to determine the turbine internal efficiency of a two-turbine unit operating in a two-stage regenerative steam power plant. He measures the pressure and temperature at the high-pressure turbine inlet and the two regenerative bleed steam stations. The pressures and temperatures are 14 MPa and 520°C, 4.0 MPa and 360°C, and 200 kPa saturated, respectively. The low-pressure turbine receives the nonextracted steam at 4.0 MPa and 360°C. The condenser is at 36°C. Determine (a) the low-pressure and high-pressure turbine internal efficiencies; (b) the quality of steam entering the condenser; (c) the total turbine work per kg; (d) the *T-s* diagram; (e) the cycle second-law efficiency.

15.36 An existing steam power plant has the following turbine test data: inlet pressure = 2.1 MPa, inlet temperature = 260°C, exhaust pressure = 140 kPa, mechanical efficiency = 95%, steam flow = 226.0 kg/s, condensate temperature = 70°C, and turbine power = 1×10^5 kW. The old boiler failed and is to be replaced by a new boiler having exit steam conditions of 4.9 MPa and 320°C. This unit will drive an additional turbine with an internal efficiency of 90%, which exhausts at 2.1 MPa. Part of the exhaust is reheated to the test conditions above, including flow rate, and the remainder heats feedwater to 205°C. Determine (a) total plant power; (b) new total steam flow rate in kg/s; (c) old turbine internal efficiency; (d) overall thermal efficiency.

15.37 A regenerative Rankine cycle, producing 250 MW, has two feedwater heaters, a closed one for the first turbine extraction and an open one for the second turbine extraction. Steam enters the turbine at 7.5 MPa and 500°C and expands to 1500 kPa, where the first extraction stage occurs. The remaining steam expands to 500 kPa, where the second extraction stage occurs. The remainder expands through the turbine and exhausts at 7.5 kPa. The closed feedwater heater drains through a trap to the open heater. Determine (a) the cycle thermal efficiency; (b) the steam flow rate entering the turbine; (c) the steam flow rate to each of the heaters.

15.38 Recalculate Problem 15.37, this time including a turbine isentropic efficiency of 88% and a pump efficiency of 80%.

15.39 A reheat-regenerative Rankine cycle, producing 250 MW, has two feedwater heaters, a closed one for the first turbine extraction and an open one for the second turbine extraction. When a fraction steam is extracted for the first stage of feed heating, the remainder is reheated to 450°C. Steam enters the turbine at 7.5 MPa and 500°C and expands to 1500 kPa, where the first extraction stage occurs. The remaining steam expands to 500 kPa, where the second extraction stage occurs. The remainder expands through the turbine and exhausts at 7.5 kPa. The closed feedwater heater drains through a trap to the open heater. Determine: (a) the cycle thermal efficiency; (b) the steam flow rate entering the turbine; (c) the steam flow rate to each of the heaters.

15.40 Recalculate Problem 15.39, including a turbine isentropic efficiency of 88% and a pump efficiency of 80%.

15.41 A reheat-regenerative Rankine cycle uses steam at 8.4 MPa and 560°C entering the high-pressure turbine. The cycle includes one steam-extraction stage for regenerative feedwater heating, the remainder at this point being reheated to 540°C. The condenser temperature is 35°C. Determine (a) the *T-s* diagram for the cycle; (b) optimum extraction pressure; (c) fraction of steam extracted; (d) turbine work in kJ/kg; (e) pump work in kJ/kg; (f) overall thermal efficiency.

15.42 A steam power plant operates on a reheat-regenerative cycle where there is one regenerative state and one reheat stage that occur at the same pressure. The steam enters the turbine at 8.5 MPa and 540°C and expands isentropically to 2.2 MPa, where steam is extracted for feed heating; the remainder is reheated to 500°C. The steam exiting the turbine is at 7 kPa and 240 m/s. There is a 10°C temperature rise in circulating water temperature through the condenser. The water enters the condenser at 20°C. The power plant produces 60 000 kW. Determine (a) the steam flow rate; (b) the cooling-water flow rate; (c) the change of available energy in the condenser; (d) the T-s diagram.

15.43 A turbine operates in a reheat-regenerative cycle with one reheat stage and two stages of regenerative heating. The heater nearest the condenser is open, while the second heater is closed (shell-and-tube) with its steam drains flowing back to the previous heater. Steam is generated at 4.9 MPa and reaches the turbine at 4.6 MPa and 390°C. Reheating occurs at 760 kPa to a temperature of 370°C. Extractions for feed heating occur at 760 kPa and 100 kPa. The condenser pressure is 7.0 kPa. In the shell-and-tube heater the temperature difference between the condensed steam and the feedwater leaving is 5°C. Determine, per kg of steam (a) the T-s diagram; (b) the initial temperature of steam leaving the boiler (the process is adiabatic); (c) the temperature of the feedwater leaving the last heater; (d) the mass fractions of steam extracted for feed heating; (e) the quality of the steam entering the condenser for a turbine efficiency of 80%.

15.44 A manufacturing facility uses a steam power plant to generate electricity and provide heating steam to the work spaces. The steam plant operates on the regenerative cycle, with one stage of feed heating. Steam enters the turbine at 5.5 MPa and 500°C, expands to 660 kPa, where extraction for feedwater heating occurs. The remaining steam expands through the turbine to 150 kPa, where extraction for space heating occurs ($y_2 = y_1$). The remaining steam expands to 15 kPa, where it is condensed. The returns from the space heating enter the condenser as a saturated liquid at the condenser temperature. Determine (a) the utilization factor; (b) the T-s diagram; (c) the turbine work per kg; (d) for a 50 000-kW load, the mass flow of steam entering the turbine; (e) the heat supplied to the working areas for the conditions in (d).

15.45 A university decides to invest in a cogeneration facility, providing 20 MW of power and steam for heating in the campus buildings. A preliminary design proposal suggests that steam be generated at 2500 kPa and 300°C. In addition a turbine may be purchased that has two extraction stages, the first occurring at 1000 kPa for building heating and the second at 300 kPa for regenerative heating with an open feedwater heater. The condenser pressure is 7.5 kPa. The returns from the buildings may be considered to be saturated at the condenser temperature. The buildings require 3000 kW of heat for the worst-case condition. The turbine's isentropic efficiency is 80% at these steam conditions. The pump efficiency is assumed to be 100%. Determine (a) the steam generator capacity in kg/s of steam produced and in the heat rate required; (b) the mass flow rate of steam extracted for building heating; (c) the cycle's utilization factor.

15.46 A cogeneration system consists of a steam generator that produces 30 kg/s of steam at 7.5 MPa and 450°C, a turbine with extraction stages at 1500 kPa for process heating and at 500 kPa for open regenerative heating, and a condenser operating at 7.5 kPa. The process heating requires 10 kg/s of steam. The returns to the condenser are at 60°C and atmospheric pressure. The turbine isentropic efficiency is 85%. Determine (a) the process heating load in kW; (b) the net power developed by the cycle; (c) the utilization factor.

15.47 The exhaust from a diesel engine is used to generate steam for process heating. The engine produces 10 MW, and the exhaust enters the steam generator at 450°C and exits at 150°C. Saturated steam is generated at 500 kPa with the returns, at 50°C and atmospheric pressure, being pumped back to the waste heat boiler. The exhaust flow rate from the diesel is 35 kg/s. Consider the exhaust to have properties similar to air. Determine (a) the steam flow rate; (b) the amount of process heat; (c) the utilization factor.

15.48 A power plant uses cogeneration and a bottoming cycle to raise the plant's utilization factor. From the steam generator 10 kg/s of steam enters the steam turbine at 5 MPa and 450°C and expands to 200 kPa. The steam exits the turbine, one-half going to process heating and the remainder to a heat exchanger, acting as the boiler in an R 12 bottoming cycle. The refrigerant leaves the boiler at 2 MPa and 100°C, enters an R 12 turbine, and expands to 750 kPa when it enters the condenser. A pump returns the R 12 to the boiler. The condensate returns from the process heating, at 1 atm and 60°C, mixes with the condensate from boiler and is pumped into the steam generator. Determine (a) the rate of heat transfer in the steam generator; (b) the process heat rate; (c) the net power produced; (d) the utilization factor.

15.49 A steam generator may be considered to be a constant-pressure combustion chamber followed by a heat exchanger where the heat from the combustion gases is transferred to water, creating steam. Consider such a steam generator where the combustion gases, with properties similar to air, enter the heat exchanger at 1500°K and are cooled to 500°K. Twenty-five kg/s of water enters the heat exchanger at 10 MPa and 175°C and leaves as a superheated vapor at 10 MPa and 500°C. $T_0 = 300$°K and $p_0 = 100$ kPa. Determine (a) the availability change of the combustion gas in kW; (b) the availability change of the water in kW; (c) the irreversibility rate in kW; (d) the second-law efficiency.

15.50 A processing plant requires 7200 kg/h of saturated steam or slightly superheated steam at 0.4 MPa, which is extracted from a steam turbine. The turbine receives 5 kg/s of steam at 7.5 MPa and 450°C. The saturated-liquid condensate from the process heater is pumped back into the discharge line of the feed pump. The condenser pressure is 10 kPa. Determine (a) the net power produced; (b) the utilization factor; (c) the enthalpy of the water entering the steam generator.

15.51 A combined gas turbine–steam power plant has a compression ratio of 15 in the gas turbine portion. Fifty kg/s of air enters the compressor at 300°K and 100 kPa. The maximum allowable temperature is 1700°K, and the gases leave the steam generator at 450°K. Steam is produced at 7.5 MPa and 400°C, enters the turbine, and is condensed at 10 kPa. Assume the gases have properties similar to air. Determine (a) the steam mass flow rate; (b) the net power produced; (c) the overall thermal efficiency.

15.52 Recalculate problem 15.51 using a compressor isentropic efficiency of 82%, steam and gas turbine isentropic efficiencies of 85%, and a pump efficiency of 80%.

15.53 A combined gas turbine–steam power plant produces 500 MW of net power. The pressure ratio of the gas turbine unit is 16, with air entering at 300°K and 100 kPa. The maximum inlet temperature to the turbine is 1750°K. The minimum gas temperature from the steam generator is 450°K. Steam is generated at 7.5 MPa and 450°C. The turbine has one open feedwater heater regenerative stage at 500 kPa. The condenser pressure is 10 kPa. Assume the gases have properties similar to air. Determine (a) the air and steam mass flow rates; (b) the cycle efficiency; (c) the availability of the gas leaving the steam generator; (d) the availability of the gas leaving the gas turbine relative to inlet air temperature and pressure. What fraction of this was used in the steam cycle?

15.54 In a binary vapor power cycle, using steam and mercury, saturated mercury vapor enters the mercury turbine at 1250 kPa ($h_g = 361.2$ kJ/kg, $s_g = 0.5024$ kJ/kg-K) and exhausts at 14 kPa (h_f = 36.0, h_{fg} = 294.2, s_f = 0.0923, s_{fg} = 0.5491). The condenser–steam generator produces a saturated steam at 3.8 MPa. The steam turbine exhausts at 7 kPa. Determine (a) the mercury turbine work per kg mercury; (b) the steam turbine work per kg steam; (c) the mercury flow rate if the steam flow rate is 25.0 kg/s; (d) the thermal efficiency for the total cycle.

15.55 An 8-stage impulse turbine receives steam at 7.0 MPa and 550°C. The stage efficiency is 80%, and the isentropic enthalpy drop per stage is 160 kJ/kg. Determine (a) the exit pressure; (b) the exit quality; (c) the reheat factor; (d) the condition curve; (e) the second-law efficiency.

15.56 The turbine in Problem 15.55 operates under different steam conditions. The steam enters at 5.5 MPa and 480°C and exhausts at 14 kPa. Determine the turbine work per kg.

15.57 Steam enters a turbine at 1.4 MPa and 320°C. The turbine internal efficiency is 70%, and the load requirement is 800 kW. The exhaust is to the back pressure system, maintained at 175 kPa. Find the steam flow rate.

15.58 A turboelectric plant has steam enter the high-pressure turbine at 16.5 MPa and 570°C. The steam expands to 3.5 MPa, where the extraction occurs, and the remainder is reheated to 540°C. The steam enters the low-pressure turbine and exhausts at 14 kPa. Additional regenerative feedwater heating occurs at 1.2 MPa. The feedwater and condensate leave each heater at the saturated steam temperature. The condensate from the condenser is subcooled to 5°C. The turbine internal efficiency is 80%, and the generator efficiency is 96%. Determine (a) steam flow rate entering the turbine if the generator load is 25 MW; (b) overall thermal efficiency; (c) heat rejected; (d) heat added; (e) the cycle second-law efficiency.

15.59 A supercritical power plant produces 800 MW when operating on a reheat-regenerative cycle with three stages of regenerative heating and one state of reheating. The extraction pressures for feedwater heating are 5.5 MPa, 600 kPa, and 100 kPa. Reheating occurs at 5.5 MPa. The inlet steam conditions to the turbine are 30 MPa and 540°C, and the reheat temperature is 520°C. The turbine exhaust pressure is 7 kPa. Determine (a) the T-s diagram; (b) the cycle efficiency; (c) the steam flow rate; (d) the circulating-water flow rate if there is a 10°C temperature rise in the condenser from 20°C; (e) the total power required for pumping.

15.60 A steam turbine carrying a full load of 50.0 MW uses 71.7 kg/s of steam. The turbine efficiency is 75%, and it exhausts steam at 25 mm Hg (absolute) with an enthalpy of 2210 kJ/kg. What are the temperature and pressure of the steam entering the turbine?

15.61 Steam is admitted to the cylinder of an engine in such a manner that the average pressure is 840 kPa. The diameter of the piston is 25.4 cm, and the length of the stroke is 30.5 cm. (a) Determine the work that can be done during one revolution, assuming that the steam is admitted successively to each side (top and bottom) of the piston. (b) What is the power produced when the engine is running at 300 rpm?

15.62 A flashed steam geothermal power plant is located where underground hot water is available as a saturated liquid at 700 kPa. The wellhead pressure is 600 kPa. The flashed steam enters a turbine at 500 kPa and expands to 13 kPa, when it is condensed. The flow rate from the well is 29.6 kg/s. T_0 = 27°C. Determine (a) the mass flow rate of steam; (b) the power produced; (c) the cooling-water flow if water is available at 30°C and a 10°C rise is allowed through the condenser; (d) the efficiency, where energy input is the available energy of the geothermal water.

15.63 The same as Problem 15.62 except now the turbine has an internal isentropic efficiency of 80%. In addition to determining (a)–(d), determine the cycle second-law efficiency.

15.64 The earth's temperature has been found to increase 4°C for every 30 m over a geothermal revservoir. A well is drilled to a depth of 1000 m. There is a 75-kPa pressure drop from the well to the turbine inlet, including the pressure loss in the separator. The condenser pressure is 14 kPa. Determine (a) the inlet pressure; (b) the work per kg of wellhead steam–water mixture; (c) the unit's efficiency (defined in Problem 15.62) if $T_0 = 27°C$.

15.65 A combined gas turbine–steam power plant is to be used for the generation of electric power. The combined unit must produce 600 MW. There are two stages for the compressor with ideal intercooling at the optimum interstage pressure and two stages for the turbine with reheating to the same turbine inlet temperature. The compressor unit receives air at 100 kPa and 290°K and operates with a pressure ratio of 9. The turbine inlet temperature is 1220°K, with reheating ocurring at 340 kPa. The turbine exhausts to the steam generator, and the products of combustion are cooled to 150°C. The steam generator produces steam at 5.5 MPa and 450°C. The steam turbine exhausts at 13 kPa. All expansion and compression processes are isentropic. Determine (a) the net gas turbine work per kg air; (b) the net steam turbine work per kg steam; (c) the overall thermal efficiency; (d) the airflow required; (e) the fuel/air ratio if $h_{RP} = 43\ 200$ kJ/kg fuel; (f) the fuel flow rate; (g) the cost in dollars per kWh of electricity produced if the fuel costs \$0.45/kg; (h) the second-law efficiency.

15.66 A plant modification to an existing 20 000-kW gas turbine unit is proposed. The existing unit has the following test data: inlet air = 300°K, 100 kPa; pressure ratio = 8; compressor efficiency = 90%; fuel/air ratio = 0.0165 kg fuel/kg air; turbine efficiency = 90%; and heating value of fuel = 43 950 kJ/kg. The proposal suggests that the addition of a steam generator using the energy of the turbine exhaust will provide significant increased power output and raise the overall plant thermal efficiency. The steam leaves the generator at 4.8 MPa and 370°C and enters the turbine. The expansion process is adiabatic with a turbine efficiency of 90% and an exhaust pressure of 13 kPa. The gas turbine products of combustion must leave the steam generator at 160°C to prevent condensation on the tube and stack surfaces. Determine (a) the available energy in the gas turbine exhaust entering the steam generator ($T_0 = 300°K$); (b) the overall unit thermal efficiency, old and proposed; (c) the total power output under proposed conditions; (d) the percentage of the available energy in (a) that was used.

PROBLEMS (English Units)

*15.1 In a Rankine cycle, steam enters the turbine at 800 psia and 800°F, which exhausts at 1 psia. Show the cycle on a T-s diagram and find (a) the quality of the steam entering the condenser; (b) the turbine work in Btu/lbm; (c) the pump work in Btu/lbm; (d) the heat supplied in Btu/lbm; (e) the heat rejected in Btu/lbm; (f) the net work of the cycle in Btu/lbm; (g) the thermal efficiency of the cycle.

*15.2 In an ideal Rankine cycle steam enters the turbine at 1000 psia and 1000°F. The condenser pressure is 1 psia. The net power produced by the cycle is 100 MW. Determine (a) the heat transfer to the water in the steam generator; (b) the cycle thermal efficiency; (c) the mass flow rate of cooling water required if it enters the condenser at 70°F and leaves at 90°F.

*15.3 Recalculate Problem *15.2 with turbine and pump efficiencies of 85% and 80%, respectively.

*15.4 A Rankine cycle produces 100 MW of power with a condenser pressure of 1 psia and an inlet turbine temperature of 1000°F. Determine the cycle thermal efficiency and the steam flow rate required for an inlet turbine pressure of (a) 2000 psia; (b) 200 psia.

*15.5 A Rankine cycle is characterized by turbine inlet conditions of 1500 psia and 1000°F. The condenser pressure is 1 psia. The heat transfer to the steam in the boiler occurs at the rate of 7.0×10^6 Btu/sec. The cooling water in the condenser increases in temperature from 70° to 85°F. Determine (a) the net power produced; (b) the cooling-water flow rate in gal/min; (c) the cycle thermal efficiency.

*15.6 Recalculate Problem *15.5 for boiler pressures of 4000 psia and 150 psia, all other conditions being the same. Compare the results to the original problem.

*15.7 Concentrating solar collectors are used to provide the heat source for a Rankine cycle using water as the working substance. The design specifications require a power output of 13,500 hp. Commercially available collectors allow steam to be generated at 400 psia and 600°F. The cycle low pressure is assumed to be 1 psia. Determine (a) the cycle thermal efficiency; (b) the steam flow rate.

*15.8 A Rankine cycle produces 13.5 hp of net power, operates with R 12 as the working fluid, and uses solar collectors as the source for heat addition. The collectors receive solar radiation at a rate of 2.5 Btu/min-ft² of collector surface area. The turbine receives saturated vapor at 300 psia, and the condenser operates at 100 psia. Determine the minimum area of collector surface required.

*15.9 It is possible to construct a Rankine-cycle power plant using warm water near the ocean's surface as a heat source and cold water from the ocean's depth as a heat sink. For a particular plant operating with this configuration, the working substance is ammonia, the surface water is 80°F, and the cold water is 40°F. The ammonia is a saturated vapor entering the turbine and is 3°F less than the seawater temperature entering the heat exchanger and 3°F more than the seawater temperature leaving the turbine. The cycle is to produce 50 MW of power. Determine (a) the cycle thermal efficiency; (b) the cycle efficiency when considering the power requirements of pumps to move the seawater, which amount to 15 MW.

*15.10 An adiabatic steam turbine receives steam at 2000 psia and expands it to a condenser pressure of 1 psia. The turbine isentropic efficiency is 85%. Determine the turbine inlet steam temperature such that the steam quality leaving the turbine is 92%.

*15.11 A Rankine-cycle power plant produces 100 MW of power and is characterized by a steam generator outlet condition of 1500 psia and 1000°F and a condenser pressure of 1 psia. The condensate leaving the condenser is subcooled by 5°F. Because of frictional and nonadiabatic effects in the piping leading to the turbine, the turbine inlet conditions are 1400 psia and 950°F. The pump discharge pressure is 1650 psia, and the turbine and pump internal efficiencies are 85%. Determine (a) the cycle thermal efficiency; (b) the steam flow rate required; (c) the heat transfer from the steam pipe connecting the boiler and turbine.

*15.12 The steam turbine in a Rankine cycle receives steam at 800 psia and 900°F and expands it to 2 psia and a quality of 95%. Heat loss from the turbine amounts to 43 Btu/lbm of steam passing through the turbine. Determine (a) the turbine isentropic efficiency; (b) the irreversibility per unit mass if the turbine casing temperature is 400°F and $T_0 = 77$°F.

*15.13 A Rankine-cycle power plant is to be designed with a maximum turbine inlet temperature of 1050°F and a minimum condenser pressure of 1 psia. The manufacturer guarantees a turbine isentropic efficiency of 85% and a pump efficiency of 80%. The manufacturer will make this guarantee for the turbine only if the exit steam condition from the turbine is 92% or greater. Determine the steam generator pressure that allows this.

*15.14 In a reheat Rankine cycle, steam first enters the turbine at 1500 psia and 1100°F. It expands to an unknown pressure and is extracted and reheated to 1000°F. Steam reenters the turbine and expands to 1 psia with a quality of 92%. Determine (a) the reheat pressure; (b) the cycle thermal efficiency.

*15.15 A reheat Rankine cycle is characterized by turbine inlet conditions of 1500 psia and 1000°F. The steam expands until it is a saturated vapor and is reheated to 900°F. It reenters the turbine and expands to a condenser pressure of 28.5 in. Hg vacuum. Heat is transferred to the steam in the boiler at 7.0×10^6 Btu/sec. The cooling water in the condenser increases in temperature from 70° to 85°F. Determine (a) the net power produced; (b) the cooling-water flow rate in lbm/sec; (c) the heat transfer to the steam in the reheat process; (d) the cycle thermal efficiency.

*15.16 A supercritical reheat Rankine cycle has three stages of reheat. The steam state entering the high-pressure turbine is 4000 psia and 1100°F. The steam expands to 800 psia and is reheated to 900°F, reenters and expands to 150 psia, and is reheated to 800°F. It reenters the turbine and exhausts at 1 psia. Determine (a) the quality or degrees of superheat of the steam entering the condenser; (b) the net work; (c) the cycle thermal efficiency.

*15.17 In the conceptual design stages of a power plant, consideration is given to a steam generator operating at 4000 psia and a maximum temperature of 1100°F. The condenser pressure is 0.7 psia. Should reheat be used in the cycle, and if so, how many stages of reheating would be needed? The steam leaving the turbine should not be superheated.

*15.18 A 30,000-kW turbogenerator operates in a reheat cycle. It receives steam at 600 psia and 700°F, and extraction for reheating occurs at 120 psia and 400°F. Returning to the turbine at 100 psia and 600°F, the steam completes the expansion to 1 psia. The combined steam rate is 10.35 lbm/kWh, and the efficiency of the generator is 95%. Determine (a) the thermal efficiency; (b) the enthalpy and quality (or temperature) of the actual exhaust; (c) steam consumption per hr at the rated load.

*15.19 In an ideal reheat cycle, steam enters the high-pressure turbine at 800 psia and 800°F, leaves at 100 psia, is reheated to 800°F, passes through the low-pressure turbine, and exhausts to a condenser at 1 psia. Determine (a) the moisture content of the steam entering the condenser; (b) the pump work in Btu/lbm; (c) the work of the high-pressure turbine; (d) the total turbine work; (e) the heat supplied; (f) the heat rejected; (g) the net work of the cycle; (h) the thermal efficiency of the cycle; (i) the *T-s* diagram.

*15.20 A regenerative Rankine cycle operates with one closed feedwater heater. The condensate from the heater passes through a steam trap and enters the condenser. The turbine inlet steam conditions are 1500 psia, 1100°F, and 3.6×10^5 lbm/hr. The steam expands isentropically to 100 psia, where extraction occurs for feedwater heating. The remaining steam expands to 1 psia. Determine (a) the cycle thermal efficiency; (b) the mass flow rate of steam to the feedwater heater; (c) the net power produced.

*15.21 Recalculate Problem *15.20, this time including a turbine isentropic efficiency of 85% and a pump efficiency of 80%.

*15.22 In some cycles drain pumps receive the condensate from the shell side of the closed feedwater heater and pump it back into the outlet piping from the tube side of the heater. Consider the following situation: Extraction steam enters the shell side of a heater at 200 psia and 450°F and condenses to a saturated liquid. The feedwater enters the tube side at 1500 psia and 200°F and leaves at 1500 psia and 350°F. A drain pump with an isentropic efficiency of 80% returns the condensate to the discharge line from the heater. Determine the temperature of the water in the discharge line after receiving the drain pump return.

*15.23 A regenerative Rankine cycle, producing 250 MW, has two feedwater heaters, a closed one for the first turbine extraction and an open one for the second turbine extraction. Steam enters the turbine at 1000 psia and 1000°F and expands to 200 psia, where the first extraction stage occurs. The remaining steam expands to 50 psia, where the second extraction stage occurs. The remainder expands through the turbine and exhausts at 1 psia. The closed feedwater heater drains through a trap to the open heater. Determine (a) the cycle thermal efficiency; (b) the steam flow rate entering the turbine; (c) the steam flow rate to each of the heaters.

*15.24 Recalculate Problem *15.23, this time including a turbine isentropic efficiency of 88% and a pump efficiency of 80%.

*15.25 A reheat-regenerative Rankine cycle, producing 250 MW, has two feedwater heaters, closed for the first turbine extraction and open for the second turbine extraction. When a fraction steam is extracted for the first stage of feedwater heating, the remainder is reheated to 900°F. Steam enters the turbine at 1000 psia and 1000°F and expands to 200 psia, where the first extraction stage occurs. The remaining steam expands to 50 psia, where the second extraction stage occurs. The remainder expands through the turbine and exhausts at 1 psia. The closed feedwater heater drains through a trap to the open heater. Determine (a) the cycle thermal efficiency; (b) the steam flow rate entering the turbine; (c) the steam flow rate to each of the heaters.

*15.26 Recalculate Problem *15.25, including a turbine isentropic efficiency of 88% and a pump efficiency of 80%.

*15.27 Steam is generated in an ideal reheat–regenerative cycle at 1500 psia and 1000°F. After expansion in the turbine to the optimum pressure for a single extraction, the steam is removed from the turbine, one portion to be used for feedwater heating and the remainder to be reheated to 900°F at the constant pressure. Condenser pressure is 1 psia. Determine (a) the extraction pressure; (b) the mass fraction of steam extracted; (c) the heat added to the cycle; (d) the heat rejected by the cycle; (e) the net cycle work; (f) the cycle thermal efficiency; (g) the T-s diagram.

*15.28 Steam is generated at 500 psia and 600°F. At the throttle of the turbine, the steam is at 475 psia and 580°F. A single extraction for feedwater heating occurs at 130 psia, and condensation occurs at 102°F. Determine the net work and thermal efficiency of the cycle. There is a small transfer of heat during the passage of the steam from the boiler to the turbine.

*15.29 In an ideal regenerative cycle, steam is generated at 400 psia and 700°F. Steam is extracted for feedwater heating at 110, 30, and 5.99 psia. Condensation occurs at 90°F. Determine (a) the mass fraction extracted at each point; (b) the net work; (c) the cycle efficiency.

*15.30 A turbogenerator, with extractions for feedwater heating at 180, 411.85, and 6.0 psia, delivers 75,000 kW when the rate of steam flow is 900,000 lbm/hr. At the throttle inlet, $p = 600$ psia and $T = 650°F$. At the exhaust, the pressure is 2.447 in. Hg absolute; the actual temperature of the feedwater entering the boiler is 360°F. The generator efficiency is 96%. Determine (a) the mass fractions; (b) the turbine efficiency; (c) the exit quality of steam.

*15.31 A turbogenerator receives steam at 500 psia and 600°F and discharges it at 1.0 psia. At 150 psia, the steam is withdrawn, and part of it is used for heating feedwater while the remainder passes through a reheater. The actual temperature of the steam as it leaves the turbine is 420°F. Upon reentering the turbine, the steam is at 140 psia and 600°F. At 20 psia and an actual temperature of 250°F, more steam is extracted. The actual temperatures of the water leaving the feedwater heaters are 225°F and 355°F. Determine (a) the steam mass fractions; (b) the cycle thermal efficiency.

*15.32 In a mercury-steam cycle, the saturated mercury vapor leaves the boiler at 60 psia ($T = 836.1°F$), and it is exhausted to the condenser-boiler at 2 in. Hg absolute. Saturated steam is generated in the condenser–boiler at 360 psia, and the steam turbine exhausts at 1 psia. Saturated liquid mercury at 2 in. Hg absolute is pumped into the boiler, and saturated water at 1 psia is pumped into the condenser–boiler. The properties of mercury at 60 psia are $h_f = 30$ Btu/lbm, $h_{fg} = 118.6$ Btu/lbm, $s_v = 0.1277$ Btu/lbm-R; at 2 in. Hg absolute, $h_f = 15.85$ Btu/lbm, $h_{fg} = 126.95$ Btu/lbm, $s_f = 0.02323$ Btu/lbm-R, $s_{fg} = 0.1385$ Btu/lbm-R. Determine (a) the net work of the mercury cycle; (b) the net work of the steam cycle; (c) lbm of mercury needed per lbm of steam; (d) the thermal efficiency of the binary vapor cycle.

*15.33 A university decides to invest in a cogeneration facility, providing 20 MW of power and steam for heating in the campus buildings. A preliminary design proposal suggests that steam be generated at 400 psia and 550°F. In addition a turbine may be purchased that has two extraction stages, the first occurring at 150 psia for building heating and the second at 50 psia for regenerative heating with an open feedwater heater. The condenser pressure is 1 psia. The returns from the buildings may be considered to be saturated at the condenser temperature. The buildings require 1.08×10^7 Btu/hr of heat for the worst-case condition. The turbine's isentropic efficiency is 80% at these steam conditions. The pump efficiency is assumed to be 100%. Determine (a) the steam generator capacity in lbm/sec of steam produced and in the heat rate required; (b) the mass flow rate of steam extracted for building heating; (c) the cycle's utilization factor.

*15.34 A cogeneration system consists of a steam generator that produces 60 lbm/sec of steam at 1000 psia and 900°F; a turbine with extraction stages at 200 psia for process heating and at 75 psia for open regenerative heating, and a condenser operating at 1 psia. The process heating requires 20 lbm/sec of steam. The returns to the condenser are at 140°F and atmospheric pressure. The turbine isentropic efficiency is 85%. Determine (a) the process heating load in Btu/hr; (b) the net power developed by the cycle; (c) the utilization factor.

*15.35 The exhaust from a diesel engine is used to generate steam for process heating. The engine produces 13,500 hp, and the exhaust enters the steam generator at 900°F and exits at 300°F. Saturated steam is generated at 75 psia with the returns, at 100°F and atmospheric pressure, being pumped back to the waste heat boiler. The exhaust flow rate from the diesel is 77 lbm/sec. Consider the exhaust to have properties similar to air. Determine (a) the steam flow rate; (b) the amount of process heat; (c) the utilization factor.

*15.36 A power plant uses cogeneration and a bottoming cycle to raise the plant's utilization factor. From the steam generator 79,200 lbm/hr of steam enters the steam turbine at

800 psia and 900°F and expands to 30 psia. The steam exits the turbine, one-half going to process heating and the remainder to a heat exchanger, acting as the boiler in an R 12 bottoming cycle. The refrigerant leaves the boiler at 300 psia and 200°F, enters an R 12 turbine, and expands to 100 psia when it enters the condenser. A pump returns the R 12 to the boiler. The condensate returns from the process heating, at 1 atm and 140°F, mixes with the condensate from boiler and is pumped into the steam generator. Determine (a) the rate of heat transfer in the steam generator; (b) the process heat rate; (c) the net power produced; (d) the utilization factor.

*15.37 A steam generator may be considered to be a constant-pressure combustion chamber followed by a heat exchanger where the heat from the combustion gases is transferred to water, creating steam. Consider such a steam generator where the combustion gases, with properties similar to air, enter the heat exchanger at 2700°R and are cooled to 900°R. At a rate of 180,000 lbm/hr water enters the heat exchanger at 1500 psia and 350°F and leaves as a superheated vapor at 1500 psia and 1000°F. $T_0 = 77$°F and $p_0 = 1$ atm. Determine (a) the availability change of the combustion gas in Btu/min; (b) the availability change of the water in Btu/min; (c) the irreversibility rate in Btu/min; (d) the second-law efficiency.

*15.38 A processing plant requires 16,000 lbm/hr of saturated steam or slightly super-heated steam at 60 psia, which is extracted from a steam turbine. The turbine receives 48,000 lbm/hr of steam at 1000 psia and 900°F. The saturated liquid condensate from the process heater is pumped back into the discharge line of the feed pump. The condenser pressure is 1 psia. Determine (a) the net power produced; (b) the utilization factor; (c) the enthalpy of the water entering the steam generator.

*15.39 A combined gas turbine–steam power plant has a compression ratio of 15 in the gas turbine portion. One hundred lbm/sec of air enters the compressor at 77°F and 1 atm. The maximum allowable temperature is 3000°R, and the gases leave the steam generator at 800°R. Steam is produced at 1000 psia and 800°F, enters the turbine, and is condensed at 2 psia. Assume the gases have properties similar to air. Determine (a) the steam mass flow rate; (b) the net power produced; (c) the overall thermal efficiency.

*15.40 Recalculate problem *15.39 using a compressor isentropic efficiency of 82%, steam and gas turbine isentropic efficiencies of 85%, and a pump efficiency of 80%.

*15.41 A combined gas turbine–steam power plant produces 500 MW of net power. The pressure ratio of the ideal-gas turbine unit is 16, with air entering at 77°F and 1 atm. The maximum inlet temperature to the turbine is 3100°R. The minimum gas temperature from the steam generator is 800°R. Steam is generated at 1000 psia and 900°F. The turbine has one open feedwater heater regenerative stage at 75 psia. The condenser pressure is 2 psia. Assume the gases have properties similar to air. Determine (a) the air and steam mass flow rates; (b) the cycle efficiency; (c) the availability of the gas leaving the steam generator; (d) the availability of the gas leaving the gas turbine relative to inlet air temperature and pressure. What fraction of this was used in the steam cycle?

*15.42 A combined gas turbine–steam power plant is to be used for the generation of electric power. The combined unit must produce 600 MW. There are two stages for the compressor with ideal intercooling at the optimum interstage pressure and two stages for the turbine with reheating to the same turbine inlet temperature. The compressor unit receives air at 14.6 psia and 530°R and operates with a pressure ratio of 9. The turbine inlet temperature is 2200°R, with reheating occurring at 50 psia. The turbine exhausts to the steam generator, and the products of combustion are cooled to 300°F.

The steam generator produces steam at 700 psia and 700°F. The steam turbine exhausts at 2 psia. All expansion and compression processes are adiabatic, with $\eta_c = 0.85$ and $\eta_t = 0.90$. Determine (a) the net gas turbine work per lbm of air; (b) the net steam turbine work per lbm of steam; (c) the overall thermal efficiency; (d) the airflow required; (e) the fuel/air ratio of $h_{RP} = 18{,}600$ Btu/lbm fuel; (f) the fuel flow rate; (g) the cost in dollars per kWh of electricity produced if the fuel costs \$0.07/lbm.

*15.43 A plant modification to an existing 40,000-kW gas turbine unit is proposed. The existing unit has the following test data: inlet air = 540°R, 14.7 psia; pressure ratio = 8; compressor efficiency = 90%; fuel/air ratio = 0.0165 lbm fuel/lbm air; turbine efficiency = 90%; and heating value of fuel = 18,900 Btu/lbm. The proposal suggests that the addition of a steam generator using the energy of the turbine exhaust will provide significantly increased power output and raise the overall plant thermal efficiency. The steam leaves the generator at 700 psia and 700°F and enters the turbine. The expansion process is adiabatic, with a turbine efficiency of 90% and an exhaust pressure of 2 psia. The gas turbine products of combustion must leave the steam generator at 320°F to prevent condensation on the tube and stack surfaces. Determine (a) the available energy in the gas turbine exhaust entering the steam generator ($T_0 = 100°F$); (b) the overall unit thermal efficiency, old and proposed; (c) the total power output under proposed conditions; (d) the percentage of the available energy in (a) that was used.

*15.44 Perform a heat balance for a four-stage reheat–regenerative cycle according to heat balance guidelines for pressure drops and equipment efficiency. Boiler pressure is 2000 psia and 1000°F; determine the optimum bleed pressures with the deaerating heater at the pressure nearest to but greater than 35 psia. The feed pump follows the deaerating heater. The condenser pressure is 1 psia, and the heater drains are cascaded. Reheating occurs at the first extraction. The plant produces 500 MW. Determine (a) the mass fractions; (b) the cycle efficiency; (c) the exit steam state from the turbine; (d) the oil required per day; (e) the dollars per kWh if oil costs 9.5¢/lbm.

*15.45 Simplify the reheat power plant shown in Figure 15.35 by eliminating all the heaters between the condensate pump and the first heater. Let the condenser operate at 90°F. Assume the high-pressure turbine exhausts at 550 psia. Calculate the mass flows if the power required is 236 MW. Use equipment efficiency guidelines as per actual heat balance discussion.

COMPUTER PROBLEMS

C15.1 Use STMCYCLE.TK or develop a spreadsheet template or computer program to investigate the effect of varying the steam generator pressure on the thermal efficiency of an ideal Rankine cycle. For a steam generator superheater outlet temperature of 500°C and a condensing temperature of 35°C, vary the steam generator pressure from 100 kPa to 30 MPa. Plot the results.

C15.2 Use STMCYCLE.TK or develop a spreadsheet template or computer program to investigate the effect of varying the steam generator superheater outlet temperature on the thermal efficiency of an ideal Rankine cycle. For a steam generator pressure of 7 MPa and a condensing temperature of 35°C, vary the superheater temperature from 300°C to 600°C. Plot the results.

C15.3 Use STMCYCLE.TK or develop a spreadsheet template or computer program to investigate the effect of varying the condensing temperature on the thermal efficiency of an ideal Rankine cycle. For steam generator outlet conditions of 7 MPa and 500°C, vary the condensing temperature from 25°C to 100°C. Plot the results.

C15.4 Modify STMCYCLE.TK to permit the analysis of an ideal regenerative Rankine cycle with a single open feedwater heater. For steam generator outlet conditions of 7 MPa and 550°C and a condensing temperature of 35°C, use the modified model to determine the bleed pressure that will result in the highest cycle thermal efficiency. Do the results confirm that the optimum heater location is where the feedwater exit temperature is halfway between the boiler saturated steam temperature and the condenser temperature?

C15.5 Modify STMCYCLE.TK to permit the analysis of an ideal reheat Rankine cycle with a single reheat stage. For steam generator outlet conditions of 7 MPa and 550°C and a condensing temperature of 35°C, use the modified model to determine the reheat pressure that will result in the highest cycle thermal efficiency.

C15.6 A 600-MW power plant has a reheat-regenerative cycle that operates with throttle conditions of 2500 psia and 1000°F; reheat at 500 psia to 1000°F; closed feedwater heaters with a TTD of 10°F at 500 psia, 225 psia, 25 psia, and 7 psia; an open feedwater heater at 90 psia; and turbine exhaust at 1 psia. The drain from each closed heater is routed to the next lower heater through a trap, except that the drain from the 7-psia heater is pumped to the open heater. The steam generator efficiency is 90%, and the turbine-generator efficiency is 85%. Oil with a higher heating value of 18,000 Btu/lbm is burned. Determine (a) the turbine-generator inlet steam flow; (b) the fuel oil consumed per hr.

C15.7 The exhaust pressure for an ideal steam turbine is 1 psia. The throttle temperature is varied from 600°F to 1000°F. Use STMCYCLE.TK to determine the maximum permitted throttle pressure if the maximum moisture content in the exhaust steam is 10%. Plot the results.

C15.8 Use STMCYCLE.TK to model a combined-cycle gas turbine–steam power plant similar to Figure 15.29. The gas turbine exhaust flow is 60 kg/s at 525°C. The steam turbine throttle temperature is 475°C. The gas turbine exhaust is cooled to a temperature 50°C above the steam generator saturation temperature. Vary the steam generator pressure and determine the pressure that will produce the maximum steam turbine power.

16

Refrigeration and Air Conditioning Systems

Refrigeration and air conditioning are distinct but interconnected fields of engineering. Figure 16.1 illustrates the relationship between the two as we note that the function of cooling is a necessary aspect of air conditioning. Cooling is necessary in many areas, such as industrial plants for process control. Air conditioning covers more than cooling the air, the primary focus of this chapter, but addresses matters of air quality, distribution, and human comfort. In this chapter we will examine

- Various types of refrigerants and reasons for their selection;
- Refrigeration cycles, including vapor-compression systems and absorption systems;
- The reciprocating compressors used in the vapor-compression systems;
- Controls for refrigeration systems;
- Systems for liquefying gases;
- Air conditioning fundamentals using the psychrometric chart;
- Various air conditioning heating and cooling processes.

Refrigeration in the engineering sense means maintaining a system at a temperature less than the temperature of the surroundings. This will not occur naturally, so a device must be developed that will maintain this condition.

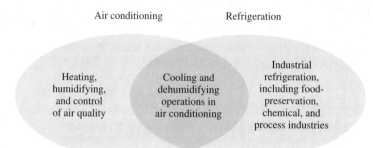

Figure 16.1 Applications of air conditioning and refrigeration.

As we saw in dealing with Carnot cycles in Chapter 7, a reversed Carnot engine will remove heat from a low-temperature reservoir and deliver this energy, plus the work necessary to transfer the heat, to a high-temperature reservoir. The refrigerated system in this case is the low-temperature reservoir.

16.1 REVERSED CARNOT CYCLE

Figure 16.2 illustrates the physical representation of the reversed Carnot cycle and the T-S diagram representing the cycle. This is a mechanical-compression system. The performance ratio of a refrigeration system is not an efficiency, but rather the coefficient of performance, COP, and is defined as the heat supplied to the cycle at a low

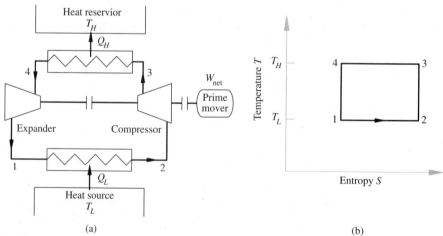

Figure 16.2 (a) The physical layout of a refrigeration system operating on a reversed Carnot cycle. (b) The T-S diagram for a reversed Carnot cycle.

temperature (the desired effect) divided by the net work (what it costs):

$$COP = \frac{Q_{in}}{W_{net}} \qquad (16.1)$$

For the reversed Carnot cycle, this is

$$COP = \frac{T_L \Delta S}{(T_H - T_L)\Delta S} = \frac{T_L}{T_H - T_L} \qquad (16.2)$$

Since the reversed Carnot cycle is the best that can be achieved, the COP for the reversed Carnot cycle is often used as a basis of comparison with actual COP values.

16.2 REFRIGERANT CONSIDERATIONS

The next question is what type of working substance, or refrigerant, can be used in the refrigeration system. There are many choices available, as Table 16.1 illustrates, but it is desirable to have the compressor intake pressure equal to or greater than atmospheric pressure, so that air is not likely to leak into the refrigeration system. It is necessary, then, to use a substance with a boiling temperature less than the surrounding temperature at atmospheric pressure. If the boiling points at atmospheric pressure of various substances are investigated, many will have temperatures low enough to serve as refrigerants. Ammonia, for instance, boils at $-33.3°C$ and R 12 at $-29.7°C$, both at atmospheric pressure. If a two-phase substance were used, the T-s diagram for the reversed Carnot vapor-compression system would look similar to Figure 16.3. In this case the heat supplied from the refrigerated region would evapo-

TABLE 16.1   PHYSICAL PROPERTIES OF REFRIGERANTS

Refrigerant number	Chemical formula	Molecular weight	Boiling point 1 atm (K)	Critical temperature (K)	Critical pressure (MPa)	Latent heat at boiling point (kJ/kgmol)	Current use	Safety group
729	Air	28.97	78.8	132.6	3.77	—	Gas cycle	1
13	$CClF_3$	104.47	191.7	302.0	3.87	15 503	Yes	1
744	CO_2	44.01	194.6	304.1	7.38	25 306	No	1
13B1	$CBrF_3$	148.9	215.4	340.1	3.96	17 679	Yes	1
22	$CHClF_2$	86.48	232.4	369.1	4.98	20 425	Yes	1
717	NH_3	17.03	239.8	406.1	11.42	23 328	Yes	2
12	CCl_2F_2	120.93	243.4	385.1	4.11	19 969	Yes	1
114	$CClF_2CClF_2$	170.93	276.7	418.9	3.27	23 442	Yes	1
21	$CHCl_2F$	102.93	282.1	451.6	5.17	24 918	No	1
11	CCl_3F	137.38	296.7	471.1	4.38	25 022	Yes	1
113	CCl_2FCClF_2	187.39	320.7	487.3	3.41	27 493	Yes	1

Source: Data extracted from ASHRAE *Guide and Data Book,* 1965–66, by permission.

Safety group 1: negligible toxicity, nonflammable. Safety group 2: toxic, flammable, or both.

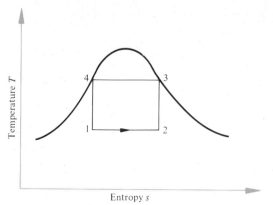

Figure 16.3 The T-s diagram for a two-phase substance operating on the reversed Carnot cycle.

rate the fluid until state 2 was reached. The compressor would compress the wet mixture isentropically to a high temperature until it was a saturated vapor. Heat rejection would occur at constant temperature as the fluid was condensed to a saturated liquid, and then the liquid would enter the constant-entropy expander, producing work, until it reached the pressure and temperature at state 1.

16.3 STANDARD VAPOR-COMPRESSION CYCLE

There are several disadvantages to the aforementioned system. First, reciprocating compressors should not operate in the saturated-mixture region, since lubricating oil may be washed away in the compression process and controlling the amount of liquid in the mixture is difficult. Second, the work performed by the expander is very small in comparison to the compressor work, and the cost for such an expander would be

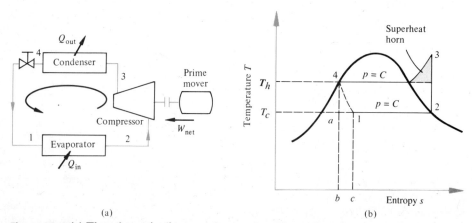

Figure 16.4 (a) The schematic diagram of a simple refrigeration system. (b) The T-s diagram for the system.

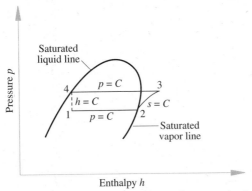

Figure 16.5 The *p-h* diagram of a standard vapor-compression system.

unnecessarily expensive. The ideal, or standard, vapor-compression system, as illustrated in Figure 16.4, overcomes these problems in two ways. First, the refrigerant receives heat until it is a saturated vapor at state 2, and, second, the expansion process is an irreversible throttling process for which only an expansion valve is needed. As in the case of power-producing cycles using two-phase substances, the heat addition, state 1 to state 2, and heat rejection, state 3 to state 4, are constant-pressure processes. The superheat horn, shown in Figure 16.4(b), illustrates the additional work required of dry compression as compared to wet compression. The area *abc*1 represents the loss of refrigerating effect due to the irreversible throttling process compared with the constant-entropy expansion process. In refrigeration practice, a *p-h* diagram is often used to analyze cycles rather than the *T-s* diagram. Figure 16.5 shows a *p-h* diagram of a standard vapor-compression system.

The first means of refrigeration used by humans was the melting of ice blocks, so when more sophisticated means of producing a refrigeration effect were developed, it was natural to express this in terms of familiar units, that of melting ice. A common unit of refrigeration is the cooling effect of 1 ton (2000 lbm) of ice melting at a constant rate in one day. The latent heat of fusion of ice is 334.9 kJ/kg (144 Btu/lbm), and so the refrigeration produced by 907.18 kg of ice (2000 lbm, or 1 ton) is 3.0384×10^5 kJ. One ton of refrigeration is thus the removal of 3.0384×10^5 kJ (288,000 Btu) in a 24-h period. Hence

$$1 \text{ ton} = 3.516 \text{ kW} = 200 \text{ Btu/min} \qquad (16.3)$$

In the appendix tables, abridged charts and tables are given for several refrigerants. R 12 properties are provided in TK Solver models.

Example 16.1

A standard vapor-compression refrigeration system produces 20 tons of refrigeration using R 12 as a refrigerant while operating between a condenser temperature of 41.6°C and an evaporator temperature of −25°C. Determine the refrigerating effect in kilojoules per kilogram, the circulating rate of R 12 in kilograms per second, the power required, the COP, the heat rejected in kilowatts, and the volume flow rate of refrigerant at compressor inlet conditions.

Solution

Given: A standard vapor-compression refrigeration system with known condenser and evaporator temperatures and with a known cooling load.

Find: The system energy characteristics, power required, and heat rejected, as well as the COP and the R 12 mass flow rate.

Sketch and Given Data:

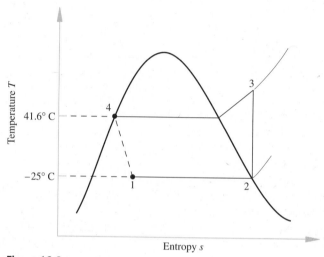

Figure 16.6

Assumptions:

1. The vapor-compression system is closed, but each component may be treated as an open system.
2. The changes in kinetic and potential energies may be neglected.
3. The condenser and evaporator processes are constant-pressure.
4. The expansion process is a throttling process.
5. Saturated vapor enters the compressor, and saturated liquid leaves the condenser.

Analysis: Before determining the energy terms associated with each component and for the entire system, find the enthalpy values around the cycle. From assumption 5 we know that $h_4 = h_f = 76.17$ kJ/kg at $41.6°C$ and $h_2 = h_g = 176.35$ kJ/kg at $-25°C$. Process 1-4 is a throttling process; hence $h_1 = h_4 = 76.17$ kJ/kg. Process 2-3 is isentropic. However, the pressure at state 3 is not explicitly given. Note that it must be equal to the saturated pressure at state 4, as process 3-4 is constant-pressure. Thus, $p_3 = p_4 = p_{sat}$ at $41.6°C = 1.0$ MPa. The value of entropy at state 2 is known, $s_2 = s_g = 0.7121$ kJ/kg-K, and because the entropy is constant from state 2 to state 3, this is also the value of s_3. Entering the superheated R 12 tables, find the enthalpy by interpolating to be $h_3 = 213.46$ kJ/kg.

The refrigerating effect is the heat added per unit mass; hence

$$q_{in} = h_2 - h_1 = 176.35 - 76.17 = 100.18 \text{ kJ/kg}$$

The system capacity provides information regarding the mass flow rate required.

$$(20 \text{ tons}) (3.516 \text{ kW/ton}) = \dot{m}(h_2 - h_1) = \dot{m}(100.18 \text{ kJ/kg})$$

$$\dot{m} = 0.7019 \text{ kg/s}$$

The power is found from a first-law analysis of the compressor, subject to the assumptions.

$$\dot{W} = \dot{m}(h_2 - h_3) = (0.7019 \text{ kg/s})(176.35 - 213.46 \text{ kJ/kg})$$

$$\dot{W} = -26.05 \text{ kW}$$

$$COP = \frac{q_{in}}{w_{net}} = \frac{100.18 \text{ kJ/kg}}{37.11 \text{ kJ/kg}} = 2.7$$

The heat rejected occurs in the condenser; from a first-law analysis

$$\dot{Q}_{out} = -\dot{m}(h_3 - h_4)$$

$$\dot{Q}_{out} = -(0.7019 \text{ kg/s})(213.46 - 76.17 \text{ kJ/kg}) = -96.36 \text{ kW}$$

The specific volume at compressor inlet conditions is equal to the saturated-vapor specific volume at $-25°C$ or $0.131\ 166$ m^3/kg. The volume flow rate is

$$\dot{V} = \dot{m}v_2 = (0.7019 \text{ kg/s})(0.131\ 166 \text{ m}^3/\text{kg})$$

$$\dot{V} = 0.092 \text{ m}^3/\text{s} = 92 \text{ liters/s}$$

Comments:

1. Knowing the temperature allows the determination of the pressure, and vice versa, when the refrigerant is in the saturated region. This assists us when analyzing the condenser and evaporator.
2. The physical dimensions of a compressor depend on the volume flow rate it must handle at inlet conditions, an important specification for the system. ▪

16.4 VAPOR-COMPRESSION SYSTEM COMPONENTS

Heat Exchangers

The two heat exchangers in the vapor-compression system often differ in type. The condenser is often a shell-and-tube heat exchanger with the refrigerant on the shell side and the cooling fluid, often water, passing through the tubes. For air-cooled condensers, a cross-flow heat exchanger is commonly found, where air blows across finned tubes through which the refrigerant flows. Refer to Figure 6.8 for these types. The most frequently encountered evaporator is a direct-expansion evaporator, where the boiling, cold refrigerant is inside a tube and air or another fluid passes over the tube. These tubes are often finned to provide greater surface area for heat transfer.

Expansion Valves

The primary type of expansion valve is the thermostatic expansion valve. It has two functions, to expand the liquid refrigerant from high pressure and warm temperature to low pressure and low temperature, and to supply the correct amount of refrigerant to the evaporator. The refrigerant leaving the evaporator is slightly superheated to about 6° to 10°F above saturation due to the control process used by the thermostatic expansion valve. The process across the expansion valve is irreversible, a throttling process.

A thermostatic expansion valve is schematically illustrated in Figure 16.7. Any throttling valve can reduce the pressure and temperature, but the thermostatic expansion valve goes one step further. It also regulates the degree of superheat leaving the evaporator coil. The refrigerant enters the evaporator as a liquid-vapor mixture. The liquid content of this liquid-vapor mixture vaporizes as it receives heat from the surroundings. Eventually, all of the liquid refrigerant evaporates. As further heat is transferred from the surroundings, the temperature of the refrigerant rises. In the case illustrated in Figure 16.7, the temperature rises 10°F above the boiling temperature, which allows for 10°F of superheat. The bulb is filled with a liquid, typically the same refrigerant that is used in the system. For a temperature of 10°F, the saturated pressure in the bulb capillary tube and on top of the diaphragm of the expansion valve is 29.3 psia. This is pressure p_3. The pressure of the refrigerant is the saturated pressure at 0°F or 23.8 psia, pressure p_2. The restraining spring is set to provide the balance, or $p_1 = 4.5$ psia, which allows for 10°F superheat.

Suppose that the heat transfer from the surroundings decreases as a result of the goods cooling down. The superheat will decrease in the outlet of the evaporator coils. As the superheat decreases, the pressure p_3 decreases. Now $(p_1 + p_2) > p_3$, so the valve closes slightly to a new position and less refrigerant enters the evaporator. When the

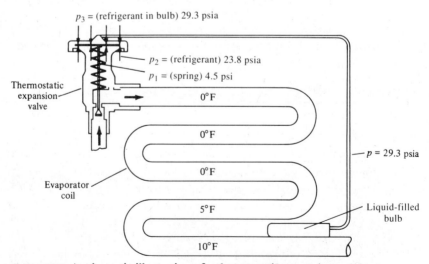

Figure 16.7 A schematic illustration of a thermostatic expansion valve.

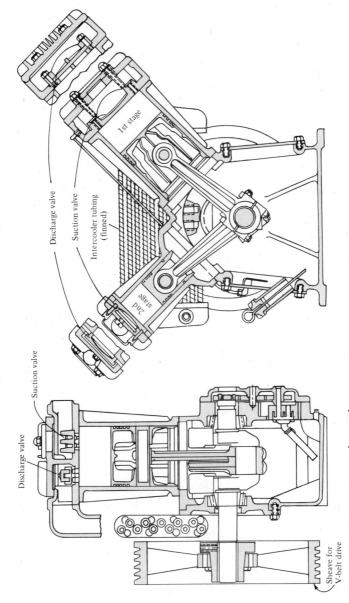

Figure 16.8 A two-stage reciprocating compressor.

Discharge valve

Suction valve

Intercooler tubing (finned)

1st stage

2nd stage

Suction valve

Discharge valve

Sheave for V-belt drive

superheat in the evaporator rises above 10°F, the expansion valve will open wider, allowing more refrigerant to flow, in an effort to maintain the superheat setting.

Compressors

The compressor is the heart of the refrigeration system, where the conversion of mechanical work to thermal energy allows the movement of heat from a low temperature to a high temperature. The four common types of compressors are reciprocating, screw, centrifugal, and vane. The most common is the reciprocating compressor, consisting of a piston moving back and forth with suction and discharge valves that are pressure-actuated flexible disks; they deflect to open and close, based on the pressure differential across them. Figure 16.8 on page 613 illustrates a two-stage reciprocating compressor.

16.5 COMPRESSORS WITHOUT CLEARANCE

All reciprocating compressors have a clearance volume between the top of the piston and the top of the cylinder, where the exhaust and intake valves are located. Some compressors are double-acting, which means they compress in both stroke directions. We will consider the compressor to be single-acting; if double-acting is desired, multiply the result by two. We will also consider the clearance volume to be zero, which means that all the gas in the cylinder is pushed out when the piston is at the top of its stroke. From these considerations, the analytical development of compressors with clearance will be made.

Figure 16.9 illustrates a p-V diagram for this cycle. From state 0 to state 1, gas intake occurs at constant pressure until the piston reaches bottom dead center at state 1; the gas is compressed polytropically from state 1 to state 2 until the pressure is that of the gas in the discharge line; the exhaust valve opens, and the gas is discharged at constant pressure from state 2 to state 3. Since no gas is left, the pressure is undefined.

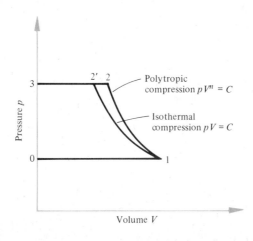

Figure 16.9 A p-V diagram for a single-acting reciprocating compressor without clearance.

As soon as the piston moves an infinitesimal amount, the intake valve opens, and gas is drawn in again from state 0 to state 1. Note that there will be a difference between the work necessary to compress the gas from state 1 to state 2 and the total work of the cycle. The line from state 1 to state 2' illustrates the path of isothermal compression. The enclosed area is less, so the cycle work is less for isothermal compression.

Let us calculate the cycle work using the ideal-gas laws for each process and add the terms together, which yields

$$W_{0-1} = p_1(V_1 - V_0) = p_1 V_1$$

$$W_{1-2} = \frac{p_2 V_2 - p_1 V_1}{1 - n}$$

$$W_{2-3} = p_2(V_3 - V_2) = -p_2 V_2 \qquad (16.4)$$

$$W_{3-0} = 0$$

$$W_{\text{cycle}} = \sum W = \frac{n}{n-1}(p_1 V_1 - p_2 V_2)$$

The cycle work equation may be further arranged to eliminate V_2 and have an expression for the work in terms of p_1, V_1, and p_2.

Since the process from state 1 to state 2 is polytropic,

$$\frac{V_2}{V_1} = \left(\frac{p_2}{p_1}\right)^{-1/n}$$

and

$$\frac{p_2 V_2}{p_1 V_1} = \left(\frac{p_2}{p_1}\right)^{(n-1)/n} \qquad (16.5)$$

Substituting equation (16.5) into equation (16.4) yields

$$W_{\text{cycle}} = \frac{n}{n-1} p_1 V_1 \left[1 - \left(\frac{p_2}{p_1}\right)^{(n-1)/n}\right] \qquad (16.6)$$

Equation (16.6) is valid for compressors without clearance. We calculate the cycle work for the isothermal case. The result is

$$W_{\text{cycle}} = -p_1 V_1 \ln\left(\frac{p_2}{p_1}\right) \qquad (16.7)$$

which is valid for compressors without clearance when the compression is isothermal.

16.6 RECIPROCATING COMPRESSORS WITH CLEARANCE

Let us extend our knowledge of reciprocating compressors without clearance so it can be applied to actual compressors, compressors with clearance. In these, the piston does not move to the top of the cylinder, and some space is left around the valves, called the *clearance volume.* This volume is usually expressed as a percentage of the total displacement volume; it is called the *clearance percentage, c,* and is defined as

$$c = \frac{\text{clearance volume}}{\text{displacement volume}} = \frac{V_3}{V_{PD}} \tag{16.8}$$

Typically, the value of c falls between 3 and 10%.

Figure 16.10 illustrates a p-V diagram for a compressor with clearance. Starting at state 1, the gas is compressed polytropically to state 2; at state 2 the exhaust valve opens, and from state 2 to state 3 the gas is discharged at constant pressure; at state 3 the piston is at the top of its stroke; as it moves down, the exhaust valve closes and the trapped gas expands, doing work on the piston until state 4 is reached; at state 4 the pressure in the cylinder is low enough for gas to be drawn in through the intake valve until state 1 is reached and the cycle is complete.

To calculate the cycle work, we note that the area 1234 is equal to the cycle work and

$$\text{area}_{1234} = \text{area}_{123'4'} - \text{area}_{433'4'}$$

where areas 123'4' and 433'4' may be calculated as if they were cycle works for a compressor without clearance using equation (16.6). The cycle work becomes, on substitution of the work expressions,

$$W_{\substack{\text{cycle}\\1234}} = \frac{n}{n-1} p_1 V_1 \left[1 - \left(\frac{p_2}{p_1} \right)^{(n-1)/n} \right] - \frac{m}{m-1} p_4 V_4 \left[1 - \left(\frac{p_3}{p_4} \right)^{(m-1)/m} \right] \tag{16.9}$$

For this development, $p_3 = p_2$, and $p_4 = p_1$. Since the expansion work is small compared with the compression work, the error involved in setting $m = n$ is very small. With these assumptions and the pressure equalities, equation (16.9) becomes

$$W_{\text{cycle}} = \frac{n}{n-1} p_1 (V_1 - V_4) \left[1 - \left(\frac{p_2}{p_1} \right)^{(n-1)/n} \right] \tag{16.10}$$

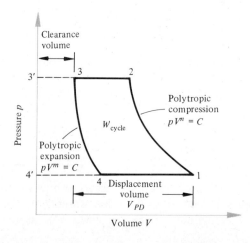

Figure 16.10 A p-V diagram for a reciprocating compressor with clearance.

Equation (16.10) is the cycle work for a gas compressor with clearance. The difference between equation (16.10) and equation (16.6) is the volume term, $(V_1 - V_4)$. This term represents the amount of gas drawn into the cylinder at T_1 and p_1. As can be seen from Figure 16.10, the smaller the clearance volume, the greater the volume of gas that can be drawn into the compressor.

16.7 COMPRESSOR PERFORMANCE FACTORS

Thus far we have been able to compute the work for a compressor in which the processes were reversible. In an actual compressor irreversibilities occur, and the ideal work is not equal to the actual work that must be supplied to the compressor. To account for these irreversibilities, we introduce several compressor performance factors.

The *compression efficiency, η_{cn},* is an indication of how closely the actual compression process approaches the ideal process:

$$\eta_{cn} = \frac{\text{theoretical work}}{\text{indicated work}} \text{(isentropic or isothermal)} \tag{16.11}$$

If the theoretical process is based on isentropic compression, the compression efficiency is called the *isentropic* or *adiabatic compression efficiency.* If the theoretical process is isothermal, it is called the *isothermal compression efficiency.* The *indicated work* is the work that the gas indicates was done on it. This is not the work supplied by the motor; mechanical losses diminish the amount of the work the motor delivers. To account for the loss, the compressor has a mechanical efficiency, η_m, which is defined as

$$\eta_m = \frac{\text{indicated work}}{\text{shaft work}} \tag{16.12}$$

where the shaft work is the work delivered by the power source. If the two efficiencies are multiplied, the overall efficiency, or *compressor efficiency, η_c,* is obtained:

$$\eta_c = \eta_{cn}\eta_m = \frac{\text{theoretical work}}{\text{shaft work}} \tag{16.13}$$

The compressor efficiency shows how well the compressor uses the energy supplied. Note that the theoretical work is the minimum energy that must be supplied, whereas the shaft work is the available energy that must actually be supplied.

Example 16.2
A refrigeration compressor has a displacement volume of 0.05 m³ and a clearance percentage of 5%. It operates at 500 rpm and receives R 12 at 100 kPa and −20°C. The compression is isentropic with a discharge pressure of 700 kPa. Determine the power required and the R 12 discharged in cubic meters per second. Perform the analysis using tables and the ideal-gas equations of state, where for R 12 $c_p = 0.5862$ kJ/kg-K, $R = 0.067$ kJ/kg-K, and $k = 1.129$.

Solution

Given: A reciprocating compressor with specified inlet and discharge conditions, process, and substance.

Find: The power required and volume flow rate at discharge using table properties and assuming the refrigerant to be an ideal gas.

Sketch and Given Data:

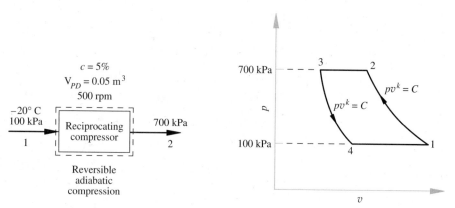

Figure 16.11

Assumptions:

1. The compression and expansion processes in the compressor are reversible adiabatic.
2. The changes in kinetic and potential energies may be neglected.
3. The compressor operates at steady-state.

Analysis: Consider the R 12 to be a pure substance; the property values for the tables at the given conditions are $h_1 = 179.861$ kJ/kg, $s_1 = 0.7401$ kJ/kg-K, and $v_1 = 0.167\,701$ m³/kg. The process from state 1 to state 2 is isentropic; hence $s_2 = s_1$ and the pressure is 700 kPa. Interpolating in the tables yields $h_2 = 215.66$ kJ/kg, $s_2 = 0.7401$ kJ/kg-K, and $v_2 = 0.028\,266$ m³/kg.

Find the mass in the cylinder; knowing the mass per revolution and the number of revolutions per minute, the mass flow rate may be determined. The volume that the refrigerant flows into is less than the displacement volume and must be determined using the expression for volumetric efficiency derived in Chapter 13. From equation (13.19) we find

$$\eta_v = 1 + c - c\left(\frac{p_2}{p_1}\right)^{1/k} = 1 + 0.05 - (0.05)\left(\frac{700}{100}\right)^{1/1.129} = 0.77$$

From equation (13.18)

$$\eta_v = 0.77 = \frac{V_{act}}{V_{PD}} = \frac{V_{act}}{0.05}$$

$$V_{act} = 0.0385 \text{ m}^3/\text{rev}$$

The volume flow rate is

$$\dot{V}_{act} = \frac{(500 \text{ rpm})(0.0385 \text{ m}^3/\text{rev})}{(60 \text{ s/min})} = 0.3208 \text{ m}^3/\text{s}$$

The mass flow rate is

$$\dot{m} = \frac{\dot{V}}{v_1} = \frac{(0.3208 \text{ m}^3/\text{s})}{(0.167\ 701 \text{ m}^3/\text{kg})} = 1.913 \text{ kg/s}$$

The power is found from the first-law analysis to be

$$\dot{W} = \dot{m}(h_1 - h_2)$$

$$\dot{W} = (1.913 \text{ kg/s})(179.861 - 215.66 \text{ kJ/kg}) = -68.5 \text{ kW}$$

The volume flow rate at discharge is

$$\dot{V}_{dis} = \dot{m}v_2 = (1.913 \text{ kg/s})(0.028\ 266 \text{ m}^3/\text{kg}) = 0.054 \text{ m}^3/\text{s}$$

Consider R 12 to be an ideal gas. Use equation (16.10) to determine the power required and ideal-gas equations of state.

$$\dot{W} = \frac{k}{k-1} p_1(V_1 - V_4)\ (N)\left[1 - \left(\frac{p_2}{p_1}\right)^{(k-1)/k}\right]$$

$$\dot{W} = \left(\frac{1.129}{0.129}\right)(100 \text{ kN/m}^2)(0.0385 \text{ m}^3)\frac{(500 \text{ rpm})}{(60 \text{ s/min})}\left[1 - \left(\frac{700}{100}\right)^{0.129/1.129}\right]$$

$$\dot{W} = -69.9 \text{ kW}$$

The volume flow rate may be found from the ideal-gas equation of state, but the temperature at state 2 must be known. Using the relationship for a reversible adiabatic process yields

$$T_2 = T_1 \left(\frac{p_2}{p_1}\right)^{(k-1)k} = (253\,°\text{K})\left(\frac{700}{100}\right)^{0.129/1.129} = 316\,°\text{K}$$

$$\dot{V} = \frac{\dot{m}RT_2}{p_2} = \frac{(1.913 \text{ kg/s})(0.067 \text{ kJ/kg-K})(316\,°\text{K})}{(700 \text{ kN/m}^2)}$$

$$\dot{V} = 0.058 \text{ m}^3/\text{s}$$

Comment: The ideal-gas equations calculate the power as 2% higher and the discharge volume flow rate as 7.4% higher when compared to using the property tables.

16.8 ACTUAL VAPOR-COMPRESSION CYCLE

Actual vapor-compression cycles deviate somewhat from the standard cycle described above. The refrigerant gas entering the compressor suction is commonly superheated; subcooling of the liquid refrigerant entering the expansion valve can occur; pressure drops occur in the system piping and across the evaporator and condenser; and friction and other losses result in a compression process that is not isentropic. Figure 16.12 compares an actual cycle with a standard cycle on a *p-h* diagram.

These deviations affect cycle performance in different ways. Piping pressure drops and compressor losses reduce performance and are generally kept to a minimum consistent with economic considerations. Subcooling of the liquid refrigerant generally has a desirable effect on cycle performance. It reduces flashing across the expansion valve and increases the refrigerating effect in the evaporator. Superheating of the compressor suction gas increases compressor work. Some superheating is necessary to insure proper operation of the thermostatic expansion valve and prevent liquid refrigerant flooding back to the compressor suction.

Refrigeration Cycle Analysis Using TK Solver

The use of TK Solver models R12SAT.TK and R12SHT.TK to determine the properties of R 12 was introduced in Chapter 4. The model R12CYCLE.TK contains the superheated and saturated R 12 property functions and permits the analysis of R12 vapor-compression cycles. The model permits changing evaporating temper-

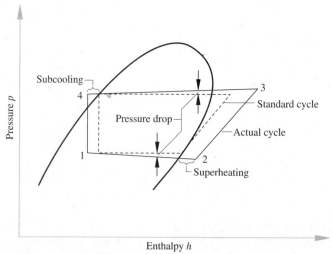

Figure 16.12 Comparison of actual cycle and standard cycle.

perature, condensing temperature, superheating, subcooling, and pressure drops in
the system and evaluating the effects.

Example 16.3

A vapor-compression system operating on R 12 has the same condensing and evaporating temperatures as the standard cycle analyzed in Example 16.1 (41.6°C and
−25°C). The superheat at the compression suction and the subcooling at the condenser outlet are 10°K. The pressure drops are 10 kPa across the evaporator and the
condenser. Use R12CYCLE.TK to calculate the refrigerating effect and COP. Compare the results to those from Example 16.1 for the standard cycle.

Solution

Given: An R 12 vapor-compression cycle with specified conditions.

Find: The refrigerating effect and COP.

Sketch and Given Data:

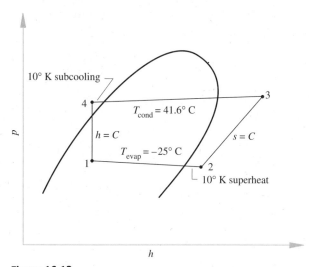

Figure 16.13

Assumptions:

1. The system is closed, but each component may be treated as an open system.
2. The changes in kinetic and potential energies may be neglected.
3. The expansion process is constant-enthalpy.
4. The compression process is isentropic.

Analysis: Enter the given data into the Variable Sheet of R12CYCLE.TK.

```
═ St Input ═════════════════════ VARIABLE SHEET ═══════════════
        — Name — Output — Unit — Comment ─────────────────────

                                     ***R 12 Cycle Performance***

           COP      2.6982
           Wcomp    40.153    kJ/kg
           Qevap    108.34    kJ/kg
           Qcond    148.49    kJ/kg
   -25     Tevap              degC    Temperature (degR, degF, degC, degK)
   41.6    Tcond              degC    Temperature (degR, degF, degC, degK)
   5       SHT                degK    Superheat at Point 2 (degR, degK)
   5       SC                 degK    Subcooling at Point 4 (degR, degK)
   10      Delp12             kPa     Pressure Drop across Evaporator
   10      Delp34             kPa     Pressure Drop across Condenser

                                       Point 1 - Evaporator Inlet
           P1       123.6     kPa     Pressure (psia, MPa, kPa)
           T1       -25       degC    Temperature (degR, degF, degC, degK)
           v1       .047118   m3/kg   Specific Volume (ft3/lbm, m3/kg)
           h1       71.317    kJ/kg   Enthalpy (BTU/lbm, kJ/kg)
           s1       .2889     kJ/kg-K Entropy (B/lbm-R, kJ/kg-K)
           x        .35559            Quality

                                       Point 2 - Compressor Suction
           P2       113.6     kPa     Pressure (psia, MPa, kPa)
           T2       -20       degC    Temperature (degR, degF, degC, degK)
           v2       .14683    m3/kg   Specific Volume (ft3/lbm, m3/kg)
           h2       179.66    kJ/kg   Enthalpy (BTU/lbm, kJ/kg)
           s2       .73088    kJ/kg-K Entropy (B/lbm-R, kJ/kg-K)

                                       Point 3 - Compressor Discharge
           P3       1008.6    kPa     Pressure (psia, MPa, kPa)
           T3       62.596    degC    Temperature (degR, degF, degC, degK)
           v3       .019468   m3/kg   Specific Volume (ft3/lbm, m3/kg)
           h3       219.81    kJ/kg   Enthalpy (BTU/lbm, kJ/kg)
           s3       .73088    kJ/kg-K Entropy (B/lbm-R, kJ/kg-K)

                                       Point 4 - Condenser Outlet
           P4       998.63    kPa     Pressure (psia, MPa, kPa)
           T4       36.6      degC    Temperature (degR, degF, degC, degK)
           v4       .00078944 m3/kg   Specific Volume (ft3/lbm, m3/kg)
           h4       71.317    kJ/kg   Enthalpy (BTU/lbm, kJ/kg)
           s4       .26119    kJ/kg-K Entropy (B/lbm-R, kJ/kg-K)
```

Comments:

1. The refrigerating effect is about 8% greater than in the standard cycle.
2. The COP is only slightly lower than for the standard cycle.
3. R12CYCLE.TK assumes that all superheating results in useful cooling. This may not be true in many real installations.

16.9 MULTISTAGE VAPOR-COMPRESSION SYSTEMS

When the ratio of the compressor discharge pressure to suction pressure, p_2/p_1, becomes high, the performance of the compressor is affected. The volumetric efficiency is reduced, compressor efficiency falls off, and high discharge temperatures

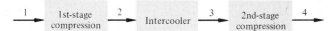

Figure 16.14 A schematic diagram of multistage compression with intercooling.

can result. High ratios of compressor pressure occur in refrigeration cycles when the evaporating temperature is much lower than the condensing temperature. Compressing the gas in two or more stages can solve many of these problems. Splitting a 25:1 pressure ratio into two 5:1 steps improves volumetric efficiency and reduces the power requirements. The use of an intercooler between stages will lower the gas discharge temperature and also improve performance.

The following derivation for a two-stage compressor uses ideal-gas equations of state, but the results are applicable to nonideal gases, such as refrigerant vapors.

Figure 16.14 illustrates a two-stage compressor with an intercooler between the first and second stages. Ideally, the intercooler will bring the temperature of the gas leaving the intercooler down to the ambient temperature. Figure 16.15 illustrates the p-V and T-s diagrams for the compressor. To accomplish this temperature drop, the intercooler may be water-jacketed. For two-stage compressors, the intercooler may consist of a parallel set of finned pipes connecting the low-pressure discharge header to the high-pressure intake header. The air from the fluted vanes on the compressor flywheel blows over the tubes, cooling the compressed gas within the tubes.

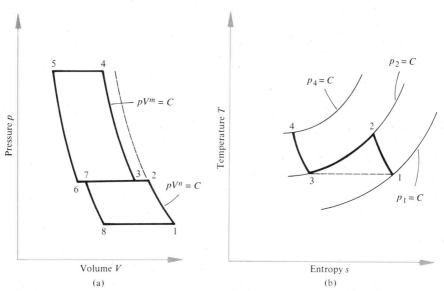

Figure 16.15 (a) The p-V diagram for a two-stage reciprocating compressor. (b) The T-s diagram for two-stage compression with ideal intercooling.

The work for the first- and second-stage cylinders may be calculated from equation (16.10):

$$W_1 = \frac{n}{n-1} p_1(V_1 - V_8)\left[1 - \left(\frac{p_2}{p_1}\right)^{(n-1)/n}\right]$$ (16.14)

The work for the second stage is

$$W_2 = \frac{m}{m-1} p_2(V_3 - V_6)\left[1 - \left(\frac{p_4}{p_2}\right)^{(m-1)/m}\right]$$ (16.15)

Experience with gas compressors has shown that $m = n$. The total work is the sum of the work for the two stages:

$$W_{total} = \frac{n}{n-1} p_1(V_1 - V_8)\left[1 - \left(\frac{p_2}{p_1}\right)^{(n-1)/n}\right]$$
$$+ \frac{n}{n-1} p_2(V_3 - V_6)\left[1 - \left(\frac{p_4}{p_2}\right)^{(n-1)/n}\right]$$ (16.16)

For steady flow through the compressor, the mass entering the first stage enters the second stage,

$$\dot{W}_{total} = \frac{n}{n-1} \dot{m}RT_1\left[1 - \left(\frac{p_2}{p_1}\right)^{(n-1)/n}\right] + \frac{n}{n-1} \dot{m}RT_3\left[1 - \left(\frac{p_4}{p_2}\right)^{(n-1)/n}\right]$$ (16.17)

For an ideal compressor, $T_3 = T_1$. Let us now find the value of p_2 that will minimize the total work. Taking the first derivative of $\dot{W}_{total}$ with respect to the variable p_2 and setting it equal to zero results in the following relationship:

$$p_2 = \sqrt{p_1 p_4}$$ (16.18)

When the value of the pressure for the intercooler is determined as in equation (16.18), the work is equal in all stages and the total work is a minimum. The second derivative of the total work, expressed as the negative of equation (16.17), is positive; thus the work is a minimum.

For a three-stage compressor, the pressure of the low-pressure intercooler, p_2, may be found in a similar manner to be

$$p_2 = \sqrt[3]{(p_1)^2 p_4}$$ (16.19)

and the pressure of the high-pressure intercooler, $p_{2'}$, is

$$p_{2'} = \sqrt[3]{(p_1)(p_4)^2}$$ (16.20)

where p_1 is the intake pressure and p_4 is the final discharge pressure.

The basic principle is that when the pressure ratio across each stage is the same, the compressor work is a minimum.

Example 16.4

A two-stage compressor receives 0.25 m³/s of ammonia at 100 kPa and 10°C and discharges it at 1000 kPa. An intercooler between the stages cools the ammonia back to 10°C. The value of n for the compression is 1.26. Determine the minimum power and the maximum temperature required for compression, the power and the maximum temperature for one stage of compression to the same pressure, the volumetric efficiency based on a clearance of 5%, and the heat removed in the intercooler.

Solution

Given: A two-stage compressor with known inlet conditions, discharge pressure, and compression process.

Find: The minimum power required for compression, the power for one stage of compression, the maximum temperature and volumetric efficiency in each case, and the heat removed in the intercooler for the two-stage compression.

Sketch and Given Data:

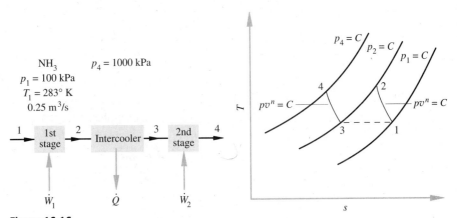

Figure 16.16

Assumptions:

1. Ammonia may be treated as an ideal gas with constant specific heats.
2. The changes in kinetic and potential energies may be neglected.
3. The compressor is an open steady-state system.

Analysis: The minimum power consumption occurs when the interstage pressure is at its optimum. Hence,

$$p_2 = \sqrt{(100 \text{ kPa})(1000 \text{ kPa})} = 316 \text{ kPa}$$

The power for one stage of compression is

$$\dot{W}_1 = \frac{n}{n-1} p_1 \dot{V}_1 \left[1 - \left(\frac{p_2}{p_1} \right)^{(n-1)/n} \right]$$

$$\dot{W}_1 = \frac{1.26}{0.26} (100 \text{ kN/m}^2)(0.25 \text{ m}^3/\text{s}) \left[1 - \left(\frac{316}{100} \right)^{0.26/1.26} \right] = -32.5 \text{ kW}$$

Since the pressure ratios are the same and $p_1 \dot{V}_1 = p_3 \dot{V}_3$, the work of each stage is equal to one another; hence the total power required is

$$\dot{W}_{total} = 2 \times \dot{W}_1 = -65.0 \text{ kW}$$

For a single-stage compressor operating to a discharge pressure of 1000 kPa, the power is

$$\dot{W}_{single} = \frac{n}{n-1} p_1 \dot{V}_1 \left[1 - \left(\frac{p_2}{p_1} \right)^{(n-1)n} \right]$$

$$\dot{W}_{single} = \frac{1.26}{0.26} (100 \text{ kN/m}^2)(0.25 \text{ m}^3/\text{s}) \left[1 - \left(\frac{1000}{100} \right)^{0.26/1.26} \right] = -73.7 \text{ kW}$$

The maximum temperature in each case is found by using the polytropic relationships between temperature and pressure.

$$T_{max} \text{ (2-stage)} = T_1 \left(\frac{p_2}{p_1} \right)^{(n-1)n} = (283^\circ \text{K}) \left(\frac{316}{100} \right)^{0.26/1.26} = 358.8^\circ \text{K}$$

$$T_{max} \text{ (1-stage)} = T_1 \left(\frac{p_2}{p_1} \right)^{(n-1)/n} = (283^\circ \text{K}) \left(\frac{1000}{100} \right)^{0.26/1.26} = 455.1^\circ \text{K}$$

Performing a first-law analysis on the intercooler, an open steady-state system, subject to the assumptions, yields

$$\dot{Q} + \dot{m} h_2 = \dot{m} h_3$$

The mass flow rate may be found from the volumetric flow rate by using the ideal-gas equation of state.

$$\dot{m} = \frac{p_1 \dot{V}_1}{RT_1} = \frac{(100 \text{ kN/m}^2)(0.25 \text{ m}^3/\text{s})}{(0.4882 \text{ kJ/kg-K})(283^\circ \text{K})} = 0.181 \text{ kg/s}$$

Using the ideal-gas equation of state for enthalpy yields

$$\dot{Q} = \dot{m}(h_3 - h_2) = \dot{m} c_p (T_3 - T_2)$$

For an ideal intercooler, $T_3 = T_1$; hence

$$\dot{Q} = (0.181 \text{ kg/s})(2.089 \text{ kJ/kg-K})(283 - 358.8^\circ \text{K}) = -28.7 \text{ kW}$$

Use equation (13.19) to determine the volumetric efficiency for each stage of the two-stage compressor. The volumetric efficiency for the first stage is

$$\eta_v = 1 + c - c \left(\frac{p_2}{p_1} \right)^{1/k}$$

$$\eta_v = 1 + 0.05 - (0.05)\left(\frac{316}{100}\right)^{1/1.26} = 0.925$$

The volumetric efficiency for the second stage is the same, as the pressure ratio is the same.

The volumetric efficiency for one stage of compression to 1000 kPa is

$$n = 1 + c - c\left(\frac{p_2}{p_1}\right)^{1/k}$$

$$n = 1 + 0.05 - (0.05)\left(\frac{1000}{100}\right)^{1/1.26} = 0.739$$

Comment: Two stages of compression significantly reduce the work required compared to a single compression stage. In this example the work for a single stage was 13.4% greater than for the two stages. Volumetric efficiency is much higher for the two-stage compressor. Multistage compressors are designed to achieve minimum work at their normal operating conditions. When these compressors operate at other than the design conditions, the work is less than that required for one stage but will not be the minimum value.   ▬

In refrigeration systems, the entire system is staged or cascaded, not just the compressor stage. This permits the attaining of lower temperatures for refrigeration with a fixed amount of compressor work.

The advantages will be noted as the system is developed. The optimum interstage pressure, p_i, is found to be

$$p_i = (p_h p_l)^{1/2} \tag{16.21}$$

where p_h is the maximum system pressure and p_l is the minimum system pressure. Figure 16.17 illustrates the schematic diagram and the *T-s* and *p-h* diagrams for the cycle.

The evaporator of the high-pressure system serves as the condenser for the low-pressure system. The *T-s* diagram shows the increased refrigerating effect, area 1*acb*, due to the cascading. The dotted line 5 to *a* is an extension of the throttling process from the high pressure to the pressure of the evaporator. Also, the decrease in work compared to that of a single stage is shown by area 3*d*76.

In this drawing the fluids are assumed to be the same, so the same *T-s* diagram may be used. As long as a closed cascade condenser is used, the fluids in the high-pressure and low-pressure systems may be different, in which case a correct *T-s* diagram should be used for each fluid. When the same fluid is used throughout the system, a direct-contact heat exchanger, instead of the closed cascade condenser, is more common. The *T-s* and *p-h* diagrams remain the same, but the physical diagram is illustrated by Figure 16.18. The flash gas from the heat exchanger goes to the high-pressure compressor intake, and the saturated liquid goes to the low-pressure throttling valve. This type of heat exchanger is more efficient, since all tube resistance to heat transfer is eliminated. The open system also provides some control due to transient changes in the system by allowing the proportion of fluid in each loop to

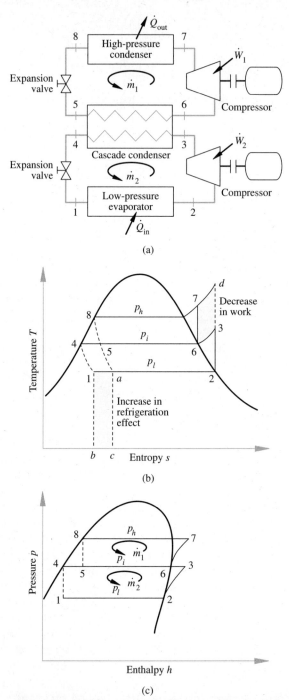

Figure 16.17 (a) The schematic diagram of a two-stage cascade refrigeration system. (b) The *T-s* diagram for the system. (c) The *p-h* diagram for the system.

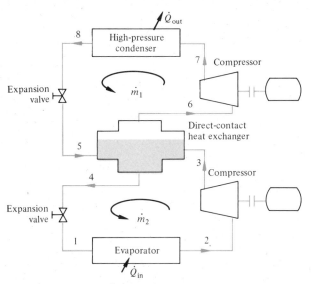

Figure 16.18 An open or direct contact cascade heat exchanger in a two-stage cascade system.

alter slightly. The intermediate pressure, p_i, will vary slightly, depending on conditions.

While the subject of mass flow rates is under discussion, how can the ratio of mass in one loop to that in the other be determined? By performing an energy balance on the cascade condenser, whether it be closed or direct contact, whether using the same fluid or different fluids.

$$\dot{m}_2 h_3 + \dot{m}_1 h_5 = \dot{m}_2 h_4 + \dot{m}_1 h_6$$

$$\frac{\dot{m}_1}{\dot{m}_2} = \frac{h_3 - h_4}{h_6 - h_5} \tag{16.22}$$

The desired refrigerating effect determines the flow rate in the low-pressure loop. For instance, if X tons of refrigeration is desired,

$$\dot{m}_2(h_2 - h_1) = 3.516X \text{ kW}$$

$$\dot{m}_2 = \frac{(3.516 \text{ kW/ton})X}{(h_2 - h_1 \text{ kJ/kg})} \text{ kg/s} \tag{16.23}$$

and $\dot{m}_1$ may be calculated from equation (16.22).

Example 16.5

A 30-ton, multistage refrigeration system uses ammonia as the refrigerant. The evaporator operates at $-30°C$, and a direct-contact cascade condenser is used for the intermediate stage. The high-pressure condenser operates at a saturated-ammonia temperature of $40°C$. Determine the individual and total power required and the COP for the cascaded system and for a single-stage refrigeration system operating between the same temperature limits.

Solution

Given: A two-stage cascade refrigeration system with known low- and high-temperature conditions and the system load capacity.

Find: The total power required and the COP, contrasted with a single-stage vapor-compression system.

Sketch and Given Data:

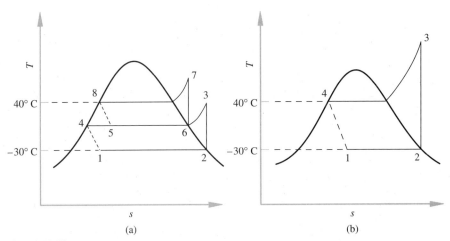

Figure 16.19

Assumptions:

1. The changes in kinetic and potential energies may be neglected.
2. The compression processes are isentropic.
3. Saturated liquids leave the condensers; saturated vapors enter the compressors.
4. The condensation process is at constant pressure.

Analysis: It is necessary to calculate the flow rates in the high- and low-pressure loops. To do this we must find the intermediate pressure. Use Tables A.9 and A.10 for ammonia properties.

$$p_i = \sqrt{p_h p_l} = \sqrt{(1554.3)(119.5)} = 431 \text{ kPa}$$

where $\qquad p_h = 1554.3 \text{ kPa} = p_{\text{sat}}$ at $40°C$

and $\qquad p_l = 119.5 \text{ kPa} = p_{\text{sat}}$ at $-30°C$.

Determine enthalpies for the various states describing the cycle: $h_2 = 1404.6$ kJ/kg; $s_3 = s_2$, $h_3 = 1574.3$ kJ/kg; $h_6 = 1443.5$ kJ/kg; $s_7 = s_6$, $h_7 = 1628.1$ kJ/kg; $h_8 = 371.7$ kJ/kg; $h_5 = 371.7$ kJ/kg; $h_4 = 181.5$ kJ/kg; $h_1 = 181.5$ kJ/kg.

Determine the mass flow rate in the low-pressure loop by equation (16.23):

$$\dot{m}_2 = \frac{(30 \text{ tons}) (3.516 \text{ kW/ton})}{(1404.6 - 181.5 \text{ kJ/kg})} = 0.0862 \text{ kg/s}$$

From equation (16.22)

$$\dot{m}_1 = \dot{m}_2 \frac{h_3 - h_4}{h_6 - h_5}$$

$$\dot{m}_1 = (0.0862 \text{ kg/s}) \frac{(1574.3 - 181.5 \text{ kJ/kg})}{(1443.5 - 371.7 \text{ kJ/kg})} = 0.112 \text{ kg/s}$$

The power of the low-pressure unit is found by an energy balance on the isentropic compressor:

$$\dot{W}_{LP} = \dot{m}_2(h_3 - h_2) = (0.0862 \text{ kg/s})(1574.3 - 1404.6 \text{ kJ/kg}) = 14.6 \text{ kW}$$

and, similarly, for the high-pressure compressor,

$$\dot{W}_{HP} = \dot{m}_1(h_7 - h_6) = (0.112 \text{ kg/s})(1628.1 - 1443.5 \text{ kJ/kg}) = 20.6 \text{ kW}$$

The total power is $\dot{W}_{\text{total}} = 35.2$ kW.

If one stage of compression is used, then cycle enthalpies are $h_1 = 1404.6$ kJ/kg, $h_2 = 1805.1$ kJ/kg, and $h_3 = h_4 = 371.7$ kJ/kg, and the mass flow rate determined by equation (16.23) is

$$\dot{m} = \frac{(30 \text{ tons})(3.516 \text{ kW/ton})}{(1404.6 - 371.7 \text{ kJ/kg})} = 0.1021 \text{ kg/s}$$

$$\dot{W} = \dot{m}(h_2 - h_1) = (0.1021 \text{ kg/s})(1805.1 - 1404.6 \text{ kJ/kg}) = 40.9 \text{ kW}$$

The COP for one stage of compression is

$$q_{\text{in}} = h_1 - h_4 = 1404.6 - 371.7 = 1032.9 \text{ kJ/kg}$$

$$w = h_2 - h_1 = 1805.1 - 1404.6 = 400.5 \text{ kJ/kg}$$

$$\text{COP}_1 = \frac{q_{\text{in}}}{w} = 2.58$$

and the COP for two stages of compression is

$$q_{\text{in}} = h_2 - h_1 = 1404.6 - 181.5 = 1223.1 \text{ kJ/kg}$$

$$w = (h_3 - h_2) + (h_7 - h_6) = 354.3 \text{ kJ/kg}$$

$$\text{COP}_2 = \frac{q_{\text{in}}}{w} = 3.45$$

Comment: The use of cascading significantly reduces the work required by 16.2% in this example. Furthermore, the coefficient of performance increases 33% compared to a single compression stage.

16.10 MULTIEVAPORATORS WITH ONE COMPRESSOR

In refrigeration systems it is not practical, or necessary, to have a separate refrigeration compressor for each evaporator. One compressor may be used to receive the refrigerant from several evaporators operating at different pressures, hence temperatures. Figure 16.20(a) illustrates a compressor receiving refrigerant from two evaporators.

The compressor intake pressure must be at the pressure of the lowest-temperature evaporator, in this case the evaporator set at 0°F. Such a situation would exist when an evaporator is used for maintaining frozen foods. In this case the pressure corresponding to 0°F is 23.8 psia. The other evaporator could maintain a space at 40°F. This would correspond to refrigerated dairy and vegetable produce, where the temperature of the refrigerant in the evaporator coils should be above 32°F to prevent dehydration of the produce. If the refrigerant in the coils is below 32°F, ice will form on the coils due to the condensing of water in the air. This lowers the relative humidity, accelerating the dehydration of the produce. To prevent this, a back-pressure valve is located on the higher-temperature box evaporator's exit piping. The back-pressure regulator maintains a higher pressure in the evaporator coils than exists at the compressor suction. The pressure at the compressor inlet must be low enough to maintain the correct temperature in the lowest-temperature space. If

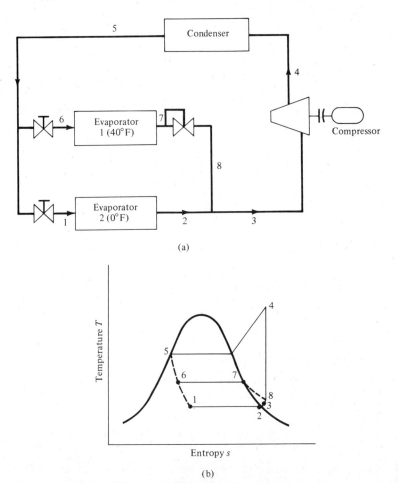

(a)

(b)

Figure 16.20 (a) A multievaporator refrigeration system with one compressor. (b) The T-s diagram for the system.

pressure in the high-pressure evaporator rises above the set point, the valve will open, allowing some refrigerant to leave the evaporator and flow to the compressor inlet. The saturated refrigerant temperature in any evaporator coil may be adjusted by regulating the back pressure; the higher the pressure, the higher the saturated temperature. Figure 16.20(b) illustrates the *T-s* diagram for the multievaporator compressor.

Example 16.6

A refrigeration system using R 12 has two evaporators operating at different temperatures and one compressor. One evaporator provides 5 tons of cooling at 40°F, and the second provides 7 tons of cooling at 0°F. The condenser pressure is 150 psia. Determine refrigerant flow through each evaporator and the COP for the system.

Solution

Given: A multievaporator system with two evaporators and their tonnage and one compressor with known evaporator and condenser states.

Find: The COP for the system and the refrigerant flow rate through each evaporator.

Sketch and Given Data:

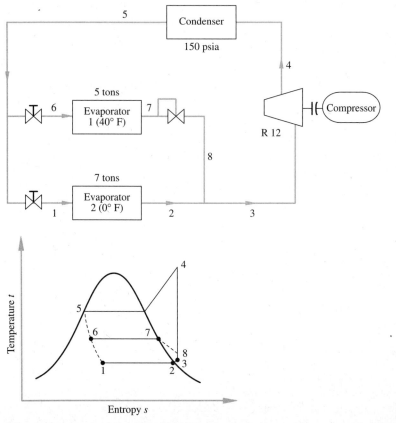

Figure 16.21

Assumptions:

1. The changes in kinetic and potential energies may be neglected.
2. Refrigerant leaves the condenser as a saturated liquid; it leaves the evaporators as a saturated vapor.
3. The compression processes are isentropic; the condensation and evaporation processes are at constant pressure.

Analysis: Determine the enthalpy states around the cycle: $h_5 = 33.4$ Btu/lbm; $h_1 = h_5 = 33.4$ Btu/lbm; $h_6 = h_5 = 33.4$ Btu/lbm; $h_2 = h_g$ at $0°F = 77.3$ Btu/lbm; $h_7 = h_g$ at $40°F = 81.4$ Btu/lbm; $h_8 = h_7 = 81.4$ Btu/lbm.

To find the flow rates through the evaporators, perform a first-law analysis. For the 40°F evaporator

$$\dot{Q} = \dot{m}_1(h_7 - h_6)$$

$$(5 \text{ tons})(200 \text{ Btu/min-ton}) = \dot{m}_1 (81.4 - 33.4 \text{ Btu/lbm})$$

$$\dot{m}_1 = 20.83 \text{ lbm/min}$$

For the 0°F evaporator

$$\dot{Q} = \dot{m}_2(h_2 - h_1)$$

$$(7 \text{ tons})(200 \text{ Btu/min-ton}) = \dot{m}_2(77.3 - 33.4 \text{ Btu/lbm})$$

$$\dot{m}_2 = 31.89 \text{ lbm/min}$$

Perform a first-law analysis at the junction of the combined flows to the compressor.

$$\dot{m}_2h_2 + \dot{m}_1h_8 = (\dot{m}_1 + \dot{m}_2)h_3$$

$$(31.89 \text{ lbm/min})(77.3 \text{ Btu/lbm}) + (20.83 \text{ lbm/min})(81.4 \text{ Btu/lbm})$$

$$= (52.72 \text{ lbm/min})(h_3 \text{ Btu/lbm})$$

$$h_3 = 78.9 \text{ Btu/lbm}$$

Thus, the refrigerant is slightly superheated. Extrapolating between saturated and superheated tables yields $s_3 = 0.1719$ Btu/lbm-R. State 4 may be determined from $p_4 = 150$ psia: $s_4 = 0.1719$; thus, $h_4 = 93.1$.

The power is

$$\dot{W} = (\dot{m}_1 + \dot{m}_2)(h_4 - h_3)$$

$$\dot{W} = (52.72 \text{ lbm/min})(93.1 - 78.9 \text{ Btu/lbm}) = 748.6 \text{ Btu/min} = 17.6 \text{ hp}$$

and the COP is

$$\text{COP} = \frac{(12 \text{ tons})(200 \text{ Btu/min-ton})}{(748.6 \text{ Btu/min})} = 3.2$$

Capacity Control

Any refrigeration system must be designed to handle the maximum refrigeration load anticipated. Typically, this occurs when warm cargo or stores are loaded and the temperature of the stores must be lowered rapidly. The total refrigertion load is the sum of energy to be removed from the goods plus the heat leakage into the refrigerated space from the surroundings. As the goods temperature decreases and reaches the desired storage temperature, the system load is equal to the heat infiltration from the surroundings, considerably less load than the initial load.

While a six-cylinder compressor may be necessary to provide the refrigerant flow and, thus, the cooling capacity at full load, only two cylinders are sufficient for the final-temperature maintenance load. The compressor can operate at the final-temperature maintenance load with all six cylinders, but it would short-cycle – start and stop with short running times. Also, the six cylinders require more power than the two cylinders because of irreversible losses.

To solve this problem, two methods are commonly used, namely, use of two-speed motors and of commercial cylinder unloaders. The capacity of a compressor is a direct function of its speed — reducing the compressor speed reduces its capacity in a linear relationship, while half-speed provides one-half the refrigerant flow. However, the driving power is not linear. When power is reduced by 50%, it will still allow sufficient refrigerant flow for about 70% of the cooling load. Speed reduction does not solve the problem of irreversible losses, nor does it gradually reduce flow. Automatic cylinder unloaders solve these problems. The automatic unloader is a hydraulic unit operating on oil pressure from the lubricating oil pump. As the suction pressure decreases, the differential in pressure between the oil and the compressor intake increases. This causes a mechanism to "unload" a pair of cylinders. Automatic cylinder unloaders hold the suction valves open to stop further compression. Typically, compressor cylinders are unloaded in pairs, or in banks, by having different set points on the hydraulic mechanism. As the differential in pressure increases, other cylinders are sequentially unloaded. The cylinders are unloaded in pairs or banks to maintain even load on the crankshaft. The last two cylinders in the example, or the last bank in general, are not provided with unloading mechanisms. As the suction pressure continues to decrease, the compressor will be shut down when the low-pressure setting is reached on the low-pressure cutout control. A further advantage of the unloaders occurs when the unit starts. The unloaded cylinders remain unloaded until the compressor oil pressure reactivates them. Since a stopped compressor has no oil pressure, the starting load on the motor is significantly reduced.

16.11 CAPILLARY TUBE SYSTEMS

Many small refrigeration systems such as those found in drinking fountains, refrigerators, and small freezers use a capillary tube as the expansion device. These systems can operate on the reciprocating vapor-compression cycle. The systems are designed to be compact and to require little routine maintenance.

The system is illustrated schematically in Figure 16.22(a). The compressor and motor are hermetically sealed into one unit. This prevents refrigerant leaks at the shaft seals and also simplifies the lubrication. The operating noise level is also reduced since the motor and compressor are completely enclosed. An obvious disadvantage is that repairs are difficult because of the inaccessibility of the moving parts.

The capillary tube has a small diameter, 0.030 to 0.050 in.; its length may be from 5 to 20 ft. The pressure drop determines the length of tubing — the longer the tubing,

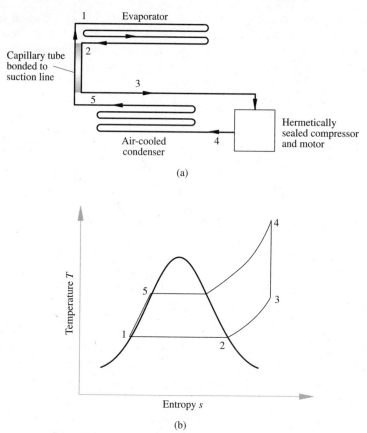

Figure 16.22 (a) A capillary tube refrigeration system. (b) The T-s diagram for the system.

the greater the pressure drop. This means the evaporator pressure, hence temperature, will be lower. Because of the very small diameter involved, flashing of the refrigerant in the tube must be avoided if proper flow of refrigerant is to continue. To assure subcooling of the refrigerant, the capillary tubing is bonded to the cold compressor suction line. The refrigerant in the capillary tube remains a liquid by being subcooled, and the vapor leaving the evaporator is superheated. This process increases the work of the compressor, and even though the refrigerating effect is improved by subcooling, there is a slight decrease in unit performance.

Figure 16.22(b) illustrates the T-s diagram for a capillary tube system. This is the ideal system with no subcooling leaving the condenser and no superheating leaving the evaporator. The bonding length must assure that state 1 is liquid, and in the ideal case a saturated liquid. The enthalpy rise from state 2 to state 3 must equal the enthalpy drop from state 5 to state 1. Ideally, the subcooling-superheating heat exchanger should cause no change in the cycle COP. Actually, due to compressor irreversibilities, the work increases significantly, more than offsetting increase in the refrigerating effect.

16.12 ABSORPTION REFRIGERATION SYSTEMS

The largest operating expense in a vapor-compression refrigeration system is due to work: 100% of the available energy is used to transfer heat from a low temperature to a high temperature. However, the work is transformed into heat and rejected from the system in the condenser. To overcome this use of available energy, the property of absorption of gases by certain fluids may be used to transfer heat from a low to a high temperature. In the chemical reaction in the absorption process, the heat of reaction is released. Since the vapor is condensed in another fluid, the latent heat of the vapor as well as the heat of reaction must be removed. There are several types of absorption refrigeration systems, among them ammonia–water, water–lithium bromide, and water–lithium chloride. We shall look at the ammonia–water system in detail. The other two use water as a refrigerant, which is practical for air conditioning where a temperature below 0°C is not needed. The ammonia–water system, however, is capable of temperatures below 0°C and can achieve tempertures as low as those of the ammonia vapor-compression system.

In this case the ammonia is the refrigerant, R, and the water is the carrier, C. The ammonia vapor is absorbed by liquid water. Figure 16.23 illustrates a schematic diagram for a simple ammonia-water refrigeration system. Some of the complexities

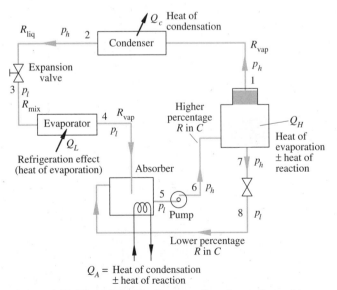

Figure 16.23 The schematic diagram of an absorption refrigeration system.

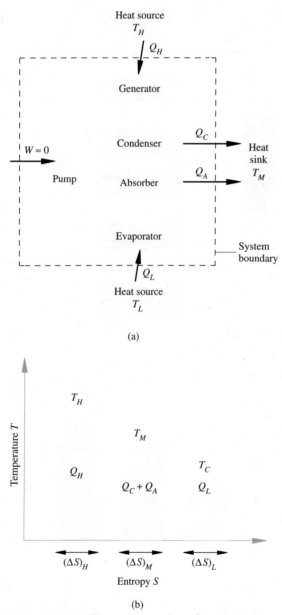

Figure 16.24 (a) The representation of an absorption refrigeration system as heat sources and sinks. (b) The T-S diagram for the system.

encountered in an actual system, such as a rectifier and heat exchanger modifications, are not included.

In Figure 16.23, a pump replaces the compressor for changing the pressure level of the system. The work of the pump is very small, so the use of 100% available energy is also very small. We start with the vapor leaving the generator at state 1. It is dry saturated ammonia vapor at pressure p_h. The ammonia is condensed in the condenser, leaving as a saturated liquid, state 2; it then passes through a throttling valve, where the pressure is reduced to p_i, leaving at state 3. The ammonia passes through the evaporator, picking up heat from the surroundings, and leaves as a saturated vapor at state 4. This cold vapor then enters the absorber, where it mixes with a hot aqueous solution and is condensed and absorbed. For ammonia the heat of reaction is positive, so a heat exchanger must be located in the absorber to cool the hot aqueous solution, improving its absorbing capability and removing the latent heat of condensation and reaction. This aqueous solution, with a high percentage of ammonia (R) in the water (C), leaves the absorber at state 5 and enters the pump, which it leaves at a high pressure p_h at state 6. The high-pressure cool mixture enters the generator, where heat is added to drive the ammonia from solution. The ammonia leaves as a saturated vapor at state 1. Some of the hot liquid, which now has a low percentage of ammonia, leaves the generator at state 7 and is reduced in pressure as it passes through a valve, leaving at state 8. The hot aqueous solution now enters the absorber.

It is possible to calculate both the maximum possible and the actual COP for an absorption refrigeration system. The reversed Carnot cycle cannot be used as a basis of comparison in this instance, because the unit does not use a compressor. The refrigeration is accomplished, ideally, by an exchange of heat isothermally. Although this does not actually happen, we can note that condensation and evaporation are isothermal processes.

Figure 16.24(a) represents the system schematically as various heat sources and sinks. It is possible to represent isothermal heat flow as

$$Q = T\Delta S \tag{16.24}$$

and Figure 16.24(b) represents these on a T-S diagram.

COP for Absorption Refrigeration

We know that the COP is the refrigerating effect divided by the cost of producing that effect. This remains unchanged.

$$\text{COP} = \frac{Q_L}{Q_H} \tag{16.25}$$

However, it is possible for the reversible case to reduce this to an equation involving temperature only. An energy balance on the system yields the following, with work neglected:

$$Q_L + Q_H = Q_C + Q_A = Q_M \tag{16.26}$$

The second law of thermodynamics, when applied to a reversible cycle, tells us that the entropy production is zero, or that

$$(\Delta S)_H + (\Delta S)_L + (\Delta S)_M = 0 \tag{16.27}$$

Using the reversible equation for heat transfer [equation (16.24)], equation (16.27) becomes

$$\frac{Q_H}{T_H} + \frac{Q_L}{T_L} = \frac{Q_M}{T_M} \tag{16.28}$$

where the direction of the heat flow has been accounted for. We then divide equation (16.26) by T_M,

$$\frac{Q_H}{T_M} + \frac{Q_L}{T_M} = \frac{Q_M}{T_M} \tag{16.29}$$

and equate equations (16.28) and (16.29):

$$\frac{Q_H}{T_H} + \frac{Q_L}{T_L} = \frac{Q_H}{T_M} - \frac{Q_L}{T_M} \tag{16.30}$$

Simplifying,

$$Q_H\left(\frac{1}{T_H} - \frac{1}{T_M}\right) = Q_L\left(\frac{1}{T_M} - \frac{1}{T_L}\right)$$

and solving for Q_L/Q_H,

$$\text{COP}_{\text{ideal}} = \frac{Q_L}{Q_H} = \frac{T_L}{T_M - T_L}\frac{T_H - T_M}{T_H} \tag{16.31}$$

Thus the ideal COP may be determined by knowing the temperature of the various heat-transfer terms. The actual COP is determined by equation (16.25) after analyzing the actual heat transferred in the absorption system.

Gas-Liquid Equilibrium

Before analyzing the ammonia absorption system, we must first discover what tools are available for the analysis. The system is a mixture of ammonia and water vapor. The system states where the ammonia and water are in equilibrium may be analyzed by using Appendix Figure B.3. This chart was developed for analyzing ammonia liquid equilibrium properties. The conditions exist wherever ammonia is in contact with water: in the absorber, generator, condenser, throttling valve, and evaporator. Figure 16.25 illustrates the various properties and specifically denotes them for a temperature of 54.4°C and a pressure of 138 kPa. The concentration fraction of the ammonia in liquid form, x', is

$$x' = \frac{\text{kilograms liquid NH}_3}{\text{kilograms mixture}} \tag{16.32}$$

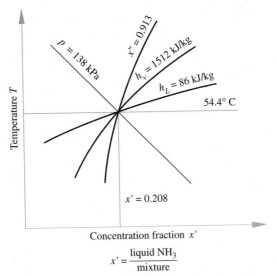

Figure 16.25 The T-x' diagram for an ammonia-water mixture at 54.4°C and 138 kPa.

and the concentration fraction of ammonia vapor is

$$x'' = \frac{\text{kilograms vapor } NH_3}{\text{kilograms mixture}} \qquad (16.33)$$

The enthalpies are of the liquid mixture, h_L, and of the vapor, h_V. Thus, once the temperature and the pressure are known, the other properties. h_V, h_L, x', x'', may be found directly from the chart.

We must be able to determine the enthalpy when two streams of different concentrations are mixed, adiabatically and with heat transfer. These processes occur in the heat exchangers. Figure 16.26(a) is a schematic of a mixing chamber, and Figures 16.26(b) and 16.26(c) show the h-x diagram for adiabatic and nonadiabatic mixing.

The concentration fraction, x, is the percentage of ammonia present in the entering stream. A conservation of mass could be determined for the ammonia, water, or mixture. Let us perform it for the ammonia:

$$(NH_3)_1 + (NH_3)_2 = (NH_3)_3$$

$$\dot{m}_1 x_1 + \dot{m}_2 x_2 = \dot{m}_3 x_3$$

$$\dot{m}_1 + \dot{m}_2 = \dot{m}_3$$

$$x_3 = \frac{\dot{m}_1 x_1 + \dot{m}_2 x_2}{\dot{m}_1 + \dot{m}_2} \qquad (16.34)$$

Note that the concentration of ammonia at state 3 is independent of heat-transfer considerations; it is determined by a mass balance on the heat exchanger. The

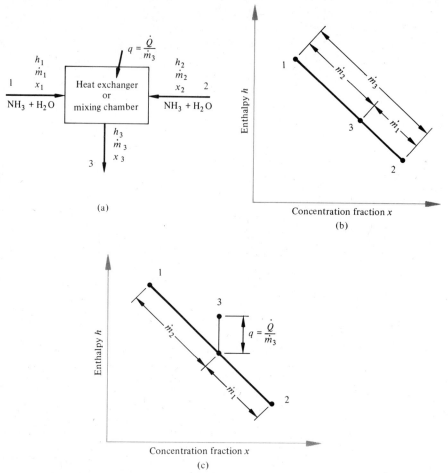

Figure 16.26 (a) The various energy and mass flows into and out of a heat exchanger. (b) The h-x diagram for an adiabatic mixing chamber. (c) The h-x diagram for a mixing chamber with heat transfer.

enthalpy does depend on the heat-transfer process. An energy balance on the exchanger yields

$$\dot{m}_1 h_1 + \dot{m}_2 h_2 + \dot{Q} = \dot{m}_3 h_3$$

$$\dot{m}_1 + \dot{m}_2 = \dot{m}_3$$

$$h_3 = \frac{\dot{m}_1 h_1 + \dot{m}_2 h_2}{\dot{m}_1 + \dot{m}_2} + \frac{\dot{Q}}{\dot{m}_3} \qquad (16.35)$$

and if the exchanger is adiabatic, $\dot{Q} = 0$, and

$$h_3 = \frac{\dot{m}_1 h_1 + \dot{m}_2 h_2}{\dot{m}_1 + \dot{m}_2} \qquad (16.36)$$

The enthalpy at state 3 with heat transfer is equal to the adiabatic enthalpy, equation (16.36), plus an increase of $\dot{Q}/\dot{m}_3$. Since the mass fraction is the same, this is represented vertically. Figures 16.26(b) and 16.26(c) demonstrate the equations.

Example 16.7

An ammonia-water absorption refrigeration system has the following operating conditions: generator temperature, 104.4°C; condenser pressure, 1103 kPa; evaporator pressure, 207 kPa; evaporator temperature, −1.1°C; absorber temperature, 26.6°C. Determine per ton of refrigeration the heat transfer in the generator, condenser, and absorber and the COP.

Solution

Given: An ammonia-water absorption refrigeration system with known operating temperatures and pressures.

Find: The heat transferred in system components and the unit COP.

Sketch and Given Data:

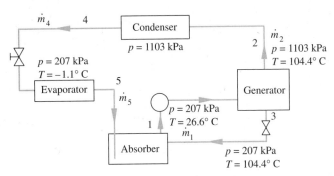

Figure 16.27

Assumptions:

1. The changes in kinetic and potential energies may be neglected.
2. Each component in the closed refrigeration system may be considered an open steady-state system.
3. Equilibrium is assumed for all system states.

Analysis: Use Figure B.3(a) to locate the enthalpies at each point.

   Point 1: $T = 26.6°C$, $p = 207$ kPa, $x' = 0.410$, $h_L = -140$ kJ/kg

   Point 2: $T = 104.4°C$, $p = 1103$ kPa, $x'' = 0.920$, $h_V = 1580$ kJ/kg

   Point 3: $T = 104.4°C$, $p = 1103$ kPa, $x' = 0.310$, $h_L = 280$ kJ/kg

   Point 4: $p = 1103$ kPa, $x_{4'} = x_{2''} = 0.920$

Since p and x' are known, find on the chart $T = 30.5°C$ and $h_L = 84$ kJ/kg.

Point 5: $p = 207$ kPa, $T = -1.1°C$, $x = x_{4'} = x_{2''} = 0.920$

At point 5, the mixture contains 92.0 % ammonia and 8.0% water. The water cannot be vaporized and must be drained off. This is called *purge liquid*. In an actual system a line exists to drain this liquid. To find the enthalpy at 5, it is necessary to find the mass of ammonia and of water, and then to compute the enthalpy of the mixture. For the purge liquid, at $p = 207$ kPa, $T = -1.1°C$, $x_{5'} = 0.620$, and $h_L = -232$ kJ/kg.

For the purge vapor at $p = 207$ kPa, $T = -1.1°C$, $x_{5''} = 0.999^+$, and $h_V = 1284$ kJ/kg.

The ammonia mass leaving the evaporator is

$$1(x_5) = m_{p1}x_{5'} + (1 - m_{p1})x_{5''}$$

where m_{p1} is the mass of the purge liquid in 1 kg of mixture,

$$1(0.920) = m_{p1}(0.620) + (1 - m_{p1})(0.999)$$

$$m_{p1} = 0.208$$

The enthalpy may be found by using equation (16.36).

$$h_5 = (0.208)(-232 \text{ kJ/kg}) + (1 - 0.208)(1284 \text{ kJ/kg}) = 967 \text{ kJ/kg}$$

All the necessary enthalpies are known, so the refrigerant flow rate per ton of refrigeration may be found.

$$\dot{m}_2 = \dot{m}_4 = \dot{m}_5 = \frac{3.516 \text{ kW/ton}}{(h_5 - h_4)} = \frac{3.516 \text{ kW/ton}}{(967 - 84 \text{ kJ/kg})} = 0.003\,982 \text{ kg/s-ton}$$

To determine $\dot{m}_1$ and $\dot{m}_3$, make a total mass balance and an ammonia balance around the generator,

Total mass balance $\dot{m}_1 = \dot{m}_2 + \dot{m}_3$

Ammonia balance $\dot{m}_1 x_1 = \dot{m}_2 x_2 + \dot{m}_3 x_3$

Substitute and solve for $\dot{m}_3$:

$$(0.003\,982 + \dot{m}_3 \text{ kg/s-ton})(0.410) = (0.003\,982 \text{ kg/s-ton})(0.920) + \dot{m}_3(0.310)$$

$$\dot{m}_3 = 0.020\,309 \text{ kg/s-ton}$$

$$\dot{m}_1 = \dot{m}_2 + \dot{m}_3 = 0.003\,982 + 0.020\,309 = 0.024\,291 \text{ kg/s-ton}$$

An energy balance on the generator will yield the heat added, $\dot{Q}_H$.

$$\dot{Q}_H + \dot{m}_1 h_1 = \dot{m}_2 h_2 + \dot{m}_3 h_3$$

$$\dot{Q}_H + (0.024\,291 \text{ kg/s-ton})(-140 \text{ kJ/kg}) = (0.003\,982 \text{ kg/s-ton})(1580 \text{ kJ/kg})$$

$$+ (0.020\,309 \text{ kg/s-ton})(280 \text{ kJ/kg})$$

$$\dot{Q}_H = 15.37 \text{ kW/ton}$$

The heat rejected in the absorber, $\dot{Q}_A$, is found by an energy balance on the absorber.

$$\dot{Q}_A + \dot{m}_1 h_1 = \dot{m}_3 h_3 + \dot{m}_s h_5$$

$$\dot{Q}_A + (0.024\ 291\ \text{kg/s-ton})(-140\ \text{kJ/kg}) = (0.020\ 309\ \text{kg/s-ton})(280\ \text{kJ/kg})$$
$$+ (0.003\ 982\ \text{kg/s-ton})(967\ \text{kJ/kg})$$

$$\dot{Q}_A = 12.94\ \text{kW/ton}$$

The heat rejected in the condenser, $\dot{Q}_C$, is found by an energy balance on the condenser.

$$\dot{Q}_C + \dot{m}_4 h_4 = \dot{m}_2 h_2$$

$$\dot{Q}_C + (0.003\ 982\ \text{kg/s-ton})(84\ \text{kJ/kg}) = (0.003\ 982\ \text{kg/s-ton})(1580\ \text{kJ/kg})$$

$$\dot{Q}_C = 5.95\ \text{kW/ton}$$

The COP for the system is

$$\text{COP} = \frac{\dot{Q}_L}{\dot{Q}_H} = \frac{3.516}{15.37} = 0.228$$

Comment: A check on the accuracy for the heat flows may be made by equating energy supplied and rejected for the cycle.

$$\dot{Q}_H + \dot{Q}_L = \dot{Q}_A + \dot{Q}_C$$

$$15.37 + 3.516 = 12.94 + 5.95$$

$$18.89\ \text{kW/ton} = 18.89\ \text{kW/ton}$$

Three-Fluid Absorption System

The other absorption systems, such as water-lithium bromide, are analyzed in a similar manner. One interesting variation on the absorption system is the three-fluid system developed by Von Platen and Munters. In this system no pump is needed, since the entire system is at uniform total pressure. On what is conventionally the high-pressure side, ammonia supports a total pressure of 1240 kPa. On the conventional low-pressure side, the evaporator, ammonia, supports only 210 kPa, and hydrogen supports 1030 kPa. A liquid seal separates the two sides, and the pressure drop of the ammonia across the liquid seal is equivalent to throttling. By the use of Dalton's law of partial pressures, refrigeration may occur even though the total system pressure is constant. The heat supply is a gas flame to the generator. This system is used commercially for small gas refrigerators and is currently marketed by Dometic under license of Electrolux of Sweden.

16.13 HEAT PUMP

The heat pump is a reversible refrigerator. It can be used for home heating in cool weather and home cooling in warm weather. Basically it is a vapor-compression refrigeration unit with an external heat sink or source, depending on the outdoor temperature. The compressor is designed to meet the heating or cooling requirements of the house, whichever is appropriate for the weather.

Let us consider Figure 16.4(a). This is the basic vapor-compression refrigeration system. In the winter it is desirable to use the heat rejected in the condenser for heating. Typically, a hot-air heating system is used in homes with heat pumps. The air-cooled condenser is located in that system. The evaporator is located outdoors and absorbs heat from the water in a well or pond, if available, or from the outside air. The COP for this mode of operation is

$$\text{COP}_h = \frac{\text{desired effect}}{\text{cost of effect}} = \frac{q_{\text{out}}}{w_{\text{net}}} = \frac{h_3 - h_4}{h_3 - h_2} \qquad (16.37)$$

In the summer, a time for home cooling, the evaporator is located in the house and the condenser outdoors. The heat from the living quarters enters the refrigeration system in the evaporator. The condenser rejects the system heat to the water or air outdoors. With some extra piping and a couple of valves, the condenser and evaporator could reverse roles depending on the desired operating mode. The COP for the cooling mode of operation is

$$\text{COP}_c = \frac{\text{desired effect}}{\text{cost of effect}} = \frac{q_{\text{in}}}{w_{\text{net}}} = \frac{h_2 - h_1}{h_3 - h_2} \qquad (16.38)$$

The heat pump capacity, motor size, and refrigerant supply must be evaluated by considering both modes of operation. The one with the greater demand determines the system size. Because of the larger capacity demanded by heating in many climates, the utilization of the heat pump has been popular in those areas where the winter temperature is not extremely low. The popularity of year-round air conditioning has prompted an increase in heat pump application, even in areas of low winter temperatures, such as New England. Another reason is that the cost of fuel oil has risen to a point where the heat pump becomes economically competitive with oil burners for home heating. Problem 16.31 yields some interesting answers in evaluating the economics of heat pump operations.

A tremendous benefit may be found by using solar heating in conjunction with the heat pump for home heating. Consider the COP_h for the reversed Carnot cycle,

$$\text{COP}_h = \frac{T_H}{T_H - T_L}$$

The solar collector is able to raise T_L, and by doing so the COP_h increases significantly. Instead of operating between 1.6° and 48.9°C, the heat pump may only have to operate between 32.2° and 48.9°C. The COP_h increases from 6.8 to 19.3. Obviously, much less work is required for the same heat output.

16.14 LOW TEMPERATURE AND LIQUEFACTION

The refrigeration process dealing with the production of very low temperatures is *cryogenics.* Exactly where the cryogenic temperature limit occurs is not rigorously defined; however, several authorities suggest that temperatures below 173 – 123°K be called cryogenic. To achieve these temperatures, the cascade refrigeration system discussed earlier is commonly used. In the cascade system different refrigerants are used in each loop so the evaporator temperature in the lowest-temperature loop will be sufficiently low for whatever purpose is desired, for example, gas liquefaction. Different refrigerants are used, so the compressor intake pressure in each stage is above or near atmospheric pressure; this prevents air infiltration and also means that the specific volume at the intake will not be great enough to cause a significant increase in the work required for compression.

The gases in the cryogenic region behave similarly to vapors we analyzed in Chapter 4. At room temperature and pressure, gases are very greatly superheated, which is why the ideal-gas equations of state may be used. An important thermodynamic phenomenon, the Joule-Thomson effect, is frequently used in cryogenic and liquefaction operations.

A throttling process produces no change in enthalpy, and hence for an ideal gas the temperature remains constant:

$$h = u + pv = c_v T + RT$$

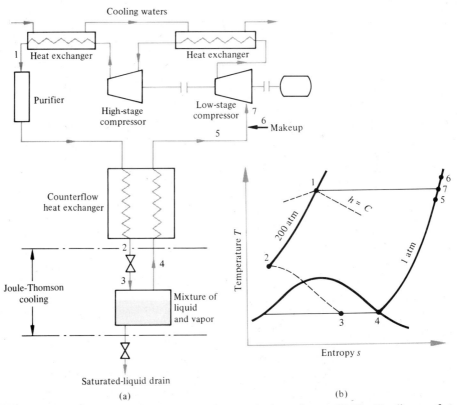

Figure 16.28 (a) A schematic diagram of the liquefaction of gases. (b) The T-s diagram for the liquefaction process.

In a real gas, however, the throttling process does produce a temperature change, either up or down. The Joule-Thomson coefficient, μ, is defined as

$$\mu \equiv \left(\frac{\partial T}{\partial p}\right)_h \tag{16.39}$$

A positive value of μ indicates that the temperature decreases as the pressure decreases, and a cooling effect is thus observed. This is true for almost all gases at ordinary pressures and temperatures. The exceptions are hydrogen gas, helium, and neon, which have a temperature increase with a pressure decrease; hence $\mu < 0$. Even for these gases there is a temperature above which the Joule-Thomson coefficient changes from negative to positive. At this *inversion temperature* $\mu = 0$.

The utilization of the Joule-Thomson coefficient in the liquefaction of gases was developed by Linde and Hampson independently. By looking at Figure 10.4(b), we can see that if the gas were at a sufficiently low temperature and high pressure, the throttling process would bring the gas into the saturated-mixture region. Here the vapor and liquid could be separated. Such a system was devised by Linde and Hampson, and it is illustrated schematically with a *T-s* diagram in Figure 16.28.

The gas is compressed isothermally in two stages (state 7 to state 1); it is then purified and cooled at constant pressure in a very efficient counterflow heat exchanger (state 1 to state 2); the gas is throttled (state 2 to state 3); and some is liquefied because of the additional Joule-Thomson cooling. The remainder at state 4, a saturated vapor, passes through the counterflow heat exchanger to state 5; makeup gas is added at state 6; and the mixture enters the compressor at state 7.

There are other variations of the liquefaction cycles and refrigeration cycles in general. However, the method of analysis, primarily a first-law analysis, is the same in all cases. Actual refrigeration systems, with their associated control systems, are beyond the scope of this text and must remain an area for further study.

16.15 AIR CONDITIONING AND REFRIGERATION

One of the most common uses of refrigeration is in air conditioning systems. As humans, we find certain environmental conditions make us feel more comfortable than others. Hence, home and work air conditioning and heating systems have been developed. The engineering speciality associated with this technology is heating, ventilating, and air conditioning (HVAC).

Air conditioning systems are designed to provide a satisfactory air state for human comfort by controlling temperature and humidity. We intuitively know that temperature, humidity, and air motion affect the way we react with the environment. Convective heat transfer to or from the body is directly proportional to the temperature difference between the body and the surrounding air and the velocity of the air. In addition, our bodies perspire as a means of cooling, and the evaporation of water from the skin is a function of the humidity of the air surrounding the skin surface.

The human body tries to maintain a constant temperature of 98.6°F, or 37°C. It may be viewed as a very sophisticated electrochemical-mechanical engine that receives energy (food and oxygen), converts a portion into chemical work and mechanical work, and rejects the rest as heat and waste. Our bodies feel comfortable in environments where the heat may be rejected comfortably. The amount of heat that we reject depends on our activity level. When we sleep, about 87 W (297 Btu/hr) is dissipated; 115 W (392 Btu/hr) when we are awake but sitting; and 440 W (1500 Btu/hr) for strenuous physical activity.

To determine what is viewed as reasonable comfort, a series of studies have been undertaken by the American Society of Heating, Refrigerating, and Air Conditioning Engineering (ASHRAE). Comfort charts, indicating desirable temperature, humidity, and air-motion conditions as well as clothing type, have been developed. The charts indicate the range of conditions that most people will find comfortable. Systems for HVAC are designed with these in mind.

In Chapter 11 we analyzed gas-vapor mixtures, particularly air and water vapor mixtures using the ideal-gas laws. In air conditioning analysis, it is more common to use psychrometric charts that account for actual mixture behavior than to use the ideal-gas law analysis.

16.16 PSYCHROMETRIC CHART

The wet- and dry-bulb temperatures are typically used to determine the relative humidity in conjunction with a psychrometric chart (Figures B.4(a) and (b)) and may be used in place of the equations developed in Chapter 11. Either method is correct, but the chart tends to be more convenient. For pressures other than atmospheric, a correction factor must be used. Many charts indicate what these corrections are, but it is sufficient, at this stage, to know that corrections must be applied.

The psychrometric charts are widely used in air conditioning applications, as well as in problems of drying substances. The chart represents the dry-bulb temperature as the abscissa, humidity ratio as the ordinate, and the wet-bulb temperature, specific volume, relative humidity, vapor pressures, and enthalpy as the other variables. Figure 16.29 illustrates how to read the chart for these variables. The enthalpy on the chart is per kilogram of dry air. The enthalpy term is the sum of the dry air and water vapor enthalpies in the mixture. Thus, $h = h_a + \omega h_f$, where h is the enthalpy given by the chart. The enthalpy of air and water may have different reference temperatures; however, the change of enthalpy is most often used, and the discrepancy, slight in the first place, is very much minimized.

The enthalpy deviation lines plotted on the chart are the difference between the enthalpies of unsaturated moist air and the saturated air at the same wet-bulb temperature. Thus, the enthalpy at the wet-bulb temperature is found and the deviation subtracted from it. This correction is typically small and can be ignored in most situations.

Let us use the psychrometric chart in Figure B.4(a) and find the various properties of atmospheric air with a dry-bulb temperature of 35 °C and a wet-bulb temperature of 21 °C. The dry-bulb temperature is on the chart's abscissa. The wet-bulb temperatures are read along the left-hand curved line, and the lines slope diagonally down to the right. Find the intersection of the dry-bulb 35 °C line and the diagonal wet-bulb temperature line of 21 °C. This point of intersection defines the air state. The other properties are read as follows: the relative humidity is 29%; the enthalpy is

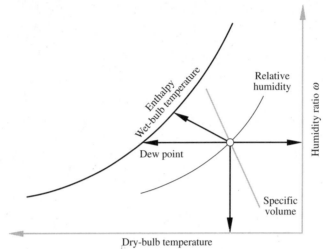

Figure 16.29 The psychrometric chart.

61 kJ/kg air; the specific volume is 0.885 m³/kg air; the humidity ratio is 0.010 kg water/kg air. The dew point is found by moving horizontally to the saturation temperature line. The temperature value, 14°C, is the dew point.

16.17 AIR CONDITIONING PROCESSES

When air is conditioned, its temperature and humidity are affected, often at the same time. The controls, the system, and all the related equipment used in an actual air conditioner will not be covered here. We will be concerned with the thermodynamic aspects of adjusting the air humidity.

Dehumidifying and Reheating

Let us consider removing moisture from air until the desired humidity is attained. Available chemical dehumidifiers, such as silica gel, are hygroscopic and absorb water. In most instances, however, the air is cooled below its dew point until the desired amount of moisture is removed. The remaining air, now cold and humid, can be reheated to achieve the desired condition. Two methods are used to cool the air. In one, chilled water is sprayed into the air; a fine spray gives a large heat-transfer surface. The air leaves as saturated air at the low water temperature. The second is the direct method employed in room air conditioners, in which air is passed over the cool or cold evaporator coils of a refrigeration unit.

From a thermodynamic viewpoint we are concerned only with the state properties entering and leaving the dehumidifier. In determining how much the air must be cooled in a dehumidification process, we must find the vapor pressure of the water at the desired final conditions. The total pressure is constant during the heating process after dehumidification, so the vapor pressure will be constant also. The vapor pressure leaving the dehumidifier must be the vapor pressure desired at the final temperature.

Most air conditioning thermodynamic analyses are based on the first law. An energy balance determines the energy requirements, and conservation of mass determines the mass of water added or removed in a process.

Example 16.8

Moist air enters a dehumidifier-reheater air conditioning unit at 0.47 m³/s, 26°C, and 80% relative humidity. It is to be cooled to saturation and then reheated to 26°C and 50% relative humidity. Determine the tons of refrigeration and the kilowatts of heating required.

Solution

Given: A cooling-dehumidifying air conditioner with air inlet and exit conditions and total flow rate.

Find: The heating and cooling requirements of the system.

Sketch and Given Data:

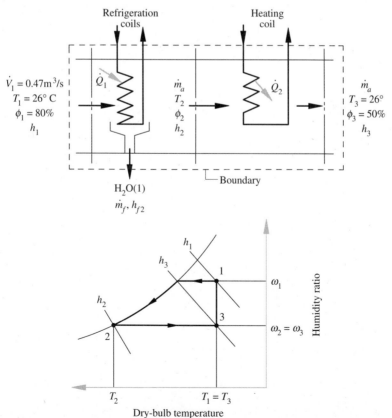

Figure 16.30

Assumptions:

1. The control volume operates at a steady state.
2. The changes in kinetic and potential energies may be neglected.
3. There is no work crossing the boundary.
4. No heat is transferred to the surroundings.
5. The total pressure remains constant.
6. The air leaving the cooling coil is saturated.

Analysis: In this example, the final conditions are known; from this we can determine the properties at point 2. The energy balance on the cooling coil yields

$$\dot{Q}_1 + \dot{m}_a h_1 = \dot{m}_f h_{f2} + \dot{m}_a h_2$$

The flow rate must be determined by finding the specific volume from the psychrometric chart: $h_1 = 69$ kJ/kg, $\omega_1 = 0.0168$ kg water/kg air, and $v_{a1} = 0.87$ m³/kg.

$$\dot{V}_1 = \dot{m}_a v_{a1}$$

$$(0.47 \text{ m}^3/\text{s}) = (\dot{m}_a (0.87 \text{ m}^3/\text{kg})$$

$$\dot{m}_a = 0.54 \text{ kg/s}$$

At state 2, $h_2 = 40.5$ kJ/kg; $\omega_2 = 0.0105$ kg water/kg air; $h_{f2} = h_f$ at T_2, $T_2 = T_{\text{dew}} = 15°C$, and $h_{f2} = 61.95$ kJ/kg.

$$\dot{Q}_1 = \dot{m}_a(h_2 - h_1) + \dot{m}_a(\omega_1 - \omega_2)h_{f2}$$

$$\dot{Q}_1 = (0.54 \text{ kg/s})(40.5 - 69 \text{ kJ/kg})$$

$$+ (0.54 \text{ kg/s air})(0.0168 - 0.0105 \text{ kg water/kg air})(61.95 \text{ kJ/kg water})$$

$$\dot{Q}_1 = -15.2 \text{ kW}$$

$$\text{Refrigeration} = \frac{+15.2 \text{ kW}}{3.516 \text{ kW/ton}} = 4.3 \text{ tons}$$

Perform an energy balance on the heater to find $\dot{Q}_2$.

$$\dot{m}_a h_2 + \dot{Q}_2 = \dot{m}_a h_3$$

$$\dot{Q}_2 = \dot{m}_2(h_3 - h_2)$$

From the psychrometric chart at state 3, $h_3 = 52.6$ kJ/kg.

$$\dot{Q}_2 = (0.54 \text{ kg/s})(52.6 - 40.5 \text{ kJ/kg}) = 6.53 \text{ kW}$$

Comments:

1. The water condensed is usually determined by multiplying the change in specific humidity by the air flow rate.
2. In practice, it is common to neglect the energy of the liquid water leaving the dehumidifier since it is small in comparison to the energy in the air-water mixture.

Humidifying and Heating

Air is conditioned in the winter as well, by adding moisture and heat. This is important in home heating; very dry air causes wooden furniture to shrink and is responsible for such adverse physiological effects as dry mucous membranes; more fuel must be consumed because of the poor heat-conduction properties of dry air. The specific heat and thermal conductivity are affected directly by the moisture in the air. When the air is dry, the temperature must be higher than for moister air to make us feel comfortable.

Example 16.9

Air enters a heater-humidifier at 10°C and 10% relative humidity. The air mixture leaves the heater-humidifier at a temperature of 24°C and a relative humidity of 50%. The water enters the humidifier at 15°C, and the air flow rate entering is 0.75 kg air/s. Determine the amount of heat and water added.

Solution

Given: An air heater-humidifier receives air at known conditions with a given flow rate, and heats and humidifies it to specified exit conditions.

Find: The heat and water required to meet the specifications.

Sketch and Given Data:

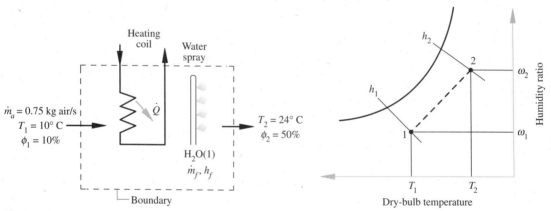

Figure 16.31

Assumptions:

1. The control volume operates at a steady state.
2. The changes in kinetic and potential energies may be neglected.
3. There is no work crossing the boundary.
4. No heat is transferred to the surroundings.
5. The total pressure remains constant.

Analysis: Perform a first-law analysis on the heater-humidifier.

$$\dot{Q} + \dot{m}_a h_1 + \dot{m}_f h_f = \dot{m}_a h_2$$

The enthalpy of the water is equal to the saturated liquid enthalpy at 15°C, $h = 61.95$ kJ/kg. The water added is equal to the mass flow rate of air times the change in absolute humidity, or

$$\dot{m}_f = \dot{m}_a(\omega_2 - \omega_1)$$

$$\dot{m}_f = (0.75 \text{ kg/s})(0.0094 - 0.0007 \text{ kg water/kg air}) = 0.006\ 52 \text{ kg/s}$$

From the psychrometric chart, $h_1 = 12$ kJ/kg air and $h_2 = 47.9$ kJ/kg air. Solving for the heat from the first-law equation yields

$$\dot{Q} = \dot{m}_a(h_2 - h_1) - \dot{m}_f h_f$$

$$\dot{Q} = (0.75 \text{ kg/s})(47.9 - 12 \text{ kJ/kg}) - (0.006\ 52 \text{ kg/s})(61.95 \text{ kJ/kg})$$

$$\dot{Q} = 26.52 \text{ kW}$$

Comment: The amount of heating and humidification need not be separated, as the final state includes the effects of water added to the mixture enthalpy and the raising of the water's temperature to the final mixture temperature.   ▬

Using TK Solver to Analyze Air Conditioning Processes

In Chapter 11, TK Solver models PSYCHRO.TK and PSYCHRO2.TK were introduced to determine the properties of air–water vapor mixtures and related processes. PSYCHRO.TK can be used instead of the psychrometric chart, and PSYCHRO2.TK can be used to analyze processes such as heating, cooling, humidification, and dehumidification.

Example 16.10
Air at 50°F and 90% relative humidity passes across a heating coil, exiting at 90°F. Use PSYCHRO.TK to determine the relative humidity of the exiting air and the heat added in British thermal units per pound mass.

Solution

Given: Air of known dry-bulb temperature and relative humidity is heated to a given dry-bulb temperature.

Find: The exit relative humidity and heat added.

Sketch and Given Data: See Figure 16.32 on page 655.

Assumptions:

1. The control volume operates at a steady state.
2. The changes in kinetic and potential energies may be neglected.
3. There is no work crossing the boundary.
4. No heat is transferred to the surroundings.

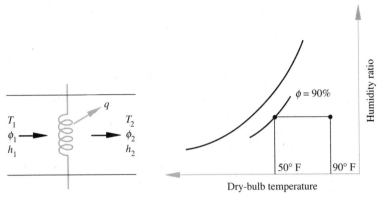

Figure 16.32

5. The total pressure remains constant.
6. There is no change in humidity ratio across the coil.

Analysis: Load PSYCHRO2.TK into TK Solver, and enter the inlet and outlet temperatures, inlet relative humidity, and zero for the change in humidity ratio. The solved Variable Sheet should look like this:

```
========================= VARIABLE SHEET =========================
St  Input — Name — Output —— Unit —— Comment ———————————————————

                                          ***Psychrometric Process***

                                          Inlet Conditions
    50      DB1                  degF     Dry-Bulb Temperature
            WB1    48.489        degF     Wet-Bulb Temperature
            DP1    47.181        degF     Dew Point Temperature
    90      RH1                  %        Relative Humidity
            W1     .0069301      lbm/lbm  Humidity Ratio
            h1     19.507        BTU/lbm  Total Enthalpy
            V1     12.998        ft3/lbm  Specific Volume

                                          Outlet Conditions
    90      DB2                  degF     Dry-Bulb Temperature
            WB2    64.473        degF     Wet-Bulb Temperature
            DP2    47.181        degF     Dew Point Temperature
            RH2    22.991        %        Relative Humidity
            W2     .0069301      lbm/lbm  Humidity Ratio
            h2     29.232        BTU/lbm  Total Enthalpy
            V2     14.017        ft3/lbm  Specific Volume

                                          Process Variables
    14.696  Pb                   psia     Barometric Pressure
            delDB  40            degF     Change in DB Temperature
    0       delW                 lbm/lbm  Change in Humidity Ratio
            delh   9.7247        BTU/lbm  Change in Enthalpy
```

Comment: Although the humidity ratio remains constant across the coil, the relative humidity decreases. ▬

Adiabatic Mixing of Air Streams

Very often in air conditioning systems airstreams are adiabatically mixed, as illustrated in Figure 16.33. This situation occurs in buildings where part of the air is recirculated and part is fresh outside air. This allows the cooling or heating unit size to be smaller than it would be if all the air had to be conditioned from the surroundings to the desired inside temperature and humidity. Exceptions to this are laboratories and hospitals, where zoning requirements dictate that all the air be continuously supplied from the surroundings and none recirculated, for obvious reasons. In adiabatic mixing we apply equations for conservation of energy and mass (air and water vapor) to find the mixture enthalpy and humidity ratio at the exit state. No work is done, the mixing is adiabatic, and kinetic and potential energies are negligible.

Thus, for air

$$\dot{m}_{a1} + \dot{m}_{a2} = \dot{m}_{a3}$$

and for water vapor

$$\dot{m}_{v1} + \dot{m}_{v2} = \dot{m}_{v3}$$

which when combined with the humidity ratio becomes

$$\dot{m}_{a1}\omega_1 + \dot{m}_{a2}\omega_2 = \dot{m}_{a3}\omega_3$$

This last equation may be solved to determine the humidity ratio at the final state.

The absence of heat transfer, work, and changes of kinetic and potential energies reduces the first-law equation to

$$\dot{m}_{a1}h_1 + \dot{m}_{a2}h_2 = \dot{m}_{a3}h_3$$

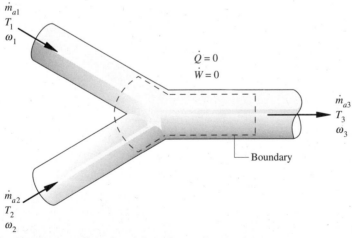

Figure 16.33 Adiabatic mixing of two airstreams.

which may be solved for h_3. The enthalpy and humidity ratio at state 3 define the state.

Evaporative Cooling

In relatively hot and dry climates air cooling can be accomplished by evaporating water into the air. The evaporation process is endothermic, requiring energy from the air and lowering its dry-bulb temperature. The total enthalpy of the air–water vapor mixture remains constant as the amount of water in the air increases. This is a low-cost, low-energy way to reduce the air's temperature. Homes in such climates, in addition to traditional air conditioning units, will often have evaporative cooling units. The air is circulated by a fan across a mesh that contains water; the water evaporates as the air passes across it. This lowers the temperature of the air and can provide sufficient cooling, at low cost, for certain conditions. When traditional air conditioning is used as well, the evaporative unit lowers the high temperature in the house, and hence reduces the power required.

Cooling Towers

As we learned in Chapter 9, all energy eventually reaches the dead state, that is, the state of the surroundings. The atmosphere is the ultimate surrounding for all processes on earth. In a heat-transfer process, the final energy will exist at the temperature of the surroundings, the air. We learned that in power plant operations a condenser is a necessary part of the cycle. Usually water is circulated through the

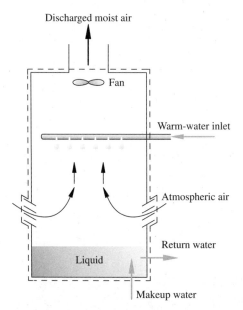

Discharged moist air

Fan

Warm-water inlet

Atmospheric air

Return water

Liquid

Makeup water

Figure 16.34 An evaporative cooling tower, also known as a wet cooling tower.

condenser, picks up the rejected heat from the system mass, and is discharged. This arrangement works very well when cooling water is in sufficient supply. In many areas, however, cooling water is not available or the environmental effect, thermal pollution, is too damaging to permit such circulation of water.

There are several methods to overcome this problem: dry cooling towers, where air is passed over tubes containing the system mass; cooling ponds; and evaporative cooling towers. In the evaporative cooling tower, circulating water is used in the condenser, but air is used to cool this circulating water. The water is partially evaporated, humidifying the air, and thus cooling the remaining circulating water. Figure 16.34 presents a schematic diagram of an evaporative cooling tower. The atmospheric air, not saturated, enters the cooling tower at the bottom, flowing counter to the direction of the warm circulating water. The circulating water evaporates, saturating the air and decreasing its own energy and therefore its temperature. By the time it has reached the bottom of the tower, the remaining water is sufficiently cool to return to the condenser. Makeup water must be added to replenish that which evaporated. The temperature and, even more so, the relative humidity of the inlet air are important in determining how much air is required to cool the water. Should the inlet air be saturated, no evaporative cooling could occur.

Example 16.11

An evaporative cooling tower is used by a large power-generating facility located in the Midwest. The circulating-water flow rate is 2000 gal/min, and the water enters at $100°F$. The air inlet conditions are $T_{db} = 80°F$ and $T_{wb} = 60°F$. The exit air is at $90°F$ and 90% relative humidity. Determine the air flow rate required to cool the water to $80°F$ and the quantity of water evaporated. The makeup water is available at $80°F$.

Solution

Given: A cooling tower receives air and a specified amount of water at known conditions. The water outlet temperature and the air outlet state are specified.

Find: The air flow rate required to provide the cooling and the water lost to evaporation.

Sketch and Given Data: See Figure 16.35 on page 659.

Assumptions:

1. The control volume operates at a steady state.
2. The changes in kinetic and potential energies may be neglected.
3. There is no work crossing the boundary.
4. No heat is transferred to the surroundings.
5. The total pressure remains constant.

Analysis: The first-law analysis of the cooling tower subject to the assumptions yields

$$\dot{m}_a h_{a_1} + \dot{m}_{fc} h_{f_1} + \dot{m}_{fm} h_{f_2} = \dot{m}_{fc} h_{f_3} + \dot{m}_a h_{a_2}$$

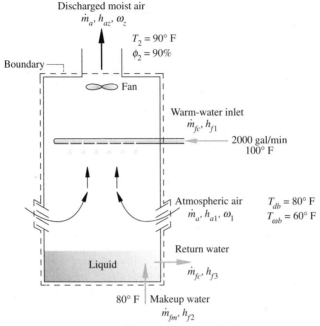

Discharged moist air
$\dot{m}_a,\ h_{az},\ \omega_z$

$T_2 = 90°\ F$
$\phi_2 = 90\%$

Boundary

Fan

Warm-water inlet
$\dot{m}_{fc},\ h_{f1}$

2000 gal/min
100° F

Atmospheric air
$\dot{m}_a,\ h_{a1},\ \omega_1$

$T_{db} = 80°\ F$
$T_{\omega b} = 60°\ F$

Return water
$\dot{m}_{fc},\ h_{f3}$

Liquid

80° F   Makeup water
$\dot{m}_{fm},\ h_{f2}$

Figure 16.35

where $\dot{m}_a$ = air mass flow rate, $\dot{m}_{fc}$ = circulating-water flow rate = 2000 gal/min, h_{f_1} = inlet circulating-water enthalpy = 67.92 Btu/lbm, h_{f_2} = inlet makeup-water enthalpy = 47.74 Btu/lbm, $\dot{m}_{fm}$ = makeup water flow rate, h_{a_1} = inlet air enthalpy = 26.5 Btu/lbm, h_{f_3} = outlet circulating-water enthalpy = 47.74 Btu/lbm, h_{a_2} = outlet air enthalpy = 52.2 Btu/lbm, ω_2 = outlet humidity ratio = 0.028 lbm vapor/lbm air, and ω_1 = inlet humidity ratio = 0.0066 lbm vapor/lbm air. The water enthalpy may be found from the steam tables and the air enthalpy from the psychrometric chart.

The mass of makeup water due to evaporation may be expressed as

$$\dot{m}_{fm} = \dot{m}_a(\omega_2 - \omega_1)$$

Substitute in the first-law equation.

$\dot{m}_a(26.5\ \text{Btu/lbm}) + (2000\ \text{gal/min})(8.34\ \text{lbm/gal})(67.92\ \text{Btu/lbm})$

$\qquad + \dot{m}_a(0.028 - 0.0066\ \text{lbm vapor/lbm air})(47.74\ \text{Btu/lbm water})$

$\qquad = (2000\ \text{gal/min})(8.34\ \text{lbm/gal})(47.74\ \text{Btu/lbm}) + \dot{m}_a(52.2\ \text{Btu/lbm})$

$\dot{m}_a = 13{,}640\ \text{lbm/min}$

The water evaporated is

$\dot{m}_f = (13{,}640\ \text{lbm air/min})(0.028 - 0.0066\ \text{lbm vapor/lbm air})$

$\dot{m}_f = 291.9\ \text{lbm/min} = 35\ \text{gal/min}$

Comments:

1. The water evaporated is less than 2% of the water supplied. The remainder is saved and recirculated.
2. Actual plants have other considerations, such as the loss of chemicals in the cooling process, that pose certain environmental impacts.   ▬

CONCEPT QUESTIONS

1. What factors render the Carnot cycle impractical for refrigeration systems using a pure substance and operating in the mixture region?

2. A refrigerator and a heat pump operate on the reversed Carnot cycle and between the same temperature limits. Which will have the higher COP?

3. Two refrigerators operate on the reversed Carnot cycle and receive heat at the same low temperature. One refrigerator operates in a warmer environment than the other. Which will consume more work for the same amount of heat received?

4. What contributes to the internal irreversibilities in the ideal vapor-compression refrigeration cycle?

5. A proposal has been made to improve the COP of the ideal vapor-compression refrigeration cycle by using a turbine in place of the throttling valve to decrease the refrigerant pressure. Discuss the merits of doing this.

6. It has been suggested that water be used as the working substance in a vapor-compression refrigeration system if the temperature of the evaporator is above freezing. Would this be feasible?

7. In improving the performance of the standard vapor-compression cycle, subcooling of the refrigerant leaving the condenser was explained. Discuss the limitations to the amount of subcooling possible.

8. In Chapter 8 we saw that for a heat engine the enclosed area on a T-s diagram represents the net work. Does the enclosed area of the T-s diagram of an actual vapor-compression cycle represent the net work? Of an ideal cycle?

9. When selecting a refrigerant for a given application, what are the important qualities?

10. Contrast the advantages and disadvantages of heat pumps.

11. Why are cascade refrigeration systems used?

12. What device allows a single refrigeration compressor to handle several evaporators operating at different pressures?

13. Explain how a thermostatic expansion valve controls the flow of refrigerant through the evaporator coils.

14. Explain the operation of an absorption refrigeration system.

15. Is the definition of the COP for an absorption refrigeration system consistent with that of a vapor-compression system? Why?

16. Why is fluid in the generator of the ammonia-absorption system heated and cooled in the absorber?

17. Why must gases be compressed to high pressures in the liquefaction process?

18. Referring to the psychrometric chart, at what conditions are the wet- and dry-bulb temperatures the same?

19. Illustrate how the dew point is found on the psychrometric chart, given two other properties.

20. What are the functions that an air conditioning system performs besides heating or cooling the air?

21. How does humidity affect the way humans react to an air dry-bulb temperature?

22. Explain humidification and dehumidification processes.

23. Why is air sometimes humidified when it is heated?

24. Why is air sometimes reheated after it has been cooled?

25. How does evaporative cooling work?

26. Explain the operation of a natural-draft evaporative cooling tower.

PROBLEMS (SI)

Refrigeration Cycle

16.1 A reversed Carnot cycle is used for heating and cooling. The work supplied is 10 kW. If the COP = 3.5 for cooling, determine (a) T_2/T_1; (b) the refrigerating effect (tons); (c) the COP for heating.

16.2 Where is the greatest gain in COP derived in a Carnot refrigeration cycle — changing the condenser or the evaporator temperature?

16.3 A reversed Carnot cycle uses R 12 as the working fluid. The refrigerant enters the condenser as a saturated vapor at 30°C and leaves as a saturated liquid. The evaporator temperature is a constant −10°C. Determine per unit mass (a) the compressor work; (b) the turbine work; (c) the heat input; (d) the COP.

16.4 A standard vapor-compression refrigeration system uses ammonia as the refrigerant. The ammonia leaves the evaporator at −20°C, and the condenser pressure is 1000 kPa. The flow rate is 20 kg/min. Determine (a) the tons of refrigeration; (b) the power required; (c) the COP.

16.5 A standard vapor-compression refrigeration system uses R 12 as the refrigerant. The R 12 leaves the evaporator at −20°C, and the condenser pressure is 1000 kPa. The flow rate is 20 kg/min. Determine (a) the tons of refrigeration; (b) the power required; (c) the COP.

16.6 A standard vapor-compression refrigeration system using R 12 operates with an evaporator temperature of −30°C and a condenser exit temperature of 50°C. The compressor requires 75 kW. Determine the capacity in tons.

16.7 A refrigeration compressor receives 1.2 m³/min of R 12 at −10°C from the evaporator of a standard vapor-compression refrigeration system. The condenser pressure is 1.2 MPa. Determine (a) the mass flow rate; (b) the power required; (c) the tons of refrigeration; (d) the COP.

16.8 A vapor-compression refrigeration system has an evaporator operating at −30°C, a condenser with a pressure of 900 kPa, and subcooling of 5°C of the R 12 leaving the condenser. Determine the percentage of increase in refrigerating effect because of the subcooling.

16.9 In a vapor-compression refrigeration system, the R 12 leaves the evaporator at 150 kPa and 0°C, enters the compressor, and is compressed isentropically to 1.2 MPa. The discharge from the condenser is subcooled by 5°C. The refrigerant flow rate is 40 kg/min. Determine (a) the tons of refrigeration; (b) the COP; (c) the increase in the refrigerating effect compared to the standard cycle.

16.10 A vapor-compression refrigeration system uses R 12 and has an evaporator pressure of 100 kPa and a condenser pressure of 900 kPa. The temperature of the refrigerant entering the compressor is − 10°C and entering the condenser is 90°C. The refrigerant leaving the condenser is subcooled 5°C, and the compressor power is 5 kW. Determine (a) the tons of refrigeration; (b) the COP; (c) the change of availability across the compressor.

16.11 A 10-ton vapor-compression refrigeration system uses R 12. The vapor enters the compressor with 10°C of superheat and exits the condenser with 5°C of subcooling. The evaporator pressure is 150 kPa, and the condenser pressure is 1000 kPa. The compressor isentropic efficiency is 80%. Determine (a) the power; (b) the COP; (c) the change of availability across the condenser.

16.12 An actual vapor-compression refrigerator has a compressor with inlet conditions of − 20°C and 130 kPa and exit conditions of 60°C and 900 kPa. The refrigerant is R 12. The liquid refrigerant enters the expansion valve at 850 kPa and exits at 160 kPa. Determine the COP.

16.13 Refrigerant 12 enters an adiabatic compressor as a saturated vapor at − 15°C and discharges at 1.0 MPa. If the compressor efficiency is 70%, determine the actual work.

16.14 An actual vapor-compression refrigerator has a 1.5-kW motor driving the compressor. The compressor efficiency is 70% and has inlet pressure and temperature of 50 kPa and − 20°C and a discharge pressure of 900 kPa. The condenser exit pressure is 800 kPa, and the expansion valve exit pressure is 60 kPa. Determine (a) the flow rate; (b) the COP; (c) the tons of refrigeration for R 12.

16.15 A Carnot engine operates between temperature reservoirs of 817°C and 25°C and rejects 25 kW to the low-temperature reservoir. The Carnot engine drives the compressor of an ideal vapor-compression refrigerator, which operates within pressure limits of 190 kPa and 1.2 MPa. The refrigerant is ammonia. Determine (a) the COP; (b) the refrigerant flow rate.

16.16 A 1-ton vapor-compression refrigeration system using R 12 as the refrigerant has the following data for the cycle: evaporating pressure = 123 kPa, condensing pressure = 600 kPa, temperature of R 12 entering the throttling valve = 20°C; and temperature of refrigerant leaving the evaporator = −7°C. Assuming isentropic compression, determine (a) the refrigerant flow rate; (b) the power required; (c) the heat rejected; (d) the second-law efficiency.

16.17 A vapor-compression refrigeration system uses R 12 as the refrigerant. The gaseous refrigerant leaves the compressor at 1200 kPa and 80°C. The heat loss during compression is 14 kJ/kg. The refrigerant enters the expansion valve at 32°C. The liquid leaves the evaporator and enters the compressor as a saturated vapor at − 15°C. The unit must produce 50 tons of refrigeration. Determine (a) the R 12 flow rate; (b) the compressor power; (c) the COP.

16.18 A 15-ton refrigeration system is used to make ice. The water is available at 10°C. Refrigerant 12 is used with saturated temperature limits of −25°C and 75°C. Determine (a) the COP; (b) the refrigerant flow rate; (c) the power required; (d) the maximum kg of ice manufactured per day.

16.19 An ideal vapor-compression refrigeration system uses a subcooling-superheating heat exchanger with 10°C of superheat added. The evaporator operates at −30°C, the condenser at 1.4 MPa. There is a cooling requirement of 50 tons. If the fluid is R 12, determine (a) the mass flow rate; (b) the COP; (c) the refrigerating effect; (d) the degrees of subcooling; (e) the power required; (f) the second-law efficiency.

16.20 A two-stage cascade vapor-compression refrigeration system uses R 12 and produces 15 tons of refrigeration. A direct-contact heat exchanger is the cascade condenser. The low-temperature evaporator pressure is 100 kPa, and the condenser pressure is 1000 kPa. The intermediate pressure is optimum. Determine (a) the mass flow rate in the upper and lower stages; (b) the COP; (c) the irreversibility rate in the cascade condenser for $T_o = 298°K$ and $p_o = 100$ kPa.

16.21 The same as Problem 16.20, assuming the isentropic compressor efficiency is 85% for the compressors.

16.22 A two-stage cascade vapor-compression refrigeration system uses R 12 and has an evaporator temperature of −20°C, a cascade condenser pressure of 400 kPa, and a condenser pressure of 1200 kPa. The flow rate through the evaporator is 1.3 kg/s. Determine (a) the mass flow rate through the condenser; (b) the COP; (c) the total power.

16.23 A two-stage cascade vapor-compression refrigeration system uses ammonia in the low-temperature loop and R 12 in the high-temperature loop. The evaporator operates at −30°C, and the ammonia discharges to the closed cascade condenser at 600 kPa. The R 12 must be 10°C lower than the saturated ammonia temperature to provide adequate cooling. The R 12 compressor has a pressure ratio of 5:1. The ammonia provides 15 tons of cooling. Determine (a) the pressure of the R 12 in the cascade heat exchanger and the condenser; (b) the mass flow rates of ammonia and R 12; (c) the total power; (d) the COP.

16.24 A compressor serves two evaporators, one providing 5 tons of cooling at −20°C and the other 10 tons of cooling at 0°C. The system is a vapor-compression refrigeration system with the condenser pressure at 1.2 MPa. Determine (a) the refrigerant flow through each evaporator; (b) the R 12 state entering the compressor; (c) the compressor power.

16.25 A vapor-compression refrigeration system uses a subcooling-superheating heat exchanger located after the evaporator to subcool the refrigerant entering the expansion valve. The refrigerant leaving the evaporator is superheated in the process. Assume the refrigerant leaves the evaporator as a saturated vapor and the condenser as a saturated liquid and that no pressure drops occur in the heat exchangers. The evaporator temperature is −10°C, the condenser pressure is 1000 kPa, and the flow rate is 20 kg/min. Determine (a) the compressor power; (b) the tons of refrigeration; (c) the COP.

16.26 A heat pump uses a vapor-compression cycle with R 12 and provides 15 kW of heat. The condenser pressure is 800 kPa, and the evaporator pressure is 300 kPa. Determine (a) the mass flow rate of R 12; (b) the COP; (c) the power required.

16.27 A vapor-compression heat pump system uses R 12. The compressor receives 0.75 m³/min at 250 kPa and 0°C and discharges it at 900 kPa and 60°C. The refrigerant leaves the condenser as a saturated liquid. Determine (a) the heat output in kW; (b) the power required; (c) the COP.

16.28 A home uses a vapor-compression heat pump with R 12 as the means for providing heat. The house requires 14 kW of heat when the inside temperature is 20°C and the outside temperature is 0°C. The minimum temperature difference between the heat exchangers and air is 10°C. (a) Specify the evaporator and condenser pressures. (b) Determine for the specified cycle its COP, the mass flow rate of R 12, and the compressor power required.

16.29 A vapor-compression heat pump uses R 12 and provides 10 kW of heating. The evaporator pressure is 200 kPa, and the refrigerant enters the compressor at 0°C. The compressor's isentropic efficiency is 80%, and the condenser pressure is 900 kPa. The electricity to drive the compressor comes from a power plant with an efficiency of 40%. Determine (a) the compressor power; (b) the ratio of heat used in the power plant to produce the electricity to the heat output of the heat pump.

16.30 The air-cooled condenser of a vapor-compression heat pump system heats 50 m³/min of air from 20° to 35°C at a constant pressure of 100 kPa. The R 12 exits the condenser at 1000 kPa and 35°C. The evaporator temperature is −10°C. Determine (a) the refrigerant mass flow rate; (b) the compressor power; (c) the COP.

16.31 A vapor-compression system, using ammonia, is used for home heating. Air is to be heated from 7° to 30°C at atmospheric pressure. The living requirements demand an air circulation rate of 0.45 m³/s. The evaporator temperature is 0°C, and the condenser pressure is 1.6 MPa. Determine (a) the refrigerant flow rate; (b) the COP; (c) the power required; (d) the cost, if electricity costs $0.05/kW-h; (e) the cost of fuel, at $0.27/liter, if the heating value of the fuel is 34 900 kJ/liter.

16.32 A vapor-compression refrigeration system using ammonia as the working substance is required to provide 30 tons of refrigeration at −16°C. To accomplish this, the ammonia operates between pressure limits of 144 kPa and 1400 kPa. Determine (a) the quality of ammonia entering the evaporator; (b) the mass flow rate; (c) the change of available energy across the condenser if $T_0 = 26°C$; (d) the volume flow rate at the compressor intake; (e) the bore (D) and stroke (L)(assume $L = D$) of a double-acting, reciprocating compressor at 600 rpm. Assume that the fluid is completely discharged from the cylinder each revolution.

16.33 A two-stage cascade refrigeration system uses ammonia as the working substance. The mass flow rate in the high-pressure loop is 0.10 kg/s. The condenser saturated temperature is 36°C, and the evaporator temperature is −40°C. The cascade condenser is direct contact. Determine (a) the cascade condenser pressure for minimum work; (b) the refrigerating effect in tons; (c) the COP; (d) the power required; (e) the second-law efficiency.

16.34 A vapor-compression system, using R 12, is to be used for home heating. The maximum heating demand is increasing the temperature of 0.50 m³/s of air at 5°C and atmospheric pressure to 30°C. The minimum condenser temperature must be 10°C greater than the maximum air temperature. The evaporator operates at −5°C. Determine (a) the refrigerant flow rate; (b) the power required; (c) the COP; (d) the cycle second-law efficiency.

16.35 A vapor-compression system, used for home heating, operates between 38°C in the saturated condenser and −5°C in the evaporator. The fluid is R 12. The home owner decides to increase the evaporator temperature to improve the efficiency of operation. For a heating load of 10 kW, investigate the variation of COP with evaporator temperatures of −5°C, 0°C, and +5°C. Do you agree with the home owner?

16.36 A vapor-compression system uses ammonia for heating and cooling. The compressor can handle 6 liters/s of ammonia at −4°C, discharging it at 1.8 MPa. Determine (a) the

COP for heating and cooling; (b) the maximum cooling load possible; (c) the maximum heating load possible.

16.37 A refrigeration unit uses the ammonia-absorption system for cooling. The unit is characterized by the following conditions: generator temperature = 98.8°C, condenser pressure = 7034 kPa, evaporator pressure = 207 kPa, evaporator temperature = 1.6°C, and absorber temperature = 23.8°C. Determine (a) the COP; (b) the heat required per ton of refrigeration; (c) the heat rejected in the condenser.

16.38 An ammonia-absorption refrigeration unit is used for air conditioning a large office building. The cooling requirements are 200 tons at −1.1°C. The vapor leaving the generator is at 115.5°C and 1241 kPa. The evaporator pressure is 138 kPa, and the absorber temperature is 23.8°C. Determine (a) the mass flow of ammonia required; (b) the COP; (c) the heat required; (d) the natural gas required if the heat release is 33 500 kJ/m^3.

16.39 For the liquefaction of gases, the following properties are given for nitrogen. Refer to Figure 16.28(a). $T_5 = T_6 = T_7 = 270°K$, $h_5 = h_6 = h_7$ kJ/kg; $h_1 = 405$ kJ/kg at 100 atm; $T_2 = 150°K$, $h_2 = 200$ kJ/kg; $T_3 = 78°K$; $h_4 = 230$ kJ/kg. Determine (a) the quality of the mixture at state 3 if $h_f = 35$ kJ/kg; (b) the kg/min required at state 7 for 1 kg of liquefied nitrogen.

16.40 A heat pump uses ammonia as the refrigerant and must deliver 30 kW to the building. The discharge pressure is 1.8 MPa, and the suction pressure is 300 kPa at 15°C. The compression is isentropic. Determine (a) the ammonia flow rate in liters/s at inlet conditions; (b) the power required; (c) COP; (d) the discharge temperature.

16.41 An air turbine receives air at 1400 kPa and 310°K and expands it to 140 kPa. The turbine internal efficiency is 70%, and the power produced is 65 kW. The air exhausting from the turbine will be used for refrigeration. The refrigerated space is to be maintained at 260°K. What is the maximum possible refrigeration in tons?

Air Conditioning

16.42 Using the psychrometric chart determine (a) the humidity ratio and specific enthalpy of air at 30°C dry-bulb and 60% relative humidity; (b) the dew point for air at 30°C dry-bulb and 25°C wet-bulb; (c) the relative humidity and humidity ratio for air at 40°C dry-bulb and 25°C wet-bulb.

16.43 A tank 1 m in diameter and 2 m long contains air at 40°C dry-bulb and 50% relative humidity. The surrounding temperature lowers at night, and the air in the tank then has a dry-bulb temperature of 25°C. How much, if any, water condenses?

16.44 Using the psychrometric chart, determine (a) the specific enthalpy and specific volume for air with a relative humidity of 60% and a dry-bulb temperature of 35°C; (b) the humidity ratio and the relative humidity, given a wet-bulb temperature of 25°C and a specific volume of 0.9 m^3/kg; (c) the wet- and dry-bulb temperatures, given a relative humidity of 70% and a humidity ratio of 0.018 kg vapor/kg air.

16.45 Twenty-five kg/min of air enters a dehumidifier at 25°C, 1 atm, and 60% relative humidity. The air is cooled to 15°C, with water and air exiting separately at this state. Determine (a) the mass flow rate of water leaving; (b) the heat transfer.

16.46 A flow of 0.5 kg/s of air at 1 atm, 30°C, and 70% relative humidity enters a dehumidifier. The air exits at 15°C and 95% relative humidity, and water exits at 15°C. Refrigerant enters the dehumidifier coils with an enthalpy of 75 kJ/kg and leaves with an enthalpy of 190 kJ/kg. Determine the refrigerant flow rate required.

16.47 A small basement dehumidifier receives 1 m³/min of air at 20°C and 70% relative humidity. The dehumidifier removes 2 liters of water in a 24-h period. What is the humidity ratio of the air leaving the dehumidifier?

16.48 An air conditioning system dehumidifies and then reheats the air. The system operates with 75 m³/min of air entering the dehumidifier at 27°C and 70% relative humidity. The air leaves the reheater at 22°C and 50% relative humidity. Determine (a) the temperature of the air leaving the dehumidifier before it is reheated; (b) the flow rate of the condensed water; (c) the tons of cooling required; (d) the reheat in kW.

16.49 Outside air entering at 55 m³/min, 10°C, and 70% relative humidity is reheated and humidified so the air condition entering the heating system is 30°C and 35% relative humidity. The humidification occurs with water at 15°C. Determine (a) the water flow rate required; (b) the heat transfer in kW.

16.50 An air conditioning system consists of a heating section followed by a humidifying section that supplies saturated steam at 100°C. One m³/s of air enters the heating section at 10°C and 70% relative humidity. It leaves the humidifying section at 22°C and 50% relative humidity. Determine (a) the temperature of the air when it leaves the heating section; (b) the flow rate of steam required; (c) the heat required.

16.51 Air at 50 m³/min, 15°C, and 60% relative humidity enters the heating section of a heater/humidifier air conditioning system. The air is heated and passes over a mesh, where water at 20°C is supplied. The air leaves the unit at 25°C and 40% relative humidity. Determine (a) the temperature of the air leaving the heating section; (b) the flow rate of water required; (c) the heat added.

16.52 An air conditioned classroom receives air at 15°C. The air leaving via the exit duct is at 26°C and 50% relative humidity. The people in the room may be viewed as adding 0.075 kg/min of water vapor at 32°C. The heat addition from the people in the room and the lights and surroundings is estimated to be 9.4 kW. Determine (a) the inlet volume flow rate; (b) the relative humidity; (c) the humidity ratio.

16.53 Air at 0.5 kg/s, 35°C, and 25% relative humidity enters an adiabatic evaporative cooler. The cooler receives water at 20°C, and the air leaves saturated at 25°C. Determine the water flow rate.

16.54 Two airstreams mix adiabatically. The first stream at 35°C and 60% relative humidity with a mass flow rate of 1 kg/s mixes with the second at 20°C and 50% relative humidity with a flow rate of 2 kg/s. Determine the exit air temperature and relative humidity.

16.55 It is desired to have 500 m³/min of air at 22°C and 50% relative humidity. This is obtained by adiabatically mixing 300 m³/min of air at 25°C and 65% relative humidity with another airstream. Specify the volume flow rate, temperature, and relative humidity of that airstream.

16.56 Air at 35°C and 40% relative humidity enters an adiabatic mixing chamber with a flow rate of 0.3 kg/s and mixes with saturated air at 7°C with a flow rate of 0.5 kg/s. Determine the relative humidity and temperature of the exit air.

16.57 Air is to be dehumidified with a silica-gel absorber. The initial air condition is 24°C with a humidity ratio of 0.018 kg water vapor/kg dry air; it is to leave with a humidity ratio of 0.005. The silica gel reduces the humidity ratio to 0.001, so a portion of the initial air bypasses the silica gel and mixes with the air leaving the gel. Determine the mass fraction bypassed.

16.58 Air enters a drier at 21°C and 25% relative humidity and leaves at 66°C and 50% relative humidity. Determine the air flow rate at 1 atm pressure if 5.44 kg/h of water is evaporated from wet material in the drier.

16.59 A continuous drier is designed to produce, in 24 h, 20 000 kg of product containing 5% water. The product enters the drier with 35% water. The air used for drying has an inlet temperature of 20°C and a relative humidity of 45% and is preheated to 65°C before reaching the product. The air leaves saturated at 43°C. Determine (a) the air flow rate in m³/s; (b) the heat required in the preheater.

16.60 A compressor receives an air–water vapor mixture at 96 kPa and 10°C with a vapor pressure of 1.0 kPa, and compresses it adiabatically to 207 kPa and 65°C. Determine (a) the work required per kg; (b) the relative humidity initially and finally.

16.61 A building has a heat loss of 300 kW. It must be heated by fresh air. The space is to be maintained at 24°C and 45% relative humidity, with an outside temperature of 2°C saturated. Humidification is achieved with 208 kPa saturated steam. Determine (a) the cold-air flow rate in m³/s; (b) the steam required in kg/s; (c) the hot-air flow rate.

16.62 Air is at 37°C dry-bulb and 15% relative humidity. The air is to be cooled by evaporative cooling until the relative humidity is 60%. Determine (a) the final temperature; (b) for 0.5 m³/s of air, the water required.

16.63 In the winter months a building is heated with conditioned air at 25°C dry-bulb and 45% relative humidity. Atmospheric air is at 2°C dry-bulb and 50% relative humidity. Water is supplied at 65°C. The air flow rate is 5 m³/s. Determine (a) the water added; (b) the heat supplied.

16.64 An air conditioning unit removes 35 kW of heat from a building. The unit handles 1.92 m³/s of air, 0.29 m³/s of which is outside air at 35°C dry-bulb and 24°C wet-bulb, while the remainder is recirculated air from the building. The air in the building is maintained at 25°C dry-bulb and 40% relative humidity. The refrigeration coil surface temperature is 10°C. (a) At what condition does the air enter the building? (b) How many kg/s of water are condensed by the air conditioning unit?

16.65 A heating system for an office bulding uses an adiabatic saturation air washer followed by a heating coil. A mixture of 1.2 m³/s of outside air at 5°C dry-bulb and 50% relative humidity and 6 m³/s of return air at 20°C dry-bulb and 55% relative humidity enter the saturator. The mixture leaves with a temperature 1°C less than saturation. The heating coil heats the mixture at 38°C. For the mixture leaving the heating coil, determine (a) the relative humidity; (b) the specific volume; (c) the heat supplied in the coil.

16.66 Water with a flow rate of 126 kg/s enters a natural-draft cooling tower at 43°C. The water is cooled to 29°C by air entering at 38°C dry-bulb and 24°C wet-bulb. The air leaves saturated at 40°C. Determine (a) the air supply in m³/s; (b) the makeup water required.

16.67 A natural-draft cooling tower receives water at 42°C and cools it to 26°C. Three hundred m³/s of atmospheric air is available at 20°C and 40% relative humidity. The exit air is saturated at 32°C. Makeup water is available at 26°C. Determine the flow rate of water that can be cooled.

16.68 A cooling tower is to be installed to supply cooling water. Water enters the tower at a rate of 190 kg/s at 46°C. Water must leave the tower with a flow rate of 190 kg/s at 29°C. The air enters at 24°C, 50% relative humidity, and 1 atm pressure, and leaves saturated at 31°C. Determine (a) the air flow rate in m³/s; (b) the makeup-water flow rate.

16.69 An air–water vapor mixture enters an adiabatic device with a pressure of 150 kPa, a temperature of 40°C, and an unknown relative humidity. The air flow rate is 0.2

kg/min. The mixture leaves the device at 30°C, 150 kPa, and 80% relative humidity. Water at 30°C is sprayed into the air for cooling. Determine the water required for 1 h of operation.

16.70 At a rate of 15 kg/s, air at 25°C and 30% relative humidity enters a device where heat is added and/or removed, and water at 20°C is removed. The exit-air dry-bulb temperature is 15°C, and the wet-bulb temperature is 5°C. Determine (a) the heat flow (denote direction); (b) the water flow rate.

Compressors

16.71 A two-stage compressor receives 0.151 kg/s of helium at 140 kPa and 300°K and delivers it at 7.0 MPa. The compression is polytropic with $n = 1.5$. The intercooler is ideal. Determine (a) the power required; (b) the intercooler pressure; (c) the maximum temperature; (d) the temperature for one stage of compression; (e) the heat transferred in the intercooler.

16.72 An air compressor is tested, and it is found that the electric motor used 37.3 kW when the compressor handled 0.189 m³/s of air at 101.4 kPa and 300°K and discharged it at 377.1 kPa. Determine (a) the overall adiabatic efficiency; (b) the overall isothermal efficiency.

16.73 Calculate the volumetric efficiency of a single-cylinder, double-acting compressor with a bore and stroke of 0.45 × 0.45 m. The compressor is tested at 150 rpm and found to deliver a gas from 101.3 kPa and 300°K to 675 kPa at a rate of 0.166 m³/s when $n = 1.33$ for expansion and compression processes.

16.74 For Problem 16.73 calculate the clearance percentage and estimate the clearance volume in m³.

16.75 A reciprocating compressor with a 3% clearance receives air at 100 kPa and 300°K and discharges it at 1.0 MPa. The expansion and compression are polytropic, with $n = 1.25$. There is a 5% pressure drop through the inlet and discharge valves. The air is warmed to 38°C by the cylinder walls by the end of the intake stroke. Determine (a) the theoretical and actual volumetric efficiency; (b) the work per kg; (c) the percentage of the work needed to overcome the throttling losses.

16.76 A reciprocating compressor has a 5% clearance with a bore and stroke of 25 × 30 cm. The compressor operates at 500 rpm. The air enters the cylinder at 27°C and 95 kPa and discharges at 2000 kPa. If $n = 1.3$ for compression and expansion processes, determine (a) the volumetric efficiency; (b) the volume of air handled at inlet conditions in m³/s; (c) the power required; (d) the mass of air discharged in kg/s; (e) the mass of air left at top dead center; (f) the p-V diagram.

16.77 A reciprocating double-acting, single-cylinder air compressor operates at 220 rpm with a piston speed of 200 m/s. The air is compressed isentropically from 96.5 kPa and 289°K to 655 kPa. The compressor clearance is 5.4%, and the air flow rate is 0.4545 kg/s. Determine for $n = 1.35$ (a) the volumetric efficiency; (b) the piston displacement; (c) the power; (d) the bore and stroke if $L = D$.

16.78 Derive expressions for the optimum intercooler pressures for a three-stage compressor with two stages of intercooling.

16.79 A small freezer is to have a cooling load of 1.1 tons at −25°C. The compressor and motor must be selected. The following conditions may be assumed: refrigerant = R 12, reciprocating compressor operating at 600 rpm, discharge pressure = 800 kPa, liquid receiver temperature = 20°C, $\eta_{cn} = 0.80$, and $\eta_m = 0.80$. Determine (a) the compressor displacement in liters; (b) the power required.

16.80 A single-stage, single-acting compressor has a 20-cm bore and a 25-cm stroke and rotates at 200 rpm. The compressor intake receives dry saturated ammonia vapor at −18°C and compresses it adiabatically to 1000 kPa. The actual shaft work of the compressor is 20% more than the reversible, isentropic compression work. The volumetric efficiency is 88%. After leaving the compressor and the condenser, the liquid ammonia is at 20°C. Determine (a) the ammonia flow rate; (b) the power input to the compressor; (c) the heat removed in the condenser.

16.81 A single-acting compressor with twin cylinders of 30 × 30 cm receives saturated ammonia vapor at 226 kPa and discharges it at 1200 kPa. Saturated liquid ammonia enters the expansion valve. Ice is to be manufactured at −9°C, and the water is available at 26°C. The compressor runs at 150 rpm, and the volumetric efficiency is 80%. Assuming the specific heat of ice to be 2.1 kJ/kg-K, determine (a) the ammonia flow rate; (b) the mass of ice manufactured; (c) the compressor power.

PROBLEMS (English Units)

Refrigeration Cycle

*16.1 A reversed Carnot cycle uses R 12 as the working fluid. The refrigerant enters the condenser as a saturated vapor at 90°F and leaves as a saturated liquid. The evaporator temperature is a constant −10°F. Determine per unit mass (a) the compressor work; (b) the turbine work; (c) the heat input; (d) the COP.

*16.2 A standard vapor-compression refrigeration system uses ammonia as the refrigerant. The ammonia leaves the evaporator at −15°F, and the condenser pressure is 150 psia. The flow rate is 45 lbm/min. Determine (a) the tons of refrigeration; (b) the power required; (c) the COP.

*16.3 A standard vapor-compression refrigeration system uses R 12 as the refrigerant. The R 12 leaves the evaporator at −15°F, and the condenser pressure is 150 psia. The flow rate is 45 lbm/min. Determine (a) the tons of refrigeration; (b) the power required; (c) the COP.

*16.4 A refrigeration compressor receives 42 ft³/min of R 12 at −10°F from the evaporator of a standard vapor-compression refrigeration system. The condenser pressure is 200 psia. Determine (a) the mass flow rate; (b) the power required; (c) the tons of refrigeration; (d) the COP.

*16.5 A vapor-compression refrigeration system has an evaporator operating at −20°F, a condenser with a pressure of 175 psia, and subcooling of 10°F of the R 12 leaving the condenser. Determine the percentage of increase in refrigerating effect because of the subcooling.

*16.6 In a vapor-compression refrigeration system, the R 12 leaves the evaporator at 20 psia and 20°F, enters the compressor, and is compressed isentropically to 200 psia. The discharge from the condenser is subcooled by 10°F. The refrigerant flow rate is 100 lbm/min. Determine (a) the tons of refrigeration; (b) the COP; (c) the increase in the refrigerating effect compared to the standard cycle.

*16.7   A vapor-compression refrigeration system uses R 12 and has an evaporator pressure of 15 psia and a condenser pressure of 150 psia. The temperature of the refrigerant entering the compressor is $0°F$ and entering the condenser is $180°F$. The refrigerant leaving the condenser is subcooled $10°F$, and the compressor power is 6.0 hp. Determine (a) the tons of refrigeration; (b) the COP; (c) the change of availability across the compressor.

*16.8   A 10-ton vapor-compression refrigeration system uses R 12. The vapor enters the compressor with $20°F$ of superheating and exits the condenser with $10°F$ of subcooling. The evaporator pressure is 15 psia, and the condenser pressure is 175 psia. The compressor isentropic efficiency is 80%. Determine (a) the power; (b) the COP; (c) the change of availability across the condenser.

*16.9   In a standard refrigerating cycle the evaporator temperature is $-10°F$, and the condenser temperature is $120°F$. Saturated vapor enters the compressor, and there is no subcooling of the liquid refrigerant. Consider ammonia and R 12, and select the refrigerant that gives the higher COP.

*16.10   A standard vapor-compression refrigerating system uses a subcooling-superheating heat exchanger with $10°F$ of superheating. The evaporator operates at $-20°F$, the condenser at 240 psia. There is a cooling requirement of 50 tons. If the fluid is R 12, determine (a) the mass flow rate; (b) the COP; (c) the refrigerating effect; (d) the degrees of subcooling; (e) the power required.

*16.11   A two-stage cascade refrigeration system uses ammonia as the working substance. The mass flow rate in the high-pressure loop is 600 lbm/hr. The condenser saturated temperature is $100°F$ and the evaporator temperature is $-40°F$. The cascade condenser is direct contact. Determine (a) the cascade condenser pressure for minimum work; (b) the refrigerating effect in tons; (c) the COP; (d) the power required.

*16.12   A two-stage cascade vapor-compression refrigeration system uses R 12 and produces 15 tons of refrigeration. A direct-contact heat exchanger is the cascade condenser. The low-temperature evaporator's temperature is $-20°F$, and the condenser pressure is 175 psia. The intermediate pressure is optimum. Determine (a) the mass flow rate in the upper and lower stages; (b) the COP; (c) the irreversibility rate in the cascade condenser for $T_0 = 77°F$ and $p_0 = 14.7$ psia.

*16.13   The same as Problem *16.12, assuming the isentropic compressor efficiency is 85% for the compressors.

*16.14   A two-stage cascade vapor-compression refrigeration system uses R 12 and has an evaporator temperature of $-10°F$, a cascade condenser pressure of 60 psia, and a condenser pressure of 200 psia. The flow rate through the evaporator is 1.3 lbm/sec. Determine (a) the mass flow rate through the condenser; (b) the COP; (c) the total power.

*16.15   A two-stage cascade vapor-compression refrigeration system uses ammonia in the low-temperature loop and R 12 in the high-temperature loop. The evaporator operates at $-25°F$, and the ammonia discharges to the closed cascade condenser at 50 psia. The R 12 must be $20°F$ lower than the saturated ammonia temperature to provide adequate cooling. The R 12 compressor has a pressure ratio of $5:1$. The ammonia provides 15 tons of cooling. Determine (a) the pressure of the R 12 in the cascade heat exchanger and the condenser; (b) the mass flow rates of ammonia and R 12; (c) the total power; (d) the COP.

*16.16   A compressor serves two evaporators, one providing 5 tons of cooling at $-20°F$ and the other 10 tons of cooling at $10°F$. The system is a vapor-compression refrigeration system with the condenser pressure at 200 psia. Determine (a) the refrigerant flow through each evaporator; (b) the R 12 state entering the compressor; (c) the compressor power.

*16.17 A vapor-compression refrigeration system uses a subcooling-superheating heat exchanger located after the evaporator to subcool the refrigerant entering the expansion valve. The refrigerant leaving the evaporator is superheated in the process. Assume the refrigerant leaves the evaporator as a saturated vapor and the condenser as a saturated liquid and that no pressure drops occur in the heat exchangers. The evaporator temperature is 0°F, the condenser pressure is 200 psia, and the flow rate is 20 lbm/min. Determine (a) the compressor power; (b) the tons of refrigeration; (c) the COP.

*16.18 A heat pump uses the vapor-compression cycle with R 12 and provides 850 Btu/min of heat. The condenser pressure is 125 psia, and the evaporator pressure is 43 psia. Determine (a) the mass flow rate of R 12; (b) the COP; (c) the power required.

*16.19 A vapor-compression heat pump system uses R 12. The compressor receives 26 ft³/min at 25 psia and 30°F and discharges it at 150 psia and 180°F. The refrigerant leaves the condenser as a saturated liquid. Determine (a) the heat output in Btu/hr; (b) the power required; (c) the COP.

*16.20 A home uses a vapor-compression heat pump with R 12 as the means for providing heating. The house requires 48,000 Btu/hr of heat when the inside temperature is 70°F and the outside temperature is 30°F. The minimum temperature difference between the heat exchangers and air is 20°F. (a) Specify the evaporator and condenser pressures. (b) Determine for the specified cycle its COP, the mass flow rate of R 12, and the compressor power required.

*16.21 A vapor-compression heat pump uses R 12 and provides 34,000 Btu/hr of heating. The evaporator pressure is 30 psia, and the refrigerant enters the compressor at 20°F. The compressor's isentropic efficiency is 80%, and the condenser pressure is 125 psia. The electricity to drive the compressor comes from a power plant with an efficiency of 40%. Determine (a) the compressor power; (b) the ratio of heat used in the power plant to produce the electricity to the heat output of the heat pump.

*16.22 The air-cooled condenser of a vapor-compression heat pump system heats 1750 ft³/min of air from 70° to 95°F at a constant pressure of 1 atm. The R 12 exits the condenser at 150 psia and 95°F. The evaporator temperature is 10°F. Determine (a) the refrigerant mass flow rate; (b) the compressor power; (c) the COP.

*16.23 A vapor-compression refrigeration system using ammonia as the working substance is required to provide 30 tons of refrigeration at 0°F. To accomplish this, the ammonia operates between pressure limits of 20.88 psia and 210 psia. Determine (a) the quality of ammonia entering the evaporator; (b) the mass flow rate; (c) the change of available energy across the condenser if $T_0 = 80°F$; (d) the volume flow rate at the compressor intake; (e) the bore (D) and stroke (L) ($L = D$) of a double-acting, reciprocating compressor at 600 rpm. Assume that the fluid is completely discharged from the cylinder each revolution.

*16.24 A vapor-compression refrigeration system uses one compressor for two evaporators, one operating at 50°F and providing 10 tons of cooling and the other operating at 10°F and providing 5 tons of cooling. Determine for R 12 and a condenser operating at 100°F (a) the mass flow rate; (b) the COP; (c) the power required; (d) the second-law efficiency.

*16.25 A capillary tube vapor-compression refrigeration system operates on R 12 with an evaporator temperature of 0°F and a condenser temperature of 120°F. The system provides 1 ton of cooling. Determine (a) the mass flow rate; (b) the power; (c) the COP; (d) the temperature of R 12 entering the compressor.

*16.26 The same operating system, limits, and refrigerant as Problem *16.25. The compressor/motor has an efficiency of 50%. Determine (a) the COP; (b) the power required. (c) The system operates 6 hr/day, and the cost of electricity is \$0.15/kWh. Find the expense per day of operating the unit.

Air Conditioning

*16.27 Using the psychrometric chart determine (a) the humidity ratio and specific enthalpy of air 90°F dry-bulb and 60% relative humidity; (b) the dew point for air at 90°F dry-bulb and 80°F wet-bulb; (c) the relative humidity and humidity ratio for air at 105°F dry-bulb and 75°F wet-bulb.

*16.28 A tank 1 ft in diameter and 6 ft long contains air at 105°F dry-bulb and 50% relative humidity. The surrounding temperature lowers at night, and the air in the tank then has a dry-bulb temperature of 80°F. How much, if any, water condenses?

*16.29 Using the psychrometric chart, determine (a) the specific enthalpy and specific volume for air with a relative humidity of 60% and a dry-bulb temperature of 100°F; (b) the humidity ratio and the relative humidity, given a wet-bulb temperature of 75°F and a specific volume of 14.5 ft³/lbm. (c) the wet- and dry-bulb temperatures, given a relative humidity of 70% and a humidity ratio of 0.018 lbm vapor/lbm air.

*16.30 Twenty-five lbm/min of air enters a dehumidifier at 75°F, 1 atm, and 60% relative humidity. The air is cooled to 60°F, with water and air exiting separately at this state. Determine (a) the mass flow rate of water leaving; (b) the heat transfer.

*16.31 Sixty-six lbm/min of air at 1 atm, 85°F, and 70% relative humidity enters a dehumidifier. The air exits at 60°F and 95% relative humidity, and water exits at 60°F. Refrigerant enters the dehumidifier coils with an enthalpy of 175 Btu/lbm and leaves with an enthalpy of 440 Btu/lbm. Determine the refrigerant flow rate required.

*16.32 A small basement dehumidifier receives 35 ft³/min of air at 70°F and 70% relative humidity. The dehumidifier removes 2 qt of water in a 24-h period. What is the humidity ratio of the air leaving the dehumidifier?

*16.33 An air conditioning system dehumidifies and then reheats the air. The system operates with 2625 ft³/min of air entering the dehumidifier at 80°F and 70% relative humidity. The air leaves the reheater at 70°F and 50% relative humidity. Determine (a) the temperature of the air leaving the dehumidifier before it is reheated; (b) the flow rate of the condensed water; (c) the tons of cooling required; (d) the reheat in Btu/min.

*16.34 Outside air entering at 1925 ft³/min, 50°F, and 70% relative humidity is reheated and humidified so the air state entering the heating system is 85°F and 35% relative humidity. The humidification occurs with water at 65°F. Determine (a) the water flow rate required; (b) the heat transfer in Btu/min.

*16.35 An air-conditioning system consists of a heating section followed by a humidifying section that supplies saturated steam at 212°F. Twenty-one hundred ft³/min of air enters the heating section at 50°F and 70% relative humidity. It leaves the humidifying section at 70°F and 50% relative humidity. Determine (a) the temperature of the air when it leaves the heating section; (b) the flow rate of steam required; (c) the heat required.

*16.36 Air at 60°F and 60% relative humidity enters at 1750 ft³/min the heating section of a heater/humidifier air conditioning system. The air is heated and passes over a mesh, where water at 70°F is supplied. The air leaves the unit at 80°F and 40% relative humidity. Determine (a) the temperature of the air leaving the heating section; (b) the flow rate of water required; (c) the heat added.

*16.37 An air-conditioned classroom receives air at 60°F. The air leaving via the exit duct is at 80°F and 50% relative humidity. The people in the room may be viewed as adding 0.165 lbm/min of water vapor at 95°F. The heat addition from the people in the room and the lights and surroundings is estimated to be 33,000 Btu/hr. Determine the inlet volume flow rate, relative humidity, and humidity ratio.

*16.38 Air at 1.1 lbm/s, 95°F, and 25% relative humidity enters an adiabatic evaporative cooler. The cooler receives water at 70°F, and the air leaves saturated at 80°F. Determine the water flow rate.

*16.39 Two airstreams mix adiabatically. The first stream at 95°C amd 60% relative humidity with a mass flow rate of 1 lbm/sec mixes with the second at 70°F and 50% relative humidity with a flow rate of 2 lbm/sec. Determine the exit air temperature and relative humidity.

*16.40 It is desired to have 17,500 ft³/min of air at 70°F and 50% relative humidity. This is obtained by adiabatically mixing 10,500 ft³/min of air at 80°F and 65% relative humidity with another airstream. Specify the volume flow rate, temperature, and relative humidity of that airstream.

*16.41 Air at 95°F and 40% relative humidity enters an adiabatic mixing chamber with a flow rate of 40 lbm/min and mixes with saturated air at 45°F with a flow rate of 65 lbm/min. Determine the relative humidity and temperature of the exit air.

*16.42 In the winter months, a building is heated with conditioned air at 78°F dry-bulb and 45% relative humidity. Atmospheric air is at 36°F dry-bulb and 50% relative humidity. Water is supplied at 150°F. The air flow rate is 1×10^4 ft³/min. Determine (a) the water added in lbm/min; (b) the heat supplied.

*16.43 An air-conditioning unit removes 1.2×10^5 Btu/hr of heat from a building. The unit handles 4000 ft³/min of air, 600 ft³/min of which is outside air at 95°F dry-bulb and 75°F wet-bulb, while the remainder is recirculated air from the building. The air in the building is maintained at 78°F dry-bulb and 40% relative humidity. The refrigeration coil surface temperature is 50°F. (a) At what condition does the air enter the building? (b) How many lbm/hr of water are condensed by the air conditioning unit?

*16.44 A cooling tower must cool 9000 gal/min of water from 84° to 68°F with inlet air conditions of 70°F dry-bulb and 60°F wet-bulb. The air exit condition is saturated at 80°F. Determine (a) the volumetric flow rate of air required; (b) the gal/min of water makeup required.

*16.45 A cooling tower receives 5×10^5 ft³/min of atmospheric air at 14.7 psia, 70°F dry-bulb, and 40% relative humidity, and discharges it at 95°F dry-bulb and 80% relative humidity. Water enters at 110°F and leaves at 76°F. Determine (a) the mass flow rate of air entering the cooling tower; (b) the mass flow rate of water entering; (c) the mass flow rate of water evaporated.

*16.46 A natural-draft cooling tower receives water at 110°F and cools it to 80°F. Atmospheric air is available at 10,500 ft³/sec, 70°F, and 40% relative humidity. The exit air is saturated at 90°F. Makeup water is available at 80°F. Determine the flow rate of water that can be cooled.

Compressors

*16.47   Air at 230 ft³/min is compressed from 14.7 psia and 70°F to 50 psia by a 14 × 14-in. single-stage, double-acting compressor operating at 100 rpm. The expansion and compression processes are reversible adiabatically. Determine (a) the discharge temperature; (b) the power required.

*16.48   A 15 × 15-in. single-stage, double-acting air compressor has a 5% clearance, runs at 150 rpm, and receives air at 14.5 psia and 74°F and discharges it at 50 psia. The compression and discharge processes are polytropic with $n = 1.35$. Determine (a) the volume and mass flow rates at inlet conditions; (b) the volume flow rate at discharge; (c) the power required.

*16.49   A gas compressor handles 350 ft³/min of a gas at 14.7 psia and 80°F and discharges it at 98 psia. The piston bore and stroke are 18 × 18 in.; the single cylinder is double-acting; the compressor operates at 150 rpm; and $n = 1.32$. Determine (a) the volumetric efficiency; (b) the clearance percentage.

*16.50   A two-stage, double-acting air compressor operates at 165 rpm. The low-pressure cylinder has a bore and stroke of 12 × 14 in., and the high-pressure cylinder a bore and stroke of 8 × 14 in. Additionally, 240 ft³/min of air enters at 14 psia and 110°F and leaves at 105 psia with ideal intercooling. The isothermal efficiency is 75%. Determine (a) the power; (b) the volumetric efficiency; (c) the volumetric efficiency if the piston rods are included and have a 2-in. diameter.

*16.51   A reciprocating air compressor with 6% clearance receives 150 ft³/min of air at 14.5 psia and 120°F. The discharge pressure is 45 psia, and the overall adiabatic efficiency is 68%. Determine the power required.

COMPUTER PROBLEMS

C16.1   Use R12CYCLE.TK to calculate the COP of a standard vapor-compression cycle operating under various evaporating and condensing temperatures. For condensing temperatures of 25°C, 35°C, and 45°C, calculate the COP for evaporating temperatures between −40° and 10°C and plot the results.

C16.2   Using R12CYCLE.TK, investigate the effects of superheating of the compressor suction vapor. For a condensing temperature of 100°F and evaporating temperatures of −10° and 40°F, vary the superheat between 0° and 20°R. Discuss your conclusions.

C16.3   Using R12CYCLE.TK, investigate the effects of subcooling of the condenser outlet. For a condensing temperature of 100°F and an evaporating temperature of −10°F, vary the subcooling between 0° and 20°R. Discuss your conclusions.

C16.4   Using R12CYCLE.TK, investigate the effects of pressure drops on the cycle. For a condensing temperature of 40°C and an evaporating temperature of −25°C, vary the evaporator and condenser pressure drops between 0 and 20 kPa. Discuss your conclusions.

C16.5   Modify R12CYCLE.TK to include equations to model an ideal reciprocating compressor. Based on a given compressor displacement, the equations should calculate the volumetric efficiency, refrigerant mass flow, compressor power, and refrigeration capacity. For a compressor with 5% clearance and 2 m³/min displacement, calculate the above variables for evaporating temperatures between −50° and 20°C and for a condensing temperature of 40°C. Plot the results.

C16.6 Use the modified version of R12CYCLE.TK developed for Problem C16.5 to analyze a heat pump installation. Calculate the BTU/hr delivered to the building and the COP for outside air temperatures between 0° and 50°F. Assume the building temperature is 70°F, the evaporating temperature is 10°F below the outside temperature, and the condensing temperature is 10°F above the building temperature. Comment on the results.

C16.7 An air–water vapor mixture has a dry-bulb temperature of 30°C and a humidity ratio of 0.015 kg vapor/kg air. Using PSYCHRO.TK, calculate the wet-bulb and dew point temperatures for a barometric pressure of 101 kPa. Repeat for a barometric pressure of 80 kPa.

C16.8 An airflow of 0.5 kg/s at 15°C and 20% relative humidity is being humidified by the addition of 0.004 kg/s of saturated steam at 101 kPa. Use PSYCHRO2.TK to determine the dry-bulb and wet-bulb temperatures of the exiting air.

17

Fluid Flow in Nozzles and Turbomachinery

This chapter explores the thermodynamic aspects of fluid flow, including property variations of the fluid while undergoing a process. The process may be reversible or irreversible, the fluid compressible or incompressible. Thermodynamics and fluid dynamics overlap a great deal, and the purpose here is not to present a course in fluid dynamics, but rather to analyze those situations that require thermodynamic understanding. All actual fluid flow is irreversible and three-dimensional, but we could not proceed very far in an analysis if we held to these restrictions. A large number of flow conditions may be approximated by assuming one-dimensional, steady-state, steady-flow conditions. These are the primary restrictions we will use in this chapter: the flow is one-dimensional; the mass flow rate through a passage is constant with time; and the property variations at a point do not change with time. These may seem overly restrictive, but in many situations the actual variations from these conditions are small and may be neglected.

In this chapter we will investigate

- The conservation of momentum;
- Various fluid properties such as acoustic velocity, Mach number, and stagnation properties;
- The flow of fluids in nozzles and diffusers;

- Why the conservation of energy describes the ideal shape of nozzles and diffusers;
- Normal shock waves;
- The effect of supersaturation on the flow of a vapor in a nozzle;
- How wind turbines convert the wind's energy into mechanical work;
- Energy transfer in turbomachinery.

17.1 CONSERVATION OF MASS

The conservation of mass for steady, one-dimensional flow is

$$\dot{m} = \frac{A\mathrm{v}}{v} \tag{17.1}$$

where $\dot{m}$ is the mass flow rate, v is the specific volume, A is the flow area, and v is the average fluid velocity. For steady flow, the mass flow rate is the same at one plane as at another plane; thus,

$$\dot{m} = \frac{A_1\mathrm{v}_1}{v_1} = \frac{A_2\mathrm{v}_2}{v_2} = \cdots = \frac{A_i\mathrm{v}_i}{v_i} \tag{17.2}$$

It will prove convenient to express the mass flow rate as a differential equation. Taking the logarithm of equation (17.1) and then the derivative yields

$$0 = \frac{dA}{A} + \frac{d\mathrm{v}}{\mathrm{v}} - \frac{dv}{v} \tag{17.3}$$

17.2 CONSERVATION OF MOMENTUM

In developing the equation for the change of momentum and force acting on a control volume, let us recall Newton's second law of motion. This states that the rate of change of momentum in a direction is equal to the sum of external forces acting on the body in the same direction.

$$\sum \mathbf{F} = \frac{d}{dt}(m\mathbf{v}) \tag{17.4}$$

There are two general types of forces that $\sum \mathbf{F}$ comprises, body and surface. Body forces are functions of the system size (volume or mass) and include forces from force fields. The greater the system size is, the greater the total force acting on the system as a result of a magnetic field. Should acceleration be important, as in the case in turbomachines, then centrifugal and Coriolis forces must be included. Surface forces are exerted on the system surface (control volume surface) by the surroundings. These forces may be represented by the viscous pressure tensor, which has normal and shearing components.

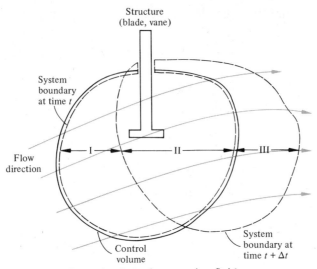

Figure 17.1 Control volume in a moving fluid.

Figure 17.1 illustrates a control volume with an object in it, such as a blade, vane, or strut. Considering only one component (direction) of equation (17.4) for Cartesian coordinates yields

$$\sum F_x = \frac{d}{dt}(mv_x) \tag{17.5}$$

Considering the right-hand side of equation (17.5), it is desirable to change systems from constant mass to constant volume. To accomplish this consider that at time t,

$$(mv_x)_t = \Sigma(v_x m)_{\mathrm{I}t} + (v_x m)_{\mathrm{II}t} + \Sigma(v_x m)_{\mathrm{III}t} \tag{17.6}$$

and at time $t + \Delta t$,

$$(mv_x)_{t+\Delta t} = \Sigma(v_x m)_{\mathrm{I}t+\Delta t} + (v_x m)_{\mathrm{II}t+\Delta t} + \Sigma(v_x m)_{\mathrm{III}t+\Delta t} \tag{17.7}$$

Subtract equation (17.6) from (17.7) and divide by Δt:

$$\frac{(mv_x)_{t+\Delta t} - (mv_x)_t}{\Delta t} = \frac{(mv_x)_{\mathrm{II}t+\Delta t} - (mv_x)_{\mathrm{II}t}}{\Delta t} + \sum v_x \frac{m_{\mathrm{I}t+\Delta t} - m_{\mathrm{I}t}}{\Delta t}$$

$$+ \sum v_x \frac{m_{\mathrm{III}t+\Delta t} - m_{\mathrm{III}t}}{\Delta t}$$

A summation of velocities $\Sigma \, v_x$ is used because there are a variety of x velocities in the fluid; each stream line would have its own. When the limit of Δt approaching zero is taken, the summations become integrals and mass differences become differentials.

$$\frac{d(mv_x)}{dt} = \frac{d}{dt}(mv_x)_{\mathrm{II}} - \int v_x \frac{dm_{\mathrm{I}}}{dt} + \int v_x \frac{dm_{\mathrm{III}}}{dt}$$

$$\frac{d(mv_x)}{dt} = \frac{d}{dt}(mv_x)_{CV} - \int v_x \dot{m}_{\mathrm{in}} + \int v_x \dot{m}_{\mathrm{out}}$$

Equation (17.5) becomes

$$\sum F_x = \frac{d}{dt}(mv_x)_{CV} + \int v_x \dot{m}_{out} - \int v_x \dot{m}_{in} \qquad (17.8)$$

Equation (17.8) is not in terms of measurable properties. Referring to equations (3.7) and (3.8), equation (17.8) becomes

$$\sum F_x = \frac{d}{dt}\int_{CV} \rho v_x d\mathcal{V} + \int_A (\rho v_x v_n dA)_{out} - \int_A (\rho v_x v_n dA)_{in} \qquad (17.9)$$

In vector notation equation (17.9) may be written

$$\sum \mathbf{F} = \frac{d}{dt}\int_{CV} \rho \mathbf{v}\, d\mathcal{V} + \int_A \rho(\mathbf{v} \cdot d\mathbf{A})\mathbf{v} \qquad (17.10)$$

For one-dimensional steady flow, equation (17.9) becomes

$$\sum F_x = \rho_2 A_2 v_{x_2}^2 - \rho_1 A_1 v_x^2 \qquad (17.11)$$

Moment of Momentum

Centrifugal forces are important in the analysis of rotating machinery. In this instance not only are forces of interest, but the moment of these forces — or torque — is important. By looking at equation (17.10) we may see that the moment of forces is the same as the moment of momentum. Since the expressions for momentum have measurable fluid quantities, they are often used. A torque is produced when a force is acting at a distance. When a wrench is used to loosen or tighten a nut, a force is applied (your hand) normal to the wrench and at a distance r (wrench length) from the nut. In vector notation this is the vector product, or

$$\sum \mathbf{F} \times \mathbf{r} = \frac{d}{dt}\int_{CV} \rho \mathbf{v} \times \mathbf{r}\, d\mathcal{V} + \int \rho(\mathbf{v} \cdot d\mathbf{A})\mathbf{v} \times \mathbf{r} \qquad (17.12)$$

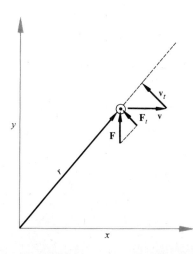

Figure 17.2 Point at a distance r acted on by force **F** and fluid velocity **v**.

Consider Figure 17.2, which illustrates a force acting at a distance r in the xy plane. The torque (moment of momentum) is in the z direction (out of the page). Equation (17.12) in Cartesian coordinates for this example is

$$\sum rF_t = \frac{d}{dt} \int_{CV} \rho r v_t \, d\mathcal{V} + \int_A (\rho v_n \, dA \; r v_t)_{\text{out}} - \int_A (\rho v_n \, dA \; r v_t)_{\text{in}} \quad (17.13)$$

17.3 ACOUSTIC VELOCITY

An important velocity in the study of fluid mechanics is the *acoustic* or *sonic velocity*. This is the velocity at which a small pressure wave moves through a fluid. When we talk, for instance, we are generating pressure waves of varying intensities. These waves move through the air at a certain velocity, the acoustic velocity, commonly called the speed of sound. This is about 336 m/s at standard atmospheric conditions. The speed of sound is finite; for example, take the case where a gun is fired at some distance, say several hundred yards. The flash or movement of the gun is seen, then the sound of gunfire is heard. The speed of light is 2.998×10^8 m/s, exceedingly greater than the acoustic velocity of 336 m/s. For larger pressure waves, such as shock waves, the propagation may be several times the acoustic velocity. The analysis of shock waves is typically included in a course on compressible fluid flow.

Let us consider Figure 17.3, which illustrates a small pressure wave, caused by the piston and moving through a compressible fluid. The pressure wave must be weak enough for the property changes to be differentially small. If the wave is too great, the criterion of differential property change is not met, and the following derivation is not valid. The wave, generated by the piston movement in Figure 17.3, moves at the acoustic velocity, a, into the stationary gas. If we place ourselves on the pressure wave, it seems as if the fluid is approaching from the right, the stationary region, at the acoustic velocity and leaving at a different velocity because of the pressure change across the wave front. The mass must be conserved entering and leaving the control surface. The sum of the forces acting on the control surface may be written by using

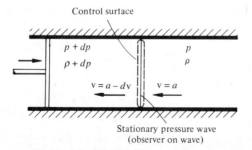

Stationary pressure wave
(observer on wave)

Figure 17.3 A small pressure wave in the stationary coordinate system.

equation (17.11) or by considering the difference in forces acting on the control surface.

$$F_{net} = pA - (p + dp)A$$

or

$$F_{net} = \dot{m}v_{x_2} - \dot{m}v_{x_1}$$

where shear forces and friction are negligible;

$$\dot{m} = \rho A v = \rho A a$$

Equating the net force across the control surface yields

$$pA - (p + dp)A = Aa[(a - dv) - a]\rho$$

which reduces to

$$dp = \rho a\, dv \qquad\qquad (17.14)$$

The continuity equation across the control surface is

$$\rho a A = (\rho + d\rho)(a - dv)A$$

We neglect second-order terms as being very much less than first-order terms. Thus,

$$dv = a\,\frac{d\rho}{\rho} \qquad\qquad (17.15)$$

Substituting equation (17.15) into equation (17.14) yields

$$a = \left(\frac{dp}{d\rho}\right)^{1/2} \qquad\qquad (17.16)$$

For small-pressure waves, the compression process is essentially isentropic. For an ideal gas, the isentropic process is

$$pv^k = C \quad \text{or} \quad p = C\rho^k$$

We differentiate

$$dp = kpv\, d\rho$$

$$\left(\frac{\partial p}{\partial \rho}\right)_s = kpv = kRT$$

Therefore, for English units,

$$a = (g_c kRT)^{1/2} \text{ ft/sec} \qquad\qquad (17.17a)$$

and for SI units,

$$a = (kRT)^{1/2} \text{ m/s} \qquad\qquad (17.17b)$$

In Chapter 10 we found that physical coefficients were defined in terms of partial derivatives. We defined the coefficient of isothermal compressibility, β_T, in equation

(10.49). Another coefficient, the coefficient of adiabatic compressibility, β_s, is

$$\beta_s = -\frac{1}{v}\left(\frac{\partial v}{\partial p}\right)_s \qquad (17.18)$$

where the minus sign allows β_s to be always positive. This coefficient is tabulated for liquids and gases. The acoustic velocity may be defined in terms of β_s as follows:

$$a = \left(\frac{1}{\rho\beta_s}\right)^{1/2} \qquad (17.19)$$

where consistent units must be used.

17.4 STAGNATION PROPERTIES

Let us consider a stream of fluid flowing through an insulated pipe. The energy of this fluid at any plane is $(h + \text{k.e.})$, the sum of the enthalpy and kinetic energy. If we follow the fluid back to its starting point, where the velocity is zero, we will find that the energy of the fluid is h_0, where the subscript zero stands for zero velocity. Thus, since the energy is constant throughout the adiabatic duct,

$$h_0 = h + \frac{v^2}{2} \text{ J/kg} \qquad (17.20a)$$

$$h_0 = h + \frac{v^2}{2g_c\mathscr{J}} \text{ Btu/lbm} \qquad (17.20b)$$

Let us work with equations (17.20a) and (17.20b) for the time being. We substitute for the enthalpy, $h = c_p T$; thus,

$$T_0 = T + \frac{v^2}{2c_p}$$

$$\frac{T_0}{T} = 1 + \frac{v^2}{2c_p T} \qquad (17.21a)$$

$$\frac{T_0}{T} = 1 + \frac{v^2}{2g_c\mathscr{J}c_p T} \qquad (17.21b)$$

Thus, the temperature at zero velocity may be expressed in terms of the fluid temperature and velocity at any plane. This zero velocity condition is the *stagnation condition*. We may determine the temperature at the stagnation state by specifying that the fluid decelerate adiabatically. If a second condition, that the deceleration be isentropic, is invoked, then the isentropic stagnation pressure may be found. For an isentropic process for an ideal gas

$$\frac{T_2}{T_1} = \left(\frac{p_2}{p_1}\right)^{(k-1)/k}$$

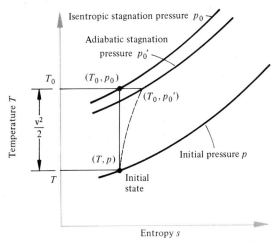

Figure 17.4 A *T-s* diagram illustrating stagnation temperature and pressure.

Thus,

$$\frac{p_0}{p} = \left(1 + \frac{v^2}{2c_p T}\right)^{k/(k-1)} \tag{17.22a}$$

$$\frac{p_0}{p} = \left(1 + \frac{v^2}{2g_c \mathscr{J} c_p T}\right)^{k/(k-1)} \tag{17.22b}$$

Equation (17.22) gives the isentropic stagnation pressure in terms of the fluid pressure, temperature, and velocity. The stagnation temperature, T_0, is the same for the adiabatic and isentropic deceleration. Figure 17.4 shows a *T-s* diagram with the various states. The irreversibilities associated with deceleration in the adiabatic case result in an entropy increase. The irreversibilities are also manifested in the adiabatic stagnation pressure, $p_{0'}$, being less than the maximum pressure, the isentropic stagnation pressure, p_0.

17.5 MACH NUMBER

The *Mach number, M*, is defined as the ratio of the actual velocity, v, divided by the local speed of sound, a, in the fluid:

$$M = \frac{v}{a} \tag{17.23}$$

Equations (17.17) and (17.23) may be substituted into equation (17.21) to yield the temperature in terms of the Mach number.

$$\frac{T_0}{T} = 1 + \frac{k-1}{2} M^2 \tag{17.24}$$

Example 17.1

A fluid is flowing through a device with a velocity of 350 m/s at 350°C and 1000 kPa. Determine the isentropic stagnation temperature, pressure, and enthalpy of the fluid if it is air and if it is steam.

Solution

Given: Steam flowing steadily through a device at known velocity, temperature, and pressure.

Find: The stagnation temperature, pressure, and enthalpy of the air and steam.

Sketch and Given Data:

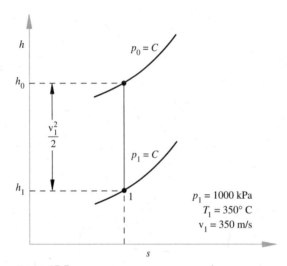

Figure 17.5

Assumptions:

1. The steam flows steadily.
2. The stagnation state is the isentropic stagnation state.
3. The initial and stagnation states are equilibrium states.

Analysis: The stagnation enthalpy and temperature for air may be found from equation (17.20a).

$$h_0 = h_1 + \frac{v_1^2}{2}$$

$$c_p T_0 = c_p T_1 + \frac{v_1^2}{2(1000 \text{ J/kJ})}$$

$$T_0 = 623°\text{K} + \frac{(350 \text{ m/s})^2}{2(1000 \text{ J/kJ})(1.0047 \text{ kJ/kg-K})} = 684°\text{K}$$

hence the stagnation enthalpy is

$$h_0 = c_p T_0 = (1.0047 \text{ kJ/kg-K})(684°\text{K}) = 687.2 \text{ kJ/kg}$$

The stagnation pressure may be found from Equation (17.22a).

$$p_0 = p_1 \left(1 - \frac{v^2}{2c_p T_1}\right)^{k/(k-1)}$$

$$p_0 = (1000 \text{ kPa}) \left(1 + \frac{(350 \text{ m/s})^2}{2(1004.7 \text{ J/kg-K})(623°\text{K})}\right)^{1.4/0.4} = 1386 \text{ kPa}$$

Determine the enthalpy and entropy of the steam's initial state using the steam tables: $h_1 = 3157.95$ kJ/kg and $s_1 = 7.3002$ kJ/kg-K. From equation (17.20a) the stagnation enthalpy may be determined:

$$h_0 = h_1 + \frac{v_1^2}{2} = 3157.95 \text{ kJ/kg} + \frac{(350 \text{ m/s})^2}{2(1000 \text{ J/kJ})} = 3219.2 \text{ kJ/kg}$$

For the isentropic stagnation state, $s_0 = s_1$. Using the steam tables and knowing the value for entropy and enthalpy allows the determination of the stagnation temperature and pressure: $T_0 = 380.7°$C and $p_0 = 1235$ kPa.

Comment: The determination of the stagnation state using the tables manually is time consuming and in some instances nearly impossible. The tables are not designed for using enthalpy and entropy as independent input variables, but for using temperature and pressure. The TK Solver model SHTSTM.TK accepts entropy and enthalpy inputs and thus readily determines the required information. Graphical techniques are also useful. The Mollier diagram gives the values of stagnation pressure and temperature from the enthalpy and entropy, the chart's coordinates. ■

17.6 FIRST-LAW ANALYSIS

Consider the control volume shown in Figure 17.6. Fluid flows steadily into and out of the control volume, but no work is done. The conservation of energy for the control volume is

$$\dot{Q} + \dot{m}[(\text{p.e.})_1 + h_1 + (\text{k.e.})_1] = \dot{m}[(\text{p.e.})_2 + h_2 + (\text{k.e.})_2]$$

We divide by $\dot{m}$, putting the equation on the basis of a unit mass within the control volume. This permits us to use the constant-mass expression for the heat flow. On the differential basis the equation becomes

$$\delta q = dh + d(\text{k.e.}) + d(\text{p.e.}) \qquad (17.25)$$

and from $h = u + pv$

$$dh = du + p \, dv + v \, dp$$

For a unit mass

$$\delta q = du + p \, dv$$

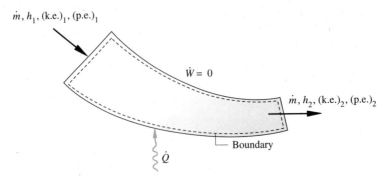

$\dot{m}, h_1, (\text{k.e.})_1, (\text{p.e.})_1$

$\dot{W} = 0$

$\dot{m}, h_2, (\text{k.e.})_2, (\text{p.e.})_2$

Boundary

$\dot{Q}$

Figure 17.6 A control with a steady flow of fluid and heat transfer.

also

$$d(\text{k.e.}) = \text{v} \, d\text{v}$$

$$d(\text{p.e.}) = g \, dz$$

Substituting into equation (17.25) yields

$$\frac{dp}{\rho} + \text{v} \, d\text{v} + g \, dz = 0 \tag{17.26}$$

Equation (17.26) is *Euler's equation for steady flow.* Bernoulli's equation in fluid mechanics is the same, but it is not written on a differential basis. Euler's equation was developed from thermodynamic or energy arguments, whereas in fluid mechanics the equation is typically developed from principles of mechanics or momentum considerations. This demonstrates, in part, the close relationship of thermodynamics and fluid mechanics and indicates the breadth of thermodynamic analysis.

17.7 NOZZLES

A nozzle is a device with two purposes: to convert thermal energy into kinetic energy and to direct the mass flow exiting from it in a specified direction. This is accomplished by means of a variable-area duct.

Let us talk about an adiabatic, variable-area duct as shown in Figure 17.7. For steady flow, the energy at any plane is constant since $q = w = \Delta\text{p.e.} = 0$ and is equal to the sum of the kinetic energy and the enthalpy.

$$\text{Total energy} = h + \text{k.e.}$$

However, this is equal to the stagnation enthalpy, h_0.

$$h_0 = h + \frac{\text{v}^2}{2} \text{ J/kg}$$

$$h_0 = h + \frac{\text{v}^2}{2g_c\mathscr{J}} \text{ Btu/lbm}$$

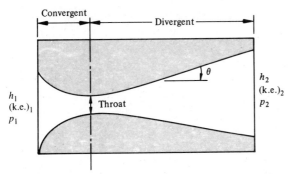

Figure 17.7 An adiabatic convergent-divergent nozzle.

We solve for the velocity:

$$v = \sqrt{2(h_0 - h)} \text{ m/s} \qquad (17.27a)$$

$$v = \sqrt{2g_c \mathcal{J}(h_0 - h)} = 223.8\sqrt{h_0 - h} \text{ ft/sec} \qquad (17.27b)$$

Thus, the velocity at any plane may be expressed in terms of the enthalpy at that plane and the stagnation enthalpy.

What happens if the flow has an initial velocity, v_1, as illustrated in the figure? The first law for an open system with no heat transfer or work interactions is

$$h_1 + (\text{k.e.})_1 = h_2 + (\text{k.e.})_2$$

$$h_1 + \frac{v_1^2}{2} = h_2 + \frac{v_2^2}{2} \qquad (17.28a)$$

$$v_2 = [2(h_1 - h_2) + v_1^2]^{1/2} \text{ m/s}$$

$$v_2 = [2g_c \mathcal{J}(h_1 - h_2) + v_1^2]^{1/2} \text{ ft/sec} \qquad (17.28b)$$

This is far more awkward to use than equation (17.27). Note that $h_1 + (\text{k.e.})_1 = h_0$, and if h_0 is used, the solution for v_2 is greatly simplified. In most instances it is easier to solve for h_0 initially from the inlet conditions, and use the stagnation enthalpy in the subsequent calculations.

One purpose of a nozzle is to convert thermal energy, enthalpy, into kinetic energy. Figure 17.7 shows a convergent (area decreasing) section followed by a divergent (area increasing) section. Since both sections must increase the velocity, let us find out why the area changes as it does.

Combine the first law and the continuity equation to find an expression for the area change in terms of the Mach number.

$$\frac{dA}{A} + \frac{dv}{v} + \frac{d\rho}{\rho} = 0 \qquad \text{continuity} \qquad (17.29)$$

$$\frac{dp}{\rho} + v\, dv = 0 \qquad \text{Euler's equation or first law} \qquad (17.30)$$

The potential energy term is not included in equation (17.30). Solve for ρ from equation (17.30) and substitute this value into equation (17.29), solving for dA/A.

$$\frac{dA}{A} = \frac{d\mathrm{v}}{\mathrm{v}}(M^2 - 1) \qquad (17.31)$$

Flow regions are given names according to the magnitude of the Mach number. When M is less than one, the flow is *subsonic;* when M equals one the flow is *sonic;* when M is greater than one the flow is *supersonic.* Here we are interested in subsonic, sonic, and supersonic flows. The change in velocity, $d\mathrm{v}$, must always be positive, as that is the purpose of a nozzle. Let us find how the area must change in these three flow regions for this to be true. According to equation (17.31) for M less than one, the area change must be negative for $d\mathrm{v}$ to be positive. This is the convergent portion of the nozzle. At M equal to one, the area change is zero, or an undefined condition exists. At the throat, M is one and the area is a minimum; thus the derivative of the area is zero. When M is greater than one, the area change must be positive for a positive $d\mathrm{v}$; therefore, the divergent portion of the nozzle increases the cross-sectional area.

If a liquid, such as water, were flowing through a nozzle, would we design a convergent-divergent nozzle? No. Let us examine the continuity equation (17.29). The liquids are essentially incompressible; hence $d\rho$ is zero. Thus, the convergent portion of the nozzle is satisfied, since the decreasing area results in increasing velocity. The divergent portion is impossible, however, since both the area and the velocity cannot increase simultaneously. In compressible fluids, the density change compensates for the area and velocity changes. For incompressible fluids, the density does not vary, and so only a convergent nozzle is used. This means that the fluid velocity cannot exceed $M = 1$. This seems a little strange, but the acoustic velocity in an ideal incompressible fluid is infinite, so the possible velocity change is great. For water, the acoustic velocity is around 1525 m/s, about 4.5 times that of air, so in terms of exit velocities, both the compressible and incompressible fluid nozzles permit a wide range of exit velocities.

Critical Pressure Ratio

At the throat conditions for compressible flow, special properties exist if the flow is at $M = 1$. These are the *critical properties,* denoted as p^* and T^*, the critical pressure and temperature. From equation (17.24) for $M = 1$, we find that

$$\frac{T_0}{T^*} = \frac{k+1}{2} \qquad (17.32)$$

and

$$\frac{p_0}{p^*} = \left(\frac{k+1}{2}\right)^{k/(k-1)} \qquad (17.33)$$

If equation (17.33) is inverted, the ratio of p^*/p_0 is called the *critical pressure ratio*.

$$\frac{p^*}{p_0} = \left(\frac{2}{k+1}\right)^{k/(k-1)} \tag{17.34}$$

Thus, the critical pressure, p^*, is a function of k only. The following lists the values we will typically use.

$k = 1.2$	$p^* = 0.564p_0$	Saturated steam
$k = 1.3$	$p^* = 0.545p_0$	Low-pressure superheated steam and supersaturated steam
$k = 1.4$	$p^* = 0.528p_0$	Air and diatomic gases at 25°C
$k = 1.67$	$p^* = 0.487p_0$	Monatomic gases

When may we use equations (17.32)–(17.34)? When the flow is isentropic and the pressure that the throat senses is less than p^*. Fortunately, the flow may be considered isentropic to the throat almost all the time. The irreversibilities in actual nozzle flow occur in the divergent portion of the nozzle. Since the area is a minimum at the throat, the condition at the throat determines the mass flow rate through the nozzle. Once critical conditions are achieved at the throat, the density and velocity are constant, so the flow rate is constant. This is called a *choked flow condition*. The nozzle cannot pass a greater flow. Pressure waves travel at the speed of sound in the divergent portion of the nozzle. Any change in the downstream pressure will travel upstream at the speed of sound, and since the flow is supersonic, the signal to change the mass flow cannot reach the throat. The following example illustrates many of the concepts we have been discussing.

Example 17.2

Air, which may be considered an ideal gas with constant specific heat, is flowing isentropically through a nozzle. It enters the nozzle with a velocity of 100 m/s, a pressure of 600 kPa, and a temperature of 600°K. The exit pressure is 150 kPa. Tabulate and plot the area, specific volume, velocity, and temperature for 50 kPa increments, including the throat conditions. In addition, determine the Mach number of the air at the nozzle exit and the force on the nozzle control volume. The flow rate is 1 kg/s.

Solution

Given: A nozzle receives air at known inlet and exit conditions and flows isentropically through it.

Find: The variation of area, velocity, specific volume, and temperature for 50-kPa pressure increments across the nozzle; the force on the nozzle and the exit Mach number.

Sketch and Given Data:

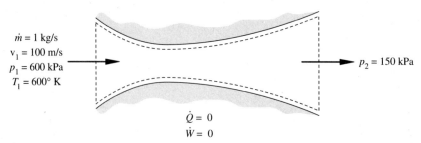

$\dot{m} = 1$ kg/s
$v_1 = 100$ m/s
$p_1 = 600$ kPa
$T_1 = 600°$ K

$p_2 = 150$ kPa

$\dot{Q} = 0$
$\dot{W} = 0$

Figure 17.8

Assumptions:

1. Air is an ideal gas with constant specific heats.
2. The flow is steady and isentropic through the nozzle.
3. The change of potential energy may be neglected.

Analysis: Calculate the stagnation properties and then velocities, areas, and specific volumes.

$$h_0 = h_1 + \frac{v_1^2}{2} = (1.0047 \text{ kJ/kg-K})(600°\text{K}) + \frac{(100 \text{ m/s})^2}{2(1000 \text{ J/kJ})} = 607.82 \text{ kJ/kg}$$

$$T_0 = h_0/c_p = \frac{(607.82 \text{ kJ/kg})}{(1.0047 \text{ kJ/kg-K})} = 604.97°\text{K}$$

$$\frac{p_0}{p} = \left(\frac{T_0}{T}\right)^{k/(k-1)} = \left(\frac{604.97}{600}\right)^{1.4/0.4}$$

$$p_0 = (600 \text{ kPa})(1.0293) = 617.6 \text{ kPa}$$

$$p^* = 0.528p_0 = 326 \text{ kPa}$$

$$v_1 = \frac{RT_1}{p_1} = \frac{(0.287 \text{ kJ/kg-K})(600°\text{K})}{(600 \text{ kN/m}^2)} = 0.287 \text{ m}^3/\text{kg}$$

$$A_1 = \frac{v_1\dot{m}}{v_1} = \frac{(0.287 \text{ m}^3/\text{kg})(1 \text{ kg/s})}{(100 \text{ m/s})} = 28.7 \text{ cm}^2$$

Find T_2. The process 1-2 is isentropic.

$$T_2 = T_1\left(\frac{p_2}{p_1}\right)^{(k-1)/k} = (600°\text{K})\left(\frac{550}{600}\right)^{0.286} = 585.25°\text{K}$$

$$v_2 = \sqrt{2(1004.7 \text{ kJ/kg-K})(604.97 - 585.25°\text{K})} = 199 \text{ m/s}$$

$$v_2 = \frac{RT_2}{p_2} = \frac{(0.287 \text{ m}^3/\text{kg})(585.25°\text{K})}{(550°\text{K})} = 0.305 \text{ m}^3/\text{kg}$$

$$A_2 = \frac{v_2\dot{m}}{\mathsf{v}_2} = \frac{(0.305 \text{ m}^3/\text{kg})(1 \text{ kg/s})}{199 \text{ m/s}} = 15.3 \text{ cm}^3$$

Continue the process, yielding the results in Table 17.1 and Figure 17.9.

TABLE 17.1

p (kPa)	v (m³/kg)	v (m/s)	T (K)	A (cm²)
600	0.287	100	600	28.7
550	0.305	199	585.25	15.3
500	0.327	267	569.5	12.2
450	0.352	324	552.6	10.9
400	0.383	377	534.3	10.2
350	0.422	427	514.3	9.9
326	0.444	450	503.9	9.8
300	0.471	476	492.1	9.9
250	0.536	526	467.1	10.2
200	0.629	579	438.2	10.9
150	0.772	636	403.6	12.1

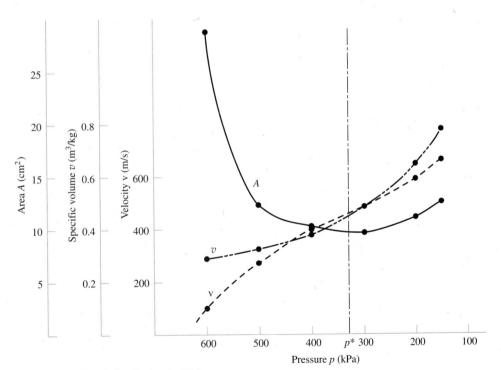

Figure 17.9 Graph for Example 17.2.

Determine the exit Mach number by finding the acoustic velocity at the exit and dividing that into the exit velocity.

$$a = \sqrt{kRT}$$

$$a = [(1.4)(287 \text{ J/kg-K})(403.6°\text{K})]^{1/2} = 402.7 \text{ m/s}$$

$$M = \frac{v}{a} = \frac{636}{402.7} = 1.58$$

The area must be in square meters for the units to balance.

$$F_{net} = \rho_2 A_2 v_2^2 - \rho_1 A_1 v_1^2 = \frac{A_2 v_2^2}{v_2} - \frac{A_1 v_1^2}{v_1}$$

The area must be in square meters for the units to balance.

$$F_{net} = \frac{(1.21 \times 10^{-3} \text{ m}^2)(636 \text{ m/s})^2}{(0.772 \text{ m}^3/\text{kg})} - \frac{(2.87 \times 10^{-3} \text{ m}^2)(100 \text{ m/s})^2}{(0.287 \text{ m}^3/\text{kg})} = 534 \text{ N}$$

Comment: The plot of the area does not indicate how long the nozzle should be to effect these area changes. In Figure 17.7 the angle of divergence is indicated by 2θ. The angle is approximately $6-12°$ for turbine nozzles. For angles greater than this, flow irreversibilities occur, limiting the efficiency of the nozzle. A trade-off is made in nozzle design between irreversibilities due to overexpansion, which occurs if 2θ is large, and increased frictional irreversibilities due to boundary-layer thickness at small angles of divergence. The figure of $6-12°$ is empirical, based on testing and design experience. No theoretical angle is determined from fluid mechanics. ▪

Example 17.3

Repeat Example 17.2, using the TK Solver model AIR.TK to compute and tabulate the specific volume, velocity, temperature, and area as pressure is varied in 50-kPa increments. Compare the results with the ideal-gas solution in Table 17.1.

Solution

Given: A nozzle with given inlet and exit conditions.

Find: The specific volume, velocity, temperature, and area as pressure is varied in 50-kPa increments.

Sketch and Given Data: See Figure 17.8.

Assumptions:

1. The flow is steady and isentropic through the nozzle.
2. The change in potential energy may be neglected.

Analysis: Load AIR.TK into TK Solver. Enter the continuity and first-law equations into the Rule Sheet. Enter the inlet conditions and zero for DELs. Use the List Solver with p2 as an input, and v2, V2, T2, and A2 saved as outputs. The results are displayed using the Table Sheet.

```
============================== RULE SHEET ==============================
S Rule ─────────────────────────────────────────────────────────────
    "Calculation Units are SI: degK, kPa, m3/kg, kJ/kg, kJ/kg-K
    P1=PvT(v1,T1)
    P2=PvT(v2,T2)
    DELh=DELh(T1,T2)
    DELu=DELu(T1,T2)
    DELs=DELs(T1,p1,T2,p2)

    mdot=A2*V2/v2           "Continuity Equation
    (V2^2-V1^2)/2=-DELh*1000    "First Law, q=w=0
=========================== VARIABLE SHEET ===========================
St Input ─ Name ─ Output ── Unit ── Comment ─────────────────────────

                                    ***Air Properties***

                                    POINT 1
      600    P1                 kPa      Pressure (kPa, MPa, psia)
      600    T1                 degK     Temperature (degK, degC, degR, degF)
             v1     .28747      m3/kg    Specific Volume (m3/kg, ft3/lbm)
      100    V1                 m/s
             A1                 m3

                                    POINT 2
L     326    P2                 kPa      Pressure (kPa, MPA, psia)
L            T2     506.27      degK     Temperature (degK, degC, degR, degF)
L            v2     .44607      m3/kg    Specific Volume (m3/kg, ft3/lbm)
L            V2     452.89      m/s
L            A2     .00098494   m2

                                    CHANGE (POINT 2 - POINT 1)
             DELh   -97.553     kJ/kg    Enthalpy (kJ/kg, BTU/lbm)
             DELu   -70.649     kJ/kg    Internal Energy (kJ/kg, BTU/lbm)
      0      DELs               kJ/kg-K  Entropy (kJ/kg-K, B/lbm-R)

      1      mdot               kg/s
```

```
   p2      v2       V2       T2        A2

   600     .28747   100      600       28.747
   550     .30609   199.97   585.71    15.307
   500     .32786   268.32   570.42    12.219
   450     .35371   326.13   553.94    10.846
   400     .38501   378.87   536.05    10.162
   350     .42384   429.13   516.42    9.8768
   326     .44607   452.89   506.27    9.8494
   300     .47355   478.6    494.64    9.8945
   250     .53992   528.7    470.04    10.212
   200     .63395   581.04   441.6     10.911
   150     .77984   637.91   407.49    12.225
```

Comments:

1. The results calculated using the ideal-gas relationships in Example 17.2 compare very favorably to the results computed using AIR.TK.
2. Only minor additions were necessary to permit AIR.TK to analyze isentropic nozzle flow.

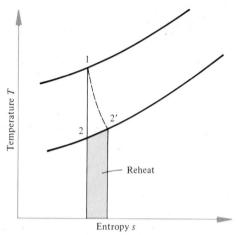

Figure 17.10 A T-s diagram illustrating reheat caused by irreversibilities in flow through a nozzle.

Nozzle Efficiency

The purpose of the nozzle is to convert enthalpy into kinetic energy. Actually, irreversibilities in flow prevent the ideal conversion from occurring. The nozzle efficiency, η_n, relates the actual and ideal conversion:

$$\eta_n = \frac{(\text{k.e.})_{2_{\text{act}}}}{(\text{k.e.})_{2_s}} \tag{17.35}$$

$$\eta_n = \frac{h_0 - h_{2'}}{(h_0 - h_2)_s} \tag{17.36}$$

Figure 7.10 illustrates a T-s diagram for a nozzle. The shaded area represents the reheat. This energy irreversibly increases the exit enthalpy to $2'$. The reheat is equal to $(h_{2'} - h_2)$. Typical values of nozzle efficiency are from 94 to 99%.

Example 17.4
Air enters a nozzle at 1.5 kg/s, 600°K, and 500 kPa with negligible velocity and expands to 101 kPa. The nozzle efficiency is 95%. Determine the air's irreversibility across the nozzle and the actual exit area and velocity. $T_0 = 300°K$.

Solution

Given: An adiabatic nozzle with a known efficiency and with specified air inlet and exit conditions.

Find: The exit velocity of the air, the nozzle area, and the irreversibility of the air.

Sketch and Given Data:

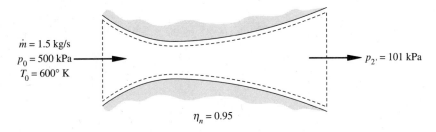

$\dot{m} = 1.5$ kg/s
$p_0 = 500$ kPa
$T_0 = 600°$ K

$p_{2'} = 101$ kPa

$\eta_n = 0.95$

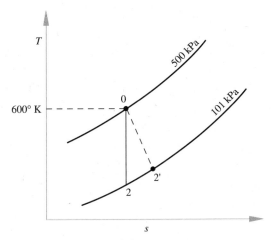

Figure 17.11

Assumptions:

1. For the conditions given, air may be considered an ideal gas with constant specific heat.
2. Air flows steadily through the nozzle. No work is done, the nozzle is adiabatic, and the change of potential energy is zero.

Analysis: Calculate the isentropic exit conditions and use the nozzle efficiency to determine the actual exit conditions.

$$T_2 = T_1 \left(\frac{p_2}{p_1}\right)^{(k-1)/k} = (600°\text{K})\left(\frac{101}{500}\right)^{0.286} = 379.9°\text{K}$$

$$\eta_n = \frac{h_0 - h_{2'}}{(h_0 - h_2)_s} = \frac{T_0 - T_{2'}}{(T_0 - T_2)_s} = 0.95 = \frac{600°\text{K} - T_{2'}}{(600 - 379.9°\text{K})}$$

$$T_{2'} = 390.9°\text{K}$$

The velocity is determined from equation (17.27a).

$$v_{2'} = \sqrt{2(h_0 - h_{2'})}$$

$$v_{2'} = \sqrt{2(1004.7 \text{ J/kg-K})(600 - 390.9°\text{K})} = 648.2 \text{ m/s}$$

The exit area is determined from the conservation of mass:

$$A_{2'} = \frac{\dot{m}v_{2'}}{v_{2'}}$$

The specific volume is determined from the ideal-gas equation of state.

$$v_{2'} = \frac{RT_{2'}}{p_{2'}} = \frac{(0.287 \text{ kJ/kg-K})(390.9°\text{K})}{(101 \text{ kN/m}^2)} = 1.111 \text{ m}^3/\text{kg}$$

$$A_{2'} = \frac{(1.5 \text{ kg/s})(1.111 \text{ m}^3/\text{kg})}{(648.2 \text{ m/s})} = 2.57 \times 10^{-3} \text{ m}^2$$

The irreversibility is found from $\dot{I} = \dot{m}T_0(s_{2'} - s_0)$. The entropy increase across the nozzle is

$$(s_{2'} - s_0) = c_p \ln\left(\frac{T_{2'}}{T_0}\right) - R \ln\left(\frac{p_{2'}}{p_0}\right)$$

$$(s_{2'} - s_0) = (1.0047) \ln\left(\frac{390.9}{600}\right) - (0.287) \ln\left(\frac{101}{500}\right) = 0.0286 \text{ kJ/kg-K}$$

$$\dot{I} = \dot{m}T_0(s_{2'} - s_0) = (1.5 \text{ kg/s})(300°\text{K})(0.0286 \text{ kJ/kg-K}) = 12.87 \text{ kW}$$

Comment: In an ideal nozzle with an isentropic expansion, there is no dissipation of the thermal in the conversion to kinetic energy. The nozzle efficiency allows one to determine entropy increase and hence the irreversibility.   ■

17.8 SUPERSATURATION

In analyzing the flow of steam through a nozzle we have dealt with equilibrium flow. The properties of steam were considered to be continuous, and an isentropic path joined the end states. In actual steam flow through a nozzle, the steam may or may not be in equilibrium during its expansion in the nozzle. A criterion for judging whether or not the steam is in metastable equilibrium is to determine whether condensation (the steam path's crossing the saturated vapor line) could have occurred for equilibrium flow. The condensation process takes time. Several molecules of steam must come together to form the nucleus of a droplet. Once the droplet is formed, other steam molecules are readily attracted. The time required is small, but the velocity through a nozzle is high, so the time that the steam is in the nozzle is also small.

In equilibrium flow, the energy released by the condensing molecules provides for increasing kinetic energy of the steam as it passes through the nozzle. If the steam does not condense, the kinetic energy increase must come from another source, the molecular translational energy, which is denoted on the macroscopic level as tem-

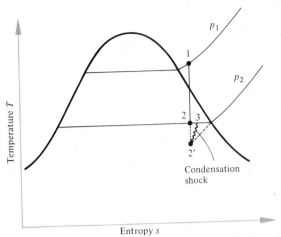

Figure 17.12 A *T-s* diagram for steam, illustrating equilibrium path 1-2 and nonequilibrium path 1-2′.

perature. There is a reduction in the temperature level compared with equilibrium steam, and steam in this condition is called *supersaturated.* The supersaturated state is a *metastable,* or *nonequilibrium,* state. If it is a nonequilibrium state, the properties cannot be determined from equilibrium tables of properties. This condition refers to steam not lying on the equilibrium surface.

Figure 17.12 illustrates the *T-s* diagram for steam and denotes the paths for an equilibrium process from state 1 to state 2 and for the nonequilibrium process state 1 to state 2′. How may the properties at state 2′ be found? By extending the constant-pressure line until it reaches the same value of entropy as state 1. Condensation has not occurred, so we may relate state 1 and state 2′ by $pv^k = C$. The values of steam at state 1 are known, so the value of v for state 2′ may be calculated knowing $p_2 = p_{2'}$. If we need to know the velocity, we cannot use the typical change of enthalpy involving the temperature. Although it is permissible to use pv relationships, $pv \neq RT$ in the supersaturated region. The molecular forces of attraction are great, so the ideal-gas equation of state is not indicative of the mixture behavior. The change of kinetic energy is still equal to the change of enthalpy, but we must develop an equation to express the enthalpy change in terms of pressure and specific volume. For isentropic flow

$$dh = v\, dp$$

$$\int_1^{2'} dh = h_{2'} - h_1 = \frac{k}{k-1}\, p_1 v_1 \left[1 - \left(\frac{p_{2'}}{p_1} \right)^{(k-1)/k} \right]$$

Thus

$$(\text{k.e.})_{2'} - (\text{k.e.})_1 = h_{2'} - h_1$$

$$\frac{v_{2'}^2 - v_1^2}{2} = \frac{k}{k-1}\, p_1 v_1 \left[1 - \left(\frac{p_{2'}}{p_1} \right)^{(k-1)/k} \right] \qquad (17.37a)$$

where the velocity is in meters per second, and

$$\frac{v_{2'}^2 - v_1^2}{2g_c} = \frac{k}{k-1} p_1 v_1 \left[1 - \left(\frac{p_{2'}}{p_1} \right)^{(k-1)/k} \right] \qquad (17.37b)$$

where the velocity is in feet per second.

The diagram in Figure 17.12 also shows *condensation shock.* This occurs when the droplet nuclei start to form and then grow very rapidly. This rapid growth appears like a shock, a sudden change, from metastable conditions to equilibrium conditions at state 3. If further expansion occurs after state 3, the expansion may be assumed to be equilibrium. This is because the droplets are present and form a nucleus for the vapor molecules to condense on. It is because no nuclei were present initially that supersaturated conditions existed. In cloud seeding, nuclei are injected into a water-vapor mixture to assist the droplet formation.

When is it necessary to account for supersaturated flow in a nozzle and why? We have seen that the throat conditions determine the mass flow rate through the nozzle. The supersaturated steam has a significantly lower specific volume than the equilibrium value, so if supersaturated flow existed at the throat, the mass flow rate through the nozzle would be greater than the equilibrium flow rate. Studies have shown that condensation shock does not occur in the convergent portion of the nozzle, but rather in the divergent portion. Since condensation shock is an irreversibility, it must be accounted for by the nozzle efficiency. As a guideline, if supersaturated flow can exist at the throat, we will assume it does exist. Condensation shock occurs in the divergent portion of the nozzle, and equilibrium flow exists at the nozzle exit. The following example compares equilibrium and supersaturated flow of steam in a nozzle.

Example 17.5

Steam expands in a nozzle from an initial condition of 400 psia and 460°F to 20 psia. The nozzle efficiency is 95%. Isentropic flow may be assumed to exist to the throat. The throat area is 0.80 in.² Determine the mass flow rate for equilibrium and supersaturated flow and the exit area.

Solution

Given: A nozzle expands steam from known initial conditions to known exit conditions. The nozzle efficiency and throat area are specified.

Find: The mass flow rate, assuming equilibrium conditions for the steam and supersaturated conditions, and the exit area.

Sketch and Given Data:

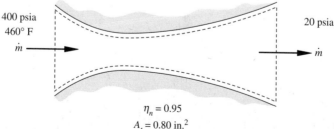

400 psia
460° F

$\dot{m}$

20 psia

$\dot{m}$

$\eta_n = 0.95$
$A_t = 0.80$ in.²

Figure 17.13

Assumptions:

1. The nozzle is adiabatic, no work is done, and the change in potential energy is zero.
2. Steady flow exists.
3. Equilibrium conditions for steam exist at the nozzle exit for both cases.
4. The flow to the throat may be considered isentropic for supersaturated and equilibrium flow.
5. The specific heat ratio, k, is 1.3 for supersaturated flow.

Analysis: Determine the equilibrium steam properties at the nozzle inlet and exit conditions. Use the steam tables or STEAM.TK.

For the inlet state $p_1 = 400$ psia, $T_1 = 460°F$, $h_1 = 1216.5$ Btu/lbm, $v_1 = 1.1977$ ft³/lbm, $s_1 = 1.4977$ Btu/lbm-R.

Find the isentropic exit state and then use the nozzle efficiency to determine the actual exit conditions. $p_2 = 20$ psia, $s_2 = s_1$, and $h_2 = 995.5$ Btu/lbm.

$$\eta_n = 0.95 = \frac{h_1 - h_{2'}}{h_1 - h_2} = \frac{1216.5 - h_{2'}}{1216.5 - 995.5}$$

$$h_{2'} = 1006.5 \text{ Btu/lbm}$$

$$v_{2'} = 16.96 \text{ ft}^3/\text{lbm}$$

The exit velocity is found by using equation (17.27b).

$$v_{2'} = 223.8\sqrt{h_1 - h_{2'}}$$

$$v_{2'} = 223.8\sqrt{1216.5 - 1006.5} = 3243 \text{ ft/sec}$$

To calculate the exit area, the mass flow rate must be determined. Determine the steam conditions for equilibrium flow to the throat.

$$p^* = 0.545 p_0 = (0.545)(400) = 218 \text{ psia}$$

$$s^* = s_1 = 1.4977 \qquad h^* = 1165.5 \text{ Btu/lbm}$$

$$v^* = 2.02 \text{ ft}^3/\text{lbm}$$

$$v^* = 223.8\sqrt{1216.5 - 1165.5} = 1598.3 \text{ ft/sec}$$

$$\dot{m} = \frac{A^* v^*}{v^*} = \frac{(0.80 \text{ in.}^2)(1598.3 \text{ ft/sec})}{(2.02 \text{ ft}^3/\text{lbm})(144 \text{ in.}^2/\text{ft}^2)} = 4.395 \text{ lbm/sec}$$

For supersaturated flow at the throat, use equation (17.37b) to determine the velocity.

$$\frac{v_t^2}{2g_c} = \frac{k}{k-1} p_1 v_1 \left[1 - \left(\frac{p_t}{p_1}\right)^{(k-1)/k}\right]$$

$$\frac{v_t^2}{2g_c} = \frac{1.3}{0.3}(400 \text{ lbf/in.}^2)(144 \text{ in.}^2/\text{ft}^2)(1.1977 \text{ ft}^3/\text{lbm})\left[1 - \left(\frac{218}{400}\right)^{0.3/1.3}\right]$$

$$v_t = 1585.6 \text{ ft/sec}$$

$$v_t = v_1 \left(\frac{p_1}{p_t}\right)^{1/k} = (1.1977 \text{ ft}^3/\text{lbm}) \left(\frac{400}{218}\right)^{1/1.3} = 1.9104 \text{ ft}^3/\text{lbm}$$

From the conservation of mass

$$\dot{m} = \frac{Av_t}{v_t} = \frac{(0.80 \text{ in.}^2)(1585.6 \text{ ft/sec})}{(144 \text{ in.}^2/\text{ft}^2)(1.9104 \text{ ft}^3/\text{lbm})} = 4.611 \text{ lbm/sec}$$

The error in mass flow rate, comparing equilibrium conditions to supersaturated conditions, is

$$\% \text{ error} = \frac{4.611 - 4.395}{4.395} \times 100 = 4.9\%$$

The exit area may be found from the conservation of mass:

$$A_{2'} = \frac{\dot{m}v_{2'}}{v_{2'}}$$

$$A_{2'} = \frac{(4.611 \text{ lbm/sec})(16.96 \text{ ft}^3/\text{lbm})(144 \text{ in.}^2/\text{ft}^2)}{(3243 \text{ ft/sec})} = 3.47 \text{ in.}^2$$

Comment: Although this problem asks that supersaturated flow be calculated, in general if supersaturated flow can exist at a nozzle throat, assume that it does and base the nozzle calculations on that. As this example indicates, a nearly 5% difference in flow rate is significant.   ▬

17.9 DIFFUSER

The diffuser has been mentioned previously, and its purpose is opposite to that of the nozzle. It converts kinetic energy into thermal energy; more specifically, it compresses a fluid to a higher pressure by decelerating the flow. The ideal process for this is isentropic. From equation (17.31), $dA/A = (dv/v)(M^2 - 1)$, we see that for supersonic flow to exist the area must converge to decrease the velocity. For subsonic flow, the area must increase for the velocity to decrease. At $M = 1$, the area and velocity changes are zero, as in the case of the nozzle. Thus, a supersonic diffuser would look similar to Figure 17.14(a). Figure 17.14(b) shows an h-s diagram for the diffuser when the exit velocity is negligibly small. For the isentropic compression $p_{02} = p_{01}$, by the definition of the stagnation state. Actually, the compression is adiabatic (the dotted path), and the stagnation pressure, p_{02}, cannot be achieved. Note that the stagnation enthalpy or temperature is only a function of the process, being adiabatic, so $h_{02} = h_{02'}$.

The same equations apply for a diffuser as for a nozzle. The first law for the diffuser, an energy balance, is

$$(\text{k.e.})_1 - (\text{k.e.})_2 = h_2 - h_1$$

Since we want the diffuser to increase the fluid pressure, the diffuser efficiency, η_d, would relate how this was performed:

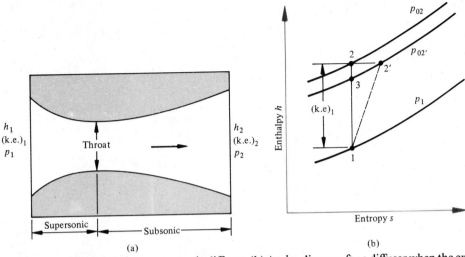

Figure 17.14 (a) An adiabatic supersonic diffuser. (b) An *h-s* diagram for a diffuser when the exit velocity is negligibly small.

$$\eta_d = \frac{\text{change in kinetic energy to achieve } p_{02'} \text{ ideally}}{\text{change in kinetic energy to achieve } p_{02'} \text{ actually}} \qquad (17.38)$$

$$\eta_d = \frac{(\text{k.e.})_1 - (\text{k.e.})_3}{(\text{k.e.})_1 - (\text{k.e.})_{2'}} = \frac{(h_1 - h_3)_s}{h_1 - h_{2'}} \qquad (17.39)$$

Supersonic diffuser efficiencies tend to be lower than nozzle efficiencies because of shock wave formation, especially at other than design conditions.

Example 17.6

Air enters a diffuser at 1.5 kg/s, 100 kPa, and 300°K with a Mach number of 4. It is decelerated in a convergent/divergent diffuser with an efficiency of 80% to negligible exit velocity. Determine the actual exit pressure and irreversibility.

Solution

Given: An adiabatic diffuser with a known efficiency decelerates air from known inlet conditions.

Find: The actual exit pressure and the irreversibility across the diffuser.

Sketch and Given Data:

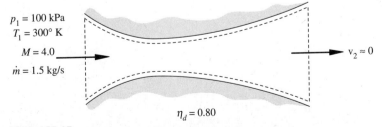

$p_1 = 100$ kPa
$T_1 = 300°$ K
$M = 4.0$
$\dot{m} = 1.5$ kg/s

$v_2 \approx 0$

$\eta_d = 0.80$

Figure 17.15

Assumptions:

1. The diffuser is adiabatic, no work is done, and the change in potential energy is zero.
2. Steady flow exists.
3. Air may be considered an ideal gas with constant specific heats.

Analysis: The inlet velocity may be determined by finding the value for the acoustic velocity and multiplying it by the Mach number.

$$a = (kRT)^{1/2} = [(1.4)(287.02 \text{ J/kg-K})(300°\text{K})]^{1/2} = 347.2 \text{ m/s}$$

$$v = Ma = 4a = 1388.8 \text{ m/s}$$

From the first-law analysis of the diffuser

$$\cancel{\dot{Q}} + \dot{m}\,[h_1 + (\text{k.e.})_1 + \cancel{(\text{p.e.})_1}] = \cancel{\dot{W}} + \dot{m}\,[h_2 + (\text{k.e.})_2 + \cancel{(\text{p.e.})_2}]$$

which reduces to

$$(\Delta\text{k.e.})_{\text{act}} = \Delta h_{\text{act}}$$

$$\frac{v_1^2 - v_2^2}{2} = h_{2'} - h_1$$

$$\frac{1388.8^2 - 0}{2(1000)} = h_{2'} - h_1 = 964.38 \text{ kJ/kg}$$

$$h_{2'} - h_1 = c_p\,(T_{2'} - T_1) = (1.0047 \text{ kJ/kg-K})(T_{2'} - 300°\text{K}) = 964.38 \text{ kJ/kg}$$

$$T_{2'} = 1260°\text{K}$$

From the diffuser efficiency

$$\eta_d = \frac{(h_1 - h_3)_s}{h_{1'} - h_{2'}} = 0.8 = \frac{(h_1 - h_3)_s}{-964.38 \text{ kJ/kg}}$$

$$(h_1 - h_3)_s = -771.5 \text{ kJ/kg} = c_p(T_1 - T_3) = (1.0047 \text{ kJ/kg-K})(300°\text{K} - T_3)$$

$$T_3 = 1068°\text{K}$$

Pressure p_3 is equal to the actual discharge pressure $p_{02'}$.

$$\frac{p_3}{p_1} = \left(\frac{T_3}{T_1}\right)^{k/(k-1)} = \left(\frac{1068}{300}\right)^{1.4/0.4} = 85.13$$

$$p_3 = p_{02} = (100 \text{ kPa})(85.13) = 8513 \text{ kPa}$$

The irreversibility is

$$\dot{I} = \dot{m}T_0(s_2 - s_1)$$

$$s_2 - s_1 = c_p \ln\left(\frac{T_{2'}}{T_1}\right) - R \ln\left(\frac{p_{02'}}{p_1}\right)$$

$$s_2 - s_1 = (1.0047) \ln\left(\frac{1260}{300}\right) - (0.287) \ln\left(\frac{8513}{100}\right) = +0.1663 \text{ kJ/kg-K}$$

$$\dot{I} = (1.5 \text{ kg/s})(300°\text{K})(0.1663 \text{ kJ/kg-K}) = 74.8 \text{ kW}$$

Comment: The irreversibilities in the diffuser cause a significant increase in the irreversibility function, decreasing the air's availability. ■

17.10 SHOCK WAVES

We have looked at isentropic flow through a nozzle and have used nozzle efficiency to correct for flow irreversibilities within the nozzle. A nozzle is designed to operate at certain conditions, but often in turbines the operating conditions are other than the design conditions. What happens then? Figure 17.16 illustrates a convergent-divergent nozzle and the pressure distribution in it for various back pressures for steady-flow conditions. Let p_{b_i} be the back pressure for any line i, $i = 1, 2, \ldots, 6$. If the back pressure is equal to the inlet pressure, no flow occurs, and the pressure is uniform through the nozzle, as shown by line 1 in Figure 17.16. If the back pressure is decreased to p_{b_2}, there is flow through the nozzle. The pressure decreases through the converging portion to the throat; the flow is subsonic, so the divergent portion of the nozzle acts as a diffuser, decelerating the fluid with the resultant rise in pressure. When the back pressure is lowered to p_{b_3}, the flow at the throat reaches sonic velocity,

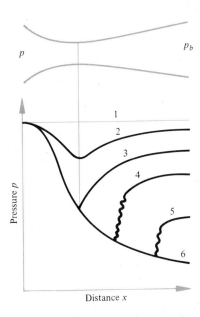

Figure 17.16 Pressure distribution in a convergent-divergent nozzle for various back pressures.

and the divergent portion of the nozzle acts as a diffuser and decelerates the flow. For steam, the value of p_{b_3}/p_0 must be about 0.82 for this condition to be achieved. As the back pressure is lowered further, the pressure decreases through the convergent portion of the nozzle, sonic velocity is achieved at the throat, and the pressure continues to decrease in the divergent portion of the nozzle, as the flow is supersonic. Lines 4 and 5 illustrate this phenomenon. At the design back pressure the flow through the entire nozzle is indicated by line 6.

Frequently the back pressure is higher than design, such as p_{b_4} or p_{b_5}. The exit pressure wave propagates into the fluid stream, indicating that the pressure for the supersonic flow is too low; it must increase. There is a discontinuity in pressure, and a *shock wave* patches the two flow regimes. The shock wave is a compression wave, where the kinetic energy of the fluid irreversibility increases the pressure of the fluid by decelerating in a very, very short distance. This is called a *normal shock wave,* since it is normal to the fluid flow. The flow on one side of the shock wave is supersonic; the flow on the other side is subsonic. The divergent portion of the nozzle acts as a diffuser for the subsonic flow in the remaining length of the nozzle.

The disturbance caused by the shock wave cannot be felt at the throat because after the throat the flow is supersonic, whereas the disturbance propagates at the sonic velocity and cannot reach the throat.

Is the mass flow rate affected by the changes in back pressure? Only until p_{b_3} is reached. After that, any lowering of the back pressure does not affect the properties at the throat. The nozzle cannot pass any greater mass flow, so it is in a choked condition. If we want to assure a given mass flow rate to a device under varying pressures, a nozzle can be used as a metering device as long as the back pressures are less than p_{b_3}.

17.11 FLOW ACROSS A NORMAL SHOCK WAVE

As discussed, under conditions that cause an abrupt deceleration from supersonic flow to subsonic flow, a shock wave is formed. A normal shock wave is one that is perpendicular, normal, to the flow. Figure 17.17 illustrates the control volume encompassing a normal shock wave. The flow is assumed steady, no heat or work leaves the control volume, and the effects of potential energy are negligible. The sketch shows a divergent portion of a nozzle, yet the areas on the x and y sides of the shock wave can be considered equal. This assumption is made possible by the shock wave being about 10^{-5} cm in thickness. Also, the forces acting on the wall because of the shock wave are very small and may be neglected when compared to the force change in the direction of flow.

Whether a shock wave occurs or not, the various conservation laws and the second law must remain valid. For the conservation of mass

$$\dot{m} = \rho_x A_x v_x = \rho_y A_y v_y$$

$$A_x = A_y$$

$$\rho_x v_x = \rho_y v_y$$

(17.40)

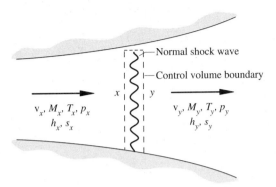

Figure 17.17 Control volume enclosing a normal shock wave.

For the conservation of energy

$$\cancel{\dot{Q}}^{0}+\cancel{\dot{m}}(h_x+\frac{v_x^2}{2}+\cancel{gz_x})=\cancel{\dot{W}}^{0}+\cancel{\dot{m}}(h_y+\frac{v_y^2}{2}+\cancel{gz_y})$$

$$h_x+\frac{v_x^2}{2}=h_y+\frac{v_y^2}{2} \tag{17.41}$$

$$h_{ox}=h_{oy} \tag{17.42}$$

For the conservation of momentum

$$A(p_x-p_y)=\rho_y A_y v_y^2-\rho_x A_x v_x^2$$

$$p_x-p_y=\rho_y v_y^2-\rho_x v_x^2 \tag{17.43}$$

For the second law

$$s_y-s_x=\frac{\Delta \dot{S}_{prod}}{\dot{m}} \tag{17.44}$$

Examination of the above equations shows that the stagnation enthalpy remains constant across the shock wave. We also note that the entropy must increase across the shock wave. The effect of the entropy increase is not manifested in the stagnation enthalpy, or temperature, but rather in a decrease in the stagnation pressure. This will be apparent by examining an h-s diagram.

Figure 17.18 illustrates an h-s diagram of the steady flow of a fluid in a constant area duct. If the flow is adiabatic, when the equations for conservation of mass and energy are combined with equation of state information for a fluid and the resulting equation is plotted on an h-s diagram, a Fanno line results. For frictionless flow if the fluid is heated or cooled, when the equations for conservation of mass and momentum are combined with equation of state information for a fluid and the resulting equation is plotted on an h-s diagram, a Rayleigh line results. These are shown in Figure 17.18. The point of maximum entropy on each line, points a and b, corre-

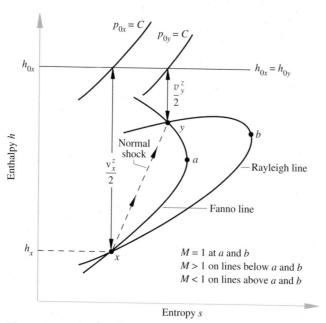

Figure 17.18 An *h-s* diagram illustrating Fanno and Rayleigh lines and direction of normal shock in changing from state *x* to state *y*.

spond to $M = 1$. Below these points are supersonic flows, and above these points are subsonic flows on both curves.

Since mass, energy, and momentum must be conserved simultaneously, the state points must lie on both the Fanno and Rayleigh lines. The points of intersection of the two lines represent states *x* and *y* as indicated. The dotted line joining the states represents the shock wave. Note that the entropy must increase along the shock wave; hence the flow must be from the *x* region to the *y* region. Notice that the stagnation pressure decreases, as p_{0y} is less than p_{0x}.

Flow of Gases with Constant Specific Heats

Often it is useful to have equations that relate the area at a given location to the area A^* that would be required for sonic flow at the same mass flow rate and where the stagnation state is the same.

$$\rho A v = \rho^* A^* v^*$$

where the * values are at $M = 1$. Solve for A/A^* and reduce the expression using the ideal-gas equation of state and the relationship for Mach number.

$$\frac{A}{A^*} = \left(\frac{1}{M}\right)\left(\frac{p^*}{p}\right)\left(\frac{T}{T^*}\right)^{1/2}$$

Use equations (17.21a) and (17.22a) to determine

$$\frac{A}{A^*} = \frac{1}{M}\left[\left(\frac{2}{k+1}\right)\left(1+\frac{k-1}{2}M^2\right)\right]^{(k+1)/2(k-1)} \tag{17.45}$$

Equations (17.21a), (17.22a), and (17.45) allow T/T_0, p/p_0, and A/A^* to be determined as a function of Mach number. Table 17.2(a) illustrates this for $k = 1.4$. In addition, a table may be created for flow across a normal shock wave.

First, the equations for flow across a shock wave may be determined. From the conservation of energy

$$h_{0x} = h_{0y}$$

Use equations (17.24) and (17.42) to obtain

$$\frac{T_{0x}}{T_x} = 1 + \frac{k-1}{2}M_x^2$$

$$\frac{T_{0y}}{T_y} = 1 + \frac{k-1}{2}M_y^2$$

$$T_{0x} = T_{0y}$$

$$\frac{T_y}{T_x} = \frac{1 + \dfrac{k-1}{2}M_x^2}{1 + \dfrac{k-1}{2}M_y^2} \tag{17.46}$$

From the conservation of momentum

$$p_x + \rho_x v_x^2 = p_y + \rho_y v_y^2$$

$$p = \rho RT \qquad a = \sqrt{kRT}$$

$$\frac{p_y}{p_x} = \frac{1 + kM_x^2}{1 + kM_y^2} \tag{17.47}$$

From the conservation of mass

$$\rho_x v_x = \rho_y v_y$$

$$p = \rho RT \qquad a = \sqrt{kRT}$$

$$\frac{p_y}{p_x} = \sqrt{\frac{T_y}{T_x}}\frac{M_x}{M_y} \tag{17.48}$$

Combine equations (17.46) and (17.47) to yield

$$M_y^2 = \frac{M_x^2 + \dfrac{2}{k-1}}{\dfrac{2k}{k-1}M_x^2 - 1} \tag{17.49}$$

The ratio of the stagnation pressure change across the shock wave is determined by using equations (17.22a) and (17.47) with the equation of state.

$$\frac{p_{0y}}{p_{0x}} = \frac{M_x}{M_y}\left[\frac{1+\frac{k-1}{2}M_y^2}{1+\frac{k-1}{2}M_x^2}\right]^{(k+1)/2(k-1)} \tag{17.50}$$

There is no area change across the shock wave; thus, combining equations (17.50) and (17.45) yields

$$\frac{A_x^*}{A_y^*} = \frac{p_{0y}}{p_{0x}} \tag{17.51}$$

Table 17.2 gives the various normal shock functions.

TABLE 17.2   ONE-DIMENSIONAL COMPRESSIBLE FLOW FUNCTIONS FOR AN IDEAL GAS WITH $k = 1.4$

	(a) Isentropic flow functions			(b) Normal shock functions				
M	T/T_0	p/p_0	A/A^*	M_x	M_y	p_y/p_x	T_y/T_x	p_{0y}/p_{0x}
0	1.000 00	1.000 00	∞	1.00	1.000 00	1.0000	1.0000	1.000 00
0.10	0.998 00	0.993 03	5.8218	1.10	0.911 77	1.2450	1.0649	0.998 92
0.20	0.992 06	0.972 50	2.9635	1.20	0.842 17	1.5133	1.1280	0.992 80
0.30	0.982 32	0.939 47	2.0351	1.30	0.785 96	1.8050	1.1909	0.979 35
0.40	0.968 99	0.895 62	1.5901	1.40	0.739 71	2.1200	1.2547	0.958 19
0.50	0.952 38	0.843 02	1.3398	1.50	0.701 09	2.4583	1.3202	0.929 78
0.60	0.932 84	0.784 00	1.1882	1.60	0.668 44	2.8201	1.3880	0.895 20
0.70	0.910 75	0.720 92	1.094 37	1.70	0.640 55	3.2050	1.4583	0.855 73
0.80	0.886 52	0.656 02	1.038 23	1.80	0.616 50	3.6133	1.5316	0.812 68
0.90	0.860 58	0.591 26	1.008 86	1.90	0.595 62	4.0450	1.6079	0.767 35
1.00	0.833 33	0.528 28	1.000 00	2.00	0.577 35	4.5000	1.6875	0.720 88
1.10	0.805 15	0.468 35	1.007 93	2.10	0.561 28	4.9784	1.7704	0.674 22
1.20	0.776 40	0.412 38	1.030 44	2.20	0.547 06	5.4800	1.8569	0.628 12
1.30	0.747 38	0.360 92	1.066 31	2.30	0.534 41	6.0050	1.9468	0.583 31
1.40	0.718 39	0.314 24	1.1149	2.40	0.523 12	6.5533	2.0403	0.540 15
1.50	0.689 65	0.272 40	1.1762	2.50	0.512 99	7.1250	2.1375	0.499 02
1.60	0.661 38	0.235 27	1.2502	2.60	0.503 87	7.7200	2.2383	0.460 12
1.70	0.633 72	0.202 59	1.3376	2.70	0.495 63	8.3383	2.3429	0.423 59
1.80	0.606 80	0.174 04	1.4390	2.80	0.488 17	8.9800	2.4512	0.389 46
1.90	0.580 72	0.149 24	1.5552	2.90	0.481 38	9.6450	2.5632	0.357 73
2.00	0.555 56	0.127 80	1.6875	3.00	0.475 19	10.333	2.6790	0.328 34
2.10	0.531 35	0.109 35	1.8369	4.00	0.434 96	18.500	4.0469	0.138 76
2.20	0.508 13	0.093 52	2.0050	5.00	0.415 23	29.000	5.8000	0.061 72
2.30	0.485 91	0.079 97	2.1931	10.00	0.387 57	116.50	20.388	0.003 04
2.40	0.464 68	0.068 40	2.4031	∞	0.377 96	∞	∞	0.0

Example 17.7

A convergent-divergent nozzle with a throat area of 4 cm² and an exit area of 8 cm² receives air at 700 kPa and 300°K. Air may be considered an ideal gas with $k = 1.4$. Choked flow exists. Determine the mass flow rate and the exit pressure for a standing shock wave at the nozzle exit and where the area is 6.22 cm².

Solution

Given: A convergent-divergent nozzle receives air at known conditions with isentropic flow before and after a shock wave.

Find: The mass flow rate of air through the nozzle and the exit pressure for stated shock wave locations.

Sketch and Given Data:

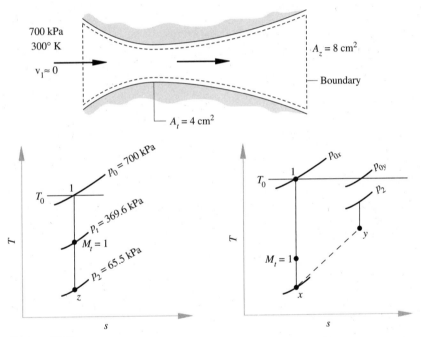

Figure 17.19

Assumptions:

1. Air is an ideal gas with constant specific heats.
2. The nozzle is adiabatic, the change in potential energy is zero, and no work is done.
3. Steady-state choked flow exists.
4. The flow is reversible adiabatic before and after the shock wave.

Analysis: Determine the mass flow rate for choked flow at the nozzle throat by finding the critical properties and then use the conservation of mass equation.

$$p^* = 0.528p_0 = (0.528)(700 \text{ kPa}) = 369.6 \text{ kPa}$$

$$T^* = (300°\text{K}) \left(\frac{369.6}{700} \right)^{0.4/1.4} = 250°\text{K}$$

$$v^* = \frac{RT^*}{p^*} = \frac{(0.287 \text{ kJ/kg-K})(250°\text{K})}{369.6 \text{ kPa}} = 0.1941 \text{ m}^3/\text{kg}$$

$$a^* = (kRT)^{1/2} = [1.4(287 \text{ kJ/kg-K})(250°\text{K})]^{1/2} = 316.9 \text{ m/s}$$

$$M^* = 1 \qquad \therefore \quad v^* = a^* = 316.9 \text{ m/s}$$

$$\dot{m} = \frac{A^* v^*}{v^*} = \frac{(0.0004 \text{ m}^2)(316.9 \text{ m/s})}{(0.1941 \text{ m}^3/\text{kg})}$$

$$\dot{m} = 0.653 \text{ kg/s}$$

Determine the Mach number just before the standing shock wave at the nozzle exit.

$$\frac{A_2}{A^*} = \frac{8 \text{ cm}^2}{4 \text{ cm}^2} = 2.0$$

From Table 17.2a $M = 2.2$ for this value of area ratio and where the flow must be supersonic.

$$\frac{p}{p_0} = 0.093 \ 52$$

$$p = (0.093 \ 52)(700) = 65.5 \text{ kPa}$$

This is the value of pressure just before the shock wave, or p_x. From Table 17.2b for $M_x = 2.2$, $M_y = 0.54706$ and $p_y/p_x = 5.48$.

$$p_y = (65.5 \text{ kPa})(0.547 \ 06) = 35.9 \text{ kPa}$$

Determine the Mach number before and after the shock wave where the area is 6.22 cm². The area ratio is

$$\frac{A}{A^*} = \frac{6.22 \text{ cm}^2}{4 \text{ cm}^2} = 1.555$$

From Table 17.2a $M = M_x = 1.9$. Using the following relationship and letting the nozzle exit be denoted as state 2 yields

$$\frac{A_2}{A_y^*} = \left(\frac{A_2}{A_x^*} \right) \left(\frac{A_x^*}{A_y^*} \right) = \left(\frac{A_2}{A_x^*} \right) \left(\frac{p_{0y}}{p_{0x}} \right)$$

For $M_x = 1.9$ $p_{0y}/p_{0x} = 0.767 \ 35$

$$\frac{A_2}{A_y^*} = \left(\frac{8 \text{ cm}^2}{4 \text{ cm}^2} \right) (0.767 \ 35) = 1.53$$

From Table 17.2a for the subsonic region the Mach number at the exit is $M_2 = 0.47$ and $p_2/p_{0y} = 0.883$.

$$p_2 = \left(\frac{p_2}{p_{0y}}\right)\left(\frac{p_{0y}}{p_{0x}}\right) p_{0x} = (0.883)(0.767)(700) = 474 \text{ kPa}$$

Comment: The mass flow rate through the nozzle is the same in both cases, as the flow is choked. Use of the isentropic flow functions and the normal shock functions allows simplification of the problem solution. The flow after the shock wave at $A = 6.22 \text{ cm}^2$ is subsonic, and the divergent portion acts as a diffuser.   ▬

Using TK Solver to Analyze Nozzle Flow

The TK Solver model CDNOZZLE.TK permits the convenient analysis of the flow of ideal gases in convergent-divergent nozzles. The model was developed using the basic equations (first law, continuity, ideal gas, isentropic process, sonic velocity) and two normal shock relationships (equations [17.47] and [17.49]). Because CDNOZZLE.TK is based on basic relationships, it is more flexible than techniques that require the use of special tables or equations.

Example 17.8
For the nozzle and inlet conditions given in Example 17.7, use CDNOZZLE.TK to determine at what area a normal shock will form for an exit pressure of 400 kPa.

Solution

Given: A convergent-divergent nozzle receives air at known conditions with isentropic flow before and after a normal shock.

Find: The area at which the shock forms for a given exit pressure.

Sketch and Given Data: See Figure 17.19 on page 709.

Assumptions:

1. Air is an ideal gas with constant specific heats.
2. The nozzle is adiabatic, the change in potential energy is zero, and no work is done.
3. Steady-state choked flow exists.
4. The flow is reversible adiabatic before and after the normal shock.

Analysis: Load CDNOZZLE.TK into TK Solver and enter the given data into the Variable Sheet. After solution, the Variable Sheet will appear as follows:

```
═══════════════════════════════ VARIABLE SHEET ═══════════════════════════════
St Input ── Name ──── Output ── Unit ──── Comment ──────────────────────────

                                           ***Convergent-Divergent Nozzle***

   1.0047   cp                   kJ/kg-K   Specific Heat
    .287    R                    kJ/kg-K   Gas Constant
   1.4      k                              Specific Heat Ratio
            mdot       .6533     kg/s      Mass Flow Rate

   0        Vinlet               m/s       Inlet Velocity
            Vthroat    317.02    m/s       Throat Velocity
            Vexit      167.67    m/s       Exit Velocity
            Vx         531.35    m/s       Velocity Upstream of Shock
            Vy         189.14    m/s       Velocity Downstream of Shock

  27        Tinlet               degC      Inlet Temperature
            Tthroat    -23.017   degC      Throat Temperature
            Texit      13.009    degC      Exit Temperature
            Tx         -113.51   degC      Temperature Upstream of Shock
            Ty         9.1958    degC      Temperature Downstream of Shock

            vinlet     .12306    m3/kg     Inlet Specific Volume
            vthroat    .19411    m3/kg     Throat Specific Volume
            vexit      .20532    m3/kg     Exit Specific Volume
            vx         .59646    m3/kg     Specific Volume Upstream of Shock
            vy         .21232    m3/kg     Specific Volume Downstream of Shock

            Ainlet               cm2       Inlet Area
   4        Athroat              cm2       Throat Area
   8        Aexit                cm2       Exit Area
            Ashock     7.3335    cm2       Area of Shock

  700       Pinlet               kPa       Inlet Pressure
            Pthroat    369.84    kPa       Throat Pressue
  400       Pexit                kPa       Exit Pressure
            Px         76.817    kPa       Pressure Upstream of Shock
            Py         381.65    kPa       Pressure Downstream of Shock

            Minlet     0                   Inlet Mach Number
            Mthroat    1                   Throat Mach Number
            Mexit      .49447              Exit Mach Number
            Mx         2.098               Mach Number Upstream of Shock
            My         .56156              Mach Number Downstream of Shock

            K1         .52834              Guess Constant (Pthroat)
            K2         .67533              Guess Constant (Pexit)
            K3         .2077               Guess Constant (Px)
            K4         .95413              Guess Constant (Py)
```

Comment: The iteration and back-solving capability of TK Solver permits the same model to be used for a variety of problems.

17.12 FLOW MEASUREMENT

The principles we have covered may be applied to flow-measuring devices, such as the venturi meter, shown in Figure 17.20. The fluid pressure decreases as the fluid accelerates to the throat—the pressure at the throat, p_2, being a minimum. The

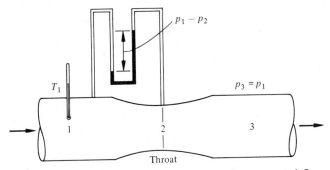

Figure 17.20 A schematic diagram illustrating a venturi flow-measuring device.

velocity may be determined by knowing the pressure change and the flow rate, determined from $\dot{m} = \rho A v$. The throat contraction is small, so the pressure change is also small. If the flow is compressible, the velocity change may be found as follows. If a density change occurs and must be accounted for, the equation $pv^k = C$ can be used to determine the functional relationship in the following integral.

The conservation of energy applied between state 1 and state 2 yields for isentropic flow

$$h_1 + (\text{k.e.})_1 = h_2 + (\text{k.e.})_2$$

$$\frac{v_2^2}{2} = \frac{v_1^2}{2} + \int_1^2 v\, dp$$

where $dh = T\, ds + v\, dp = v\, dp$ for isentropic flow. For incompressible flow, $v = C$.

$$v_2 = [v_1^2 + 2v_1(p_2 - p_1)]^{1/2}\ \text{m/s} \tag{17.52a}$$

$$v_2 = [v_1^2 + 2g_c v_1(p_2 - p_1)]^{1/2}\ \text{ft/sec} \tag{17.52b}$$

The velocity, specific volume, and area being known, the mass flow rate may be calculated.

There are other flow-measuring devices, such as orifices, but we will not discuss these or the empirical coefficients that must be included in any study of flow-measuring devices. Most fluid dynamics texts cover this area in greater detail.

17.13 WIND POWER

The use of wind power as a power source appears attractive for several reasons. First, wind power is a renewable resource, for which the required technology has already been developed. Second, no air, water, or thermal pollution is associated with it. Third, weather modification due to wind utilization is negligible. The drawbacks to wind-power utilization are relatively low efficiencies, high capital costs, noise, aesthetic problems, and reliability of equipment.

Wind power is not a reliable energy source, the output being a function of the wind velocity. Since wind-velocity fluctuations do not normally coincide with power-requirement variations, most wind-power systems must include energy storage systems.

There are several types of wind-driven machines, operating on several different principles. Although each type has its advantages, the one type that stands out as the most promising is the horizontal-axis, two-bladed propeller wind turbine, illustrated in Figure 17.21.

A wind turbine should have the following characteristics: the ability to maintain optimum alignment with the wind; a low starting torque; the ability to endure high winds; and, if used to drive a generator, a high rotational speed.

The propeller windmill is almost always either a two- or three-bladed design; the two-bladed design is more widely used because it is strong, simple, and less expensive. The horizontal-axis windmill can be positioned so that the blades lie upwind or downwind of the tower. The downwind design is usually preferred for larger machines, where a tail vane is not practical. On smaller models, a tail vane keeps the blades pointing into the wind. Large windmills are usually steered by a pilot wind vane, coupled to the drive gear, which operates to keep the windmill in constant alignment with the wind. The pilot wind vane is more sensitive to wind shifts than is

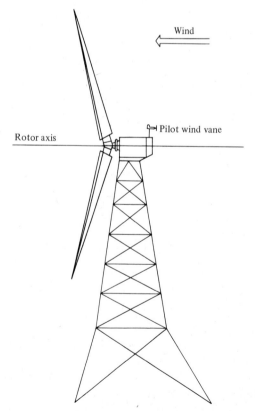

Figure 17.21 A two-bladed horizontal windmill.

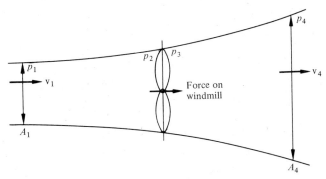

Figure 17.22 Windmill propeller in a flow field.

the large windmill. At the cutout wind speed, the blades are turned edgewise to the wind to protect the machine against damage due to high winds.

Let us develop the equations governing the windmill. Figure 17.22 illustrates a windmill propeller located in a moving fluid.

When the fluid between planes 1 and 4 is isolated, the only force acting is that exerted by the fluid on the propeller. The force acting on the windmill is equal to the pressure drop across the blades.

$$F = (p_2 - p_3)A \tag{17.53}$$

The force is also equal to

$$F = \dot{m}(v_1 - v_4)$$

If we let v be the mean velocity across the blades and ρ be the air density,

$$F = \rho v A(v_1 - v_4) \tag{17.54}$$

Combining equations (17.53) and (17.54) yields

$$p_2 - p_3 = \rho v(v_1 - v_4) \tag{17.55}$$

Applying the first law to the flow between planes 1 and 2 yields

$$u_1 + p_1 v_1 + \tfrac{1}{2}v_1^2 = u_2 + p_2 v_2 + \tfrac{1}{2}v_2^2 \tag{17.56}$$

The temperature is constant; hence $u_1 = u_2$ and

$$p_1 + \tfrac{1}{2}\rho_1 v_1^2 = p_2 + \tfrac{1}{2}\rho_2 v_2^2 \tag{17.57a}$$

Similarly, for planes 3 and 4,

$$p_3 + \tfrac{1}{2}\rho_3 v_3^2 = p_4 + \tfrac{1}{2}\rho_4 v_4^2 \tag{17.57b}$$

For a windmill operating in an unconfined fluid, pressures p_1 and p_4 are equal. Since this is so, equations (17.56) and (17.57) may be combined, on the assumption that the density is constant and by virtue of the conservation of mass $v_3 = v_2$.

$$p_2 - p_3 = \tfrac{1}{2}\rho(v_1^2 - v_4^2) \tag{17.58}$$

Combining equations (17.55) and (17.58) yields

$$v = \frac{v_1 + v_4}{2} \tag{17.59}$$

The mean velocity across the propeller blades is equal to the average of the upstream and downstream velocities, measured at some distance from the windmill. Thus, the velocity drop through the propeller is the same ahead as behind.

The windmill efficiency is the ratio of the power output to the total power available in the airstream of area A and velocity v. This was derived in Chapter 3.

$$\text{Power available} = \frac{\dot{m}v_1^2}{2}$$

$$\text{Power output} = \frac{(v_1^2 - v_4^2)Av\rho}{2}$$

$$\eta = \frac{(v_1 + v_4)(v_1^2 - v_4^2)}{2v_1^3} \tag{17.60}$$

The maximum efficiency is found by differentiating η with respect to (v_4/v_1) and setting the result equal to zero. Let

$$y = \frac{v_4}{v_1}$$

Then

$$\eta = \frac{(v_1 + yv_1)(v_1^2 - y^2v_1^2)}{2v_1^3} = \frac{(1 + y)(1 - y^2)}{2} \tag{17.61}$$

$$\frac{d\eta}{dy} = 0 = 3y^2 + 2y - 1$$

The only physically possible solution is $y = \frac{1}{3}$. This results in a value of $v_4/v_1 = \frac{1}{3}$, which when substituted in equation (17.61) yields a maximum efficiency of 59.3%. This is the maximum percentage of energy that can be used from the available energy, the inlet wind's kinetic energy.

17.14 ENERGY TRANSFER IN A TURBOMACHINE

The word *turbomachine* is derived from two Latin words, *turbo*, a spinning object, and *machina*, a solid device. If these are combined, the result is a "machine that spins or rotates." The rotating portion, or *rotor*, interacts with the fluid that encompasses it. A turbomachine is, therefore, a device that imparts energy to (does work on) or receives energy from (has work done by) a fluid. See Figure 17.23 for a simplified sketch noting the rotor; fixed portion, or *stator*; and fluid flow direction. Thermodynamically, the effect of the energy transfer is manifested as a change in the stagnation enthalpy.

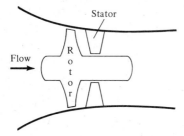

Figure 17.23 Schematic of rotor and stator.

A one-dimensional model will be used for the enclosed machines to be considered here. Often this will restrict our consideration to one portion of the machine at a time: a single stage or blade. The flow from a turbomachine blade is definitely unsteady as viewed from an absolute reference system. However, slightly upstream and downstream of the rotor (assuming many blades) the flow may often be approximated as steady. By proper choice of the flow passage and use of a relative reference plane, steady planar flow can often be used to perform preliminary analysis of the energy transfer in a turbomachine.

Turbomachine Classification

Classification of turbomachines is made by various means: direction of energy, fluid flow, fluid medium, and mode of energy transfer. Let us consider first the flow in turbomachines, which will range from radial, as shown in Figure 17.24(a), where the flow is essentially normal to the rotation axis, to axial through various stages of mixed directional flow, as shown in Figure 17.24(b), to axial as shown in Figure 17.24(c).

As noted earlier, the fluid may do work on or receive work from the rotor, and this gives a primary means of classification. In the case of pumps, fans, blowers, and compressors the work is done on the fluid, resulting in an increase in stagnation enthalpy; for turbines the work is done by the fluid on the rotor, and the stagnation enthalpy decreases. The fluid medium is also commonly used to classify turbines, for example, gas turbine, steam turbine, and water turbine.

Stagnation enthalpy changes are the most inclusive way to describe the energy transfer. However, it is not conventional for incompressible flow cases. Thus, a distinction in classification comes about: incompressible flow machines where the Mach number is typically less than 0.3 throughout, and compressible where the Mach number is high. For incompressible machines, pressure, velocity, and head are the fluid parameters used.

Efficiency

The efficiency of a device is typically a measure of how well the energy transformation is completed. This is translated into comparing the actual fluid properties, that is, the actual energy transfer, to the theoretical fluid properties if the process were ideal. The measurement of the properties will be taken at the inlet and outlet of the device, and specific values will be assumed to be average properties. It will prove useful to

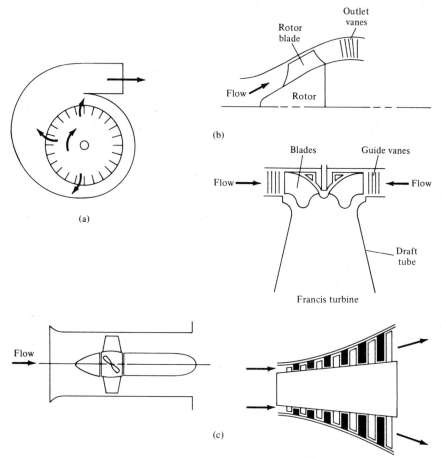

Figure 17.24 (a) A radial-flow fan. (b) Mixed-flow machines. (c) An axial-flow machine.

develop efficiency expressions for compressible and incompressible fluids. There are several aspects to efficiency, with the bottom line being cost — cost of improved design, cost of operation, and cost of manufacture. Inefficient designs developed in the era of inexpensive energy are no longer possible, and increased focus is on optimum operating states. This affects the designer, the manufacturer, and the operator. In the past the most significant efficiency improvements dealt with the cycle, not the components. The cause of turbomachine inefficiencies in many instances is due to secondary flows within the turbomachine, and design modifications can be made to minimize the effect of these flows.

Equipment function also affects efficiency. Often the focus is on how well a particular energy transformation occurs — this is the first-law efficiency and the one with which we are most familiar. The second-law efficiency adds a more difficult question — that of matching system and source energy-level requirements. The focus is on the least net entropy increase or the least net loss of availability.

Compressible Fluid Efficiencies

Consider the flow through a turbine from state 1 to state 2′. This is illustrated on an *h-s* diagram in Figure 17.25. The total fluid energy at these states is represented by the stagnation enthalpies h_{01} and $h_{02'}$. If the process is adiabatic and the change in potential energy is neglected, equation (3.61) shows that the indicated turbine work is equal to $w_t = h_{01} - h_{02'}$. The ideal process is a reversible adiabatic expansion from the same initial condition to a final state characterized by a stagnation enthalpy h_{02}. The efficiency, η_t, is

$$\eta_t = \frac{\text{indicated work}}{\text{ideal work}} = \frac{h_{01} - h_{02'}}{h_{01} - h_{02}} \tag{17.62}$$

We note that in Figure 17.25 the final-state static pressure, p_2, is indicated as well as the stagnation pressures, p_{02} and $p_{02'}$. If the discharge kinetic energy is not used, the discharge enthalpy, not a stagnation enthalpy, should be used for determining the work, and the turbine efficiency is

$$\eta_t = \frac{h_{01} - h_{2'}}{h_{01} - h_2} \tag{17.63}$$

The work for a compressor is equal to the change in the stagnation enthalpy across the unit, $w_c = h_{02'} - h_{01}$, or for the reversible adiabatic case, $w_c = h_{02} - h_{01}$. The *h-s* diagram for the compressor is illustrated in Figure 17.26. Should the compressor be cooled, then the ideal process cannot be considered reversible adiabatic but will typically be isothermal. The compressor efficiency, η_c, is defined as

$$\eta_c = \frac{\text{ideal work for compression}}{\text{actual work for compression}} = \frac{h_{02} - h_{01}}{h_{02'} - h_{01}} \tag{17.64}$$

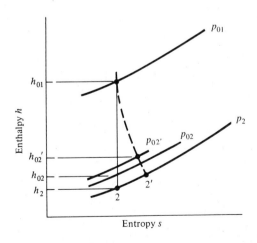

Figure 17.25 An *h-s* diagram for flow through a turbine.

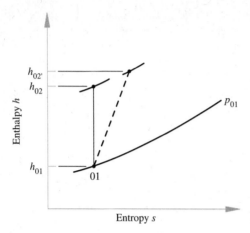

Figure 17.26 An h-s diagram for a compressor.

For an ideal gas it is possible to obtain expressions for the turbine and compressor work in terms of pressures. If $dh = c_p\, dT$ and c_p is relatively constant across the unit, the following simplification results:

$$T_{01} - T_{02} = T_{01}\left[1 - \frac{T_{02}}{T_{01}}\right] = T_{01}\left[1 - \left(\frac{p_{02}}{p_{01}}\right)^{(k-1)/k}\right]$$

$$T_{01} - T_{02} = T_{01}\left[1 - \left(\frac{p_{02}}{p_{01}}\right)^{(k-1)/k}\right] \qquad (17.65)$$

If the initial state (01) and the total pressure ratio are known, the ideal work may be calculated [$w_t = c_p(T_{01} - T_{02})$]. This ideal work is sometimes called the *ideal specific energy transfer*. A similar expression may be developed for the compressor.

Incompressible Fluid Efficiencies

The fluid efficiency for the incompressible case will simplify the calculations because the density is constant. We will be investigating the *useful* or *available* energy the fluid possesses after passing through a turbomachine.

Starting with the first law, equation (3.55a),

$$q - w + \left(\frac{p}{\rho} + u + \frac{V^2}{2} + qz\right)_{in} = \left(\frac{p}{\rho} + u + \frac{V^2}{2} + gz\right)_{out}$$

which written as a differential is

$$\delta q - \delta w = d\left(\frac{p}{\rho}\right) + du + d\left(\frac{V^2}{2}\right) + d(gz) \qquad (17.66)$$

For a unit mass

$$\delta q = du + p\, dv = du + p\, d\left(\frac{1}{\rho}\right)$$

which is the first law for constant-mass, reversible processes, allowing the expression $\delta w = p\, dv$. Expanding equation (17.66) yields

$$\delta q - \delta w = p\, d\left(\frac{1}{\rho}\right) + \frac{1}{\rho}\, dp + du + d\left(\frac{V^2}{2}\right) + d(gz)$$

$$\delta q - \delta w = \delta q + \frac{1}{\rho}\, dp + d\left(\frac{V^2}{2}\right) + d(gz)$$

Hence

$$-\delta w = \frac{1}{\rho}\, dp + d\left(\frac{V^2}{2}\right) + d(gz) \qquad (17.67)$$

If $\rho = C$, then

$$\frac{\Delta p}{\rho} + \Delta\left(\frac{V^2}{2}\right) + \Delta(gz) = -w$$

This represents the fluid's useful work. The same could be obtained by starting with equation (3.55a) and assuming a constant-temperature process $[u(T) = C]$. This is not necessary, but the results confirm our intuitive feeling that any temperature change in fluid may not be converted into work in a rotative device as it could in the compressible case. The specific energy given or delivered by the rotor (rotor specific energy), ΔE_r, will be developed by calculating changes in relative and absolute fluid velocities across the rotor.

The efficiency for a pump, called the hydraulic pump efficiency, e_p, is defined as

$$e_p = \frac{\text{useful specific energy transferred to the fluid}}{\text{rotor specific energy}}$$

$$e_p = \frac{\Delta(p/\rho + V^2/2 + gz)}{\Delta E_r} \qquad (17.68)$$

In a similar fashion the hydraulic turbine efficiency, e_t, is defined as

$$e_t = \frac{\text{rotor specific energy}}{\text{specific energy ideally available from fluid}}$$

$$e_t = \frac{\Delta E_r}{(p/\rho + V^2/2 + gz)} \qquad (17.69)$$

Velocity Diagrams

Energy transfer between the rotor and fluid is related to the fluid's change of angular momentum, which in turn depends on the change of the tangential component of the fluid velocity. The changes of the radial and axial components do not contribute to rotor torque. This requires us to develop an understanding of the manner in which the velocity vectors change through a turbomachine, and to develop a procedure for representing the vector components. The local rotor velocity, $\mathbf{u}$, with a magnitude defined as $2\pi rN$, where r is the local radius and N is the rotor rotational speed in revolutions per second, relates the absolute fluid velocity, $\mathbf{V}$, to the relative fluid velocity, $\mathbf{v}$. The reference frame for the absolute velocities is referenced to the stator. Based on these definitions

$$\mathbf{V} = \mathbf{v} + \mathbf{u} \qquad (17.70)$$

Equation (17.70) forms the basis for construction of the velocity diagrams, which will be helpful in determining changes in angular momentum. The absolute velocity may also be written in terms of its radial, $\mathbf{V}_r$, tangential, $\mathbf{V}_t$, and axial, $\mathbf{V}_a$, components

$$\mathbf{V} = \mathbf{V}_t + \mathbf{V}_r + \mathbf{V}_a \qquad (17.71)$$

In order to illustrate the significance of equation (17.70), we will consider three simple turbomachine rotors.

Examples of Turbomachines

Radial Fan   Consider the radial-flow fan rotor shown in Figure 17.27. The absolute inlet velocity, $\mathbf{V}_1$, is radial ($\mathbf{V}_{r1}$) and so the tangential component, $\mathbf{V}_{t1}$, is zero. For purposes of this discussion, it will be assumed that the fluid angles are equal to the blade angles. Later we will see that this normally is not the case, since there may be an angle of incidence at the inlet and an angle of deviation at the exit even for the design case. Equation (17.70) states that the vector sum of $\mathbf{v}_1$ and $\mathbf{u}_1$ must equal $\mathbf{V}_1$.

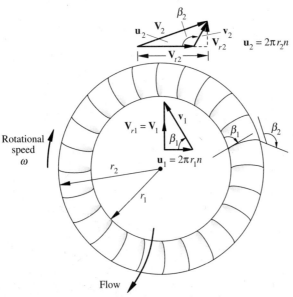

Figure 17.27 Constant-width radial-flow fan rotor with forward-curved vanes.

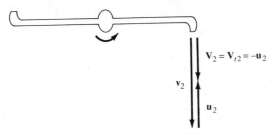

Figure 17.28 Lawn sprinkler rotor with exit velocity diagrams.

The inlet diagram is drawn to satisfy this condition. A similar application of equation (17.70) at discharge results in the discharge velocity diagram shown in Figure 17.27. Note that u_2 is greater than u_1 because the radius is greater, and for the same reason V_{r2} is less than V_{r1} for a constant-width rotor. The absolute velocity at exit now has a component in both the tangential direction, V_{t2}, and the radial direction, V_{r2}. As noted earlier, it is the tangential component that is related to rotor torque. Therefore, the important change that is observed is an increase in the tangential component of the absolute velocity in the direction of rotation. This indicates that energy has been transferred from the rotor to the fluid.

   Lawn Sprinkler   Now it will be useful to compare this rotor with the simple lawn sprinkler shown in Figure 17.28. The sprinkler makes use of a high relative exit velocity that is opposite in direction to the rotor speed. Again, using equation (17.70), the outlet velocity diagram for the operating condition $|u_2| = \frac{1}{2}|v_2|$ must appear as shown in Figure 17.28. The tangential component of the absolute velocity is initially zero, $V_{t1} = 0$, and changes from inlet to outlet. However, in this case the tangential component of absolute velocity has changed from 0 to a value opposite in direction to the rotor velocity. That is, it has decreased. Hence, energy has been extracted from the fluid as it passed through the rotor (sprinkler arm), and thus we call it a turbine.
   Suppose that, while rotating at a speed consistent with u_2, the nozzle tip suddenly breaks off so that the arm rotates as shown in Figure 17.29. Intuition tells us that in this configuration, the machine will be slowed by friction and soon will stop rotating. Why? The relative velocity vector is now radial. Application of equation (17.70) shows that the absolute velocity now has a tangential component that is in the direction of the rotor rotation. This machine is now a pump, and in the absence of external energy, dissipation will quickly slow the sprinkler arm to zero velocity.

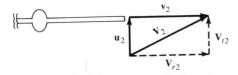

Figure 17.29 Lawn sprinkler rotor shortly after losing the nozzle.

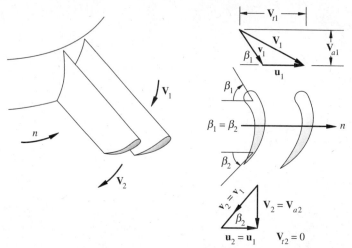

Figure 17.30 Flow through an impulse turbine stage (oblique and end views) with velocity diagrams.

Impulse Turbine   As the last illustration of the use of equation (17.70), consider an axial-flow impulse turbine stage with the simple rotor partially shown in Figure 17.30. The fluid is directed by a nozzle onto a rotor, and the rotor is designed to operate with no pressure change. Bernoulli's equation implies, for the ideal case, that there is no change in the relative velocities if the pressure change through the rotor is zero; hence, $|v_1| = |v_2|$. Also, for an axial-flow machine at a given radius, $u_1 = u_2 = u$. As we did with the fan, assume the fluid follows the blade angles. Furthermore, let us operate the rotor in a manner that will use the maximum amount of available energy. This requires that the discharge kinetic energy be a minimum; hence V_2 is equal to the axial component ($V_2 = V_{a2}$), and the tangential component, V_{t2}, is 0. Equation (17.70) dictates that the exit diagram must appear as shown in Figure 17.30. Now we have a situation opposite to that which we found for the fan and lawn sprinkler rotor, where the absolute inlet tangential velocity component was zero. However, we are interested in changes only in the tangential component, and the diagram illustrates that the change in V_t is opposite to that of u; hence, the device is a turbine.

17.15 FLUID-ROTOR ENERGY TRANSFER

Until this point we have dealt in a qualitative manner when looking at turbine-fluid energy interaction. The concept of velocity diagrams has been covered, and now we will extend our capabilities to that of a general turbomachine.

Euler Turbine Equation

Figure 17.31 illustrates a rotor of a turbomachine, where the fluid enters at the hub of radius r_1 with a velocity V_1, and exits at the rim of radius r_2 with a velocity V_2. The

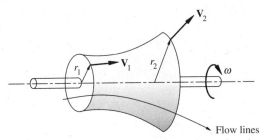

Figure 17.31 Control volume for a general turbo-machine.

mass flow rate is steady, there is no gain or loss (leakage) in the control volume, and the fluid state properties at a point in the control volume are constant with time. Accordingly, the energy flux terms across the control surface (heat, work) are also constant with time.

Let us analyze each velocity component (radial, axial, tangential) to see its effect on machinery performance and whether or not is must be included in our analysis. A change in the axial velocity will produce a net force in the axial direction. This force will tend to move the rotor unless measures are taken to resist this force or thrust. Thrust bearings are installed to prevent axial movement of the turbine rotor. The change of velocity in the radial direction will produce a net force in this direction that acts on the journal bearings supporting the rotor. For well-designed bearings there will be little effect on rotor torque, and the changes in radial and axial velocities can be neglected. The change of velocity of the tangential component will produce a net force on the tangential direction. This affects the rotor torque; hence, we will analyze this velocity component in greater detail.

If we apply equation (17.13) to this case for steady flow, it reduces to

$$\sum rF_t = \dot{m}(r_2 V_{t2} - r_1 V_{t1}) \tag{17.72}$$

where

$$\dot{m} = \int_A \rho V_n \, dA$$

Equation (17.72) represents the torque, τ, where $\tau = \Sigma rF_t$. The rate of energy transfer may be expressed in terms of the torque and the rotor angular velocity, ω. Thus,

$$E = \omega\tau \tag{17.73}$$

A sign convention must be adopted for the velocities. The product of ωV is positive when they are both in the clockwise direction. Should one be counterclockwise, the product would be negative. Substituting equation (17.73) into equation (17.72) yields

$$E = \dot{m}(\omega r_2 \, V_{t2} - \omega r_1 \, V_{t1}) = \dot{m}(u_2 \, V_{t2} - u_1 \, V_{t1}) \tag{17.74}$$

where $\omega r = \mathbf{u}$, the rotor velocity at radius r. Dividing by $\dot{m}$ allows one to find the rotor energy transfer per unit mass.

$$e = u_2 \, V_{t2} - u_1 \, V_{t1} \tag{17.75}$$

If u and V_t are in the same direction, the product is positive. This results from the derivation of equation (17.75) from a vector equation, and for this scalar form to be valid, the direction must be accounted for. Equations (17.72), (17.74), and (17.75) are all forms of the Euler turbine equation. It is very apparent that the fluid-rotor energy transfer of primary importance is accounted for by the variation in the rotor linear velocity, $\mathbf{u}$, and the fluid's tangential velocity component, $\mathbf{V}_t$.

It is necessary to correlate equation (17.75) with thermodynamic sign convention. If the system is considered to be the fluid and work is done on the fluid by the rotor (pump, compressor), the energy transfer is

$$e_p = u_2 V_{t2} - u_1 V_{t1} > 0 \tag{17.76}$$

while for a turbine, the fluid does work on the rotor and the sign must be reversed, or

$$e_t = u_1 V_{t1} - u_2 V_{t2} > 0 \tag{17.77}$$

The previous Euler turbine equations express the fluid energy in terms of the fluid tangential and linear rotor velocities. It is possible and useful to have the energy in terms of absolute ($\mathbf{V}$), linear rotor ($\mathbf{u}$), and relative ($\mathbf{v}$) velocities. Figure 17.32 illustrates the ideal velocity diagrams for the inlet and exit conditions of the generalized turbomachine illustrated in Figure 17.31. Consider the exit velocity diagram.

$$V_{r2}^2 = V_2^2 - V_{t2}^2$$

or

$$V_{r2}^2 = v_2^2 - (u_2 - V_{t2})^2$$

Equating the two yields

$$u_2 V_{t2} = \tfrac{1}{2}[V_2^2 + u_2^2 - v_2^2]$$

In a similar fashion

$$u_1 V_{t1} = \tfrac{1}{2}[V_1^2 + u_1^2 - v_1^2]$$

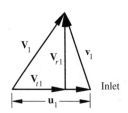

Inlet

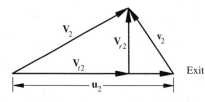

Exit

Figure 17.32 Ideal velocity diagrams for a generalized turbomachine.

and substituting into equation (17.75) yields

$$e = \tfrac{1}{2}[(V_2^2 - V_1^2) + (u_2^2 - u_1^2) + (v_1^2 - v_2^2)] \tag{17.78}$$

The terms on the right-hand side of equation (17.78) may be physically interpreted as follows: the first term, $\tfrac{1}{2}(V_2^2 - V_1^2)$, is the change in the fluid's kinetic energy in passing through the machine. In both a turbine and pump it is desirable to have V_2^2 small, though this may not be the actual case. In a turbine V_2^2 represents work that has not been transferred to the rotor, and in a pump it represents the incomplete conversion of kinetic energy to enthalpy (manifested by a fluid pressure increase).

The second term, $\tfrac{1}{2}(u_2^2 - u_1^2)$, represents the change of fluid energy as it moves in the radial direction, known as the *centrifugal effect*. The centrifugal force caused by the variation of velocity in the radial direction causes the fluid pressure to change. For example, when the discharge valve of a centrifugal pump is closed, the mass flow is zero, the absolute and relative velocities are zero, and the pressure at discharge is produced only by the centrifugal effect.

The third term, $\tfrac{1}{2}(v_1^2 - v_2^2)$, represents the Bernoulli effect, or the conversion of velocity head into pressure head.

The conversion of kinetic energy (velocity) into enthalpy (pressure) occurs in three ways: (1) the change in the absolute kinetic energy across the turbomachine casing; (2) the change of pressure within the rotor due to centrifugal forces; and (3) the change of pressure within the rotor due to a change in relative velocity within the rotor.

Degree of Reaction

The *degree of reaction, R,* specifies the portion of the total energy transfer across a rotor related to changes of static pressure (enthalpy) across the rotor. This may be defined in terms of equations (17.78) and (17.75) as

$$R = \frac{\tfrac{1}{2}[(u_2^2 - u_1^2) + (v_1^2 - v_2^2)]}{u_2 V_{t2} - u_1 V_{t1}} \tag{17.79}$$

The degree of reaction for a given machine may be negative or positive and may vary throughout the machine.

Energy Analysis of Selected Turbomachines

Let us consider the three machines analyzed previously: the radial fan, the lawn sprinkler, and the impulse turbine. These devices have a wide spectrum of operating characteristics.

Lawn Sprinkler   From equation (17.79) and with the aid of the velocity diagram in Figure 17.28, for the operating condition $|u_2| = \tfrac{1}{2}(v_2)$ we have

$$V_{t1} = u_1 = v_1 = 0$$
$$V_2 = V_{t2} = -u_2 \quad \text{and} \quad v_2 = |2u_2|$$

$$R = \frac{\frac{1}{2}[u_2^2 - v_2^2]}{-u_2^2} = \frac{\frac{1}{2}[u_2^2 - 4u_2^2]}{-u_2^2} = +1.5$$

The rotor-specific energy transfer may be calculated from equation (17.77):

$$e = +u_2^2$$

This is an example of a turbine — the work is positive. Note that if the operating speed is changed, the degree of reaction will be different. As u_2 approaches v_2, the degree of reaction approaches 1.0, as does the theoretical efficiency.

Radial Fan   Radial flow is the primary direction of flow in many turbomachines, such as pumps and compressors, and particularly so in the case of industrial fans. Since the devices are common, let us examine in a little more detail the characteristics of these flow machines as demonstrated by the radial-flow fan. Figure 17.27 illustrates the fan with inlet and discharge velocity diagrams. A variety of factors could be held constant; however, at this point it will be beneficial to determine the machine performance as a function of the blade exit angle.

To this end let us specify that (1) $u_2 = 3u_1$; (2) the radial velocity is constant; (3) there is no inlet whirl (angular momentum is zero); (4) the inlet blade angle, β_1, is 63.43°; (5) the exit blade angle, β_2, is variable. From geometry

$$V_1 = u_1 \tan(63.43) = 2u,$$

$$v_1^2 = 4u_1^2 + u_1^2 = 5u_1^2$$

and from equation (17.75) the rotor energy transfer per unit mass is

$$e = u_2 V_{t2} - u_1 V_{t1} = u_2 V_{t2} \tag{17.80}$$

since $V_{t1} = 0$.

From the velocity diagram, Figure 17.33, we have

$$x = V_{r2} \cot(180 - \beta_2) = -V_{r2} \cot \beta_2$$

$$V_{t2} = u_2 + x = u_2 - V_{r2} \cot \beta_2$$

$$v_2^2 = x^2 + V_{r2}^2 = 4u_1^2(1 + \cot^2 \beta_2)$$

and e becomes

$$e = u_2^2 - u_2 V_{r2} \cot \beta_2$$

$$u_2 = 3u_1 \qquad V_{r2} = 2u_1$$

$$e = 3u_1^2(3 - 2 \cot \beta_2) \tag{17.81}$$

The equation for the degree of reaction (17.79) reduces to

$$R = \frac{\frac{1}{2}[9u_1^2 - u_1^2 - 4u_1^2 - 4u_1^2 \cot^2 \beta_2 + 5u_1^2]}{3u_1^2(3 - 2 \cot \beta_2)}$$

$$R = \frac{9 - 4 \cot^2 \beta_2}{6(3 - 2 \cot \beta_2)} \tag{17.82}$$

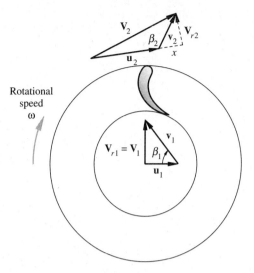

Figure 17.33 Velocity diagram for a radial fan.

If we arbitrarily let $u_1^2 = 1.0$ and plot the values of R and e, Figure 17.34 is the result. We note that there is a sign change for e at 33°. In this case the value of e is negative for exit blade angles less than 33° and positive for values greater than 33°. The only variable is the exit blade angle, and the machine operates as a turbine (e negative) or a compressor or pump (e positive). This may seem somewhat contradictory, but we defined e in equations (17.76) and (17.77) to allow for the work expressions to be positive whenever dealing with a given device.

It is interesting to note the change in the velocity diagrams as β_2 increases. As β_2 increases from 33° to 146°, the absolute velocity at discharge increases. The result is less energy conversion into static pressure. The extreme case exists for values of β_2 above 146° where there is a static head decrease, which signifies that a portion of the inlet pressure head (enthalpy) was converted into kinetic energy.

Impulse Turbine   The axial-flow impulse turbine is illustrated in Figure 17.30, with the velocity diagrams repeated in Figure 17.35.

The degree of reaction is $R = 0$, since there is no static pressure change across the rotor. The pressure change occurs in the nozzle before the impulse stage. In this case several simplifications occur in the velocity diagram. Since there is no change in static pressure, $|v_2| = |v_1|$, though there is a change in direction; the linear velocity is constant, $u_2 = u_1$; and the energy transfer occurs solely because of a change in the absolute velocities, $V_2 < V_1$. The h-s diagram for this process is illustrated in Figure 17.36. In the nozzle the stagnation enthalpy remains constant (process 1-2). The stagnation enthalpy located by point a is constant, $T_{01} = T_{02}$. The fluid enters the moving blades at state 2, exiting at state 3'. There is energy transfer from the fluid to the rotor, reducing the stagnation enthalpy (temperature) until state 3' is reached. The new stagnation state is characterized by T_{03}, and $p_{03'}$. The kinetic energy exiting the nozzle is equal to $(h_{01} - h_2)$, and the kinetic energy leaving the rotor is equal to

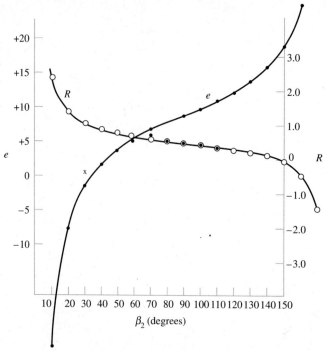

Figure 17.34 The specific energy transfer and degree of reaction for a radial fan as a function of β_2.

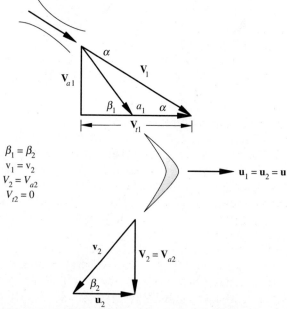

Figure 17.35 Velocity diagram for an ideal impulse turbine.

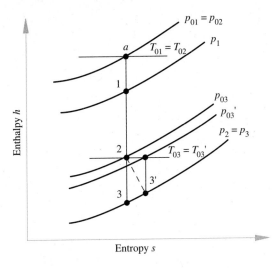

Enthalpy h

Entropy s

Figure 17.36 An h-s diagram for an impulse turbine.

$(h_{03} - h_{3'})$. The actual expansion process has irreversibilities associated with the flow, and hence $s_{3'} > s_3$.

State 3 is ideally characterized by an absolute velocity in the axial direction only $(V_{t3} = 0)$. However, for the fluid to exit the blade, $V_{a3} > 0$.

Examining the rotor specific energy transfer, using the notation of Figure 17.35 we obtain

$$e = u_1 V_{t1} - u_2 V_{t2} = u_1 V_{t1}$$

for the ideal case, where $V_{t2} = 0$. Thus, only the change in the tangential component of the absolute velocity contributes to work being done on the runner. Any nonzero value of V_{t2} reduces e, since it results in an increase in V_2 and thus the kinetic energy leaving the blade. It can be shown that $V_{t2} = 0$ when $u_1/V_1 = \cos \alpha/2$. This ratio of blade speed to entering steam speed is an important parameter in turbine design.

Theoretical Head-Capacity Diagrams

The Euler equation can be used to determine the head-capacity relationships for ideal incompressible flow conditions as follows. Consider the velocity diagram in Figure 17.37 for a radial-flow impeller. For radial flow with $V_{t1} = 0$,

$$\Delta E_r = u_2 V_{t2} = u_2(u_2 - V_{r2} \cot \beta_2) \tag{17.83}$$

The volume flow rate, $\dot{V}$, is

$$\dot{V} = V_{r2} A_2$$

where A_2 is normal to V_{r2}, and hence

$$\Delta E_r = u_2 \left(u_2 - \frac{\dot{V}}{A_2} \cot \beta_2 \right) \tag{17.84}$$

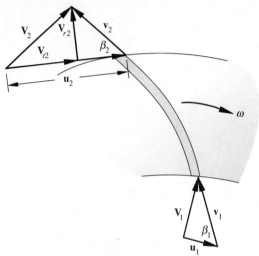

Figure 17.37 A radial-flow impeller.

If a pump or fan is running at constant speed, the following values are constant: u_2, A_2, β_2. This means the head is a function of flow only. Thus,

$$\Delta E_r = a + b\dot{V} \tag{17.85}$$

where $a = (u_2)^2$ and

$b = (u_2/A_2) \cot \beta_2$.

There are three conditions that we wish to examine as $\cot \beta_2$ varies from positive $(0 < \beta_2 < 90°)$ to negative $(\beta > 90°)$, with zero being $\beta_2 = 90°$. If $0 < \beta_2 < 90$, the blades are backward-curved; if $\beta_2 = 90°$, the blades are radial; and if $\beta_2 > 90°$, the blades are forward-curved. Figure 17.38 illustrates the head-capacity characteristics, as developed from equation (17.85). The shutoff head is the same for all designs $(u_2^2/2)$. Examination of velocity diagrams, Figure 17.39, along with Figure 17.40 leads to some interesting conclusions. At first glance it would seem desirable to have forward-curved blades, because the head increases with flow. There are two reasons why this is not desirable. First, the high ΔE_r occurs because v_2 is very large, and the high velocity cannot be efficiently transformed into pressure. Second, from the operational and control viewpoints the forward-curved blades lead to unstable operation. Power increases dramatically as flow increases, and it would be easy to overload the driving device. Radial blades also have the disadvantage of operational instability but are sometimes attractive in high-speed machines because the blade stress is low. Subsequently, the backward-curved blades are the ones most often found. The exit velocity is comparatively low, so the conversion of kinetic energy into pressure (enthalpy) does not affect overall machine performance as much as in the forward-curved case, and the machine is operationally stable.

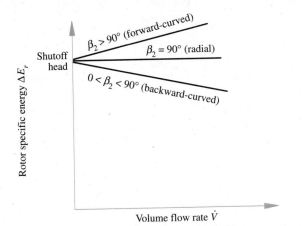

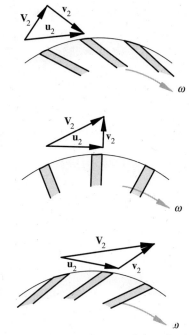

Figure 17.38 Theoretical head-capacity diagram.

Figure 17.39 Velocity diagrams for various vane curvatures.

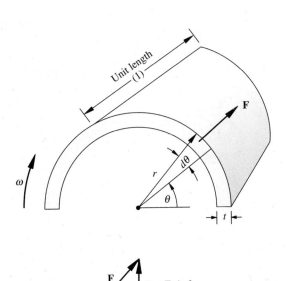

Figure 17.40 Diagram for hoop stress.

Limitations on Energy Transfer in Axial-Flow Machines

From examination of the rotor specific energy transfer for the impulse turbine, we note that $e \propto u^2$. From this we can imagine that by increasing the turbine speed, more energy can be transferred. Unfortunately, there are material limitations that prevent us from ever increasing the velocity.

Consider Figure 17.40, which illustrates a rotating ring. It may be imagined that the ring is acted on by an internal pressure (centrifugal force), which leads us to evaluate the hoop stress.

The centrifugal force is $F = mr\omega$:

$$F = \int_0^\pi \rho t r \, d\theta \, (1) \, \frac{u^2}{r}$$

However, we are interested in the force acting in the vertical direction, F_y:

$$F_y = \rho \int_0^\pi tr \frac{u^2}{r} \sin \theta \, d\theta = 2\rho t u^2 \tag{17.86}$$

The area resisting this force is $2\sigma t(1)$.

$$A = 2\sigma t \tag{17.87}$$

Equating (17.86) and (17.87) yields

$$u^2 = \frac{\sigma}{\rho} \tag{17.88}$$

Thus, the rotational velocity of the rotor may be related to material properties. Certainly the previous development is not generally applicable because parts, such as blades, are attached to the rotor. Also, temperature effects have been neglected. However, the trend is clearly demonstrated by considering the following properties of 0.8% carbon steel:

$$\rho = 7848 \text{ kg/m}^3$$

$$\sigma = 827 \text{ MPa tension}$$

$$u^2 = \frac{827 \, 000 \, 000}{7848} = 105 \, 377$$

$$u = 324.6 \text{ m/s}$$

Consider an ideal impulse turbine with the nozzle angle $\alpha = 0$ operating at an optimum blade speed/steam speed ratio.

$$\frac{u_1}{V_1} = \frac{\cos \alpha}{2}$$

$$\cos \alpha = \cos (0°) = 1$$

Thus

$$2u_1 = V_1 = V_{t1}$$

and

$$V_2 = V_{t2} = 0$$

and substituting in the Euler turbine equation (17.77) yields

$$e = 2u_1^2$$

Substituting for $u_1 = 324.6$ m/s

$$e = 2(105\ 377) = 210.7 \text{ kJ/kg}$$

Since even the most cursory examination of a steam or gas turbine shows that the available energy (isentropic enthalpy drop across the unit) is much greater, multiple turbine staging is a requirement.

CONCEPT QUESTIONS

1. What is meant by moment of momentum?
2. What is sound, and how is it generated?
3. Can sound waves travel in a vacuum?
4. What is acoustic velocity?
5. Why is the acoustic velocity in water greater than in air?
6. What is stagnation temperature? Stagnation pressure?
7. Are the isentropic and stagnation temperatures the same?
8. Are the isentropic and stagnation pressures the same?
9. Define the Mach number.
10. Is the Mach number for a plane flying 500 mph at an elevation of 1000 ft the same as for a plane flying 500 mph at an elevation of 40,000 ft? Why?
11. What are the purposes of a nozzle?
12. Why do nozzles handling liquids have only a convergent portion?
13. What is the critical pressure ratio?
14. What are the limitations on the angle of divergence in a convergent-divergent nozzle?
15. Describe the initial conditions necessary for supersaturated flow in a nozzle.
16. What is a condensation shock?
17. Can a normal shock wave be formed in the convergent portion of a convergent-divergent nozzle? Why?
18. What do the intersections of the Fanno and Rayleigh lines represent?
19. Can the Mach number of a fluid be greater than 1 after a normal shock? Why?
20. Under what conditions does the divergent portion of a nozzle accelerate flow? Decelerate flow?
21. A diffuser serves what purpose?

22. Explain the operation of a venturi meter.

23. What is a turbomachine?

24. Describe the general shape of head-capacity diagrams for forward-curved, radial, and backward-curved vanes in a radial-flow pump.

25. Why do material limitations result in multiple staging in turbines?

PROBLEMS (SI)

17.1 Helium is flowing in a pipeline at a velocity of 1000 m/s and with a pressure of 120 kPa and a temperature of 300 K. Determine the stagnation temperature and the isentropic stagnation pressure.

17.2 Helium is flowing in a duct with a velocity of 350 m/s and pressure and temperature of 100 kPa and 25°C. Determine (a) the Mach number; (b) the stagnation pressure and temperature.

17.3 Air flows in a 200-cm² duct at a velocity of 180 m/s, a pressure of 119 kPa, and a temperature of 28°C. Determine the temperature, pressure, and area required when the flow becomes sonic.

17.4 A thermocouple is inserted into a duct where the air velocity is 200 m/s, and reads 82°C. What is the actual air temperature?

17.5 Find the isentropic stagnation temperature and pressure for the following fluids flowing through a duct at 2.5 MPa, 350°C, and 450 m/s: (a) helium; (b) nitrogen; (c) steam.

17.6 Oxygen expands in an adiabatic nozzle from inlet conditions of 1100°K and negligible velocity to an exit temperature of 400°K. Determine the exit Mach number.

17.7 Determine the acoustic velocity of R 12 at 1 MPa and 50°C.

17.8 Air expands in a reversible adiabatic manner from 2.0 MPa and 100°C to 500 kPa. Determine the ratio of initial to final acoustic velocity.

17.9 A hypersonic aircraft is designed to fly at Mach 20 at an elevation where the atmospheric temperature is 217°K. Determine the stagnation temperature on the leading edge of the aircraft's wing.

17.10 Air in a convergent nozzle has the following stagnation conditions: $p_0 = 200$ kPa, $T_0 = 370$°K. If the back pressure is 100 kPa, find the exit Mach number and velocity.

17.11 Air is flowing through a nozzle at a rate of 3 kg/s. Inlet conditions are 1400 kPa, 890°K, and 100 m/s, and the exit pressure is 300 kPa. For increments of 100 kPa, plot the variation of area, specific volume, and temperature through the nozzle. Use the air tables.

17.12 Air is flowing through a nozzle with inlet conditions of 600 kPa and 1200°K and an exit condition of 150 kPa. The nozzle efficiency is 96%. If the throat area is 6.45×10^{-4} m², determine (a) the exit velocity; (b) the exit Mach number; (c) the flow rate.

17.13 At 630 kPa and 1200°K, 27.2 kg/s of helium flows through the inlet nozzles of a gas turbine. The exit pressure is 280 kPa. The throat diameter of each circular nozzle is 1.5 cm. Determine (a) the critical pressure; (b) the minimum number of nozzles required; (c) the force on the nozzles.

17.14 Steam flows isentropically through a nozzle from v = 200 m/s, T = 370°C, and p = 3.5 MPa. Determine (a) the isentropic stagnation pressure; (b) the exit specific volume; (c) the exit velocity; (d) the critical temperature if the exit pressure is 700 kPa.

17.15 The high-pressure turbine in a power plant receives 25.0 kg/s of steam at 420°C and

3.5 MPa. The steam exits the first nozzle block at 2.0 MPa and enters the blades. Determine (a) the enthalpy of steam exiting the nozzle; (b) the velocity of steam leaving the nozzle; (c) the total force on the nozzle block; (d) the minimum number of nozzles required with a 1-cm exit diameter.

17.16 Steam at 3.8 MPa and 260°C enters a nozzle with a throat area of 4.5 cm². The nozzle discharges at 1.0 MPa. Determine (a) the mass flow rate; (b) the exit steam quality; (c) the exit area; (d) the specific volume at the throat.

17.17 Methane flows through an ideal nozzle with the following inlet conditions: static pressure is 700 kPa, static temperature is 300°K, and velocity is 125 m/s. The nozzle discharges into a static pressure of 550 kPa. Determine (a) the exit static temperature; (b) the exit specific volume; (c) the exit velocity.

17.18 Steam enters an adiabatic nozzle at 1.4 MPa, 260°C, and negligible velocity, and expands to 140 kPa with a quality of 97%. Determine (a) the steam exit conditions; (b) the nozzle efficiency.

17.19 A circular convergent-divergent nozzle handles 0.15 kg/s of air from 20°C and 1.0 MPa. The nozzle has an efficiency of 95%. The discharge pressure is 100 kPa. Determine (a) the exit velocity; (b) the exit temperature; (c) the length from the throat for $\theta = 5°$.

17.20 A gaseous mixture of 75% CO_2 and 25% He, on a molal basis, flows through a 30-cm-I.D. pipe with a velocity of 150 m/s at a pressure of 600 kPa and a temperature of 115°C. The mixture velocity is to be increased to 350 m/s by reducing the pipe diameter. During the process the static temperature remains constant at 115°C by heat transfer. Determine (a) the heat transfer to or from the gas; (b) the static pressure where the velocity is 350 m/s; (c) the diameter where the velocity is 350 m/s.

17.21 An ideal gas in a rocket has the following conditions: nozzle inlet chamber pressure = 2800 kPa, nozzle exit pressure = 28 kPa, k for the gas = 1.2, molecular weight = 21.0, nozzle inlet chamber temperature = 2500°K. Determine (a) the critical pressure ratio; (b) the velocity at the throat; (c) the exit temperature; (d) the exit velocity; (e) the ratio of exit area to throat area.

17.22 Air flows through a test section and undergoes a normal shock where the upstream conditions are $M_x = 1.8$, $p_x = 500$ kPa, and $T_x = 300°K$. Determine (a) p_y; (b) p_{0x}; (c) T_{0x}; (d) the change in specific entropy across the shock wave.

17.23 A quantity of air, 0.6 kg/s, flows steadily through a convergent-divergent nozzle from inlet conditions of 400 kPa, 500°K, and 200 m/s to an exit plane pressure of 280 kPa. A normal shock wave occurs in the divergent section where $M_x = 2.0$. The flow is isentropic before and after the shock. Determine (a) the stagnation temperature, T_{0x}; (b) the stagnation pressure, p_{0x}; (c) the pressure p_x; (d) the pressure p_y; (e) the stagnation pressure, p_{0y}; (f) the stagnation temperature, T_{0y}; (g) the exit area; (h) the throat area.

17.24 Compute the entropy production across the shock wave in Problem 17.23.

17.25 A supersonic wind tunnel is created by locating the test section at the exit of a convergent-divergent nozzle. The inlet air is at 1.2 MPa, 320°K, and negligible velocity. The test-section area is 0.12 m², and the desired Mach number is 2.0. (a) Determine the air pressure, temperature, and velocity. (b) Discuss the effect of moisture in the air.

17.26 Air flows through a convergent-divergent nozzle with inlet conditions of 1 MPa, 300°K, and negligible velocity. A normal shock wave stands at the nozzle exit, where $M = 2.2$ just before the shock wave. Determine the pressure, temperature, Mach number, velocity, and stagnation pressure after the shock wave.

17.27 Determine the change of specific entropy across the shock wave in Problem 17.26.

17.28 Air flows through a convergent-divergent nozzle with inlet conditions of 2 MPa, 350°K, and negligible velocity. The exit area is twice the throat area. Determine the exit pressure such that a normal shock wave occurs at the exit plane.

17.29 Air enters a diffuser at 90 kPa, 260°K, 260 m/s, and a flow rate of 12 kg/s. The diffuser efficiency is 95%. Determine the diffuser exit temperature, pressure, and area if the exit velocity is 70 m/s.

17.30 A Pitot tube is installed in a 76-mm-I.D. pipe to determine the water flow rate by measuring the velocity head. The water temperature is 20°C. The manometer reads 38 mm Hg. Determine the water flow rate in kg/s.

17.31 Dry saturated steam enters a diffuser at 14 kPa and an unknown velocity. It exits at 42 kPa with negligible velocity. Determine (a) the discharge temperature for isentropic flow; (b) the discharge temperature if $\eta_d = 0.80$.

17.32 Carbon monoxide enters a diffuser at 100 kPa and 60°C with a Mach number of 3.0. The diffuser efficiency is 85%, the flow rate is 10 kg/s, and the exit velocity is negligible. Determine (a) the exit pressure; (b) the exit temperature; (c) the minimum area.

17.33 A venturi flow meter is used in a plant to measure water flowing through a 30-cm pipe. The throat diameter of the venturi is 23 cm, and the pressure drop is 15 cm of water. Determine the mass flow rate.

17.34 Air is flowing through an 80-mm-I.D. pipe at 100 kPa and 50°C. The venturi diameter is 40 mm, and a pressure drop of 280 mm of water is recorded across the venturi. Determine the volume flow rate of air.

17.35 Air with a velocity of 225 m/s enters an isentropic diffuser with an inlet area of 0.15 m². The inlet static temperature and pressure are 255°K and 63 kPa. The exit velocity is 100 m/s. Determine (a) the air flow rate; (b) the static and total exit temperatures; (c) the static and total exit pressures; (d) the exit area.

17.36 A wind turbine with a 61-m blade diameter produces 2000 kW when the wind velocity is 40 km/h. The air temperature is 25°C, and the pressure is 101 kPa. Determine the wind turbine efficiency.

17.37 For the turbine in Problem 17.36 determine the velocity leaving the blades.

17.38 A wind turbine has a start-up wind speed of 17 km/h and cuts out at a wind velocity of 57 km/h. The wind turbine has a blade diameter of 30 m. Consider the wind turbine to have a constant efficiency of 40%. For a 24-h period, the following wind speeds, in km/h, are available. The air density is constant at 1.181 kg/m³.

Time period	Average wind speed
0000–0400	10
0400–0800	18
0800–1200	25
1200–1600	20
1600–2000	33
2000–2400	30

Determine the power produced.

17.39 Ten kg/s of air flows across an impulse blade in an isentropic process. It enters the blade with a velocity of 457 m/s, a pressure of 110 kPa, and a temperature of 93°C, and at an angle of 18°. The blade velocity is 244 m/s, and the air leaves at an angle of 45° relative to the blade. Determine (a) the velocity diagram; (b) the power produced by the turbine.

17.40 Steam flows through a turbine in an isentropic manner with inlet conditions of 800 kPa and 650°C and exits at 100 kPa. The steam enters the turbine blade at an angle of 20°, the blade exit angle is 50°, and the blade/speed ratio is 0.5. Determine the ideal turbine efficiency.

17.41 An axial-flow fan operates at 1500 rpm. The blade inlet and exit angles are 30° and 60°, respectively, and guide vanes give the flow entering the first stage an angle of 30°. The ratio of blade tip to hub diameter is 1.375, and the hub diameter is 0.8 m. The air enters at 25°C and 1 atm and may be considered incompressible. Determine (a) the discharge velocity diagram; (b) the torque; (c) the power.

17.42 Water with a flow rate of 9.45 kg/s enters a mixed, axial-flow pump with a uniform velocity. The impeller hub diameter is 2.5 cm, and the overall diameter is 10 cm. The flow leaves at 3 m/s relative to the radial blades. The impeller rotates at 3400 rpm. Determine (a) the impeller exit width; (b) the torque required; (c) the power required.

PROBLEMS (ENGLISH UNITS)

*17.1 A convergent nozzle receives 10,000 lbm/hr of steam at 500 psia, 600°F, and 300 ft/sec and discharges it through an exit area of 0.5 in.^{2} at 225 psia and 1700 ft/sec. Determine the minimum force necessary to hold the nozzle in position.

*17.2 A thermocouple is inserted into a duct where the air velocity is 650 ft/sec and reads 180°F. What is the actual air temperature?

*17.3 Find the isentropic stagnation temperature and pressure for the following fluids flowing through a duct at 400 psia, 650°F, and 1500 ft/sec: (a) helium; (b) nitrogen; (c) steam.

*17.4 Oxygen expands in an adiabatic nozzle from inlet conditions of 2000°R and negligible velocity to an exit temperature of 720°R. Determine the exit Mach number.

*17.5 Determine the acoustic velocity of R 12 at 150 psia and 120°F.

*17.6 Air expands in a reversible adiabatic manner from 300 psia and 200°F to 60 psia. Determine the ratio of initial to final acoustic velocity.

*17.7 A hypersonic aircraft is designed to fly at Mach 20 at an elevation where the atmospheric temperature is −70°F. Determine the stagnation temperature on the leading edge of the aircraft's wing.

*17.8 Three lbm/sec of air is flowing through a nozzle. Inlet conditions are 200 psia, 1600°R, and 200 ft/sec, and the exit pressure is 40 psia. For increments of 20 psia, plot the variation of area, specific volume, and temperature through the nozzle. Use the air tables.

*17.9 A nozzle has a minimum area of 1 in.^{2} and an efficiency of 95%. It receives air at 50 psia, 1240°F, and 250 ft/sec and discharges it at 30 psia. Determine (a) the flow rate; (b) the discharge stagnation enthalpy. (c) Sketch the nozzle shape.

*17.10 At 90 psia and 2100°R, 3600 lbm/min of helium flows through the inlet nozzles of a gas turbine. The exit pressure is 40 psia. The throat diameter of each circular nozzle is

0.5 in. Determine (a) the critical pressure; (b) the minimum number of nozzles required; (c) the force on the nozzles.

*17.11 Air flows through a test section and undergoes a normal shock where the upstream conditions are $M_x = 1.8$, $p_x = 75$ psia, and $T_x = 540°R$. Determine (a) p_y; (b) p_{0x}; (c) T_{0x}; (d) the change in specific entropy across the shock wave.

*17.12 A quantity of air, 1.3 lbm/sec, flows steadily through a convergent-divergent nozzle from inlet conditions of 60 psia, 900°R, and 650 ft/sec to an exit plane pressure of 40 psia. A normal shock wave occurs in the divergent section where $M_x = 2.0$. The flow is isentropic before and after the shock. Determine (a) the stagnation temperature, T_{0x}; (b) the stagnation pressure, p_{0x}; (c) the upstream shock pressure, p_x; (d) the downstream shock pressure, p_y; (e) the stagnation pressure, p_{0y}; (f) the stagnation temperature, T_{0y}; (g) the exit area; (h) the throat area.

*17.13 Compute the entropy production across the shock wave in Problem *17.12.

*17.14 A supersonic wind tunnel is created by locating the test section at the exit of a convergent-divergent nozzle. The inlet air is at 175 psia, 575°R, and negligible velocity. The test-section area is 1.0 ft², and the desired Mach number is 2.0. (a) Determine the air pressure, temperature, and velocity. (b) Discuss the effect of moisture in the air.

*17.15 Air flows through a convergent-divergent nozzle with inlet conditions of 150 psia, 80°F, and negligible velocity. A normal shock wave stands at the nozzle exit, where $M = 2.2$ just before the shock wave. Determine the pressure, temperature, Mach number, velocity, and stagnation pressure after the shock wave.

*17.16 Determine the change of specific entropy across the shock wave in Problem *17.15.

*17.17 Air flows through a convergent-divergent nozzle with inlet conditions of 300 psia, 170°F, and negligible velocity. The exit area is twice the throat area. Determine the exit pressure such that a normal shock wave occurs at the exit plane.

*17.18 Air enters a diffuser at 13.2 psia, 10°F, 850 ft/sec, and a flow rate of 25 lbm/sec. The diffuser efficiency is 95%. Determine the diffuser exit temperature, pressure, and area if the exit velocity is 230 ft/sec.

*17.19 Dry saturated steam enters a diffuser at 2 psia and an unknown velocity. It exits at 6 psia with negligible velocity. Determine (a) the discharge temperature for isentropic flow; (b) the discharge temperature if $\eta_d = 0.80$.

*17.20 A Pitot tube is installed in a 3-in.-I.D. pipe to determine the water flow rate by measuring the velocity head. The water temperature is 65°F. The manometer reads 1.5 in. Hg. Determine the water flow rate in lbm/sec.

*17.21 A Pelton wheel turbine is shown schematically below, where a jet of water strikes the buckets tangentially and the exit angle is 165°. Derive (a) an expression for the torque and power produced on the rotor; (b) the u/V ratio to produce the maximum power.

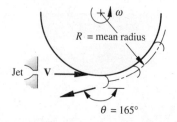

*17.22 Consider the Francis turbine illustrated below. The available head of water is 330 ft, and the turbine runs at 120 rpm. The water enters and leaves the impeller tangent to the blade tips, and the absolute velocity is radial at the turbine exit. (a) Develop the inlet and exit velocity diagrams. (b) Determine β_1 and β_2, the torque, and power.

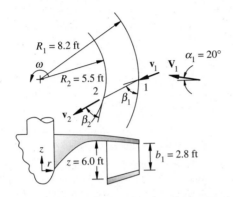

*17.23 A centrifugal pump, whose impeller is illustrated below, operates at 1750 rpm and receives water with no angular momentum. The water enters and leaves tangent to the blades. (a) Determine the inlet and exit velocity diagrams. (b) Let $\beta_1 = 30°$; evaluate the volume flow in gal/min. (c) Find the torque and power for $\beta_2 = 70°$, $90°$, $110°$.

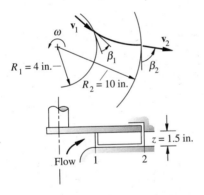

*17.24 A lawn sprinkler, as shown below, receives water at 25 psia with a flow of 1 gal/min. The sprinkler arm has a radius of 8 in., and the jet discharges at a speed of 56 ft/sec relative to the sprinkler arm at an angle $30°$ above the horizontal. The torque required to overcome bearing friction is 8.8 in.-lbf. Determine the rpm of the sprinkler.

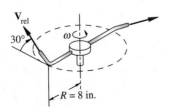

COMPUTER PROBLEMS

C17.1 Develop a TK Solver model, spreadsheet template, or computer program to design a nozzle with a circular cross section to isentropically expand a jet engine's exhaust from 110 kPa and 800°C to 40 kPa. Assume the gas $k = 1.33$ and $R = 0.285$ kJ/kg-K. The inlet velocity is negligible, and the flow rate is 25 kg/s. The output should be a table of nozzle area, nozzle diameter, temperature, velocity, Mach number, and pressure in 5-kPa increments.

C17.2 Use the TK Solver model CDNOZZLE.TK to analyze a nozzle designed to expand 1 lbm/sec of helium from 200 psia and 200°F to 14.7 psia. For certain exit pressures above design, a normal shock wave will form in the divergent portion of the nozzle. For a range of shock locations from the throat to the exit, determine the exit pressure and Mach number, the pressure and Mach number before the shock and after the shock, and the exit velocity. Display the results in a table.

C17.3 Develop a TK Solver model, spreadsheet template, or computer program to produce normal shock tables similar to Table 17.2(b) for $k = 1.2$, 1.3, and 1.5.

C17.4 Air at 101.3 kPa and 25°C flows through a 0.36-m circular duct. A Pitot tube is used to measure the velocity profile across the duct. By summing the product of the velocity times the flow area for different radii, the volume flow rate may be determined. The Pitot tube static pressures at various radii are given below. Develop a TK Solver model, spreadsheet template, or computer program to compute the volume and mass flow rates.

Radius (m)	Pitot tube static reading (mm air)
0	58.4
0.047	51.0
0.060	47.0
0.098	42.3
0.114	38.6
0.127	34.3
0.139	31.0
0.150	27.2
0.161	20.9
0.170	15.4
0.180	0.0

C17.5 The rotor energy transfer, e, for an ideal impulse turbine can be shown to be

$$e = \frac{2u(V_1 \cos \alpha - u)}{g_c} \text{ ft-lb}_f$$

Using this relationship, develop a TK Solver model, spreadsheet template, or computer program to compute the ideal turbine efficiency. For nozzle angles (α) of 10° and 30°, and blade speeds (u) from 0 to approximately twice the optimum ($V_1 \cos \alpha$), compute the efficiency and plot it versus the blade speed/steam speed ratio (u/V_1).

18

Heat Transfer and Heat Exchangers

This chapter introduces the concept of heat transfer and applies it to some special cases. The field of heat transfer is quite extensive, and this chapter is not meant to replace study in the field, but rather to introduce you to the various modes of heat transfer. In equilibrium thermodynamics we have used heat flow in the analysis of problems, but we have not discussed how the heat transfer occurs physically. We have dealt with different thermodynamic states and have observed the property changes, and from this deduced the heat transfer. Heat flow is a transient problem, so its analysis will involve more than investigation of equilibrium states. The laws of heat transfer do obey the first and second laws of thermodynamics; energy is conserved, and heat must flow from hot to cold. The purpose of including rudiments of heat-exchanger thermal analysis is that heat exchangers have formed an essential element in many of the systems we have analyzed—power plants, refrigeration units, gas compressors—and by the time this basic course in thermal analysis is finished, you should have an understanding of the selection process for heat exchangers. In this chapter you will

- Investigate the different modes of heat transfer and the laws governing heat-transfer behavior within the modes;
- Combine the modes indicative of actual heat transfer to or from devices;
- Discover why insulation may in certain circumstances increase heat transfer;
- Analyze a variety of heat exchangers found in many thermal systems.

18.1 MODES OF HEAT TRANSFER

There are three modes of heat transfer—*conduction, radiation,* and *convection.* We define them as follows:

1. Conduction is the heat transfer within a medium. In solids, particularly metals, conduction is due to the drift of free electrons and phonon vibration. At low temperatures, phonon vibration, the vibration of crystalline structure, is the primary mechanism for conduction, and at higher temperatures electron drift is the primary mechanism. Regardless of the mechanism, energy is transferred from one atom or molecule to another, resulting in a flow of energy within a medium. In gas the mechanism for conduction is primarily molecular collision. The conduction is dependent on the pressure and temperature, which act in obvious ways to increase the chance of molecular collisions. In liquids the mechanism for conduction is the combination of electron drift and molecular collision. Conduction in liquids is temperature, not pressure, dependent.

2. Radiation is thermal energy flow, via electromagnetic waves, between two bodies separated by a distance. Electromagnetic waves, which are a function of body-surface temperature, transfer heat and thus constitute thermal radiation.

3. Convection is the heat transfer between a solid surface and a fluid. This is a mixed mode, in that at the solid-fluid interface heat is transferred by conduction, molecular collisions between the solid and fluid molecules. As a result of these collisions the temperature in the fluid changes, the density varies, and bulk fluid motion occurs. The high- and low-temperature fluid elements mix, and heat is transferred between the solid and fluid by convection.

18.2 LAWS OF HEAT TRANSFER

The laws for radiation and conduction heat transfer are based on experiments and, as with all laws, cannot be proved; but because nothing contradicts them, they are assumed to be valid. The expression for convective heat flow is not a law, but an empirical equation.

Conduction

The law for conduction heat transfer is *Fourier's law,* named for the man who proposed it. It states that the conductive heat flow, q_λ, is a product of (1) the thermal conductivity of the material, λ; (2) the area normal to the heat flow, A; and (3) the temperature gradient, dT/dx, across the area.

The thermal conductivity is a property of the material, like specific heat, and is a measure of how well a material conducts heat. Fourier's law for one-dimensional flow in rectangular coordinates is

$$q_\lambda = -\lambda A \frac{dT}{dx} \quad \text{W} \quad [\text{Btu/hr}] \tag{18.1}$$

The units on each term are

$$\lambda: \qquad W/m \cdot K \qquad [Btu/hr\text{-}ft\text{-}°F]$$

$$A: \qquad m^2 \qquad [ft^2]$$

$$dT/dx: \quad K/m \qquad [R/ft]$$

The reason for the minus sign on equation (18.1) is that we want the heat flow to be positive. Since the temperature gradient, dT/dx, must be negative to cause heat to flow in the $+x$ direction — hot to cold, hence decreasing — the minus sign corrects for this. The values for the thermal conductivity depend on the molecular structure and are highest for a solid phase, the most compact phase structure, less for the liquid phase, and still less for the vapor phase. Table A.22 lists the properties of many materials.

Before equation (18.1) can be solved, we must know whether the heat flow is constant with time, steady-state, or whether it varies with time, transient or unsteady-state. We will consider only steady-state heat transfer in this chapter. Let us consider a plane wall, shown in Figure 18.1, with constant temperatures on each surface. The wall is a distance L thick. The variables T and x may be separated in equation (18.1).

$$q_\lambda \, dx = -\lambda A \, dT$$

We integrate from $x = 0$ and $T = T_h$ to $x = L$ and $T = T_c$:

$$\int_0^L q_\lambda \, dx = -\int_{T_h}^{T_c} \lambda A \, dT = \int_{T_c}^{T_h} \lambda A \, dT$$

Since A, λ, and q_λ are constant,

$$q_\lambda = \lambda A \frac{T_h - T_c}{L} = \frac{T_h - T_c}{R_\lambda} \tag{18.2}$$

where

$$R_\lambda = \frac{L}{\lambda A} \tag{18.3}$$

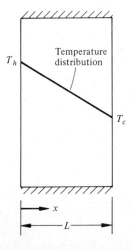

Figure 18.1 Temperature distribution in a plane wall for steady-state heat transfer.

is the *conductive resistance*. This illustrates the similarity between electric current flow being equal to a voltage potential divided by an electrical resistance, and heat flow being equal to a temperature potential divided by a thermal resistance. We were able to integrate equation (18.1) because the problem is steady-state and thus constant; it is a plane wall, so A is constant; and the thermal conductivity does not vary with temperature, so it is constant. The thermal conductivity does vary with temperature, as might be suspected, but we will assume it to be constant and compensate for the variation by using values averaged over the temperature range.

Radiation

The law for radiative heat transfer was discovered by two men: J. Stefan determined the law experimentally, and L. Boltzmann derived it theoretically from statistical mechanics. The law is that the radiant heat flow, q_r, for a blackbody is proportional to the surface area, A, times the absolute surface temperature to the fourth power. The Stefan-Boltzmann constant, σ, makes the proportionality an equation. Thus,

$$E_b = \sigma A T^4 \text{ W} \qquad [\text{Btu/hr}] \tag{18.4}$$

where E_b is the blackbody emissive power and

$$\sigma = 5.67 \times 10^{-8} \text{ W/m}^2 \cdot \text{K}^4 = 0.1714 \times 10^{-8} \text{ Btu/hr-ft}^2\text{-R}^4$$

$$A: \text{m}^2 \, [\text{ft}^2]$$

$$T: \text{K} \, [\text{R}]$$

A blackbody, or black surface, is one that absorbs all the radiation incident upon it.

Equation (18.4) describes the emission of radiation from a black surface; however, it does not indicate what the net radiative heat transfer, q_r, will be between two surfaces. Consider that surface 1 at temperature T_1 is completely enclosed by another black surface, surface 2, at temperature T_2. The net radiant heat transfer is

$$q_r = \sigma A_1 (T_1^4 - T_2^4) \tag{18.5}$$

where all temperatures are absolute. Unfortunately, real bodies, surfaces, are not perfect absorbers and radiators but emit, for the same surface temperature, a fraction of the blackbody radiation. This fraction is called the emittance, ϵ:

$$\epsilon = \frac{\text{actual surface radiation at } T}{\text{black surface radiation at } T} \tag{18.6}$$

The actual surfaces are called *gray surfaces*. Thus, the net rate of heat transfer between a gray surface at temperature T_1 to a surrounding black surface at temperature T_2 is

$$q_r = \sigma A_1 \epsilon_1 (T_1^4 - T_2^4) \tag{18.7}$$

Table A.22 lists emittances for various surfaces.

The restriction of a total black enclosure may be eliminated by using the modulus $\mathcal{F}_{1\text{-}2}$, which accounts for the relative geometries of the surfaces (not all radiation leaving state 1 necessarily reaches state 2) and the surface emittances. Thus, equation (18.7) becomes

$$q_r = \sigma A_1 \mathscr{F}_{1\text{-}2}(T_1^4 - T_2^4) \tag{18.8}$$

Radiant heat transfer frequently occurs with other modes of heat transfer, and the use of a radiative resistance R_r is helpful. Let us also define q_r to be

$$q_r = \frac{T_1 - T_{2'}}{R_r} \tag{18.9}$$

where $T_{2'}$ is an arbitrary reference temperature. The radiative resistance is found by combining equations (18.8) and (18.9) to obtain

$$R_r = \frac{T_1 - T_{2'}}{\sigma A_1 \mathscr{F}_{1\text{-}2}(T_1^4 - T_2^4)} \tag{18.10}$$

Convection

Convection heat transfer is a combination of conduction, fluid flow, and mixing. There are two types of convection: free convection, where density changes cause the bulk fluid motion, and forced convection, where a pressure difference causes the bulk fluid motion. This motion is often created by a fan or pump. The expression for convective heat flow is not a law but an empirical equation. The convective heat flux, q_c, is the product of three terms.

1. The solid-fluid surface area, A.
2. The temperature difference between the solid surface, T_s, and the fluid temperature far from the surface, T_∞.
3. The average unit convective coefficient, $\overline{h}_c$.

$$q_c = \overline{h}_c A(T_s - T_\infty) \text{ W} \qquad [\text{Btu/hr}] \tag{18.11}$$

Equation (18.11) rigorously defines $\overline{h}_c$, but this is the equation conventionally used to describe convective heat flow. The units on the terms are

$$\overline{h}_c: \quad \text{W}/(\text{m}^2 \cdot \text{K}) \qquad [\text{Btu/hr-ft}^2\text{-}°\text{F}]$$

$$A: \quad \text{m}^2 \qquad [\text{ft}^2]$$

$$T: \quad \text{K} \qquad [°\text{F}]$$

Equation (18.11) may be expressed in terms of a convective resistance R_c.

$$q_c = \frac{T_s - T_\infty}{R_c} \tag{18.12}$$

Comparing equations (18.12) and (18.11) yields an expression for R_c.

$$R_c = \frac{1}{\overline{h}_c A} \tag{18.13}$$

TABLE 18.1 TYPICAL UNIT CONVECTIVE
COEFFICIENT VALUES

Mode	$\bar{h}_c$(W/m²-K)
Free convection, air	5–25
Forced convection, air	10–200
Free convection, water	20–100
Forced convection, water	50–10 000

In calculating the heat flux the term that may be difficult to determine is $\bar{h}_c$. This term relates the fluid's physical properties and the fluid velocity over the solid surface. These affect the rate at which thermal energy may enter or leave the fluid. Table 18.1 illustrates the wide range of values $\bar{h}_c$ may have. The tremendous variation in values of $\bar{h}_c$ makes its selection very important in convective heat-transfer analysis.

Figure 18.2 illustrates the forced convection velocity and temperature profiles of a cold fluid moving over a heated plate. Several observations may be made. The velocity decreases as it approaches the solid surface, reaching zero in the fluid layer next to the surface. The heat transfer from the solid to the fluid is by conduction, since the fluid layer has zero velocity. This must equal the heat transfer by convection into the rest of the fluid, or

$$\frac{q_c}{A} = -\lambda_f \frac{\partial T}{\partial y}\bigg|_{y=0} = \bar{h}_c(T_s - T_\infty) \tag{18.14}$$

where λ_f is the thermal conductivity of the fluid. From this expression we see that the heat flux, q_c/A, is directly proportional to the thermal conductivity of the fluid and to the temperature gradient at the wall. The temperature gradient is directly proportional to the fluid velocity, with higher velocities allowing higher temperature gradients.

Convection heat transfer involves fluid velocity, fluid properties, and the overall convective coefficient. It is possible to relate these properties, but first dimensionless

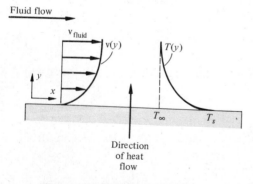

Figure 18.2 Velocity and temperature profiles in convective heat transfer.

parameters are needed. The dimensionless number for velocity is the Reynolds number. For flow through a tube or duct the Reynolds number, Re_D, is defined as

$$Re_D = \frac{v\rho D}{\mu} = \frac{vD}{\nu} \qquad (18.15)$$

where v is the average fluid velocity, ρ is the fluid density, D is the inside diameter, μ is the fluid dynamic viscosity, and ν is the fluid kinematic viscosity, $\mu\rho$.

Often the duct is not circular but rectangular, or an annulus formed by a tube within a tube. In these cases the characteristic dimension is not the tube diameter but an equivalent diameter, called the hydraulic diameter, D_H, defined as

$$D_H = 4\left(\frac{\text{cross-sectional area for fluid flow}}{\text{wetted perimeter}}\right)$$

Thus, for flow through a duct with a width a and height b

$$D_H = 4\left[\frac{a \cdot b}{2(a+b)}\right] = \frac{2ab}{a+b}$$

For flow through an annulus with inside diameter D_1 and outside diameter D_2

$$D_H = 4\left[\frac{\pi(D_2^2 - D_1^2)}{4\pi(D_2 + D_1)}\right] = D_2 - D_1$$

Low Reynolds numbers, up to about 2300, are indicative of laminar pipe flow. From 2300 to 6000 the laminar pipe flow begins a transition to turbulent pipe flow. The flow is usually completely turbulent at 6000. This becomes a matter for concern when different correlations are used for laminar and turbulent flow. Turbulent flow has a flatter velocity profile than laminar flow, allowing a greater average velocity and a greater temperature at the wall.

The Prandtl number, Pr, is a dimensionless number relating fluid properties.

$$Pr = \frac{\mu c_p}{\lambda} = \frac{\nu}{\alpha} \qquad (18.16)$$

where α is the thermal diffusivity,

$$\alpha = \frac{\lambda}{\rho c_p}$$

The Nusselt number, Nu, is the dimensionless form of the convective coefficient.

$$Nu = \frac{\bar{h}_c D}{\lambda_f} \qquad (18.17)$$

A great number of tests have been run to correlate these dimensionless numbers. There are many correlations, and the following ones are typical. There are two flow regimes, laminar and turbulent. For laminar flow in short tubes, Seider and Tate developed the expression

$$Nu = 1.86 \, (Re)^{1/3}(Pr)^{1/3}\left(\frac{D}{L}\right)^{1/3}\left(\frac{\mu}{\mu_s}\right)^{0.14} \qquad (18.18)$$

In this expression L is the pipe length, and all fluid properties are evaluated at the average fluid temperature except for μ_s. This term is evaluated at the wall surface temperature. The unit convective coefficient is dependent on the fluid viscosity, which varies with temperature. This additional viscosity term accounts for the temperature variation.

In turbulent flow the following equation may be used when the temperature difference between the tube or duct surface and the average fluid temperature is not greater than 5.5°C for liquids or 55.5°C for gases,

$$Nu = 0.023(Re)^{0.8}(Pr)^n \qquad (18.19)$$

where $n = 0.4$ for heating and $n = 0.3$ for cooling.

In those cases where the previous temperature limits are exceeded or the fluid's viscosity is greater than that of water, use

$$Nu = 0.027(Re)^{0.8}(Pr)^{1/3} \left(\frac{\mu}{\mu_s}\right)^{0.14} \qquad (18.20)$$

In this case all properties are evaluated at the average fluid temperature except μ_s, which is evaluated at the wall temperature.

How are these equations used? Table A.23 lists properties of various substances. Equations (18.18)–(18.20) are used to solve for $\bar{h}_c$, which is used in equation (18.12), where T_∞ is the average fluid temperature, to calculate the heat flux.

Example 18.1

Water enters a 3-cm-diameter tube with a velocity of 50 m/s and a temperature of 20°C and is heated. Calculate the average unit convective coefficient and the Reynolds number.

Solution

Given: Water enters a tube with a known diameter, temperature, and velocity.

Find: The Reynolds number and unit convective coefficient.

Sketch and Given Data:

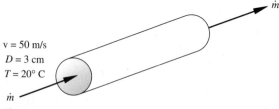

v = 50 m/s
D = 3 cm
T = 20° C

Figure 18.3

Assumptions:

1. Water flows steadily through the tube.
2. Water's properties may be evaluated at 20°C.

Analysis: The Reynolds number may be evaluated from equation (18.15),

$$\text{Re} = \frac{vD}{\nu}$$

From Table A.23 the kinetic viscosity, Prandtl number, and thermal conductivity are $\nu = 1.006 \times 10^{-6}$ m²/s, Pr = 7.0, and $\lambda = 0.597$ W/m-K.

$$\text{Re} = \frac{(50 \text{ m/s})(0.03 \text{ m})}{(1.006 \times 10^{-6} \text{ m}^2/\text{s})} = 1.491 \times 10^6 \quad \text{(turbulent)}$$

Because the flow is turbulent, equation (18.19) may be used to determine the Nusselt number and then the unit convective coefficient.

$$\text{Nu} = 0.023(\text{Re})^{0.8}(\text{Pr})^{0.4}$$

$$\text{Nu} = (0.023)(1.491 \times 10^6)^{0.8}(7.0)^{0.4}$$

$$\text{Nu} = 4350 = \frac{\bar{h}_c D}{\lambda}$$

$$4350 = \frac{(\bar{h}_c \text{ W/m}^2\text{-K})(0.03 \text{ m})}{(0.597 \text{ W/m-K})}$$

$$\bar{h}_c = 8.656 \times 10^4 \text{ W/m}^2\text{-K} = 86.6 \text{ kW/m}^2\text{-K}$$

Comments:

1. When determining the Nusselt number, the flow regime must first be established by finding the Reynolds number.
2. When a fluid is heated, the average fluid temperature across the heat exchanger is used in determining the fluid's physical properties.   ▮

18.3 COMBINED MODES OF HEAT TRANSFER

Most engineering applications involve a combination of heat-transfer modes. Figure 18.4(a) shows a plane wall. Heat is transferred by the fluid at a high temperature to the wall by convection and radiation. Within a medium, the plane wall, heat is transferred by conduction, q_λ. At the outside wall surface, heat is again transferred by convection to the surrounding fluid at a low temperature. At low temperatures radiation does not have an appreciable effect. It is possible to express the total heat flow, q, in terms of the extreme temperatures, T_h and T_l, and the resistances to heat that occur between the two temperature extremes.

On the furnace side, the total heat flow, q, is equal to

$$q = q_r + q_{ch} = \frac{T_h - T_{wh}}{R_1} \tag{18.21}$$

where

$$R_1 = \frac{(R_{ch})(R_r)}{R_{ch} + R_r}$$

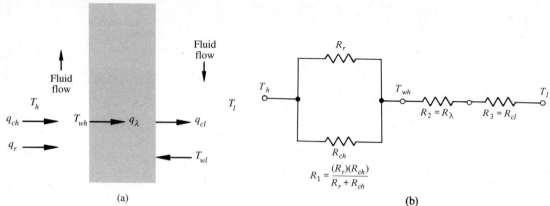

Figure 18.4 (a) A plane wall with convective and radiative heat transfer at the surface. (b) The thermal circuit for the wall.

In the wall, the total heat flow is equal to the conductive heat flow, q_λ; that is,

$$q = q_\lambda = \frac{T_{wh} - T_{wl}}{R_\lambda} \qquad (18.22)$$

and at the outside surface

$$q = q_{cl} = \frac{T_{wl} - T_l}{R_{cl}} \qquad (18.23)$$

Figure 18.4(b) shows the thermal circuit for the wall.

$$q = \frac{T_h - T_{wh}}{R_1}$$

$$q = \frac{T_{wh} - T_{wl}}{R_2}$$

$$q = \frac{T_{wl} - T_l}{R_3}$$

We multiply each equation by its resistance, add, and solve the resulting equation for q:

$$q = \frac{T_h - T_l}{R_1 + R_2 + R_3} = \frac{\Delta T}{\Sigma_{i=1}^3 R_i} \qquad (18.24)$$

Another coefficient, the *overall coefficient of heat transfer, U,* is frequently referred to in heat-transfer analyses. The equation for heat flow using the overall coefficient of heat transfer is

$$q = UA\,\Delta T \qquad (18.25)$$

and comparing equations (18.24) and (18.25) yields

$$UA = \frac{1}{\Sigma_{i=1}^3 R_i} \qquad (18.26)$$

18.4 CONDUCTION THROUGH A COMPOSITE WALL

It is possible to have conduction through more than one material. For instance, an actual furnace wall is composed of different materials in series and parallel, such as illustrated in Figure 18.5. The material might be chrome ore, followed by three types

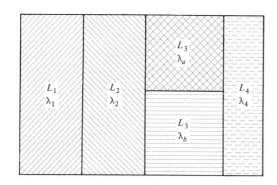

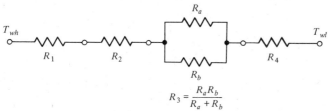

$$R_3 = \frac{R_a R_b}{R_a + R_b}$$

Figure 18.5 Composite wall with thermal circuit.

of fire bricks, followed by the steel casing. For a steady state, the heat flow may be written in terms of the sum of thermal resistances:

$$q = \frac{T_{wh} - T_{wl}}{\Sigma_{i=1}^{4} R_i} \tag{18.27}$$

where

$$R_3 = \frac{R_a R_b}{R_a + R_b}$$

18.5 CONDUCTION IN CYLINDRICAL COORDINATES

Thus far we have dealt only with plane walls; the area has been constant in the direction perpendicular to the heat flow. In cylindrical coordinates, such as heat transfer through a pipe, the area varies with radial distance, and the conduction equation must account for this variation. Figure 18.6 illustrates a cylinder of length l.

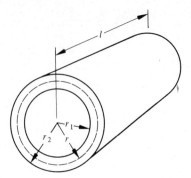

Figure 18.6 A hollow cylinder.

We will assume that heat flows only in the radial direction, so Fourier's law is written as

$$q_\lambda = -\lambda A \frac{dT}{dr}$$

The area is normal to the heat flow, $A = 2\pi r l$; hence,

$$q_\lambda = -2\pi r l\lambda \frac{dT}{dr} \tag{18.28}$$

For steady state, q_λ is constant; we may separate the variables and integrate from $r = r_1$ and $T = T_1$ to $r = r_2$ and $T = T_2$.

$$\int_{r_1}^{r_2} q_\lambda \frac{dr}{r} = \int_{T_2}^{T_1} 2\pi\lambda l \, dT$$

$$q_\lambda = \frac{2\pi\lambda l(T_1 - T_2)}{\ln(r_2/r_1)} \tag{18.29}$$

Equation (18.29) may be expressed in terms of thermal resistance, R_λ, as follows:

$$q_\lambda = \frac{T_1 - T_2}{R_\lambda}$$

where

$$R_\lambda = \frac{\ln(r_2/r_1)}{2\pi\lambda l}$$

Equation (18.29) was developed for inside temperature greater than the outside temperature, with the heat flow in the positive r direction. Should the outside temperature be greater, the heat flow would be negative, indicating flow opposite to the direction assumed in the derivation of equation (18.29). A more common occurrence is when the pipe is insulated. Typically steam pipes are insulated to prevent heat loss, and refrigerant pipes are insulated to prevent heat gain. Figure 18.7 illustrates a

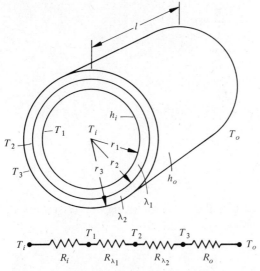

Figure 18.7 The schematic diagram and thermal circuit for an insulated pipe.

cross-sectional sketch of an insulated pipe. The heat flow, q, may be determined for each section.

$$q = q_{ci} = h_i 2\pi r_1 l(T_i - T_1) = \frac{T_i - T_1}{R_i} \tag{18.30}$$

where

$$R_i = \frac{1}{2\pi r_1 l h_i}$$

$$q = q_{\lambda_1} = -2\pi l \lambda_1 r \frac{dT}{dr}$$

$$q = q_{\lambda_1} = \frac{2\pi l \lambda_1 (T_1 - T_2)}{\ln(r_2/r_1)} = \frac{T_1 - T_2}{R_{\lambda_1}} \tag{18.31}$$

and where

$$R_{\lambda_1} = \frac{\ln(r_2/r_1)}{2\pi l \lambda_1}$$

Similarly,

$$q = q_{\lambda_2} = \frac{2\pi l \lambda_2 (T_2 - T_3)}{\ln(r_3/r_2)} = \frac{T_2 - T_3}{R_{\lambda_2}} \tag{18.32}$$

where

$$R_{\lambda_2} = \frac{\ln(r_3/r_2)}{2\pi l \lambda_2}$$

For the outside surface,

$$q = q_{co} = 2\pi r_3 h_o l (T_3 - T_o) = \frac{T_3 - T_o}{R_o} \tag{18.33}$$

where

$$R_o = \frac{1}{2\pi r_3 h_o l}$$

The heat flow can now be calculated by summing the resistances and knowing the overall temperature difference:

$$q = \frac{T_i - T_o}{\sum_{i=1}^4 R_i}$$

Example 18.2

Saturated steam at 500°K flows in a 20-cm-I.D., 21-cm-O.D. pipe. The pipe is covered with 8 cm of insulation that has a thermal conductivity of 0.10 W/m-K. The pipe's conductivity is 52 W/m-K. The ambient temperature is 300°K. The unit convective coefficient on the inside is 18 000 W/m²-K and on the outside is 12 W/m²-K. Determine the heat loss from 4 m of pipe. Calculate the overall coefffficient of heat transfer based on the outside area.

Solution

Given: A pipe is covered with insulation, and the temperatures inside the pipe and of the outside air are specified. In addition the thermal properties of the insulation and pipe are known as are the convective coefficients inside and outside.

Find: The overall coefficient of heat transfer and the total heat transfer from the pipe.

Sketch and Given Data: See Figure 18.8 on page 757.

Assumptions:

1. Steady-state conditions exist.
2. The properties are uniform for each material.

Analysis: It is easiest to calculate the heat loss per unit length of pipe, $l = 1$ m, and then correct for the total length asked for in the problem.

The heat transfer may be found by initially calculating the resistance for each mode of heat transfer and summing the resistances. The overall temperature differ-

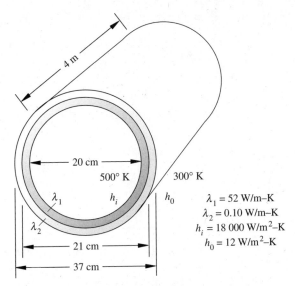

Figure 18.8

ence is divided by this summation to determine the heat transfer per unit length of pipe.

$$q = \frac{\Delta T}{\Sigma_{i=1}^{4} R_i}$$

$$R_1 = \frac{1}{2\pi l r_1 h_i} = \frac{1}{2\pi(1 \text{ m})(0.01 \text{ m})(18\ 000 \text{ W/m}^2\text{-K})} = 0.000\ 088\ ^\circ\text{K/W}$$

$$R_2 = \frac{\ln (r_2/r_1)}{2\pi \lambda_1 l} = \frac{\ln \left(\dfrac{0.105}{0.1}\right)}{2\pi(52 \text{ W/m-K})(1 \text{ m})} = 0.00149\ ^\circ\text{K/W}$$

$$R_3 = \frac{\ln (r_3/r_2)}{2\pi \lambda_2 l} = \frac{\ln \left(\dfrac{0.185}{0.105}\right)}{2\pi(0.10 \text{ W/m-K})(1 \text{ m})} = 0.9014\ ^\circ\text{K/W}$$

$$R_4 = \frac{1}{2\pi l r_3 h_o} = \frac{1}{2\pi(1 \text{ m})(0.185 \text{ m})(12 \text{ W/m}^2\text{-K})} = 0.0716\ ^\circ\text{K/W}$$

$$\Sigma R_i = 0.9733\ ^\circ\text{K/W}$$

$$q = \frac{(500 - 300\ ^\circ\text{K})}{(0.9733\ ^\circ\text{K/W})} = 205.2 \text{ W}$$

The heat transfer for 4 m of pipe is

$$Q = (4 \text{ m})(205.2 \text{ W/m}) = 820.8 \text{ W}$$

From equation (18.26) the overall coefficient of heat transfer may be determined.

$$UA = \frac{1}{\Sigma R_i} = \frac{1}{0.9733} = 1.026 \text{ W/K}$$

$$A = \pi dl$$

$$U = \frac{1.026 \text{ W/K}}{\pi(0.37 \text{ m})(1 \text{ m})} = 0.8826 \text{ W/m}^2\text{-K}$$

Comment: In this example the length of pipe is reasonably short, so the use of a unit length is not as necessary as when the length is long. Under the latter circumstance the numbers in the equation become so large as to be incomprehensible. ▄

18.6 CRITICAL INSULATION THICKNESS

We might intuitively believe that adding insulation to a pipe would decrease the heat flow from the pipe. While this is usually the case, it is not always true for small-diameter pipes and wires. Consider a cylinder with an inside radius r_i at temperature T_i, an outside radius r_o, and a thermal conductivity λ in a room where the ambient temperature is T_o. The two resistances to heat transfer are the conductive resistance from r_i to r_o and the convective resistance at the outside surface. Therefore,

$$q = \frac{2\pi\lambda l(T_i - T_o)}{\ln(r_o/r_i) + (\lambda/\bar{h}_o r_o)} \tag{18.34}$$

Thus, for a fixed inside radius and temperature potential the heat flow will be a function of the two thermal resistances. As r_o increases, one term increases logarithmically, while the other decreases linearly. The sum of these two terms is the total resistance, so a given change in radius may increase or decrease the denominator of equation (18.34). It is possible to develop an expression for the maximum heat transfer as a function of the outside radius. The outside radius that yields this maximum heat flow is the *critical radius.* For wires, it may be desirable to maximize the heat flow from the wire so as to minimize line voltage drops. This may be accomplished by adding enough insulation to the wire so that the outside radius is the critical radius. This is called the *critical insulation thickness.* If the object is to prevent heat loss, the outside radius must be greater than the critical radius. Economics determines how much insulation will be added in this case. This typically occurs in refrigeration systems.

To find the critical radius, we take the derivative of q with respect to r_o in equation (18.34) and set it equal to zero. Solving for the value of r_o gives the critical radius, r_{oc}:

$$r_{oc} = \frac{\lambda}{h_o} \tag{18.35}$$

Thus, the critical radius is a function of the outside convective coefficient and the thermal conductivity of the insulation. The critical radius is usually less than about 2.5 cm, so for pipes or wires with a radius greater than the critical radius, the critical insulation thickness is not physically obtainable.

18.7 HEAT EXCHANGERS

Heat exchangers are commonly used in most thermal systems. They allow the transfer of heat from one fluid to another. In direct-contact heat exchangers the same substance, at two different states, is mixed. More common is the heat exchanger in which one fluid flows through tubes and the other flows over the tubes. There are a number of tube configurations; we shall analyze a few, but the principles of analysis apply to all.

Why analyze heat exchangers at all? In the thermal system we will need a heat exchanger that allows the transfer of a given amount of heat. The thermal analysis will determine the heat exchanger size. There are two subsequent steps, mechanical design and manufacturing design, which are not within the scope of this text. After the thermal analysis, we can use manufacturers' catalogs to select the heat exchanger that satisfies our requirements.

It is desirable to have an equation that will denote the heat transferred in a heat exchanger in terms of the overall coefficient of heat transfer, U, the total tube surface area, A, and some temperature difference between the inlet and outlet states, $\overline{\Delta T}$. Consider a parallel-flow, tube-in-tube heat exchanger, with its temperature distribution, shown in Figure 18.9.

Let us examine a differential length of heat exchanger, with differential area dA. The heat transferred across this differential area can be expressed in three equivalent ways: the heat lost by the hot fluid, the heat gained by the cold fluid, or the heat transferred in the heat exchanger. Thus,

$$q = -\dot{m}_h c_{ph}\, dT_h = \dot{m}_c c_{pc}\, dT_c = U\, dA(T_h - T_c) \qquad (18.36)$$

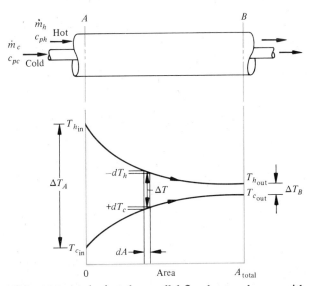

Figure 18.9 A tube-in-tube parallel-flow heat exchanger with temperature-area diagram.

In equation (18.36) the negative sign allows the heat loss by the hot fluid to be positive in accordance with the other terms denoting a positive quantity; ΔT at any dA is the difference between the hot and cold fluid temperatures at that differential area. Let

$$\dot{m}_h c_{ph} = C_h \tag{18.37}$$

$$\dot{m}_c c_{pc} = C_c \tag{18.38}$$

Rearrange equation (18.36) as follows:

$$\frac{U\,dA}{C_h} = \frac{-dT_h}{T_h - T_c}$$

$$\frac{U\,dA}{C_c} = \frac{dT_c}{T_h - T_c}$$

Adding these two equations, we obtain

$$\frac{U\,dA}{C_h} + \frac{U\,dA}{C_c} = \frac{dT_c - dT_h}{T_h - T_c}$$

Let

$$\theta = T_h - T_c$$

Then

$$\frac{U\,dA}{C_h} + \frac{U\,dA}{C_c} = \frac{-d\theta}{\theta} \tag{18.39}$$

Let us assume that U is constant, or an effective value, across the entire heat exchanger and integrate equation (18.39) across the heat exchanger, from the A end to the B end:

$$\left(\frac{U}{C_h} + \frac{U}{C_c}\right)\int_A^B dA = -\int_A^B \frac{d\theta}{\theta}$$

$$A_A = 0 \qquad \theta_A = \Delta T_A$$

$$A_B = A \qquad \theta_B = \Delta T_B$$

$$\left(\frac{UA}{C_h} + \frac{UA}{C_c}\right) = \ln\left(\frac{\Delta T_A}{\Delta T_B}\right) \tag{18.40}$$

$$q = UA\,\overline{\Delta T} = C_h(T_{h_{in}} - T_{h_{out}}) = C_c(T_{c_{out}} - T_{c_{in}})$$

$$UA = \frac{C_h(T_{h_{in}} - T_{h_{out}})}{\overline{\Delta T}} = \frac{C_c(T_{c_{out}} - T_{c_{in}})}{\overline{\Delta T}}$$

We substitute into equation (18.40) and solve for $\overline{\Delta T}$:

$$\overline{\Delta T} = \text{LMTD} = \frac{\Delta T_A - \Delta T_B}{\ln\,(\Delta T_A/\Delta T_B)} \tag{18.41}$$

LMTD is the log mean temperature difference, and thus the total heat transferred in a tube-in-tube heat exchanger for parallel flow is

$$q = UA \text{ LMTD} \qquad (18.42)$$

Equation (18.42) is usually used to calculate the total surface area, hence the tube length, of the heat exchanger. The heat transferred may be calculated from the energy gain or loss of one of the fluids. Before we apply these tools of analysis, let us examine two other heat exchanger types, the counterflow and the shell-and-tube.

Figure 18.10 illustrates a counterflow heat exchanger with the fluid temperature distributions. Figure 18.11 illustrates a shell-and-tube heat exchanger with the fluid temperature distributions. In the shell-and-tube and parallel-flow heat exchangers, it is not possible to have $\Delta T_A = \Delta T_B$ unless there is no heat transfer. In the counterflow heat exchanger, however, it is possible to have heat exchange with $\Delta T_A = \Delta T_B$. This means the LMTD is undefined. How to find the heat exchanged? Physically, $\Delta T_A = \Delta T_B$ means that the temperature difference throughout the heat exchanger is constant, or

$$q = UA \, \overline{\Delta T} = UA \, \Delta T_A \qquad (18.43)$$

Let us examine the counterflow heat exchanger in a little more detail. Note that as the area becomes greater, the difference between the hot- and cold-fluid temperature distribution becomes smaller, and, in the limit of infinite area, the lines are coincident. This means that the hot fluid is cooled to the cold-fluid inlet temperature and the cold fluid is heated to the hot-fluid inlet temperature. This complete exchange of heat is possible only in the counterflow heat exchanger. What thermodynamic principle makes this possible? The heat exchanger that has the more reversible heat exchange is more efficient and transfers more heat for a given area of tube surface. Constant-temperature heat exchange is the reversible case, so the heat ex-

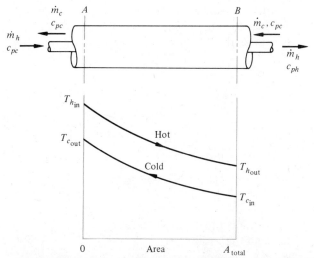

Figure 18.10 A tube-in-tube counterflow heat exchanger with a temperature-area diagram.

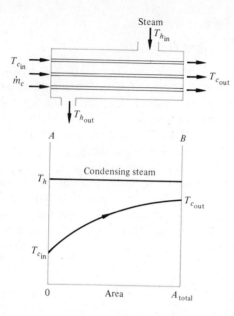

Figure 18.11 A shell-and-tube heat exchanger with temperature-area diagram for the case of condensing steam on the shell side.

changer that can, in the limit, achieve this is thermodynamically more efficient. In the limit the counterflow exchangers transfer heat at constant temperature. No other heat exchanger approaches this state, so the counterflow is more efficient. An example illustrates this point.

Example 18.3

It is necessary to cool 5000 lbm/hr of oil, $c_p = 0.8$ Btu/lbm-F, from 250°F to 150°F. Water, $c_p = 1.0$ Btu/lbm-F, is available with a flow rate of 4500 lbm/hr at a temperature of 50°F. The overall coefficient of heat transfer for a tube-in-tube heat exchanger is 15 Btu/hr-ft²-F. Determine the length of 1-in.-I.D. tubing required for counterflow and for parallel flow.

Solution

Given: A tube-in-tube heat exchanger with specified hot and cold fluids and their flow rates and temperatures, and the overall coefficient of heat.

Find: The length of tubing required.

Sketch and Given Data: See Figure 18.12 on page 763.

Assumptions:

1. Heat transfer is steady-state and steady-flow.
2. The water flows through the inside tube.
3. The heat exchanger is adiabatic.

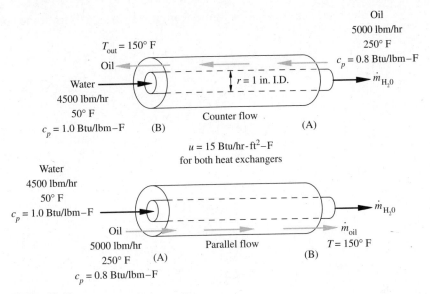

Figure 18.12

Analysis: For any heat exchanger, $q = UA\,\overline{\Delta T}$. In a counterflow tube-in-tube heat exchanger, $\overline{\Delta T}$ = LMTD. To find the LMTD we first must determine the water exit temperature. Use a first-law analysis, which yields

$$[\dot{m}(h_o - h_i)]_{\text{H}_2\text{O}} = [\dot{m}(h_i - h_o)]_{\text{oil}}$$

Using the relationship that $\Delta h \approx c_p\,\Delta T$,

$$[\dot{m}c_p\,\Delta T]_{\text{H}_2\text{O}} = [\dot{m}c_p\,\Delta T]_{\text{oil}}$$

$$\Delta T_{\text{H}_2\text{O}} = \frac{(5000\text{ lbm oil/hr})(0.8\text{ Btu/lbm oil–F})(100°\text{F})}{(4500\text{ lbm water/hr})(1.0\text{ Btu/lbm water–F})}$$

$$\Delta T_{\text{H}_2\text{O}} = 88.88°\text{F}$$

$$(T_{\text{H}_2\text{O}})_{\text{out}} = 138.88°\text{F}$$

Referring to Figure 18.13,

$$\Delta T_B = 150 - 50 = 100°\text{F}$$

$$\Delta T_A = 250 - 138.88 = 111.12°\text{F}$$

$$\text{LMTD} = \frac{\Delta T_A - \Delta T_B}{\ln\left(\dfrac{\Delta T_A}{\Delta T_B}\right)} = \frac{111.12 - 100}{\ln\left(\dfrac{111.12}{100}\right)} = 105.46°\text{F}$$

The total heat transfer is found from a first-law analysis of the control volume that the oil flows through.

$$\dot{Q} + \dot{m}\,(h + \cancel{k.e.} + \cancel{p.e.})_{\text{in}} = \cancel{\dot{W}} + \dot{m}\,(h + \cancel{k.e.} + \cancel{p.e.})_{\text{out}}$$

$$\dot{Q} = \dot{m}\,(h_o - h_i) = \dot{m}\,c_p(T_o - T_i)$$

$$\dot{Q} = \left(4500\,\frac{\text{lbm oil}}{\text{hr}}\right)\left(0.8\,\frac{\text{Btu}}{\text{lbm oil-F}}\right)(150 - 250^\circ\text{F})$$

$$\dot{Q} = -400{,}000\ \text{Btu/hr}$$

This value of heat transfer is q.

$$q = UA\ \text{LMTD}$$

$$A = \frac{(400{,}000\ \text{Btu/hr})}{(15\ \text{Btu/hr-ft}^2\text{-F})(105.46^\circ\text{F})} = 252.8\ \text{ft}^2$$

$$A = \pi dL = 252.8\ \text{ft}^2$$

$$L = \frac{(252.8\ \text{ft}^2)}{\pi(\tfrac{1}{12}\ \text{ft})} = 965.8\ \text{ft}$$

For the parallel-flow configuration,

$$\Delta T_A = 250 - 50 = 200^\circ\text{F}$$

$$\Delta T_B = 150 - 138.88 = 11.12^\circ\text{F}$$

$$\text{LMTD} = \frac{200 - 11.12}{\ln\left(\dfrac{200}{11.12}\right)} = 65.36^\circ\text{F}$$

The area is

$$A = \frac{q}{(U)(\text{LMTD})} = \frac{(400{,}000\ \text{Btu/hr})}{(15\ \text{Btu/hr-ft}^2\text{-F})(65.36^\circ\text{F})}$$

$$A = 407.9\ \text{ft}^2$$

$$A = \pi dL = 407.9\ \text{ft}^2$$

$$L = \frac{(407.9\ \text{ft}^2)}{\pi(\tfrac{1}{12}\ \text{ft})} = 1558.3\ \text{ft}$$

Comment: The parallel-flow configuration area requires a 61% increase compared to the counterflow configuration for the same amount of heat transferred.   ▬

Correction Factor

Thus far the heat exchangers have had a simple geometry. When the construction becomes more complex, $\overline{\Delta T} \neq \text{LMTD}$, and the LMTD must be modified by a correction factor F.

$$\overline{\Delta T} = F \, \text{LMTD} \tag{18.44}$$

$$q = UA \, \overline{\Delta T} \tag{18.45}$$

All that remains is to find a means for calculating the correction factor. Figures 18.13 and 18.14 show the correction factors for more complex shell-and-tube constructions illustrated schematically by Figure 18.15 and 18.16. By evaluating the two dimensionless numbers, P and Z, the correction factor, F, may be found from the appropriate chart. There are other charts for other exchangers, but we will consider only the shell-and-tube type, since that is the most common, or at least one of the most common, type of heat exchangers.

The first dimensionless parameter, P, is

$$P = \frac{T_{t_{\text{out}}} - T_{t_{\text{in}}}}{T_{s_{\text{in}}} - T_{t_{\text{in}}}} \tag{18.46}$$

where the subscripts t and s denote the shell-and-tube conditions. This is a measure of the transfer effectiveness: the numerator is an indication of the actual heat transferred, while the denominator indicates the maximum possible heat transfer.

The second dimensionless parameter, Z, is

$$Z = \frac{\dot{m}_t c_{pt}}{\dot{m}_s c_{ps}} = \frac{T_{s_{\text{in}}} - T_{s_{\text{out}}}}{T_{t_{\text{out}}} - T_{t_{\text{in}}}} \tag{18.47}$$

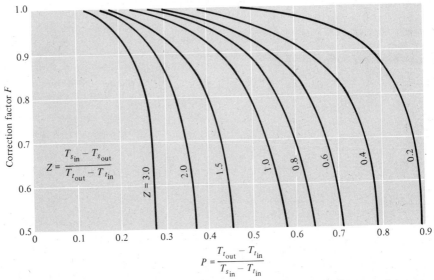

Figure 18.13 Correction factor for a heat exchanger with one shell pass and two, or a multiple of two, tube passes. (*Source:* R. A. Bowman, A. C. Mueller, and W. M. Nagle, "Mean Temperature Differences in Design," *Trans. ASME* **62**:283–294 (1940). Reproduced with permission.)

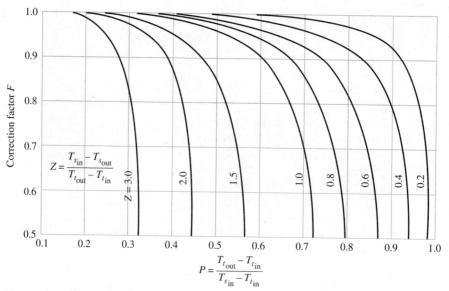

Figure 18.14 Correction factor for a heat exchanger with two shell passes and two, or a multiple of two, tube passes. (*Source:* R. A. Bowman, A. C. Mueller, and W. M. Nagle, "Mean Temperature Differences in Design," *Trans. ASME* 62:283–294 (1940). Reproduced with permission.)

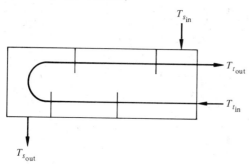

Figure 18.15 Heat exchanger with one shell pass and two, or a multiple of two, tube passes.

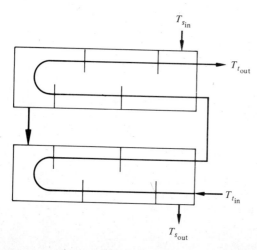

Figure 18.16 Heat exchanger with two shell passes and two tube passes.

and is a measure of the relative heat capacity of the fluids. If the terms in equation (18.47) are cross-multiplied, the result is the conservation of energy applied to the two fluids in the heat exchanger.

Example 18.4

A two-shell pass, four-tube pass, counterflow heat exchanger is used for cooling oil with a specific heat of 3.55 kJ/kg-K and a flow rate of 2.52 kg/s. The oil enters the tube side of the heat exchanger at 125°C and leaves at 50°C. Water enters the shell side at 20°C with a flow rate of 3.15 kg/s. The overall coefficient of heat transfer is 116 W/m²-K. Determine the heat transferred, the LMTD, the correction factor, F, and the heat-transfer surface area.

Solution

Given: A heat exchanger with known configuration and coefficient of heat transfer and defined hot- and cold-fluid states.

Find: The heat transferred, the LMTD, the heat exchanger correction factor, and the surface area required for heat transfer.

Sketch and Given Data:

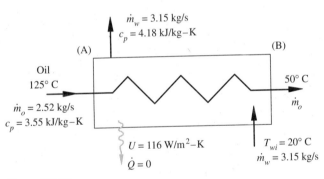

Figure 18.17

Assumptions:

1. Conditions are steady-state, with steady flow of oil and water.
2. The properties of the liquids are constant over the heat exchanger.
3. The heat exchanger is adiabatic: no heat is transferred to the surroundings.

Analysis: From the conservation of energy, we know that the heat transferred from the oil is received by the water. This allows us to determine the water exit temperature and the amount of heat that is transferred:

$$\dot{Q} = \dot{m}_o(h_i - h_o)_{oil} = (2.52 \text{ kg/s})(3.55 \text{ kJ/kg-K})(125 - 50°\text{K})$$
$$\dot{Q} = 670.95 \text{ kW}$$

$$\dot{m}_o(h_i - h_o)_{oil} = \dot{m}_w(h_o - h_i)_w$$

$$\dot{m}_o c_{po}(T_i - T_o)_{oil} = \dot{m}_w c_{pw}(T_o - T_i)_w$$

$$(2.52 \text{ kg/s})(3.55 \text{ kJ/kg-K})(125 - 50°\text{K}) = (3.15 \text{ kg/s})(4.18 \text{ kJ/kg-K})(\Delta T)_w$$

$$\Delta T_w = 50.9°\text{C}$$

$$(T_o)_w = 20 + 50.9 = 70.9°\text{C}$$

The LMTD is

$$\Delta T_A = 125 - 70.9 = 54.1°\text{C}$$

$$\Delta T_B = 50 - 20 = 30°\text{C}$$

$$\text{LMTD} = \frac{\Delta T_A - \Delta T_B}{\ln\left(\dfrac{\Delta T_A}{\Delta T_B}\right)} = \frac{54.1 - 30}{\ln\left(\dfrac{54.1}{30}\right)} = 40.87°\text{C}$$

The coefficients P and Z are calculated from equations (18.46) and (18.47):

$$P = \frac{T_{t_{out}} - T_{t_{in}}}{T_{s_{in}} - T_{t_{in}}} = \frac{50 - 125}{20 - 125} = 0.714$$

$$Z = \frac{T_{s_{in}} - T_{s_{out}}}{T_{t_{out}} - T_{t_{in}}} = \frac{20 - 70.9}{50 - 125} = 0.678$$

The correction factor, F, from Figure 18.14 is 0.882.

The average temperature difference across the heat exchanger is

$$\overline{\Delta T} = F\,\text{LMTD} = (0.882)(40.87) = 36.04°\text{C}$$

The area may be solved for using equation (18.43).

$$q = UA\,\overline{\Delta T}$$

$$A = \frac{q}{U\,\overline{\Delta T}} = \frac{(670\,950 \text{ W})}{(116 \text{ W/m}^2\text{-K})(36.04°\text{K})} = 160.5 \text{ m}^2$$

Comment: The A and B ends of the heat exchanger are arbitrary and may be interchanged; the LMTD will be the same, regardless.

Heat Exchanger Effectiveness

With the use of the correction factors we are able to compensate for complex constructions and calculate the heat transferred for a given flow condition. What if the conditions change or we wish to use the exchanger for another purpose? Will we be able to determine, from what we know about the existing performance, future performance? No. However, this is soon to be remedied.

Let us define heat exchanger effectiveness $\mathscr{E}$.

$$\mathscr{E} = \frac{\text{actual heat transfer}}{\text{maximum possible heat transfer}}$$

$$q_{\text{act}} = C_h(T_{h_{\text{in}}} - T_{h_{\text{out}}}) = C_c(T_{c_{\text{out}}} - T_{c_{\text{in}}}) \tag{18.48}$$

$$q_{\text{max}} = C_{\text{min}}(T_{h_{\text{in}}} - T_{c_{\text{in}}})$$

where c_{min} is the smaller value of c_h and c_c. The maximum possible heat exchange would occur in an infinite-area counterflow heat exchanger. Thus,

$$\mathscr{E} = \frac{C_h(T_{h_{\text{in}}} - T_{h_{\text{out}}})}{C_{\text{min}}(T_{h_{\text{in}}} - T_{c_{\text{in}}})} \tag{18.49}$$

or

$$\mathscr{E} = \frac{C_c(T_{c_{\text{out}}} - T_{c_{\text{in}}})}{C_{\text{min}}(T_{h_{\text{in}}} - T_{c_{\text{in}}})} \tag{18.50}$$

Equations (18.49) and (18.50) are equal, and the choice of which to use is up to you. Generally, you use the one where the c's cancel.

$$\mathscr{E}C_{\text{min}}(T_{h_{\text{in}}} - T_{c_{\text{in}}}) = C_h(T_{h_{\text{in}}} - T_{h_{\text{out}}}) = C_c(T_{c_{\text{out}}} - T_{c_{\text{in}}}) \tag{18.51}$$

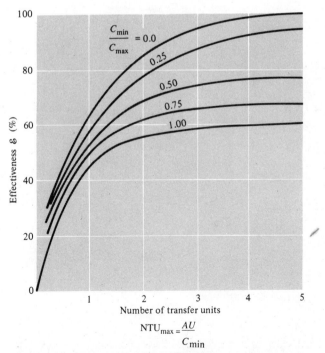

Figure 18.18 Effectiveness for heat exchanger with one shell pass and a multiple of two tube passes. (*Source:* Adapted by permission from W. M. Kays and A. L. London, *Compact Heat Exchangers,* New York: McGraw-Hill, 1958.)

Once the effectiveness and the inlet conditions are known, the outlet conditions may be calculated from equation (18.51). Figures 18.18 and 18.19 are charts used to determine the heat exchanger effectiveness for shell-and-tube heat exchangers.

There are two dimensionless quantities that must be determined before the effectiveness may be found. The ratio C_{min}/C_{max} is one quantity; it is the ratio of the flow specific heats. The second quantity, NTU_{max}, is

$$NTU_{max} = \frac{AU}{C_{min}} \qquad (18.52)$$

where AU is the value for the new condition. The value $(AU)_{old}$ may be determined from $q = UA\,\Delta T$ at the original operating conditions. Generally U will change with a change in operating conditions because of variation in the convective coefficient, and this change must be determined before the effectiveness can be found. In this text we will simplify this step—unless directed otherwise, we will assume that UA is a constant for the heat exchanger.

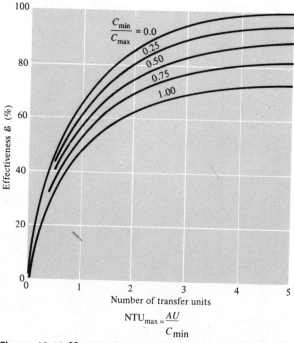

Figure 18.19 Heat exchanger effectiveness for two shell passes and a multiple of two tube passes. (*Source:* Adapted by permission from W. M. Kays and A. L. London, *Compact Heat Exchangers*, New York: McGraw-Hill, 1958.)

Example 18.5

A single-shell pass, four-tube pass, counterflow heat exchanger operates under the following conditions. Flue gases enter the heat exchanger at 260°C and leave at 150°C with an average specific heat of $c_p = 1.05$ kJ/kg-K and a flow rate of 0.51 kg/s. Water enters the exchanger at 125°C and a flow rate of 0.38 kg/s. A change in operating conditions occurs; a feedwater heater must be bypassed, so the water enters at 65°C and a flow rate of 0.31 kg/s. The specific heat of water may be assumed constant at $c_p = 4.18$ kJ/kg-K. Determine the new outlet water conditions.

Solution

Given: A heat exchanger with known operating conditions that change in a specified manner.

Find: The new outlet conditions for the water being heated.

Sketch and Given Data:

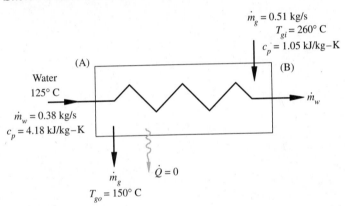

Figure 18.20

Assumptions:

1. The old and new conditions are steady-state.
2. The heat exchanger is adiabatic.
3. The flue gas supply and states remain the same.
4. Water is on the tube side, flue gas on the shell side.

Analysis: Determine the ratio of c_{min} to c_{max}.

$$C_g = (\dot{m}c_p)_g = (0.51 \text{ kg/s})(1.05 \text{ kJ/kg-K}) = 0.5355 \text{ kW/K}$$

$$C_w = (\dot{m}c_p)_w = (0.38 \text{ kg/s})(4.18 \text{ kJ/kg-K}) = 1.5884 \text{ kW/K}$$

$$\frac{C_{min}}{C_{max}} = \frac{0.5355}{1.5884} = 0.337$$

From the conservation of energy equation, we know that the heat transferred from the gas is received by the water. The outlet water condition at the initial state may be determined as follows:

$$(\dot{m}(h_o - h_i))_w = (\dot{m}(h_i - h_o))_g$$

$$[\dot{m}c_p(T_o - T_i)]_w = [\dot{m}c_p(T_i - T_o)]_g$$

$$[(0.38 \text{ kg/s})(4.18 \text{ kJ/kg-K})(T_0 - 125\,^\circ\text{C})]_w = [(0.51 \text{ kg/s})(1.05 \text{ kJ/kg-K})(260$$

$$- 150\,^\circ\text{C})]$$

$$(T_o)_w = 162.1\,^\circ\text{C}$$

$$P = \frac{T_{t_{out}} - T_{t_{in}}}{T_{s_{in}} - T_{t_{in}}} = \frac{162.1 - 125}{260 - 125} = 0.275$$

$$Z = \frac{T_{s_{in}} - T_{s_{out}}}{T_{t_{out}} - T_{t_{in}}} = \frac{260 - 150}{162.1 - 125} = 2.96$$

From Figure 18.13 the correction factor, F, is 0.75.

$$\Delta T_A = 150 - 125 = 25\,^\circ\text{C}$$

$$\Delta T_B = 260 - 162.1 = 97.9\,^\circ\text{C}$$

$$\text{LMTD} = \frac{\Delta T_A - \Delta T_B}{\ln\left(\dfrac{\Delta T_A}{\Delta T_B}\right)} = \frac{25 - 97.9}{\ln\left(\dfrac{25}{97.9}\right)} = 53.4\,^\circ\text{C}$$

$$\overline{\Delta T} = F\,\text{LMTD} = (0.75)(53.4) = 40.05\,^\circ\text{C}$$

$$UA = \frac{q}{\overline{\Delta T}} = \frac{(0.51 \text{ kg/s})(1.05 \text{ kJ/kg-K})(260 - 150\,^\circ\text{C})}{(40.05\,^\circ\text{C})}$$

$$UA = 1.470 \text{ kW/K}$$

$$\frac{UA}{C_{min}} = \frac{1.470}{0.5355} = 2.74$$

Determine the effectiveness from Figure 18.18: $\mathscr{E} = 0.81$. Once the effectiveness is known, the water outlet conditions at the new operating conditions may be found.

$$\mathscr{E}C_{min}(T_{h_{in}} - T_{c_{in}}) = C_c(T_{c_{out}} - T_{c_{in}})$$

$$(0.81)(0.5355 \text{ kW/K})(260 - 65\,^\circ\text{C}) = (0.31 \text{ kg/s})(4.18 \text{ kJ/kg-K})\Delta T_c$$

$$\Delta T_c = 65.2 = \Delta T_w$$

$$(T_o)_w = 65 + 65.2 = 130.2\,^\circ\text{C}$$

Comment: Often when the operating conditions change, the fluids on both the shell and tube sides vary. The value of UA may be assumed constant if the variance is not too great. The area, of course, does not change, but U changes if the convective coefficients vary significantly from one operating condition to another. ▬

CONCEPT QUESTIONS

1. How does conductive heat transfer occur?
2. What is the mechanism that radiative heat transfer uses to transmit thermal energy?
3. Is the equation for convective heat transfer a law?
4. Describe thermal conductivity. Is it higher for copper than for stone? Why?
5. What is blackbody radiation?
6. What is free, or natural, convection?
7. How is forced convection different than natural convection?
8. What is the Reynolds number? Why is it used?
9. Describe the relationship between the Nusselt number and the convective coefficient.
10. Describe the overall coefficient of heat transfer in terms of thermal resistances.
11. Why does adding insulation — thermal resistance — to wire sometimes increase the heat transfer from the wire?
12. Describe the ideal, tube-in-tube, counterflow, and parallel-flow heat exchangers.
13. What is a shell-and-tube heat exchanger?
14. Are actual shell-and-tube heat exchangers more like parallel-flow or like counterflow heat exchangers?

PROBLEMS (SI)

18.1 An experiment is undertaken to determine the thermal conductivity of an unknown material. The material is 5 cm thick and has a diameter of 15 cm. It is placed on a hot plate of equal diameter where the surface temperature is maintained at 60°C. The outer surface temperature is 24°C, and the power used by the hot plate is 45 W. Determine the thermal conductivity of the material.

18.2 A wall with a thermal conductivity of 0.30 W/m-K is maintained at 40°C. The heat flow through the wall is 250 W. The wall surface area is 1.5 m², and its thickness is 1 cm. Determine the temperature of the other surface.

18.3 The sum of the resistances for the outside wall of a house is 8.0°K-m²/W. For a temperature difference of 25°C across the wall and a total surface area of 150 m², determine the heat flow through the wall.

18.4 A dry-ice storage chest is a wooden box lined with glass fiber insulation 5 cm thick. The wooden box is 2 cm thick and is cubical, 60 cm on an edge. The inside surface temperature is −76°C, and the outside surface temperature is 18°C. Determine (a) the wood-insulation interface temperature; (b) the heat gain per day.

18.5 The surface of a furnace wall is at a temperature of 1200°C. The outside wall temperature is 38°C. The furnace wall construction has 15 cm of refractory material, $\lambda = 1.73$ W/m-K, and the outside wall is 1-cm steel, $\lambda = 44$ W/m-K. What thickness of refractory brick must be used between the refractory material and the wall if the heat loss is not to exceed 0.7 kW/m²? The thermal conductivity of the brick is 0.34 W/m-K.

18.6 A furnace is constructed with 20 cm of firebrick, $\lambda = 1.36$ W/m-K, 10 cm of insulating brick, $\lambda = 0.26$ W/m-K, and 20 cm of building brick, $\lambda = 0.69$ W/m-K. The inside surface temperature is 650°C, and the outside air temperature is 32°C. The heat loss from the furnace wall is 0.56 kW/m². Determine (a) the unit convective coefficient for the air; (b) the temperature 25 cm in from the outside surface.

18.7   A composite furnace wall has an inside wall temperature of 1100°C and an outside wall temperature of 38°C. Three types of bricks are available as follows:

Brick	λ (W/m·K)	Thickness (cm)	Maximum allowable temperature (°C)
1	1.55	11.5	—
2	0.21	7.5	950
3	0.70	10.0	150

The heat loss must not exceed 0.73 kW/m². Determine (a) the minimum wall thickness; (b) the actual heat loss for (a).

18.8   The oven of a stove must have sufficient insulation so that the outside surface temperature of the stove is not greater than 50°C. To accomplish this, insulation, $\lambda = 0.11$ W/m-K, is used between the inside and outside metal surfaces. The room temperature is 20°C, and the outside unit convective coefficient is 8 W/m²-K. Neglecting the resistance of the metal, what is the minimum thickness of insulation required, if the inside temperature reaches 315°C?

18.9   For the composite wall illustrated in Figure 18.5, the following values apply: $L_1 = 20$ cm, $\lambda_1 = 75$ W/m-K; $L_2 = 25$ cm, $\lambda_2 = 60$ W/m-K; $L_3 = 30$ cm, $\lambda_a = 20$ W/m-K, $\lambda_b = 60$ W/m-K; $L_4 = 50$ cm, $\lambda_4 = 50$ W/m-K. One surface is maintained at 400°C while the other is maintained at 100°C. Determine the heat flow and the temperature at the L_3/L_4 interface.

18.10   Referring to Figure 18.4(a), the fluid on the left is air at 800°C, the fluid on the right is water at 80°C, and the wall is 30 cm thick with a thermal conductivity of 0.50 W/m-K. The left-hand side convective coefficient is 100 W/m²-K, and the right-hand side convective coefficient is 1000 W/m²-K. The gas radiates as a blackbody. Determine (a) the overall circuit resistance; (b) the heat flow; (c) the wall temperatures.

18.11   The filament of a 75-W light bulb may be considered a blackbody radiating into a black enclosure at 70°C. The filament diameter is 0.10 mm, and the length is 5 cm. Considering only radiation, determine the filament temperature.

18.12   Water with a flow rate of 0.4 kg/s and a temperature of 10°C enters a 2.5-cm tube. A constant wall temperature of 50°C is maintained. Determine (a) the convective heat transfer coefficient; (b) for a tube 5.0 m long, the water exit temperature.

18.13   Air enters a heating duct with cross-sectional dimensions of 7.5 × 15 cm. The air enters the 5-m-long duct with a temperature of 0°C, and the duct surface is maintained at 67°C. If the air exit temperature is to be 20°C, what is the air flow rate?

18.14   Engine oil flows at 0.3 m/s through a 2.5-cm-diameter tube. The oil temperature is 160°C, and the tube surface temperature is 150°C. Determine the unit convective coefficient for tube lengths of (a) 2 m; (b) 4 m.

18.15   Water with a flow rate of 2.6 kg/s is heated from 10°C to 24°C as it passes through a 5-cm pipe. The inside pipe surface temperature is 95°C. Determine the pipe length required.

18.16 A dry saturated vapor enters 40 m of an 11.5-cm-O.D., 10-cm-I.D. pipe with a flow rate of 0.12 kg/s. The pipe is covered with 5 cm of 85% magnesia insulation. The surface temperature of the insulation is constant at 38°C, and the surrounding air temperature is 27°C. The unit outside convective coefficient is 10 W/m²-K. The latent heat of the vapor is 140 kJ/kg. Determine (a) the total heat loss in 40 m; (b) the quality of the vapor exiting the pipe.

18.17 A 15-cm-diameter steam line carries saturated steam at 6 MPa and is located in a tunnel with stagnant air with a temperature of 50°C. The outside unit convective coefficient is 8.5 W/m²-K. The steam line is covered with 3 cm of 85% magnesia insulation. It is decided to reduce the heat loss by one-half. How much 85% magnesia insulation is required? Neglect thermal resistance of pipe and inside surface.

18.18 A 5-cm-O.D. steel tube is used to transport saturated steam at 480 kPa. The tube is insulated with a 5-cm layer of 85% magnesia insulation followed by 3.5 cm of cork. The cork surface temperature is 24°C. Find the heat loss per m of pipe, neglecting pipe resistance.

18.19 Exhaust gases, cooling from 370°C to 175°C, are used to heat 18.9 kg/s of oil from 10°C to 100°C in a parallel-flow heat exchanger. The gas side area is 2000 m². The average specific heat of the gases is 1.09 kJ/kg-K, and the average specific heat for the oil is 2.5 kJ/kg-K. Determine (a) the overall coefficient of heat transfer based on the gas side area; (b) the LMTD; (c) the heat transferred.

18.20 The stack gas from a chemical operation contains noxious vapors that must be condensed by lowering their temperature from 315°C to 38°C. The gas volume flow rate is 0.7 m³/s. Water is available at 10°C at 1.26 kg/s. A shell-and-tube exchanger with two shell passes and four tube passes will be used, with water flowing through the tubes. The gas has a specfic heat of 1.11 kJ/kg-K and a gas constant of 0.26 kJ/kg-K. Determine (a) the required surface area if $U = 60$ W/m²-K; (b) the LMTD.

18.21 A 22.7-kg/s flow of air enters a preheater at 28°C and leaves at 150°C; 23.7 kg/s of exhaust gases, $c_p = 1.09$ kJ/kg-K, enters at 315°C. The overall coefficient of heat transfer is 710 W/m²-K. Determine (a) the exit exhaust gas temperature; (b) the surface area for parallel flow; (c) the surface area for counterflow; (d) the LMTD.

18.22 Crude oil, $c_p = 1.92$ kJ/kg-K, flows at the rate of 0.315 kg/s through the inside of a concentric, double-pipe heat exchanger and is heated from 32°C to 96°C. Another hydrocarbon, $c_p = 2.5$ kJ/kg-K, enters at 240°C. The overall coefficient of heat transfer is found to be 4400 W/m²-K. Determine for a minimum temperature difference of 17°C between the fluids (a) the LMTD for parallel flow and counterflow; (b) the surface area for parallel flow and counterflow.

18.23 A shell-and-tube heat exchanger must transfer 205 kW to a solution with a specific heat of 3.26 kJ/kg-K and change its temperature from 65°C to 93°C. Steam is available at 250 kPa. The unit convective coefficient is 3400 W/m²-K for the inside and 7300 W/m²-K for the outside. The thermal conductivity for the tubes is 111 W/m-K, and the tubes have 4.0 cm O.D. and 3.0 cm I.D. and are 3 m long. Determine the number of tubes required.

18.24 Exhaust steam with a flow of 63 kg/s at 7 kPa enters a single-pass surface condenser with an outside surface area of 7500 m². The steam quality is 90%, and the outlet condensate temperature is subcooled 5°C. The cooling-water flow rate is 4413 kg/s entering at 20°C. Find the overall coefficient of heat transfer.

18.25 A one-shell-pass, six-tube-pass heat exchanger is used as an economizer on a steam generator. The flue gas, $c_p = 1.09$ kJ/kg-K, enters at 350°C and leaves at 205°C with a flow rate of 58.0 kg/s. The feedwater enters at 175°C with a flow rate of 63 kg/s. A change of operating conditions occurs; the water flow is now 25.2 kg/s, entering at 138°C. The new gas flow rate is 23.8 kg/s, but the gas temperature remains the same. Determine (a) the old surface area required if $U = 170$ W/m²-K; (b) the effectiveness; (c) the new water outlet temperature.

18.26 A two-shell-pass, four-tube-pass heat exchanger is used as a feedwater heater and has an effectiveness of 80%. Water with a flow rate of 12.6 kg/s enters at 115°C, and air enters at 290°C with a flow rate of 25.2 kg/s. Determine (a) the air exit temperature; (b) the water exit temperature.

18.27 Liquid R 12 at a temperature of 15°C leaves the condenser through a 2.5-cm-O.D. copper tube and goes to the throttling valve in a refrigeration system. The question arises: should the line be insulated? There are three possible solutions: insulate with 9.5 mm of foam rubber, $\lambda = 0.16$ W/m-K; insulate with 15 mm of foam rubber, $\lambda = 0.16$ W/m-K; and do not insulate. The outside convective coefficient is estimated to be 7.0 W/m²-K, and the ambient temperature is 29°C. What is your recommendation and why?

18.28 Water is flowing through a 20-mm-I.D., 25-mm-O.D. brass tube, at a rate of 0.38 kg/s. The water enters at 10°C and leaves at 65°C. Heat is supplied by saturated steam condensing at 200 kPa on the outside tube surface. The outside unit convective coefficient, h_0, is 5600 W/m²-K; the inside unit convective coefficient, h_i, is 1700 W/m²-K. Determine (a) the overall coefficient of heat transfer, U, based on the outside tube area; (b) the length of tubing required.

18.29 A new plant process requires 50.4 kg/s of air to be heated from 4°C to 32°C. Saturated water at 280 kPa is available for heating the air and has a supply capacity of 8.82 kg/s. An old heat exchanger is suggested for use in the new process. Records show the following: dry saturated steam flow with no subcooling, 1.94 kg/s at 2000 kPa; airflow, 96.5 kg/s, exiting at 65°C. Determine (a) the original LMTD; (b) the effectiveness; (c) whether the heat exchanger can be used in the new process.

18.30 A counterflow shell-and-tube heat exchanger cools 1.6 liters/s of oil ($c_p = 2.5$ kJ/kg-K, $\rho = 720$ kg/m³) from 95°C to 50°C. Cooling water enters the tubes at 15°C and leaves at 38°C. The overall coefficient of heat transfer is 450 W/m²-K. Fouling of the tubes occurs, resulting in one-quarter of the tubes being blocked. The conditions remain the same except for the outlet temperatures. Determine the oil and water exit temperatures under the new conditions.

18.31 By traveling through a brass tube, 2.5 cm O.D. and 2.2 cm I.D., which is surrounded by steam at 55 kPa and 90% quality, 9.4 kg/s of water is heated from 15°C to 60°C. Assume the inside unit convective coefficient is 2000 W/m²-K, and the outside unit convective coefficient is 7950 W/m²-K. Determine (a) the length of tube required; (b) the steam supply in kg/h if no subcooling occurs.

PROBLEMS (ENGLISH UNITS)

*18.1 A boiler furnace wall must have a heat loss no greater than 700 Btu/hr-ft² and is made of a material with a thermal conductivity of 0.60 Btu/hr-ft-F. The inner wall surface temperature is 2000°F, and the outer surface temperature is 800°F. What wall thickness is required?

*18.2 A composite wall is composed of 8 in. of fire-clay brick, 6 in. of chrome brick, and 4 in. of common brick. The inside wall surface temperature is 2000°F, and the outside surface temperature is 300°F. Determine (a) the heat loss per ft² of wall area; (b) the temperatures at the brick interfaces; (c) the temperature 8 in. in from the outer surface.

*18.3 Compare the different heat flux resulting from conduction through 1 in. of the following materials when subjected to a 30°F temperature difference: copper, steel, wood, glass fiber.

*18.4 Hot water flows through a 4.5-in.-O.D. steel pipe that is insulated with 2 in. of magnesia. The temperature drop across the insulation is 125°F. Find the heat loss per 100 ft of pipe.

*18.5 Steam-generating tubes in a boiler have a 2-in.-O.D. and a 0.125-in. thickness. The boiling water within the tube receives 6700 Btu/hr-ft² when the overall temperature drop is 1000°F. Determine the percentage of temperature decrease in the metal tube.

*18.6 Saturated steam at 200 psia flows through a 6-in. steel pipe (6.625 in. O.D., 6.065 in. I.D.). The surrounding temperature is 80°F, and inside and outside unit convective coefficients are 1500 and 2.9 Btu/hr-ft²-F, respectively. Determine the thickness of 85% magnesia required to reduce the heat loss to 10% of the bare pipe case.

*18.7 An air conditioning duct is rectangular in cross section (15 × 30 in.) and has air at 15 psia and 40°F flowing through it with a velocity of 24 ft/sec. Determine the convective coefficient.

*18.8 A 1.8 × 10⁵ lbm/hr flow of air enters a preheater at 80°F and leaves at 300°F; 1.88 × 10⁵ lbm/hr of exhaust gases, $c_p = 0.26$ Btu/lbm-R, enters at 600°F. The overall coefficient of heat transfer is 125 Btu/hr-ft²-F. Determine (a) the exit exhaust gas temperature; (b) the surface area for parallel flow; (c) the surface area for counterflow; (d) the LMTD.

*18.9 Crude oil, $c_p = 0.56$ Btu/lbm-R, flows at the rate of 2500 lbm/hr through the inside of a concentric, double-pipe heat exchanger and is heated from 90°F to 200°F. Another hydrocarbon, $c_p = 0.60$ Btu/lbm-R, enters at 460°F. The overall coefficient of heat transfer is found to be 75 Btu/hr-ft²-F. Determine for a minimum temperature difference of 30°F between the fluids (a) the LMTD for parallel flow and counterflow; (b) the surface area for parallel flow and counterflow.

*18.10 Exhaust steam with a flow of 500,000 lbm/hr at 1 psia enters a single-pass surface condenser with an outside surface area of 75,000 ft². The steam quality is 90%, and the outlet condensate temperature is subcooled 10°F. The cooling-water flowrate is 70,000 gal/min entering at 70°F. Find the overall coefficient of heat transfer.

*18.11 A one-shell-pass, six-tube-pass heat exchanger is used as an economizer on a steam generator. The flue gas, $c_p = 0.26$ Btu/lbm-F, enters at 650°F and leaves at 400°F with a flow rate of 4.6 × 10⁵ lbm/hr. The feedwater enters at 350°F with a flow rate of 5 × 10⁵ lbm/hr. A change of operating conditions occurs; the water flow is now 2 × 10⁵ lbm/hr, entering at 280°F. The new gas flow rate is 1.89 × 10⁵ lbm/hr, but the gas temperature remains the same. Determine (a) the old surface area required if $U = 30$ Btu/hr-ft²-F; (b) the effectiveness; (c) the new water outlet temperature.

COMPUTER PROBLEMS

C18.1 Develop a TK Solver model, spreadsheet template, or computer program to calculate the heat transmitted from an insulated cylinder using equation 18.34. For three different combinations of h_o and Γ, compute the heat transmitted for a range of radii smaller and larger than the critical radius. Plot q versus r_o.

C18.2 Consider water being heated or cooled as it flows through a tube with a constant temperature along its length. Develop a TK Solver model, spreadsheet template, or computer program to analyze convective heat transfer to or from the water. Use Equations 18.18 and 18.20 to compute the Nusselt numbers for the laminar and turbulent regions. Calculate the film coefficient, heat transferred per unit length and outlet temperature for a range of flows from laminar to turbulent.

C18.3 Develop a TK Solver model, spreadsheet template, or computer program to calculate the heat lost from an insulated pipe to the surrounding air. Consider convection from the fluid in the pipe to the inside pipe surface, conduction through the pipe wall and the insulation, and convection and radiation to the surrounding air from the outside surface. Investigate the effect of varying the insulation thickness on the pipe and insulation temperatures and heat transmission. Investigate the error of not including radiative heat transfer from the outside surface to the surrounding air.

References

Badger, P. H. *Equilibrium Thermodynamics.* Boston: Allyn Bacon, 1967.

Callen, H. B. *Thermodynamics.* New York: Wiley, 1960.

Çengel, Y. A., and Boles, M. A. *Thermodynamics: An Engineering Approach.* New York: McGraw-Hill, 1989.

Cohen, H.; Roger, G. F. C.; and Saravanamuttoo, H. I. H. *Gas Turbine Theory,* 2nd ed. New York: Wiley, 1972.

Dodge, D. F. *Chemical Engineering Thermodynamics.* New York: McGraw-Hill, 1944.

Faires, V. M. *Thermodynamics,* 5th ed. New York: Macmillan, 1970.

Jordan, R. C., and Priester, G. B. *Refrigeration and Air Conditioning.* Englewood Cliffs, N.J.: Prentice Hall, 1956.

Keenan, J. H. *Thermodynamics.* New York: Wiley, 1941.

Moran, M. J., and Shapiro, H. N. *Fundamentals of Engineering Thermodynamics,* 2nd ed. New York: Wiley, 1992.

Obert, E. F. *Concepts of Thermodynamics.* New York: McGraw-Hill, 1960.

Obert, E. F. *Internal Combustion Engines,* 3rd ed. New York: International Textbook Company, 1968.

Shapiro, A. H. *The Dynamics and Thermodynamics of Compressible Fluid Flow.* New York: Ronald Press, 1953.

Shepherd, D. G. *Principles of Turbomachinery.* New York: Macmillan, 1956.

Stoecker, W. F., and Jones, J. W. *Refrigeration and Air Conditioning,* 2nd ed. New York: McGraw-Hill, 1982.

Taylor, C. F. *Internal Combustion Engine in Theory and Practice,* 2nd ed. Cambridge, Mass.: M.I.T. Press, 1966.

Van Wylen, G. J., and Sonntag, R. E. *Fundamentals of Classical Thermodynamics,* 3rd ed. New York: Wiley, 1986.

List of Symbols

A	area	n	moles
a	acceleration	n	polytropic exponent
a	acoustic velocity	Nu	Nusselt number
$\mathscr{A}$	availability	p.e.	specific potential energy
A.E.	available energy	P.E.	potential energy
C	constant	p	pressure
c_p	constant-pressure specific heat, mass basis	p_r	reduced pressure
$\bar{c}_p$	constant-pressure specific heat, mole basis	Pr	Prandtl number
		Q	total heat flow
		q	specific heat flow
c_v	constant-volume specific heat, mass basis	$\dot{Q}$	heat-transfer rate
$\bar{c}_v$	constant-volume specific heat, mole basis	R	specific gas constant
		$\bar{R}$	universal gas constant
COP	coefficient of performance	Re	Reynolds number
d	diameter	r	radius
E	total energy	r	compression ratio
e	specific energy	$r_{a/f}$	air/fuel ratio
$\mathscr{E}$	heat exchanger effectiveness	r_c	cutoff ratio
F	force	r_p	pressure ratio
G	total Gibbs function	S	total entropy
g	specific Gibbs function	s	specific entropy, mass basis
g	local gravitational acceleration	$\bar{s}$	specific entropy, mole basis
g_c	dimensional constant	T	absolute temperature
H	total enthalpy	U	total internal energy
h	specific enthalpy	U	overall heat transfer coefficient
$\bar{h}$	specific enthalpy, mole basis	U.E.	unavailable energy
$\bar{h}_c$	unit convective coefficient	u	specific internal energy, mass basis
K	chemical equilibrium constant	$\bar{u}$	specific internal energy, mole basis
k	ratio of specific heats		
k.e.	specific kinetic energy	V	total volume
K.E.	kinetic energy	v	specific volume
L	length	v	velocity
l	length	v_r	reduced specific volume
ln	natural logarithm	W	work
M	Mach number	$\dot{W}$	rate of work (power)
M	molecular weight	w	specific work
m	mass	x	quality of wet vapor
$\dot{m}$	mass flow rate	Z	compressibility factor

z	elevation	θ	angle
α	thermal diffusivity	λ	thermal conductivity
β	adiabatic compressibility	μ	Joule-Thomson coefficient
Δ	a change in	μ	dynamic viscosity
δ	infinitesimal change	ν	kinematic viscosity
ϵ	strain, unit	ρ	density
η	efficiency	σ	surface tension
η_{th}	thermal efficiency	σ	Stefan-Boltzmann constant
η_c	compressor efficiency	t	time
η_d	diffuser efficiency	τ	torque
η_m	mechanical efficiency	ϕ	entropy function
η_n	nozzle efficiency	ϕ	relative humidity
η_p	propulsion efficiency	ω	angular velocity
η_t	turbine internal efficiency	ω	specific humidity
η_v	volumetric efficiency	ψ	specific flow availability

Appendices

Appendix A

TABLE A.1 GAS CONSTANTS AND SPECIFIC HEATS AT LOW PRESSURES AND 25°C (77°F)

Gas	M (lbm/pmol)	c_p (Btu/lbm-R)	c_p (kJ/kg-K)	c_v (Btu/lbm-R)	c_v (kJ/kg-K)	k	R (ft-lbf/lbm-R)	R (kJ/kg-K)
Acetylene C_2H_2	26.036	0.4048	1.6947	0.3285	1.3753	1.232	59.35	0.3195
Air	28.97	0.24	1.0047	0.1714	0.7176	1.4	53.34	0.287
Ammonia NH_3	17.032	0.499	2.089	0.382	1.5992	1.304	90.73	0.4882
Argon Ar	39.95	0.1244	0.5208	0.0747	0.3127	1.666	38.68	0.2081
Carbon dioxide CO_2	44.01	0.2016	0.844	0.1565	0.6552	1.288	35.11	0.1889
Carbon monoxide CO	28.01	0.2487	1.0412	0.1778	0.7444	1.399	55.17	0.2968
Chlorine Cl_2	70.914	0.1144	0.4789	0.0864	0.3617	1.324	21.79	0.1172
Ethane C_2H_6	30.068	0.4186	1.7525	0.3526	1.4761	1.187	51.39	0.2765
Ethylene C_2H_4	28.052	0.3654	1.5297	0.2946	1.2333	1.24	55.09	0.2964
Helium He	4.003	1.241	5.1954	0.745	3.1189	1.666	386.04	2.077
Hydrazine N_2H_4	32.048	0.393	1.6453	0.33	1.3815	1.195	48.22	0.2594
Hydrogen H_2	2.016	3.419	14.3136	2.434	10.190	1.4	766.54	4.125
Methane CH_4	16.043	0.5099	2.1347	0.3861	1.6164	1.321	96.33	0.5183
Neon Ne	20.183	0.246	1.0298	0.1476	0.6179	1.666	76.57	0.4120
Nitrogen N_2	28.016	0.2484	1.0399	0.1775	0.7431	1.399	55.16	0.2968
Oxygen O_2	32	0.2194	0.9185	0.1573	0.6585	1.395	48.29	0.2598
Propane C_3H_8	44.094	0.3985	1.6683	0.3535	1.4799	1.127	35.05	0.1886
Sulfur dioxide SO_2	64.07	0.1487	0.6225	0.1177	0.4927	1.263	24.12	0.1298
Water vapor H_2O	18.016	0.4454	1.8646	0.3352	1.4033	1.329	85.77	0.4615
Xenon Xe	131.3	0.0378	0.1582	0.0227	0.0950	1.666	11.77	0.0633

TABLE A.2   PROPERTIES OF AIR AT LOW PRESSURES (SI Units)

T (K)	h (kJ/kg)	p_r	u (kJ/kg)	v_r	ϕ (kJ/kg-K)
100	99.76	0.029 90	71.06	2230	1.4143
110	109.77	0.041 71	78.20	1758.4	1.5098
120	119.79	0.056 52	85.34	1415.7	1.5971
130	129.81	0.074 74	92.51	1159.8	1.6773
140	139.84	0.096 81	99.67	964.2	1.7515
150	149.86	0.123 18	106.81	812.0	1.8206
160	159.87	0.154 31	113.95	691.4	1.8853
170	169.89	0.190 68	121.11	594.5	1.9461
180	179.92	0.232 79	128.28	515.6	2.0033
190	189.94	0.281 14	135.40	450.6	2.0575
200	199.96	0.3363	142.56	396.6	2.1088
210	209.97	0.3987	149.70	351.2	2.1577
220	219.99	0.4690	156.84	312.8	2.2043
230	230.01	0.5477	163.98	280.0	2.2489
240	240.03	0.6355	171.15	251.8	2.2915
250	250.05	0.7329	178.29	227.45	2.3325
260	260.09	0.8405	185.45	206.26	2.3717
270	270.12	0.9590	192.59	187.74	2.4096
280	280.14	1.0889	199.78	171.45	2.4461
290	290.17	1.2311	206.92	157.07	2.4813
300	300.19	1.3860	214.09	144.32	2.5153
310	310.24	1.5546	221.27	132.96	2.5483
320	320.29	1.7375	228.45	122.81	2.5802
330	330.34	1.9352	235.65	113.70	2.6111
340	340.43	2.149	242.86	105.51	2.6412
350	350.48	2.379	250.05	98.11	2.6704
360	360.58	2.626	257.23	91.40	2.6987
370	370.67	2.892	264.47	85.31	2.7264
380	380.77	3.176	271.72	79.77	2.7534
390	390.88	3.481	278.96	74.71	2.7796
400	400.98	3.806	286.19	70.07	2.8052
410	411.12	4.153	293.45	65.83	2.8302
420	421.26	4.522	300.73	61.93	2.8547
430	431.43	4.915	308.03	58.34	2.8786
440	441.61	5.332	315.34	55.02	2.9020
450	451.83	5.775	322.66	51.96	2.9249
460	462.01	6.245	329.99	49.11	2.9473
470	472.25	6.742	337.34	46.48	2.9693
480	482.48	7.268	344.74	44.04	2.9909
490	492.74	7.824	352.11	41.76	3.0120
500	503.02	8.411	359.53	39.64	3.0328
510	513.32	9.031	366.97	37.65	3.0532
520	523.63	9.684	374.39	35.80	3.0733
530	533.98	10.372	381.88	34.07	3.0930
540	544.35	11.097	389.40	32.45	3.1124
550	554.75	11.858	396.89	30.92	3.1314

TABLE A.2 PROPERTIES OF AIR AT LOW PRESSURES (SI Units) *(Continued)*

T (K)	h (kJ/kg)	p_r	u (kJ/kg)	v_r	ϕ (kJ/kg-K)
560	565.17	12.659	404.44	29.50	3.1502
570	575.57	13.500	411.98	28.15	3.1686
580	586.04	14.382	419.56	26.89	3.1868
590	596.53	15.309	427.17	25.70	3.2047
600	607.02	16.278	434.80	24.58	3.2223
610	617.53	17.297	442.43	23.51	3.2397
620	628.07	18.360	450.13	22.52	3.2569
630	638.65	19.475	457.83	21.57	3.2738
640	649.21	20.64	465.55	20.674	3.2905
650	659.84	21.86	473.32	19.828	3.3069
660	670.47	23.13	481.06	19.026	3.3232
670	681.15	24.46	488.88	18.266	3.3392
680	691.82	25.85	496.65	17.543	3.3551
690	702.52	27.29	504.51	16.857	3.3707
700	713.27	28.80	512.37	16.205	3.3861
710	724.01	30.38	520.26	15.585	3.4014
720	734.20	31.92	527.72	15.027	3.4156
730	745.62	33.72	536.12	14.434	3.4314
740	756.44	35.50	544.05	13.900	3.4461
750	767.30	37.35	552.05	13.391	3.4607
760	778.21	39.27	560.08	12.905	3.4751
770	789.10	41.27	568.10	12.440	3.4894
780	800.03	43.35	576.15	11.998	3.5035
790	810.98	45.51	584.22	11.575	3.5174
800	821.94	47.75	592.34	11.172	3.5312
810	832.96	50.08	600.46	10.785	3.5449
820	843.97	52.49	608.62	10.416	3.5584
830	855.01	55.00	616.79	10.062	3.5718
840	866.09	57.60	624.97	9.724	3.5850
850	877.16	60.29	633.21	9.400	3.5981
860	888.28	63.09	641.44	9.090	3.6111
870	899.42	65.98	649.70	8.792	3.6240
880	910.56	68.98	658.00	8.507	3.6367
890	921.75	72.08	666.31	8.233	3.6493
900	932.94	75.29	674.63	7.971	3.6619
910	944.15	78.61	682.98	7.718	3.6743
920	955.38	82.05	691.33	7.476	3.6865
930	966.64	85.60	699.73	7.244	3.6987
940	977.92	89.28	708.13	7.020	3.7108
950	989.22	93.08	716.57	6.805	3.7227
960	1000.53	97.00	725.01	6.599	3.7346
970	1011.88	101.06	733.48	6.400	3.7463
980	1023.25	105.24	741.99	6.209	3.7580
990	1034.63	109.57	750.48	6.025	3.7695
1000	1046.03	114.03	759.02	5.847	3.7810
1020	1068.89	123.12	775.67	5.521	3.8030

TABLE A.2 PROPERTIES OF AIR AT LOW PRESSURES (SI Units) *(Continued)*

T (K)	h (kJ/kg)	p_r	u (kJ/kg)	v_r	ϕ (kJ/kg-K)
1040	1091.85	133.34	793.35	5.201	3.8259
1060	1114.85	143.91	810.61	4.911	3.8478
1080	1137.93	155.15	827.94	4.641	3.8694
1100	1161.07	167.07	845.34	4.390	3.8906
1120	1184.28	179.71	862.85	4.156	3.9116
1140	1207.54	193.07	880.37	3.937	3.9322
1160	1230.90	207.24	897.98	3.732	3.9525
1180	1254.34	222.2	915.68	3.541	3.9725
1200	1277.79	238.0	933.40	3.362	3.9922
1220	1301.33	254.7	951.19	3.194	4.0117
1240	1324.89	272.3	969.01	3.037	4.0308
1260	1348.55	290.8	986.92	2.889	4.0497
1280	1372.25	310.4	1004.88	2.750	4.0684
1300	1395.97	330.9	1022.88	2.619	4.0868
1320	1419.77	352.5	1040.93	2.497	4.1049
1340	1443.61	375.3	1059.03	2.381	4.1229
1360	1467.50	399.1	1077.17	2.272	4.1406
1380	1491.43	424.2	1095.36	2.169	4.1580
1400	1515.41	450.5	1113.62	2.072	4.1753
1420	1539.44	478.0	1131.90	1.9808	4.1923
1440	1563.49	506.9	1150.23	1.8942	4.2092
1460	1587.61	537.1	1168.61	1.8124	4.2258
1480	1611.80	568.8	1187.03	1.7350	4.2422
1500	1635.99	601.9	1205.47	1.6617	4.2585
1550	1696.63	691.4	1251.78	1.4948	4.2983
1600	1757.55	791.2	1298.35	1.3485	4.3369
1650	1818.70	902.0	1345.17	1.2197	4.3745
1700	1907.39	1025.1	1392.18	1.1065	4.4112
1750	1941.63	1160.5	1439.38	1.0056	4.4469
1800	2003.36	1310.3	1486.76	0.9164	4.4817
1850	2065.27	1475.0	1534.33	0.8370	4.5156
1900	2127.37	1655.4	1582.09	0.7659	4.5487
1950	2182.72	1830.4	1624.91	0.7084	4.5777
2000	2252.06	2067.9	1678.07	0.6449	4.6127
2500	2883.91	5521	2166.41	0.3019	4.8946
3000	3526.54	12 490	2665.55	0.16015	5.1288
3500	4177.22	25 123	3172.71	0.09289	5.3294

Source: Adapted from J. H. Keenan and J. Kaye, *Gas Tables,* New York: John Wiley & Sons, 1945.

TABLE A.3 PRODUCTS—400% THEORETICAL AIR—AT LOW PRESSURES

T (K)	h (kJ/kg)	p_r	u (kJ/kg)	v_r	ϕ (kJ/kg·K)
200	201.2	0.3279	143.8	11782	6.2919
220	221.4	0.4588	158.2	9262.8	6.3884
240	241.7	0.6236	172.8	7433	6.4764
260	262.0	0.8275	187.4	6069	6.5577
280	282.3	1.0756	201.9	5028	6.6330
300	302.6	1.3738	216.5	4218	6.7033
320	323.1	1.7279	231.2	3577	6.7692
340	343.5	2.144	245.9	3063	6.8311
360	364.0	2.629	260.6	2645	6.8897
380	384.6	3.190	275.5	2301	6.9454
400	405.1	3.836	290.3	2014.4	6.9981
420	425.8	4.572	305.2	1774.4	7.0486
440	446.6	5.408	320.2	1571.5	7.0968
460	467.4	6.354	335.3	1398.4	7.1430
480	488.3	7.418	350.4	1249.9	7.1876
500	509.2	8.611	365.7	1121.6	7.2304
520	530.3	9.944	381.0	1010.2	7.2717
540	551.5	11.428	396.4	912.7	7.3116
560	572.8	13.075	412.0	827.4	7.3504
580	594.1	14.897	427.6	752.1	7.3877
600	615.6	16.909	443.3	685.5	7.4241
620	637.1	19.125	459.1	626.3	7.4596
640	658.8	21.56	475.0	573.4	7.4940
660	680.5	24.22	491.0	526.2	7.5275
680	702.4	27.14	507.2	484.0	7.5600
700	724.4	30.32	523.4	445.9	7.5919
720	746.5	33.80	539.7	411.5	7.6230
740	768.6	37.57	556.2	380.5	7.6533
760	790.9	41.66	572.7	352.4	7.6831
780	813.3	46.11	589.4	326.8	7.7122
800	835.8	50.91	606.1	303.6	7.7407
820	858.4	56.10	623.0	282.3	7.7686
840	881.1	61.72	640.0	262.9	7.7959
860	903.9	67.76	657.0	245.2	7.8227
880	926.8	74.26	674.1	228.9	7.8490
900	949.8	81.24	691.4	214.0	7.8749
920	972.8	88.74	708.7	200.26	7.9002
940	996.0	96.78	726.1	187.61	7.9252
960	1019.2	105.39	743.6	175.94	7.9496
980	1042.6	114.60	761.2	165.18	7.9736
1000	1066.0	124.45	778.9	155.21	7.9973
1050	1124.9	152.05	823.5	133.40	8.0549
1100	1184.3	184.33	868.5	115.27	8.1101
1150	1244.2	221.90	914.0	100.11	8.1633
1200	1304.5	265.4	960.0	87.36	8.2148
1250	1365.3	315.3	1006.4	76.56	8.2643

TABLE A.3   PRODUCTS—400% THEORETICAL AIR—AT LOW PRESSURES
(Continued)

T (K)	h (kJ/kg)	p_r	u (kJ/kg)	v_r	ϕ (kJ/kg-K)
1300	1426.4	372.6	1053.1	67.37	8.3123
1350	1487.8	438.0	1100.2	59.54	8.3587
1400	1542.8	512.3	1147.6	52.79	8.4037
1450	1611.7	596.2	1195.4	46.98	8.4472
1500	1674.1	690.9	1243.4	41.94	8.4896
1550	1736.7	797.3	1291.7	37.55	8.5306
1600	1799.7	916.4	1340.3	33.73	8.5707
1650	1862.9	1049.4	1389.2	30.37	8.6096
1700	1926.4	1197.4	1438.3	27.42	8.6475
1750	1990.0	1361.7	1487.6	24.83	8.6843
1800	2053.9	1543.5	1537.1	22.53	8.7204
1850	2118.0	1744.5	1586.8	20.49	8.7555
1900	2182.3	1965.7	1636.8	18.67	8.7899

Source: Adapted from J. H. Keenan and J. Kaye, *Gas Tables,* New York: John Wiley & Sons, 1945.

TABLE A.4   PRODUCTS—200% THEORETICAL AIR—AT LOW PRESSURES

T (K)	h (kJ/kg)	p_r	u (kJ/kg)	v_r	ϕ (kJ/kg-K)
200	202.4	0.3200	145.0	12073	6.2882
220	222.8	0.4491	159.6	9463	6.3854
240	243.3	0.6124	174.4	7571	6.4746
260	263.9	0.8152	189.2	6161	6.5567
280	284.5	1.0630	204.0	5088	6.6331
300	305.1	1.3620	218.9	4255	6.7043
320	325.8	1.1788	233.9	3597	6.7712
340	346.6	2.140	248.9	3070	6.8341
360	367.4	2.632	264.0	2642	6.8936
380	388.3	3.204	279.1	2291	6.9501
400	409.2	3.864	294.3	1999.3	7.0039
420	430.3	4.621	309.6	1755.7	7.0553
440	451.4	5.483	325.0	1550.1	7.1044
460	472.7	6.462	340.5	1375.2	7.1515
480	494.0	7.566	356.1	1225.5	7.1969
500	515.4	8.808	371.7	1096.4	7.2406
520	536.8	10.202	387.5	984.6	7.2829
540	558.5	11.758	403.4	887.2	7.3235
560	580.2	13.490	419.4	801.9	7.3632
580	602.0	15.413	435.4	726.9	7.4014
600	624.0	17.542	451.6	660.8	7.4386
620	646.0	19.894	467.9	602.0	7.4747
640	668.1	22.48	484.3	549.8	7.5099
660	690.4	25.33	500.8	503.3	7.5442
680	712.8	28.46	517.4	461.6	7.5776

TABLE A.4 PRODUCTS—200% THEORETICAL AIR—AT LOW PRESSURES
(Continued)

T (K)	h (kJ/kg)	p_r	u (kJ/kg)	v_r	ϕ (kJ/kg·K)
700	735.2	31.87	534.2	424.2	7.6101
720	757.8	35.62	551.0	390.5	7.6420
740	780.5	39.68	568.0	360.2	7.6731
760	803.4	44.12	585.1	332.8	7.7034
780	826.3	48.94	602.3	307.9	7.7334
800	849.4	54.17	619.6	285.3	7.7625
820	872.5	59.83	637.0	264.7	7.7911
840	895.8	65.97	654.5	246.0	7.8192
860	919.1	72.60	672.1	228.8	7.8467
880	942.6	79.75	689.8	213.2	7.8736
900	966.2	87.44	707.6	198.81	7.9001
920	989.8	95.73	725.5	185.64	7.9261
940	1013.6	104.63	743.6	173.53	7.9517
960	1037.4	114.19	761.6	162.39	7.9768
980	1061.4	124.44	779.9	152.11	8.0015
1000	1085.4	135.43	798.2	142.63	8.0256
1050	1146.0	166.34	844.4	121.93	8.0848
1100	1207.0	202.71	891.0	104.82	8.1417
1150	1268.6	245.3	938.2	90.57	8.1964
1200	1330.6	294.8	985.9	78.64	8.2492
1250	1393.1	352.0	1034.0	68.57	8.3002
1300	1456.0	418.0	1082.5	60.06	8.3496
1350	1519.2	493.6	1131.4	52.83	8.3973
1400	1582.9	580.0	1180.7	46.62	8.4436
1450	1646.9	678.1	1230.4	41.30	8.4886
1500	1711.2	789.4	1280.3	36.71	8.5322
1550	1775.8	914.9	1330.6	32.72	8.5746
1600	1840.7	1056.2	1381.1	29.26	8.6159
1650	1906.0	1214.6	1432.0	26.24	8.6560
1700	1971.4	1391.7	1483.1	23.596	8.6951
1750	2037.2	1589.2	1534.5	21.274	8.7333
1800	2103.1	1808.7	1586.0	19.225	8.7704
1850	2169.3	2052.2	1637.8	17.415	8.8068
1900	2235.7	2321.4	1689.9	15.812	8.8422
1950	2302.3	2618	1742.1	14.385	8.8768
2000	2369.0	2946	1794.5	13.114	8.9106
2100	2503.2	3700	1899.9	10.966	8.9760
2200	2638.0	4603	2006.0	9.234	9.0388

Source: Adapted from J. H. Keenan and J. Kaye, *Gas Tables,* New York: John Wiley & Sons, 1945.

TABLE A.5 SATURATED STEAM TEMPERATURE TABLE (SI Units)

T (°C)	p (kPa)	Specific volume (m³/kg)		Internal energy (kJ/kg)			Enthalpy (kJ/kg)			Entropy (kJ/kg-K)		
		Sat. liquid, v_f	Sat. vapor, v_g	Sat. liquid, u_f	Evap., u_{fg}	Sat. vapor, u_g	Sat. liquid, h_f	Evap., h_{fg}	Sat. vapor, h_g	Sat. liquid, s_f	Evap., s_{fg}	Sat. vapor, s_g
0	0.6106	0.000 999 96	206.42	-0.017 892	2375.2	2375.2	-0.017 281	2501.3	2501.3	-0.003 101	9.1572	9.1541
5	0.872 47	0.001 000 1	147.09	20.431	2361.9	2382.3	20.432	2490.2	2510.6	0.071 787	8.9528	9.0246
10	1.2287	0.001 000 5	106.3	41.104	2348.2	2389.3	41.105	2478.8	2520.0	0.145 92	8.7545	8.9005
15	1.7071	0.001 001 1	77.846	61.944	2334.4	2396.3	61.946	2467.2	2529.2	0.219 16	8.5624	8.7815
20	2.3414	0.001 002	57.726	82.907	2320.3	2403.2	82.909	2455.5	2538.4	0.291 43	8.3761	8.6676
25	3.1728	0.001 003 2	43.313	103.95	2306.1	2410.1	103.96	2443.5	2547.5	0.362 65	8.1957	8.5583
30	4.2505	0.001 004 5	32.862	125.05	2291.8	2416.9	125.06	2431.5	2556.6	0.432 78	8.0208	8.4536
35	5.633	0.001 006 1	25.196	146.18	2277.5	2423.6	146.18	2419.4	2565.6	0.5018	7.8513	8.3532
40	7.389	0.001 007 9	19.511	167.31	2263.1	2430.4	167.32	2407.2	2574.5	0.569 71	7.6871	8.2568
50	12.355	0.001 012 1	12.028	209.55	2234.1	2443.6	209.56	2382.7	2592.2	0.7022	7.3733	8.0755
60	19.946	0.001 017 1	7.6696	251.69	2205.0	2456.7	251.71	2358.0	2609.7	0.830 41	7.0779	7.9083
70	31.196	0.001 022 7	5.0418	293.73	2175.9	2469.6	293.76	2333.1	2626.9	0.9546	6.7991	7.7537
80	47.404	0.001 029	3.4067	335.68	2146.5	2482.2	335.73	2308.0	2643.7	1.0751	6.5354	7.6105
90	70.169	0.001 036	2.36	377.58	2117.0	2494.6	377.66	2282.5	2660.2	1.1922	6.2853	7.4774
100	101.33	0.001 043 6	1.6722	419.5	2087.1	2506.6	419.61	2256.5	2676.1	1.3062	6.0473	7.3535
110	143.38	0.001 051 8	1.2095	461.5	2056.7	2518.2	461.65	2230.0	2691.6	1.4175	5.8201	7.2376
120	198.7	0.001 060 6	0.891 22	503.64	2025.8	2529.4	503.85	2202.7	2706.5	1.5263	5.6026	7.1289
130	270.34	0.001 07	0.667 96	545.97	1994.2	2540.1	546.26	2174.5	2720.7	1.6328	5.3937	7.0265

140	361.64	0.001 08	0.508 41	588.54	1961.8	2550.3	588.93	2145.2	2734.2	1.7373	5.1924	6.9297
150	476.3	0.001 090 8	0.392 44	631.4	1928.4	2559.8	631.92	2114.8	2746.8	1.8399	4.9979	6.8378
160	618.38	0.001 102 2	0.3068	674.57	1894.1	2568.7	675.25	2083.2	2758.4	1.9408	4.8093	6.7501
170	792.32	0.001 114 4	0.242 64	718.08	1858.7	2576.7	718.96	2050.0	2769.0	2.0401	4.626	6.6661
180	1002.9	0.001 127 4	0.193 91	761.94	1822.0	2584.0	763.07	2015.4	2778.4	2.1379	4.4474	6.5853
190	1255.2	0.001 141 3	0.156 44	806.17	1784.0	2590.2	807.6	1979.0	2786.6	2.2343	4.2729	6.5072
200	1554.7	0.001 156 2	0.127 29	850.79	1744.7	2595.4	852.59	1940.8	2793.3	2.3295	4.1018	6.4312
210	1907.3	0.001 172 3	0.104 37	895.82	1703.7	2599.5	898.05	1900.5	2798.6	2.4235	3.9337	6.3572
220	2319.1	0.001 189 6	0.086 154	941.29	1661.1	2602.4	944.05	1858.2	2802.2	2.5166	3.768	6.2845
230	2796.6	0.001 208 3	0.071 55	987.26	1616.7	2603.9	990.63	1813.4	2804.0	2.6088	3.6041	6.2129
240	3346.6	0.001 228 6	0.059 735	1033.8	1570.2	2604.0	1037.9	1766.0	2803.9	2.7005	3.4415	6.142
250	3976.2	0.001 250 9	0.050 097	1081.0	1521.5	2602.5	1086.0	1715.7	2801.6	2.7918	3.2795	6.0714
260	4692.8	0.001 275 3	0.042 175	1129.0	1470.2	2599.1	1135.0	1662.1	2797.1	2.8831	3.1175	6.0006
270	5504.3	0.001 302 2	0.035 613	1177.9	1415.9	2593.9	1185.1	1604.8	2789.9	2.9745	2.9546	5.9291
280	6418.7	0.001 332 1	0.030 14	1228.0	1358.4	2586.4	1236.6	1543.3	2779.8	3.0665	2.7899	5.8565
290	7444.6	0.001 365 7	0.025 54	1279.5	1296.8	2576.3	1289.6	1476.8	2766.4	3.1595	2.6224	5.7819
300	8591.	0.001 403 8	0.021 648	1332.6	1230.7	2563.3	1344.7	1404.6	2749.3	3.2538	2.4507	5.7045
310	9867.7	0.001 447 5	0.018 327	1387.7	1159.1	2546.8	1402.0	1325.6	2727.6	3.3499	2.2732	5.6232
320	11 286	0.001 498 8	0.015 469	1445.1	1080.7	2525.8	1462.0	1238.4	2700.4	3.4484	2.0879	5.5362
330	12 858	0.001 560 4	0.012 983	1505.3	993.87	2499.2	1525.4	1140.7	2666.1	3.55	1.8913	5.4413
340	14 599	0.001 637 5	0.010 789	1569.8	894.97	2464.7	1593.7	1028.6	2622.2	3.6572	1.6775	5.3347
350	16 527	0.001 740 2	0.008 807 2	1641.0	777.48	2418.5	1669.8	894.27	2564.0	3.7744	1.4351	5.2094
360	18 668	0.001 894 6	0.006 939 6	1725.5	626.0	2351.5	1760.9	720.17	2481.1	3.9133	1.1374	5.0507
370	21 052	0.002 219 5	0.004 931 5	1848.8	381.89	2230.7	1895.5	438.99	2334.5	4.1202	0.68256	4.8028
374.14	22 090	0.003 155	0.003 155	2029.6	0	2029.6	2099.3	0	2099.3	4.4298	0	4.4298

TABLE A.6 SATURATED STEAM PRESSURE TABLE (SI Units)

p (kPa)	T (°C)	Specific volume (m³/kg) Sat. liquid, v_f	Sat. vapor, v_g	Internal energy (kJ/kg) Sat. liquid, u_f	Evap., u_{fg}	Sat. vapor, u_g	Enthalpy (kJ/kg) Sat. liquid, h_f	Evap., h_{fg}	Sat. vapor, h_g	Entropy (kJ/kg·K) Sat. liquid, s_f	Evap., s_{fg}	Sat. vapor, s_g
1	6.9677	0.001 000 2	129.23	28.543	2356.5	2385.1	28.544	2485.8	2514.3	0.101 06	8.874	8.9751
2	17.481	0.001 001 5	67.009	72.335	2327.4	2399.7	72.337	2461.4	2533.8	0.255 15	8.4692	8.7244
3	24.064	0.001 002 9	45.667	100.01	2308.8	2408.8	100.01	2445.8	2545.8	0.349 39	8.229	8.5784
4	28.946	.001 004 2	34.802	120.6	2294.9	2415.5	120.6	2434.1	2554.7	0.418 08	8.0572	8.4753
5	32.861	.001 005 4	28.194	137.14	2283.6	2420.8	137.14	2424.6	2561.7	0.472 41	7.9232	8.3956
7.5	40.28	.001 008	19.239	168.49	2262.2	2430.7	168.5	2406.5	2575	0.573 48	7.678	8.2515
10	45.799	.001 010 2	14.675	191.81	2246.3	2438.1	191.82	2393	2584.8	0.647 08	7.5028	8.1499
20	60.058	.001 017 1	7.6502	251.94	2204.9	2456.8	251.96	2357.9	2609.8	0.831 15	7.0762	7.9074
30	69.099	.001 022 2	5.2299	289.94	2178.5	2468.4	289.97	2335.4	2625.3	0.943 57	6.8236	7.7672
40	75.862	.001 026 3	3.994	318.33	2158.7	2477	318.37	2318.4	2636.8	1.0257	6.6428	7.6685
50	81.322	.001 029 9	3.2408	341.22	2142.6	2483.9	341.27	2304.6	2645.9	1.0908	6.5016	7.5923
60	85.931	.001 033 1	2.7324	360.53	2129	2489.6	360.6	2292.9	2653.5	1.1449	6.3855	7.5304
70	89.937	.001 036	2.3653	377.32	2117.2	2494.5	377.39	2282.7	2660	1.1915	6.2868	7.4783
80	93.49	.001 038 6	2.0875	392.21	2106.6	2498.8	392.29	2273.5	2665.8	1.2323	6.2009	7.4332
90	96.69	.001 041	1.8697	405.63	2097	2502.6	405.72	2265.2	2670.9	1.2688	6.1248	7.3936
100	99.608	.001 043 3	1.6942	417.86	2088.2	2506.1	417.97	2257.6	2675.5	1.3018	6.0564	7.3582
120	104.78	.001 047 4	1.4286	439.58	2072.6	2512.2	439.71	2243.9	2683.6	1.3598	5.9373	7.2971
140	109.29	.001 051 2	1.2368	458.52	2058.9	2517.4	458.67	2231.9	2690.5	1.4097	5.8359	7.2456
160	113.3	.001 054 6	1.0915	475.37	2046.6	2522	475.54	2221.1	2696.6	1.4536	5.7474	7.201
180	116.91	.001 057 8	0.977 62	490.59	2035.4	2526	490.78	2211.2	2702	1.4929	5.6689	7.1618
200	120.21	.001 060 8	0.885 82	504.5	2025.1	2529.6	504.72	2202.1	2706.8	1.5285	5.5982	7.1267
250	127.4	.001 067 5	0.718 77	534.96	2002.4	2537.4	535.23	2181.9	2717.1	1.6054	5.4471	7.0525

300	133.51	.001 073 4	0.605 86	560.9	1982.9	2543.8	561.22	2164.3	2725.5	1.6698	5.3221	6.9919
350	138.85	.001 078 8	0.524 27	583.62	1965.5	2549.2	584	2148.7	2732.7	1.7254	5.2152	6.9406
400	143.6	.001 083 8	0.462 47	603.93	1949.9	2553.8	604.36	2134.4	2738.8	1.7745	5.1217	6.8961
450	147.89	.001 088 4	0.413 99	622.34	1935.5	2557.9	622.83	2121.4	2744.2	1.8185	5.0383	6.8568
500	151.82	.001 092 8	0.374 89	639.24	1922.3	2561.5	639.78	2109.2	2749	1.8584	4.9631	6.8215
600	158.82	.001 100 8	0.315 67	669.45	1898.2	2567.7	670.11	2087	2757.1	1.929	4.8313	6.7603
700	164.94	.004 108 1	0.272 86	696.01	1876.8	2572.8	696.79	2067	2763.8	1.99	4.7182	6.7082
800	170.4	.001 114 9	0.240 43	719.83	1857.2	2577.1	720.72	2048.7	2769.4	2.044	4.6188	6.6628
900	175.35	.001 121 2	0.214 97	741.48	1839.2	2580.7	742.49	2031.7	2774.2	2.0925	4.53	6.6226
1000	179.88	.001 127 2	0.194 44	761.39	1822.5	2583.9	762.52	2015.8	2778.3	2.1367	4.4496	6.5863
1200	187.96	.001 138 4	0.163 34	797.11	1791.9	2589	798.47	1986.5	2785	2.2147	4.3082	6.5229
1400	195.04	.001 148 7	0.140 85	828.62	1764.4	2593	830.23	1959.9	2790.2	2.2824	4.1862	6.4686
1600	201.38	.001 158 3	0.1238	856.96	1739.1	2596.1	858.82	1935.3	2794.2	2.3425	4.0785	6.4209
1800	207.12	.001 167 5	0.110 43	882.81	1715.7	2598.5	884.91	1912.3	2797.3	2.3965	3.9818	6.3783
2000	212.39	.001 176 3	0.099 638	906.63	1693.7	2600.3	908.98	1890.6	2799.6	2.4458	3.8939	6.3397
2500	223.96	.001 196 8	0.079 989	959.43	1643.8	2603.2	962.42	1840.7	2803.2	2.5532	3.7029	6.2561
3000	233.86	.001 215 9	0.066 694	1005.2	1599	2604.1	1008.8	1795.4	2804.2	2.6443	3.5412	6.1855
3500	242.56	.001 234 2	0.057 079	1045.8	1557.9	2603.8	1050.1	1753.4	2803.5	2.7239	3.4	6.1239
4000	250.35	.001 251 7	0.049 79	1082.7	1519.7	2602.4	1087.7	1713.9	2801.5	2.7951	3.2738	6.0689
4500	257.43	.001 268 8	0.044 067	1116.6	1483.6	2600.2	1122.3	1676.2	2798.5	2.8596	3.1592	6.0188
5000	263.93	.001 285 5	0.039 45	1148.1	1449.2	2597.3	1154.5	1640	2794.6	2.919	3.0536	5.9726
7500	290.51	.001 367 5	0.025 326	1282.1	1293.6	2575.7	1292.4	1473.3	2765.7	3.1642	2.6138	5.778
10 000	310.98	.001 452 1	0.018 029	1393.2	1151.7	2544.9	1407.7	1317.5	2725.2	3.3595	2.2555	5.6149
12 500	327.81	.001 545 8	0.013 499	1491.8	1013.8	2505.6	1511.2	1163.2	2674.3	3.5274	1.9355	5.4629
15 000	342.17	.001 657	0.010 344	1584.5	871.4	2455.9	1609.3	1001.7	2611.1	3.6815	1.6279	5.3094
17 500	354.68	.001 803	0.007 928	1678.2	712.49	2390.7	1709.7	819.68	2529.4	3.8354	1.3056	5.1409
20 000	365.72	.002 037 7	0.005 847	1786.9	507.58	2294.5	1827.7	583.77	2411.5	4.0154	0.913 75	4.9291
22 000	373.68	.002 651 1	0.003 761 2	1951.5	164.15	2115.7	2009.9	188.57	2198.4	4.2985	0.291 53	4.59
22 090	374.14	.003 155	0.003 155	2029.6	0	2029.6	2099.3	0	2099.3	4.4298	0	4.4298

TABLE A.7 SUPERHEATED STEAM VAPOR TABLE (SI UNITS)

p = 10 kPa (45.799°C)

T (°C)	v (m³/kg)	u (kJ/kg)	h (kJ/kg)	s (kJ/kg·K)
50	14.867	2443	2591.6	8.1759
100	17.197	2515.2	2687.1	8.4497
150	19.514	2587.6	2782.7	8.6891
200	21.827	2660.9	2879.2	8.9038
250	24.138	2735.6	2976.9	9.0995
300	26.447	2811.6	3076.1	9.2798
350	28.756	2889.1	3176.7	9.4475
400	31.065	2968.2	3278.9	9.6047
500	35.681	3131.3	3488.1	9.8934
600	40.298	3301.1	3704.1	10.155
700	44.914	3477.9	3927	10.396
800	49.53	3661.7	4157	10.62
900	54.145	3852.6	4394.1	10.831
1000	58.761	4050.8	4638.4	11.03
1100	63.377	4256.2	4889.9	11.22
1200	67.992	4468.9	5148.8	11.401
1300	72.608	4688.9	5415	11.575

p = 50 kPa (81.322°C)

T (°C)	v (m³/kg)	u (kJ/kg)	h (kJ/kg)	s (kJ/kg·K)
100	3.4186	2511	2681.9	7.6958
150	3.8901	2585.3	2779.8	7.9411
200	4.3566	2659.5	2877.4	8.1582
250	4.821	2734.6	2975.7	8.355
300	5.2844	2810.9	3075.1	8.5359
350	5.7473	2888.6	3175.9	8.7039
400	6.2097	2967.8	3278.3	8.8613
500	7.1341	3131	3487.7	9.1503
600	8.058	3300.9	3703.8	9.4122
700	8.9816	3477.7	3926.8	9.6532
800	9.9051	3661.5	4156.8	9.8774
900	10.829	3852.5	4393.9	10.088
1000	11.752	4050.7	4638.3	10.287
1100	12.675	4256.1	4889.8	10.477
1200	13.598	4468.8	5148.7	10.658
1300	14.522	4688.8	5414.9	10.833

p = 100 kPa (99.608°C)

T (°C)	v (m³/kg)	u (kJ/kg)	h (kJ/kg)	s (kJ/kg·K)
100	1.6956	2505.4	2675	7.3609
150	1.9368	2582.5	2776.2	7.6145
200	2.1727	2657.8	2875.1	7.8346
250	2.4064	2733.4	2974.1	8.0328
300	2.6391	2810	3073.9	8.2145
350	2.8711	2887.9	3175	8.383
400	3.1028	2967.3	3277.5	8.5406
500	3.5656	3130.6	3487.2	8.83
600	4.028	3300.6	3703.4	9.092
700	4.4901	3477.5	3926.5	9.3331
800	4.9521	3661.4	4156.6	9.5573
900	5.4139	3852.3	4393.7	9.768
1000	5.8757	4050.5	4638.1	9.9672
1100	6.3375	4256	4889.7	10.157
1200	6.7992	4468.7	5148.6	10.338
1300	7.2609	4688.7	5414.8	10.513

p = 250 kPa (127.40°C)

T (°C)	v (m³/kg)	u (kJ/kg)	h (kJ/kg)	s (kJ/kg·K)
150	0.7644	2573.3	2764.4	7.1697
200	0.862 24	2652.4	2867.9	7.4002
250	0.957 61	2729.8	2969.2	7.603
300	1.0518	2807.4	3070.3	7.787
350	1.1455	2885.9	3172.2	7.9569
400	1.2387	2965.6	3275.3	8.1154
500	1.4246	3129.5	3485.6	8.4056
600	1.61	3299.8	3702.3	8.6682
700	1.7952	3476.8	3925.6	8.9095
800	1.9802	3660.8	4155.8	9.134
900	2.1652	3851.8	4393.1	9.3447
1000	2.35	4050.1	4637.6	9.5441
1100	2.5348	4255.6	4889.3	9.7338
1200	2.7196	4468.3	5148.2	9.9153
1300	2.9044	4688.4	5414.5	10.09

p = 500 kPa (151.82°C)

T (°C)	v (m³/kg)	u (kJ/kg)	h (kJ/kg)	s (kJ/kg·K)
200	0.425 13	2642.7	2855.3	7.0597
250	0.474 56	2723.5	2960.8	7.271
300	0.522 72	2802.9	3064.2	7.4592
350	0.570 21	2882.4	3167.5	7.6315
400	0.617 32	2962.9	3271.5	7.7914
500	0.710 9	3127.6	3483	8.0833
600	0.804 02	3298.3	3700.3	8.3467
700	0.896 89	3475.6	3924.1	8.5885
800	0.989 61	3659.8	4154.6	8.8132
900	1.0822	3851	4392.1	9.0241
1000	1.1748	4049.4	4636.8	9.2236
1100	1.2673	4254.9	4888.6	9.4135
1200	1.3598	4467.8	5147.7	9.5951
1300	1.4522	4687.9	5414	9.7695

p = 1000 kPa (179.88°C)

T (°C)	v (m³/kg)	u (kJ/kg)	h (kJ/kg)	s (kJ/kg·K)
200	0.206 02	2621	2827.1	6.6929
250	0.232 83	2710	2942.9	6.925
300	0.258 06	2793.5	3051.6	7.1228
350	0.282 54	2875.4	3157.9	7.3002
400	0.3066	2957.3	3263.9	7.4633
500	0.354 05	3123.8	3477.8	7.7585
600	0.401 02	3295.4	3696.5	8.0235
700	0.447 74	3473.3	3921.1	8.2662
800	0.4943	3657.9	4152.2	8.4916
900	0.540 77	3849.4	4390.2	8.7029
1000	0.587 18	4047.9	4635.1	8.9026
1100	0.633 54	4253.7	4887.2	9.0927
1200	0.679 86	4466.6	5146.5	9.2744
1300	0.726 16	4686.9	5413	9.449

p = 1500 kPa (198.29°C)

T (°C)	v	u	h	s
200	0.132 43	2595.9	2794.5	6.4505
250	0.152 06	2695.4	2923.5	6.7093
300	0.169 77	2783.6	3038.3	6.9183
350	0.186 62	2868.1	3148.1	7.1014
400	0.203 01	2951.7	3256.2	7.2677
500	0.2351	3119.9	3472.5	7.5664
600	0.266 68	3292.5	3692.6	7.8331
700	0.298 02	3471	3918	8.0767
800	0.3292	3656	4149.8	8.3027
900	0.360 29	3847.7	4388.2	8.5144
1000	0.391 31	4046.5	4633.5	8.7144
1100	0.422 28	4252.4	4885.8	8.9046
1200	0.453 22	4465.5	5145.3	9.0865
1300	0.484 13	4685.8	5412	9.2612

p = 2500 kPa (223.96°C)

T (°C)	v	u	h	s
250	0.087 05	2662.2	2879.8	6.4076
300	0.098 973	2762.4	3009.8	6.6446
350	0.1098	2852.9	3127.5	6.8409
400	0.120 11	2940	3240.3	7.0145
500	0.139 92	3112.1	3461.9	7.3204
600	0.159 21	3286.7	3684.7	7.5906
700	0.178 24	3466.4	3912	7.8362
800	0.197 12	3652.1	4144.9	8.0633
900	0.2159	3844.5	4384.2	8.2759
1000	0.234 61	4043.6	4630.2	8.4764
1100	0.253 27	4249.8	4883	8.6671
1200	0.271 91	4463.2	5143	8.8492
1300	0.290 51	4683.7	5410	9.0242

p = 5000 kPa (263.93°C)

T (°C)	v	u	h	s
300	0.045 337	2698.5	2925.2	6.2085
350	0.051 976	2810.6	3070.4	6.4511
400	0.057 822	2908.7	3197.8	6.6474
500	0.068 512	3091.9	3434.4	6.9743
600	0.078 601	3271.9	3664.9	7.2538
700	0.088 407	3454.7	3896.8	7.5044
800	0.098 055	3642.5	4132.8	7.7345
900	0.1076	3836.3	4374.3	7.949
1000	0.117 09	4036.5	4621.9	8.1509
1100	0.126 52	4243.5	4876.1	8.3426
1200	0.135 92	4457.5	5137.1	8.5255
1300	0.1453	4678.5	5405	8.701

p = 10 000 kPa (310.98°C)

T (°C)	v	u	h	s
350	0.022 43	2700.3	2924.6	5.9448
400	0.026 427	2835.9	3100.2	6.2156
500	0.032 747	3049	3376.5	6.5982
600	0.038 276	3241.5	3624.3	6.8989
700	0.043 484	3431.1	3865.9	7.1601
800	0.048 521	3623.1	4108.3	7.3966
900	0.053 456	3819.8	4354.4	7.6151
1000	0.058 323	4022.1	4605.4	7.8198
1100	0.063 143	4230.8	4862.2	8.0134
1200	0.067 929	4446	5125.3	8.1979
1300	0.072 687	4668.1	5395	8.3745

p = 15 000 kPa (342.17°C)

T (°C)	v	u	h	s
350	0.011 667	2534	2709	5.4668
400	0.015 668	2744.1	2979.1	5.8846
500	0.020 773	3002.5	3314.1	6.3496
600	0.024 82	3210	3582.3	6.6753
700	0.028 504	3407.1	3834.6	6.9482
800	0.032 008	3603.6	4083.7	7.1912
900	0.035 407	3803.3	4334.4	7.4139
1000	0.038 736	4007.8	4588.8	7.6214
1100	0.042 018	4218	4848.3	7.8171
1200	0.045 264	4434.6	5113.5	8.003
1300	0.048 485	4657.7	5385	8.1808

p = 20 000 kPa (365.72°C)

T (°C)	v	u	h	s
400	0.009 971 7	2623.8	2823.2	5.5595
500	0.014 748	2951.9	3246.9	6.1497
600	0.018 082	3177.5	3539.1	6.5049
700	0.021 011	3382.7	3802.9	6.7903
800	0.023 751	3583.9	4058.9	7.0401
900	0.026 381	3786.7	4314.3	7.267
1000	0.028 942	3993.4	4572.2	7.4774
1100	0.031 455	4205.3	4834.4	7.675
1200	0.033 932	4423.1	5101.8	7.8625
1300	0.036 383	4647.3	5375	8.0414

p = 25 000 kPa

T (°C)	v	u	h	s
400	0.006 192 2	2461.7	2616.5	5.1932
500	0.011 104	2896.7	3174.3	5.9731
600	0.014 035	3144	3494.9	6.3632
700	0.016 513	3358	3770.8	6.6618
800	0.018 795	3564.1	4034	6.9186
900	0.020 966	3770	4294.2	7.1498
1000	0.023 066	3979	4555.6	7.3631
1100	0.025 117	4192.6	4820.5	7.5628
1200	0.027 133	4411.7	5090	7.7517
1300	0.029 123	4636.9	5365	7.9318

p = 30 000 kPa

T (°C)	v	u	h	s
500	0.008 652 1	2836.1	3095.7	5.808
600	0.011 335	3109.7	3449.8	6.2394
700	0.013 514	3333	3738.5	6.5519
800	0.015 491	3544.2	4008.9	6.8159
900	0.017 355	3753.4	4274	7.0514
1000	0.019 148	3964.6	4539	7.2676
1100	0.020 892	4179.8	4806.6	7.4693
1200	0.0226	4400.2	5078.2	7.6598
1300	0.024 282	4626.5	5354.9	7.8411

p = 35 000 kPa

T (°C)	v	u	h	s
500	0.006 883 2	2769.2	3010.2	5.6475
600	0.009 407 8	3074.7	3404	6.1278
700	0.011 372	3307.9	3705.9	6.4548
800	0.013 131	3524.2	3983.8	6.726
900	0.014 776	3736.6	4253.8	6.966
1000	0.016 35	3950.1	4522.4	7.1851
1100	0.017 874	4167.1	4792.6	7.3889
1200	0.019 363	4388.8	5066.5	7.5809
1300	0.020 825	4616.1	5344.9	7.7633

TABLE A.8 COMPRESSED LIQUID TABLE (SI UNITS)

T (°C)	v (m³/kg)	u (kJ/kg)	h (kJ/kg)	s (kJ/kg·K)	v (m³/kg)	u (kJ/kg)	h (kJ/kg)	s (kJ/kg·K)	v (m³/kg)	u (kJ/kg)	h (kJ/kg)	s (kJ/kg·K)
	p = 5 MPa (263.99°C)				p = 10 MPa (311.06°C)				p = 15 MPa (342.24°C)			
Sat.	0.001 285 9	1147.8	1154.2	2.9202	0.001 452 4	1393.0	1407.6	3.3596	0.001 658 1	1585.6	1610.5	3.6848
0	0.000 997 7	0.04	5.04	0.0001	0.000 995 2	0.09	10.04	0.0002	0.000 992 8	0.15	15.05	0.004
20	0.000 999 5	83.65	88.65	0.2956	0.000 997 2	83.36	93.33	0.2945	0.000 995 0	83.06	97.99	0.2934
40	0.001 005 6	166.95	171.97	0.5705	0.001 003 4	166.35	176.38	0.5686	0.001 001 3	165.76	180.78	0.5666
60	0.001 014 9	250.23	255.30	0.8285	0.001 012 7	249.36	259.49	0.8258	0.001 010 5	248.51	263.67	0.8232
80	0.001 026 8	333.72	338.85	1.0720	0.001 024 5	332.59	342.83	1.0688	0.001 022 2	331.48	346.81	1.0656
100	0.001 041 0	417.52	422.72	1.3030	0.001 038 5	416.12	426.50	1.2992	0.001 036 1	414.74	430.28	1.2955
120	0.001 057 6	501.80	507.09	1.5233	0.001 054 9	500.08	510.64	1.5189	0.001 052 2	498.40	514.19	1.5145
140	0.001 076 8	586.76	592.15	1.7343	0.001 073 7	584.68	595.42	1.7292	0.001 070 7	582.66	598.72	1.7242
160	0.001 098 8	672.62	678.12	1.9375	0.001 095 3	670.13	681.08	1.9317	0.001 091 8	667.71	684.09	1.9260
180	0.001 124 0	759.63	765.25	2.1341	0.001 119 9	756.65	767.84	2.1275	0.001 115 9	753.76	770.50	2.1210
200	0.001 153 0	848.1	853.9	2.3255	0.001 148 0	844.5	856.0	2.3178	0.001 143 3	841.0	858.2	2.3104
220	0.001 186 6	938.4	944.4	2.5128	0.001 180 5	934.1	945.9	2.5039	0.001 174 8	929.9	947.5	2.4953
240	0.001 226 4	1031.4	1037.5	2.6979	0.001 218 7	1026.0	1038.1	2.6872	0.001 211 4	1020.8	1039.0	2.6771
260	0.001 274 9	1127.9	1134.3	2.8830	0.001 264 5	1121.1	1133.7	2.8699	0.001 255 0	1114.6	1133.4	2.8576
280					0.001 321 6	1220.9	1234.1	3.0548	0.001 308 4	1212.5	1232.1	3.0393
300					0.001 397 2	1328.4	1342.3	3.2469	0.001 377 0	1316.6	1337.3	3.2260
320									0.001 472 4	1431.1	1453.2	3.4247
340									0.001 631 1	1567.5	1591.9	3.6546

T	p = 20 MPa (365.81 °C)				p = 30 MPa				p = 50 MPa			
Sat.	0.002 036	1785.6	1826.3	4.0139								
0	0.000 990 4	0.19	20.01	0.0004	0.000 985 6	0.25	29.82	0.0001	0.000 976 6	0.20	49.03	0.0014
20	0.000 992 8	82.77	102.62	0.2923	0.000 988 6	82.17	111.84	0.2899	0.000 980 4	81.00	130.02	0.2848
40	0.000 999 2	165.17	185.16	0.5646	0.000 995 1	164.04	193.89	0.5607	0.000 987 2	161.86	211.21	0.5527
60	0.001 008 4	247.68	267.85	0.8206	0.001 004 2	246.06	276.19	0.8154	0.000 996 2	242.98	292.79	0.8052
80	0.001 019 9	330.40	350.80	1.0624	0.001 015 6	328.30	358.77	1.0561	0.001 007 3	324.34	374.70	1.0440
100	0.001 033 7	413.39	434.06	1.2917	0.001 029 0	410.78	441.66	1.2844	0.001 020 1	405.88	456.89	1.2703
120	0.001 049 6	496.76	517.76	1.5102	0.001 044 5	493.59	524.93	1.5018	0.001 034 8	487.65	539.39	1.4857
140	0.001 067 8	580.69	602.04	1.7193	0.001 062 1	576.88	608.75	1.7098	0.001 051 5	569.77	622.35	1.6915
160	0.001 088 5	665.35	687.12	1.9204	0.001 082 1	660.82	693.28	1.9096	0.001 070 3	652.41	705.92	1.8891
180	0.001 112 0	750.95	773.20	2.1147	0.001 104 7	745.59	778.73	2.1024	0.001 091 2	735.69	790.25	2.0794
200	0.001 138 8	837.7	860.5	2.3031	0.001 130 2	831.4	865.3	2.2893	0.001 114 6	819.7	875.5	2.2634
220	0.001 169 3	925.9	949.3	2.4870	0.001 159 0	918.3	953.1	2.4711	0.001 140 8	904.7	961.7	2.4419
240	0.001 204 6	1016.0	1040.0	2.6674	0.001 192 0	1006.9	1042.6	2.6490	0.001 170 2	990.7	1049.2	2.6158
260	0.001 246 2	1108.6	1133.5	2.8459	0.001 230 3	1097.4	1134.3	2.8243	0.001 203 4	1078.1	1138.2	2.7860
280	0.001 296 5	1204.7	1230.6	3.0248	0.001 275 5	1190.7	1229.0	2.9986	0.001 241 5	1167.2	1229.3	2.9537
300	0.001 359 6	1306.1	1333.3	3.2071	0.001 330 4	1287.9	1327.8	3.1741	0.001 286 0	1258.7	1323.0	3.1200
320	0.001 443 7	1415.7	1444.6	3.3979	0.001 399 7	1390.7	1432.7	3.3539	0.001 338 8	1353.3	1420.2	3.2868
340	0.001 568 4	1539.7	1571.0	3.6075	0.001 492 0	1501.7	1546.5	3.5426	0.001 403 2	1452.0	1522.1	3.4557
360	0.001 822 6	1702.8	1739.3	3.8772	0.001 626 5	1626.6	1675.4	3.7494	0.001 483 8	1556.0	1630.2	3.6291
380					0.001 869 1	1781.4	1837.5	4.0012	0.001 588 4	1667.2	1746.6	3.8101

Source: Reproduced with permission from J. H. Keenan, F. G. Keyes, P. G. Hill, and J. G. Moore, *Steam Tables*, New York: John Wiley & Sons, 1969.

TABLE A.9 SATURATED AMMONIA TABLE (SI Units)

T (°C)	p (kPa)	Specific volume (m³/kg)			Enthalpy (kJ/kg)			Entropy (kJ/kg·K)		
		Sat. liquid, v_f	Evap., v_{fg}	Sat. vapor, v_g	Sat. liquid, h_f	Evap., h_{fg}	Sat. vapor, h_g	Sat. liquid, s_f	Evap., s_{fg}	Sat. vapor, s_g
−50	40.88	0.001 424	2.6239	2.6254	−44.3	1416.7	1372.4	−0.1942	6.3502	6.1561
−48	45.96	0.001 429	2.3518	2.3533	−35.5	1411.3	1375.8	−0.1547	6.2696	6.1149
−46	51.55	0.001 434	2.1126	2.1140	−26.6	1405.8	1379.2	−0.1156	6.1902	6.0746
−44	57.69	0.001 439	1.9018	1.9032	−17.8	1400.3	1382.5	−0.0768	6.1120	6.0352
−42	64.42	0.001 444	1.7155	1.7170	−8.9	1394.7	1385.8	−0.0382	6.0349	5.9967
−40	71.77	0.001 449	1.5506	1.5521	0.0	1389.0	1389.0	0.0000	5.9589	5.9589
−38	79.80	0.001 454	1.4043	1.4058	8.9	1383.3	1392.2	0.0380	5.8840	5.9220
−36	88.54	0.001 460	1.2742	1.2757	17.8	1377.6	1395.4	0.0757	5.8101	5.8858
−34	98.05	0.001 465	1.1582	1.1597	26.8	1371.8	1398.5	0.1132	5.7372	5.8504
−32	108.37	0.001 470	1.0547	1.0562	35.7	1365.9	1401.6	0.1504	5.6652	5.8156
−30	119.55	0.001 476	0.9621	0.9635	44.7	1360.0	1404.6	0.1873	5.5942	5.7815
−28	131.64	0.001 481	0.8790	0.8805	53.6	1354.0	1407.6	0.2240	5.5241	5.7481
−26	144.70	0.001 487	0.8044	0.8059	62.6	1347.9	1410.5	0.2605	5.4548	5.7153
−24	158.78	0.001 492	0.7373	0.7388	71.6	1341.8	1413.4	0.2967	5.3864	5.6831
−22	173.93	0.001 498	0.6768	0.6783	80.7	1335.6	1416.2	0.3327	5.3188	5.6515
−20	190.22	0.001 504	0.6222	0.6237	89.7	1329.3	1419.0	0.3684	5.2520	5.6205
−18	207.71	0.001 510	0.5728	0.5743	98.8	1322.9	1421.7	0.4040	5.1860	5.5900
−16	226.45	0.001 515	0.5280	0.5296	107.8	1316.5	1424.4	0.4393	5.1207	5.5600
−14	246.51	0.001 521	0.4874	0.4889	116.9	1310.0	1427.0	0.4744	5.0561	5.5305
−12	267.95	0.001 528	0.4505	0.4520	126.0	1303.5	1429.5	0.5093	4.9922	5.5015
−10	290.85	0.001 534	0.4169	0.4185	135.2	1296.8	1432.0	0.5440	4.9290	5.4730
−8	315.25	0.001 540	0.3863	0.3878	144.3	1290.1	1434.4	0.5785	4.8664	5.4449
−6	341.25	0.001 546	0.3583	0.3599	153.5	1283.3	1436.8	0.6128	4.8045	5.4173
−4	368.90	0.001 553	0.3328	0.3343	162.7	1276.4	1439.1	0.6469	4.7432	5.3901

Temp.										
−2	398.27	0.001 559	0.3094	0.3109	171.9	1269.4	1441.3	0.6808	4.6825	5.3633
0	429.44	0.001 566	0.2879	0.2895	181.1	1262.4	1443.5	0.7145	4.6223	5.3369
2	462.49	0.001 573	0.2683	0.2698	190.4	1255.2	1445.6	0.7481	4.5627	5.3108
4	497.49	0.001 580	0.2502	0.2517	199.6	1248.0	1447.6	0.7815	4.5037	5.2852
6	534.51	0.001 587	0.2335	0.2351	208.9	1240.6	1449.6	0.8148	4.4451	5.2599
8	573.64	0.001 594	0.2182	0.2198	218.3	1233.2	1451.5	0.8479	4.3871	5.2350
10	614.95	0.001 601	0.2040	0.2056	227.6	1225.7	1453.3	0.8808	4.3295	5.2104
12	658.52	0.001 608	0.1910	0.1926	237.0	1218.1	1455.1	0.9136	4.2725	5.1861
14	704.44	0.001 616	0.1789	0.1805	246.4	1210.4	1456.8	0.9463	4.2159	5.1621
16	752.79	0.001 623	0.1677	0.1693	255.9	1202.6	1458.5	0.9788	4.1597	5.1385
18	803.66	0.001 631	0.1574	0.1590	265.4	1194.7	1460.0	1.0112	4.1039	5.1151
20	857.12	0.001 639	0.1477	0.1494	274.9	1186.7	1461.5	1.0434	4.0486	5.0920
22	913.27	0.001 647	0.1388	0.1405	284.4	1178.5	1462.9	1.0755	3.9937	5.0692
24	972.19	0.001 655	0.1305	0.1322	294.0	1170.3	1464.3	1.1075	3.9392	5.0467
26	1033.97	0.001 663	0.1228	0.1245	303.6	1162.0	1465.6	1.1394	3.8850	5.0244
28	1098.71	0.001 671	0.1156	0.1173	313.2	1153.6	1466.8	1.1711	3.8312	5.0023
30	1166.49	0.001 680	0.1089	0.1106	322.9	1145.0	1467.9	1.2028	3.7777	4.9805
32	1237.41	0.001 689	0.1027	0.1044	332.6	1136.4	1469.0	1.2343	3.7246	4.9589
34	1311.55	0.001 698	0.0969	0.0986	342.3	1127.6	1469.9	1.2656	3.6718	4.9374
36	1389.03	0.001 707	0.0914	0.0931	352.1	1118.7	1470.8	1.2969	3.6192	4.9161
38	1469.92	0.001 716	0.0863	0.0880	361.9	1109.7	1471.5	1.3281	3.5669	4.8950
40	1554.33	0.001 726	0.0815	0.0833	371.7	1100.5	1472.2	1.3591	3.5148	4.8740
42	1642.35	0.001 735	0.0771	0.0788	381.6	1091.2	1472.8	1.3901	3.4630	4.8530
44	1734.09	0.001 745	0.0728	0.0746	391.5	1081.7	1473.2	1.4209	3.4112	4.8322
46	1829.65	0.001 756	0.0689	0.0707	401.5	1072.0	1473.5	1.4518	3.3595	4.8113
48	1929.13	0.001 766	0.0652	0.0669	411.5	1062.2	1473.7	1.4826	3.3079	4.7905
50	2032.62	0.001 777	0.0617	0.0635	421.7	1052.0	1473.7	1.5135	3.2561	4.7696

Source: National Bureau of Standards Circular no. 142, *Tables of Thermodynamic Properties of Ammonia,* Extracted by permission.

TABLE A.10 SUPERHEATED AMMONIA TABLE (SI Units)

p (kPa) (Sat. temp., °C)		−20	−10	0	10	20	30	40	50	60	70	80	100
									Temperature (°C)				
50 (−46.54)	v	2.4474	2.5481	2.6482	2.7479	2.8473	2.9464	3.0453	3.1441	3.2427	3.3413	3.4397	
	h	1435.8	1457.0	1478.1	1499.2	1520.4	1541.7	1563.0	1584.5	1606.1	1627.8	1649.7	
	s	6.3256	6.4077	6.4865	6.5625	6.6360	6.7073	6.7766	6.8441	6.9099	6.9743	7.0372	
75 (−39.18)	v	1.6233	1.6915	1.7591	1.8263	1.8932	1.9597	2.0261	2.0923	2.1584	2.2244	2.2903	
	h	1433.0	1454.7	1476.1	1497.5	1518.9	1540.3	1561.8	1583.4	1605.1	1626.9	1648.9	
	s	6.1190	6.2028	6.2828	6.3597	6.4339	6.5058	6.5756	6.6434	6.7096	6.7742	6.8373	
100 (−33.61)	v	1.2110	1.2631	1.3145	1.3654	1.4160	1.4664	1.5165	1.5664	1.6163	1.6659	1.7155	1.8145
	h	1430.1	1452.2	1474.1	1495.7	1517.3	1538.9	1560.5	1582.2	1604.1	1626.0	1648.0	1692.6
	s	5.9695	6.0552	6.1366	6.2144	6.2894	6.3618	6.4321	6.5003	6.5668	6.6316	6.6950	6.8177
125 (−29.08)	v	0.9635	1.0059	1.0476	1.0889	1.1297	1.1703	1.2107	1.2509	1.2909	1.3309	1.3707	1.4501
	h	1427.2	1449.8	1472.0	1493.9	1515.7	1537.5	1559.3	1581.1	1603.0	1625.0	1647.2	1691.8
	s	5.8512	5.9389	6.0217	6.1006	6.1763	6.2494	6.3201	6.3887	6.4555	6.5206	6.5842	6.7072
150 (−25.23)	v	0.7984	0.8344	0.8697	0.9045	0.9388	0.9729	1.0068	1.0405	1.0740	1.1074	1.1408	1.2072
	h	1424.1	1447.3	1469.8	1492.1	1514.1	1536.1	1558.0	1580.0	1602.0	1624.1	1646.3	1691.1
	s	5.7526	5.8424	5.9266	6.0066	6.0831	6.1568	6.2280	6.2970	6.3641	6.4295	6.4933	6.6167
200 (−18.86)	v		0.6199	0.6471	0.6738	0.7001	0.7261	0.7519	0.7774	0.8029	0.8282	0.8533	0.9035
	h		1442.0	1465.5	1488.4	1510.9	1533.2	1555.5	1577.7	1599.9	1622.2	1644.6	1689.6
	s		5.6863	5.7737	5.8559	5.9342	6.0091	6.0813	6.1512	6.2189	6.2849	6.3491	6.4732
250 (−13.67)	v		0.4910	0.5135	0.5354	0.5568	0.5780	0.5989	0.6196	0.6401	0.6605	0.6809	0.7212
	h		1436.6	1461.0	1484.5	1507.6	1530.3	1552.9	1575.4	1597.8	1620.3	1642.8	1688.2
	s		5.5609	5.6517	5.7365	5.8165	5.8928	5.9661	6.0368	6.1052	6.1717	6.2365	6.3613
300 (−9.23)	v			0.4243	0.4430	0.4613	0.4792	0.4968	0.5143	0.5316	0.5488	0.5658	0.5997
	h			1456.3	1480.6	1504.2	1527.4	1550.3	1573.0	1595.7	1618.4	1641.1	1686.7
	s			5.5493	5.6366	5.7186	5.7963	5.8707	5.9423	6.0114	6.0785	6.1437	6.2693
350 (−5.35)	v			0.3605	0.3770	0.3929	0.4086	0.4239	0.4391	0.4541	0.4689	0.4837	0.5129
	h			1451.5	1476.5	1500.7	1524.4	1547.6	1570.7	1593.6	1616.5	1639.3	1685.2
	s			5.4600	5.5502	5.6342	5.7135	5.7890	5.8615	5.9314	5.9990	6.0647	6.1910
400 (−1.89)	v			0.3125	0.3274	0.3417	0.3556	0.3692	0.3826	0.3959	0.4090	0.4220	0.4478
	h			1446.5	1472.4	1497.2	1521.3	1544.9	1568.3	1591.5	1614.5	1637.6	1683.7
	s			5.3803	5.4735	5.5597	5.6405	5.7173	5.7907	5.8613	5.9296	5.9957	6.1228
450 (1.26)	v			0.2752	0.2887	0.3017	0.3143	0.3266	0.3387	0.3506	0.3624	0.3740	0.3971
	h			1441.3	1468.1	1493.6	1518.2	1542.2	1565.9	1589.3	1612.6	1635.8	1682.2
	s			5.3078	5.4042	5.4926	5.5752	5.6532	5.7275	5.7989	5.8678	5.9345	6.0623

		20	30	40	50	60	70	80	100	120	140	160	180
500	v	0.2698	0.2813	0.2926	0.3036	0.3144	0.3251	0.3357	0.3565	0.3771	0.3975		
(4.14)	h	1489.9	1515.0	1539.5	1563.4	1587.1	1610.6	1634.0	1680.7	1727.5	1774.7		
	s	5.4314	5.5157	5.5950	5.6704	5.7425	5.8120	5.8793	6.0079	6.1301	6.2472		
600	v	0.2217	0.2317	0.2414	0.2508	0.2600	0.2691	0.2781	0.2957	0.3130	0.3302		
(9.29)	h	1482.4	1508.6	1533.8	1558.5	1582.7	1606.6	1630.4	1677.7	1724.9	1772.4		
	s	5.322	5.4102	5.4923	5.5697	5.6436	5.7144	5.7826	5.9129	6.0363	6.1541		
700	v	0.1874	0.1963	0.2048	0.2131	0.2212	0.2291	0.2369	0.2522	0.2672	0.2821		
(13.81)	h	1474.5	1501.9	1528.1	1553.4	1578.2	1602.6	1626.8	1674.6	1722.4	1770.2		
	s	5.2259	5.3179	5.4029	5.4826	5.5582	5.6303	5.6997	5.8316	5.9562	6.0749		
800	v	0.1615	0.1696	0.1773	0.1848	0.1920	0.1991	0.2060	0.2196	0.2329	0.2459	0.2589	
(17.86)	h	1466.3	1495.0	1522.2	1548.3	1573.7	1598.6	1623.1	1671.6	1719.8	1768.0	1816.4	
	s	5.1387	5.2351	5.3232	5.4053	5.4827	5.5562	5.6268	5.7603	5.8861	6.0057	6.1202	
900	v		0.1488	0.1559	0.1627	0.1693	0.1757	0.1820	0.1942	0.2061	0.2178	0.2294	
(21.54)	h		1488.0	1516.2	1543.0	1569.1	1594.4	1619.4	1668.5	1717.1	1765.7	1814.4	
	s		5.1593	5.2508	5.3354	5.4147	5.4897	5.5614	5.6968	5.8237	5.9442	6.0594	
1000	v		0.1321	0.1388	0.1450	0.1511	0.1570	0.1627	0.1739	0.1847	0.1954	0.2058	0.2162
(24.91)	h		1480.6	1510.0	1537.7	1564.4	1590.3	1615.6	1665.4	1714.5	1763.4	1812.4	1861.7
	s		5.0889	5.1840	5.2713	5.3525	5.4292	5.5021	5.6392	5.7674	5.8888	6.0047	6.1159
1200	v			0.1129	0.1185	0.1238	0.1289	0.1338	0.1434	0.1526	0.1616	0.1705	0.1792
(30.96)	h			1497.1	1526.6	1554.7	1581.7	1608.0	1659.2	1709.2	1758.9	1808.5	1858.2
	s			5.0629	5.1560	5.2416	5.3215	5.3970	5.5379	5.6687	5.7919	5.9091	6.0214
1400	v			0.0944	0.0995	0.1042	0.1088	0.1132	0.1216	0.1297	0.1376	0.1452	0.1528
(36.28)	h			1483.4	1515.1	1544.7	1573.0	1600.2	1652.8	1703.9	1754.3	1804.5	1854.7
	s			4.9534	5.0530	5.1434	5.2270	5.3053	5.4501	5.5836	5.7087	5.8273	5.9406
1600	v				0.0851	0.0895	0.0937	0.0977	0.1053	0.1125	0.1195	0.1263	0.1330
(41.05)	h				1502.9	1534.4	1564.0	1592.3	1646.4	1698.5	1749.7	1800.5	1851.2
	s				4.9584	5.0543	5.1419	5.2232	5.3722	5.5084	5.6355	5.7555	5.8699
1800	v				0.0739	0.0781	0.0820	0.0856	0.0926	0.0992	0.1055	0.1116	0.1177
(45.39)	h				1490.0	1523.5	1554.6	1584.1	1639.8	1693.1	1745.1	1796.5	1847.7
	s				4.8693	4.9715	5.0635	5.1482	5.3018	5.4409	5.5699	5.6914	5.8069
2000	v				0.0648	0.0688	0.0725	0.0760	0.0824	0.0885	0.0943	0.0999	0.1054
(49.38)	h				1476.1	1512.0	1544.9	1575.6	1633.2	1687.6	1740.4	1792.4	1844.1
	s				4.7834	4.8930	4.9902	5.0786	5.2371	5.3793	5.5104	5.6333	5.7499

Source: National Bureau of Standards Circular no. 142, *Tables of Thermodynamic Properties of Ammonia.* Extracted by permission.

Note: Units are v, m^3/kg; h, kJ/kg; s, kJ/kg-K.

TABLE A.11 SATURATED REFRIGERANT-12 TABLE (SI Units)

T (°C)	p (MPa)	Specific volume (m³/kg)			Enthalpy (kJ/kg)			Entropy (kJ/kg·K)		
		Sat. liquid, v_f	Evap., v_{fg}	Sat. vapor, v_g	Sat. liquid, h_f	Evap., h_{fg}	Sat. vapor, h_g	Sat. liquid, s_f	Evap., s_{fg}	Sat. vapor, s_g
−90	0.0028	0.000 608	4.414 937	4.415 545	−43.243	189.618	146.375	−0.2084	1.0352	0.8268
−85	0.0042	0.000 612	3.036 704	3.037 316	−38.968	187.608	148.640	−0.1854	0.9970	0.8116
−80	0.0062	0.000 617	2.137 728	2.138 345	−34.688	185.612	150.924	−0.1630	0.9609	0.7979
−75	0.0088	0.000 622	1.537 030	1.537 651	−30.401	183.625	153.224	−0.1411	0.9266	0.7855
−70	0.0123	0.000 627	1.126 654	1.127 280	−26.103	181.640	155.536	−0.1197	0.8940	0.7744
−65	0.0168	0.000 632	0.840 534	0.841 166	−21.793	179.651	157.857	−0.0987	0.8630	0.7643
−60	0.0226	0.000 637	0.637 274	0.637 910	−17.469	177.653	160.184	−0.0782	0.8334	0.7552
−55	0.0300	0.000 642	0.490 358	0.491 000	−13.129	175.641	162.512	−0.0581	0.8051	0.7470
−50	0.0391	0.000 648	0.382 457	0.383 105	−8.772	173.611	164.840	−0.0384	0.7779	0.7396
−45	0.0504	0.000 654	0.302 029	0.302 682	−4.396	171.558	167.163	−0.0190	0.7519	0.7329
−40	0.0642	0.000 659	0.241 251	0.241 910	−0.000	169.479	169.479	−0.0000	0.7269	0.7269
−35	0.0807	0.000 666	0.194 732	0.195 398	4.416	167.368	171.784	0.0187	0.7027	0.7214
−30	0.1004	0.000 672	0.158 703	0.159 375	8.854	165.222	174.076	0.0371	0.6795	0.7165
−25	0.1237	0.000 679	0.130 487	0.131 166	13.315	163.037	176.352	0.0552	0.6570	0.7121
−20	0.1509	0.000 685	0.108 162	0.108 847	17.800	160.810	178.610	0.0730	0.6352	0.7082
−15	0.1826	0.000 693	0.090 326	0.091 018	22.312	158.534	180.846	0.0906	0.6141	0.7046
−10	0.2191	0.000 700	0.075 946	0.076 646	26.851	156.207	183.058	0.1079	0.5936	0.7014
−5	0.2610	0.000 708	0.064 255	0.064 963	31.420	153.823	185.243	0.1250	0.5736	0.6986
0	0.3086	0.000 716	0.054 673	0.055 389	36.022	151.376	187.397	0.1418	0.5542	0.6960

5	0.3626	0.000 724	0.046 761	0.047 485	40.659	148.859	189.518	0.1585	0.5351	0.6937
10	0.4233	0.000 733	0.040 180	0.040 914	45.337	146.265	191.602	0.1750	0.5165	0.6916
15	0.4914	0.000 743	0.034 671	0.035 413	50.058	143.586	193.644	0.1914	0.4983	0.6897
20	0.5673	0.000 752	0.030 028	0.030 780	54.828	140.812	195.641	0.2076	0.4803	0.6879
25	0.6516	0.000 763	0.026 091	0.026 854	59.653	137.933	197.586	0.2237	0.4626	0.6863
30	0.7449	0.000 774	0.022 734	0.023 508	64.539	134.936	199.475	0.2397	0.4451	0.6848
35	0.8477	0.000 786	0.019 855	0.020 641	69.494	131.805	201.299	0.2557	0.4277	0.6834
40	0.9607	0.000 798	0.017 373	0.018 171	74.527	128.525	203.051	0.2716	0.4104	0.6820
45	1.0843	0.000 811	0.015 220	0.016 032	79.647	125.074	204.722	0.2875	0.3931	0.6806
50	1.2193	0.000 826	0.013 344	0.014 170	84.868	121.430	206.298	0.3034	0.3758	0.6792
55	1.3663	0.000 841	0.011 701	0.012 542	90.201	117.565	207.766	0.3194	0.3582	0.6777
60	1.5259	0.000 858	0.010 253	0.011 111	95.665	113.443	209.109	0.3355	0.3405	0.6760
65	1.6988	0.000 877	0.008 971	0.009 847	101.279	109.024	210.303	0.3518	0.3224	0.6742
70	1.8858	0.000 897	0.007 828	0.008 725	107.067	104.255	211.321	0.3683	0.3038	0.6721
75	2.0874	0.000 920	0.006 802	0.007 723	113.058	99.068	212.126	0.3851	0.2845	0.6697
80	2.3046	0.000 946	0.005 875	0.006 821	119.291	93.373	212.665	0.4023	0.2644	0.6667
85	2.5380	0.000 976	0.005 029	0.006 005	125.818	87.047	212.865	0.4201	0.2430	0.6631
90	2.7885	0.001 012	0.004 246	0.005 258	132.708	79.907	212.614	0.4385	0.2200	0.6585
95	3.0569	0.001 056	0.003 508	0.004 563	140.068	71.658	211.726	0.4579	0.1946	0.6526
100	3.3440	0.001 113	0.002 790	0.003 903	148.076	61.768	209.843	0.4788	0.1655	0.6444
105	3.6509	0.001 197	0.002 045	0.003 242	157.085	49.014	206.099	0.5023	0.1296	0.6319
110	3.9784	0.001 364	0.001 098	0.002 462	168.059	28.425	196.484	0.5322	0.0742	0.6064
112	4.1155	0.001 792	0.000 005	0.001 797	174.920	0.151	175.071	0.5651	0.0004	0.5655

Source: Copyright 1955 and 1956, E. I. du Pont de Nemours & Company, Inc. Reprinted by permission.

TABLE A.12 SUPERHEATED REFRIGERANT-12 TABLE (SI Units)

T (°C)	v (m³/kg)	h (kJ/kg)	s (kJ/kg·K)	v (m³/kg)	h (kJ/kg)	s (kJ/kg·K)	v (m³/kg)	h (kJ/kg)	s (kJ/kg·K)
	p = 0.05 MPa (−45.16°C)			p = 0.10 MPa (−30.09°C)			p = 0.15 MPa (−20.15°C)		
−20.0	0.341 857	181.042	0.7912	0.167 701	179.861	0.7401			
−10.0	0.356 227	186.757	0.8133	0.175 222	185.707	0.7628	0.114 716	184.619	0.7318
0.0	0.370 508	192.567	0.8350	0.182 647	191.628	0.7849	0.119 866	190.660	0.7543
10.0	0.384 716	198.471	0.8562	0.189 994	197.628	0.8064	0.124 932	196.762	0.7763
20.0	0.398 863	204.469	0.8770	0.197 277	203.707	0.8275	0.129 930	202.927	0.7977
30.0	0.412 959	210.557	0.8974	0.204 506	209.866	0.8482	0.134 873	209.160	0.8186
40.0	0.427 012	216.733	0.9175	0.211 691	216.104	0.8684	0.139 768	215.463	0.8390
50.0	0.441 030	222.997	0.9372	0.218 839	222.421	0.8883	0.144 625	221.835	0.8591
60.0	0.455 017	229.344	0.9565	0.225 955	228.815	0.9078	0.149 450	228.277	0.8787
70.0	0.468 978	235.774	0.9755	0.233 044	235.285	0.9269	0.154 247	234.789	0.8980
80.0	0.482 917	242.282	0.9942	0.240 111	241.829	0.9457	0.159 020	241.371	0.9169
90.0	0.496 838	248.868	1.0126	0.247 159	248.446	0.9642	0.163 774	248.020	0.9354

T (°C)	v (m³/kg)	h (kJ/kg)	s (kJ/kg·K)	v (m³/kg)	h (kJ/kg)	s (kJ/kg·K)	v (m³/kg)	h (kJ/kg)	s (kJ/kg·K)
	p = 0.20 MPa (−12.53°C)			p = 0.25 MPa (−6.25°C)			p = 0.30 MPa (−0.86°C)		
0.0	0.088 608	189.669	0.7320	0.069 752	188.644	0.7139	0.057 150	187.583	0.6984
10.0	0.092 550	195.878	0.7543	0.073 024	194.969	0.7366	0.059 984	194.034	0.7216
20.0	0.096 418	202.135	0.7760	0.076 218	201.322	0.7587	0.062 734	200.490	0.7440
30.0	0.100 228	208.446	0.7972	0.079 350	207.715	0.7801	0.065 418	206.969	0.7658
40.0	0.103 989	214.814	0.8178	0.082 431	214.153	0.8010	0.068 049	213.480	0.7869
50.0	0.107 710	221.243	0.8381	0.085 470	220.642	0.8214	0.070 635	220.030	0.8075
60.0	0.111 397	227.735	0.8578	0.088 474	227.185	0.8413	0.073 185	226.627	0.8276
70.0	0.115 055	234.291	0.8772	0.091 449	233.785	0.8608	0.075 705	233.273	0.8473
80.0	0.118 690	240.910	0.8962	0.094 398	240.443	0.8800	0.078 200	239.971	0.8665
90.0	0.122 304	247.593	0.9149	0.097 327	247.160	0.8987	0.080 673	246.723	0.8853
100.0	0.125 901	254.339	0.9332	0.100 238	253.936	0.9171	0.083 127	253.530	0.9038
110.0	0.129 483	261.147	0.9512	0.103 134	260.770	0.9352	0.085 566	260.391	0.9220

p = 0.40 MPa (8.15°C)

p = 0.50 MPa (15.59°C)

p = 0.60 MPa (22.00°C)

t	v	h	s	v	h	s	v	h	s
20.0	0.045 836	198.762	0.7199	0.035 646	196.935	0.6999			
30.0	0.047 971	205.428	0.7423	0.037 464	203.814	0.7230	0.030 422	202.116	0.7063
40.0	0.050 046	212.095	0.7639	0.039 214	210.656	0.7452	0.031 966	209.154	0.7291
50.0	0.052 072	218.779	0.7849	0.040 911	217.484	0.7667	0.033 450	216.141	0.7511
60.0	0.054 059	225.488	0.8054	0.042 565	224.315	0.7875	0.034 887	223.104	0.7723
70.0	0.056 014	232.230	0.8253	0.044 184	231.161	0.8077	0.036 285	230.062	0.7929
80.0	0.057 941	239.012	0.8448	0.045 774	238.031	0.8275	0.037 653	237.027	0.8129
90.0	0.059 846	245.837	0.8638	0.047 340	244.932	0.8467	0.038 995	244.009	0.8324
100.0	0.061 731	252.707	0.8825	0.048 886	251.869	0.8656	0.040 316	251.016	0.8514
110.0	0.063 600	259.624	0.9008	0.050 415	258.845	0.8840	0.041 619	258.053	0.8700
120.0	0.065 455	266.590	0.9187	0.051 929	265.862	0.9021	0.042 907	265.124	0.8882
130.0	0.067 298	273.605	0.9364	0.053 430	272.923	0.9198	0.044 181	272.231	0.9061

p = 0.70 MPa (27.66°C)

p = 0.80 MPa (32.74°C)

p = 0.90 MPa (37.37°C)

t	v	h	s	v	h	s	v	h	s
40.0	0.026 761	207.580	0.7148	0.022 830	205.924	0.7016	0.019 744	204.170	0.6982
50.0	0.028 100	214.745	0.7373	0.024 068	213.290	0.7248	0.020 912	211.765	0.7131
60.0	0.029 387	221.854	0.7590	0.025 247	220.558	0.7469	0.022 012	219.212	0.7358
70.0	0.030 632	228.931	0.7799	0.026 380	227.766	0.7682	0.023 062	226.564	0.7575
80.0	0.031 843	235.997	0.8002	0.027 477	234.941	0.7888	0.024 072	233.856	0.7785
90.0	0.033 027	243.066	0.8199	0.028 545	242.101	0.8088	0.025 051	241.113	0.7987
100.0	0.034 189	250.146	0.8392	0.029 588	249.260	0.8283	0.026 005	248.355	0.8184
110.0	0.035 332	257.247	0.8579	0.030 612	256.428	0.8472	0.026 937	255.593	0.8376
120.0	0.036 458	264.374	0.8763	0.031 619	263.613	0.8657	0.027 851	262.839	0.8562
130.0	0.037 572	271.531	0.8943	0.032 612	270.820	0.8838	0.028 751	270.100	0.8745
140.0	0.038 673	278.720	0.9119	0.033 592	278.055	0.9016	0.029 639	277.381	0.8923
150.0	0.039 764	285.946	0.9292	0.034 563	285.320	0.9189	0.030 515	284.687	0.9098

TABLE A.12 SUPERHEATED REFRIGERANT-12 TABLE (SI Units) (Continued)

T (°C)	p = 1.00 MPa (41.69°C)			p = 1.20 MPa (49.31°C)			p = 1.40 MPa (56.09°C)		
	v (m³/kg)	h (kJ/kg)	s (kJ/kg·K)	v (m³/kg)	h (kJ/kg)	s (kJ/kg·K)	v (m³/kg)	h (kJ/kg)	s (kJ/kg·K)
50.0	0.018 366	210.162	0.7021	0.014 483	206.661	0.6812			
60.0	0.019 410	217.810	0.7254	0.015 463	214.805	0.7060	0.012 579	211.457	0.6876
70.0	0.020 397	225.319	0.7476	0.016 368	222.687	0.7293	0.013 448	219.822	0.7123
80.0	0.021 341	232.739	0.7689	0.017 221	230.398	0.7514	0.014 247	227.891	0.7355
90.0	0.022 251	240.101	0.7895	0.018 032	237.995	0.7727	0.014 997	235.766	0.7575
100.0	0.023 133	247.430	0.8094	0.018 812	245.518	0.7931	0.015 710	243.512	0.7785
110.0	0.023 993	254.743	0.8287	0.019 567	252.993	0.8129	0.016 393	251.170	0.7988
120.0	0.024 835	262.053	0.8475	0.020 301	260.441	0.8320	0.017 053	258.770	0.8183
130.0	0.025 661	269.369	0.8659	0.021 018	267.875	0.8507	0.017 695	266.334	0.8373
140.0	0.026 474	276.699	0.8839	0.021 721	275.307	0.8689	0.018 321	273.877	0.8558
150.0	0.027 275	284.047	0.9015	0.022 412	282.745	0.8867	0.018 934	281.411	0.8738
160.0	0.028 068	291.419	0.9187	0.023 093	290.195	0.9041	0.019 535	288.946	0.8914

T (°C)	p = 1.60 MPa (62.19°C)			p = 1.80 MPa (67.75°C)			p = 2.00 MPa (72.88°C)		
	v (m³/kg)	h (kJ/kg)	s (kJ/kg·K)	v (m³/kg)	h (kJ/kg)	s (kJ/kg·K)	v (m³/kg)	h (kJ/kg)	s (kJ/kg·K)
70.0	0.011 208	216.650	0.6959	0.009 406	213.049	0.6794			
80.0	0.011 984	225.177	0.7204	0.010 187	222.198	0.7057	0.008 704	218.859	0.6909
90.0	0.012 698	233.390	0.7433	0.010 884	230.835	0.7298	0.009 406	228.056	0.7166
100.0	0.013 366	241.397	0.7651	0.011 526	239.155	0.7524	0.010 035	236.760	0.7402
110.0	0.014 000	249.264	0.7859	0.012 126	247.264	0.7739	0.010 615	245.154	0.7624
120.0	0.014 608	257.035	0.8059	0.012 697	255.228	0.7944	0.011 159	253.341	0.7835
130.0	0.015 195	264.742	0.8253	0.013 244	263.094	0.8141	0.011 676	261.384	0.8037
140.0	0.015 765	272.406	0.8440	0.013 772	270.891	0.8332	0.012 172	269.327	0.8232
150.0	0.016 320	280.044	0.8623	0.014 284	278.642	0.8518	0.012 651	277.201	0.8420
160.0	0.016 864	287.669	0.8801	0.014 784	286.364	0.8698	0.013 116	285.027	0.8603
170.0	0.017 398	295.290	0.8975	0.015 272	294.069	0.8874	0.013 570	292.822	0.8781
180.0	0.017 923	302.914	0.9145	0.015 752	301.767	0.9046	0.014 013	300.598	0.8955

	p = 2.50 MPa (84.21°C)			p = 3.00 MPa (93.97°C)			p = 3.50 MPa (102.58°C)		
90.0	0.006 595	219.562	0.6823						
100.0	0.007 264	229.852	0.7103	0.005 231	220.529	0.6770			
110.0	0.007 837	239.271	0.7352	0.005 886	232.068	0.7075	0.004 324	222.121	0.6750
120.0	0.008 351	248.192	0.7582	0.006 419	242.208	0.7336	0.004 959	234.875	0.7078
130.0	0.008 827	256.794	0.7798	0.006 887	251.632	0.7573	0.005 456	245.661	0.7349
140.0	0.009 273	265.180	0.8003	0.007 313	260.620	0.7793	0.005 884	255.524	0.7591
150.0	0.009 697	273.414	0.8200	0.007 709	269.319	0.8001	0.006 270	264.846	0.7814
160.0	0.010 104	281.540	0.8390	0.008 083	277.817	0.8200	0.006 626	273.817	0.8023
170.0	0.010 497	289.589	0.8574	0.008 439	286.171	0.8391	0.006 961	282.545	0.8222
180.0	0.010 879	297.583	0.8752	0.008 782	294.422	0.8575	0.007 279	291.100	0.8413
190.0	0.011 250	305.540	0.8926	0.009 114	302.597	0.8753	0.007 584	299.528	0.8597
200.0	0.011 614	313.472	0.9095	0.009 436	310.718	0.8927	0.007 878	307.864	0.8775

	p = 4.00 MPa (110.32°C)		
120.0	0.003 736	224.863	0.6771
130.0	0.004 325	238.443	0.7111
140.0	0.004 781	249.703	0.7386
150.0	0.005 172	259.904	0.7630
160.0	0.005 522	269.492	0.7854
170.0	0.005 845	278.684	0.8063
180.0	0.006 147	287.602	0.8262
190.0	0.006 434	296.326	0.8453
200.0	0.006 708	304.906	0.8636
210.0	0.006 972	313.380	0.8813
220.0	0.007 228	321.774	0.8985
230.0	0.007 477	330.108	0.9152

Source: Copyright 1955 and 1956, E. I. du Pont de Nemours & Company, Inc. Reprinted by permission.

TABLE A.13 PROPERTIES OF AIR AT LOW PRESSURES (English Units)

T (°R)	h (Btu/lbm)	p_r	u (Btu/lbm)	v_r	φ (Btu/lbm-R)	T (°R)	h (Btu/lbm)	p_r	u (Btu/lbm)	v_r	φ (Btu/lbm-R)
360	85.97	0.3363	61.29	396.6	0.503 69	1460	358.63	50.34	258.54	10.743	0.847 04
380	90.75	0.4061	64.70	346.6	0.516 63	1480	363.89	53.04	262.44	10.336	0.850 62
400	95.53	0.4858	68.11	305.0	0.528 90	1500	369.17	55.86	266.34	9.948	0.854 16
420	100.32	0.5760	71.52	270.1	0.540 58	1520	374.47	58.78	270.26	9.578	0.857 67
440	105.11	0.6776	74.93	240.6	0.551 72	1540	379.77	61.83	274.20	9.226	0.861 13
460	109.90	0.7913	78.36	215.33	0.562 35	1560	385.08	65.00	278.13	8.890	0.864 56
480	114.69	0.9182	81.77	193.65	0.572 55	1580	390.40	68.30	282.09	8.569	0.867 94
500	119.48	1.0590	85.20	174.90	0.582 33	1600	395.74	71.73	286.06	8.263	0.871 30
520	124.27	1.2147	88.62	158.58	0.591 73	1620	401.09	75.29	290.04	7.971	0.874 62
537	128.10	1.3593	91.53	146.34	0.599 45	1640	406.45	78.99	294.03	7.691	0.877 91
540	129.06	1.3860	92.04	144.32	0.600 78						
560	133.86	1.5742	95.47	131.78	0.609 50	1660	411.82	82.83	298.02	7.424	0.881 16
580	138.66	1.7800	98.90	120.70	0.617 93	1680	417.20	86.82	302.04	7.168	0.884 39
600	143.47	2.005	102.34	110.88	0.626 07	1700	422.59	90.95	306.06	6.924	0.887 58
620	148.28	2.249	105.78	102.12	0.633 95	1720	428.00	95.24	310.09	6.690	0.890 74
640	153.09	2.514	109.21	94.30	0.641 59	1740	433.41	99.69	314.13	6.465	0.893 37
660	157.92	2.801	112.67	87.27	0.649 02	1760	438.83	104.30	318.18	6.251	0.896 97
680	162.73	3.111	116.12	80.96	0.656 21	1780	444.26	109.08	322.24	6.045	0.900 03
700	167.56	3.446	119.58	75.25	0.663 21	1800	449.71	114.03	326.32	5.847	0.903 08
720	172.39	3.806	123.04	70.07	0.670 02	1820	455.17	119.16	330.40	5.658	0.906 09
740	177.23	4.193	126.51	65.38	0.676 65	1840	460.63	124.47	334.50	5.476	0.909 08
760	182.08	4.607	129.99	61.10	0.683 12	1860	466.12	129.95	338.61	5.302	0.912 03
780	186.94	5.051	133.47	57.20	0.689 42	1880	471.60	135.64	342.73	5.134	0.914 97
800	191.81	5.526	136.97	53.63	0.695 58	1900	477.09	141.51	346.85	4.974	0.917 88
820	196.69	6.033	140.47	50.35	0.701 60	1920	482.60	147.59	350.98	4.819	0.920 76
840	201.56	6.573	143.98	47.34	0.707 47	1940	488.12	153.87	355.12	4.670	0.923 62

860	206.46	7.149	147.50	44.57	0.713 23
880	211.35	7.761	151.02	42.01	0.718 86
900	216.26	8.411	154.57	39.64	0.724 38
920	221.18	9.102	158.12	37.44	0.729 79
940	226.11	9.834	161.68	35.41	0.735 09
960	231.06	10.610	165.26	33.52	0.740 30
980	236.02	11.430	168.83	31.76	0.745 40
1000	240.98	12.298	172.43	30.12	0.750 42
1020	245.97	13.215	176.04	28.59	0.755 36
1040	250.95	14.182	179.66	27.17	0.760 19
1060	255.96	15.203	183.29	25.82	0.764 96
1080	260.97	16.278	186.93	24.58	0.769 64
1100	265.99	17.413	190.58	23.40	0.774 26
1120	271.03	18.604	194.25	22.30	0.778 80
1140	276.08	19.858	197.94	21.27	0.783 26
1160	281.14	21.18	201.63	20.293	0.787 67
1180	286.21	22.56	205.33	19.377	0.792 01
1200	291.30	24.01	209.05	18.514	0.796 28
1220	296.41	25.53	212.78	17.700	0.800 50
1240	301.52	27.13	216.53	16.932	0.804 66
1260	306.65	28.89	220.28	16.205	0.808 76
1280	311.79	30.55	224.05	15.518	0.812 80
1300	316.94	32.39	227.83	14.868	0.816 80
1320	322.11	34.31	231.63	14.253	0.820 75
1340	327.29	36.31	235.43	13.670	0.824 64
1360	332.48	38.41	239.25	13.118	0.828 48
1380	337.68	40.59	243.08	12.593	0.832 29
1400	342.90	42.88	246.93	12.095	0.836 04
1420	348.14	45.26	250.79	11.622	0.839 75
1440	353.37	47.75	254.66	11.172	0.843 41

1960	493.64	160.37	359.28	4.527	0.926 45
1980	499.17	167.07	363.43	4.390	0.929 26
2000	504.71	174.00	367.61	4.258	0.932 05
2020	510.26	181.16	371.79	4.130	0.934 81
2040	515.82	188.54	375.98	4.008	0.937 56
2060	521.39	196.16	380.18	3.890	0.940 26
2080	526.97	204.02	384.39	3.777	0.942 96
2100	532.55	212.1	388.60	3.667	0.945 64
2150	546.54	233.5	399.17	3.410	0.952 22
2200	560.59	256.6	409.78	3.176	0.958 68
2250	574.69	281.4	420.46	2.961	0.965 01
2300	588.82	308.1	431.16	2.765	0.971 23
2350	603.00	336.8	441.91	2.585	0.977 32
2400	617.22	367.6	452.70	2.419	0.983 31
2450	631.48	400.5	463.54	2.266	0.989 19
2500	645.78	435.7	474.40	2.125	0.994 97
2550	660.12	473.3	485.31	1.9956	1.000 64
2600	674.49	513.5	496.26	1.8756	1.006 23
2650	688.90	556.3	507.25	1.7646	1.011 72
2700	703.35	601.9	518.26	1.6617	1.017 12
2750	717.83	650.4	529.31	1.5662	1.022 44
2800	732.33	702.0	540.40	1.4775	1.027 67
2850	746.88	756.7	551.52	1.3951	1.032 82
2900	761.45	814.8	562.66	1.3184	1.037 88
2950	776.05	876.4	573.84	1.2469	1.042 88
3000	790.68	941.4	585.04	1.1803	1.047 79
3500	938.40	1829.3	698.48	0.7087	1.093 83
4000	1088.26	3280	814.06	0.4518	1.133 34
4500	1239.86	5521	931.39	0.3019	1.169 05
5000	1392.87	8837	1050.12	0.209 59	1.201 29
6000	1702.29	20120	1291.00	0.110 47	1.257 69
6500	1858.44	28974	1412.87	0.083 10	1.282 68

Source: Reproduced with permission of authors and publisher from J. H. Keenan and J. Kaye, *Gas Tables*, New York: John Wiley & Sons, 1945.

TABLE A.14 SATURATED STEAM TEMPERATURE TABLE (English Units)

T (°F)	p (psia)	Specific volume (ft³/lbm) Sat. liquid, v_f	Sat. vapor, v_g	Internal energy (Btu/lbm) Sat. liquid, u_f	Evap., u_{fg}	Sat. vapor, u_g	Enthalpy (Btu/lbm) Sat. liquid, h_f	Evap., h_{fg}	Sat. vapor, h_g	Entropy (Btu/lbm-R) Sat. liquid, s_f	Evap., s_{fg}	Sat. vapor, s_g
32.018	0.088 732	0.016 018	3300.5	0.038 893	1021.1	1021.1	0.039 156	1075.3	1075.3	−0.000 645	2.1869	2.1863
40	0.121 85	0.016 019	2442.1	7.8309	1016	1023.8	7.8313	1071.1	1078.9	0.015 226	2.1435	2.1587
50	0.178 41	0.016 026	1700.9	17.7	1009.5	1027.2	17.701	1065.6	1083.4	0.034 91	2.0908	2.1257
60	0.256 87	0.016 038	1204.3	27.659	1002.9	1030.5	27.66	1060.1	1087.8	0.054 333	2.0399	2.0943
70	0.364 06	0.016 054	865.8	37.68	996.14	1033.8	37.681	1054.5	1092.1	0.073 465	1.9908	2.0642
80	0.508 42	0.016 076	631.46	47.741	989.35	1037.1	47.743	1048.8	1096.5	0.092 286	1.9433	2.0356
90	0.700 24	0.016 102	466.78	57.826	982.51	1040.3	57.828	1043	1100.8	0.110 78	1.8975	2.0082
100	0.951 95	0.016 132	349.44	67.92	975.63	1043.6	67.922	1037.2	1105.1	0.128 95	1.8532	1.9821
110	1.2783	0.016 166	264.72	78.013	968.73	1046.7	78.017	1031.3	1109.4	0.146 79	1.8104	1.9572
120	1.6969	0.016 204	202.8	88.099	961.81	1049.9	88.104	1025.5	1113.6	0.1643	1.7691	1.9334
130	2.228	0.016 246	157	98.173	954.88	1053.1	98.179	1019.6	1117.8	0.1815	1.7291	1.9106
140	2.8951	0.016 292	122.77	108.23	947.93	1056.2	108.24	1013.7	1121.9	0.198 39	1.6904	1.8888
150	3.7255	0.016 341	96.901	118.28	940.97	1059.3	118.29	1007.8	1126.1	0.214 98	1.653	1.8679
160	4.7497	0.016 393	77.166	128.31	934	1062.3	128.32	1001.8	1130.1	0.231 29	1.6167	1.848
170	6.0026	0.016 449	61.968	138.33	927	1065.3	138.35	995.81	1134.2	0.247 33	1.5815	1.8288
180	7.5228	0.016 508	50.16	148.34	919.97	1068.3	148.37	989.77	1138.1	0.263 11	1.5473	1.8104
190	9.3537	0.016 57	40.909	158.35	912.91	1071.3	158.38	983.69	1142.1	0.278 65	1.5141	1.7928
200	11.543	0.016 635	33.602	168.36	905.81	1074.2	168.4	977.54	1145.9	0.293 95	1.4819	1.7758
210	14.143	0.016 703	27.787	178.37	898.65	1077	178.42	971.33	1149.7	0.309 04	1.4505	1.7595
220	17.21	0.016 774	23.126	188.4	891.44	1079.8	188.45	965.04	1153.5	0.323 92	1.4199	1.7438
230	20.808	0.016 848	19.363	198.43	884.16	1082.6	198.5	958.65	1157.2	0.338 61	1.39	1.7286

240	25.002	0.016 925	16.307	208.49	876.81	1085.3	208.56	952.18	1160.7	0.353 12	1.3609	1.714
250	29.864	0.017 005	13.808	218.56	869.38	1087.9	218.66	945.59	1164.2	0.367 45	1.3324	1.6999
260	35.473	0.017 088	11.753	228.67	861.85	1090.5	228.78	938.88	1167.7	0.381 61	1.3046	1.6862
270	41.909	0.017 175	10.052	238.8	854.21	1093	238.94	932.04	1171	0.395 63	1.2773	1.673
280	49.259	0.017 264	8.6384	248.97	846.47	1095.4	249.13	925.06	1174.2	0.409 49	1.2506	1.6601
290	57.617	0.017 357	7.4562	259.18	838.62	1097.8	259.36	917.93	1177.3	0.423 21	1.2244	1.6477
300	67.078	0.017 453	6.4627	269.42	830.63	1100.1	269.64	910.64	1180.3	0.4368	1.1987	1.6355
320	89.726	0.017 655	4.9125	290.03	814.26	1104.3	290.33	895.53	1185.9	0.4636	1.1486	1.6122
340	118.08	0.017 873	3.7877	310.82	797.3	1108.1	311.21	879.67	1190.9	0.489 92	1.1	1.59
360	153.09	0.018 107	2.9581	331.8	779.69	1111.5	332.32	862.97	1195.3	0.515 81	1.0528	1.5686
380	195.8	0.018 36	2.3366	352.98	761.36	1114.3	353.65	845.36	1199	0.5413	1.0068	1.5481
400	247.3	0.018 633	1.8645	374.37	742.27	1116.6	375.22	826.74	1202	0.566 43	0.9617	1.5281
420	308.79	0.018 929	1.5012	395.98	722.33	1118.3	397.06	807.02	1204.1	0.591 26	0.917 42	1.5087
440	381.52	0.019 253	1.2182	417.85	701.45	1119.3	419.2	786.09	1205.3	0.615 82	0.873 76	1.4896
460	466.83	0.019 607	0.9953	440	679.53	1119.5	441.69	763.81	1205.5	0.640 19	0.830 53	1.4707
480	566.1	0.019 997	0.817 92	462.48	656.43	1118.9	464.58	740.02	1204.6	0.664 43	0.787 54	1.452
500	680.82	0.020 428	0.675 38	485.39	631.99	1117.4	487.96	714.5	1202.5	0.688 64	0.744 54	1.4332
520	812.51	0.020 911	0.559 78	508.79	606	1114.8	511.94	687.02	1199	0.712 92	0.701 29	1.4142
540	962.81	0.021 453	0.465 18	532.82	578.2	1111	536.64	657.25	1193.9	0.737 37	0.657 48	1.3948
560	1133.4	0.022 072	0.387 09	557.62	548.23	1105.8	562.24	624.79	1187	0.762 13	0.612 74	1.3749
580	1326.2	0.022 786	0.322 02	583.34	515.67	1099	588.94	589.1	1178	0.787 34	0.566 63	1.354
600	1543.1	0.023 627	0.267 25	610.19	479.92	1090.1	616.94	549.49	1166.4	0.813 14	0.518 55	1.3317
620	1786.6	0.024 645	0.2206	638.37	440.16	1078.5	646.51	504.95	1151.5	0.839 72	0.467 69	1.3074
640	2059.4	0.025 93	0.180 23	668.54	394.68	1063.2	678.42	453.48	1131.9	0.8677	0.412 39	1.2801
660	2365	0.027 666	0.144 43	701.92	340.3	1042.2	714.02	391.4	1105.4	0.898 25	0.349 58	1.2478
680	2708	0.030 354	0.111 11	741.88	268.99	1010.9	757.09	309.46	1066.5	0.934 73	0.271 54	1.2063
700	3094.8	0.036 758	0.074 486	804.11	144.73	948.84	825.16	166.34	991.5	0.993	0.143 44	1.1364
705.44	3203.6	0.050 53	0.050 53	872.6	0	872.6	902.5	0	902.5	1.058	0	1.058

TABLE A.15 SATURATED STEAM PRESSURE TABLE (English Units)

p (psia)	T (°F)	Specific volume (ft³/lbm)		Internal energy (Btu/lbm)			Enthalpy (Btu/lbm)			Entropy (Btu/lbm-R)		
		Sat. liquid, v_f	Sat. vapor, v_g	Sat. liquid, u_f	Evap., u_{fg}	Sat. vapor, u_g	Sat. liquid, h_f	Evap., h_{fg}	Sat. vapor, h_g	Sat. liquid, s_f	Evap., s_{fg}	Sat. vapor, s_g
1	101.64	0.016 137	333.59	69.577	974.5	1044.1	69.58	1036.2	1105.8	0.1319	1.8461	1.978
2	125.99	0.016 229	173.76	94.134	957.66	1051.8	94.14	1022	1116.1	0.174 64	1.745	1.9196
3	141.39	0.016 298	118.73	109.63	946.97	1056.6	109.64	1012.9	1122.5	0.200 71	1.6851	1.8858
4	152.89	0.016 356	90.651	121.18	938.96	1060.1	121.19	1006	1127.2	0.219 72	1.6424	1.8621
5	162.16	0.016 405	73.541	130.48	932.48	1063	130.49	1000.5	1131	0.234 78	1.609	1.8437
6	169.98	0.016 449	61.993	138.31	927.01	1065.3	138.33	995.82	1134.2	0.2473	1.5815	1.8288
8	182.79	0.016 525	47.354	151.13	918	1069.1	151.16	988.08	1139.2	0.267 46	1.538	1.8054
10	193.14	0.016 59	38.429	161.49	910.68	1072.2	161.52	981.76	1143.3	0.283 47	1.5039	1.7874
14.696	212	0.016 716	26.807	180.31	897.27	1077.6	180.35	970.12	1150.5	0.311 92	1.4445	1.7564
15	212.96	0.016 724	26.298	181.34	896.52	1077.9	181.39	969.47	1150.9	0.313 47	1.4413	1.7548
20	227.89	0.016 832	20.094	196.31	885.71	1082	196.37	960.01	1156.4	0.335 52	1.3963	1.7318
25	240	0.016 925	16.308	208.48	876.81	1085.3	208.56	952.18	1160.7	0.353 11	1.3609	1.714
30	250.26	0.017 007	13.749	218.83	869.18	1088	218.92	945.41	1164.3	0.367 82	1.3317	1.6995
35	259.21	0.017 082	11.901	227.87	862.44	1090.3	227.98	939.41	1167.4	0.3805	1.3068	1.6873
40	267.17	0.017 15	10.501	235.93	856.38	1092.3	236.06	933.99	1170	0.391 68	1.285	1.6767
45	274.37	0.017 213	9.4034	243.24	850.85	1094.1	243.38	929.01	1172.4	0.4017	1.2656	1.6673
50	280.94	0.017 273	8.5183	249.93	845.74	1095.7	250.09	924.39	1174.5	0.410 79	1.2482	1.6589
60	292.64	0.017 382	7.1775	261.87	836.53	1098.4	262.07	916.02	1178.1	0.426 81	1.2176	1.6444
70	302.86	0.017 481	6.2085	272.36	828.33	1100.7	272.58	908.52	1181.1	0.440 66	1.1915	1.6321
80	311.97	0.017 572	5.4744	281.74	820.9	1102.6	282	901.68	1183.7	0.4529	1.1685	1.6214

90	320.22	0.017 658	4.8984	290.26	814.08	1104.3	290.55	895.37	1185.9	0.463 88	1.1481	1.612
100	327.76	0.017 738	4.4339	298.08	807.75	1105.8	298.41	889.47	1187.9	0.473 86	1.1296	1.6035
120	341.21	0.017 887	3.7302	312.09	796.25	1108.3	312.49	878.68	1191.2	0.4915	1.0972	1.5887
140	352.99	0.018 023	3.2214	324.43	785.94	1110.4	324.9	868.92	1193.8	0.506 78	1.0692	1.576
160	363.51	0.018 15	2.8359	335.51	776.52	1112	336.04	859.95	1196	0.520 31	1.0447	1.565
180	373.04	0.018 27	2.5333	345.59	767.82	1113.4	346.2	851.6	1197.8	0.532 47	1.0227	1.5552
200	381.78	0.018 383	2.2893	354.87	759.7	1114.6	355.55	843.75	1199.3	0.543 55	1.0027	1.5463
250	400.95	0.018 647	1.845	375.4	741.34	1116.7	376.26	825.83	1202.1	0.567 62	0.959 57	1.5272
300	417.34	0.018 889	1.5443	393.1	725.03	1118.1	394.15	809.71	1203.9	0.587 97	0.923 27	1.5112
350	431.73	0.019 116	1.3269	408.77	710.2	1119	410.01	794.9	1204.9	0.605 69	0.891 75	1.4974
400	444.6	0.019 331	1.1622	422.92	696.5	1119.4	424.35	781.09	1205.4	0.621 45	0.863 78	1.4852
450	456.29	0.019 539	1.0328	435.86	683.67	1119.5	437.49	768.05	1205.5	0.635 68	0.838 52	1.4742
500	467.02	0.019 739	0.928 49	447.85	671.56	1119.4	449.67	755.64	1205.3	0.648 71	0.815 43	1.4641
600	486.21	0.020 126	0.770 34	469.55	648.99	1118.5	471.78	732.29	1204.1	0.671 95	0.774 19	1.4461
700	503.09	0.020 499	0.655 95	488.96	628.09	1117.1	491.62	710.4	1202	0.692 39	0.737 89	1.4303
800	518.21	0.020 865	0.569 19	506.67	608.4	1115.1	509.76	689.57	1199.3	0.710 74	0.705 17	1.4159
900	531.95	0.021 227	0.501 03	523.07	589.62	1112.7	526.61	669.53	1196.1	0.727 51	0.675 19	1.4027
1000	544.58	0.021 588	0.445 99	538.42	571.54	1110	542.42	650.08	1192.5	0.743 01	0.647 34	1.3903
1200	567.18	0.022 316	0.362 37	566.74	536.86	1103.6	571.69	612.37	1184.1	0.771 13	0.596 37	1.3675
1400	587.07	0.023 067	0.3016	592.7	503.43	1096.1	598.67	575.59	1174.3	0.796 38	0.5499	1.3463
1600	604.88	0.023 857	0.2552	616.94	470.63	1087.6	624	539.12	1163.1	0.819 55	0.506 44	1.326
1800	621.04	0.024 704	0.218 37	639.89	437.96	1077.8	648.11	502.46	1150.6	0.841 14	0.464 95	1.3061
2000	635.83	0.025 633	0.188 21	662.05	404.74	1066.8	671.54	464.91	1136.4	0.861 72	0.424 38	1.2861
2500	668.15	0.028 594	0.1307	717.07	314.08	1031.1	730.3	361.31	1091.6	0.912 06	0.320 37	1.2324
3000	695.32	0.034 34	0.084 349	784.51	185.39	969.9	803.57	213.15	1016.7	0.974 34	0.184 55	1.1589
3203.6	705.44	0.050 53	0.050 53	872.6	0	872.6	902.5	0	902.5	1.058	0	1.058

TABLE A.16 SUPERHEATED STEAM VAPOR TABLE (English Units)

T (°F)	v (ft³/lbm)	u (Btu/lbm)	h (Btu/lbm)	s (Btu/lbm-R)	v (ft³/lbm)	u (Btu/lbm)	h (Btu/lbm)	s (Btu/lbm-R)	v (ft³/lbm)	u (Btu/lbm)	h (Btu/lbm)	s (Btu/lbm-R)
	p = 1.0 psia (101.64°F)				p = 5.0 psia (162.16°F)				p = 10.0 psia (193.14°F)			
200	392.54	1077.3	1149.9	2.0512	78.153	1076	1148.3	1.8718	38.845	1074.2	1146.1	1.7927
250	422.44	1094.5	1172.6	2.0843	84.223	1093.6	1171.5	1.9056	41.942	1092.4	1170	1.8275
300	452.3	1111.8	1195.5	2.1152	90.253	1111.1	1194.6	1.937	44.995	1110.3	1193.5	1.8595
350	482.13	1129.2	1218.4	2.1444	96.26	1128.7	1217.8	1.9664	48.025	1128.1	1216.9	1.8892
400	511.95	1146.8	1241.5	2.172	102.25	1146.4	1241	1.9942	51.041	1145.9	1240.4	1.9172
450	541.77	1164.6	1264.8	2.1982	108.24	1164.3	1264.4	2.0205	54.047	1163.9	1263.9	1.9437
500	571.57	1182.5	1288.3	2.2233	114.22	1182.3	1288	2.0456	57.046	1182	1287.5	1.9689
600	631.17	1219	1335.8	2.2702	126.16	1218.8	1335.6	2.0926	63.034	1218.6	1335.2	2.016
700	690.77	1256.3	1384.2	2.3136	138.1	1256.2	1384	2.1361	69.011	1256	1383.7	2.0596
800	750.35	1294.5	1433.3	2.3542	150.02	1294.4	1433.2	2.1767	74.983	1294.2	1433	2.1002
900	809.94	1333.5	1483.4	2.3923	161.95	1333.4	1483.2	2.2148	80.951	1333.3	1483.1	2.1383
1000	869.52	1373.4	1534.3	2.4283	173.87	1373.3	1534.2	2.2508	86.916	1373.2	1534.1	2.1744
1200	988.67	1456	1638.9	2.4952	197.71	1455.9	1638.8	2.3178	98.843	1455.8	1638.7	2.2413
1400	1107.8	1542.2	1747.2	2.5566	221.55	1542.2	1747.1	2.3792	110.77	1542.1	1747.1	2.3027

T (°F)	v (ft³/lbm)	u (Btu/lbm)	h (Btu/lbm)	s (Btu/lbm-R)	v (ft³/lbm)	u (Btu/lbm)	h (Btu/lbm)	s (Btu/lbm-R)	v (ft³/lbm)	u (Btu/lbm)	h (Btu/lbm)	s (Btu/lbm-R)
	p = 14.696 psia (211.93°F)				p = 25 psia (240.0°F)				p = 50 psia (280.94°F)			
250	28.428	1091.2	1168.5	1.7835	16.561	1088.6	1165.2	1.7211				
300	30.532	1109.5	1192.5	1.8159	17.835	1107.6	1190.1	1.7549	8.7724	1102.8	1183.9	1.6721
350	32.611	1127.5	1216.2	1.846	19.081	1126.1	1214.4	1.7858	9.4284	1122.6	1209.9	1.705
400	34.676	1145.5	1239.8	1.8742	20.311	1144.4	1238.4	1.8144	10.065	1141.8	1234.9	1.7349
450	36.73	1163.5	1263.4	1.9008	21.531	1162.7	1262.3	1.8413	10.691	1160.6	1259.5	1.7626
500	38.778	1181.6	1287.1	1.9261	22.744	1181	1286.2	1.8668	11.309	1179.2	1283.9	1.7886
600	42.862	1218.4	1334.9	1.9734	25.157	1217.9	1334.3	1.9143	12.532	1216.7	1332.6	1.8368
700	46.936	1255.8	1383.5	2.017	27.561	1255.5	1383	1.9581	13.744	1254.6	1381.7	1.8809
800	51.004	1294.1	1432.8	2.0576	29.958	1293.8	1432.4	1.9988	14.95	1293.1	1431.4	1.9218
900	55.068	1333.2	1482.9	2.0958	32.352	1332.9	1482.6	2.037	16.152	1332.3	1481.8	1.9602
1000	59.13	1373.1	1533.9	2.1319	34.743	1372.9	1533.7	2.0732	17.352	1372.4	1533	1.9964
1200	67.25	1455.7	1638.6	2.1988	39.521	1455.6	1638.4	2.1402	19.747	1455.2	1637.9	2.0636
1400	75.365	1542	1747	2.2603	44.295	1541.9	1746.8	2.2017	22.138	1541.6	1746.4	2.1251

T (°F)	p = 100 psia (327.76°F)				p = 200 psia (381.78°F)				p = 300 psia (417.34°F)			
350	4.5921	1114.9	1199.8	1.6188								
400	4.937	1136	1227.4	1.6517	2.3611	1122.8	1210.2	1.5594				
450	5.2675	1156.1	1253.6	1.6813	2.5493	1146.3	1240.6	1.5938	1.6367	1134.9	1225.8	1.5361
500	5.5894	1175.7	1279.1	1.7085	2.7258	1168	1268.9	1.6239	1.7675	1159.5	1257.6	1.5701
600	6.2179	1214.2	1329.3	1.758	3.0595	1209	1322.3	1.6767	2.0053	1203.5	1314.9	1.6267
700	6.8349	1252.7	1379.2	1.8029	3.3798	1248.9	1374	1.7232	2.2274	1244.9	1368.6	1.6751
800	7.4455	1291.6	1429.4	1.8443	3.693	1288.6	1425.3	1.7655	2.4419	1285.6	1421.1	1.7184
900	8.0522	1331.1	1480.1	1.8829	4.0021	1328.7	1476.8	1.8047	2.6519	1326.3	1473.5	1.7582
1000	8.6564	1371.4	1531.6	1.9193	4.3086	1369.4	1528.8	1.8415	2.8592	1367.3	1526.1	1.7954
1200	9.8601	1454.4	1636.9	1.9867	4.9166	1452.9	1634.9	1.9094	3.2687	1451.4	1632.8	1.8638
1400	11.06	1541	1745.7	2.0484	5.5207	1539.8	1744.1	1.9713	3.6743	1538.6	1742.5	1.926

T (°F)	p = 400 psia (444.6°F)				p = 600 psia (486.21°F)				p = 800 psia (518.21°F)			
450	1.1743	1121.8	1208.7	1.4892								
500	1.2851	1150	1245.1	1.5281	0.794 76	1127.5	1215.7	1.4586				
600	1.477	1197.7	1307	1.5894	0.946 26	1185	1290	1.5323	0.677 93	1170.5	1270.9	1.4863
700	1.6508	1240.9	1363	1.6398	1.0732	1232.2	1351.4	1.5875	0.783 28	1223	1338.9	1.5476
800	1.8161	1282.5	1416.9	1.6842	1.1899	1276.1	1408.2	1.6343	0.876 29	1269.4	1399.1	1.5972
900	1.9767	1323.8	1470.1	1.7247	1.3013	1318.7	1463.2	1.6762	0.963 34	1313.5	1456.1	1.6406
1000	2.1344	1365.3	1523.2	1.7623	1.4096	1361.1	1517.6	1.7147	1.047	1356.8	1511.8	1.6801
1200	2.4448	1449.8	1630.8	1.8311	1.6208	1446.8	1626.7	1.7846	1.2087	1443.6	1622.6	1.751
1400	2.7511	1537.3	1741	1.8936	1.8279	1534.9	1737.8	1.8476	1.3662	1532.4	1734.7	1.8146
1600	3.0551	1628.2	1854.3	1.9513	2.0326	1626.2	1851.8	1.9056	1.5214	1624.1	1849.3	1.8729

T (°F)	p = 1000 psia (544.58°F)				p = 1500 psia (596.2°F)				p = 2000 psia (635.83°F)			
600	0.514 03	1153.9	1249	1.445	0.283 65	1099.3	1178	1.3444				
700	0.608 41	1213	1325.6	1.5141	0.3719	1184.3	1287.6	1.4434	0.248 96	1148.4	1240.5	1.3785
800	0.687 75	1262.4	1389.6	1.567	0.435 17	1243.5	1364.3	1.5068	0.307 42	1222.3	1336	1.4576
900	0.7604	1308.1	1448.8	1.6121	0.4893	1294.1	1429.9	1.5569	0.353 17	1279.1	1409.8	1.5139
1000	0.829 42	1352.5	1506	1.6525	0.539 01	1341.3	1490.9	1.6001	0.393 55	1329.7	1475.4	1.5603
1200	0.961 51	1440.5	1618.4	1.7245	0.631 78	1432.5	1607.9	1.675	0.466 85	1424.5	1597.2	1.6384
1400	1.0893	1530	1731.5	1.7886	0.719 92	1523.8	1723.6	1.7406	0.535 24	1517.5	1715.6	1.7055
1600	1.2146	1622.1	1846.8	1.8473	0.805 59	1617	1840.6	1.8002	0.601 08	1611.9	1834.3	1.766
1800	1.3385	1717.4	1965.1	1.902	0.889 75	1713	1960	1.8553	0.665 39	1708.7	1954.9	1.8217

TABLE A.16 SUPERHEATED STEAM VAPOR TABLE (English Units) (Continued)

T (°F)	p = 2500 psia (668.15°F)				p = 3000 psia (695.32°F)				p = 4000 psia			
	v (ft³/lbm)	u (Btu/lbm)	h (Btu/lbm)	s (Btu/lbm-R)	v (ft³/lbm)	u (Btu/lbm)	h (Btu/lbm)	s (Btu/lbm-R)	v (ft³/lbm)	u (Btu/lbm)	h (Btu/lbm)	s (Btu/lbm-R)
700	0.169 98	1102.2	1180.8	1.3105	0.111 42	1041.9	1103.7	1.2329	0.105 17	1099	1176.8	1.2767
800	0.229 39	1198	1304.1	1.4127	0.175 97	1170	1267.7	1.3691	0.146 04	1207	1315.1	1.3826
900	0.271 04	1263.1	1388.5	1.4771	0.215 89	1245.9	1365.7	1.4439	0.174 31	1278.6	1407.6	1.4482
1000	0.306 08	1317.7	1459.3	1.5273	0.247 62	1305.1	1442.6	1.4984				
1200	0.367 84	1416.3	1586.4	1.6088	0.301 79	1407.9	1575.5	1.5837	0.219 19	1391	1553.2	1.5417
1400	0.424 41	1511.2	1707.5	1.6775	0.350 51	1504.9	1699.5	1.654	0.258 12	1492.1	1683.2	1.6154
1600	0.478 37	1606.7	1828	1.7389	0.396 55	1601.6	1821.7	1.7163	0.294 28	1591.2	1809	1.6795
1800	0.530 77	1704.3	1949.9	1.7952	0.441 03	1700	1944.8	1.7732	0.328 84	1691.2	1934.6	1.7376

TABLE A.17 COMPRESSED LIQUID TABLE (English Units)

T (°F)	p = 500 psia (467.13°F) v (ft³/lbm)	u (Btu/lbm)	h (Btu/lbm)	s (Btu/lbm-R)	p = 1000 psia (544.75°F) v (ft³/lbm)	u (Btu/lbm)	h (Btu/lbm)	s (Btu/lbm-R)	p = 1500 psia (596.39°F) v (ft³/lbm)	u (Btu/lbm)	h (Btu/lbm)	s (Btu/lbm-R)
Sat.	0.019 748	447.70	449.53	0.649 04	0.021 591	538.39	542.38	0.743 20	0.023 461	604.97	611.48	0.808 24
32	0.015 994	0.00	1.49	0.000 00	0.015 967	0.03	2.99	0.000 05	0.015 939	0.05	4.47	0.000 07
50	0.015 998	18.02	19.50	0.035 99	0.015 972	17.99	20.94	0.035 92	0.015 946	17.95	22.38	0.035 84
100	0.016 106	67.87	69.36	0.129 32	0.016 082	67.70	70.68	0.129 01	0.016 058	67.53	71.99	0.128 70
150	0.016 318	117.66	119.17	0.214 57	0.016 293	117.38	120.40	0.214 10	0.016 268	117.10	121.62	0.213 64
200	0.016 608	167.65	169.19	0.293 41	0.016 580	167.26	170.32	0.292 81	0.016 554	166.87	171.46	0.292 21
250	0.016 972	217.99	219.56	0.367 02	0.016 941	217.47	220.61	0.366 28	0.016 910	216.96	221.65	0.365 54
300	0.017 416	268.92	270.53	0.436 41	0.017 379	268.24	271.46	0.435 52	0.017 343	267.58	272.39	0.434 63
350	0.017 954	320.71	322.37	0.502 49	0.017 909	319.83	323.15	0.501 40	0.017 865	318.98	323.94	0.500 34
400	0.018 608	373.68	375.40	0.566 04	0.018 550	372.55	375.98	0.564 72	0.018 493	371.45	376.59	0.563 43
450	0.019 420	428.40	430.19	0.627 98	0.019 340	426.89	430.47	0.626 32	0.019 264	425.44	430.79	0.624 70
500					0.020 36	483.8	487.5	0.6874	0.020 24	481.8	487.4	0.6853
550									0.021 58	542.1	548.1	0.7469

T (°F)	p = 2000 psia (636.00°F) v (ft³/lbm)	u (Btu/lbm)	h (Btu/lbm)	s (Btu/lbm-R)	p = 3000 psia (695.52°F) v (ft³/lbm)	u (Btu/lbm)	h (Btu/lbm)	s (Btu/lbm-R)	p = 5000 psia v (ft³/lbm)	u (Btu/lbm)	h (Btu/lbm)	s (Btu/lbm-R)
Sat.	0.025 649	662.40	671.89	0.862 27	0.034 310	783.45	802.50	0.973 20				
32	0.015 912	0.06	5.95	0.000 08	0.015 859	0.09	8.90	0.000 09	0.015 755	0.11	14.70	-0.000 01
50	0.015 920	17.91	23.81	0.035 75	0.015 870	17.84	26.65	0.035 55	0.015 773	17.67	32.26	0.035 08
100	0.016 034	67.37	73.30	0.128 39	0.015 987	67.04	75.91	0.127 77	0.015 897	66.40	81.11	0.126 51
200	0.016 527	166.49	172.60	0.291 62	0.016 476	165.74	174.89	0.290 46	0.016 376	164.32	179.47	0.288 18
300	0.017 308	266.93	273.33	0.433 76	0.017 240	265.66	275.23	0.432 05	0.017 110	263.25	279.08	0.428 75
400	0.018 439	370.38	377.21	0.562 16	0.018 334	368.32	378.50	0.559 70	0.018 141	364.47	381.25	0.555 06
450	0.019 191	424.04	431.14	0.623 13	0.019 053	421.36	431.93	0.620 11	0.018 803	416.44	433.84	0.614 51
500	0.020 14	479.8	487.3	0.6832	0.019 944	476.2	487.3	0.6794	0.019 603	469.8	487.9	0.6724
560	0.021 72	551.8	559.8	0.7565	0.021 382	546.2	558.0	0.7508	0.020 835	536.7	556.0	0.7411
600	0.023 30	605.4	614.0	0.8086	0.022 74	597.0	609.6	0.8004	0.021 91	584.0	604.2	0.7876
640					0.024 75	654.3	668.0	0.8545	0.023 34	634.6	656.2	0.8357
680					0.028 79	728.4	744.3	0.9226	0.025 35	690.6	714.1	0.8873
700									0.026 76	721.8	746.6	0.9156

Source: Reproduced with permission from J. H. Keenan, G. Keyes, G. Hill, and J. G. Moore, *Steam Tables*, New York: John Wiley & Sons, 1969.

TABLE A.18 SATURATED AMMONIA TABLE (English Units)

p (psia)	T (°F)	Volume (ft³/lbm) Liquid, v_f	Volume (ft³/lbm) Vapor, v_g	Enthalpy (Btu/lbm) Liquid, h_f	Enthalpy (Btu/lbm) Evap., h_{fg}	Enthalpy (Btu/lbm) Vapor, h_g	Entropy (Btu/lbm-R) Liquid, s_f	Entropy (Btu/lbm-R) Vapor, s_g
5.0	−63.11	0.022 72	49.31	−24.5	612.8	588.3	−0.0599	1.4857
10.0	−41.34	0.023 19	25.81	−1.4	598.5	597.1	−0.0034	1.4276
15.0	−27.29	0.023 50	17.67	13.6	588.8	602.4	0.0318	1.3938
15.98	−25.0	0.023 57	16.66	16.0	587.2	603.2	0.0374	1.3886
18.30	−20.0	0.023 69	14.68	21.4	583.4	605.0	0.0497	1.3774
19.7	−17.20	0.023 75	13.70	24.4	581.6	606.0	0.0560	1.3710
20.0	−16.64	0.023 78	13.50	25.0	581.2	606.2	0.0578	1.3700
20.88	−15.0	0.023 81	12.97	26.7	580.0	606.7	0.0618	1.3664
21.0	−14.78	0.023 82	12.90	27.0	579.8	606.8	0.0623	1.3659
23.74	−10.0	0.023 93	11.50	32.1	576.4	608.5	0.0738	1.3558
24.7	−8.40	0.023 97	11.086	33.8	575.2	609.0	0.0772	1.3525
25.0	−7.96	0.023 98	10.96	34.3	574.8	609.1	0.0787	1.3515
26.92	−5.0	0.024 06	10.23	37.5	572.6	610.1	0.0857	1.3454
28.0	−3.40	0.024 10	9.853	39.3	571.4	610.7	0.0895	1.3421
30.0	−0.57	0.024 18	9.236	42.3	569.3	611.6	0.0962	1.3364
30.42	0	0.024 19	9.116	42.9	568.9	611.8	0.0975	1.3352
34.27	5.0	0.024 32	8.150	48.3	565.0	613.3	0.1092	1.3253
34.7	5.52	0.024 33	8.067	48.9	564.6	613.5	0.1104	1.3243
35.0	5.89	0.024 34	7.991	49.3	564.3	613.6	0.1113	1.3236
37.7	9.07	0.024 43	7.452	52.8	561.8	614.6	0.1187	1.3175
38.51	10.0	0.024 46	7.304	53.8	561.1	614.9	0.1208	1.3157
40.0	11.66	0.024 51	7.047	55.6	559.7	615.4	0.1246	1.3125
43.14	15.00	0.024 60	6.562	59.2	557.1	616.3	0.1323	1.3062
48.21	20.0	0.024 74	5.910	64.7	553.1	617.8	0.1437	1.2969
50.0	21.67	0.024 79	5.710	66.5	551.7	618.2	0.1475	1.2939
53.73	25.00	0.024 88	5.334	70.2	548.9	619.1	0.1551	1.2879
60.0	30.21	0.025 04	4.805	75.9	544.6	620.5	0.1668	1.2787
73.32	40.0	0.025 33	3.971	86.8	536.2	623.0	0.1885	1.2618
100.0	56.05	0.025 84	2.952	104.7	521.8	626.5	0.2237	1.2356
107.6	60.0	0.025 97	2.751	109.2	518.1	627.3	0.2322	1.2294
120.0	66.02	0.026 18	2.476	116.0	512.4	628.4	0.2452	1.2201
124.3	68.00	0.026 25	2.393	118.3	510.5	628.8	0.2494	1.2170
128.8	70.0	0.026 32	2.312	120.5	508.6	629.1	0.2537	1.2140
130.0	70.53	0.026 34	2.291	121.1	508.1	629.2	0.2548	1.2132
140.0	74.79	0.026 49	2.132	126.0	503.9	629.9	0.2638	1.2068
150.0	78.81	0.026 64	1.994	130.6	499.9	630.5	0.2724	1.2009
153.0	80.0	0.026 68	1.955	132.0	498.7	630.7	0.2749	1.1991
166.4	85.0	0.026 87	1.801	137.8	493.6	631.4	0.2854	1.1918
170.0	86.29	0.026 92	1.764	139.3	492.3	631.6	0.2881	1.1900
174.8	88.00	0.026 98	1.716	141.2	490.6	631.8	0.2917	1.1875
180.0	89.78	0.027 06	1.667	143.3	488.7	632.0	0.2954	1.1850
180.6	90.0	0.027 07	1.661	143.5	488.5	632.0	0.2958	1.1846
186.6	92.00	0.027 15	1.609	145.8	486.4	632.2	0.3000	1.1818

TABLE A.18 SATURATED AMMONIA TABLE (English Units) (Continued)

p (psia)	T (°F)	Volume (ft³/lbm) Liquid, v_f	Volume (ft³/lbm) Vapor, v_g	Enthalpy (Btu/lbm) Liquid, h_f	Enthalpy (Btu/lbm) Evap., h_{fg}	Enthalpy (Btu/lbm) Vapor, h_g	Entropy (Btu/lbm-R) Liquid, s_f	Entropy (Btu/lbm-R) Vapor, s_g
190.0	93.13	0.027 20	1.581	147.2	485.2	632.4	0.3024	1.1802
200.0	96.34	0.027 32	1.502	150.9	481.8	632.7	0.3090	1.1756
205.0	97.90	0.027 38	1.466	152.7	480.1	632.8	0.3122	1.1734
210.0	99.43	0.027 45	1.431	154.6	478.4	633.0	0.3154	1.1713
214.7	100.90	0.027 51	1.400	156.2	476.9	633.1	0.3180	1.1690
220.0	102.42	0.027 58	1.367	158.0	475.2	633.2	0.3216	1.1671
247.0	110.00	0.027 90	1.217	167.0	466.7	633.7	0.3372	1.1566
250.0	110.8	0.027 92	1.202	168.0	465.8	633.8	0.3388	1.1555
260.0	113.42	0.028 16	1.155	171.1	462.8	633.9	0.3441	1.1518
266.2	115.0	0.028 13	1.128	173.0	460.9	633.9	0.3474	1.1497
275.0	117.22	0.028 23	1.091	175.6	458.4	634.0	0.3519	1.1466
280.0	118.45	0.028 29	1.072	177.1	456.9	634.0	0.3545	1.1449
286.4	120.0	0.028 36	1.047	179.0	455.0	634.0	0.3576	1.1427
300.0	123.21	0.028 51	0.999	182.9	451.1	634.0	0.3642	1.1383

Source: National Bureau of Standards Circular no. 142, *Tables of Thermodynamic Properties of Ammonia.* Extracted by permission.

TABLE A.19 SUPERHEATED AMMONIA TABLE (English Units)

	Absolute pressure, psia (saturation temperature)							
T (°F)	v (ft³/lbm)	h (Btu/lbm)	s (Btu/lbm-R)	v (ft³/lbm)	h (Btu/lbm)	s (Btu/lbm-R)	v (ft³/lbm)	h (Btu/lbm)	s (Btu/lbm-R)
	p = 15 psia (−27.29°F)			p = 20 psia (−16.64°F)			p = 25 psia (−7.96°F)		
Sat.	17.67	602.4	1.3938	13.50	606.2	1.3700	10.96	609.1	1.3515
−20	18.01	606.4	1.4031	—	—	—	—	—	—
−10	18.47	611.9	0.4154	13.74	610.0	1.3784	—	—	—
0	18.92	617.2	1.4272	14.09	615.5	1.3907	11.19	613.8	1.3616
10	19.37	622.5	0.4386	14.44	621.0	0.4025	11.47	619.4	0.3738
20	19.82	627.8	0.4497	14.78	626.4	0.4138	11.75	625.0	0.3855
30	20.26	633.0	0.4604	15.11	631.7	0.4248	12.03	630.4	0.3967
40	20.70	638.2	0.4709	15.45	637.0	0.4356	12.30	635.8	0.4077
50	21.14	643.4	1.4812	15.78	642.3	1.4460	12.57	641.2	1.4183
60	21.58	648.5	0.4912	16.12	647.5	0.4562	12.84	646.5	0.4287
70	22.01	653.7	0.5011	16.45	652.8	0.4662	13.11	651.8	0.4388
80	22.44	658.9	0.5108	16.78	658.0	0.4760	13.37	657.1	0.4487
90	22.88	664.0	0.5203	17.10	663.2	0.4856	13.64	662.4	0.4584
100	23.31	669.2	1.5296	17.43	668.5	1.4950	13.90	667.7	1.4679
110	23.74	674.4	0.5388	17.76	673.7	0.5042	14.17	673.0	0.4772
120	24.17	679.6	0.5478	18.08	678.9	0.5133	14.43	678.2	0.4864
130	24.60	684.8	0.5567	18.41	684.2	0.5223	14.69	683.5	0.4954

			Absolute pressure, psia (saturation temperature)					
	v	h	s	v	h	s	v	h	s
T (°F)	(ft³/lbm)	(Btu/lbm)	(Btu/lbm-R)	(ft³/lbm)	(Btu/lbm)	(Btu/lbm-R)	(ft³/lbm)	(Btu/lbm)	(Btu/lbm-R)
	p = 15 psia (−27.29°F)			p = 20 psia (−16.64°F)			p = 25 psia (−7.96°F)		
140	25.03	690.0	0.5655	18.73	698.4	0.5312	14.95	688.8	0.5043
150	25.46	695.3	1.5742	19.05	694.7	1.5399	15.21	694.1	1.5131
160	25.88	700.5	0.5827	19.37	700.0	0.5485	15.47	699.4	0.5217
170	26.31	705.8	0.5911	19.70	705.3	0.5569	15.73	704.7	0.5303
180	26.74	711.1	0.5995	20.02	710.6	0.5653	15.99	710.1	0.5387
190	27.16	716.4	0.6077	20.34	715.9	0.5736	16.25	715.4	0.5470
200	27.59	721.7	1.6158	20.66	721.2	1.5817	16.50	720.8	1.5552
220	28.44	732.4	0.6318	21.30	732.0	0.5978	17.02	731.6	0.5713
240	—	—	—	21.94	742.8	0.6135	17.53	742.5	0.5870
260	—	—	—	—	—	—	18.04	753.4	0.6025
	p = 30 psia (−0.57°F)			p = 35 psia (5.89°F)			p = 40 psia (11.66°F)		
Sat.	9.236	611.6	1.3364	7.991	613.6	1.3236	7.047	615.4	1.3125
10	9.492	617.8	1.3497	8.078	616.1	1.3289	—	—	—
20	9.731	623.5	0.3618	8.287	622.0	0.3413	7.203	620.4	1.3231
30	9.966	629.1	0.3733	8.493	627.7	0.3532	7.387	626.3	0.3353
40	10.20	634.6	0.3845	8.695	633.4	0.3646	7.568	632.1	0.3470
50	10.43	640.1	1.3953	8.895	638.9	1.3756	7.746	637.8	1.3583
60	10.65	645.5	0.4059	9.093	644.4	0.3863	7.922	643.4	0.3692
70	10.88	650.9	0.4161	9.289	649.9	0.3967	8.096	648.9	0.3797
80	11.10	656.2	0.4261	9.484	655.3	0.4069	8.268	654.4	0.3900
90	11.33	661.6	0.4359	9.677	660.7	0.4168	8.439	659.9	0.4000
100	11.55	666.9	1.4456	9.869	666.1	1.4265	8.609	665.3	1.4098
110	11.77	672.2	0.4550	10.06	671.5	0.4360	8.777	670.7	0.4194
120	11.99	677.5	0.4642	10.25	676.8	0.4453	8.945	676.1	0.4288
130	12.21	682.9	0.4733	10.44	682.2	0.4545	9.112	681.5	0.4381
140	12.43	688.2	0.4823	10.63	687.6	0.4635	9.278	686.9	0.4471
150	12.65	693.5	1.4911	10.82	692.9	1.4724	9.444	692.3	1.4561
160	12.87	698.8	0.4998	11.00	698.3	0.4811	9.609	697.7	0.4648
170	13.08	704.2	0.5083	11.19	703.7	0.4897	9.774	703.1	0.4735
180	13.30	709.6	0.5168	11.38	709.1	0.4982	9.938	708.5	0.4820
190	13.52	714.9	0.5251	11.56	714.5	0.5066	10.10	714.0	0.4904
200	13.73	720.3	1.5334	11.75	719.9	1.5148	10.27	719.4	1.4987
220	14.16	731.1	0.5495	12.12	730.7	0.5311	10.59	730.3	0.5150
240	14.59	742.0	0.5653	12.49	741.7	0.5469	10.92	741.3	0.5309
260	15.02	753.0	0.5808	12.86	752.7	0.5624	11.24	752.3	0.5465
280	15.45	764.1	0.5960	13.23	763.7	0.5776	11.56	763.4	0.5617
300	—	—	—	—	—	—	11.88	774.6	0.5766

TABLE A.19 SUPERHEATED AMMONIA TABLE (English Units) *(Continued)*

T (°F)	v (ft³/lbm)	h (Btu/lbm)	s (Btu/lbm-R)	v (ft³/lbm)	h (Btu/lbm)	s (Btu/lbm-R)	v (ft³/lbm)	h (Btu/lbm)	s (Btu/lbm-R)

T (°F)	v (ft³/lbm)	h (Btu/lbm)	s (Btu/lbm-R)	v (ft³/lbm)	h (Btu/lbm)	s (Btu/lbm-R)	v (ft³/lbm)	h (Btu/lbm)	s (Btu/lbm-R)
	p = 50 psia (21.67°F)			*p* = 60 psia (30.21°F)			*p* = 70 psia (37.70°F)		
Sat.	5.710	618.2	1.2939	4.805	620.5	1.2787	4.151	622.4	1.2658
30	5.838	623.4	1.3046	—	—	—	—	—	—
40	5.988	629.5	0.3169	4.933	626.8	1.2913	4.177	623.9	1.2688
50	6.135	635.4	1.3286	5.060	632.9	1.3035	4.290	630.4	1.2816
60	6.280	641.2	0.3399	5.184	639.0	0.3152	4.401	636.6	0.2937
70	6.423	646.9	0.3508	5.307	644.9	0.3265	4.509	642.7	0.3054
80	6.564	652.6	0.3613	5.428	650.7	0.3373	4.615	648.7	0.3166
90	6.704	658.2	0.3716	5.547	656.4	0.3479	4.719	654.6	0.3274
100	6.843	663.7	1.3816	5.665	662.1	1.3581	4.822	660.4	1.3378
110	6.980	669.2	0.3914	5.781	667.7	0.3681	4.924	666.1	0.3480
120	7.117	674.7	0.4009	5.897	673.3	0.3778	5.025	671.8	0.3579
130	7.252	680.2	0.4103	6.012	678.9	0.3873	5.125	677.5	0.3676
140	7.387	685.7	0.4195	6.126	684.4	0.3966	5.224	683.1	0.3770
150	7.521	691.1	1.4286	6.239	689.9	1.4058	5.323	688.7	1.3863
160	7.655	696.6	0.4374	6.352	695.5	0.4148	5.420	694.3	0.3954
170	7.788	702.1	0.4462	6.464	701.0	0.4236	5.518	699.9	0.4043
180	7.921	707.5	0.4548	6.576	706.5	0.4323	5.615	705.5	0.4131
190	8.053	713.0	0.4633	6.687	712.0	0.4409	5.711	711.0	0.4217
200	8.185	718.5	1.4716	6.798	717.5	1.4493	5.807	716.6	1.4302
210	8.317	724.0	0.4799	6.909	723.1	0.4576	5.902	722.2	0.4386
220	8.448	729.4	0.4880	7.019	728.6	0.4658	5.998	727.7	0.4469
240	8.710	740.5	0.5040	7.238	739.7	0.4819	6.187	738.9	0.4631
260	8.970	751.6	0.5197	7.457	750.9	0.4976	6.376	750.1	1.4789
280	9.230	762.7	1.5350	7.675	762.1	1.5130	6.563	761.4	0.4943
300	9.489	774.0	0.5500	7.892	773.3	0.5281	6.750	772.7	0.5095
	p = 80 psia (44.40°F)			*p* = 90 psia (50.47°F)			*p* = 100 psia (56.05°F)		
Sat.	3.655	624.0	1.2545	3.266	625.3	1.2445	2.952	626.5	1.2356
50	3.712	627.7	1.2619	—	—	—	—	—	—
60	3.812	634.3	0.2745	3.353	631.8	1.2571	2.985	629.3	1.2409
70	3.909	640.6	0.2866	3.442	638.3	0.2695	3.068	636.0	0.2539
80	4.005	646.7	0.2981	3.529	644.7	0.2814	3.149	642.6	0.2661
90	4.098	652.8	0.3092	3.614	650.9	0.2928	3.227	649.0	0.2778
100	4.190	658.7	1.3199	3.698	657.0	1.3038	3.304	655.2	1.2891
110	4.281	664.6	0.3303	3.780	663.0	0.3144	3.380	661.3	0.2999
120	4.371	670.4	0.3404	3.862	668.9	0.3247	3.454	667.3	0.3104
130	4.460	676.1	0.3502	3.942	674.7	0.3347	3.527	673.3	0.3206
140	4.548	681.8	0.3598	4.021	680.5	0.3444	3.600	679.2	0.3305
150	4.635	687.5	1.3692	4.100	686.3	1.3539	3.672	685.0	1.3401
160	4.722	693.2	0.3784	4.178	692.0	0.3633	3.743	690.8	0.3495
170	4.808	698.8	0.3874	4.255	697.7	0.3724	3.813	696.6	0.3588

	Absolute pressure, psia (saturation temperature)								
T (°F)	v (ft³/lbm)	h (Btu/lbm)	s (Btu/lbm-R)	v (ft³/lbm)	h (Btu/lbm)	s (Btu/lbm-R)	v (ft³/lbm)	h (Btu/lbm)	s (Btu/lbm-R)
	p = 80 psia (44.40°F)			p = 90 psia (50.47°F)			p = 100 psia (56.05°F)		
180	4.893	704.4	0.3963	4.332	703.4	0.3813	3.883	702.3	0.3678
190	4.978	710.0	0.4050	4.408	709.0	0.3901	3.952	708.0	0.3767
200	5.063	715.6	1.4136	4.484	714.7	1.3988	4.021	713.7	1.3854
210	5.147	721.3	0.4220	4.560	720.4	0.4073	4.090	719.4	0.3940
220	5.231	726.9	0.4304	4.635	726.0	0.4157	4.158	725.1	0.4024
230	5.315	732.5	0.4386	4.710	731.7	0.4239	4.226	730.8	0.4108
240	5.398	738.1	0.4467	4.785	737.3	0.4321	4.294	736.5	0.4190
250	5.482	743.8	1.4547	4.859	743.0	1.4401	4.361	742.2	1.4271
260	5.565	749.4	0.4626	4.933	748.7	0.4481	4.428	747.9	0.4350
280	5.730	760.7	0.4781	5.081	760.0	0.4637	4.562	759.4	0.4507
300	5.894	772.1	0.4933	5.228	771.5	0.4789	4.695	770.8	0.4660
	p = 110 psia (61.21°F)			p = 120 psia (66.02°F)			p = 140 psia (74.79°F)		
Sat.	2.693	627.5	1.2275	2.476	628.4	1.2201	2.132	629.9	1.2068
70	2.761	633.7	1.2392	2.505	631.3	1.2255	—	—	—
80	2.837	640.5	0.2519	2.576	638.3	0.2386	2.166	633.8	1.2140
90	2.910	647.0	0.2640	2.645	645.0	0.2510	2.228	640.9	0.2272
100	2.981	653.4	1.2755	2.712	651.6	1.2628	2.288	647.8	1.2396
110	3.051	659.7	0.2866	2.778	658.0	0.2741	2.347	654.5	0.2515
120	3.120	665.8	0.2972	2.842	664.2	0.2850	2.404	661.1	0.2628
130	3.188	671.9	0.3076	2.905	670.4	0.2956	2.460	667.4	0.2738
140	3.255	677.8	0.3176	2.967	676.5	0.3058	2.515	673.7	0.2843
150	3.321	683.7	1.3274	3.029	682.5	1.3157	2.569	679.9	1.2945
160	3.386	689.6	0.3370	3.089	688.4	0.3254	2.622	686.0	0.3045
170	3.451	695.4	0.3463	3.149	694.3	0.3348	2.675	692.0	0.3141
180	3.515	701.2	0.3555	3.209	700.2	0.3441	2.727	698.0	0.3236
190	3.579	707.0	03644	3.268	706.0	0.3531	2.779	704.0	0.3328
200	3.642	712.8	1.3732	3.326	711.8	1.3620	2.830	709.9	1.3418
210	3.705	718.5	0.3819	3.385	717.6	0.3707	2.880	715.8	0.3507
220	3.768	724.3	0.3904	3.442	723.4	0.3793	2.931	721.6	0.3594
230	3.830	730.0	0.3988	3.500	729.2	0.3877	2.981	727.5	0.3679
240	3.892	735.7	0.4070	3.557	734.9	0.3960	3.030	733.3	0.3763
250	3.954	741.5	1.4151	3.614	740.7	1.4042	3.080	739.2	1.3846
260	4.015	747.2	0.4232	3.671	746.5	0.4123	3.129	745.0	0.3928
270	4.076	752.9	0.4311	3.727	752.2	0.4202	3.179	750.8	0.4008
280	4.137	758.7	0.4389	3.783	758.0	0.4281	3.227	756.7	0.4088
290	4.198	764.5	0.4466	3.839	763.8	0.4359	3.275	762.5	0.4166
300	4.259	770.2	1.4543	3.895	769.6	1.4435	3.323	768.3	1.4243
320	—	—	—	—	—	—	3.420	780.0	0.4395

TABLE A.19 SUPERHEATED AMMONIA TABLE (English Units) *(Continued)*

				Absolute pressure, psia (saturation temperature)					
T (°F)	v (ft³/lbm)	h (Btu/lbm)	s (Btu/lbm-R)	v (ft³/lbm)	h (Btu/lbm)	s (Btu/lbm-R)	v (ft³/lbm)	h (Btu/lbm)	s (Btu/lbm-R)
	$p = 160$ psia (82.64°F)			$p = 180$ psia (89.78°F)			$p = 200$ psia (96.34°F)		
Sat.	1.872	631.1	1.1952	1.667	632.0	1.1850	1.502	632.7	1.1756
90	1.914	636.6	1.2055	1.668	632.2	1.1853	—	—	—
100	1.969	643.9	1.2186	1.720	639.9	1.1992	1.520	635.6	1.1809
110	2.023	651.0	0.2311	1.770	647.3	0.2123	1.567	643.4	0.1947
120	2.075	657.8	0.2429	1.818	654.4	0.2247	1.612	650.9	0.2077
130	2.125	664.4	0.2542	1.865	661.3	0.2364	1.656	658.1	0.2200
140	2.175	670.9	0.2652	1.910	668.0	0.2477	1.698	665.0	0.2317
150	2.224	677.2	1.2757	1.955	674.6	1.2586	1.740	671.8	1.2429
160	2.272	683.5	0.2859	1.999	681.0	0.2691	1.780	678.4	0.2537
170	2.319	689.7	0.2958	2.042	687.3	0.2792	1.820	684.9	0.2641
180	2.365	695.8	0.3054	2.084	693.6	0.2891	1.859	691.3	0.2742
190	2.411	701.9	0.3148	2.126	699.8	0.2987	1.897	697.7	0.2840
200	2.457	707.9	1.3240	2.167	705.9	1.3081	1.935	703.9	1.2935
210	2.502	713.9	0.3331	2.208	712.0	0.3172	1.972	710.1	0.3029
220	2.547	719.9	0.3419	2.248	718.1	0.3262	2.009	716.3	0.3120
230	2.591	725.8	0.3506	2.288	724.1	0.3350	2.046	722.4	0.3209
240	2.635	731.7	0.3591	2.328	730.1	0.3436	2.082	728.4	0.3296
250	2.679	737.6	1.3675	2.367	736.1	1.3521	2.118	734.5	1.3382
260	2.723	743.5	0.3757	2.407	742.0	0.3605	2.154	740.5	0.3467
270	2.766	749.4	0.3838	2.446	748.0	0.3687	2.189	746.5	0.3550
280	2.809	755.3	0.3919	2.484	753.9	0.3768	2.225	752.5	0.3631
290	2.852	761.2	0.3998	2.523	759.9	0.3847	2.260	758.5	0.3712
300	2.895	767.1	1.4076	2.561	765.8	1.3926	2.295	764.5	1.3791
320	2.980	778.9	0.4229	2.637	777.7	0.4081	2.364	776.5	0.3947
340	3.064	790.7	0.4379	2.713	789.6	0.4231	2.432	788.5	0.4099
	$p = 210$ psia (99.43°F)			$p = 220$ psia (102.42°F)			$p = 240$ psia (108.09°F)		
Sat.	1.431	633.0	1.1713	1.367	633.2	1.1671	1.253	633.6	1.2592
110	1.480	641.5	1.1863	1.400	639.4	1.1781	1.261	635.3	1.1621
120	1.524	649.1	0.1996	1.443	647.3	0.1917	1.302	643.5	0.1764
130	1.566	656.4	0.2121	1.485	654.8	0.2045	1.342	651.3	0.1898
140	1.608	663.5	0.2240	1.525	662.0	0.2167	1.380	658.8	0.2025
150	1.648	670.4	1.2354	1.564	669.0	1.2281	1.416	666.1	1.2145
160	1.687	677.1	0.2464	1.601	675.8	0.2394	1.452	673.1	0.2259
170	1.725	683.7	0.2569	1.638	682.5	0.2501	1.487	680.0	0.2369
180	1.762	690.2	0.2672	1.675	689.1	0.2604	1.521	686.7	0.2475
190	1.799	696.6	0.2771	1.710	695.5	0.2704	1.554	693.3	0.2577
200	1.836	702.9	1.2867	1.745	701.9	1.2801	1.587	699.8	1.2677
210	1.872	709.2	0.2961	1.780	708.2	0.2896	1.619	706.2	0.2773
220	1.907	715.3	0.3053	1.814	714.4	0.2989	1.651	712.6	0.2867
230	1.942	721.5	0.3143	1.848	720.6	0.3079	1.683	718.9	0.2959

TABLE A.19 SUPERHEATED AMMONIA TABLE (English Units) *(Continued)*

	Absolute pressure, psia (saturation temperature)								
T (°F)	v (ft³/lbm)	h (Btu/lbm)	s (Btu/lbm-R)	v (ft³/lbm)	h (Btu/lbm)	s (Btu/lbm-R)	v (ft³/lbm)	h (Btu/lbm)	s (Btu/lbm-R)
	$p = 210$ psia (99.43°F)			$p = 220$ psia (102.42°F)			$p = 240$ psia (108.09°F)		
240	1.977	727.6	0.3231	1.881	726.8	0.3168	1.714	725.1	0.3049
250	2.011	733.7	1.3317	1.914	732.9	1.3255	1.745	731.3	1.3137
260	2.046	739.8	0.3402	1.947	739.0	0.3340	1.775	737.5	0.3224
270	2.080	745.8	0.3486	1.980	745.1	0.3424	1.805	743.6	0.3308
280	2.113	751.8	0.3568	2.012	751.1	0.3507	1.835	749.8	0.3392
290	2.147	757.9	0.3649	2.044	757.2	0.3588	1.865	755.9	0.3474
300	2.180	763.9	1.3728	2.076	763.2	1.3668	1.895	762.0	1.3554
320	2.246	775.9	0.3884	2.140	775.3	0.3825	1.954	774.1	0.3712
340	2.312	787.9	0.4037	2.203	787.4	0.3978	2.012	786.3	0.3866
360	2.377	800.0	0.4186	2.265	799.5	0.4127	2.069	798.4	0.4016
380	2.442	812.0	0.4331	2.327	811.6	0.4273	2.126	810.6	0.4163
	$p = 260$ psia (113.42°F)			$p = 280$ psia (118.45°F)			$p = 300$ psia (123.21°F)		
Sat.	1.155	633.9	1.1518	1.072	634.0	1.1449	0.999	634.0	0.1383
120	1.182	639.5	1.1617	1.078	635.4	1.1473	—	—	—
130	1.220	647.8	0.1757	1.115	644.0	0.1621	1.023	640.1	1.1487
140	1.257	655.6	0.1889	1.151	652.2	0.1759	1.058	648.7	0.1632
150	1.292	663.1	1.2014	1.184	660.1	1.1888	1.091	656.9	1.1767
160	1.326	670.4	0.2132	1.217	667.6	0.2011	1.123	664.7	0.1894
170	1.359	677.5	0.2245	1.249	674.9	0.2127	1.153	672.2	0.2014
180	1.391	684.4	0.2354	1.279	681.9	0.2239	1.183	679.5	0.2129
190	1.422	691.1	0.2458	1.309	688.9	0.2346	1.211	686.5	0.2239
200	1.453	697.7	1.2560	1.339	695.6	1.2449	1.239	693.5	1.2344
210	1.484	704.3	0.2658	1.367	702.3	0.2550	1.267	700.3	0.2447
220	1.514	710.7	0.2754	1.396	708.8	0.2647	1.294	706.9	0.2546
230	1.543	717.1	0.2847	1.424	715.3	0.2742	1.320	713.5	0.2642
240	1.572	723.4	0.2938	1.451	721.8	0.2834	1.346	720.0	0.2730
250	1.601	729.7	1.3027	1.478	728.1	1.2924	1.372	726.5	1.2827
260	1.630	736.0	0.3115	1.505	734.4	0.3013	1.397	732.9	0.2917
270	1.658	742.2	0.3200	1.532	740.7	0.3099	1.422	739.2	0.3004
280	1.686	748.4	0.3285	1.558	747.0	0.3184	1.447	745.5	0.3090
290	1.714	754.5	0.3367	1.584	753.2	0.3268	1.472	751.8	0.3175
300	1.741	760.7	1.3449	1.610	759.4	1.3350	1.496	758.1	1.3257
320	1.796	772.9	0.3608	1.661	771.7	0.3511	1.544	770.5	0.3419
340	1.850	785.2	0.3763	1.712	784.0	0.3667	1.592	782.9	0.3576
360	1.904	797.4	0.3914	1.762	796.3	0.3819	1.639	795.3	0.3729
380	1.957	809.6	0.4062	1.811	808.7	0.3967	1.686	807.7	0.3878
400	2.009	821.9	1.4206	1.861	821.0	1.4112	1.732	820.1	1.4024

Source: National Bureau of Standards Circular no. 142, *Tables of Thermodynamic Properties of Ammonia.* Extracted by permission.

TABLE A.20   SATURATED REFRIGERANT-12 TABLE (English Units)

T (°F)	p (psia)	Specific volume (ft³/lbm)			Enthalpy (Btu/lbm)			Entropy (Btu/lbm-R)		
		Sat. liquid, v_f	Evap., v_{fg}	Sat. vapor, v_g	Sat. liquid, h_f	Evap., h_{fg}	Sat. vapor, h_g	Sat. liquid, s_f	Evap., s_{fg}	Sat. vapor, s_g
−130	0.412 24	0.009 736	70.7203	70.730	−18.609	81.577	62.968	−0.049 83	0.247 43	0.197 60
−120	0.641 90	0.009 816	46.7312	46.741	−16.565	80.617	64.052	−0.043 72	0.237 31	0.193 59
−110	0.970 34	0.009 899	31.7671	31.777	−14.518	79.663	65.145	−0.037 79	0.227 80	0.190 02
−100	1.4280	0.009 985	21.1541	22.164	−12.466	78.714	66.248	−0.032 00	0.218 83	0.186 83
−90	2.0509	0.010 073	15.8109	15.821	−10.409	77.764	67.355	−0.026 37	0.210 34	0.183 98
−80	2.8807	0.010 164	11.5228	11.533	−8.3451	76.812	68.467	−0.020 86	0.202 29	0.181 43
−70	3.9651	0.010 259	8.5584	8.5687	−6.2730	75.853	69.580	−0.015 48	0.194 64	0.179 16
−60	5.3575	0.010 357	6.4670	6.4774	−4.1919	74.885	70.693	−0.010 21	0.187 16	0.177 14
−50	7.1168	0.010 459	4.9637	4.9742	−2.1011	73.906	71.805	−0.005 06	0.180 38	0.175 33
−40	9.3076	0.010 564	3.8644	3.8750	0	72.913	72.913	0	0.173 73	0.173 73
−30	11.999	0.010 674	3.0478	3.0585	2.1120	71.903	74.015	0.004 96	0.167 33	0.172 29
−20	15.267	0.010 788	2.4321	2.4429	4.2357	70.874	75.110	0.009 83	0.161 19	0.171 02
−10	19.189	0.010 906	1.9628	1.9727	6.3716	69.824	76.196	0.014 62	0.155 27	0.169 89
0	23.849	0.011 030	1.5979	1.6089	8.5207	68.750	77.271	0.019 32	0.149 56	0.168 88
10	29.335	0.011 160	1.3129	1.3241	10.684	67.651	78.335	0.023 95	0.144 03	0.167 98
20	35.736	0.011 296	1.0875	1.0988	12.863	66.522	79.385	0.028 52	0.138 67	0.167 19
30	43.148	0.011 438	0.907 36	0.918 80	15.058	65.361	80.419	0.033 01	0.133 47	0.166 48
40	51.667	0.011 588	0.761 98	0.773 57	17.273	64.163	81.436	0.037 45	0.128 41	0.165 86
50	61.394	0.011 746	0.643 62	0.655 37	19.507	62.926	82.433	0.041 84	0.123 46	0.165 30
60	72.433	0.001 913	0.546 48	0.558 39	21.766	61.643	83.409	0.046 18	0.118 61	0.164 79
70	84.888	0.012 089	0.466 09	0.478 18	24.050	60.309	84.359	0.050 48	0.113 86	0.164 34
80	98.870	0.012 277	0.399 07	0.411 35	26.365	58.917	85.282	0.054 75	0.109 17	0.163 92
90	114.49	0.012 478	0.342 81	0.355 29	28.713	57.461	86.174	0.059 00	0.104 53	0.163 53
100	131.86	0.012 693	0.295 25	0.307 94	31.100	55.929	87.029	0.063 23	0.099 92	0.163 15
110	151.11	0.012 924	0.255 77	0.267 69	33.531	54.313	87.844	0.067 45	0.095 34	0.162 79

TABLE A.20 SATURATED REFRIGERANT-12 TABLE (English Units) (Continued)

T (°F)	p (psia)	Specific volume (ft³/lbm) Sat. liquid, v_f	Evap., v_{fg}	Sat. vapor, v_g	Enthalpy (Btu/lbm) Sat. liquid, h_f	Evap., h_{fg}	Sat. vapor, h_g	Entropy (Btu/lbm-R) Sat. liquid, s_f	Evap., s_{fg}	Sat. vapor, s_g
120	172.35	0.013 174	0.220 19	0.233 26	36.013	52.597	88.610	0.071 68	0.090 73	0.162 41
130	195.71	0.013 447	0.190 19	0.203 64	38.553	50.768	89.321	0.075 83	0.086 09	0.162 02
140	221.32	0.013 746	0.164 24	0.177 99	41.162	48.805	89.967	0.080 21	0.081 38	0.161 59
150	249.31	0.014 078	0.141 56	0.155 64	43.850	46.684	90.534	0.084 53	0.076 57	0.161 10
160	279.82	0.014 449	0.121 59	0.136 04	46.633	44.373	91.006	0.088 93	0.072 60	0.160 53
170	313.00	0.014 871	0.103 86	0.118 73	49.529	41.830	91.359	0.093 42	0.066 43	0.159 85
180	349.00	0.015 360	0.087 94	0.103 30	52.562	38.999	91.561	0.098 04	0.060 96	0.159 00
190	387.98	0.015 942	0.073 476	0.089 418	55.769	35.792	91.561	0.102 84	0.055 11	0.157 93
200	430.90	0.016 559	0.060 069	0.076 728	59.203	32.075	91.278	0.107 89	0.048 62	0.156 51
210	475.52	0.017 601	0.047 242	0.064 843	62.959	27.599	90.558	0.113 32	0.039 21	0.154 53
220	524.43	0.018 986	0.035 154	0.053 140	67.246	21.790	89.036	0.119 43	0.032 06	0.151 49
230	577.03	0.021 854	0.017 581	0.039 435	72.893	12.229	85.122	0.127 89	0.017 73	0.145 12
233.6 (critical)	596.9	0.028 70	0	0.028 70	78.86	0	78.86	0.1359	0	0.1359

Source: Copyright 1955 and 1956, E. I. DuPont de Nemours & Co., Inc. Reprinted by permission.

TABLE A.21 SUPERHEATED REFRIGERANT-12 TABLE (English Units)

T (°F)	5 psia (−62.33°F) v (ft³/lbm)	h (Btu/lbm)	s (Btu/lbm-R)	10 psia (−37.24°F) v (ft³/lbm)	h (Btu/lbm)	s (Btu/lbm-R)	15 psia (−20.75°F) v (ft³/lbm)	h (Btu/lbm)	s (Btu/lbm-R)
0	8.0611	78.582	0.196 63	3.9809	78.246	0.184 71	2.6201	77.902	0.177 51
20	8.4265	81.309	0.202 44	4.1691	81.014	0.190 61	2.7494	80.712	0.183 49
40	8.7903	84.090	0.208 12	4.3556	83.828	0.196 35	2.8770	83.561	0.189 31
60	9.1528	86.922	0.213 67	4.5408	86.689	0.201 97	3.0031	86.451	0.194 98
80	9.5142	89.806	0.219 12	4.7248	89.596	0.207 46	3.1281	89.383	0.200 51
100	9.8747	92.738	0.224 45	4.9079	92.548	0.212 83	3.2521	92.357	0.205 93
120	10.234	95.717	0.229 68	5.0903	95.546	0.218 09	3.3754	95.373	0.211 22
140	10.594	98.743	0.234 81	5.2720	98.586	0.223 25	3.4981	98.429	0.216 40
160	10.952	101.812	0.239 85	5.4533	101.669	0.228 80	3.6202	101.525	0.221 48
180	11.311	104.925	0.244 79	5.6341	104.793	0.233 26	3.7419	104.661	0.226 46
200	11.668	108.079	0.249 64	5.8145	107.957	0.238 13	3.8632	107.835	0.231 35
220	12.026	111.272	0.254 41	5.9946	111.159	0.242 91	3.9841	111.046	0.236 14

T (°F)	20 psia (−8.13°F) v (ft³/lbm)	h (Btu/lbm)	s (Btu/lbm-R)	25 psia (2.24°F) v (ft³/lbm)	h (Btu/lbm)	s (Btu/lbm-R)	30 psia (11.12°F) v (ft³/lbm)	h (Btu/lbm)	s (Btu/lbm-R)
20	2.0391	80.403	0.178 29	1.6125	80.088	0.174 14	1.3278	79.765	0.170 65
40	2.1373	83.289	0.184 19	1.6932	83.012	0.180 12	1.3969	82.730	0.176 71
60	2.2340	86.210	0.189 92	1.7723	85.965	0.185 91	1.4644	85.716	0.182 57
80	2.3295	89.168	0.195 50	1.8502	88.950	0.191 55	1.5306	88.729	0.188 26
100	2.4241	92.164	0.200 95	1.9271	91.968	0.197 04	1.5957	91.770	0.193 79
120	2.5179	95.198	0.206 28	2.0032	95.021	0.202 40	1.6600	94.843	0.199 18
140	2.6110	98.270	0.211 49	2.0786	98.110	0.207 63	1.7237	97.948	0.204 45
160	2.7036	101.380	0.216 59	2.1535	101.234	0.212 76	1.7868	101.086	0.209 60
180	2.7957	104.528	0.221 59	2.2279	104.393	0.217 78	1.8494	104.258	0.214 63
200	2.8874	107.712	0.226 49	2.3019	107.588	0.222 69	1.9116	107.464	0.219 57
220	2.9789	110.932	0.231 30	2.3756	110.817	0.227 52	1.9735	110.702	0.224 40
240	3.0700	114.186	0.236 02	2.4491	114.080	0.232 25	2.0351	113.973	0.229 15

T (°F)	35 psia (18.92°F) v (ft³/lbm)	h (Btu/lbm)	s (Btu/lbm-R)	40 psia (25.92°F) v (ft³/lbm)	h (Btu/lbm)	s (Btu/lbm-R)	50 psia (38.14°F) v (ft³/lbm)	h (Btu/lbm)	s (Btu/lbm-R)
40	1.1850	82.442	0.173 75	1.0258	82.148	0.171 12	0.802 48	81.540	0.166 55
60	1.2442	85.463	0.179 68	1.0789	85.206	0.177 12	0.847 13	84.676	0.172 71
80	1.3021	88.504	0.185 42	1.1306	88.277	0.182 92	0.890 25	87.811	0.178 62
100	1.3589	91.570	0.191 00	1.1812	91.367	0.188 54	0.932 16	90.953	0.184 34
120	1.4148	94.663	0.196 43	1.2309	94.480	0.194 01	0.973 13	94.110	0.189 88
140	1.4701	97.785	0.201 72	1.2798	97.620	0.199 33	1.0133	97.286	0.195 27
160	1.5248	100.938	0.206 89	1.3282	100.788	0.204 53	1.0529	100.485	0.200 51
180	1.5789	104.122	0.211 95	1.3761	103.985	0.209 61	1.0920	103.708	0.205 63
200	1.6327	107.338	0.216 90	1.4236	107.212	0.214 57	1.1307	106.958	0.210 64
220	1.6862	110.586	0.221 75	1.4707	110.469	0.219 44	1.1690	110.235	0.215 53
240	1.7394	113.865	0.226 51	1.5176	113.757	0.224 20	1.2070	113.539	0.220 32
260	1.7923	117.175	0.231 17	1.5642	117.074	0.228 88	1.2447	116.871	0.225 02

T (°F)	60 psia (48.64°F) v (ft³/lbm)	h (Btu/lbm)	s (Btu/lbm-R)	70 psia (57.40°F) v (ft³/lbm)	h (Btu/lbm)	s (Btu/lbm-R)	80 psia (66.21°F) v (ft³/lbm)	h (Btu/lbm)	s (Btu/lbm-R)
60	0.692 10	84.126	0.168 92	0.580 88	83.552	0.165 56	—	—	—
80	0.729 64	87.330	0.174 97	0.614 58	86.832	0.171 75	0.527 95	86.316	0.168 85
100	0.765 88	90.528	0.180 79	0.646 85	90.091	0.177 68	0.557 34	89.640	0.174 89
120	0.801 10	93.731	0.186 41	0.678 03	93.343	0.183 39	0.585 56	92.945	0.180 70
140	0.835 51	96.945	0.191 86	0.708 36	96.597	0.188 91	0.612 86	96.242	0.186 29
160	0.869 28	100.776	0.197 16	0.738 00	99.862	0.194 27	0.639 43	99.542	0.191 70
180	0.902 52	103.427	0.202 33	0.767 08	103.141	0.199 48	0.665 43	102.851	0.196 96
200	0.935 31	106.700	0.207 36	0.795 71	106.439	0.204 55	0.690 95	106.174	0.202 07
220	0.967 75	109.997	0.212 29	0.823 97	109.756	0.209 51	0.716 09	109.513	0.207 06
240	0.999 88	113.319	0.217 10	0.851 91	113.096	0.214 35	0.740 90	112.872	0.211 93
260	1.0318	116.666	0.221 82	0.879 59	116.459	0.219 09	0.765 44	116.251	0.216 69
280	1.0634	120.039	0.226 44	0.907 05	119.846	0.223 73	0.789 75	119.652	0.221 35

90 psia (73.79°F)

Temp			
100	0.487 49	89.175	0.172 34
120	0.513 46	92.536	0.178 24
140	0.538 45	95.879	0.183 91
160	0.562 68	99.216	0.189 38
180	0.586 29	102.557	0.194 69
200	0.609 41	105.905	0.199 84
220	0.632 13	109.267	0.204 86
240	0.654 51	112.644	0.209 76
260	0.676 62	116.040	0.214 55
280	0.698 49	119.456	0.219 23
300	0.720 16	122.892	0.223 81
320	0.741 66	126.349	0.228 30

100 psia (80.76°F)

Temp			
100	0.431 38	88.694	0.169 96
120	0.455 62	92.116	0.175 97
140	0.478 81	95.507	0.181 72
160	0.501 18	98.884	0.187 26
180	0.522 91	102.257	0.192 62
200	0.544 13	105.633	0.197 82
220	0.564 92	109.018	0.202 87
240	0.585 38	112.415	0.207 80
260	0.605 54	115.828	0.212 61
280	0.625 46	119.258	0.217 31
300	0.645 18	122.707	0.221 91
320	0.664 72	126.176	0.226 41

125 psia (96.17°F)

Temp			
100	0.329 43	87.407	0.164 55
120	0.350 86	91.008	0.170 87
140	0.370 98	94.537	0.176 86
160	0.390 15	98.023	0.182 58
180	0.408 57	101.484	0.188 07
200	0.426 42	104.934	0.193 38
220	0.443 80	108.380	0.198 53
240	0.460 81	111.829	0.203 53
260	0.477 50	115.287	0.208 40
280	0.493 94	118.756	0.213 16
300	0.510 16	122.238	0.217 80
320	0.526 19	125.737	0.222 35

150 psia (109.45°F)

Temp			
120	0.280 07	89.800	0.166 29
140	0.298 45	93.498	0.172 56
160	0.315 66	97.112	0.178 49
180	0.332 00	100.675	0.184 15
200	0.347 69	104.206	0.189 58
220	0.362 85	107.720	0.194 83
240	0.377 61	111.226	0.199 92
260	0.392 03	114.732	0.204 85
280	0.406 17	118.242	0.209 67
300	0.420 08	121.761	0.214 36
320	0.433 79	125.290	0.218 94
340	0.447 33	128.833	0.223 43

175 psia (121.19°F)

Temp			
120	—	—	—
140	0.245 95	92.373	0.168 59
160	0.261 98	96.142	0.174 78
180	0.276 97	99.823	0.180 62
200	0.291 20	103.447	0.186 20
220	0.304 85	107.036	0.191 56
240	0.318 04	110.605	0.196 74
260	0.330 87	114.162	0.201 75
280	0.343 39	117.717	0.206 62
300	0.355 57	121.273	0.211 37
320	0.367 73	124.835	0.215 99
340	0.379 63	128.407	0.220 52

200 psia (131.74°F)

Temp			
120	—	—	—
140	0.205 79	91.137	0.164 80
160	0.221 21	95.100	0.171 30
180	0.235 35	98.921	0.177 37
200	0.248 60	102.652	0.183 11
220	0.261 17	106.325	0.188 60
240	0.273 23	109.962	0.193 87
260	0.284 89	113.576	0.198 96
280	0.296 23	117.178	0.203 90
300	0.307 30	120.775	0.208 70
320	0.318 15	124.373	0.213 37
340	0.328 81	127.974	0.217 93

TABLE A.21   SUPERHEATED REFRIGERANT-12 TABLE (English Units) (Continued)

T (°F)	250 psia (150.24°F) v (ft³/lbm)	250 psia h (Btu/lbm)	250 psia s (Btu/lbm-R)	300 psia (166.18°F) v (ft³/lbm)	300 psia h (Btu/lbm)	300 psia s (Btu/lbm-R)	400 psia (192.93°F) v (ft³/lbm)	400 psia h (Btu/lbm)	400 psia s (Btu/lbm-R)
160	0.162 49	92.717	0.164 62	—	—	—	—	—	—
180	0.176 05	96.925	0.171 30	0.134 82	94.556	0.165 37	—	—	—
200	0.188 24	100.930	0.177 47	0.146 97	98.975	0.172 17	0.091 005	93.718	0.160 92
220	0.199 52	104.809	0.183 26	0.157 74	103.136	0.178 38	0.103 16	99.046	0.168 88
240	0.210 14	108.607	0.188 77	0.167 61	107.140	0.184 19	0.113 00	103.735	0.175 68
260	0.220 27	112.351	0.194 04	0.176 85	111.043	0.189 69	0.121 63	108.105	0.181 83
280	0.230 01	116.060	0.199 13	0.185 62	114.879	0.194 95	0.129 49	112.286	0.187 56
300	0.239 44	119.747	0.204 05	0.194 02	118.670	0.200 00	0.136 80	116.343	0.192 98
320	0.248 62	123.420	0.208 82	0.202 14	122.430	0.204 89	0.143 72	120.318	0.198 14
340	0.257 59	127.088	0.213 46	0.210 02	126.171	0.209 63	0.150 32	124.235	0.203 10
360	0.266 39	130.754	0.217 99	0.217 70	129.900	0.214 23	0.156 68	128.112	0.207 89
380	0.275 04	134.423	0.222 41	0.225 22	133.624	0.218 72	0.162 85	131.961	0.212 58

T (°F)	500 psia (215.10°F) v (ft³/lbm)	500 psia h (Btu/lbm)	500 psia s (Btu/lbm-R)	600 psia v (ft³/lbm)	600 psia h (Btu/lbm)	600 psia s (Btu/lbm-R)
220	0.064 207	92.397	0.156 83	—	—	—
240	0.077 620	99.218	0.166 72	0.047 488	91.024	0.153 35
260	0.087 054	104.526	0.174 21	0.061 922	99.741	0.165 66
280	0.094 923	109.277	0.180 72	0.070 859	105.637	0.173 74
300	0.101 90	113.729	0.186 66	0.078 059	110.729	0.180 53
320	0.108 29	117.997	0.192 21	0.084 333	115.420	0.186 63
340	0.114 26	122.143	0.197 46	0.090 017	119.871	0.192 27
360	0.119 92	126.205	0.202 47	0.095 289	124.167	0.197 57
380	0.125 33	130.207	0.207 30	0.100 25	128.355	0.202 62
400	0.130 54	134.166	0.211 96	0.104 98	132.466	0.207 46
420	0.135 59	138.096	0.216 48	0.109 52	136.523	0.212 13
440	0.140 51	142.004	0.220 87	0.113 91	140.539	0.216 64

Source: Copyright 1955 and 1956, E. I. duPont de Nemours & Co., Inc. Reprinted by permission.

TABLE A.22 PROPERTIES OF SELECTED MATERIALS AT 20°C (68°F)

Material	λ (W/m-K)	λ (Btu/hr·ft·°F)	c (kJ/kg-K)	c (Btu/lbm-°F)	ρ (kg/m³)	ρ (lbm/ft³)	ϵ
Metals							
Aluminum	236	136	0.896	0.214	2702	169	0.037 (polished)
Copper	399	230	0.383	0.901	8933	558	0.035 (slightly tarnished)
Gold	316	182	0.129	0.031	19 300	1205	0.446 (not polished)
Iron	81	47	0.452	0.108	7870	491	0.741 (oxidized smooth)
Lead	35	20	0.129	0.031	11 340	708	0.226 (gray oxidized)
Magnesium	156	90	1.017	0.243	1740	109	
Nickel	91	53	0.446	0.106	8900	556	0.389 (oxidized)
Silver	427	247	0.234	0.056	10 500	656	0.03 (polished)
Zinc	121	70	0.385	0.092	7140	446	0.237 (tarnished)
Alloys							
Brass (70% Cu, 30% Zn)	111	64	0.385	0.092	8522	532	0.209 (tarnished)
Cast iron ($\approx$4% C)	52	30	0.420	0.100	7272	454	0.64 (oxidized)
Stainless steel	14	8	0.461	0.110	7817	488	0.14
Steel (1% C)	43	25	0.473	0.113	7801	487	0.921 (oxidized rough)
Insulating							
Asbestos	0.113	0.065	0.816	0.195	383	23.9	
Brick, common	0.42	0.243	0.840	0.220	1800	112.4	0.911 (red rough)
Chrome brick	0.774	0.447	0.837	0.200	3011	188.0	
Cork, boards	0.042	0.024	1.880	0.499	150	9.4	
Fire-clay brick	0.347	0.200	0.963	0.230	2322	145.0	0.75
Glass fiber	0.035	0.020			220	13.7	
Magnesia (85%)	0.022	0.013			270	16.8	
Wood (oak)	0.069	0.040	2.386	0.570	528	33.0	0.840 (planed)

TABLE A.23 PHYSICAL PROPERTIES OF SELECTED FLUIDS

Temperature, T (K)	Density, ρ (kg/m³)	Specific heat, c_p (J/kg-K)	Thermal conductivity, λ (W/m-K)	Thermal diffusivity, $\alpha \times 10^6$ (m²/s)	Absolute viscosity, $\mu \times 10^6$ (N·s/m²)	Kinematic viscosity, $v \times 10^6$ (m²/s)	Prandtl number, Pr
Dry air at atmospheric pressure[a]							
273	1.252	1011	0.0237	19.2	17.456	13.9	0.71
293	1.164	1012	0.0251	22.0	18.240	15.7	0.71
313	1.092	1014	0.0265	24.8	19.123	17.6	0.71
333	1.025	1017	0.0279	27.6	19.907	19.4	0.71
353	0.968	1019	0.0293	30.6	20.790	21.5	0.71
373	0.916	1022	0.0307	33.6	21.673	23.6	0.71
473	0.723	1035	0.0370	49.7	25.693	35.5	0.71
573	0.596	1047	0.0429	68.9	39.322	49.2	0.71
673	0.508	1059	0.0485	89.4	32.754	64.6	0.72
773	0.442	1076	0.0540	113.2	35.794	81.0	0.72
1273	0.268	1139	0.0762	240	48.445	181	0.74
Water at saturation pressure[a]							
273	999.3	4226	0.558	0.131	1794	1.789	13.7
293	998.2	4182	0.597	0.143	993	1.006	7.0
313	992.2	4175	0.633	0.151	658	0.658	4.3
333	983.2	4181	0.658	0.159	472	0.478	3.00
353	971.8	4194	0.673	0.165	352	0.364	2.25
373	958.4	4211	0.682	0.169	278	0.294	1.75
473	862.8	4501	0.665	0.170	139	0.160	0.95
573	712.5	5694	0.564	0.132	92.2	0.128	0.98
Refrigerant R 12 (CCl_2F_2), saturated liquid[b]							
223	1547	875.0	0.067	5.01	4.796	0.310	6.2
233	1519	884.7	0.069	5.14	4.238	0.279	5.4
243	1490	895.6	0.069	5.26	3.770	0.253	4.8
253	1461	907.3	0.071	5.39	3.433	0.235	4.4
263	1429	920.3	0.073	5.50	3.158	0.221	4.0
273	1397	934.5	0.073	5.57	2.990	0.214	3.8
283	1364	949.6	0.073	5.60	2.769	0.203	3.6
293	1330	965.9	0.073	5.60	2.633	0.198	3.5
303	1295	983.5	0.071	5.60	2.512	0.194	3.5
313	1257	1001.9	0.069	5.55	2.401	0.191	3.5
323	1216	1021.6	0.067	5.45	2.310	0.190	3.5
Unused engine oil, saturated liquid[b]							
273	899.1	1796	0.147	911	3848	4280	471
293	888.2	1880	0.145	872	799	900	104
313	876.1	1964	0.144	834	210	240	28.7
333	864.0	2047	0.140	800	72.5	83.9	10.5
353	852.0	2131	0.138	769	32.0	37.5	4.90
373	840.0	2219	0.137	738	17.1	20.3	2.76
393	829.0	2307	0.135	710	10.3	12.4	1.75
413	816.9	2395	0.133	686	6.54	8.0	1.16
433	805.9	2483	0.132	663	4.51	5.6	0.84

[a] Source: K. Raznjevic, *Handbook of Thermodynamic Tables and Charts,* New York: McGraw-Hill Book Company, 1976.

[b] Source: E. R. G. Eckert and R. M. Drake, *Analysis of Heat and Mass Transfer,* New York: McGraw-Hill Book Company, 1972.

Appendix B

TABLE B.1 MOLLIER (ENTHALPY–ENTROPY) DIAGRAM FOR STEAM

Entropy, s, kJ/kgK

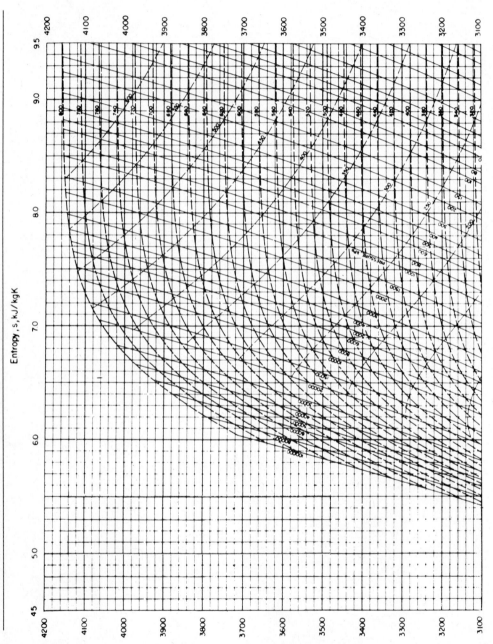

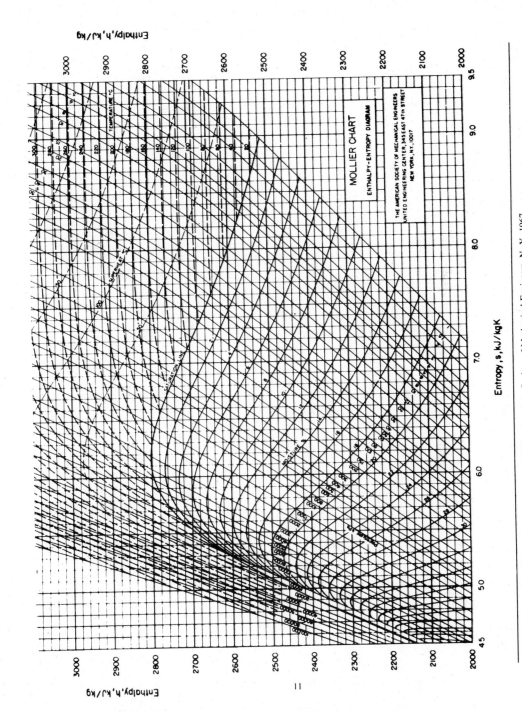

MOLLIER CHART

ENTHALPY-ENTROPY DIAGRAM

THE AMERICAN SOCIETY OF MECHANICAL ENGINEERS
UNITED ENGINEERING CENTER, 345 EAST 47th STREET
NEW YORK, N.Y., 10017

Enthalpy, h, kJ/kg

Entropy, s, kJ/kgK

TEMPERATURE °C

11

Source: Reprinted by permission from ASME Steam Tables, American Society of Mechanical Engineers, N. Y. 1967.

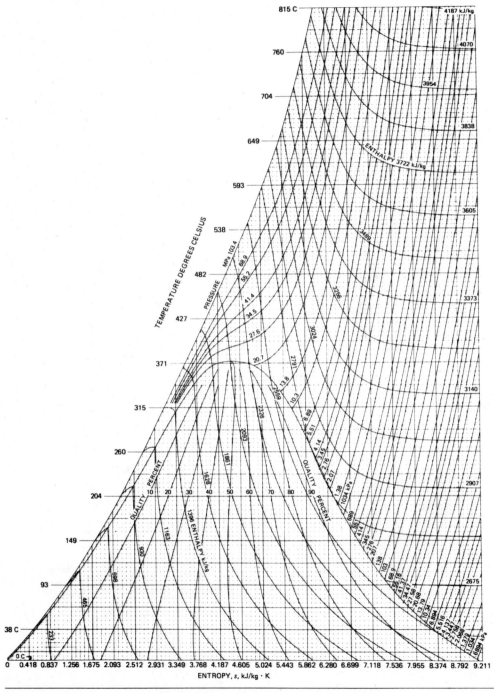

Source: Reprinted by permission from ASME Steam Tables, American Society of Mechanical Engineers, N. Y. 1967.

TABLE B.3(a)   AMMONIA–WATER EQUILIBRIUM CHART (SI Units)   *Source:* Reprinted by permission from *Refrigerating Engineering,* Vol. 58(10) 1950.

(continued overleaf)

TABLE B.3(b) AMMONIA–WATER EQUILIBRIUM CHART (ENGLISH UNITS)

Source: Reprinted by permission from *Refrigerating Engineer* 58(10) 1950.

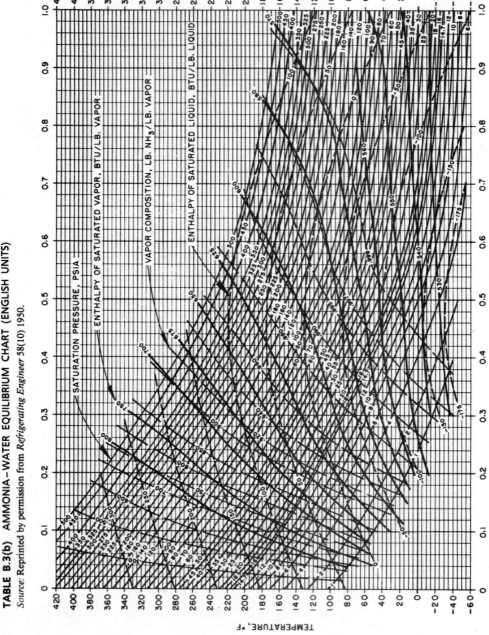

TABLE B.4(a) PSYCHROMETRIC CHART (SI Units)

Source: Reproduced by permission of Carrier Corporation.

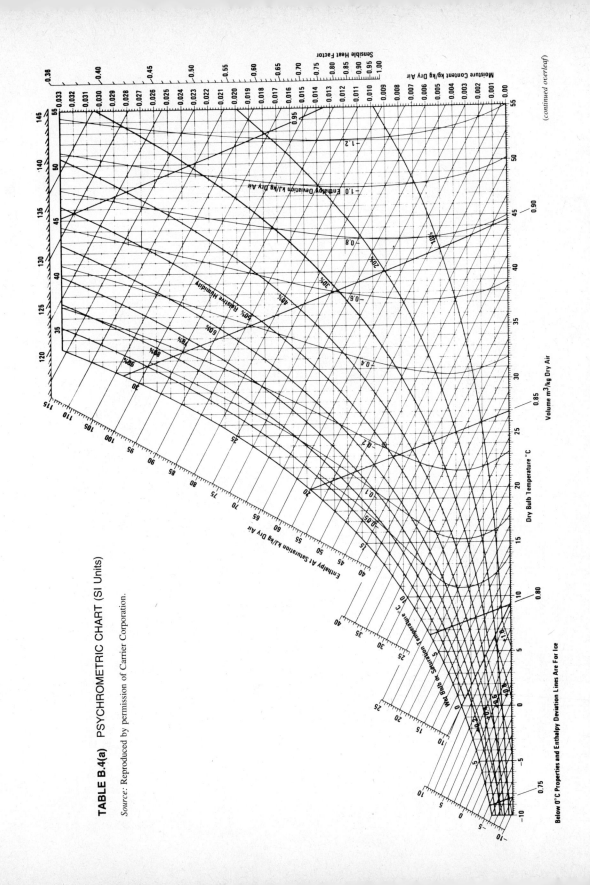

(continued overleaf)

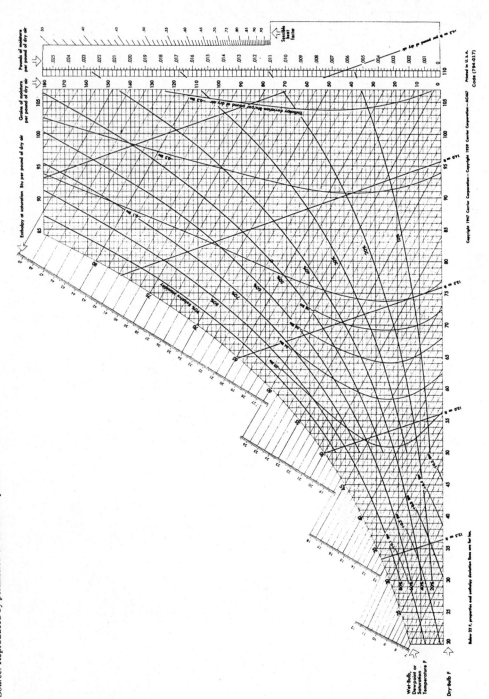

TABLE B.4(b) PSYCHROMETRIC CHART (ENGLISH UNITS)
Source: Reproduced by permission of Carrier Corporation.

Appendix C

TABLE C.1 ENTHALPIES OF FORMATION, GIBBS FUNCTION OF FORMATION, AND ABSOLUTE ENTROPY AT 25°C AND 1 ATM PRESSURE

Substance	M	$\bar{h}_f^\circ$ (kJ/kgmol)	h_f° (Btu/pmol)	$\bar{g}_f^\circ$ (kJ/kgmol)	g_f° (Btu/pmol)	$\bar{s}^\circ$ (kJ/kgmol-K)	s° (Btu/pmol-R)
Acetylene C_2H_2(g)	26.038	226 866	97,542	209 290	89,987	200.98	48.004
Ammonia NH_3(g)	17.032	−45 926	−19,746	−16 390	−7,047	192.72	46.03
Benzene C_6H_6(g)	78.108	82 976	35,676	129 732	55,780	269.38	64.34
Butane C_4H_{10}(g)	58.124	−126 223	−54,270	−17 164	−7,380	310.32	74.12
Carbon C, graphite	12.011	0	0	0	0	5.69	1.359
Carbon dioxide CO_2(g)	14.01	−393 757	−169,297	−394 631	−169,677	213.83	51.072
Carbon monoxide CO(g)	28.01	−110 596	−47,551	−137 242	−59,009	197.68	47.214
Dodecane $C_{12}H_{26}$(g)	170.328	−290 971	−125,104				
Dodecane $C_{12}H_{26}$(l)	170.328	−394 199	−169,487				
Ethane C_2H_6(g)	30.07	−84 718	−36,425	−32 095	−14,148	229.64	54.85
Ethene C_2H_4(g)	28.054	52 315	22,493	68 159	29,306	219.58	52.447
Hydrazine N_2H_4(g)	32.048	95 410	41,022	159 260	68,476	238.77	57.03
Methane CH_4(g)	16.043	−74 917	−32,211	−50 844	−21,861	186.27	44.49
Octane C_8H_{18}(g)	114.23	−208 581	−89,680	16 536	7,110	467.04	111.55
Octane C_8H_{18}(l)	114.23	−250 102	−107,532	6 614	2,844	361.03	86.23
Propane C_3H_8(g)	44.097	−103 909	−44,676	−23 502	−10,105	270.09	64.51
Water H_2O(g)	18.016	−241 971	−104,036	−228 729	−98,345	188.85	45.106
Water H_2O(l)	18.016	−286 010	−122,971	−237 327	−102,042	69.98	16.716

Source: JANAF Thermochemical Tables, Document PB 168-370, Clearinghouse for Federal Scientific and Technical Information, 1965.

TABLE C.2 IDEAL-GAS ENTHALPY AND ABSOLUTE ENTROPY AT 1 ATM PRESSURE

Nitrogen, diatomic (N_2)
$(\bar{h}_f^\circ)_{298} = 0$ kJ/kmol $= 0$ Btu/pmol
$M = 28.016$

T (K)	T (R)	$\bar{h}^\circ - \bar{h}^\circ_{298}$ (kJ/kmol)	$\bar{h}^\circ - \bar{h}^\circ_{537}$ (Btu/pmol)	$\bar{s}^\circ$ (Btu/pmol-R)	$\bar{s}^\circ$ (kJ/kmol-K)
0	0	−8669	−3,730	0	0
100	180	−5770	−2,483	38.170	159.813
200	360	−2858	−1,229	42.992	179.988
298	537	0	0	45.770	191.611
300	540	54	23	45.813	191.791
400	720	2971	1,278	47.818	200.180
500	900	5912	2,543	49.386	206.740
600	1,080	8891	3,825	50.685	212.175
700	1,260	11 937	5,135	51.806	216.866
800	1,440	15 046	6,473	52.798	221.016
900	1,620	18 221	7,839	53.692	224.757
1000	1,800	21 460	9,232	54.507	228.167
1100	1,980	24 757	10,651	55.258	231.309
1200	2,160	28 108	12,092	55.955	234.225
1300	2,340	31 501	13,552	56.604	236.941
1400	2,520	34 936	15,030	57.212	239.484
1500	2,700	38 405	16,522	57.784	241.878
1600	2,880	41 903	18,027	58.324	244.137
1700	3,060	45 430	19,544	58.835	246.275
1800	3,240	48 982	21,073	59.320	248.304
1900	3,420	52 551	22,608	59.782	250.237
2000	3,600	56 141	24,152	60.222	252.078
2100	3,780	59 748	25,704	60.642	253.836
2200	3,960	63 371	27,263	61.045	255.522
2300	4,140	67 007	28,827	61.431	257.137
2400	4,320	70 651	30,395	61.802	258.689
2500	4,500	74 312	31,970	62.159	260.183
2600	4,680	77 973	33,548	62.503	261.622
2700	4,860	81 659	35,131	62.835	263.011
2800	5,040	85 345	36,716	63.155	264.350
2900	5,220	89 036	38,304	63.465	265.647
3000	5,400	92 738	39,897	63.765	266.902
3200	5,760	100 161	43,090	64.337	269.295
3400	6,120	107 608	46,294	64.877	271.555
3600	6,480	115 031	49,509	65.387	273.689
3800	6,840	122 570	52,731	65.871	275.714
4000	7,200	130 076	55,960	66.331	277.638
4200	7,560	137 603	59,198	66.770	279.475
4400	7,920	145 143	62,442	67.189	281.228
4600	8,280	152 699	65,693	67.591	282.910
4800	8,640	160 272	68,951	67.976	284.521
5000	9,000	167 858	72,214	68.346	286.069
5200	9,360	175 456	75,483	68.702	287.559
5400	9,720	183 071	78,759	69.045	288.994
5600	10,180	190 703	82,042	69.377	290.383
5800	10,540	198 347	85,331	69.698	291.726
6000	10,800	206 008	88,627	70.008	293.023

Oxygen, diatomic (O_2)
$(\bar{h}_f^\circ)_{298} = 0$ kJ/kmol $= 0$ Btu/pmol
$M = 32.00$

T (K)	T (R)	$\bar{h}^\circ - \bar{h}^\circ_{298}$ (kJ/kmol)	$\bar{h}^\circ - \bar{h}^\circ_{537}$ (Btu/pmol)	$\bar{s}^\circ$ (Btu/pmol-R)	$\bar{s}^\circ$ (kJ/kmol-K)
0	0	−8682	−3,735	0	0
100	180	−5778	−2,486	41.395	173.306
200	360	−2866	−1,233	46.218	198.486
298	537	0	0	49.004	205.142
300	540	54	23	49.047	205.322
400	720	3029	1,303	51.091	213.874
500	900	6088	2,619	52.722	220.698
600	1,080	9247	3,978	54.098	226.455
700	1,260	12 502	5,378	55.297	231.272
800	1,440	15 841	6,815	56.361	235.924
900	1,620	19 246	8,280	57.320	239.936
1000	1,800	22 707	9,769	58.192	243.585
1100	1,980	26 217	11,279	58.991	246.928
1200	2,160	29 765	12,805	59.729	250.016
1300	2,340	33 351	14,348	60.415	252.886
1400	2,520	36 966	15,903	61.055	255.564
1500	2,700	40 610	17,471	61.656	258.078
1600	2,880	44 279	19,049	62.222	260.446
1700	3,060	47 970	20,637	62.757	262.685
1800	3,240	51 689	22,237	63.265	264.810
1900	3,420	55 434	23,848	63.749	266.835
2000	3,600	59 199	25,468	64.210	268.764
2100	3,780	62 986	27,097	64.652	270.613
2200	3,960	66 802	28,739	65.076	272.387
2300	4,140	70 634	30,388	65.483	274.090
2400	4,320	74 492	32,047	65.876	275.735
2500	4,500	78 375	33,718	66.254	277.316
2600	4,680	82 274	35,395	66.620	278.848
2700	4,860	86 199	37,084	66.974	280.329
2800	5,040	90 144	38,781	67.317	281.764
2900	5,220	94 111	40,487	67.650	283.157
3000	5,400	98 098	42,203	67.973	284.508
3200	5,760	106 127	45,657	68.592	287.098
3400	6,120	114 232	49,144	69.179	289.554
3600	6,400	122 399	52,657	69.737	291.889
3800	6,840	130 629	56,198	70.269	294.115
4000	7,200	138 913	59,762	70.776	296.236
4200	7,560	147 248	63,347	71.262	298.270
4400	7,920	155 628	66,953	71.728	300.219
4600	8,280	164 046	70,574	72.176	302.094
4800	8,640	172 502	74,212	72.606	303.893
5000	9,000	180 987	77,863	73.019	305.621
5200	9,360	189 502	81,526	73.418	307.290
5400	9,720	198 037	85,198	73.803	308.901
5600	10,180	206 593	88,879	74.175	310.458
5800	10,540	215 166	92,567	74.535	311.964
6000	10,800	223 756	96,262	74.883	313.420

TABLE C.2 IDEAL-GAS ENTHALPY AND ABSOLUTE ENTROPY AT 1 ATM PRESSURE *(Continued)*

Hydrogen, diatomic (H_2)
$(\bar{h}_f^\circ)_{298} = 0$ kJ/kmol $= 0$ Btu/pmol
$M = 2.016$

T (K)	T (R)	$\bar{h}^\circ - \bar{h}_{298}^\circ$ (kJ/kmol)	$\bar{h}^\circ - \bar{h}_{537}^\circ$ (Btu/pmol)	$\bar{s}^\circ$ (Btu/pmol-R)	$\bar{s}^\circ$ (kJ/kmol-K)
0	0	−8468	−3,643	0	0
100	180	−5293	−2,277	24.387	102.145
200	360	−2770	−1,192	28.520	119.437
298	537	0	0	31.208	130.684
300	540	54	23	31.251	130.864
400	720	2958	1,273	33.247	139.215
500	900	5883	2,531	34.806	145.738
600	1,080	8812	3,791	36.082	151.077
700	1,260	11 749	5,054	37.165	155.608
800	1,440	14 703	6,325	38.107	159.549
900	1,620	17 682	7,607	38.946	163.060
1000	1,800	20 686	8,899	39.702	166.223
1100	1,980	23 723	10,206	40.394	169.118
1200	2,160	26 794	11,527	41.033	171.792
1300	2,340	29 907	12,866	41.628	174.281
1400	2,520	33 062	14,224	42.187	176.620
1500	2,700	36 267	15,602	42.716	178.833
1600	2,880	39 522	17,003	43.217	180.929
1700	3,060	42 815	18,419	43.695	182.929
1800	3,240	46 150	19,854	44.150	184.833
1900	3,420	49 522	21,305	44.586	186.657
2000	3,600	52 932	22,772	45.004	188.406
2100	3,780	56 379	24,255	45.406	190.088
2200	3,960	59 860	25,753	45.793	191.707
2300	4,140	63 371	27,263	46.166	193.268
2400	4,320	66 915	28,787	46.527	194.778
2500	4,500	70 492	30,326	46.875	196.234
2600	4,680	74 090	31,874	47.213	197.649
2700	4,860	77 718	33,435	47.540	199.017
2800	5,040	81 370	35,006	47.857	200.343
2900	5,220	85 044	36,587	48.166	201.636
3000	5,400	88 743	38,178	48.465	202.887
3200	5,760	96 199	41,386	49.040	205.293
3400	6,120	103 738	44,629	49.586	207.577
3600	6,400	111 361	47,909	50.107	209.757
3800	6,840	119 064	51,223	50.605	211.841
4000	7,200	126 846	54,571	51.082	213.837
4200	7,560	134 700	57,949	51.540	215.753
4400	7,920	142 624	61,358	51.980	217.594
4600	8,280	150 620	64,798	52.405	219.372
4800	8,640	158 682	68,267	52.815	221.087
5000	9,000	166 808	71,762	53.211	222.744
5200	9,360	174 996	75,285	53.595	224.351
5400	9,720	183 247	78,835	53.967	225.907
5600	10,180	191 556	82,409	54.328	227.418
5800	10,540	199 924	86,009	54.679	228.886
6000	10,800	208 346	89,633	55.020	230.313

TABLE C.2 IDEAL-GAS ENTHALPY AND ABSOLUTE ENTROPY AT 1 ATM PRESSURE (Continued)

Hydrogen, monatomic (H)
$(\bar{h}_f^\circ)_{298} = 217\,986$ kJ/kmol $= 93,780$ Btu/pmol
$M = 1.008$

T (K)	T (R)	$\bar{h}^\circ - \bar{h}_{298}^\circ$ (kJ/kmol)	$\bar{h}^\circ - \bar{h}_{537}^\circ$ (Btu/pmol)	$\bar{s}^\circ$ (Btu/pmol-R)	$\bar{s}^\circ$ (kJ/kmol-K)
0	0	−6197	−2,666	0	0
100	180	−4117	−1,771	21.965	92.011
200	360	−2042	−878	25.408	106.417
298	537	0	0	27.392	114.718
300	540	38	16	27.423	114.847
400	720	2117	911	28.852	120.826
500	900	4197	1,805	29.961	125.466
600	1,080	6276	2,700	30.867	129.257
700	1,260	8351	3,593	31.632	132.458
800	1,440	10 431	4,487	32.296	135.236
900	1,620	12 510	5,382	32.881	137.684
1000	1,800	14 590	6,277	33.404	139.872
1100	1,980	16 669	7,171	33.878	141.855
1200	2,160	18 749	8,066	34.310	143.662
1300	2,340	20 824	8,959	34.708	145.328
1400	2,520	22 903	9,853	35.076	146.867
1500	2,700	24 983	10,748	35.419	148.303
1600	2,880	27 062	11,642	35.739	149.641
1700	3,060	29 142	12,537	36.041	150.905
1800	3,240	31 217	13,430	36.325	152.093
1900	3,420	33 296	14,324	36.593	153.215
2000	3,600	35 376	15,219	36.848	154.281
2100	3,780	37 455	16,114	37.090	155.294
2200	3,960	39 535	17,008	37.322	156.265
2300	4,140	41 610	17,901	37.542	157.185
2400	4,320	43 689	18,796	37.754	158.072
2500	4,500	45 769	19,690	37.957	158.922
2600	4,680	47 848	20,585	38.152	159.737
2700	4,860	49 928	21,479	38.339	160.520
2800	5,040	52 007	22,374	38.520	161.277
2900	5,220	54 082	23,267	38.694	162.005
3000	5,400	56 162	24,161	38.862	162.708
3200	5,760	60 321	25,951	39.183	164.051
3400	6,120	64 475	27,738	39.484	165.311
3600	6,400	68 634	29,527	39.768	166.499
3800	6,840	72 793	31,316	40.037	167.624
4000	7,200	76 948	33,104	40.292	168.691
4200	7,560	81 107	34,893	40.534	169.704
4400	7,920	85 266	36,682	40.765	170.670
4600	8,280	89 420	38,470	40.986	171.595
4800	8,640	93 579	40,259	41.198	172.482
5000	9,000	97 734	42,046	41.400	173.327
5200	9,360	101 893	43,835	41.595	174.143
5400	9,720	106 052	45,625	41.783	174.930
5600	10,180	110 207	47,412	41.963	175.683
5800	10,540	114 365	49,201	42.138	176.415
6000	10,800	118 524	50,990	42.306	177.118

Carbon dioxide (CO_2)

$(\bar{h}_f^{\circ})_{298} = -393\ 757$ kJ/kmol $= -169{,}297$ Btu/pmol

$M = 44.011$

T (K)	T (R)	$\bar{h}^{\circ} - \bar{h}_{298}^{\circ}$ (kJ/kmol)	$\bar{h}^{\circ} - \bar{h}_{537}^{\circ}$ (Btu/pmol)	$\bar{s}^{\circ}$ (Btu/pmol-R)	$\bar{s}^{\circ}$ (kJ/kmol-K)
0	0	−9364	−4,028	0	0
100	180	−6456	−2,777	42.758	179.109
200	360	−3414	−1,469	47.769	199.975
298	537	0	0	51.072	213.795
300	540	67	29	51.127	214.025
400	720	4008	1,724	53.830	225.334
500	900	8314	3,577	56.122	234.924
600	1,080	12 916	5,557	58.126	243.309
700	1,260	17 761	7,641	59.910	250.773
800	1,440	22 815	9,815	61.522	257.517
900	1,620	28 041	12,064	62.992	263.668
1000	1,800	33 405	14,371	64.344	269.325
1100	1,980	38 894	16,733	65.594	274.555
1200	2,160	44 484	19,138	66.756	279.417
1300	2,340	50 158	21,578	67.841	283.956
1400	2,520	55 907	24,052	68.859	288.216
1500	2,700	61 714	26,550	69.817	292.224
1600	2,880	67 580	29,074	70.722	296.010
1700	3,060	73 492	31,617	71.578	299.592
1800	3,240	79 442	34,177	72.391	302.993
1900	3,420	85 429	36,752	73.165	306.232
2000	3,600	91 450	39,343	73.903	309.320
2100	3,780	97 500	41,945	74.608	312.269
2200	3,960	103 575	44,559	75.284	315.098
2300	4,140	109 671	47,182	75.931	317.805
2400	4,320	115 788	49,813	76.554	320.411
2500	4,500	121 926	52,454	77.153	322.918
2600	4,680	128 085	55,103	77.730	325.332
2700	4,860	134 256	57,758	78.286	327.658
2800	5,040	140 444	60,421	78.824	329.909
2900	5,220	146 645	63,088	79.344	332.085
3000	5,400	152 862	65,763	79.848	334.193
3200	5,760	165 331	71,127	80.810	338.218
3400	6,120	177 849	76,513	81.717	342.013
3600	6,400	190 405	81,914	82.574	345.599
3800	6,840	202 999	87,332	83.388	349.005
4000	7,200	215 635	92,768	84.162	352.243
4200	7,560	228 304	98,219	84.901	355.335
4400	7,920	241 003	103,682	85.607	358.289
4600	8,280	253 734	109,159	86.284	361.122
4800	8,640	266 500	114,651	86.933	363.837
5000	9,000	279 295	120,155	87.557	366.448
5200	9,360	292 123	125,674	88.158	368.963
5400	9,720	304 984	131,207	88.738	371.389
5600	10,180	317 884	136,757	89.299	373.736
5800	10,540	330 821	142,322	89.841	376.004
6000	10,800	343 791	147,902	90.367	378.205

Carbon monoxide (CO)
$(\bar{h}_f^\circ)_{298} = -110\,596$ kJ/kmol $= -47,551$ Btu/pmol
$M = 28.011$

T (K)	T (R)	$\bar{h}^\circ - \bar{h}_{298}^\circ$ (kJ/kmol)	$\bar{h}^\circ - \bar{h}_{537}^\circ$ (Btu/pmol)	$\bar{s}^\circ$ (Btu/pmol-R)	$\bar{s}^\circ$ (kJ/kmol-K)
0	0	−8699	−3,730	0	0
100	180	−5770	−2,483	39.613	165.850
200	360	−2858	−1,229	44.435	186.025
298	537	0	0	47.214	197.653
300	540	54	23	47.257	197.833
400	720	2975	1,280	49.265	206.234
500	900	5929	2,551	50.841	212.828
600	1,080	8941	3,847	52.152	218.313
700	1,260	12 021	5,171	53.287	223.062
800	1,440	15 175	6,529	54.293	227.271
900	1,620	18 397	7,915	55.200	231.066
1000	1,800	21 686	9,329	56.028	234.531
1100	1,980	25 033	10,769	56.790	237.719
1200	2,160	28 426	12,229	57.496	240.673
1300	2,340	31 865	13,709	58.154	243.426
1400	2,520	35 338	15,203	58.769	245.999
1500	2,700	38 848	16,713	59.348	248.421
1600	2,880	42 384	18,234	59.893	250.702
1700	3,060	45 940	19,764	60.409	252.861
1800	3,240	49 522	21,305	60.898	254.907
1900	3,420	53 124	22,855	61.363	256.852
2000	3,600	56 739	24,410	61.807	258.710
2100	3,780	60 375	25,974	62.230	260.480
2200	3,960	64 019	27,542	62.635	262.174
2300	4,140	67 676	29,115	63.024	263.802
2400	4,320	71 340	30,694	63.397	265.362
2500	4,500	75 023	32,276	63.756	266.865
2600	4,680	78 714	33,863	64.102	268.312
2700	4,860	82 408	35,453	64.435	269.705
2800	5,040	86 115	37,048	64.757	271.053
2900	5,220	89 826	38,644	65.069	272.358
3000	5,400	93 542	40,243	65.370	273.618
3200	5,760	100 998	43,450	65.945	276.023
3400	6,120	108 479	46,669	66.487	278.291
3600	6,400	115 976	49,894	66.999	280.433
3800	6,840	123 495	53,129	67.485	282.467
4000	7,200	131 026	56,369	67.946	284.396
4200	7,560	138 578	59,618	68.387	286.241
4400	7,920	146 147	62,874	68.807	287.998
4600	8,280	153 724	66,134	69.210	289.684
4800	8,640	161 322	69,403	69.596	291.299
5000	9,000	168 929	72,675	69.967	292.851
5200	9,360	176 548	75,953	70.325	294.349
5400	9,720	184 184	79,238	70.669	295.789
5600	10,180	191 832	82,528	71.001	297.178
5800	10,540	199 489	85,822	71.322	298.521
6000	10,800	207 162	89,123	71.633	299.822

Nitric oxide (NO)
$(\bar{h}_f^\circ)_{298} = 90\ 592$ kJ/kmol $= 38{,}974$ Btu/pmol
$M = 30.008$

T (K)	T (R)	$\bar{h}^\circ - \bar{h}^\circ_{298}$ (kJ/kmol)	$\bar{h}^\circ - \bar{h}^\circ_{537}$ (Btu/pmol)	$\bar{s}^\circ$ (Btu/pmol-R)	$\bar{s}^\circ$ (kJ/kmol-K)
0	0	−9192	−3,955	0	0
100	180	−6071	−2,612	42.286	177.034
200	360	−2950	−1,269	47.477	198.753
298	537	0	0	50.347	210.761
300	540	54	23	50.392	210.950
400	720	3042	1,309	52.444	219.535
500	900	6058	2,606	54.053	226.267
600	1,080	9146	3,935	55.397	231.890
700	1,260	12 309	5,296	56.562	236.765
800	1,440	15 548	6,689	57.596	241.091
900	1,620	18 857	8,113	58.528	244.991
1000	1,800	22 230	9,563	59.377	248.543
1100	1,980	25 652	11,036	60.157	251.806
1200	2,160	29 121	12,528	60.878	254.823
1300	2,340	32 627	14,036	61.548	257.626
1400	2,520	36 166	15,559	62.175	260.250
1500	2,700	39 731	17,093	62.763	262.710
1600	2,880	43 321	18,637	63.317	265.028
1700	3,060	46 932	20,191	63.840	267.216
1800	3,240	50 559	21,751	64.335	269.287
1900	3,420	54 204	23,319	64.806	271.258
2000	3,600	57 861	24,892	65.255	273.136
2100	3,780	61 530	26,471	65.683	274.927
2200	3,960	65 216	28,057	66.092	276.638
2300	4,140	68 906	29,644	66.484	278.279
2400	4,320	72 609	31,237	66.861	279.856
2500	4,500	76 320	32,834	67.223	281.370
2600	4,680	80 036	34,432	67.571	282.827
2700	4,860	83 764	36,036	67.907	284.232
2800	5,040	87 492	37,640	68.232	285.592
2900	5,220	91 232	39,249	68.545	286.902
3000	5,400	94 977	40,860	68.849	288.174
3200	5,760	102 479	44,087	69.427	290.592
3400	6,120	110 002	47,324	69.973	292.876
3600	6,400	117 545	50,569	70.488	295.031
3800	6,840	125 102	53,820	70.976	297.073
4000	7,200	132 675	57,078	71.440	299.014
4200	7,560	140 260	60,341	71.882	300.864
4400	7,920	147 863	63,612	72.305	302.634
4600	8,280	155 473	66,886	72.709	304.324
4800	8,640	163 101	70,168	73.097	305.947
5000	9,000	170 736	73,453	73.470	307.508
5200	9,360	178 381	76,741	73.828	309.006
5400	9,720	186 042	80,037	74.173	310.449
5600	10,180	193 707	83,335	74.507	311.847
5800	10,540	201 388	86,639	74.829	313.194
6000	10,800	209 074	89,946	75.140	314.495

TABLE C.2 IDEAL-GAS ENTHALPY AND ABSOLUTE ENTROPY AT 1 ATM
PRESSURE *(Continued)*

Nitrogen dioxide (NO_2)
$(\bar{h}_f^\circ)_{298} = 33\ 723$ kJ/kmol $= 14{,}508$ Btu/pmol
$M = 46.005$

T (K)	T (R)	$\bar{h}^\circ - \bar{h}_{298}^\circ$ (kJ/kmol)	$\bar{h}^\circ - \bar{h}_{537}^\circ$ (Btu/pmol)	$\bar{s}^\circ$ (Btu/pmol-R)	$\bar{s}^\circ$ (kJ/kmol-K)
0	0	$-10\ 196$	$-4{,}387$	0	0
100	180	-6870	$-2{,}956$	48.356	202.431
200	360	-3502	$-1{,}507$	53.925	225.732
298	537	0	0	57.324	239.953
300	540	67	29	57.379	240.183
400	720	3950	1,699	60.041	251.321
500	900	8150	3,506	62.279	260.685
600	1,080	12 640	5,438	64.234	268.865
700	1,260	17 368	7,472	65.975	276.149
800	1,440	22 288	9,589	67.544	282.714
900	1,620	27 359	11,770	69.971	288.684
1000	1,800	32 552	14,004	70.278	294.153
1100	1,980	37 836	16,277	71.482	299.190
1200	2,160	43 196	18,583	72.597	303.855
1300	2,340	48 618	20,916	73.634	308.194
1400	2,520	54 095	23,272	74.604	312.253
1500	2,700	59 609	25,645	75.513	316.056
1600	2,880	65 157	28,031	76.369	319.637
1700	3,060	70 739	30,433	77.178	323.022
1800	3,240	76 345	32,845	77.943	326.223
1900	3,420	81 969	35,264	78.670	329.265
2000	3,600	87 613	37,692	79.362	332.160
2100	3,780	93 274	40,127	80.022	334.921
2200	3,960	98 947	42,568	80.653	337.562
2300	4,140	104 633	45,014	81.257	340.089
2400	4,320	110 332	47,466	81.837	342.515
2500	4,500	116 039	49,921	82.394	344.846
2600	4,680	121 754	52,380	83.930	347.089
2700	4,860	127 478	54,842	83.446	349.248
2800	5,040	133 206	57,307	83.944	351.331
2900	5,220	138 942	59,774	84.425	353.344
3000	5,400	144 683	62,224	84.890	355.289
3200	5,760	156 180	67,190	85.777	359.000
3400	6,120	167 695	72,144	86.611	362.490
3600	6,400	179 222	77,103	87.398	365.783
3800	6,840	190 761	82,067	88.144	368.904
4000	7,200	202 309	87,035	88.852	371.866
4200	7,560	213 865	92,007	89.525	374.682
4400	7,920	225 430	96,982	90.168	377.372
4600	8,280	237 003	101,961	90.783	379.946
4800	8,640	248 580	106,942	91.372	382.410
5000	9,000	260 161	111,924	91.937	384.774
5200	9,360	271 751	116,910	92.480	387.046
5400	9,720	283 340	121,896	93.003	389.234
5600	10,180	294 934	126,884	93.507	391.343
5800	10,540	306 532	131,873	93.993	393.376
6000	10,800	318 130	136,863	94.463	395.343

Water (H_2O)

$(\bar{h}_f^\circ)_{298} = -241\ 971$ kJ/kmol $= -104{,}036$ Btu/pmol

$M = 18.016$

T (K)	T (R)	$\bar{h}^\circ - \bar{h}_{298}^\circ$ (kJ/kmol)	$\bar{h}^\circ - \bar{h}_{537}^\circ$ (Btu/pmol)	$\bar{s}^\circ$ (Btu/pmol-R)	$\bar{s}^\circ$ (kJ/kmol-K)
0	0	−9904	−4,261	0	0
100	180	−6615	−2,846	36.396	152.390
200	360	−3280	−1,411	41.916	175.486
298	537	0	0	45.106	188.833
300	540	63	27	45.155	189.038
400	720	3452	1,485	47.484	198.783
500	900	6920	2,977	49.334	206.523
600	1,080	10 498	4,516	50.891	213.037
700	1,260	14 184	6,102	52.249	218.719
800	1,440	17 991	7,740	53.464	223.803
900	1,620	21 924	9,432	54.570	228.430
1000	1,800	25 978	11,176	55.592	232.706
1100	1,980	30 167	12,978	56.545	236.694
1200	2,160	34 476	14,832	57.441	240.443
1300	2,340	38 903	16,736	58.288	243.986
1400	2,520	43 447	18,691	59.092	247.350
1500	2,700	48 095	20,691	59.859	250.560
1600	2,880	52 844	22,734	60.591	253.622
1700	3,060	57 685	24,817	61.293	256.559
1800	3,240	62 609	26,935	61.965	259.371
1900	3,420	67 613	29,088	62.612	262.078
2000	3,600	72 689	31,271	63.234	264.681
2100	3,780	77 831	33,484	63.834	267.191
2200	3,960	83 036	35,723	64.412	269.609
2300	4,140	88 295	37,985	64.971	271.948
2400	4,320	93 604	40,270	65.511	274.207
2500	4,500	98 964	42,575	66.034	276.396
2600	4,680	104 370	44,901	66.541	278.517
2700	4,860	109 813	47,243	67.032	280.571
2800	5,040	115 294	49,601	67.508	282.563
2900	5,220	120 813	51,975	67.971	284.500
3000	5,400	126 361	54,362	68.421	286.383
3200	5,760	137 553	59,177	69.284	289.994
3400	6,120	148 854	64,039	70.102	293.416
3600	6,400	160 247	68,940	70.881	296.676
3800	6,840	171 724	73,877	71.622	299.776
4000	7,200	183 280	78,849	72.331	302.742
4200	7,560	194 903	83,849	73.008	305.575
4400	7,920	206 585	88,875	73.658	308.295
4600	8,280	218 325	93,926	74.281	310.901
4800	8,640	230 120	99,000	74.881	313.412
5000	9,000	241 957	104,092	75.459	315.830
5200	9,360	253 839	109,204	76.016	318.160
5400	9,720	265 768	114,336	76.553	320.407
5600	10,180	277 738	119,486	77.074	322.587
5800	10,540	289 746	124,652	77.577	324.692
6000	10,800	301 796	129,836	78.065	326.733

TABLE C.2 IDEAL-GAS ENTHALPY AND ABSOLUTE ENTROPY AT 1 ATM PRESSURE *(Continued)*

Hydroxyl (OH)
$(\bar{h}_f^\circ)_{298} = 39\ 463$ kJ/kmol $= 16{,}978$ Btu/pmol
$M = 17.008$

T (K)	T (R)	$\bar{h}^\circ - \bar{h}_{298}^\circ$ (kJ/kmol)	$\bar{h}^\circ - \bar{h}_{537}^\circ$ (Btu/pmol)	$\bar{s}^\circ$ (Btu/pmol-R)	$\bar{s}^\circ$ (kJ/kmol-K)
0	0	−9171	−3,946	0	0
100	180	−6138	−2,641	35.726	149.587
200	360	−2975	−1,280	40.985	171.591
298	537	0	0	43.880	183.703
300	540	54	23	43.925	183.892
400	720	3033	1,305	45.974	192.465
500	900	5991	2,578	47.551	199.063
600	1,080	8941	3,847	48.837	204.443
700	1,260	11 903	5,121	49.927	209.004
800	1,440	14 878	6,401	50.877	212.979
900	1,620	17 887	7,695	51.724	216.523
1000	1,800	20 933	9,005	52.491	219.732
1100	1,980	24 025	10,336	53.195	222.677
1200	2,160	27 158	11,684	53.847	225.405
1300	2,340	30 342	13,054	54.455	227.949
1400	2,520	33 568	14,441	55.027	230.342
1500	2,700	36 840	15,849	55.566	232.598
1600	2,880	40 150	17,273	56.077	234.736
1700	3,060	43 501	18,715	56.563	236.769
1800	3,240	46 890	20,173	57.025	238.702
1900	3,420	50 308	21,643	57.467	240.551
2000	3,600	53 760	23,128	57.891	242.325
2100	3,780	57 241	24,626	58.296	244.020
2200	3,960	60 752	26,136	58.686	245.652
2300	4,140	64 283	27,655	59.062	247.225
2400	4,320	67 839	29,185	59.424	248.739
2500	4,500	71 417	30,724	59.773	250.200
2600	4,680	75 015	32,272	60.110	251.610
2700	4,860	78 634	33,829	60.436	252.974
2800	5,040	82 266	35,392	60.752	254.296
2900	5,220	85 918	36,963	61.058	255.576
3000	5,400	89 584	38,540	61.355	256.819
3200	5,760	96 960	41,713	61.924	259.199
3400	6,120	104 387	44,908	62.462	261.450
3600	6,480	111 859	48,123	62.973	263.588
3800	6,840	119 378	51,358	63.458	265.618
4000	7,200	126 934	54,608	63.922	267.559
4200	7,560	134 528	57,875	64.364	269.408
4400	7,920	142 156	61,157	64.788	271.182
4600	8,280	149 816	64,453	65.195	272.885
4800	8,640	157 502	67,759	65.586	274.521
5000	9,000	165 222	71,080	65.963	276.099
5200	9,360	172 967	74,412	66.326	277.617
5400	9,720	180 736	77,755	66.676	279.082
5600	10,180	188 531	81,108	67.015	280.500
5800	10,540	196 351	84,472	67.343	281.873
6000	10,800	204 192	87,845	67.661	283.203

Source: JANAF Thermochemical Tables, Document PB 168-370, Clearinghouse for Federal Scientific and Technical Information, 1965.

TABLE C.3 ENTHALPY OF COMBUSTION (HEATING VALUE) OF VARIOUS COMPOUNDS

Fuel	Negative of higher heating value $H_2O(l)$ (kJ/kg)	Negative of lower heating value $H_2O(v)$ (kJ/kg)	Negative of higher heating value $H_2O(l)$ (Btu/lbm)	Negative of lower heating value $H_2O(v)$ (Btu/lbm)
Liquid hydrocarbon				
Methane CH_4				
Ethane C_2H_6				
Propane C_3H_8	−49 977	−45 984	−21,490	−19,773
Butane C_4H_{10}	−49 149	−45 363	−21,134	−19,506
Pentane C_5H_{12}	−48 637	−44 977	−20,914	−19,340
Hexane C_6H_{14}	−48 307	−44 728	−20,772	−19,233
Heptane C_7H_{16}	−48 065	−44 551	−20,668	−19,157
Octane C_8H_{18}	−47 886	−44 419	−20,591	−19,100
Decane $C_{10}H_{22}$	−47 637	−44 232	−20,484	−19,020
Dodecane $C_{12}H_{26}$	−47 465	−44 102	−20,410	−18,964
Acetylene C_2H_2				
Benzene C_6H_6	−41 825	−40 137	−17,985	−17,259
Hydrogen H_2				
Gaseous hydrocarbon				
Methane CH_4	−55 491	−50 005	−23,861	−21,502
Ethane C_2H_6	−51 870	−47 479	−22,304	−20,416
Propane C_3H_8	−50 346	−46 346	−21,649	−19,929
Butane C_4H_{10}	−49 519	−45 732	−21,293	−19,665
Pentane C_5H_{12}	−49 005	−45 346	−21,072	−19,499
Hexane C_6H_{14}	−48 674	−45 095	−20,930	−19,391
Heptane C_7H_{16}	−48 430	−44 916	−20,825	−19,314
Octane C_8H_{18}	−48 249	−44 781	−20,747	−19,256
Decane $C_{10}H_{22}$	−47 995	−44 593	−20,638	−19,175
Dodecane $C_{12}H_{26}$	−47 823	−44 460	−20,564	−19,118
Acetylene C_2H_2	−49 907	−48 219	−21,460	−20,734
Benzene C_6H_6	−42 260	−40 572	−18,172	−17,446
Hydrogen H_2	−141 856	−120 012	−60,998	−51,605

Source: JANAF Thermochemical Tables, Document PB 168-370, Clearinghouse for Federal Scientific and Technical Information, 1965.

TABLE C.4 NATURAL LOGARITHM OF EQUILIBRIUM CONSTANT K

For the reaction $\nu_A A + \nu_B B \rightleftharpoons \nu_C C + \nu_D D$ the equilibrium constant K is defined as

T (K)	$H_2 \rightleftharpoons 2H$	$O_2 \rightleftharpoons 2O$	$N_2 \rightleftharpoons 2N$	$H_2O \rightleftharpoons H_2 + \frac{1}{2}O_2$	$H_2O \rightleftharpoons \frac{1}{2}H_2 + OH$	$CO_2 \rightleftharpoons CO + \frac{1}{2}O_2$	$\frac{1}{2}N_2 + \frac{1}{2}O_2 \rightleftharpoons NO$
298	−164.018	−186.988	−367.493	−92.214	−106.214	−103.768	−35.052
500	−92.840	−105.643	−213.385	−52.697	−60.287	−57.622	−20.295
1000	−39.816	−45.163	−99.140	−23.169	−26.040	−23.535	−9.388
1200	−30.887	−35.018	−80.024	−18.188	−20.289	−17.877	−7.569
1400	−24.476	−27.755	−66.342	−14.615	−16.105	−13.848	−6.270
1600	−19.650	−22.298	−56.068	−11.927	−13.072	−10.836	−5.294
1800	−15.879	−18.043	−48.064	−9.832	−10.663	−8.503	−4.536
2000	−12.853	−14.635	−41.658	−8.151	−8.734	−6.641	−3.931
2200	−10.366	−11.840	−36.404	−6.774	−7.154	−5.126	−3.433
2400	−8.289	−9.510	−32.024	−5.625	−5.838	−3.866	−3.019
2600	−6.530	−7.534	−28.317	−4.654	−4.725	−2.807	−2.671
2800	−5.015	−5.839	−25.130	−3.818	−3.769	−1.900	−2.372
3000	−3.698	−4.370	−22.372	−3.092	−2.943	−1.117	−2.114
3200	−2.547	−3.085	−19.950	−2.457	−2.218	−0.435	−1.888
3400	−1.529	−1.948	−17.813	−1.897	−1.582	0.163	−1.690
3600	−0.622	−0.939	−15.911	−1.398	−1.014	0.695	−1.513
3800	0.189	−0.032	−14.212	−0.951	−0.507	1.170	−1.356
4000	0.921	0.783	−12.673	−0.548	−0.050	1.593	−1.216
4500	2.473	2.500	−9.427	0.306	0.914	2.484	−0.921
5000	3.712	3.882	−6.820	0.990	1.683	3.191	−0.686
5500	4.730	5.010	−4.679	1.554	2.312	3.765	−0.497
6000	5.577	5.950	−2.878	2.026	2.837	4.239	−0.341

Source: *JANAF Thermochemical Tables*, Document PB 168-370, Clearinghouse for Federal Scientific and Technical Information, 1965.

Appendix D

Solving Thermodynamics Problems with the Personal Computer

The personal computer has become an essential engineering problem-solving tool. Almost all chapters in this book include problems to be solved using a computer, and the disk included with the text contains thermodynamic property models and representative problem solutions. While a high-level language such as Fortran or Pascal could be used to solve these thermodynamics problems, we have elected to use spreadsheet and equation-solving software (Lotus 1-2-3 and TK Solver).

The use of these powerful general-purpose packages greatly reduces the time necessary to solve an individual problem and allows you to concentrate on engineering principles and to avoid struggling with coding the problem and getting the program to run. Input and output routines, including graphics, are part of the packages. You do not need to be a skilled programmer to get data in, process it, and produce professional-looking output.

SPREADSHEETS

The spreadsheet solutions included on the disk and listed in the text were developed using a Lotus-compatible spreadsheet. These models should run with little or no modifications on any spreadsheet that can load a Lotus .WKS file. The models use only the more common Lotus commands and avoid the use of macros and other such features, to maximize compatibility.

The example spreadsheet solutions listed in the text show both the calculated output as you would see it on the computer monitor, and the display of the cell contents as text. The cell text display shows exactly what data, equations, and labels were entered. This information will permit you to recreate the solution on any spreadsheet software.

The ability to create graphics easily is one of the benefits of using a spreadsheet. Your spreadsheet model will typically display your problem solution in a tabular format. The graphics menu gives you a number of graph types to choose from. The type you will use most commonly is the X-Y graph. The basic steps in displaying an X-Y graph are

1. Enter the Graphics menu (normally /G).
2. Select X-Y type.
3. Specify range of cells containing the *x* data (i.e., A1 . . A10).
4. Specify range of cells containing the *y* data (i.e., B1 . . . B10).

5. Enter titles and labels.
6. Specify special graph format options.
7. Display the graph on the monitor and/or print hardcopy.

TK SOLVER

TK Solver is an equation-solving software package. Many problems can be solved using either a spreadsheet or TK Solver, but typically each has advantages in a particular situation. The important advantage TK Solver has is its ability to solve systems of nonlinear simultaneous equations. Problems that are difficult or impossible to solve on a spreadsheet or that would require many hours to code using Fortran or Pascal can be easily solved with TK Solver.*

TK Solver is organized into "sheets." The most important sheets are the Rule Sheet and the Variable Sheet. The equations that must be used to solve the problem at hand are entered into the Rule Sheet. These equations can be entered in any order and need not be solved for the unknown variables. As you enter the Rules (equations), TK Solver will automatically begin creating the Variable Sheet. Each time you enter an equation with new variable names, the new variables are entered on the Variable Sheet. The basic steps in solving a problem using TK Solver are

1. Enter the equations into the Rule Sheet.
2. Enter the value of all known variables into the Input column of the Variable Sheet.
3. Press the "Solve" key (**F9**). Your answers will appear in the Output column of the Variable Sheet.

Obviously solving other than a simple problem is a bit more involved than this, but the time needed to learn to solve fairly complex problems is surprisingly short. The more-advanced functions and capabilities can be learned as you need them. Some of the features of TK Solver that you will find most useful are

1. Unit conversions (Unit Sheet).
2. Ability to solve trial-and-error problems using the Iterative Solver.
3. Ability to write your own subroutines (Function Sheet).
4. Tabular output capabilities (Table Sheet).
5. Graphical output capabilities (Plot Sheet).

On the disk included with the text are a series of TK Solver models. They can be identified by the .TK file extension (i.e., UNITS.TK). Included are models with steam, R 12, and air properties; units conversions; the psychrometric chart; and selected example problems. The property models can be used as provided, as "electronic tables," or merged into models you develop. This second capability permits the development of complicated process and system models in which fluid properties can be automatically determined.

---

* For more information about TK Solver, contact Universal Technical Systems (UTS) at 1220 Rock Street, Rockford, IL, 61101, or phone 1-815-963-2220, FAX 1-815-963-8884.

Units

The units conversion model (UNITS.TK) is part of each property model. This gives each property model extensive units display and conversion capability. To change the units, merely type over the units listed with the ones desired. When using the units capabilities of TK Solver, it is important to understand the difference between Calculation Units and Display Units. Calculation Units are used internally in the equations in the Rule and Function Sheets. Display Units are those in which the values of variables are displayed. TK Solver interprets the first unit entered to be the Calculation Unit. To check the Calculation Units that TK Solver is using for a particular variable, "dive" to the Variable Subsheet by pressing > with the cursor on the variable of interest. (If a question mark appears in front of a variable value, it usually means you made a typo entering the unit name.)

Merging Models

The fluid property models included on the disk were developed using Rule and List functions, two of the three types of user-defined function types available in the Function Sheet. All the property models were developed using SI units (degK, kPa, m3/kg, kJ/kg, kJ/kg-K) as the Calculation Units. When merging any of these models into another, all variables must initially be entered as SI, then changed to the desired Display Units. All equations that call the property functions must take into consideration the Calculation Units.

To merge a property model into a model you are developing, the procedure is as follows:

1. With your model up in TK Solver, load the other model "on top" of yours using the normal load commands (/SL or F3).
2. Edit the Variable, Rule, Unit, and other sheets to eliminate duplicate and unwanted entries.
3. Enter rules calling the property functions into the Rule Sheet. Remember to make all calls to the property functions consistent with the Calculation Units of the function.

Iterative Solver

The main equation-solving tool in TK Solver is the Direct Solver, a powerful substitution method. As powerful as the Direct Solver is, there are some problems it cannot solve. For those situations, the Iterative Solver, a Newton-Rapson method, is available. It is activated by assigning "guess" values to one or more of the unknown variables as follows:

1. Enter G (for guess) in the Status Field of the variables you wish to assign guesses.
2. Enter the guess value in the Input Field.
3. Attempt a solution (press F9).
4. If TK Solver is unsuccessful, try a different guess value.
5. If repeated guesses fail, try rearranging the equations in the Rule Sheet.

List Solving — Multiple Solutions of a Model

List Solving permits a model to be solved a number of times in succession with one or more of the input values changed. The results are saved in lists and can be displayed as tables (using the Table Sheet) or graphs (using the Plot Sheet). The basic steps are as follows:

1. Set up the problem and get it to run properly in TK Solver. The problem can require use of either the Direct or the Iterative Solver.
2. Enter **L** in the Variable Sheet Status Column of each of the following variables: (1) input variables that will change with each solution, (2) output variables that you want saved, (3) output variables that serve as guesses in an iterative solution.
3. Enter the List Sheet (=**L**). You should see all the variables you identified in step 2. The List Subsheets for the input variables and guess variables must be filled as follows: (1) Move the cursor to the variable. (2) Dive to the subsheet (type >) and fill the list. You can do this manually, or using one of the automatic list fill options (type !). TK Solver automatically enters the element numbers.
4. Enter List Solve (**F10** or /LL). TK Solver will perform the multiple solutions and store the results in the List Variable Subsheets.

Tables

Tables can be displayed only after completion of a List Solve. Data must be available in the List Subsheets. To create a table, proceed as follows:

1. Enter the Table Sheet (=**T**).
2. Enter a name for the table being created.
3. Dive to the Table Subsheet (type >).
4. Enter the name of the lists you want displayed.
5. Enter title, headings, etc.
6. Press **F8** to display the table. To print a copy rather than display it on the monitor, change the entry in the subsheet from Screen (**S**) to Printer (**P**).

Graphs

Graphs (TK Solver calls them plots) can be displayed only after completion of a List Solve. Data must be available in the List Subsheets. To create a plot, proceed as follows:

1. Enter the Plot Sheet (=**P**).
2. Enter a name for the plot being created and the plot type (line, bar, or pie) desired.
3. Dive to the Plot Subsheet (type >).
4. Enter the requested data. As a minimum, you must enter the X-Axis List and the Y-Axis List(s). Enter other information as necessary to modify and enhance the graph.

5. Press **F7** to display the graph. To print a copy on an IBM- or Epson-compatible printer, press **P** when the graph is being displayed. Version 1.1 of TK Solver contains special printer drivers. If you have specified the driver for your printer, pressing **O** will produce a higher-quality plot.

Thermodynamics Models

The disk included with your book includes a series of TK Solver models to assist in solving thermodynamics problems. The models will work in Versions 1.0 and 1.1 of TK Solver Plus and Release 2.0 of TK Solver. Brief descriptions of the models and some hints and suggestions on their use are provided here.

UNITS.TK   A Unit Sheet with most common conversions needed for thermodynamics problems. It is included as part of all the property and cycle models.

SATSTM.TK   Calculates saturated properties of steam. The property equations are contained in seven Rule Functions: Psat, Hf, Hg, Vf, Vg, Sf, and Sg. The input to all functions is the saturated temperature. Psat returns the saturated pressure when provided with the saturated temperature. Hf and Hg return the saturated liquid and vapor enthalpy; Vf and Vg the saturated liquid and vapor specific volume; and Sf and Sg the saturated liquid and vapor entropy. Since Rule Functions can be "back solved," the saturation pressure or any of the other properties can be the input parameter, but this will require the use of the Iterative Solver. The model is valid for temperatures above 0°C and below the critical point.

SHTSTM.TK   Calculates superheated properties of steam. The property equations are contained in three Rule Functions: Hsuper, Vsuper, and Ssuper. The input to all three functions is the temperature and pressure. The functions return the enthalpy, specific volume, and entropy, respectively. If properties other than the temperature and pressure are the input variables, the Iterative Solver must be used. The model is valid for pressures between 7 and 3500 kPa and temperatures below 1000°C.

STEAM.TK   When provided with two independent properties, this model will determine if steam is superheated or a mixture and then determine the steam properties. It is useful in situations when the steam condition is not obvious. STEAM.TK requires the use of the Iterative Solver and is sensitive to the initial guesses. For certain input parameters, it may be necessary to experiment if the model doesn't converge with the default values.

STMCYCLE.TK   Analyzes a simple Rankine cycle. It uses the property models SATSTM.TK and SHTSTM.TK. It can easily be modified to include component efficiencies and to analyze more complicated cycles with regenerative heaters and reheat.

STEAMFUN.TK Contains the Rule functions from SATSTM.TK and SHTSTM.TK. This model can be loaded directly into another model, providing the capability of determining saturated and superheated steam properties.

R12SAT.TK Calculates saturated properties of R 12. The property data is contained in seven List Functions (Psat, Hf, Hg, Vf, Vg, Sf, and Sg). The saturated temperature is the input to all seven functions. Psat returns the saturated pressure; Vf and Vg return the saturated liquid and vapor specific volume; Hf and Hg return the saturated liquid and vapor enthalpy; and Sf and Sg return the saturated liquid and vapor entropy. The model is valid for temperatures between $-90°C$ and the critical point.

R12SHT.TK Calculates superheated properties of R 12. The property equations are contained in three Rule Functions: P, H, and S. P returns the pressure, H the enthalpy, and S the entropy when provided with the temperature and specific volume. Note that the equations require the use of the Iterative Solver and are sensitive to the initial guess for specific volume. The model includes a technique for automatically providing an initial guess of the specific volume based on the ideal-gas law. R12SHT.TK is valid for pressures between 40 kPa and 4000 kPa and temperatures below 230°C.

R12CYCLE.TK Analyzes an R 12 vapor-compression cycle. It includes the property functions from R12SAT.TK and R12SHT.TK and allows varying the evaporating temperature, condensing temperature, superheating, subcooling, and pressure drops in the system.

R12FUN.TK Contains the Rule and List functions from R12SAT.TK and R12SHT.TK. This model can be loaded directly into another model, providing the capability of determining saturated and superheated R 12 properties.

AIR.TK Calculates the properties of air. The property equations are contained in four Rule Functions: PvT, DELh, DELu, and DELs. PvT uses the Redlich-Kwong equation to calculate the air pressure, given the temperature and specific volume. DELh, DELu, and DELs calculate the change in enthalpy, internal energy, and entropy, respectively, between two state points. Each requires the two temperatures as inputs, and DELs also requires the pressures. The Redlich-Kwong equation requires the use of the Iterative Solver for all solutions. AIR.TK is valid for temperatures between 250°K and 2000°K.

AIRCYCLE.TK Analyzes a four-point air cycle using the four Rule Functions from AIR.TK. It can be used to analyze simple Otto, diesel, Brayton, and other such cycles. Since it is based on real-gas data rather than the ideal-gas relationships, the results are more accurate. It can be easily modified to analyze more complicated versions of the basic cycles or to include other effects to more accurately represent realistic behavior. Convergence can be a problem with some combinations of input and output variables. If you experience convergence problems, analyze the cycle one process at a time. Add processes to the solution until the cycle is complete. DIESEL.TK and OTTO.TK are modified versions of AIRCYCLE.TK for the analysis of Diesel and Otto cycles.

AIRFUN.TK   Contains the Rule functions from AIR.TK. This model can be loaded directly into another model, providing the capability of determining air properties.

PSYCHRO.TK   Permits the convenient determination of the properties of air–water vapor mixtures. With the input of any two independent properties (dry-bulb temperature, wet-bulb temperature, dew point, relative humidity, humidity ratio), all the other properties can be determined.

PSYCHRO2.TK   Based on PSYCHRO.TK, analyzes processes of air–water vapor mixtures. The change in dry-bulb temperature, humidity ratio, or enthalpy can be used as an input to determine the unknown quantities.

COMBUST.TK   Solves combustion process problems involving up to three dissociation reactions: water into hydrogen and oxygen, water into hydrogen and hydroxyl, and carbon dioxide into carbon monoxide and oxygen. The data on ideal-gas enthalpies from Table C.2 and the equilibrium constants from Table C.4 are contained in List Functions.

ORSAT.TK   Analyzes the products of a combustion process. The back-solving capabilities of TK Solver permit the Orsat stack gas analysis to be determined, given the fuel analysis and excess air, or the fuel composition and excess air determined, given the stack gas analysis.

CDNOZZLE.TK   Permits the convenient analysis of the flow of ideal gases in convergent-divergent nozzles, including the formation of normal shock waves in the divergent section. CDNOZZLE.TK is based on basic relationships (first law, continuity, ideal gas, isentropic process, sonic velocity) and is therefore more flexible than techniques that require the use of special tables or equations. The Iterative Solver is required, and the model is sensitive to the initial guesses. The default values may not work in some situations, and some experimentation may be necessary.

REFERENCES

1. T. F. Irvine, Jr., and P. E. Liley, *Steam and Gas Tables with Computer Equations,* Orlando, FL: Academic Press, 1984.
2. V. Ganapathy, *Basic Programs for Steam Plant Engineers,* New York: Marcel Dekker, 1986.
3. R. C. Downing, "Refrigerant Equations," *ASHRAE Transactions* 80, part 2, no. 2313 (1974).
4. W. C. Reynolds, *Thermodynamic Properties in SI,* Stanford, CA: Department of Mechanical Engineering, Stanford University, 1979.

Answers to Selected Problems

Answers to Selected Problems

Chapter 1
1.1 1.2 x 10^10 kWh   **1.3** 28.4 days   **1.5** 62,040,000 gallons total

Chapter 2
2.1 520 kPa, 310 kPa   **2.3** 4704 N   **2.5** 6 kg/s, 90 min   **2.7** 10.34 m   **2.9** 8400 N
2.11 2.22 atm, 1688 mm Hg abs   **2.13** −2731.5°X   **2.15** 24.5 kg   **2.17** 456.9 kPa
2.19 80.1 kg   **2.21** 1180 kPa   **2.23** 294 N   **2.25** 80 kPa   **2.27** 663 m
2.29 2119.8 kPa   **2.31** 126.4 kPa   **2.33** 0.107 kg   **2.35** 38.15°C   ***2.1** 64.5 psia,
39.5 psia   ***2.3** 1080 lbf   ***2.5** 59.58 lbm   ***2.7** 95.49 psia   ***2.9** 61.7 lbf
*2.11 2.22 psia   ***2.13** 2177.1 ft   ***2.15** 282.2 psia   ***2.17** 15.4 psia   ***2.19** 0.45 lbm
*2.21 2.66 ft³/lbm   ***2.23** 71.3 ft³/lbm

Chapter 3
3.1 68.4 kJ   **3.3** 70 kg   **3.5** 7.67 m/s   **3.7** 562.5 kJ, 350 kJ   **3.9** 7.7 kJ
3.11 −481.6 kJ/kg   **3.13** −10.9 kW   **3.15** 68.6 kJ   **3.17** −0.0056 J   **3.19 (a)** 2450 kJ
(b) 2450 kJ   **(c)** 2450 kJ   **3.21** −121.1 kW   **3.23** 41 kg/day   **3.25** −22818 kW;
493.9 kg/s   **3.27** 772.2 m/s   **3.29** 490 kW; 920 kJ/kg   **3.31** 5.17 kW; 17.99 kW
3.33 27.3 kJ   **3.35** 625 kJ   **3.37** −586 kJ   **3.39** 2.74 s   **3.41** 116.7 kW; 1.38 s
3.45 140 kJ, 80 kJ, 80 kJ   **3.47** 32.9 min   **3.49** 55.6 kW; 0.265 m/s   **3.51** 7.67 m/s
3.53 1.44 kJ/kg   **3.55** 106.5 kJ   **3.57** 17357 W   **3.59** 158.4 MJ   **3.61** 24.8 kJ,
0.8 m³/kg   **3.63** −274.4 kJ, 25 kJ   **3.65** 525 MW   **3.67** 23 016 kg/s   **3.69** 14 517 kg
*3.1 −14.48 Btu   ***3.3** 464 Btu   ***3.5** −36.3 hp   ***3.7** 2934 ft/sec   ***3.9** 90,000 ft-lb_f
*3.11 −620,967 ft-lb_f   ***3.13** 7.29 sec   ***3.15** 160.7 hp; 1.4 sec   ***3.19** 140 Btu, 80 Btu,
80 Btu   **3.21** 30.8 min   ***3.23** 73.3 kW; 0.51 ft/sec   ***3.25** 25.4 ft/sec
*3.27 92.8 Btu/lbm   ***3.29** 1045 ft-lb_f   ***3.31** 8424 Btu/hr   ***3.33** 168 480 kJ
*3.35 11.2 Btu/lbm, 12.82 ft³/lbm   ***3.37** −246.9 Btu, 30.1 Btu   ***3.39** 739 MW
*3.41 1111.1 MW; 1.042 × 10^8 lbm/hr   ***3.43** 2896 gal.

Chapter 4
4.1 7.389 kPa, 0.1025 kg; 1554.33 kPa, 24.0 kg; 0.9607 MPa, 110.07 kg   **4.7** 2599.3 kJ/kg,
0.1147 m³/kg   **4.9** 122 458 kJ   **4.11** −441.3 kJ/kg, 441.3 kJ/kg, −336.3 kJ/kg
4.13 178.09 kg, 11.55 kg; 0.03045 m³, 0.46955 m³   **4.15** 1551.7 kJ/kg, 0.23813 m³/kg
4.17 16,527 kPa, −307.8 kJ/kg   **4.19** 3.123 kg/s, 0.677 kg/s   **4.21** 3481.8 kJ/kg, 524.9°C
4.23 2472 kPa, 259°C   **4.25** 0.943   **4.27** 0.9954   **4.29** 11.08 kW   **4.31** −3.17 kJ/kg
4.33 3090 kW, 0.0648 m³   **4.37** 0.7715, 0.8457   **4.39** 1.76154 m³   **4.41** 0.017016 kg
4.43 0.000477 m³, 0.499532 m³   **4.45** 0.2 m³, 0.3392   **4.47** 282 kPa   **4.49** −147.6 kJ
4.51 1170 kPa, 55.7°C   **4.53** 37.98 kJ/kg   ***4.3** 0.743%, 99.257%   ***4.5** 594 Btu
*4.7 676.5 Btu, 53.4 Btu, 38.4 Btu   ***4.9** 97,755 Btu   ***4.13** 0.288, 0.1096
*4.15 27.947 ft³   ***4.17** 0.1356 lbm   ***4.19** 0.00468 ft³, 1.0 ft³   ***4.21** 2.7855 ft³, 0.2062
*4.23 338.7 psia   ***4.25** −18.2 Btu   ***4.27** 29.6 psia, 94.5°F   ***4.29** 140.9 Btu/lbm

Chapter 5

5.1 1.111 kJ/kg-K   **5.3** 0.01925 kg, 0.71255 kg, 1.887 kPa   **5.5** 45.66 kg, 337.1 kPa
5.7 371.2 kPa, 371.0 kPa   **5.9** 309.0 kPa, 308.8 kPa   **5.11** 0.0216 m^3/kg, 0.00686 m^3/kg,
1.384 m^3/kg   **5.13** 1.35 kJ/kg-K, 1.08 kJ/kg-K, 30.79 kg/kgmol   **5.15** 862.3 kJ/kg,
654.1 kJ/kg   **5.17** 2.1 cm   **5.19** 2.41 kJ/kg-K, 1.87 kJ/kg-K, 3.11 kJ/kg-K, 2.39 kJ/kg-K
5.21 4.286 kJ/kg-K, 0.286 m^3 **5.23** 4875.5 N   **5.25** 13.39 m   **5.27** 400.5°K
5.31 26,868 kPa, 24,373 kPa   **5.33** 25.37 br/m   **5.35** 0-27260 kPa, 0-15980 kPa,
0-12220 kPa   **5.37** 472.5°K, 225.96 kPa   **5.39** 27.73 kJ, 61.1 kJ   ***5.1** 232.23 Btu/lbm,
229.98 Btu/lbm   ***5.3** 1.595 ft^3/lbm, 1.531 ft^3/lbm   ***5.5** 35.1 ft-lbf/lbm-R, 44.03 lbm/pmol
*5.7 33.66 lbm/pmol, 45.91 ft-lbf/lbm-R, 0.196 Btu/lbm-R   ***5.9** 11,703 ft
*5.11 65.37 lbm   ***5.13** 60.61 ft-lbf/lbm-R   ***5.15** 1.2 Btu/lbm-R, 11.79 ft^3
*5.17 4665.5 psia, 4182.6 psia   ***5.19** 24.86 br/m   ***5.21** 0-3954 psia, 0-2318 psia,
1772 psia   ***5.23** 850.2°R, 33.9 psia   ***5.25** 5989 ft-lbf, 13,191 ft-lbf

Chapter 6

6.1 52.4°K   **6.3** 2256 kPa   **6.5** 1032.5 kJ, 758 kPa   **6.7** 917.3 kJ/kg   **6.9** 284.6 m/s
6.11 21.2 kg/s   **6.13** 40.1 kJ, 28 kJ   **6.15** 576 K, 310 kJ, 310 kJ   **6.17** 900°C
6.19 −221.8 MW, 2628.4 kg/s   **6.21** 1911 lifts   **6.23** 268.9 m/s, 262.4 kPa   **6.25** −29.7 kJ/
kg, −53.3 kJ/kg, 248.8 K, 0.5135 m^3/kg   **6.27** 275.6 kJ, 78 kJ, 275.6 kJ, 197.6 kJ
6.29 2218 kJ, 2218 kJ, 0, 0   **6.31** 83.4 kJ, 89.4 kJ, 14 kJ, 7 kJ, 250 kPa   **6.33** 482.1°K,
−374.2 kW   **6.35** 450.1 kJ, 270.1 kJ   **6.37** 531.3 K, 16 623 kW   **6.39** 676 K, 703.4 kJ/kg,
83.94 kg/s   **6.41** 255.4 kJ   **6.43** heat in = 1916.4 kJ/kg, heat out = −1560 kJ/kg
6.45 heat in = 9491.1 kJ, net work = −2977.9 kJ   **6.47** 465.3 K   **6.49** 39.8 kJ
6.51 −426.9 kJ/kg, −103.7 kJ/kg   **6.53** −32.25 kJ, −15.75 kJ   **6.55** −6.6 kJ
6.57 net work = 337.3 kJ   **6.59** 9687 kW   **6.61** 215.7°C   **6.63** −17.6 kW
6.65 −36.7 kW   **6.67** 47.6°C   **6.69** 59.5 kg/s, −3764 kW   **6.71** 0.04 kg/s
6.73 28.97 kg/s   **6.75** 25.4%   **6.77** 669.5 m/s, 179.9°C   **6.79** −101.7 kW   **6.81** 2750 kJ,
1.44 kg   **6.83** 342.3°C, 4900 kg   ***6.1** 36.7 Btu   ***6.3** 0, −54 138 ft-lb$_f$, −69.57 Btu
*6.5 95.2%   ***6.7** 0.95 psia, −20.9 Btu   ***6.9** 0.6571, 0.7266, 0.6539 ft^3/lbm   ***6.11** 1000°R,
−142.2 Btu, −53.5 Btu   ***6.15** 133.2 Btu, 478.2 Btu   ***6.17** W_{1-2} = 0, Q_{1-2} = −354.6 Btu,
W_{2-3} = 149 Btu, Q_{2-3} = 726.1 Btu   ***6.19** −503.4 Btu, 4234.4 Btu   ***6.21** −16.8 hp, −123.5
Btu/min, 6.14 lbm/min   ***6.23** −3.6 hp, 2.92 ft^3/min   ***6.25** 1603 hp   ***6.27** 25.8 psi
*6.29 1031 Btu/sec   ***6.31** 28.9 lbm/min, −925.4 Btu/min   ***6.33** 515.5 lbm/min
*6.35 949°F   ***6.37** 51 psia   ***6.39** −668.3 hp, −20,295 Btu/min   ***6.41** −218.3 hp,
324.4 hp, 106.1 hp, 15075 Btu/min   ***6.43** 0.987 lbm, −124.8°F   ***6.45** 417.6 Btu/min, 13.4°F

Chapter 7

7.1 516.5 K, 37.5%, 429.4 kPa, −20 kJ   **7.3** 66.9%, −17.5 kJ, 33.5 kJ   **7.5** 70%, 285.7 kW,
20 liter   **7.7** 84%   **7.9** 95.5 kJ/cycle, −31.8 kJ/cycle, 530.8 kW, 0.1837 m^3, 23 kPa
7.11 632.4 K, 183.5 kJ, 116.2 kJ   **7.13** 8.4%   **7.15** 0.095 kW, 1.16 kW   **7.17** 197.7°C, 1.27,
1400 kJ   **7.19** 7.66, 1.02 kg/s   **7.21** 714 K   **7.23** 0.39 kW   **7.27** no   **7.29** 60%,
745 K   **7.31** 75%   **7.33** yes   **7.35** 2.29 kW, 1.51 kW   **7.39** 2.15 hr, $31.57
7.41 30 kW   ***7.1** 915°F, 2696 Btu/min   ***7.3** 33.2 Btu/lbm   ***7.5** −34 F   ***7.7** 60.2%,
1331 R, 30.1 Btu   ***7.9** 0.69 hp   ***7.11** 0.174, 0.5, 1.08, 2.38, 8.0 Btu   ***7.13** yes
*7.15 no   ***7.17** 9.1%   ***7.19** −2.5 °F   ***7.21** 68.2 Btu, −16.7 Btu, 68.2 Btu, 284.5 Btu,
−16.7 Btu, −284.5 Btu 75.5%   ***7.23** 2.4 hrs, $33.21   ***7.25** 3318 Btu/min

Chapter 8

8.1 −0.987 kJ/K, −0.829 kJ/K, −5.104 kJ/K   **8.3** 1.67, 1.22, 2.05   **8.5** yes   **8.7** 57.2 kJ,
0.109 m^3/kg, 0.582 m^3/kg   **8.9** 0.9   **8.11** 0.174 kg/s, −0.0133 kW/K; 0.0218 kg/s,
−0.0138 kW/K, 0.00263 kg/s, −0.0138 kW/K   **8.13** 313.6 K, 0.068 kJ/K, 26.4 kJ
8.15 7.2208 kJ/kg-K   **8.17** 2197.8 kPa, 935.4 kJ; 526.1 kJ, 1.1688 kJ/K   **8.19** 624 kPa,
17.9 kJ, 0.042 m^3 **8.21** 475.7 K, 0.0848 kg/s, −31.8 kW   **8.25** 28.76 kg/s, −8.2 kW/K,
+12.8 kW/K   **8.27** −0.0267 kW/K, +0.0286 kW/K   **8.29** 1.0 kJ/K

8.31 −0.1964 kJ/K   **8.33** 0.159 kJ/K   **8.35** 159.3 kJ   **8.37** 0.824 kW/K
8.39 0.097 kJ/K   **8.41** 2.65 kJ/kg   **8.43** 71.9°C, 14.8 kJ/K   **8.45** 0.129 kJ/K
8.47 −0.0417, +0.0439, 0.0022 kJ/K   **8.49** −0.0038 kW/K, 0.0051 kW/K, 0.0013 kW/K
8.51 468 m/s, 464 m/s   **8.53** 36.9 kJ   **8.55** no at 700 kPa; yes at 900 kPa   **8.57** −1.5,
+1.875, +0.375 kJ/K   **8.59** 0.1968 kJ/K, 0.0485 kJ/K   **8.61** B to A   **8.63** 492.3 kJ/kg,
0.778 kJ/kg-K   **8.65** 0.0296 kW/K   **8.67** 448 K, 0.526 kJ/K   __*8.1__ 0.0848 Btu/lbm-R
__*8.3__ 16%   __*8.5__ +0.036 Btu/R   __*8.7__ 0.498 Btu/sec-R   __*8.9__ 208.9 Btu/hr-R
__*8.11__ 0.07 Btu/R   __*8.13__ 0.806 Btu/min-R, −0.709 Btu/min-R   __*8.15__ 20 psia,
0.3363 Btu/R   __*8.17__ 45.5%   __*8.19__ ΔS_{prod} = 110.4 Btu/min-R   __*8.21__ 1.55 Btu/R,
−0.93 Btu/R, 0.62 Btu/R   __*8.23__ −0.027 Btu/lbm-R, +0.035 Btu/lbm-R, 0.008 Btu/lbm-R
__*8.25__ 154.3°F, 6.05 Btu/R   __*8.27__ 0.038 Btu/R   __*8.29__ −0.119 Btu/min-R, 0.192 Btu/min-R,
0.073 Btu/min-R   __*8.31__ 1534 ft/sec, 1536 ft/sec   __*8.33__ 70.6 Btu   __*8.35__ no for 100 psia;
no for 125 psia.   __*8.37__ 0.057 Btu/R   __*8.39__ ΔS_{air} = −0.090 Btu/R; ΔS_{ne} = 0.128 Btu/R
__*8.41__ −25 F, yes   __*8.43__ 0.223 Btu/lbm-R   __*8.45__ 897°F, 0.245 Btu/R

Chapter 9
9.1 814.2 kJ/kg, −9.3 kJ/kg, 115.1 kJ/kg, 2.2 kJ/kg   **9.3** 0.5 J   **9.5** 29.5 m³   **9.9** 116.1 kJ/
kg, −1.581 kJ/kg   **9.11** −2939.1 kJ/kg, −261.3 kJ/kg, −810.1 kJ/kg, −209.1 kJ/kg   **9.13** 330 K,
375 kPa, 86.4 kJ   **9.15** 143.5 kJ, 89.7 kJ, 33.6 kJ   **9.17** 3 kW   **9.19** 1.5 kW, 2.5 kW
9.21 1201.1 kJ/kg   **9.23** 422.2 kJ/kg, −9.1 kJ/kg, 73.1 kJ/kg   **9.25** 298.1 kJ/kg, 8942.4 kW
9.27 −442.1 kW, 387.7 kW   **9.29** 56.15 kW, −0.91 kW   **9.31** 2728 m/s, −1555 kW
9.33 −32.8 kW, −76.5 kW, 110.1 kW   **9.35** 4.9 kg/s, 45.1 kg/s, 1374 kW   **9.39** −2143.6 kW,
1932.5 kW, 90.2%   **9.41** −518.1 kJ/kg   **9.43** 1285.9 kJ/kg   **9.45** −87.5 kJ/kg,
−16.8 kJ/kg   **9.47** 97.2 kJ   **9.49** 22 093 kW, −15 608 kW, 9498 kW, 20.78 kW/K
9.51 1275 kW, 11452 kW, 0   **9.53** −712.43 kJ/kg, 254.55 kJ/kg, 0.5091 kJ/kg   **9.55** 0.4869 kJ/
K, 94.8%, 90.2%   **9.57** 96.5%   **9.59** 184.3 kJ   **9.61** 0.0914 kJ/K, −27.4 kJ
__*9.1__ 372.1 Btu/lbm, −1.0 Btu/lbm   __*9.3__ 23.5 ft-lb_f   __*9.5__ 1048.7 ft³   __*9.7__ 45.4 Btu/lbm,
−2.8 Btu/lbm   __*9.9__ −91.3 Btu/lbm, −1098.4 Btu/lbm, −77.9 Btu/lbm, −353.2 Btu/lbm
__*9.11__ 595 R, 50 psia, 92.3 Btu   __*9.13__ 61.7 Btu, 39 Btu, 14.6 Btu   __*9.15__ 89.2 Btu/min
__*9.17__ 1.8 kW, 1.8 kW   __*9.19__ 514.1 Btu/lbm   __*9.21__ 167.0 Btu/lbm, −3.8 Btu/lbm,
95.5 Btu/lbm   __*9.23__ 114.7 Btu/lbm, 7456 Btu/sec   __*9.25__ −477.6 Btu/sec, 420.2 Btu/sec
__*9.27__ 3210.8 Btu/min, −52.7 Btu/min   __*9.29__ 9481 ft/sec, −782.8 Btu/sec   __*9.31__ 2.3 Btu/sec,
−142.5 Btu/sec, 140.2 Btu/sec   __*9.33__ 643 lbm/min, 5847 lbm/min, 82,322 Btu/min
__*9.37__ −2138 Btu/sec, 1929.4 Btu/sec, 90.2%   __*9.39__ 44.5 Btu   __*9.41__ 301.9 Btu/lbm
__*9.43__ 1095.5 Btu/sec, 10,180 Btu/sec, 100 Btu/sec, 8984 Btu/sec, 89.1%

Chapter 10
10.7 1875.0 kJ/kg, 1100.6 kJ/kg, 143.3 kJ/kg   **10.13** 0.00328 K⁻¹; 0.0035 K⁻¹
10.15 −247.2 kJ, −249.5 kJ, 0.654 kJ/K   **10.17** 0.0142 K/kPa   **10.19** 115 kPa, 114.8 kPa,
115 kPa   **10.21** 0.434 kg, 0.436 kg, 0.438 kg   **10.23** 6885 kPa, 0.244 m³, 563 kJ
10.35 6.75 T_c, 5.3 T_c __*10.3__ 790 F, 0.052 R/psia   __*10.5__ 0.450 Btu/lbm-R, 0.64 Btu/lbm-R
__*10.7__ 728.8 psia, 840.2 psia, 722.5 psia   __*10.9__ 428 psia, 408.7 psia, 431 psia
__*10.11__ 1.006 lbm, 1.01 lbm, 1.01 lbm   __*10.13__ 1024 psia, 8.86 ft³, 119.6 Btu/lbm

Chapter 11
11.1 0.270, 0.494, 0.236; 30 kPa, 40 kPa, 30 kPa; 35.61 kg/kg mol   **11.3** 44.94 kg/kg mol,
0.185 kJ/kg-K; 0.627, 0.373   **11.5** 0.94, 0.06; 141 kPa, 6 kPa   **11.7** 0.346, 0.425, 0.229; 0.552,
0.431, 0.017   **11.9** 391.3 K, 290 kPa; 0.5537 kJ/K; 0   **11.11** 9365 kPa, 0.1671 kJ/K
11.13 0.0155 kg, 4.3 kPa, 126.9 kPa, −71.1 kJ   **11.15** 60°C, 0.0691 kg vapor/kg air, 64.1%
11.17 0.122, 0.076, 0.802; 0.777 kJ/kg-K   **11.19** −325.3 kW, 0.0808 kW/K   **11.21** 578°K,
684.2 kW   **11.23** 88%, −311.1 kW, 0.128 kW/K   **11.25** 0.604 m³, 0.162 m³; 191.3 kPa;
0.379 kJ/K   **11.27** 1.268 kJ/K, 0   **11.29** 203.7°C, 7.443 kW/K   **11.31** 60.6%
11.3 0.261 kg   **11.35** 196.6 kPa   **11.37** −5.4 kJ, −37.9 kJ   **11.39** 375 kPa, −2883.5 kJ,
−7.97 kJ/K   **11.41** 328K   __*11.1__ 0.372 lbm removed, 0.593 lbm added   __*11.3__ 594.3°R;

0.785, 0.215, 0.909, 0.091; 40.68 ft-lb$_f$/lbm-R, 37.98 lbm/pmol; 121.1 Btu/min-R   ***11.5** 584 R, 111.6 psia; 0.811 Btu/R   ***11.7** 0.122, 0.076, 0.802; 2.44 psia, 1.52 psia, 16.04 psia; 144.4 ft-lb$_f$/lbm-R   ***11.9** −417.8 hp, 3.075 Btu/min-R   ***11.11** 578°R, 490.5 Btu/sec ***11.13** 88%, 189.3 hp, 0.03 Btu/sec-R   ***11.15** 22.92 ft^3, 5.47 ft^3; 26.96 psia; 0.205 Btu/R ***11.17** 0.3027 Btu/R, 0   ***11.19** 400 F, 1.678 Btu/R   ***11.21** 63.7%   ***11.23** 0.017 lbm ***11.25** 29.2 psia   ***11.27** −0.144 Btu, −0.952 Btu   ***11.29** 56.25 psia, −78.6 Btu, −0.1209 Btu/R   ***11.31** 140 F

Chapter 12

12.1 907.9 kg, 10.5% CO_2, 3.3% O_2, 11.9% H_2O, 74.3% N_2 **12.3** 3586 kg   **12.5** 2.14 kPa, 36 kg   **12.7** 15.0 kg air/kg fuel, 88.06 kg mol air/kg mol fuel, 0.726 kg fuel/kg water, 30.42 kg/kgmol, 28.65 kg/kgmol, 0.942   **12.11** 26476 kJ/kg   **12.27** 47.7°C, 45°C   **12.29** 3662883 kJ/kg mol fuel, 45°C, 0.0125 kg/s   **12.45** 3110 K, 980 kPa   **12.47** 0.08 CO_2, 0.151 H_2O, 0.039 O_2, 0.73 N_2; 54 C; −313.1 kW; 1.61 m/s   **12.51** 9770.7 kg/hr   **12.53** 2334 K, 2336 K, 2409 K **12.55** 48 855 kJ/kg   **12.57** 59.5%   **12.59** 2214 K   **12.73** 2.09 V, 1.025 **12.75** 0.00099 kg/s   ***12.1** 48.73 lbm   ***12.3** $n = 1, x = 4$ ***12.5** $n = 14.8, x = 2.31$; −3.5% deficient air   ***12.31** −23,328 Btu/lbm, −21,133 Btu/lbm   ***12.33** 25 sec

Chapter 13

13.1 56.5%, 1057 kPa   **13.3** 713.5 kJ/kg, 44.9%   **13.5** 47.5%, 290 kJ, −152.4 kJ, −71.1 kJ **13.9** 1685 kJ/cycle, 84.25 kW, 430.8 kPa   **13.11** 1994.6 kJ/kg, 1209.2 kJ/kg   **13.13** 60%, 1404.1 kJ/kg, 1050 kPa   **13.21** 1040.3 kJ/kg, 1705.5 kJ/kg, 1308 kPa   **13.23** −202.3 kJ/kg **13.27** 0.0275 kg fuel/kg air, 0.316 kg/min, 90 kW, 8.5 kg/min, 746.3 K   **13.29** 22.5% **13.31** 330 kJ/kg, −251 kJ/kg, −388.8 kJ/kg, 0.8531 kJ/kg-K   **13.33** 11.4, 0.00435 kg/s, 34.7%, 60.4 kW, 0.257 kg/s   **13.35** 225.5 kW, 0.02424 m^3/s, 541 kPa   **13.37** 30.4%, 3.958 × 10^{-3} kg/s   **13.39** 6.8, 442.2 kPa   **13.41** 18.5 × 26.1 cm   ***13.1** 9.74, 59.7%, 11.4%, 4306 R   ***13.9** 113 hp, 62.7 psia   ***13.17** 5.9%, 52%, 1020.5 hp

Chapter 14

14.1 44.8%, 375 kJ/kg, 837.1 kJ/kg   **14.3** 255 kJ/kg, 637.6 K, 4.664 m^3/kg   **14.5** 3 **14.7** 794 K, 38.6%, 0.631 kg/s   **14.9** 80.6%, 84.2%, 25.5%   **14.19** 0.1148 kg/s, 35.8%, 824K, −1339.7 kW   **14.21** 0.0229 kg fuel/kg air   **14.23** 0.01925 kg fuel/kg air, 729 K, 51.9%, 655 K, 40.7%   **14.29** 54.8%, 0.01940 kg fuel/kg air, 688.1 kJ/kg, −230.5 kJ/kg, 247.1 kJ/kg, −83.2 kJ/kg, 53.4%   **14.31** −249.4 kJ/kg, 539.9 kJ/kg, 627.4 kJ/kg, 46.3%, 13 072 kW, −60.1 kJ/kg **14.33** 353 kg/s, 0.952, 7.06 kg/s   ***14.1** 6.7, 42%   ***14.9** 15 lbm fuel/min, 36.2%, 1479°R ***14.11** 0.01862 lbm fuel/lbm air, 1286 R, 53.9%, 1173 R, 42.4%

Chapter 15

15.1 44.2%, 968.4 kJ/kg, −303.8 kJ/kg   **15.17** 12.6%   **15.19** 42.3%   **15.21** 84%, 46.2% **15.31** 43.8%, 33.3 kg/s, 0.1434, 0.1158, 80.9%   **15.33** 908.9 kg/s, 2342 MW, 66.9 kg/s, 0.1066, 0.0879, 0.083, 42.7%   **15.35** 72.6%, 67.3%; 92.7%; 791.6 kJ/kg; 34.1%; 55.8%   **15.43** 392°C, 163.3°C, 0.1341, 0.0704, 97.2%   **15.55** 26.5 kPa, 95.2%, 1.062, 85.3%   **15.57** 2.6 kg/s **15.59** 51.2%, 610 kg/s, 18 124 kg/s, 20 MW   **15.61** 25.9 kJ/rev, 129.8 kW   **15.63** 0.80 kg/s, 2133 kW, 40 kg/s, 11.5%, 36.4%   **15.65** 477.3 kJ/kg air, 1128 kJ/kg steam, 52.1%, 895 kg/s, 0.00741 kg fuel/kg air, 0.02235 kg fuel/kg air, 26.6 kg/s, 0.072 $/kW-h, 68.3%   ***15.1** 79.3%, 506.4 Btu/lbm, −2.3 Btu/lbm, 1326.1 Btu/lbm, −822 Btu/lbm, 504.1 kJ/kg, 38.2% ***15.27** 133 psia, 0.219, 1379.4 Btu/lbm, −754.9 Btu/lbm, 624.5 Btu/lbm, 45.3%   ***15.29** 0.085, 0.075, 0.07; 410.3 Btu/lbm, 38.8%   ***15.31** 0.128, 0.098; 31.6%   ***15.43** −1.53 × 10^6 Btu/min; 33.9%, 49.5%; 58320 kW; 68%

Chapter 16

16.1 0.777, 9.95 tons, 4.48   **16.13** 43 kJ/kg   **16.15** 4.0, 0.244 kg/s   **16.17** 1.538 kg/s, 97.8 kW, 1.8   **16.19** 1.985 kg/s, 1.78, 88.6 kJ/kg, 5.4°C, 98.5 kW, 63.1%   **16.31** 0.0096 kg/s,

6.6, 1.825 kW, $2.19/day, $8.06/day   **16.33** 315.7 kPa, 26.7 tons, 2.61, 35.9 kW   **16.37** 0.277, 12.69 kW/ton, 5.25 kW/ton   **16.39** 71.7%, 4.6 kg/min   **16.41** 9.6 tons   **16.57** 23.5%
16.59 1.13 m³/s, 61.6 kW   **16.61** 7.23 m³/s, 0.0351 kg/s, 7.9 m³/s   **16.63** 0.0408 kg/s, 252.8 kW   **16.65** 20%, 0.89 m³/kg, 213.1 kW   **16.69** 0.05 kg/hr   **16.71** −519 kW, 990 kPa, 575.8 K, 1105 K, 216.4 kW   **16.73** 46.4%   **16.75** 82.5%, 79.6%; −260.8 kJ/kg; 5.5%
16.77 83.1%, 0.0641 m³/stroke, −93.5 kW, $D = L = 0.434$ m   **16.79** 41.7 liters, 1.63 kW
16.81 0.2039 kg/s, 0.484 kg/s, −49.6 kW   ***16.9** NH_3 ***16.11** 46.9 psia, 19.9 tons, 2.53, 27.67 kW   ***16.23** 0.22, 796.3 lbm/hr, −47.4 Btu/lbm, 172.1 ft³/min $L = d = 6.8$ inches
***16.25** 3.39 lbm/min, 3.0, 1.57 hp, 120 F   ***16.43** T = 59°F DB, $\phi = 72\%$; 21 lbm/hr
***16.45** 37,064 lbm/min, 33430 lbm/min, 874.7 lbm/min   ***16.47** 751.9°R, −21.6 hp
***16.49** 44%, 17.5%   ***16.51** −18.66 hp

Chapter 17
17.1 338.5 K, 162.3 kPa   **17.3** 264.2 K, 75.3 kPa, 153.3 cm²   **17.13** 306.8 kPa, 514 nozzles, 50.5 kN   **17.15** 3107.3 kJ/kg, 567.8 m/s, 14.2 kN, 76 nozzles   **17.17** 282.9 K, 0.267 m³/kg, 297.7 m/s   **17.19** 519.3 m/s, 158.8 K, 2.25 cm   **17.21** 0.564, 1040 m/s, 1160 K, 2523 m/s, 11.88   **17.31** 152°C, 176.1°C   **17.33** 88.06 kg/s   **17.35** 29.05 kg/s; 275.2 K, 280.2 K; 82.3 kPa, 87.6 kPa; 0.278 m²   **17.37** 4.37 m/s   **17.39** 874.2 kW   **17.41** 302.5 Nm, 47.5 kW   ***17.1** 120.9 lb_f   ***17.9** 0.6235 lbm/sec, 409.2 Btu/lbm   ***17.19** 305.6 F, 349.4 F   ***17.21** ½   ***17.23** 4145 gal/min; 2202.7 ft-lb_f, 733.7 hp; 2279.2 ft-lb_f, 759.2 hp; 2355.6 ft-lb_f, 750.1 hp

Chapter 18
18.1 3.53 W/m-K   **18.3** 468.7 W   **18.5** 0.535m   **18.7** 64.5 cm, 642.7 W/m²
18.9 14.57 kW/m²   **18.11** 3029 K   **18.13** 0.03 kg/s   **18.15** 2.43m   **18.17** 4.7 cm
18.19 11.7 W/m²-K, 181.7°C, 4252 kW   **18.21** 207.3°C, 27.49 m², 22.77 m², 142.56°C
18.23 7   **18.25** 754.4 m², 0.78, 178.7C   **18.29** 165.6°C, 0.704, yes   **18.31** 345.1 m, 0.854 kg/s   ***18.1** 1.03 ft   ***18.3** 82, 833, 9004, 14.4, 7.2 Btu/hr-ft²   ***18.5** 0.3%
***18.7** 4.34 Btu/hr-ft²-F   ***18.9** 135.3°F, 106.5°F; 15.17 ft², 19.28 ft²   ***18.11** 6447 ft², 62%, 334.5 F

Index